ANNALS OF THE NEW YORK ACADEMY OF SCIENCES

Volume 763

The New York Academy of Sciences
2 East 63rd Street
New York, New York 10021

THE IMIDAZOLINE RECEPTOR: PHARMACOLOGY, FUNCTIONS, LIGANDS, AND RELEVANCE TO BIOLOGY AND MEDICINE

ANNALS OF THE NEW YORK ACADEMY OF SCIENCES

Volume 763

THE IMIDAZOLINE RECEPTOR: PHARMACOLOGY, FUNCTIONS, LIGANDS, AND RELEVANCE TO BIOLOGY AND MEDICINE

Edited by Donald J. Reis, Pascal Bousquet, and Angelo Parini

The New York Academy of Sciences
New York, New York
1995

♾ *The paper used in this publication meets the minimum requirements of American National Standard for Information Sciences – Permanence of Paper for Printed Library Materials, ANSI Z39.48-1984.*

Library of Congress Cataloging-in-Publication Data

The imidazoline receptor : pharmacology, functions, ligands, and relevance to biology and medicine / edited by Donald J. Reis, Pascal Bousquet, and Angelo Parini.
p. cm. — (Annals of the New York Academy of Sciences ; v. 763)
Includes bibliographical references and index.
ISBN 0-89766-935-5 (cloth : alk. paper). — ISBN 0-89766-936-3 (paper : alk. paper)
1. Imidazoline—Receptors—Congresses. 2. Imidazoline-Derivatives—Physiological effect—Congresses. 3. Alpha adrenoceptors—Congresses. I. Reis, Donald J. II. Bousquet, Pascal. III. Parini, Angelo. IV. Series.
Q11.N5 vol. 763
[RM666.I35]
500 s—dc20
[615′.7]

95-316
CIP

PCP
Printed in the United States of America
ISBN 0-89766-935-5 (cloth)
ISBN 0-89766-936-3 (paper)
ISSN 0077-8923

ANNALS OF THE NEW YORK ACADEMY OF SCIENCES

Volume 763
July 12, 1995

THE IMIDAZOLINE RECEPTOR: PHARMACOLOGY, FUNCTIONS, LIGANDS, AND RELEVANCE TO BIOLOGY AND MEDICINE [a]

Editors
DONALD J. REIS, PASCAL BOUSQUET, AND ANGELO PARINI

Conference Organizer
DONALD J. REIS

CONTENTS

[a] This volume is the result of a conference entitled **II International Symposium on Imidazoline Receptors** held in New York, New York on July 19–20, 1994.

Part II. Imidazoline Receptors: Structure, Pharmacology, and Regulation

Part III. Imidazoline Receptor Distribution

Part IV. Signal Transduction

Part V. Endogenous Ligands

Part VI. Cellular Actions: Kidney, Pancreas, Adipocytes, Trachea

Part VII. Cellular Actions: CNS and Autonomic Nervous System

Part VIII. Cardiovascular System and Hypertension

*****

Financial assistance was received from:

- BIOPHARMA LABORATOIRES BIOPHARMACEUTIQUES DE FRANCE
- SOLVAY PHARMA DEUTSCHLAND GmbH
- BEIERSDORF-LILLY GmbH
- INSTITUT DE RECHERCHES INTERNATIONALES SERVIER
- ORION-FARMOS PHARMACEUTICALS
- AB ASTRA
- STERLING WINTHROP INC.
- LABORATORIOS LASA
- PFIZER INC.
- SMITHKLINE BEECHAM PHARMACEUTICALS

Preface

DONALD J. REIS,[a] PASCAL BOUSQUET,[b] AND ANGELO PARINI[c]

[a]*Cornell University Medical College*
New York, New York 10021

[b]*University Louis Pasteur*
67000 Strasbourg, France

[c]*Institut Louis Bugnard*
31054 Toulouse Cédex, France

The concept of the imidazoline receptor is barely over a decade old. At first it was proposed, on the basis of structure/function studies, to be a nonadrenergic receptive site, mediating the hypotensive actions of clonidine in brain.[1] Ligand binding sites soon proved the correctness of the surmise,[2] and soon established that imidazoline receptors existed in multiple forms, were expressed in a variety of organs, cells, and tissues, often in association with mitochondria,[3] and had actions independent of those on blood pressure. The identification of agmatine as an endogenous ligand(s) for imidazoline receptors[4] and the probability that other "clonidine-displacing substances"[5] exist suggest that imidazoline receptors may be a receptive component of a novel neurotransmitter, hormonal and/or paracrine system with widespread influences in the body's economy.

In response to increasing interest, the First International Symposium on Imidazoline Preferring Receptors was organized in Paris in 1992 to review the "state of the field." Some of the longer contributions were published in full length, and the remainder as abstracts, in a slim journal supplement (*Fundamental & Clinical Pharmacology*, Vol. 6/Suppl. 1, 1992). However, in the face of new and notable discoveries and larger numbers of investigators, the Second International Symposium on Imidazoline Receptors was organized and held at the Waldorf-Astoria Hotel in New York on July 19 and 20, 1994. The meeting was designated as an official satellite of the XIIth International Congress of Pharmacology held in Montreal, Canada from July 24-29, 1994. This volume represents the proceedings of that satellite. The size and quality of the contributions attest to the viability of the field.

The Editors wish to single out for particular thanks the Scientific Advisory Committee, which helped to select the most suitable work for presentation and review-free communications. The meeting would never have seen the light of day without a generous seed grant from Biopharma and subsequent generous contributions from Solvay, Beiersdorf-Lilly, Servier, and other contributors listed in the Table of Contents. Kathleen Koenig of Cornell University Medical College was the Administrative Director and was assisted by Janine Vinciguerra. Catherine Macrae of The Macrae Group facilitated the preliminary publication agreement and attended to administrative details. Nancy Marmor at Cornell interfaced with the able editorial group of the New York Academy of Sciences, including Bill Boland and his collaborators, Sheila Kane and Angela Fink, in the preparation of the volume.

REFERENCES

1. Bousquet, P., J. Feldman & J. Schwartz. 1984. Central cardiovascular effects of alpha adrenergic drugs: Differences between catecholamines and imidazolines. J. Pharmacol. Exp. Ther. **230:** 232–236.
2. Ernsberger, P., M. P. Meeley, J. J. Mann & D. J. Reis. 1987. Clonidine binds to imidazole binding sites as well as $alpha_2$-adrenoceptors in the ventrolateral medulla. Eur. J. Pharmacol. **134:** 1–13.
3. Tesson, F., C. Prip-Buus, A. Lemoine, J. P. Pegorier & A. Parini. 1991. Subcellular distribution of imidazoline-guanidinium-receptive sites in human and rabbit liver. Major localization to the mitochondrial outer membrane. J. Biol. Chem. **266:** 155–160.
4. Li, G., S. Regunathan, C. J. Barrow, J. Eshraghi, R. Cooper & D. J. Reis. 1994. Agmatine is an endogenous clonidine-displacing substance in brain. Science **263:** 966–969.
5. Atlas, D. & Y. Burstein. 1984. Isolation and partial purification of a clonidine-displacing endogenous brain substance. Eur. J. Biochem. **144:** 287–293.

Pharmacological Characteristics of Alpha-2 Adrenergic Receptor Subtypes

DAVID B. BYLUND

Department of Pharmacology
University of Nebraska Medical Center
600 South 42nd Street
Omaha, Nebraska 68198-6260

The family of adrenergic receptors mediates the actions of the endogenous catecholamines and neurohormonal messengers epinephrine and norepinephrine. Many members of this receptor family have been cloned, and their predicted amino acid sequences indicate that they are members of the G-protein-coupled receptor superfamily.

The initial subclassification of adrenergic receptors into alpha and beta types was based on their pharmacological rather than their functional characteristics,[1] as had been done by previous investigators. Subsequently, beta-adrenergic receptors were subdivided into beta-1 and beta-2 subtypes, and recently a beta-3 subtype was identified. This is also a pharmacological classification scheme based on a comparison of the relative potencies of agonists and antagonists. In contrast to the relatively simple subclassification of beta-adrenergic receptors based on pharmacological characteristics the subdivision of alpha-adrenergic receptors was more circuitous. As early as 1965, it was suggested that there were different pharmacological types of alpha-adrenergic receptors.[2] However, not until 1973 were the terms alpha-1 and alpha-2 first suggested for alpha-adrenergic receptor subtypes.[3] It was initially proposed that alpha-1 adrenergic receptors were postsynaptic and alpha-2 adrenergic receptors were presynaptic.[4,5] However, this anatomical classification of alpha-adrenergic receptors was quickly shown to be in error, because many examples of alpha-2 adrenergic receptors that did not have a presynaptic localization were found. Thus, although alpha-2 adrenergic receptors can be at pre- or postsynaptic or, indeed, at nonsynaptic locations (such as platelets), the majority of presynaptic alpha-adrenergic receptors have alpha-2 adrenergic receptor pharmacological characteristics. The pharmacological subclassification of alpha-adrenergic receptors into alpha-1 and alpha-2 adrenergic receptor subtypes was formalized in 1977.[6]

ADRENERGIC RECEPTOR TYPES

Recently we suggested that in contrast to the traditional classification of adrenergic receptors into alpha and beta, it is preferable to classify adrenergic receptors into one of three major types, alpha-1, alpha-2, and beta.[7] This pharmacological classification scheme is based on three observations. First, selective agonists were identified that have high affinity for one of the three major types but much lower affinity (1,000-

to 10,000-fold) for the other two types. By contrast, the agonists used to define the subtypes generally are only 10- to 50-fold selective. Second, each of the three major types of adrenergic receptors couples to its second messenger system through a distinct family of guanine nucleotide regulatory binding proteins (G-proteins). Alpha-1 receptors are coupled through G_q, alpha-2 receptors through G_i, and beta receptors through the G_s family. Third, from a molecular structural point of view, the predicted amino acid sequence for the three major types of receptors indicates that both beta and alpha-1 receptor types have short third intracellular loops and long COOH-terminal tails, whereas the alpha-2 receptor has a long third intracellular loop and a short COOH-terminal tail. Furthermore, alpha-1, alpha-2, and beta types all have similar predicted amino acid sequence identities to each other. Thus, there is no molecular reason to consider the alpha-1 and alpha-2 types more closely related to each other than either type is to the beta-adrenergic receptor type.[7]

ALPHA-2 ADRENERGIC RECEPTOR SUBTYPES

Heterogeneity of alpha-2 adrenergic receptors was noted in many laboratories on the basis of both functional and radioligand binding studies. Often this heterogeneity was interpreted as the result of pharmacological subtypes of the receptor. Most evidence for alpha-2 adrenergic receptor subtypes originally came from binding studies in various tissues and functional studies in cell lines and more recently in cells transfected with the genes (or cDNA) for the receptors.[7,8] On the basis of these studies, four pharmacological and three molecular alpha-2 adrenergic receptor subtypes have been defined.[8,9] The alpha-2A adrenergic receptor subtype, for which prazosin and ARC-239 have relatively low affinity, is found in the human platelet and the HT29 cell.[10] The gene for this receptor, cloned from the human,[11] was localized to chromosome 10. Therefore, this cloned human alpha-2 adrenergic receptor is known as alpha-2-C10 adrenergic receptor. Orthologous clones have been found in the rat[12] and pig.[13] The second subtype, the alpha-2B adrenergic receptor, which has been identified in neonatal rat lung[10] and the NG108 cell,[10] has been cloned from the rat (RNG),[14] human (alpha-2-C2),[15,16] and mouse[17] and has relatively high affinity for prazosin and ARC-239. A third subtype, alpha-2C adrenergic receptor, has been identified in an opossum kidney (OK) cell line[18] and cloned from that cell line.[19] This receptor has a relatively high affinity for prazosin but is pharmacologically distinct from the alpha-2B subtype.[20] A gene for an alpha-2 adrenergic receptor has also been cloned from the human kidney[21] and is located on chromosome 4. The expressed alpha-2-C4 adrenergic receptor subtype has the same pharmacological characteristics as does the alpha-2C adrenergic receptor from the opossum and the orthologues that have been cloned from the rat[12] and mouse.[22] A fourth pharmacological subtype, the alpha-2D adrenergic receptor, has been identified in the rat salivary gland[23] and bovine pineal gland.[24] This pharmacological subtype has been cloned from the rat[12] and mouse.[22] On the basis of a predicted amino acid sequence, the alpha-2D adrenergic receptor appears to be an orthologue of the human alpha-2-C10 (alpha-2A) and is therefore designated by some investigators as the alpha-2A/D adrenergic receptor to emphasize its genetic similarity to the human alpha-2-C10 as well as its pharmacological differences from the alpha-2A subtype.

TABLE 1. Classification of Alpha-2 Adrenergic Receptor Subtypes

		Pharmacological Classification		Molecular Classification		
Subtype	Species	Tissue/Cell Line	Ref.	Name	Tissue	Ref.
Alpha-2A	Human	Platelet, HT29	10	Alpha-2-C10	Genomic	11
Alpha-2A	Porcine	Brain		(Pig α_{2A})	Genomic	13
Alpha-2A	Chicken	Pineal gland	32			
Alpha-2A/D	Rat	Salivary gland	23	RG20	Genomic	12
Alpha-2A/D	Mouse			Mα_2-10H	Genomic	22
Alpha-2A/D	Bovine	Pineal gland	24			
Alpha-2B	Rat	Lung, NG108	10	RNGα_2	Kidney	14
Alpha-2B	Human			Alpha-2-C2	Genomic	15
Alpha-2B	Mouse			Mα_2-2H	Genomic	17
Alpha-2C	Opossum	Kidney, OK cell	20	OKc2	OK cell	19
Alpha-2C	Human			Alpha-2-C4	Kidney	21
Alpha-2C	Rat			RG10	Genomic	12
Alpha-2C	Mouse			Mα_2-4H	Genomic	22

TABLE 2. Comparison of pK_i Values for Pharmacologically Defined Subtypes with Those for Cloned Subtypes of Alpha-2 Adrenergic Receptors

	Correlation Coefficient			
	Alpha-2-C10	Alpha-2-C2	Alpha-2-C4	RG20
HT29 cell (Alpha-2A)	0.98	0.24	0.63	0.86[a]
Rat lung (Alpha-2B)	0.33	0.95	0.80	0.45
OK cell (Alpha-2C)	0.74	0.63	0.95	0.57
Bovine pineal (Alpha-2D)	0.73	0.57[b]	0.53[c]	0.93

NOTE: Data are from refs. 25 and 26.
[a] Alpha-2-C10 vs RG20.
[b] Bovine pineal vs rat lung.
[c] Bovine pineal vs OK cell.

The current classification scheme for alpha-2 adrenergic receptor subtypes, summarized in TABLE 1, indicates a clear relationship among the pharmacologically defined and molecularly cloned subtypes. The major evidence for this correspondence comes from radioligand binding data comparing the K_i values from inhibition experiments in the prototypic tissues for the pharmacological subtypes and those in cells transfected with the clones.[25,26] The results of these experiments, summarized in TABLE 2, show that corresponding subtypes have much higher correlation coefficients (>0.93) than do noncorresponding subtypes.

Many workers in the field doubt that additional alpha-2 adrenergic receptor subtypes will be found in addition to the four pharmacological or three genetic subtypes that have been identified to date. Some evidence, however, suggests that additional subtypes and/or heterogeneity do exist. For example, current subtypes do not appear to account for some of the observed pharmacological differences between pre- and postsynaptic receptors. Two novel alpha-2 adrenergic receptor antagonists, SK&F 104978 and SK&F 104856, appear to functionally discriminate among alpha-2 adrenergic receptors in a way that is not easily explained by their binding affinities at the known subtypes.[9] In addition, Wikberg and coworkers showed that the alpha-2 adrenergic receptor drug guanoxabenz can delineate both alpha-2A[27] and alpha-2B adrenergic receptor[28] subtypes (but not alpha-2C adrenergic receptors) into further subtypes. This, however, appears to be due to metabolism of the guanoxabenz rather than to receptor heterogeneity.

IMIDAZOLINE BINDING SITES

Several widely used alpha-2 adrenergic receptor drugs, including clonidine and idazoxan, also bind with high affinity to nonadrenergic sites, which have been designated as imidazoline binding sites.[29] These binding sites are not part of the alpha-2 adrenergic receptor family because of their very low affinity for the catecholamines epinephrine and norepinephrine. There is good evidence that these imidazoline binding sites are receptors and have important physiological effects, such as the regulation of blood pressure, and the hypotensive effects of some alpha-2 adrenergic receptor drugs in some species are mediated by these receptors. The endogenous substance for this receptor has been named CDS for clonidine-displacing substance.[30]

Recently, a potential CDS was isolated from bovine brain and identified as agmatine by mass spectroscopy.[31] This relatively simple compound (decarboxylated arginine) had not previously been detected in brain, but it has many of the expected characteristics of CDS and of an endogenous ligand for the imidazoline receptors.[31] The reported affinity (IC_{50} value to inhibit [^{3}H]*p*-aminoclonidine binding) of agmatine for the alpha-2 adrenergic receptor of rat cerebral cortex (presumably mostly the alpha-2A/D subtype) is 4 μM.[31] To discover if agmatine has a differential affinity for the various pharmacological subtypes, we determined the K_i values for agmatine for the inhibition of [^{3}H]RX821002 binding for alpha-2A (CHO cells transfected with the human alpha-2-C10 receptor), alpha-2B (CHO cells transfected with the rat RNG receptor), alpha-2C (CHO cells transfected with the OK alpha-2C receptor), and alpha-2D (bovine neuroretina) adrenergic receptor subtypes. As shown in TABLE 3, agmatine has similar and low affinities in the range of 3.4–5.7 μM for all four alpha-2 adrenergic receptor subtypes. The affinity of agmatine is significantly lower (higher K_i) than that of norepinephrine for the alpha-2 adrenergic receptor subtypes. In preliminary experiments, we assessed the activity of agmatine in a functional assay for the alpha-2A adrenergic receptor (inhibition of forskolin-stimulated cyclic AMP production in HT29 cells). Over the concentration range of 1 nM to 100 μM, agmatine

TABLE 3. Affinity of Agmatine for Alpha-2 Adrenergic Receptor Subtypes

	Agmatine	Clonidine	Norepinephrine
Alpha-2A			
K_i (nM)	4,300 ± 400	15 ± 2	116 ± 18
Slope	1.23	0.76	0.56
Alpha-2B			
K_i (nM)	3,420 ± 170	27 ± 5	244 ± 37
Slope	1.21	0.88	0.68
Alpha-2C			
K_i (nM)	5,740 ± 110	48 ± 1	24 ± 1
Slope	1.02	0.77	0.58
Alpha-2D			
K_i (nM)	4,670 ± 200	53[a]	209[a]
Slope	1.08	...	...

NOTE: K_i values are means ± SEM of three determinations.
[a] From ref. 33.

did not significantly inhibit forskolin-stimulated cyclic AMP production (agonist activity), nor did it significantly reverse the UK 14,304 inhibition of forskolin-stimulated cyclic AMP production (antagonist activity). The combined results of the binding and functional experiments indicate that agmatine has little, if any, activity at the alpha-2 adrenergic receptor.

SUMMARY

The adrenergic receptor family has three main members: alpha-1, alpha-2, and beta receptor types. Each of these three types contains three or more subtypes. There are currently four pharmacological (2A, 2B, 2C, and 2D) and three molecular (or genetic) alpha-2 adrenergic subtypes (2A/D, 2B, and 2C). Because different functions are likely to be mediated by different subtypes, much effort is being directed towards understanding the physiological roles of the various subtypes. However, little is currently known about the specific function mediated by the various subtypes.

REFERENCES

1. AHLQUIST, R. P. 1948. A study of adrenotropic receptors. Am. J. Physiol. **153:** 586–600.
2. ROSSUM, J. M. 1965. Different types of sympathomimetic α-receptors. J. Pharm. Pharmacol. **17:** 202–216.
3. DELBARRE, B. & H. SCHMITT. 1973. A further attempt to characterize sedative receptors activated by clonidine in chickens and mice. Eur. J. Pharmacol. **22:** 355–359.
4. LANGER, S. Z. 1974. Presynaptic regulation of catecholamine release. Biochem. Pharmacol. **23:** 1793–1800.

5. STARKE, K., H. MONTEL, W. GAYK & R. MERKER. 1974. Comparison of the effects of clonidine on pre- and postsynaptic adrenoceptors in the rabbit pulmonary artery. Naunyn-Schmiedebergs Arch. Pharmacol. **285:** 133-150.
6. BERTHELSON, S. & W. A. PETTINGER. 1977. A functional basis for classification of α-adrenergic receptors. Life Sci. **21:** 595-606.
7. BYLUND, D. B. 1988. Subtypes of α_2-adrenoceptors: Pharmacological and molecular biological evidence converge. Trends Pharmacol. Sci. **9:** 356-361.
8. BYLUND, D. B. 1992. Subtypes of alpha-1 and alpha-2 adrenergic receptors. FASEB J. **6:** 832-839.
9. BYLUND, D. B., D. C. EIKENBERG, J. P. HIEBLE, S. Z. LANGER, R. J. LEFKOWITZ, K. P. MINNEMAN, P. B. MOLINOFF, R. R. RUFFOLO, JR. & U. TRENDELENBURG. 1994. IV. International union of pharmacology nomenclature of adrenoceptors. Pharmacol. Rev. **46:** 121-136.
10. BYLUND, D. B., C. RAY-PRENGER & T. J. MURPHY. 1988. Alpha-2A and alpha-2B adrenergic receptor subtypes: Antagonist binding in tissues and cell lines containing only one subtype. J. Pharmacol. Exp. Ther. **245:** 600-607.
11. KOBILKA, B. K., H. MATSUI, T. S. KOBILKA, T. L. YANG-FENG, U. FRANCKE, M. G. CARON, R. J. LEFKOWITZ & J. W. REGAN. 1987. Cloning, sequencing, and expression of the gene coding for the human platelet α_2-adrenergic receptor. Science **238:** 650-656.
12. LANIER, S. M., S. DOWNING, E. DUZIE & C. J. HOMCY. 1991. Isolation of rat genomic clones encoding subtypes of the α_2-adrenergic receptor. Identification of a unique receptor subtype. J. Biol. Chem. **266:** 10470-10478.
13. GUYER, C. A., D. A. HORSTMAN, A. L. WILSON, J. D. CLARK, E. J. CRAGOE, JR. & L. E. LIMBIRD. 1990. Cloning, sequencing, and expression of the gene encoding the porcine α_2-adrenergic receptor: Allosteric modulation by Na^+, H^+, and amiloride analogs. J. Biol. Chem. **265:** 17307-17317.
14. ZENG, D., J. K. HARRISON, D. D. D'ANGELO, C. M. BARBER, A. L. TUCKER, Z. LU & K. R. LYNCH. 1990. Molecular characterization of a rat α_{2B}-adrenergic receptor. Proc. Natl. Acad. Sci. USA **87:** 3102-3106.
15. LOMASNEY, J. W., W. LORENZ, L. F. ALLEN, K. KING, J. W. REGAN, T. L. YANG-FENG, M. G. CARON & R. J. LEFKOWITZ. 1990. Expression of the α_2-adrenergic receptor subtype, the gene for which is located on chromosome 2. Proc. Natl. Acad. Sci. USA **87:** 5094-5098.
16. WEINSHANK, R. L., J. M. ZGOMBICK, M. MACCHI, N. ADHAM, H. LICHTBLAU, T. A. BRANCHEK & P. R. HARTIG. 1990. Cloning, expression, and pharmacological characterization of a human α_{2B}-adrenergic receptor. Mol. Pharmacol. **38:** 681-688.
17. CHRUSCINSKI, A. J., R. E. LINK, D. A. DAUNT, G. S. BARSH & B. K. KOBILKA. 1992. Cloning and expression of the mouse homolog of the human α_2-C2 adrenergic receptor. Biochem. Biophys. Res. Commun. **186:** 1280-1287.
18. MURPHY, T. J. & D. B. BYLUND. 1988. Oxymetazoline inhibits adenylate cyclase by activation of serotonin-1 receptors in the OK cell, an established renal epithelial cell line. Mol. Pharmacol. **34:** 1-7.
19. BLAXALL, H. S., D. R. CERUTIS, N. A. HASS, L. J. IVERSEN & D. B. BYLUND. 1994. Cloning and expression of the α_{2C}-adrenergic receptor from the OK cell line. Mol. Pharmacol. **45:** 176-181.
20. BLAXALL, H. S., T. J. MURPHY, J. C. BAKER, C. RAY & D. B. BYLUND. 1991. Characterization of the alpha-2C adrenergic receptor subtype in the opossum kidney and in the OK cell line. J Pharmacol. Exp. Ther. **259:** 323-329.
21. REGAN, J. W., T. S. KOBILKA, T. L. YANG-FENG, M. G. CARON, R. J. LEFKOWITZ & B. K. KOBILKA. 1988. Cloning and expression of human kidney cDNA for an α_2-adrenergic receptor subtype. Proc. Natl. Acad. Sci. USA **85:** 6301-6305.
22. LINK, R., D. DAUNT, G. BARSH, A. CHRUSCINSKI & B. KOBILKA. 1992. Cloning of two mouse genes encoding α_2-adrenergic receptor subtypes and identification of a single

amino acid in the mouse α_2-C10 homolog responsible for an interspecies variation in antagonist binding. Mol. Pharmacol. **42:** 16-27.

23. MICHEL, A. D., D. N. LOURY & R. L. WHITING. 1989. Differences between the α_2-adrenoceptor in rat submaxillary gland and the α_{2A}- and α_{2B}-adrenoceptor subtypes. Br. J. Pharmacol. **98:** 890-897.
24. SIMONNEAUX, V., M. EBADI & D. B. BYLUND. 1991. Identification and characterization of alpha-2D adrenergic receptors in bovine pineal gland. Mol. Pharmacol. **40:** 235-241.
25. BYLUND, D. B., H. S. BLAXALL, L. J. IVERSEN, M. G. CARON, R. J. LEFKOWITZ & J. W. LOMASNEY. 1992. Pharmacological characteristics of α_2-adrenergic receptors: Comparison of pharmacologically defined subtypes with subtypes identified by molecular cloning. Mol. Pharmacol. **42:** 1-5.
26. O'ROURKE, M. F., L. J. IVERSEN, J. W. LOMASNEY & D. B. BYLUND. 1994. Species orthologs of the alpha-2A adrenergic receptor: The pharmacologic properties of the bovine and rat receptors differ from the human and porcine receptors. J. Pharmacol. Exp. Ther. **271:** 735-740.
27. UHLÉN, S., Y. XIA, V. CHHAJLANI, C. C. FELDER & J. E. S. WIKBERG. 1992. [^{3}H]-MK-912 binding delineates two α_2-adrenoceptor subtypes in rat CNS one of which is identical with the cloned pA2d α_2-adrenoceptor. Br. J. Pharmacol. **106:** 986-995.
28. UHLÉN, S. & J. E. S. WIKBERG. 1991. Delineation of three pharmacological subtypes of α_2-adrenoceptors in the kidney. Br. J. Pharmacol. **104:** 657-664.
29. ERNSBERGER, P., M. P. MEELEY, J. J. MANN & D. J. REIS. 1987. Clonidine binds to imidazole binding sites as well as alpha$_2$-adrenoceptors in the ventrolateral medulla. Eur. J. Pharmacol. **134:** 1-13.
30. ATLAS, D. & Y. BURSTEIN. 1984. Isolation and partial purification of clonidine-displacing endogenous brain substance. Eur. J. Pharmacol. **144:** 287-293.
31. LI, G., S. REGUNATHAN, C. J. BARROW, J. ESHRAGHI, R. COOPER & D. J. REIS. 1994. Agmatine: An endogenous clonidine-displacing substance in the brain. Science **263:** 966-969.
32. BYLUND, D. B., P. K. RUDEEN, L. J. PETTERBORG & C. RAY-PRENGER. 1988. Identification of α_2-adrenergic receptors in chicken pineal gland using [^{3}H]rauwolscine. J. Neurochem. **51:** 81-86.
33. HARRISON, J. K., D. D. D'ANGELO, D. ZENG & K. R. LYNCH. 1991. Pharmacological characterization of rat α_2-adrenergic receptors. Mol. Pharmacol. **40:** 407-412.

Possible Structural and Functional Relationships between Imidazoline Receptors and α_2-Adrenoceptors

J. PAUL HIEBLE AND ROBERT R. RUFFOLO, JR.

Division of Pharmacological Sciences
SmithKline Beecham Pharmaceuticals
709 Swedeland Road
King of Prussia, Pennsylvania 19406

During the last few years substantial progress has been made in the characterization of the properties and pharmacological role of the imidazoline receptors. Perhaps the most important development was the identification of agmatine as a potential endogenous ligand for the imidazoline receptors.[1] Drugs having agonist activity at I_1 receptors are now in clinical use as antihypertensive agents, with central I_1 receptor activation postulated as their primary mechanism of action. Although the functional effects of I_2 receptor stimulation are less clearly understood, the recent availability of selective antagonists should facilitate the characterization of the physiological roles of this receptor.

On the other hand, it is still difficult to dissociate imidazoline receptors from α_2-adrenoceptors both pharmacologically and functionally. All I_1 receptor agonists, including agmatine, also have comparable affinity for the α_2-adrenoceptor, and no antagonist that selectively blocks I_1 receptors vis-à-vis α_2-adrenoceptors is available. Although there are subtle differences in the distribution of imidazoline receptors and α_2-adrenoceptors within the central nervous system, most central nuclei contain both receptor classes. In many cases, stimulation of I_1 receptors and α_2-adrenoceptors produces identical physiological responses.

Although practically no structural information is available on the imidazoline receptors, the significant number of structurally distinct agonists and antagonists with affinity for both imidazoline receptors and α_2-adrenoceptors suggests some degree of similarity at the molecular level between these two receptor classes. This assumption is supported by the identical functional role of imidazoline receptors and α_2-adrenoceptors in many organ systems.

In this report we provide a brief update on imidazoline receptor characterization and subclassification, compare the structural features required for activation of imidazoline receptors and α_2-adrenoceptors, and contrast the pharmacological profiles of agents having selectivity between the two receptor classes.

SUBCLASSIFICATION AND NOMENCLATURE OF IMIDAZOLINE RECEPTORS

The historical background of the experiments leading to the recognition of discrete imidazoline receptors was reviewed previously.[2] We briefly summarize the current

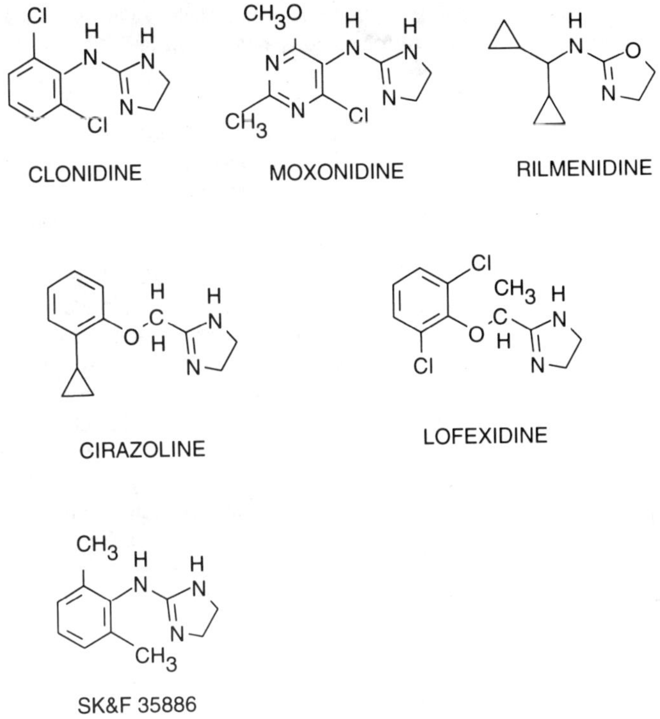

FIGURE 1. Imidazolines and oxazolines having affinity for imidazoline receptors and α_2-adrenoceptors.

consensus regarding the pharmacological subclassification of imidazoline receptors in order to define heterogeneity in imidazoline receptors and to differentiate these receptors from α_2-adrenoceptors.

Nonadrenergic binding sites recognizing molecules that contain an imidazoline ring or structurally related moiety have been referred to by a variety of names, including imidazole guanidinium receptor sites (IGRS), imidazoline preferring receptors (IPR), nonadrenergic idazoxan binding sites (NAIBS), idazoxan receptors, and imidazoline receptors. The nomenclature for these "receptors" is now becoming standardized, with sites identified by [^{3}H] *p*-aminoclonidine being designated as I_1 receptors and those identified by [^{3}H] idazoxan as I_2 receptors.[3]

I_1 receptors have high affinity for several α_2-adrenoceptor agonists of the imidazoline class including clonidine, rilmenidine, tizanidine, and moxonidine (FIG. 1). Several imidazoles that do not have affinity for α_2-adrenoceptors, including imidazole-4-acetic acid and cimetidine, can inhibit binding to I_1 receptors at relatively high concentrations.[4] Functional studies have shown that *p*-aminoclonidine, clonidine, rilmenidine, and moxonidine can act as agonists at the I_1 receptor site.[4,5] Idazoxan is also capable of inhibiting [^{3}H] *p*-aminoclonidine binding to I_1 receptors, and it was shown to be an I_1 receptor antagonist in functional experiments.

The α_2-adrenoceptor agonists of the guanidine class, such as guanabenz and guanfacine, have very low affinity for some I_1 receptors,[6] and they do not activate central I_1 receptors to lower blood pressure.[4] However, the putative endogenous I receptor agonist agmatine will inhibit [^{3}H] *p*-aminoclonidine binding to I_1 receptors, albeit at relatively high concentrations (approximately 1 μM).[1]

Subtypes of the I_1 receptor may possibly exist and perhaps are related to species differences. In membranes prepared from bovine ventrolateral medulla, the aralkylguanidine guanabenz (FIG. 2) has virtually no ability to inhibit [^{3}H] *p*-aminoclonidine binding.[4,6] However, in other tissues, such as guinea pig kidney,[7] guanabenz is at least equipotent with clonidine in displacing [^{3}H] *p*-aminoclonidine.

The binding of [^{3}H] idazoxan to I_2 receptors can be inhibited by cirazoline, naphazoline, UK 14,304, and guanabenz. Several other agents show variable affinity against this radioligand, depending on the tissue studied. This had led to proposals for further subclassification of the I_2 receptor, based on the affinity for amiloride or clonidine or on the stereoselectivity of medetomidine.[8–10] Although it was initially proposed that these differences in affinity were the result of species differences,[11] the finding of two pharmacologically distinct I_2 sites in the guinea pig[10] suggested that true receptor subtypes may exist. However, these data are complicated by the discovery of ligand-dependent interconversion of different affinity states for the I_2 receptor.[7,12] It now appears that subtypes of the I_2 receptor do exist, but their pharmacological profiles have yet to be accurately defined.

At least a portion of the I_2 binding sites may be located intracellularly, primarily on the mitochondrial membrane.[13–15] This finding may be consistent with the proposed functional effects of I_2 receptor activation, most of which can be related to regulation of intracellular ion transport.[15–17]

Several other radioligands are reported to bind to both α_2-adrenoceptors and additional nonadrenergic sites suggested to be imidazoline receptors. These include [^{3}H] rilmenidine,[18,19] [^{3}H] atipamezole,[20] and [^{3}H] cirazoline.[21] The sites labeled by these ligands appear to have pharmacological characteristics that are distinct from those defined for either I_1 or I_2 receptors. Although the I_1 receptors of bovine brain stem appear to have similar characteristics when labeled by [^{3}H] clonidine, [^{3}H] *p*-aminoclonidine, or [^{125}I] *p*-iodoclonidine,[6] a recent study suggests that [^{3}H] clonidine can identity an imidazoline receptor in human medulla having different characteristics.[22] Some of the differences observed between laboratories studying imidazoline receptors may result from different methods used to dissect the imidazoline receptor binding sites from the binding of these ligands to the α_2-adrenoceptors also present in the same tissues.

STRUCTURAL FEATURES OF AGONISTS AT IMIDAZOLINE RECEPTOR AND α_2-ADRENOCEPTOR SITES

The structural requirements for activation of α_2-adrenoceptors have been studied in much more detail than have those for activation of the imidazoline receptor.[23] The primary distinction is that the phenethylamines, including endogenous catecholamine neurotransmitters, are α_2-adrenoceptor agonists, but they are virtually inactive at I_1 receptor sites. The inability of phenethylamines to activate I_2 receptors is supported

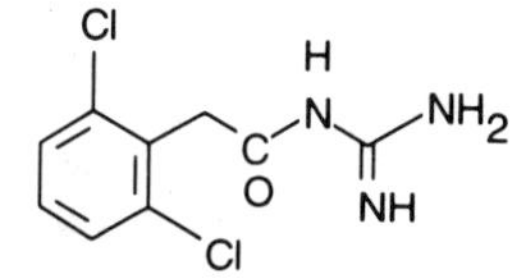

GUANFACINE

EC_{50} α_2 (ileum)= 10 nM

EC_{50} α_2 (atrium)= 91 nM

K_i α_2 (brain)= 2 nM

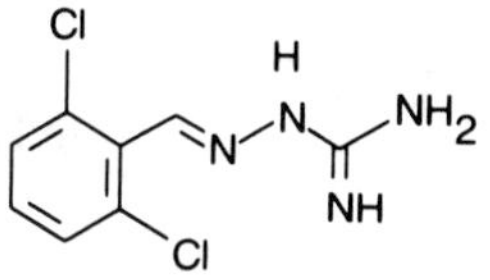

GUANABENZ

EC_{50} α_2 (ileum) = 14 nM

EC_{50} α_2 (atrium)= 100 nM

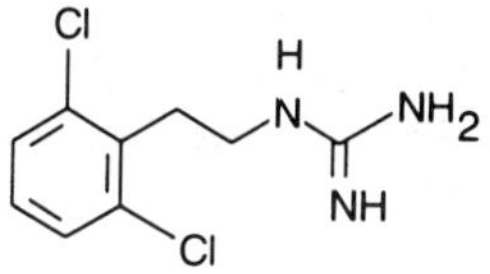

SK&F 70699

EC_{50} α_2 (ileum)= 95 nM

EC_{50} α_2 (atrium)= 190 nM

K_i α_2 (brain)= 4 nM

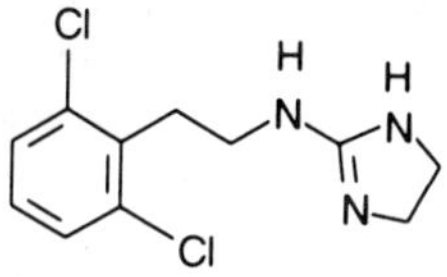

SK&F 108767

EC_{50} α_2 (atrium) = 10,000 nM

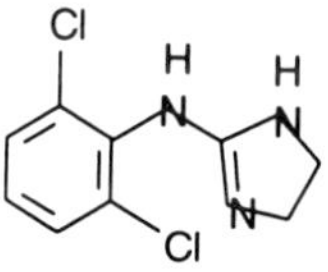

CLONIDINE

EC_{50} α_2 (ileum) = 8.5 nM

EC_{50} α_2 (atrium)= 15 nM

K_i α_2 (brain) = 0.9 nM

FIGURE 2. α_2-Adrenoceptor agonist activity of aralkylguanidines and imidazolines. α_2-Adrenoceptor agonist activity determined as inhibition of short circuit current in rat ileum[71] or inhibition of the inotropic response induced by field stimulation.[54]

primarily by their very low affinity for these sites in radioligand binding assays, inasmuch as the functional responses to I_2 receptor stimulation have been less clearly defined. In contrast, the histamine receptor antagonist cimetidine has much higher affinity for I_1 receptors vis-à-vis α_2-adrenoceptors and can activate central I_1 receptors, thereby reducing blood pressure and heart rate following local administration to the rostral ventrolateral medulla (RVL) in the anesthetized rat.[4] It has not been established if the affinity of cimetidine for the I_1 receptor results from the imidazole ring or the cyanoguanidine moiety. Cimetidine has very low affinity for the I_2 receptor.[10]

Most other compounds found to be I_1 receptor agonists can also activate the α_2-adrenoceptor. These mixed α_2/I_1 receptor agonists fall into four structural classes, the iminoimidazolidines, 2-aralkylimidazolines, oxazolines, and guanidines. Although these classes bear some structural similarity, clearly guanidines cannot be considered open-chain analogs of the iminoimidazolidines, because ring-opened or oxazoline analogs of clonidine have at least a 100-fold lower affinity for the α_2-adrenoceptor,[24] and SK&F 108767, the cyclized analog of the guanidine SK&F 70699, loses all activity at the α_2-adrenoceptor (FIG. 2).

The activity of SK&F 70699 or SK&F 108767 as an I_1 receptor agonist has not been determined, but other aralkylguanidines, such as guanfacine and guanabenz, do not activate this receptor[4] and have low affinity as inhibitors of [^{3}H] *p*-aminoclonidine or [H^3] clonidine binding in bovine ventrolateral medulla.[4,6] Indeed, the only known guanidine having affinity for the I_1 receptor in this tissue is its putative endogenous agonist agmatine. The affinity of agmatine for I_1 receptors in bovine medulla (K_i = 700 nM versus [^{3}H] *p*-aminoclonidine)[1] is low compared to that for the imidazoline and oxazoline agonists shown in FIGURE 1 (K_i < 10 nM).

Imidazolines of diverse structure have high affinity for I_1 and I_2 receptors, and based on more limited functional data, they appear able to activate these receptors. Clonidine and oxymetazoline produce nearly equal I_1 receptor-mediated reductions in blood pressure and heart rate when administered to the RVL of anesthetized rats.[4] Moxonidine produces a similar effect,[25] but it appears to have greater selectivity than does clonidine for I_1 receptors versus α_2-adrenoceptors.[6]

Cirazoline and lofexidine differ from other imidazolines in having an ether linkage between aryl and imidazoline rings. Both of these compounds have high affinity for I_1 sites.[6] This structural modification may favor interactions with I_2 receptors, because lofexidine has higher affinity than does clonidine at both of the I_2 receptor subtypes present in the guinea pig,[10] and cirazoline is generally found to be the most potent inhibitor of [^{3}H] idazoxan binding of all imidazolines tested,[11] with selectivity for I_2 vis-à-vis I_1 receptors.[25] Lofexidine is postulated to be an I_1 receptor agonist on the basis of its ability to produce centrally mediated antihypertensive action. Cirazoline stimulates catecholamine release from bovine adrenal chromaffin cells.[26] This effect was postulated to result from I_2 receptor activation, but as will be detailed, the nature of the imidazoline receptor subtype present in this tissue is controversial.

Idazoxan has been shown to be an I_1 receptor antagonist on the basis of its ability to block the sympatholytic effect of clonidine[4] or rilmenidine.[27] Based on the effects of guanine nucleotides on [^{3}H] idazoxan binding, it has been postulated that idazoxan may have partial agonist activity at the I_2 receptor site.[28] Phentolamine, a competitive antagonist of α_2-adrenoceptors, may act as an I_2 receptor agonist.[29] Naphazoline, although having moderate affinity for I_1 sites, failed to induce a sympatholytic effect when administered to the rat RVL.[4] Although functional studies have not been performed, it is possible that naphazoline, which, like idazoxan, bears a polycyclic ring system attached to the imidazoline ring, may also have affinity, but not efficacy, at the I_1 receptor.

A novel oxazoline, rilmenidine stimulates central I_1 receptors.[25,30] Like moxonidine, rilmenidine shows significant selectivity for I_1 receptors versus α_2-adrenoceptors.[6]

RESPONSES MEDIATED BY α_2-ADRENOCEPTOR AND IMIDAZOLINE RECEPTOR ACTIVATION

As just noted, many compounds have agonist activity at both α_2-adrenoceptor and imidazoline receptor sites. Likewise, in many cases, stimulation of these two receptor classes produces identical functional responses. The primary example is the centrally mediated sympatholytic effect induced by α_2- and I_1 receptor agonists. It is well established that central α_2-adrenoceptor activation using a wide variety of agonists including phenethylamines, azepines, and imidazolines results in inhibition of sympathetic outflow and a consequent decrease in blood pressure. However, Ernsberger *et al.*[4] demonstrated that the ability of a series of compounds to lower blood pressure and heart rate when administered to the RVL of anesthetized rats correlated well with their affinities for the I_1 receptor, but not with their affinities for α_2-adrenoceptors. Furthermore, the hypotensive action of clonidine was antagonized by idazoxan, a mixed I_1/α_2-adrenoceptor antagonist, but not by SK&F 86466, an α_2-adrenoceptor antagonist with no affinity for I_1 receptors. Similar results were obtained for rilmenidine.[27]

When clonidine is administered to spontaneously hypertensive rats via intravenous injection, however, the antihypertensive action appears to be mediated by α_2-adrenoceptors, inasmuch as SK&F 86466 and idazoxan are both effective antagonists.[31] Furthermore, guanabenz, which is ineffective when administered directly to the RVL,[4] produces an equivalent antihypertensive effect to that of clonidine, which is also sensitive to blockade by SK&F 86466 and idazoxan when administered intravenously. However, intracisternal administration of yohimbine was unable to block the hypotensive action of intravenous clonidine in the anesthetized normotensive rat at a dose capable of blocking clonidine-induced metabolic effects in the locus coeruleus.[32]

Intravenous administration of rilmenidine to conscious rabbits produced a sympatholytic effect, which was evaluated by measuring the firing rate of renal sympathetic nerves.[33] This action was sensitive to blockade by low doses of yohimbine, a non-imidazoline α_2-adrenoceptor antagonist having no affinity for the I_1 receptor,[34,35] and RX 821002, a methoxy analog of idazoxan that is relatively ineffective against the hypotensive action of intracisternal rilmenidine in the rabbit.[36]

Hence, central I_1 receptors and α_2-adrenoceptors both mediate sympatholytic action, and their relative contributions to the effect of a particular agonist may depend on the route of administration. This is consistent with the premise that the sympatholytic effect of phenethylamine and imidazoline agonists involves different sites of action within the central nervous system.[2]

Several other responses may be mediated by both α-adrenoceptors and imidazoline receptors. The vasoconstrictor action of 4-aminotolazoline in the rat aorta is mediated, as with most other imidazolines, via α-adrenoceptor activation.[37] However, the constrictor action of 4-isothiocyanatotolazoline in this tissue may be mediated by an imidazoline receptor, because the response can be blocked by idazoxan, but not by phentolamine, prazosin, or SK&F 104078. These antagonists have shown very low affinity for I_2 sites.[7,38] 4-Isothiocyanatotolazoline was a relatively potent inhibitor of [^{3}H] idazoxan binding to I_2 receptors in porcine renal cortex.[39] The affinity of this potential agonist and most of the antagonists at the I_1 receptor site have not been reported, preventing the subtype assignment of this response.

α_2-Adrenoceptor stimulation is known to inhibit stimulation-evoked neurotransmitter release from sympathetic nerve terminals. It has been suggested that an imidazoline receptor may also mediate this effect on the basis of the ability of certain imidazolines to inhibit stimulation-induced [^{3}H] norepinephrine overflow in rabbit blood vessels after treatment of the tissues with rauwolscine or phenoxybenzamine to eliminate an α_2-adrenoceptor-mediated effect.[40,41] Several imidazolines were found to be prejunctional α_2-adrenoceptor antagonists before treatment with rauwolscine, and they became apparent imidazoline receptor agonists following α_2-adrenoceptor blockade. The relatively low potency of rauwolscine as an antagonist of the neuroinhibitory effect of clonidine and other imidazolines was also used to suggest an imidazoline receptor-mediated effect on neurotransmission.[41] However, the pharmacological characteristics of this putative receptor do not correspond with those of either I_1 or I_2 receptors, because it does not appear to be activated by moxonidine, and the potency of rauwolscine against the neuroinhibitory effect of norepinephrine is also unexpectedly low.

Stimulation of both α_2-adrenoceptors and imidazoline receptors may be involved in inhibition of the nonadrenergic noncholinergic twitch response of rabbit iris to field stimulation. The guanidine agonist aganodine, which was an agonist at the putative prejunctional imidazoline receptor site in rabbit pulmonary artery,[41] produced rauwolscine-insensitive inhibition of neurotransmission in this tissue.[42] Oxymetazoline had a similar inhibitory effect, but in this case rauwolscine was a potent antagonist. The affinity of aganodine for I_1 or I_2 receptors has not been reported.

An I_1 receptor-mediated component to the lowering of intraocular pressure by UK 14,304[43] and moxonidine[44] has been reported. This represents yet another example in which α_2-adrenoceptor and I_1 receptor stimulation may result in the same pharmacological response; however, there are multiple mechanisms for controlling intraocular pressure, and individual mechanisms may be selectively influenced by α_2-adrenoceptor and I_1 receptor activation.[44]

In other cases, imidazoline receptor activation may produce a distinct response from that produced by activation of the α_2-adrenoceptor. Both I_1 and I_2 receptors can be detected in the kidney via radioligand binding assays.[45] Several imidazolines, including cirazoline, idazoxan, and UK 14,304, produced a small, but statistically significant inhibition of $^{22}Na^+$ influx and Na^+ dependent re-alkalization upon acidification in isolated proximal tubular cells.[17,46] In this system, the effects of these imidazolines are opposite those observed with norepinephrine, which increased ^{22}Na influx. These data suggest that I_2 receptors and α_2-adrenoceptors on these cells mediate opposite actions[17] and that both cirazoline and idazoxan are agonists at this I_2 receptor. Idazoxan produced an increase in urine output in both normally hydrated and water-loaded rats; this effect was not observed with structurally related compounds not having affinity for the I_2 receptor, suggesting that this receptor may be involved.[47] However, α_2-adrenoceptor agonists are known to increase urine output in the rat,[48,49] and idazoxan is known to have partial agonist activity at the α_2-adrenoceptor site in other systems.[50] Therefore, an α_2-adrenoceptor-mediated mechanism for this action of idazoxan remains a possibility.

In the anesthetized rat, intrarenal infusion of moxonidine, an agonist with selectivity for I_1 versus α_2-adrenoceptors[6] and no affinity for the I_2 receptor,[51] produced an increase in osmolar clearance, with no evidence for the α_2-adrenoceptor-mediated effects on free-water clearance observed with clonidine.[52] This effect was attenuated by idazoxan, but not by rauwolscine, suggesting involvement of the I_1 receptor. Similar results were observed with 2-(2,6-dimethylphenyl)imino-imidazolidine (SK&F 35886).[53] SK&F 35886 is reported to have high affinity for the I_1 receptor,[52] but its I_1 versus α_2-selectivity relative to clonidine is not known. However, both SK&F 35886[54] and moxonidine[55] have greater intrinsic activity at the α_1-adrenoceptor site than does clonidine, and the possibility of an α_1-adrenoceptor contribution to the differences in renal actions of these agonists has not been eliminated.

It has been suggested that cultured bovine adrenal chromaffin cells would represent a useful system for the study of imidazoline receptors, because these cells do not appear to possess α_2-adrenoceptors.[15] However, data reported in this system do not present a consistent profile. Some studies[6,56] demonstrate the presence of specific [^{3}H] *p*-aminoclonidine binding to these cells, with a profile consistent with that expected for an I_1 receptor. However, others[15,26] cannot detect specific binding of this radioligand, but do find specific sites for [^{3}H] idazoxan, having a pharmacological profile that is consistent with that of the I_2 receptor subtype. Diverse functional effects are also produced by imidazoline receptor activation in this system. Clonidine and other imidazolines are reported to inhibit the catecholamine release induced by nicotinic receptor activation.[29,57,58] On the other hand, using an apparently identical primary cell culture preparation, compounds inhibiting the binding of [^{3}H] idazoxan stimulated the basal release of catecholamines[26]; the endogenous imidazoline receptor agonist agmatine produced a similar increase in basal catecholamine release,[1] as does a partially purified bovine preparation presumably containing agmatine as its active pharmacologic principle.[59] It is possible that imidazoline receptors mediate both the enhancement of basal catecholamine release and the inhibition of stimulated release, although clonidine (1 μM) had no effect on basal catecholamine release.[57]

SELECTIVE BLOCKADE OF α_2-ADRENOCEPTORS AND IMIDAZOLINE RECEPTORS

Characterization of the responses to imidazoline receptor activation and the differentiation of these responses from those resulting from α_2-adrenoceptor activation require selective antagonists. Some agents showing substantial selectivity were recently identified.

Idazoxan is the only compound that has demonstrated functional I_1 receptor antagonist activity. This compound has affinity for both I_1 and I_2 receptors and α_2-adrenoceptors.[25] Efaroxan, which is structurally similar to idazoxan (FIG. 3), has even higher I_1 receptor affinity than does idazoxan,[25] with over 1,000-fold lower affinity

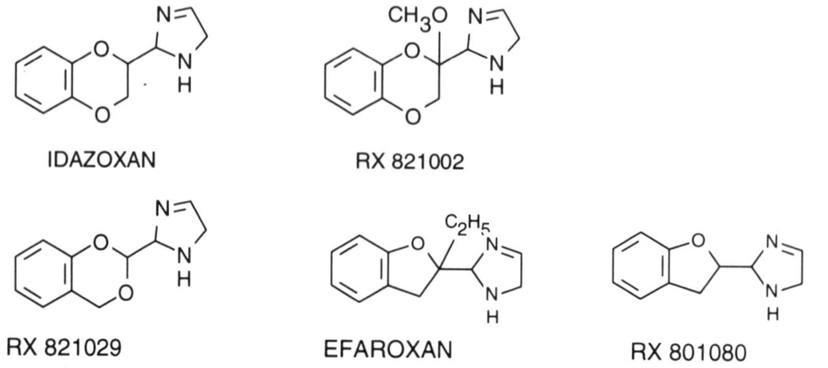

FIGURE 3. Analogs of idazoxan having selectivity for I_2 receptors or α_2-adrenoceptors.

for the I_2 receptor.[60] Efaroxan retains affinity for the α_2-adrenoceptor comparable to that of idazoxan.[60] The activity of efaroxan as a functional antagonist of I_1 receptor-mediated actions has not been reported. Although cimetidine has moderate affinity for the I_1 receptor, with essentially no affinity for either the I_2 receptor or α_2-adrenoceptors,[25] it is not clear if this agent is an agonist or antagonist at the I_1 receptor site. Central administration of cimetidine lowers blood pressure in the anesthetized rat,[4] but it has not been reported if this response is sensitive to blockade by idazoxan.

Agents such as SK&F 86466 or yohimbine block central α_2-adrenoceptors without attenuating the hypotensive action induced by I_1 receptor activation.[4,27]

Structure-activity relationships for selective interaction between I_2 receptors and α_2-adrenoceptors have been studied more extensively. However, it is not clear if idazoxan and its structural analogs are antagonists at the I_2 receptor. In one of the only responses attributable to an I_2 receptor interaction, inhibition of ^{22}Na influx into renal tubular cells, idazoxan and cirazoline produced similar effects. These data could be used as evidence for an agonist action of both idazoxan and cirazoline at the I_2 receptor site.

Minor structural changes can dramatically influence the I_2 receptor versus α_2-adrenoceptor selectivity of idazoxan analogs. Introduction of a methoxy substituent on the benzodioxan ring bearing the imidazoline ring abolishes I_2 receptor affinity while enhancing α_2-adrenoceptor affinity. This analog, RX 821002, is now widely used as a selective α_2-adrenoceptor antagonist and as a tritiated radioligand for identification of α_2-adrenoceptors.[38] Substitution at this position is also responsible for the low affinity of efaroxan for the I_2 receptor, because the corresponding benzofuran lacking the ethyl substituent (RX 801080) has high affinity for both I_2 receptors and α_2-adrenoceptors.[60]

Interestingly, methylation of one of the imidazoline nitrogen atoms reduces, but does not eliminate, affinity for the I_2 receptor.[60] This analog has no affinity for the α_2-adrenoceptor and represents the first example of an agent having selective affinity for the I_2 receptor. Another compound having selective affinity for the I_2 receptor is the 1,3-benzodioxanyl analog of idazoxan (RX 821029) which has 3 to 5-fold higher affinity for I_2 receptors, with at least a 10-fold lower affinity for the α_2-adrenoceptor.[61]

Structural modifications of cirazoline can also influence I_2 receptor versus α_2-adrenoceptor selectivity. Removal of the *ortho* cyclopropyl substituent, or its replacement by ethyl, did not substantially influence affinity for either receptor class, but potency as an α_1-adrenoceptor agonist was reduced nearly 100-fold.[62] The 3-amino derivative of cirazoline has high affinity for the I_2 receptor and has been used to prepare photoaffinity ligands for labeling of I_2 receptor proteins.[63] Introduction of a 5-methoxy substituent for the 2-ethyl analog resulted in a compound with enhanced selectivity, with at least 100-fold lower affinity for the α_2-adrenoceptor than for the I_2 receptor.[62]

Interestingly, methylation of the carbon atom bridging the phenyl and imidazoline rings totally abolished the ability to inhibit the binding of [^{3}H] idazoxan to rabbit kidney membranes.[62] However, lofexidine, a close analog differing only by the presence of chlorine atoms at positions 2 and 6 of the aromatic ring, has high affinity as an inhibitor of [^{3}H] idazoxan binding to guinea pig kidney, cortex, or ileum[10,45] or to the I_2-like sites identified by [^{3}H] rilmenidine in rat cortex.[18]

LIGAND-RECEPTOR INTERACTIONS FOR α_2-ADRENOCEPTORS AND IMIDAZOLINE RECEPTORS

The molecular structure of the α_2-adrenoceptor is well known, and recombinant receptors from several species are now available. Based on their pharmacological characteristics, four distinct α_2-adrenoceptor subtypes have been identified and designated as α_{2A}, α_{2B}, α_{2C}, and α_{2D}, with the α_{2A}- and α_{2D}-adrenoceptors likely to be species variants.[64] Nevertheless, in contrast to the large amount of data available for the β-adrenoceptor, the actual mode of interaction between agonists and antagonists and the α_2-adrenoceptor has not been studied extensively. As observed for the β-adrenoceptor, the nitrogen atom of the catecholamine agonists appears to interact with Asp^{113} of the α_{2A}-adrenoceptor, located on transmembrane helix III.[65] The aromatic hydroxyl groups may interact with Cys^{201} and Ser^{204} on transmembrane helix V.[65] Recent evidence suggests a stereoselective interaction between norepinephrine and epinephrine with Ser^{90} on transmembrane helix II, suggesting that this residue is involved in the binding of the β-hydroxyl group.[66]

Imidazolines with high affinity for the α_2-adrenoceptor, such as clonidine, have none of the structural elements required for agonist activity in the phenethylamine series. Although catechol substitution on the aromatic ring results in potent and efficacious agonists, it is not required, and introduction of a hydroxyl group to the carbon atom bridging the aromatic and imidazoline rings reduces the affinity of catecholimidazolines for the α-adrenoceptor.[67] Hence it is likely that these two chemical classes interact with the α_2-adrenoceptor in a different orientation. In contrast to the catecholamine neurotransmitters, mutation of Ser^{90} to alanine does not substantially change the affinity of the recombinant α_{2A}-adrenoceptor for either clonidine, tolazoline, or either enantiomer of α-hydroxytolazoline.[66]

None of the imidazoline receptors has yet been cloned, but several proteins with pharmacological characteristics consistent with those of the I_2 receptor have been isolated[68,69] or labeled via a photoaffinity ligand.[63] From these biochemical studies it is clear that the proteins isolated or labeled are not α_2-adrenoceptors. Based on the binding of [^{3}H] rilmenidine to rat brain and kidney, molecular modeling calculations have proposed a model of the imidazoline receptor which correlates with the ability of a series of drugs to inhibit the binding of this radioligand.[19]

[^{3}H]Clonidine binds to both I_1 receptors and α_2-adrenoceptors in bovine rostral ventrolateral medulla. Selective attack on arginine residues by phenylglyoxal inhibited the binding of clonidine to I_1 receptors, but not to α_2-adrenoceptors, whereas attack on cysteine residues with *N*-ethylmaleimide inhibited binding to α_2-adrenoceptors but not to I_1 receptors.[70] This is consistent with the primary sequence of the α_2-adrenoceptors, which have no conserved arginine residues in locations likely to interact with an extracellular ligand.

CONCLUSIONS

Although there is convincing evidence that imidazoline receptors and α_2-adrenoceptors represent distinct entities, it is clear that they share several common features. This finding is based both on common effects mediated by both receptor classes and on the identification of many agents showing high affinity for both receptor classes. The latter is probably related to the fact that nearly all agents that have been tested to date for imidazoline receptor affinity were initially prepared as α-adrenoceptor agonists or antagonists or were derived from such agents by relatively minor structural modification. No large scale effort to identify, de novo, agents selectively interacting with imidazoline receptors has been reported. Nevertheless, the ability of several structural classes, including imidazolines, imidazoles, and guanidines, to bind to both imidazoline receptors and α_2-adrenoceptors suggests a similar relative orientation of important binding elements between the receptors.

The fact that minor structural modifications can alter the ratios between I_1 receptor, I_2 receptor, and α_2-adrenoceptor affinity would suggest a relatively hindered ligand binding pocket. Like the α_2-adrenoceptor, the I_1 receptor appears to be membrane bound and may be coupled to a guanine nucleotide regulatory protein. Hence it is reasonable to assume that it will have the seven transmembrane spanning regions postulated for many other neurotransmitter and hormone receptors. Molecular characterization of ligand interactions with imidazoline receptors will not be possible until these receptors have been cloned and expressed. In the meantime, studies on the interactions of imidazoline and guanidine ligands with the α_2-adrenoceptor can provide useful background data for comparison, because it is likely that these structures will interact with both membrane-bound receptors in a similar manner.

The most widely characterized action of the I_1 receptor, centrally mediated inhibition of sympathetic outflow, is also produced by central α_2-adrenoceptor activation, but probably at a different site. Dissociation of this sympatholytic effect from that of α_2-adrenoceptor stimulation would be facilitated by the identification of an I_1 receptor antagonist lacking significant affinity for the α_2-adrenoceptor. Such an antagonist might also allow identification of other pharmacologically important responses mediated by I_1 receptor activation.

SUMMARY

Although it is now well established that imidazoline receptors and α_2-adrenoceptors are discrete entities with distinct endogenous ligands, the two receptor classes apparently have several common features. While the catecholamines stimulate α_2-adrenoceptors but not imidazoline receptors, agmatine, a guanidine analog that may be an endogenous imidazoline receptor ligand, can interact with both I_1 and I_2 imidazoline receptors as well as α_2-adrenoceptors, although, interestingly, other guanidines such as guanabenz are highly selective for α_2-adrenoceptors versus I_1 receptors. Most I_1 receptor agonists such as moxonidine, rilmenidine, and clonidine can also stimulate α_2-adrenoceptors, and the same physiological response is produced by activation of central I_1 receptors and α_2-adrenoceptors, but their anatomical locations differ. The imidazoline idazoxan is an antagonist at I_1, I_2, and α_2-receptors, but minor structural alterations of idazoxan can result in molecules with selectivity for either α_2-adrenoceptors or imidazoline receptors. The precise mode of interaction of imidazoline agonists and antagonists with the α_2-adrenoceptor is not yet understood, and structures of the imidazoline receptors are still unknown. Nevertheless, the fact that many agents can stimulate or block both receptor classes, combined with the fact that α_2-adrenoceptors and I_1 receptors can mediate identical physiological responses, suggests that many common structural features may be present.

REFERENCES

1. Li, G., S. Regunathan, C. J. Barrow, J. Eshraghi, R. Cooper & D. J. Reis. 1994. Science **263:** 966-969.
2. Hieble, J. P. & R. R. Ruffolo, Jr. 1992. Fundam. Clin. Pharmacol. **6** (Suppl. 1): 7s-13s.
3. Michel, M. C. & P. Ernsberger. 1992. Trends Pharmacol. Sci. **13:** 369-370.
4. Ernsberger, P., R. Giuliano, R. N. Willette & D. J. Reis. 1990. J. Pharmacol. Exp. Ther. **253:** 408-418.
5. Smyth, D. D., P. Li, D. E. Blandford & S. B. Penner. 1992. J. Pharmacol. Exp. Ther. **262:** 1080-1086.
6. Ernsberger, P., T. H. Damon, L. M. Graff, S. G. Schaffer & M. O. Christen. 1993. J. Pharmacol. Exp. Ther. **264:** 172-182.
7. Wikberg, J. E. S., S. Uhlen & V. Chhajlani. 1992. Pharmacol. & Toxicol. **70:** 208-219.
8. Reis, D. J., S. Regunathan, H. Wang, D. L. Feinstein & M. P. Meeley. 1992. Fundam. Clin. Pharmacol. **6** (Suppl. 1): 23s-29s.
9. Michel, M. C. & P. A. Insel. 1989. Trends Pharmacol. Sci. **10:** 342-344.
10. Wikberg, J. E. S., S. Uhlen & V. Chhajlani. 1991. Eur. J. Pharmacol. **193:** 335-340.
11. Kilpatrick, A. T., C. C. Brown & A. C. Mackinnon. 1992. Biochem. Soc. Trans. **20:** 113-118.
12. Rangel, N., S. Uhlen & J. E. S. Wikberg. 1993. Pharmacol. & Toxicol. **73:** 86-90.
13. Michel, M. C., J. W. Regan, M. A. Gerhardt, R. R. Neubig, P. A. Insel & H. J. Motulsky. 1990. Mol. Pharmacol. **37:** 65-68.
14. Tesson, F., C. Prip-Buus, A. Lemoine, J. P. Pegorier & A. Parini. 1991. J. Biol. Chem. **266:** 155-160.
15. Regunathan, S., M. P. Meeley & J. Reis. 1993. Biochem. Pharmacol. **48:** 1667-1675.
16. Regunathan, S., M. J. Evinger, M. P. Meeley & D. J. Reis. 1991. Biochem. Pharmacol. **42:** 2011-2018.
17. Coupry, I., I. Limon, F. Tesson, V. Lachaud, C. Gargalidis-Moudanos & A. Parini. 1992. Therapie **47:** 519-524.
18. King, P. R., A. L. Gundlach, B. Jarrott & W. J. Louis. 1992. Eur. J. Pharmacol. **218:** 101-108.

19. King, P. R., S. Suzuki, T. L. Nero, A. L. Gundlach & W. J. Louis. 1993. J. Hypertension **11** (Suppl. 5): S160-S161.
20. Sjoholm, B., J.-M. Savola & M. Scheinin. 1992. Fundam. Clin. Pharmacol. **6** (Suppl. 1): 56s.
21. Angel, I., H. Schoemaker, J. Allen, M. Le Rouzic, C. Pimoule, S. Arbilla & S. Z. Langer. 1992. Fundam. Clin. Pharmacol. **6** (Suppl. 1): 54s.
22. Bricca, G., H. Greney, J. Zhang, M. Dontenwill, J. Stutzmann, A. Belcourt & P. Bousquet. 1994. Eur. J. Pharmacol. **266:** 25-33.
23. Nichols, A. J. & R. R. Ruffolo, Jr. 1991. *In* α-Adrenoceptors: Molecular Biology, Biochemistry and Pharmacology. Progress in Basic Clinical Pharmacology. **8:** 75-114. Karger. Basel.
24. Jarrott, B., R. J. Summers, A. J. Culvenor & W. J. Louis. 1980. Circ. Res. **46** (Suppl. I): I15-I20.
25. Ernsberger, P. R., K. L. Westbrooks, M. O. Christen & S. G. Schafer. 1992. J. Cardiovasc. Pharmacol. **20** (Suppl. 4): S1-S10.
26. Regunathan, S., M. P. Meeley & D. J. Reis. 1991. Soc. Neurosci. Abstr. **17:** 355.
27. Gomez, R. E., P. Ernsberger, G. Feinland & D. J. Reis. 1991. Eur. J. Pharmacol. **195:** 181-191
28. Hussain, J. F., D. A. Kendall & V. G. Wilson. 1993. Br. J. Pharmacol. **109:** 831-837.
29. Ohara-Imaizumi, M. & K. Kumakura. 1992. Cell. & Mol. Neurobiol. **12:** 273-283.
30. Tibirica, E., J. Feldman, C. Mermet, L. Monassier, F. Gonon & P. Bousquet. 1991. Eur. J. Pharmacol. **209:** 213-221.
31. Hieble, J. P. & D. C. Kolpak. 1993. Br. J. Pharmacol. **110:** 1635-1639.
32. Tibirica, E., J. Feldman, C. Mermet, F. Gonon & P. Bousquet. 1990. J. Pharmacol. Exp. Ther. **256:** 606-613.
33. Szabo, B., R. Urban & K. Starke. 1993. N-S Arch. Pharmacol. **348:** 593-600.
34. Ernsberger, P., M. P. Meeley, J. J. Mann & D. J. Reis. 1986. Eur. J. Pharmacol. **134:** 1-13.
35. Bricca, G., M. Dontenwill, A. Molines, J. Feldman, A. Belcourt & P. Bousquet. 1989. Eur. J. Pharmacol. **162:** 1-9.
36. Sannajust, F., J. Rathner, M. Dubar & G. A. Head. 1992. Fund. Clin. Pharmacol. **6** (Suppl. 1): 53s.
37. Shams, G., B. V. Venkataraman, A. Hamada, D. D. Miller, P. N. Patil & D. R. Feller. 1991. Eur. J. Pharmacol. **199:** 315-323.
38. Langin, D., H. Paris & M. Lafontan. 1990. J. Pharmacol. Exp. Ther. **37:** 876-885.
39. Hussain, J. F., F. Miall, P. Patil, D. Miller, D. A. Kendall & V. G. Wilson. 1992. Fundam. Clin. Pharmacol. **6** (Suppl. 1): 54s.
40. Gothert, M. & G. J. Molderings. 1991. Naunyn-Schmiedebergs Arch. Pharmacol. **343:** 271-282.
41. Molderings, G. J., F. Hentrich & M. Gothert. 1991. Naunyn-Schmiedebergs Arch. Pharmacol. **344:** 630-638.
42. Fuder, H. & M. Selbach. 1993. Naunyn-Schmiedebergs Arch. Pharmacol. **347:** 394-401.
43. Burke, J., A. Kharlamb, T. Shan, E. Runde, E. Padillo & G. Manlapaz. 1992. Fundam. Clin. Pharmacol. **6** (Suppl. 1): 56s.
44. Ogidigben, M., T.-C. Chu & D. E. Potter. 1994. J. Pharmacol. Exp. Ther. **269:** 897-904.
45. Wikberg, J.E.S., N. Rangel, R. Muceniece, S. Uhlen & V. Chhajlani. 1992. Fundam. Clin. Pharmacol. **6** (Suppl. 1): 54s.
46. Bidet, M., P. Poujeol & A. Parini. Biochem. Biophys. Acta **1024:** 173-179.
47. Jackson, H. C., I. J. Griffin, S. D. Birkett & D. J. Nutt. 1992. Br. J. Pharmacol. **106:** 443-446.
48. Gellai, M. 1990. Am. Physiol. Soc. **259:** F1-F8.

49. Brooks, D. P., R. M. Edwards, P. D. DePalma, T. A. Fredrickson, J. P. Hieble & M. Gellai. 1991. J. Pharmacol. Exp. Ther. **259** (3): 1277-1282.
50. Limberger, N. & K. Starke. 1983. Naunyn-Schmiedebergs Arch. Pharmacol. **324:** 75-78.
51. Michel, M. C. & R. Schafers. 1992. J. Cardiovasc. Pharmacol. **20** (Suppl. 4): S24-S30.
52. Allan, D. R., S. B. Penner & D. D. Smyth. 1993. Br. J. Pharmacol. **108:** 870-875.
53. Smyth, D. D. & P. Li. 1991. Can. J. Physiol. Pharmacol. **69:** Axxxi.
54. Hieble, J. P. & R. G. Pendelton. 1979. Naunyn-Schmiedebergs Arch. Pharmacol. **309:** 217-224.
55. Armah, B. I. 1988. Arzneim-Forsch. Drug Res. **38** (II): 10.
56. Ernsberger, P., G. Feinland, M. J. Evinger, M. P. Meeley & D. J. Reis. 1989. Am. J. Hypertens. **2:** 93A.
57. Powis, D. A. & P. F. Baker. 1985. Mol. Pharmacol. **29:** 134-141.
58. Nguyen, T. T. & A. DeLean. 1987. Can. J. Physiol. Pharmacol. **65:** 823-827.
59. Regunathan, S., M. P. Meeley & J. Reis. 1991. Mol. Pharmacol. **40:** 884-888.
60. Langin, D., H. Paris, M. Dauzats & M. Lafontan. 1990. Eur. J. Pharmacol. (Mol. Pharmacol.) **188:** 261-272.
61. Hudson, A., N. Mallard, D. Nutt & C. Chapleo. 1992. Fundam. Clin. Pharmacol. **6** (Suppl. 1): 45s.
62. Ivkovic, B., V. Bakthavachalam, J. Neumeyer, S. Lanier, L. Brasili & M. Pigini. 1992. Fundam. Clin. Pharmacol. **6** (Suppl. 1): 47s.
63. Lanier, S. M., B. Ivkovic, I. Singh, J. L. Neumeyer & V. Bakthavachalam. 1993. J. Biol. Chem. **268:** 16047-16051.
64. Bylund, D. B., D. C. Eikenberg, J. P. Hieble, S. Z. Langer, R. J. Lefkowitz, K. P. Minneman, P. B. Molinoff, R. R. Ruffolo, Jr. & U. Trendelenburg. 1994. Pharmacol. Rev. **46:** 121-136.
65. Wang, C.-D., M. A. Buck & C. M. Fraser. 1991. Mol. Pharmacol. **40:** 168-179.
66. Li, Y.-O., J. P. Hieble, D. J. Bergsma, A. M. Swift, S. Ganguly & R. R. Ruffolo, Jr. 1994. Pharmacol. Comm., in press.
67. Ruffolo, R. R., Jr. & J. E. Waddell. 1983. J. Pharmacol. Exp. Ther. **224:** 559-566.
68. Limon, I., I. Coupry, S. M. Lanier & A. Parini. 1992. J. Biol. Chem. **267:** 21645-21649.
69. Wang, H., S. Regunathan, M. P. Meeley & D. J. Reis. 1992. J. Pharmacol. Exp. Ther. **42:** 792-801.
70. Shen, I. H. & P. Ernsberger. 1994. FASEB J. **8:** A355.
71. Hieble, J. P., G. P. McCafferty, D. Naselsky & W. E. Bondinell. 1990. Eur. J. Pharmacol. **183:** 1204-1205.

I_1-Imidazoline Receptors

Definition, Characterization, Distribution, and Transmembrane Signaling[a]

PAUL ERNSBERGER,[b] MARILYN E. GRAVES, LYNETTE M. GRAFF, NASSER ZAKIEH, PHONG NGUYEN, LAURA A. COLLINS, KELLY L. WESTBROOKS, AND GUINIVERE G. JOHNSON

Departments of Medicine, Pharmacology, and Neuroscience
Case Western Reserve School of Medicine
10900 Euclid Ave.
Cleveland, Ohio 44106-4982

Clonidine, an imidazolidine centrally acting antihypertensive agent, is considered the quintessential agonist for the α_2-adrenergic receptor. The imidazoline idazoxan is widely considered the α_2-antagonist of choice. Indeed, an accepted criterion for α_2-receptor-mediated actions is that they are elicited by clonidine and blocked by idazoxan or yohimbine.[1] Now open to question is the status of clonidine, idazoxan, and a host of other heterocyclic agonists and antagonists considered to be relatively specific for α-adrenergic or histamine receptors. Doubts that the diverse actions of imidazolines were entirely mediated by α-adrenergic receptors were raised early when clonidine was found to be a weak partial agonist or even an antagonist in many α_2-receptor systems. Furthermore, α-antagonists as well as α-agonists lower blood pressure when administered into brain stem ventricles.[2] For example, phentolamine, an imidazoline α-antagonist, elicits a centrally mediated fall in blood pressure by inhibiting sympathetic outflow.[2] The effect of clonidine is additive with phentolamine, ruling out an α_2 mechanism. Karppanen[3] noted interactions between clonidine and imidazole compounds such as cimetidine and imidazole-4-acetic acid and proposed that clonidine may act via brain stem "imidazole receptors."[3] The first direct evidence for a nonadrenergic action of imidazolines was obtained from microinjection directly into the rostral ventrolateral medulla (RVLM), the site of action for clonidine.[4] Cirazoline, an imidazoline α_2-antagonist, showed a clonidine-like vasodepressor action.[4] In contrast, α-methylnorepinephrine, a potent non-imidazoline α_2-agonist, had

[a] This work was supported by the National Institutes of Health (R29-HL44514 to P.E.). L. M. Graff and K. L. Westbrooks were partially supported by the Cornell Tradition undergraduate work-study program. P. Nguyen was supported in part by CWRU undergraduate work-study. G. G. Johnson was supported by the Department of Medicine First-Year Student Summer Research Program.

[b] Address for correspondence: Paul Ernsberger, PhD, Division of Hypertension, Case Western Reserve University, Cleveland, OH 44106-4982.

no effect. These findings led to the proposal that the action of clonidine within the RVLM was mediated not by α_2-adrenergic receptors but by a novel receptor specific for imidazolines.

The assumption that clonidine was specific for α_2-receptors was based in part on binding assays showing that [^{3}H]clonidine labeled α_2-receptors in brain membranes. However, [^{3}H]clonidine binding assays in brain had always used cerebral cortex membranes because of this brain region's very high density of α_2-adrenergic receptors. When [^{3}H]clonidine binding assays were conducted using membranes from the RVLM, the site of action of clonidine, it was found that [^{3}H]clonidine labeled not only α_2-adrenergic receptors but also a novel population of nonadrenergic binding sites that were insensitive to catecholamines and other adrenergic agents with a phenylethylamine structure and were specific for imidazoles and imidazolines. In cell membranes from the RVLM of bovine brain stem, [^{3}H]clonidine binds with nearly equal nanomolar affinities to both α_2-adrenergic and I_1-imidazoline sites.[5–9]

An endogenous ligand for imidazoline receptors was recently identified as agmatine, a precursor of the polyamine pathway in bacteria not previously known to be present in mammalian brain.[10] Agmatine inhibits [^{3}H]clonidine binding and is synthesized in brain by a novel enzyme, arginine decarboxylase, not previously known to exist in mammals. We recently were able to confirm and extend the initial radioligand binding results to establish that agmatine is about 30-fold selective for I_1 relative to α_2 receptors, in contrast to the sixfold selectivity reported by Li and colleagues[10] (data not shown).

In the present study, we seek to (1) distinguish the ligand selectivity profile of I_1-imidazoline sites relative to other recognition sites for imidazolines, (2) determine subcellular distribution and sensitivity to guanine nucleotides for I_1 sites, (3) determine the expression of I_1-imidazoline sites relative to α_2-adrenergic sites in various tissue and cell types and in specific regions of the brain stem, and (4) demonstrate the specific activation of a transmembrane signaling pathway by I_1-imidazoline receptors.

METHODS

Tissues and Cells. Bovine brains and adrenal glands were obtained from a local slaughterhouse and handled as previously described.[7–9] Human prostate samples were excess pathology specimens free from neoplasms and were processed by published procedures.[11] Rat pancreases were obtained from Pel-Freez (Rogers, Arkansas), and islets were isolated by collagenase digestion and settling at unit gravity. In a group of 9 cats anesthetized with pentobarbital (45 mg/kg) and ventilated with room air, 18 carotid bodies were dissected out and frozen in liquid nitrogen. Carotid bodies were also removed from 20 rabbits. No adrenergic drugs were administered to the animals at any time. PC12 pheochromocytoma cells were grown under 5% CO_2 in RPMI-1640 medium supplemented with 10% heat-inactivated horse serum (Gibco) and 5% fetal calf serum (Upstate Biotechnology) without antibiotics. Media were changed three times a week, and cells were passaged once a week. Nerve growth factor (7S; Upstate Biotechnology; 200 ng/ml) was added daily for 4 days before each experiment to induce neuronal differentiation. Published procedures were used for primary derived astrocytes[12] and mesangial cells,[13] NG108-15 neuroblastoma-glioma hybrid clonal cells,[14] freshly isolated primary rabbit tracheal epithelial cells,[15] and epithelial cells derived from proximal, distal, and collecting portions of the rabbit nephron.[16]

Membrane Isolation and [^{3}H]Clonidine and [^{125}I]PIC Binding Assays. Bovine brains and adrenal glands were processed as described elsewhere.[7–9] Kidney, prostate, and carotid body samples were processed by digestion and lysis in a hypotonic medium containing collagenase, EDTA, $ZnCl_2$, and protease inhibitors as recently described.[11] Kidney, prostate, and carotid body lysates and pancreatic islets were homogenized and crude membranes isolated and washed by the same method as that used for bovine brain. Isolated cells were treated similarly, except that initial homogenization was done with either a hand-held glass-glass homogenizer or a nitrogen cavitation device rather than a polytron. Binding assays of I_1 sites with [^{3}H]clonidine or [^{125}I]PIC were performed as described.[7–9] (See Ernsberger *et al.* [this volume] for details.)

Autoradiographic Studies with [^{125}I]PIC. Autoradiography was performed in the presence of vehicle alone, 10 μM epinephrine to block α_2-adrenergic sites, or 10 μM cimetidine to mask I_1-imidazoline sites, as earlier described.[9] (See Haxhiu *et al.* [this volume] for additional details.)

ELISA Assays for PGE_2. These assays were performed using a secondary antibody linked to acetylcholinesterase and were conducted according to the manufacturer's instructions (Cayman Chemical Co.). Culture medium was first concentrated on a C18 extraction column (Waters Sep-Pak) and eluted with methanol. In each experiment, one sample was spiked with 500 pg PGE_2 to determine recovery. All PGE_2 data were corrected for recovery.

DEFINITION OF I_1-IMIDAZOLINE BINDING SITES VERSUS OTHER RECOGNITION SITES FOR IMIDAZOLINES

We proposed the terminology of I_1- and I_2-imidazoline sites for the high-affinity nonadrenergic binding sites labeled by [^{3}H]clonidine and [^{3}H]idazoxan, respectively.[17–20] To test the hypothesis that imidazolines bind to multiple distinct recognition sites, we determined the affinities of 24 structurally diverse compounds at α_2-adrenergic and I_1-imidazoline sites and compared these values with I_2-imidazoline affinities from the literature. Each panel in FIGURE 1 compares affinities, expressed as the negative log of the K_i value (pK_i), between two different binding sites. The dashed line in each panel is the line of identity, which is where each point would fall if the two sites were identical in their affinities for the different ligands. In FIGURE 1A, there is no correlation between the affinity at α_2-adrenergic receptors in the RVLM and I_1-imidazoline affinity in the same tissue. Only one compound falls on the line of identity (1), which is clonidine itself. The non-imidazoline ligands (15-24) all fell above the diagonal, indicating selectivity for α_2-adrenergic receptors. Yohimbine (23) was the least α_2-selective of the non-imidazoline antagonists, showing only 30-fold lower affinity for I_1 than for α_2 binding sites. The imidazoles (12-14) fell below the diagonal, showing that these compounds bound preferentially to I_1. The imidazolidines (1-4) and the imidazolines (5-11) were scattered on both sides of the diagonal. There were no compounds in the lower right corner, which is the region where extremely selective high-affinity ligands for I_1 sites would be found. Seven compounds (3,7,9,10,12–14) and brain extracts containing putative endogenous ligand (25) all show a selectivity for I_1 over α_2 of between one and two log units, of which efaroxan (9) is the most potent antagonist and moxonidine (3) the most potent agonist.

A strong correlation was noted between I_1-imidazoline affinities determined in two different tissues, bovine RVLM and rat kidney (FIG. 1B). Data were obtained from both renal cortex and renal medulla using [^{3}H]clonidine or [^{3}H]*p*-aminoclonidine, with indistinguishable results (compare closed and open symbols). The high correlation coefficient (0.94) and slope (0.74) imply that differences in ligand specificity at I_1-imidazoline sites between species (bovine versus rat) or tissues (brain versus kidney) may be relatively minor. Similarly, affinity values at I_1-imidazoline sites in canine prostate were compared with those in kidney and brain stem (not shown). The affinity (-log K_i) for each of the competing drugs at I_1 sites in canine prostate was highly correlated with affinities at I_1 sites in rat kidney ($r = 0.949$; $n = 5$; $p < 0.02$) with a slope very close to 1 (1.00 ± 0.19). Affinity values for competing drugs were also highly correlated with those obtained in the bovine RVLM ($r = 0.946$; $n = 6$; $p < 0.005$) and a slope only slightly less than 1 (0.75 ± 0.13). The only significant difference between species was that cimetidine showed 10- to 30-fold higher affinity in the dog than in the rat or cow. These data suggested that the I_1 sites in the canine prostate were nearly identical to I_1 sites in rat kidney and bovine brain stem. Furthermore, affinities at human platelet I_1 sites correlated well with data from bovine RVLM ($r > 0.9$).[21] Thus, the binding properties of I_1-imidazoline receptors are highly conserved between mammalian species.

Relative binding affinities of 24 different compounds at I_1 sites were compared with results from I_2 sites in rat liver[22] (FIG. 1C) or human adipocytes[23] (FIG. 1D). The results in these two tissues from two species were similar, showing no correlation to I_1 affinity. Most of the ligands fell below the diagonal, indicating higher affinity at I_1 than at I_2 sites. Efaroxan (9) was most selective, showing 100,000-fold higher affinity for I_1 than for I_2 sites. A few compounds showed similar affinities at I_1 and I_2 sites and thus fell close to the line of identity, including bromoxidine (11), naphazoline (6), histamine (14), and SK&F 104078 (21). Many non-imidazolines showed low affinity at both I_1 and I_2 and thus fell in the lower left (16–20). Idazoxan (5) showed a modest preference for I_2 over I_1 whereas guanabenz (15) was highly selective for I_2. The fact that several compounds have similar affinities at I_1 and I_2 sites was recently noted, and this prompted the suggestion that these two subtypes may be similar or even identical.[24] However, the lack of any overall correlation with a large series of compounds and the high degree of selectivity shown by certain imidazolines argue against this thesis. A similar lack of correlation was obtained when correlating I_2 affinities determined in guinea pig brain[25] with I_1 affinities.[17]

The significant difference in ligand specificity shown in FIGURE 1 implies, but does not establish, the theory that I_1- and I_2-imidazoline sites are distinct and independent. To test if these two subtypes might coexist in bovine RVLM membranes, we characterized [^{3}H]idazoxan binding in the same membranes and under the same conditions used to study I_1-imidazoline sites (FIG. 2). Each of the α_2-adrenergic agonists completely inhibited specific [^{3}H]idazoxan binding as defined in the presence of 0.1 mM phentolamine. There was no evidence for catecholamine-insensitive [^{3}H]-

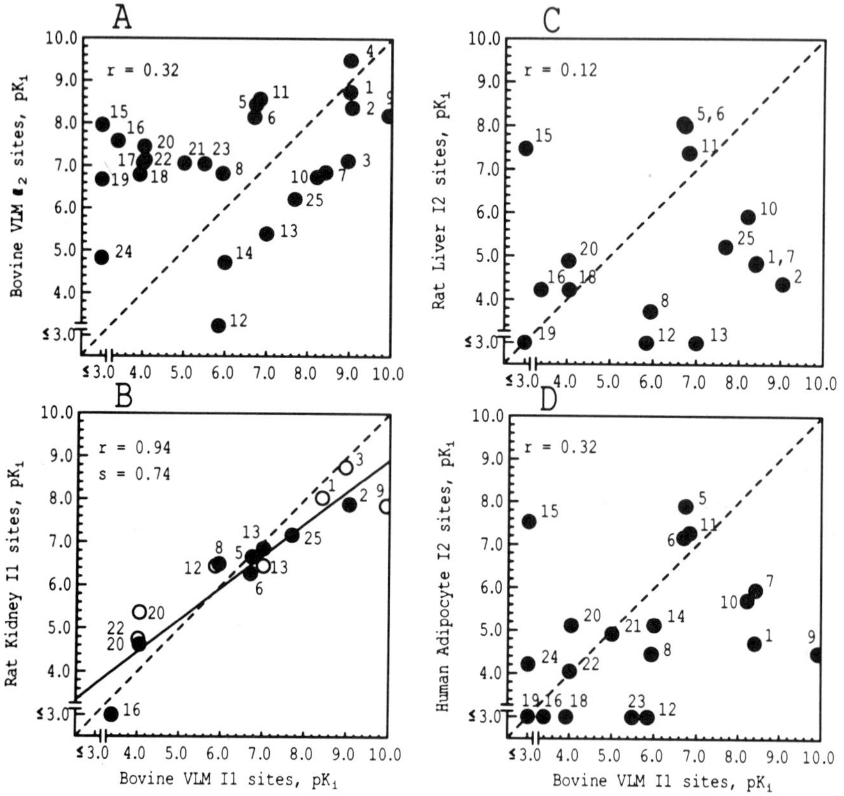

FIGURE 1. Comparison of binding affinities at α_2-adrenergic and I_1 and I_2 imidazole binding sites. Affinities of test agents for specific binding sites were expressed as the negative log of the K_i (pK_i) as determined by analysis of inhibition curves using the LIGAND program.

Key to test agents: Imidazolidines: 1 = clonidine, 2 = *p*-aminoclonidine, 3 = moxonidine, and 4 = *p*-iodoclonidine. *Imidazolines:* 5 = idazoxan, 6 = naphazoline, 7 = oxymetazoline, 8 = phentolamine, 9 = efaroxan (RX 821037), 10 = rilmenidine, and 11 = bromoxidine (UK 14,304). *Imidazoles:* 12 = imidazole-4-acetic acid, 13 = cimetidine, and 14 = histamine. *Guanidines:* 15 = guanabenz. *Phenylethylamines:* 16 = (–)-epinephrine, 17 = (–)-α-methylnorepinephrine 18 = (–)-norepinephrine, and 19 = phenylphrine. *Other:* 20 = SK&F 86466 (6-chloro-*N*-methyl-2,3,4,5-tetrahydro-1-H-3-benzazepine), 21 = SK&F 104078 (6-chloro-9-[3-methyl-2-butenyl]oxyl-3-methyl-2,3,4,5-tetrahydro-1-H-3-benzazepine), 22 = phenoxybenzamine, 23 = yohimbine, 24 = prazosin, and 25 = clonidine-displacing substance (CDS). The K_i for CDS is arbitrarily expressed as units × 10^{-6}, where 1 unit inhibits [^{3}H]clonidine binding to cerebral cortex membranes by 50%.

In each panel, the affinity of the ligands at I_1 sites in bovine RVLM labeled by any of three clonidine analogs ([^{3}H]*p*-aminoclonidine, [^{125}I]*p*-iodoclonidine, or [^{3}H]clonidine, all of which produced nearly equivalent pK_i values) was compared to binding affinity at another site. The *dashed diagonal* is the line of identity, around which all points should fall if the two binding sites were closely related. (**A**) ligand affinity at bovine RVLM I_1 sites was contrasted to affinity at α_2-adrenergic sites in the same tissue. (**B**) Ligand affinity at bovine RVLM I_1 sites is strongly correlated with affinities at I_1 sites in rat kidney in both cortex (*closed circles*) and the medulla (*open circles*). The *solid line* represents the linear regression best fit. (**C** and **D**) Ligand affinity at bovine RVLM I_1 sites was compared with affinities at I_2 sites labeled by [^{3}H]idazoxan in either rat liver[22] (**C**) or human adipocytes[23] (**D**).

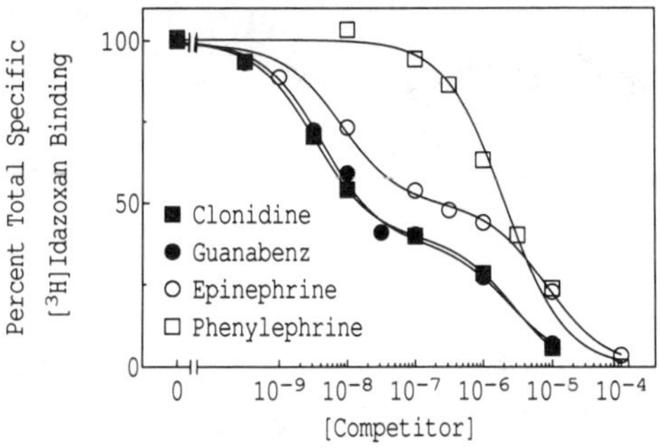

FIGURE 2. Inhibition of [^{3}H]idazoxan binding to membranes prepared from bovine RVLM by increasing concentrations of imidazolidine, guanidine, or phenylethylamine α_2-adrenergic agonists. Each point represents the mean of three assays conducted in triplicate. Curves were derived by least squares fit to two-component logistic equations (GraphPad, Motulsky, 1990). RVLM membranes (100 μg protein) were incubated with 2.0 nM [^{3}H]idazoxan and serial dilutions of the competing agent. Nonspecific binding was defined in the presence of 0.1 mM phentolamine. In separate saturation assays conducted with a series of eight concentrations of [^{3}H]idazoxan, the K_D was 3.9 ± 0.6 nM and the B_{max} was 322 ± 80 fmol/mg protein. The K_D is within the range reported for [^{3}H]idazoxan binding to α_2-adrenergic receptors, but below that generally reported for I_2 sites. The B_{max} is higher than the estimate obtained using [^{3}H]clonidine in RVLM membranes (TABLE 2), which is consistent with the ability of agonist radioligands to label a portion of α_2-adrenergic receptors which are in a high affinity state, whereas antagonist radioligands label the entire population of sites.[1]

idazoxan binding. Each ligand showed the expected order of potency for an α_2-adrenergic receptor (pIC_{50} for high affinity site ± standard error): guanabenz (8.55 ± 0.13) = clonidine (8.38 ± 0.07) > epinephrine (8.10 ± 0.09) > phenylephrine (5.69 ± 0.05). The agonists showed biphasic curves, presumably representing high affinity and low affinity states of the α_2-adrenergic receptor, as anticipated for an antagonist radioligand (low affinity site pIC_{50}s: guanabenz (5.98 ± 0.24) ≥ clonidine (5.62 ± 0.14) > epinephrine (5.09 ± 0.10)). Thus, I_2-imidazoline sites were not detected in bovine RVLM membranes. Because I_2 sites are localized in mitochondria,[26,27] the data in FIGURE 2 imply that our membrane preparation may contain a low number of mitochondria. By the same token, a prior report that I_2 sites are abundant in human RVLM membranes[28] probably reflects greater contamination by mitochondria rather than a true species difference. The lack of I_2 sites in a standardized preparation expressing I_1-imidazoline sites further supports the notion that these sites are distinct.

To test the hypothesis that I_1-imidazoline sites are localized to mitochondria, similar to I_2 sites, we examined the subcellular distribution of [^{125}I]PIC binding sites in PC12 cells using sucrose density gradient purification of plasma membranes. Purified plasma membranes contained a very high proportion of I_1 sites in parallel with the distribution of muscarinic receptor sites. Indeed, I_1-imidazoline sites were enriched severalfold by sucrose density gradient purification (B_{max} = 560 ± 120 fmol/mg protein; TABLE 2). About 15% of both I_1-imidazoline and muscarinic acetylcholine binding sites were present in the mitochondrial fraction. Similar data have been

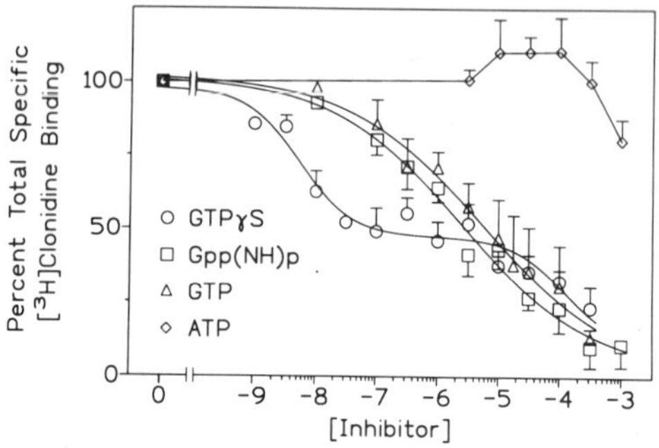

FIGURE 3. Dose-dependent inhibition of specific [^{3}H]clonidine binding to adrenomedullary chromaffin cells by guanine and adenine nucleotides. Each point represents the mean of four to six experiments, each conducted in replicate. Nonspecific binding was defined in the presence of 0.1 mM naphazoline. [^{3}H]Clonidine binding was dose dependently and completely inhibited by the nonhydrolyzable GTP analogs GTPγS and Gpp(NH)p and by GTP itself. In contrast, ATP produced a slight but significant increase in binding.

obtained in human platelet plasma membrane fractions.[21] Thus, unlike I_2 sites, I_1 sites were primarily localized to the plasma membrane and are not concentrated in mitochondria.

Guanine nucleotides have been consistently reported not to influence [^{3}H]idazoxan binding or its inhibition by competitors, indicating that I_2 sites are not coupled to G-proteins.[22–28] We sought to determine if I_1-imidazoline receptor binding might be regulated by guanine nucleotides. Adrenomedullary chromaffin cells were used as a model system, because these cells lack α_2-adrenergic receptors. As shown in FIGURE 3, guanine nucleotides potently and completely inhibited specific [^{3}H]clonidine binding to chromaffin cell membranes in the following potency order ($IC_{50} \pm$ SE): GTPγS (5.2 ± 1.2 nM) > Gpp(NH)p (2.5 ± 0.9 μM) > GTP (5.7 ± 1.4 μM) >>> ATP, which is consistent with that observed for receptors coupled to guanine nucleotide binding regulatory proteins. The reason for the biphasic inhibition by GTPγS is not readily apparent (low affinity component IC_{50} = 230 ± 68 μM), but conceivably tissue thioesterases might degrade this GTP analog. ATP did not inhibit I_1-imidazoline binding and actually increased it by about 10% ($p < 0.05$, analysis of variance). The latter finding rules out nonspecific inhibitory effects of nucleotide triphosphates. These data are consistent with previous reports showing guanine nucleotide sensitivity of I_1-imidazoline sites in bovine RVLM, PC12 cells, and canine prostate,[11,19] and they suggest that I_1-imidazoline sites may be coupled to a G-protein and might therefore belong to the superfamily of G-protein-coupled receptors.

The present data support the hypothesis that there are subtypes of imidazoline binding sites.[20,27] The properties of α_2-adrenergic receptor subtypes, I_1- and I_2-imidazoline sites, are compared in TABLE 1. Putative I_1- and I_2-imidazoline binding sites are labeled by [^{3}H]clonidine and [^{3}H]idazoxan, respectively. These subtypes differ from each other and from α_2-adrenergic receptors in their relative preferences for ligands of different chemical classes. The imidazolidines, which include clonidine, moxonidine, and other centrally acting antihypertensive agents, show high affinity

TABLE 1. Comparison of $\alpha_{2A\text{-}D}$, I_1, and I_2 Binding Sites

	$\alpha_{2A,B,C,D}$	I_1	I_2
Radioligands	[^{3}H]Rauwolscine [^{3}H]Idazoxan [^{3}H]Clonidine [^{125}I]*p*-Iodoclonidine	[^{3}H]Clonidine [^{125}I]*p*-Iodoclonidine [^{3}H]Moxonidine	[^{3}H]Idazoxan
Ligand specificity by chemical class	Imidazolidines ≥ Imidazolines > Phenylethyalmines ≫ Imidazoles[a]	Imidazolidines ≥ Imidazolines > Imidazoles ≫ Phenylethylamines[a]	Imidazolines > Imidazolidines ≫ Imidazoles Phenylethylamines[a]
Endogenous ligand	Epinephrine, agmatine?[10]	Agmatine?[10]	Agmatine?[10]
G-protein coupling	Yes[1]	Yes[a]	No[27]
Subcellular fraction	Plasma membrane[1,21]	Plasma membrane[a]	Mitochondria[26,27]; other?[22]
Signaling mechanism	Adenyl cyclase inhibition[1]; other?	Prostaglandin release[a]; K^+ channel?; other?	Monoamine oxidase inhibition?[b]
Brain distribution	Widespread[24] Neurons ≥ glia[12]	Brainstem reticular formation[a]; neurons ≫ glia[a]	Forebrain; circumventricular organs[24]; glia > neurons[29]

[a] Data from present paper.
[b] Parini and colleagues, this volume.

for α_2 and I_1 sites. I_2 sites selectively bind imidazolines relative to imidazolidines and imidazoles, whereas I_1 sites recognize all three classes of ring structure. I_1- and I_2-imidazoline sites differ in subcellular distribution, brain localization, and interactions with G-protein. Furthermore, I_1 sites are not expressed by glial astrocytes (TABLE 2), whereas these cells are the primary cell type expressing I_2 sites in brain.[29] As shown in the present study, I_1- but not I_2-imidazoline sites: (1) show high affinity for centrally acting antihypertensive drugs ($pK_i > 7$); (2) are concentrated in plasma membrane fractions, (3) are present in RVLM membranes, and (4) appear to be coupled to a G-protein.

HETEROGENEITY OF I_1-IMIDAZOLINE BINDING SITES

Several laboratories have identified nonadrenergic binding sites for [^{3}H]clonidine and other imidazolines that are low-affinity high-capacity sites distinct from I_1-imidazoline receptors as described here. For instance, Bricca and colleagues[30] reported on a site in human RVLM with an affinity for clonidine of 105 nM. Puzzlingly, the K_D values in that study estimated by saturation experiments (7 nM) and by kinetic analysis (2 nM) were almost 50-fold lower than the IC_{50} values for clonidine competi-

TABLE 2. Density of I_1-Imidazoline and α_2-Adrenergic Receptors in Different Cells and Tissues

Tissue or Cell	Radioligand	B_{max} (fmol/mg)	
		I_1	α_2
Neural			
Cow rostral ventrolateral medulla (RVLM)[a]	[^{3}H]Clonidine	52 ± 11	141 ± 26
RVLM[a]	[^{3}H]*p*-aminoclonidine	49 ± 4	111 ± 14
RVLM[a]	[^{125}I]*p*-iodoclonidine	90 ± 8	136 ± 7
Cow dorsomedial medulla	[^{3}H]Clonidine	60 ± 5	52 ± 4
Frontal cerebral cortex	[^{3}H]*p*-aminoclonidine	70 ± 12	800 ± 180
Astrocytes[a]	[^{3}H]*p*-aminoclonidine	0 ± 0	16 ± 2
NG 108-15 cells	[^{3}H]*p*-aminoclonidine	290 ± 30	261 ± 48
Chromaffin cells	[^{3}H]Clonidine	450 ± 90	0 ± 0
PC12 cells	[^{125}I[*p*-iodoclonidine	20 ± 4	0 ± 0
PC12 plasma membrane	[^{125}I]*p*-iodoclonidine	560 ± 120	0 ± 0
Cat carotid body	[^{125}I]*p*-iodoclonidine	16 ± 5	10 ± 2
Rabbit carotid body	[^{125}I]*p*-iodoclonidine	16 ± 2	23 ± 2
Epithelial			
Rat renal cortex[c]	[^{3}H]*p*-aminoclonidine	41 ± 10	65 ± 9
Rat renal medulla	[^{3}H]Clonidine	62 ± 8	27 ± 2
Rabbit mesangial cell	[^{125}I]*p*-iodoclonidine	27 ± 3	20 ± 2
Rabbit proximal tubule	[^{125}I]*p*-iodoclonidine	39 ± 7	190 ± 17
Rabbit distal tubule	[^{125}I]*p*-iodoclonidine	69 ± 10	44 ± 7
Rabbit collecting duct	[^{125}I]*p*-iodoclonidine	73 ± 8	14 ± 1
Rabbit tracheal epithelium	[^{125}I]*p*-iodoclonidine	18 ± 2	19 ± 2
Dog prostate[d]	[^{3}H]Clonidine	18 ± 2	8 ± 1
Human prostate	[^{3}H]Clonidine	113 ± 12	45 ± 4
Secretory			
Pancreatic islet	[^{125}I]*p*-iodoclonidine	39 ± 4	21 ± 4

NOTE: B_{max} values were determined by nonlinear curve-fitting with the LIGAND program from four to eight saturation experiments using a series of seven to eight concentrations of radioligand and conducted in triplicate or quadruplicate. An exception was the carotid body, which due to its small size (<1 mg) was tested at a single radioligand concentration. K_D values at I_1-imidazoline sites ranged from 0.5 to 2 nM for [^{125}I]*p*-iodoclonidine, 3 to 12 nM for [^{3}H]*p*-aminoclonidine, and 4 to 15 nM for [^{3}H]clonidine. K_D values at α_2-adrenergic sites ranged from 0.3 to 1 nM for [^{125}I]*p*-iodoclonidine, 0.5 to 13 nM for [^{3}H]*p*-aminoclonidine, and 2 to 10 nM for [^{3}H]clonidine. Brain tissue tended to show the lowest K_D values for both receptor types.

[a] From ref. 8.
[b] From ref. 12.
[c] From ref. 34.
[d] From ref. 11.

tion for [^{3}H]clonidine sites. In contrast, in the same series of experiments,[30] clonidine showed high affinity for I_1-imidazoline sites in the bovine RVLM, with results nearly superimposable on those of a prior report[5] and with nearly identical K_D values (6 versus 3 nM). Another recent report from this group examined rabbit brain stem.[31] In that study, the K_D values for [^{3}H]clonidine were 3 and 18 nM for α_2-adrenergic and I_1-imidazoline sites, respectively, which were well within the range reported for bovine RVLM. However, in competition studies, all the imidazolines had low affinity, with BDF 6143 showing the highest affinity at 320 nM, and bromoxidine, oxymetazoline, phentolamine, and idazoxan all showing about the same 2-4 μM affinity, whereas moxonidine had no detectable effect on [^{3}H]clonidine binding to nonadrenergic sites. Comparable results were obtained in human RVLM.[32]

Similarly, studies of [^{3}H]*p*-aminoclonidine binding in rat kidney showed about 100 nM affinity for clonidine and *p*-aminoclonidine, but >1 μM affinities for idazoxan and phentolamine.[33] Cimetidine, an agent active at H_2-histamine and I_1-imidazoline sites, showed almost no affinity. The non-imidazoline compound guanabenz showed the highest affinity of any agent (10 nM), in agreement with studies of low affinity [^{125}I]*p*-iodoclonidine sites ("I_3 sites") expressed by human platelets.[21] The K_D for [^{3}H]*p*-aminoclonidine was surprisingly low, about 100 nM by either saturation or kinetic analysis, and the B_{max} was 1,000 fmol/mg protein.[33] Both the K_d and the B_{max} of the nonadrenergic sites labeled by [^{3}H]*p*-aminoclonidine in rat kidney were about 25-fold higher than those previously reported for I_1-imidazoline sites labeled by [^{3}H]*p*-aminoclonidine in rat kidney.[34]

In rat cerebral cortex, which lacks high affinity I_1-imidazoline sites detectable by [^{125}I]*p*-iodoclonidine autoradiography,[9,35] [^{3}H]rilmenidine labels a low affinity (K_d = 250 nM) high capacity (B_{max} = 1,600 fmol/mg protein) site.[36] The imidazolines clonidine, idazoxan, lofexidine, naphazoline, and oxymetazoline had affinities between 3 and 20 μM at this cortical site. A similar site was characterized in human platelets[21] and in untransfected CHO cells (Piletz *et al.*, submitted). Recently full characterization of [^{3}H]clonidine binding in bovine adrenal chromaffin cells was reported.[37] Two populations of specific sites were identified, one having a high affinity for clonidine (15 nM) as well as for naphazoline, BDF 6143, moxonidine, and rilmenidine (K_i < 100 nM), whereas the second site had affinities for these same compounds ranging from 2 μM (clonidine) to 23 μM (BDF 6143). The first site resembles the I_1-imidazoline sites characterized in brain stem, but the second site appears to resemble the low affinity sites described by others.

High affinity I_1-imidazoline binding sites (K_d < 20 nM) were described by several independent groups. Human platelet I_1 sites were extensively characterized and appear identical to bovine RVLM sites.[21] Another report replicated key findings on I_1 sites in bovine RVLM.[30] Yet another report characterized binding sites in rat brain[38] that appear indistinguishable from I_1-imidazoline sites of the bovine RVLM. Another laboratory found a nonadrenergic binding site for [^{3}H]*p*-aminoclonidine with a K_D of 2 nM and a K_i for clonidine and oxymetazoline of 9 and 2 nM, respectively, in close agreement with previous reports.[39] Moreover, a study of [^{3}H]clonidine binding in rat vas deferens found a site with 18 nM affinity for idazoxan but 30 μM affinity for yohimbine,[40] a pattern consistent with I_1-imidazoline sites. Finally, high affinity I_1-imidazoline sites were independently characterized in bovine adrenomedually chromaffin cells.[37] This literature indicates that I_1 imidazoline sites can be reproducibly labeled in radioligand binding assays. Failure to detect these sites probably reflects technical problems with the assay. (See Ernsberger *et al.*, this volume.)

The data so far suggest that I_1-imidazoline sites may exist as distinct high and low affinity forms. Because low affinity I_1-imidazoline sites are insensitive to guanine nucleotides,[21,32,37] and high and low affinity I_1 sites recognize many of the same compounds, it is tempting to speculate that the low affinity I_1 site represents an uncoupled I_1-receptor. If so, high affinity I_1-imidazoline sites could be converted to low affinity sites by adding guanine nucleotide. Development of an antagonist radioligand would help resolve this issue.

DISTRIBUTION OF I_1-IMIDAZOLINE BINDING SITES

Imidazolines elicit actions independent of α-adrenergic receptors in a variety of tissues. I_1-Imidazoline binding has been characterized in detail in the RVLM, but little is known of the distribution of I_1 in different cells and tissues compared to α_2-adrenergic receptors. We labeled I_1 and α_2 sites with [^{3}H]clonidine, [^{3}H]*p*-aminoclonidine, or [^{125}I]*p*-iodoclonidine, each with comparable results. As shown in TABLE 2, I_1 as well as α_2 sites were expressed in dorsomedial medulla as well as RVLM. In cerebral cortex, α_2 sites predominate 10 : 1 over I_1 sites. Astrocytes express α_2 but not I_1, implying that in the brain I_1-imidazoline sites may be localized primarily to neurons. NG 108-15 neuroblastoma-glioma cells express both sites in abundance, and presumably I_1 expression is contributed by the neuroblastoma parent rather than the glioma parent. Chromaffin and PC12 pheochromocytoma cells lack α_2 but express I_1, mainly in the plasma membrane. The carotid body, like chromaffin cells, derives embryologically from the neural crest. In both cat and rabbit, carotid bodies express both I_1 and α_2.

In kidney, I_1 sites were expressed mainly in epithelial cells of the distal nephron and medulla, whereas α_2 sites were enriched in the proximal tubule (TABLE 2). I_1-Imidazoline sites increase in density along the nephron from the mesangium to the proximal, distal, and collecting tubule. α_2-Adrenergic receptors were most abundant in proximal tubule cells and were present in much lower density elsewhere in the nephron. The distribution of both I_1-imidazoline and α_2-adrenergic receptor sites within the kidney by autoradiography[34] is consistent with their apparent distribution along the nephron. Human as well as dog prostate expresses I_1 as well as α_2 within epithelial cell regions. Pancreatic islets express I_1 as well as α_2, suggesting that I_1 may be present in secretory as well as epithelial cell types. Another cell type specialized for secretion, the platelet, also expresses I_1 as well as α_2 sites.[21,41] These data show that I_1-imidazoline sites are expressed by many different tissues and cell types of ectodermal origin and imply widespread physiological functions for the I_1-imidazoline receptor.

As shown in TABLE 2, α_2-adrenergic receptors were expressed in the absence of I_1-imidazoline sites in glial astrocytes and frontal cerebral cortex. Conversely, I_1-imidazoline sites were expressed in the absence of α_2-adrenergic receptors in adrenal chromaffin and PC12 cells. These findings make it unlikely that imidazoline sites were a subunit of the α_2-adrenergic receptor or an associated modulatory protein. I_1-Receptors appear to be expressed independently of α_2-receptors.

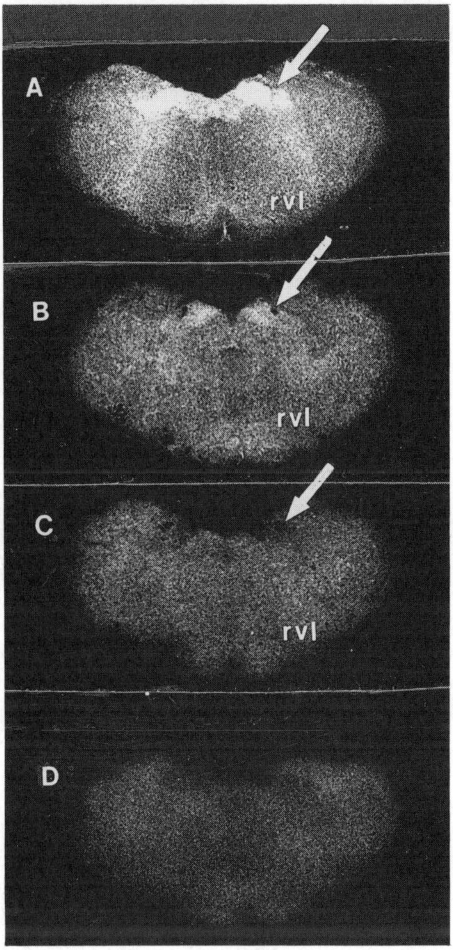

FIGURE 4. Autoradiography of [^{125}I]PIC binding sites in the medulla of a spontaneously hypertensive rat. Shown are photographic prints made under identical conditions using the autoradiograms as negatives. Light areas correspond to the highest intensity of labeling. Adjacent 15-μm sections of rat brain stem were incubated with 0.5 nM [^{125}I]*p*-iodoclonidine under one of four different conditions. The section in panel **A,** incubated with antioxidant vehicle alone, shows the total binding of [^{125}I]PIC. Binding sites were concentrated in the region of the nucleus tractus solitarii (*arrow* indicates the solitary tract), but were distributed across the reticular formation, including the RVLM (labeled rvl). The section in panel **B** was incubated with 10 μM cimetidine to selectively mask I_1-imidazoline sites and thus shows the distribution of α_2-adrenergic sites. Labeling in the rvl was inhibited by 10 μM cimetidine. The section below in panel **C** was incubated with 10 μM epinephrine to mask α_2-receptors and thus shows the distribution of I_1 binding. Note that [^{125}I]PIC labeling was almost completely eliminated by epinephrine in the area of the nucleus tractus solitarii. The labeling of the rvl region persisted in the presence of epinephrine. The section in panel **D** was incubated with 10 μM BDF-6143 to block both I_1 and α_2 binding and thus shows the distribution of nonspecific sites.

VISUALIZATION OF I_1-IMIDAZOLINE BINDING SITES

Quantitative receptor autoradiography of imidazoline and α_2-receptors has been reported in rat medulla oblongata[9,35] and the kidney.[34] Mapping of the distribution of I_1 relative to α_2 sites in rat medulla is shown in FIGURE 4. Adjacent 15 μm sections were incubated 1 hour with 0.5 nM [^{125}I]*p*-iodoclonidine and either vehicle (*top,* total binding), 0.1 mM cimetidine to mask I_1, leaving mainly α_2 (*second*), 10 μM epinephrine to mask α_2, leaving mainly I_1 (*third*), or 0.1 mM BDF-6143 to define nonspecific binding (*bottom*). Labeling in the nucleus of the solitary tract (arrows mark the solitary tract) and other dorsal nuclei is mainly α_2, because it was blocked by epinephrine but not cimetidine. I_1-Imidazoline sites (*third*) were absent in the nucleus of the solitary tract but were widely distributed across reticular areas, including rvl. I_1-imidazoline sites are also present in the midline and lateral raphe nuclei. A similar pattern of α_2 sites in dorsal nuclei and I_1 sites in ventral regions was seen in the pons (not shown). The locus coeruleus expresses mainly α_2, whereas sites were present in ventral tegmental areas, including the A5 region.[35] Thus, I_1-imidazoline sites were preferentially located in the brain stem reticular formation, implying participation in autonomic control and other integrative brain stem functions.

POSSIBLE FUNCTIONS OF NEURONAL I_1-IMIDAZOLINE RECEPTORS

The evidence implicating I_1-imidazoline sites in the vasodepressor actions of imidazolines has been reviewed elsewhere.[9,42,43] To summarize: (1) Some imidazoline α_2-antagonists lower blood pressure when microinjected into the RVLM.[4] Non-imidazolines which are potent full agonists at α_2-receptors, such as α-methylnorepinephrine and guanabenz, are 100-fold less efficacious than are the imidazolines in the RVLM.[6] (3) Binding affinity at I_1 sites in the RVLM correlates with clinical antihypertensive efficacy ($r = 0.996$, $n = 4$) and with vasodepressor activity within the RVLM ($r = 0.79$, $n = 8$), whereas α_2-affinity correlates with neither measure of efficacy ($r = -0.24$ and -0.17, respectively).[7,8] (4) The vasodepressor action of imidazolines (clonidine, rilmenidine, or moxonidine), given either intravenously or by RVLM microinjection, can be prevented or reversed by RVLM microinjection of imidazoline antagonists (idazoxan, efaroxan), whereas non-imidazoline antagonists (SK&F 86466, rauwolscine) have little effect.[6,7,42,44]

Initial tests of the effects of selective antagonists on the action of imidazolines within the RVLM have used microinjection for the delivery of both agonist and antagonist. However, local microinjection introduces high local concentrations of the drug whose effects may not be entirely representative of the actions of systemically administered agents. Therefore, we examined the ability of locally injected antagonists within the RVLM to prevent the effect of systemically administered drug. Microinjection of the selective α_2-antagonist SK&F 86466 into the RVLM in either a low dose (1 nmol) or a high dose (10 nmol) did not significantly attenuate the effect of intravenous rilmenidine (0.5 mg/kg). In contrast, the α_2/I_1 antagonist idazoxan completely abolished the effect of intravenous rilmenidine even at a low dose (1 nmol).[44] In addition, microinjection of the selective I_1-imidazoline antagonist efaroxan (10 nmol) into the RVLM completely prevented the hypotensive action of intravenous moxonidine (40 μg/kg).[42] Furthermore, Nosjean and Guyenet[45] showed that microinjection of rauwolscine (2 nmol) into the RVLM elicited a fall in blood pressure, similar to that of other α_2-antagonists. Subsequent injection of clonidine produced a further fall in blood pressure, such that the total depressor response to rauwolscine followed by clonidine was indistinguishable from the response to clonidine alone. In contrast to rauwolscine, the I_1/α_2 antagonist idazoxan completely blocked the action of clonidine after microinjection into the RVLM.[45] These studies implicate I_1-imidazoline receptors within the RVLM in the action of systemically administered imidazolines.

In the carotid body, I_1-imidazoline receptors may facilitate electrical activity elicited by hypoxia, whereas in contrast, α_2-adrenergic receptors inhibit chemosensory function.[46,47] Thus, the separation of α_2-adrenergic and I_1-imidazoline receptor-mediated effects is more distinct than in the brain stem, inasmuch as the two receptors appear to mediate opposite effects. This interpretation is supported by the contrasts between guanabenz and clonidine, two α_2-adrenergic agonists expressing nearly identical affinity and efficacy at α_2-receptors.[1] In the absence of α_2-blockade, guanabenz inhibited nearly two thirds of the chemosensory response to hypoxia. Clonidine also tended to decrease chemosensory responding, but the effect was inconsistent and facilitation was sometimes observed. After blockade of α_2-adrenergic receptors with the specific antagonist SK&F 86466, guanabenz had no effect. In contrast, α_2-blockade unmasked a stimulatory effect of clonidine.[46] These data demonstrate that clonidine acts on a receptor other than the α_2-adrenergic receptor to facilitate chemosensory discharge. These data support the hypothesis that I_1-imidazoline binding sites are functional receptors and are functionally distinct from α_2-adrenergic receptors.

Clonidine acts within the preoptic area to induce a fall in body temperature which can be mimicked by norepinephrine and blocked by the non-imidazoline α_2-antagonists rauwolscine, yohimbine,[48] or SK&F 86466.[49] Direct microinjection of clonidine into the preoptic area elicits a biphasic response.[50] The initial hypothermic response was completely blocked by rauwolscine and could also be elicited by microinjection of norepinephrine. The delayed hyperthermic response was not blocked by rauwolscine but could be blocked by indomethacin, implying that the hyperthermic action of clonidine was mediated by prostaglandin release.[50] The rauwolscine-insensitive hyperthermic response might be mediated by I_1-imidazoline receptors. Thus, in the preoptic area, α_2- and I_1-receptors may mediate opposite effects on body temperature. Interestingly, clonidine may be an effective treatment for hypothermia in Shapiro's syndrome, which is caused by a hypothalamic deficit.[51]

In ocular pharmacology, there is extensive evidence linking an I_1-imidazoline receptor mechanism to reductions in intraocular pressure. (See Campbell *et al.* and Burke *et al.*, this volume.) A centrally mediated mydriasis response is another candidate for mediation by I_1-receptors. Microinjection of 20 nmol clonidine into the cat preoptic area elicits a long-lasting mydriasis, which was not blocked by α_2-adrenergic blockade with yohimbine or phentolamine.[52] In contrast, the sedative action of clonidine microinjections was completely prevented by α_2-blockade. Furthermore, the non-imidazoline α_2-agonists norepinephrine and phenylephrine did not elicit mydriasis even in doses up to 300 nmol.[52]

Clonidine is a potent sedative in diverse animal species. α_2-Adrenergic receptors almost certainly mediate sedation. First, the alkaloid antagonists yohimbine and rauwolscine potently and completely reverse the sedation elicited by clonidine,[53] whereas higher doses were required to entirely reverse vasodepressor actions.[53] Second, profound sedation is elicited by α_2-agonists with little affinity for I_1 sites, including the antihypertensive agents α-methyldopa, guanabenz, and guanfacine. The potent α_2-agonist dexmedetomidine, which has only micromolar affinity for I_1-imidazoline sites,[21] induces marked sedation, even anesthesia, without lowering blood pressure.[54] In contrast, two clonidine analogs with reduced affinity for α_2-adrenergic receptors relative to I_1-imidazoline sites, namely, rilmenidine and moxonidine, show less tendency to produce sedation than does clonidine itself.[7] In fact, moxonidine actually improves alertness and performance on a driving simulator.[55] The sedative action of α_2-agonists may be mediated within the locus coeruleus, because local microinjection of clonidine or non-imidazoline α_2-adrenergic agonists into the locus coeruleus elicits sedation.[56] The activity of neurons in the locus coeruleus is potently inhibited by clonidine and by non-imidazoline α_2-agonists, whereas the selective I_1-agonist rilmenidine is inhibitory only at very high doses.[57] These data are consistent with the absence of I_1 sites from the rat locus coerleus.[35] Finally, clonidine interferes with respiratory reflexes through activation of α_2-adrenergic receptors, whereas doses of the selective I_1-agonist moxonidine that lower blood pressure to a similar degree lack any inhibitory effect on reflex function (Haxhiu *et al.*, this volume). Thus, unlike α_2-adrenergic receptors, I_1-imidazoline receptors do not appear to induce nonspecific depression of central nervous system function.

POSSIBLE FUNCTIONS OF NONNEURONAL I_1-IMIDAZOLINE RECEPTORS

I_1-Imidazoline binding sites were recently characterized in adrenomedullary chromaffin cells.[7,8,19] In these cells, [^{3}H]clonidine binds to I_1-imidazoline but not to α_2-adrenergic receptors (TABLE 2). Chromaffin cell I_1-receptors may be functional. Clonidine inhibits the release of catecholamines, ATP, and leu-enkephalin from adrenomedullary chromaffin cells elicited by nicotinic stimulation, and these effects are not mediated by adrenergic receptors.[58] The action of clonidine is not mimicked by non-imidazolines, including epinephrine, norepinephrine, or phenylephrine, but imidazolines such as naphazoline and oxymetazoline elicit clonidine-like effects. Clonidine's effects on adrenomedullary chromaffin cells cannot be reversed by non-imidazoline antagonists such as yohimbine and phenoxybenzamine. Phentolamine and tolazoline, α-antagonists that bind to I_1-imidazoline sites, have a clonidine-like action. Alternatively, one group reported a stimulatory effect of clonidine on catecholamine release from chromaffin cells.[10] However, high concentrations of imidazolines (>1 μM) are required for either the inhibitory or the stimulatory effects. Furthermore, the order of potency of the imidazolines differs from that at I_1-imidazoline sites. For example, phentolamine is consistently more potent than clonidine, whereas the converse is true in binding studies. Whether I_1-imidazoline receptors regulate adrenal catecholamine release remains an open question.

In the kidney, I_1-imidazoline receptors may facilitate sodium excretion. Intrarenal administration of the I_1-agonist moxonidine increases excretion of sodium and water, and these actions are blocked by idazoxan, an I_1-antagonist, but not by the selective α_2-antagonist rauwolscine.[59] Renal tubule epithelial cells of the distal tubule and collecting duct (TABLE 2) are a potential site of action for the natriuretic action of selective I_1-imidazoline agonists in the isolated perfused rat kidney. Interestingly, I_1-imidazoline binding in the kidney was upregulated early in the development of hypertension in an angiotensin II infusion model,[18] consistent with a role of renal I_1-imidazoline receptors in blood pressure regulation.

Nonrenal epithelia that express I_1-imidazoline sites, such as those of the trachea and prostate, may also be affected by I_1-imidazoline agents. In tracheal epithelial cells, there is evidence for I_1-imidazoline receptor-mediated regulation of sodium transport (Liedtke and Ernsberger, this volume). Insulin release from pancreatic islet cells is probably under control of an imidazoline receptor, and the I_1-imidazoline subtype is present and therefore a candidate for mediating this action (see elsewhere in this volume).

SIGNALING PATHWAYS ASSOCIATED WITH THE I_1-IMIDAZOLINE RECEPTOR

Despite the potential importance of the I_1-imidazoline receptor, little is known regarding the transmembrane signaling pathways it activates. Thus far, it is only known that these receptors do not affect classic signaling pathways through cyclic nucleotides or phophatidylinositol-selective phospholipase C (PI-PLC).[15,60,61] The only studies by others of signaling mechanisms coupled to I_1-receptors have reported largely negative results. In adrenomedullary chromaffin cells and in tracheal epithelial cells, clonidine in concentrations up to 0.1 mM had no effect on inositol phosphate accumulation or on basal or forskolin-stimulated cAMP levels.[15,60] Moxonidine and cimetidine were also inactive.[15] Clonidine increased cGMP in chromaffin cells, but this effect was not shared by other I_1-agonists and presumably was nonspecific.[60,61] Clonidine increases the influx of ^{45}Ca, but only after the cells have been preloaded with isotope,[60] implying an action on calcium efflux similar to that of thapsigargin. The minimal effective concentration of clonidine was 10 μM,[60] which is 1,000 times its binding affinity. Thus, stimulation of I_1-receptors has no effect on cyclic nucleotides, inositol phosphates, or calcium influx.

Given the negative results obtained so far for activation of the classic second messenger pathways by I_1-imidazoline receptors, we tested the possibility that I_1-receptors may be coupled to phospholipase A2. Receptor activation of phospholipase A2 liberates arachidonic acid, the precursor for prostaglandins.[62] Rat PC12 clonal pheochromocytoma cells, which lack α_2-adrenergic receptors (TABLE 2), were treated with nerve growth factor to induce a neuronal phenotype, which produces a 250% increase in the density of I_1-imidazoline sites.[63] Stimulation of I_1-receptors in PC12 cells by the I_1-agonist moxonidine elicits a nearly fivefold increase in release of PGE_2 from PC12 cells (FIG. 5). Maximal stimulation was obtained with 0.1 μM moxonidine, and a 10 nM concentration elicited a significant response (230 ± 90% of control; n = 4). These doses are consistent with the IC_{50} of moxonidine at I_1-imidazoline sites in PC12 cell plasma membranes (IC_{50} = 16 ± 8 nM; n = 2). The PGE_2 release elicited by 0.1 μM moxonidine was attenuated by the addition of another I_1-ligand, BDF-6143. When given alone, BDF-6143 had no effect on PGE_2 release (data not shown). These data suggest that moxonidine is indeed an I_1-imidazoline agonist and BDF-6143 is an I_1-antagonist. Cimetidine, which behaves like an I_1-agonist in binding and functional assays in the RVLM,[6] also elicits PGE_2 release which can be antagonized by BDF-6143 (FIG. 5). The calcium ionophore A23187 elicits only a modest increase in PGE_2 release, in contrast to other cell types. This suggests that phospholipase A2 is relatively insensitive to increases in intracellular calcium in PC12 cells, unlike many nonneuronal cells.[62] Prostaglandin release elicited by I_1-imidazoline receptor stimulation may be independent of calcium influx, consistent with the lack of effect of clonidine on ^{45}Ca influx.[60]

These studies suggest that the physiological actions elicited by stimulation of I_1-imidazoline receptors may be mediated by the generation of prostaglandins. The

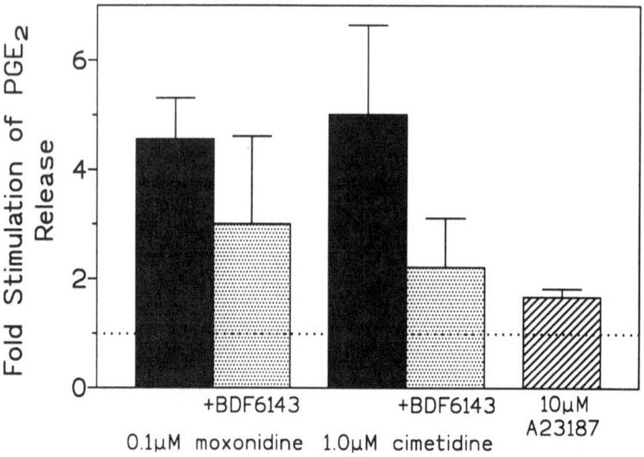

FIGURE 5. Release of prostaglandin E_2 into the medium elicited by the selective I_1-agonist moxonidine in PC12 cells. Data represent the mean of four experiments in duplicate. PC12 cells were seeded into 75-cm^2 flasks coated with rat tail collagen and 2 days later treated with 200 ng/ml NGF daily for 6 days. Confluent cultures were treated with increasing concentrations of moxonidine with or without calcium ionophore (10 μM) for 15 minutes and the medium collected, concentrated on C18 columns, and assayed for prostaglandin E_2 by ELISA. Basal release was 2.4 pg/75 cm^2 flask per 15 minutes. Bars show the effect of moxonidine (0.1 μM), cimetidine (1.0 μM), and their antagonism by BDF-6143 (10 μM). For comparison, the effect of the calcium ionophore A23187 (10 μM) is shown. Moxonidine and cimetidine stimulated prostaglandin release, and this effect was antagonized by the putative I_1-imidazoline antagonist BDF-6143.

following properties of PGE_2 release from PC12 cells in response to moxonidine implicate a specific receptor-mediated signaling pathway: (1) dose-dependence; (2) effective concentrations of agonists approximate their binding affinity; (3) inhibition by an antagonist that binds to the receptor and elicits no response on its own; and (4) the response can be elicited by structurally dissimilar agents (moxonidine and cimetidine) which bind to I_1-receptors. Interestingly, prior studies linked the vasodepressor action of clonidine to arachidonic acid metabolites. Blockade of brain prostaglandin synthesis by intracerebroventricular indomethacin attenuates the hypotensive action of clonidine[64] and the hyperthermic action within the preoptic nucleus.[50] These data support the existence of a hypothetical pathway whereby I_1-imidazoline receptor activation stimulates phospholipase A2, thereby liberating arachidonic acid, resulting in the generation of prostaglandins, which then act on neuronal receptors to lower blood pressure and elicit other responses.

SUMMARY

Data were presented showing that I_1-imidazoline sites show a unique ligand specificity that differs markedly from that of any of the α_2-adrenergic subtypes or the I_2-imidazoline sites labeled by [^{3}H]idazoxan. On the other hand, the ligand specificity of I_1-imidazoline sites is maintained across mammalian species (cow, rat, dog, and human) and between different tissues and cell types. I_1-Imidazoline sites can be further distinguished from I_2 sites because the latter, unlike I_1 sites, were not present in RVLM membranes from bovine brain stem. Furthermore, I_1-imidazoline sites were modulated by guanine nucleotides with a specificity appropriate for a receptor coupled to G-protein and were mainly localized to plasma membranes. I_1-Imidazoline sites show a unique pattern of distribution between diverse tissues and cell types and appear to be a neuroepithelial marker as well as being present in secretory cells of the pancreatic islets. The widespread distribution of I_1-imidazoline sites implies that the functional significance of this putative receptor may have been underestimated. The signaling pathway associated with the I_1-imidazoline receptor remains to be fully elucidated, but is likely that activation of phospholipase A2 leading to release of arachidonic acid and subsequent generation of prostaglandins plays a major role.

ACKNOWLEDGMENTS

We express our appreciation to Dr. Janice Douglas of CWRU Medicine for supplying renal epithelial and mesangial cells and providing support and encouragement; Dr. Diane Felsen of Cornell University Medical College, for providing dog and human prostate samples; Dr. Carole Liedtke of CWRU Pediatrics for supplying tracheal epithelial cells; Dr. Lorraine Iacovitti of Hahnemann University for providing astrocytes; and Dr. Lloyd Greene of Columbia University for providing PC12 cells.

REFERENCES

1. Bylund, D. B. & D. C. U'Prichard. 1983. Characterization of α_1- and α_2-adrenergic receptors. Int. Rev. Neurobiol. **24:** 343-431.
2. Bogaievsky, D., Y. Bogaievsky, D. Tsoucaris-Kupfer & H. Schmitt. 1974. Blockade of the central hypotensive effect of clonidine by alpha-adrenoreceptor antagonists in rats, rabbits and dogs. Clin. Exp. Pharmacol. Physiol. **1:** 527-534.
3. Karppanen, H. 1981. Interrelationships between clonidine and histaminergic mechanisms. Trends Pharmacol. Sci. **2:** 35-37.
4. Bousquet, P., J. Feldman, R. Bloch & J. Schwartz. 1984. Central cardiovascular effects of α-adrenergic drugs: Differences between catecholamines and imidazolines. J. Pharmacol. Exp. Ther. **230:** 232-236.
5. Ernsberger, P., M. P. Meeley, J. J. Mann & D. J. Reis. 1987. Clonidine binds to imidazole binding sites as well as α_2-adrenoceptors n the ventrolateral medulla. Eur. J. Pharmacol. **134:** 1-13.
6. Ernsberger, P., R. Giuliano, R. N. Willette & D. J. Reis. 1990. Role of imidazole receptors in the vasodepressor response to clonidine analogs in the rostral ventrolateral medulla. J. Pharmacol. Exp. Ther. **253:** 408-418.
7. Ernsberger, P. R., K. L. Westbrooks, M. O. Christen & S. G. Schäfer. 1992. A second generation of centrally acting antihypertensive agents act on putative I_1-imidazoline receptors. J. Cardiovasc. Pharmacol. **20** (Suppl. 4): S1-S10.
8. Ernsberger, P., T. H. Damon, L. M. Graff, M. O. Christen & S. G. Schäfer. 1993. Moxonidine, a centrally-acting antihypertensive agent, is a selective ligand for I_1-imidazoline sites. J. Pharmacol. Exp. Ther. **264:** 172-182.
9. Ernsberger, P., M. A. Haxhum, L. M. Graff, L. A. Collins, I. Dreshaj, D. L. Grove, M. E. Graves, S. G. Schäfer & M. O. Christen. 1994. A novel mechanism of action

for hypertension control: Moxonidine as a selective I_1-imidazoline agonist. Cardiovasc. Drugs Ther. **8** (Suppl. 1): 27-41.
10. LI, G., S. REGUNATHAN, C. J. BARROW, J. ESHRAGHI, R. COOPER & D. J. REIS. 1994. Agmatine: An endogenous clonidine-displacing substance in the brain. Science **263:** 966-969.
11. FELSEN, D., P. ERNSBERGER, P. M. SUTARIA, R. J. NEJAT, P. NGUYEN, M. MAY, D. S. BRESLIN, D. N. MARION & E. D. VAUGHAN, JR. 1994. Identification, localization and functional analysis of imidazoline and α-adrenergic receptors in canine prostate. J. Pharmacol. Exp. Ther. **268:** 1063-1071.
12. ERNSBERGER, P., L. IACOVITTI & D. J. REIS. 1990. Atrocytes cultured from specific brain regions differ in their expression of adrenergic binding sites. Brain Res. **517:** 202-208.
13. ERNSBERGER, P., J. ZHOU, T. H. DAMON & J. G. DOUGLAS. 1992. Angiotensin II receptor subtypes in cultured rat renal mesangial cells. Am. J. Physiol. Renal, Fluid Electrolyte Physiol. **263:** F411-F416.
14. ERNSBERGER, P., M. P. MEELEY & D. J. REIS. 1989. Neuroblastoma-glioma hybrid cells contain clonidine-displacing substance. Eur. J. Pharmacol. **174:** 135-138.
15. LIEDTKE, C. M., J. FURIN & P. ERNSBERGER. 1993. α_2-Adrenergic, but not imidazole, agonists activate NaCl cotransport in rabbit tracheal epithelial cells. Am. J. Physiol. Cell Physiol. **264:** C568-C576.
16. WELSH, C., G. DUBYAK & J. G. DOUGLAS. 1988. Relationship between phospholipase C activation and prostaglandin E_2 and cyclic adenosine monophosphate production in rabbit tubular epithelial cells. J. Clin. Invest. **81:** 710-719.
17. ERNSBERGER, P. 1991. Heterogeneity of imidazoline binding sites in brain: Proposed I_1 and I_2 subtypes. IBRO World Cong. Neurosci. **3:** 274.
18. ERNSBERGER, P., C. SIMS & J. G. DOUGLAS. 1991. I_1-imidazoline sites in rat renal medulla: Binding properties and up-regulation by subpressor angiotensin infusion. Hypertension **18:** 380.
19. ERNSBERGER, P., K. L. WESTBROOKS & L. M. GRAFF. 1991. Imidazoline binding sites labeled by clonidine analogs (I_1 sites): Putative G-protein-linked receptors. FASEB J. **5:** A1066.
20. MICHEL, M. C. & P. ERNSBERGER. 1992. Keep an eye on the I site: Imidazoline-preferring receptors. Trends Pharmacol. Sci. **13:** 369-370.
21. PILETZ, J. E. & K. SLETTEN. 1993. Nonadrenergic imidazoline binding sites on human platelets. J. Pharmacol. Exp. Ther. **267:** 1493-1502.
22. ZONNENSCHEIN, R., S. DIAMANT & D. ATLAS. 1990. Imidazoline receptors in rat liver cells: A novel receptor or a subtype of α_2-adrenoceptors. Eur. J. Pharmacol. **190:** 203-215.
23. LANGIN, D., H. PARIS & M. LAFONTAN. 1990. Binding of [^{3}H]idazoxan and of its methoxy derivative [^{3}H] RX821002 in human fat cells: [^{3}H]idazoxan but not [^{3}H] RX821002 labels additional non-α_2-adrenergic binding sites. Mol. Pharmacol. **37:** 876-885.
24. DE VOS, H., G. BRICCA, J. DE KEYSER, J.-P. DE BACKER, P. BOUSQUET & G. VAUQUELIN. 1994. Imidazoline receptors, non-adrenergic idazoxan binding sites and α_2-adrenoceptors in the human central nervous system. Neuroscience **59:** 589-598.
25. WIKBERG, J. E. S. & S. UHLÉN. 1990. Further characterization of the guinea pig cerebral cortex idazoxan receptor: Solubilization, distinction from the imidazole site, and demonstration of cirazoline as an idazoxan receptor-selective drug. J. Neurochem. **55:** 192-203.
26. TESSON, F., C. PRIP-BUUS, A. LEMOINE, J.-P. PEGORIER & A. PARINI. 1991. Subcellular distribution of imidazoline-guanidinium-receptive sites in human and rabbit liver. Major localization to the mitochondrial outer membrane. J. Biol. Chem. **266:** 155-160.
27. PARINI, A., I. COUPRY, F. TESSON, I. LIMON & C. GARGALIDIS-MOUDANOS. 1992. Imidazoline-guanadinium receptive site (IGRS): Mitochondrial distribution and regulation properties. Fundam. Clin. Pharmacol. **6** (Suppl. 1): 35s-40s.

28. BRICCA, G., H. GRENEY, M. DONTENWILL-KIEFFER, J. ZHANG, A. BELCOURT & P. BOUSQUET. 1993. Heterogeneity of the specific imidazoline binding of [^{3}H]idazoxan in the human cerebral cortex. Neurochem. Int. **22:** 153-163.
29. REGUNATHAN, S., D. L. FEINSTEIN & D. J. REIS. 1993. Expression of non-adrenergic imidazoline sites in rat cerebral cortical astrocytes. J. Neurosci. Res. **34:**681-688.
30. BRICCA, G., M. DONTENWILL, A. MOLINES, J. FELDMAN, A. BELCOURT & P. BOUSQUET. 1989. The imidazoline preferring receptor: Binding studies in bovine, rat and human brainstem. Eur. J. Pharmacol. **162:** 1-9.
31. BRICCA, G., J. ZHANG, H. GRENEY, M. DONTENWILL, J. STUTZMANN, A. BELCOURT & P. BOUSQUET. 1993. Relevance of the use of [^{3}H]clonidine to identify imidazoline receptors in rabbit brainstem. Br. J. Pharmacol. **110:** 1537-1543.
32. BRICCA, G., H. GRENEY, J. ZHANG, M. DONTENWILL, J. STUTZMANN, A. BELCOURT & P. BOUSQUET. 1994. Human brain imidazoline receptors: Further characterization with [^{3}H]clonidine. Eur. J. Pharmacol. Mol. Pharmacol. **266:** 25-33.
33. MACKINNON, A. C., M. STEWART, H. J. OVERMAN, M. SPEDDING & C. M. BROWN. 1993. [^{3}H]*p*-aminoclonidine and [^{3}H]idazoxan label different populations of imidazoline sites on rat kidney. Eur. J. Pharmacol. **232:** 79-87.
34. ERNSBERGER, P., G. FEINLAND, M. P. MEELEY & D. J. REIS. 1990. Characterization and visualization of clonidine-sensitive imidazole sites in rat kidney which recognize clonidine-displacing substance. Am. J. Hypertens. **30:** 90-97.
35. ERNSBERGER, P. & L. A. COLLINS. 1993. Autoradiography of I_1-imidazoline and α_2-adrenergic sites in rat brainstem: Reticular localization of I_1-sites. Soc. Neurosci. Abstr. **19:** 1694.
36. KING, P. R., A. L. GUNDLACH, B. JARROTT & W. J. LOUIS. 1992. α_2-Adrenoceptor and catecholamine-insensitive binding sites for [^{3}H]rilmenidine in membranes from rat cerebral cortex. Eur. J. Pharmacol. **218:** 101-108.
37. MOLDERINGS, G. J., D. MOURA, K. FINK, H. BONISCH & M. GOTHERT. 1993. Binding of [^{3}H]clonidine to I_1-imidazoline sites in bovine adrenal medullary membranes. Naunyn-Schmiedeberg's Arch. Pharmacol. **348:** 70-76.
38. KAMISAKI, Y., T. ISHIKAWA, Y. TAKAO, H. OMODANI, N. KUNO & T. ITOH. 1990. Binding of [^{3}H]*p*-aminoclonidine to two sites, α_2-adrenoceptors and imidazoline binding sites: Distribution of imidazoline binding sites in rat brain. Brain Res. **514:** 15-21.
39. MURAMATSU, I. & S. KIGOSHI. 1992. Tizanidine may discriminate between imidazoline-receptors and α_2-adrenoceptors. Jpn. J. Pharmacol. **59:** 457-459.
40. FISZMAN, M. L., M. A. LUCHELLI-FORTIS & F. J. STEFANO. 1989. Amphetamine antagonizes the presynaptic inhibitory effect of clonidine through an interaction at the level of the α_2-adrenoceptors. Gen. Pharmacol. **20:** 351-358.
41. PILETZ, J. E., A. C. ANDORN, J. R. UNNERSTALL & A. HALARIS. 1991. Binding of [^{3}H]-*p*-aminoclonidine to α_2-adrenoceptor states plus a non-adrenergic site on human platelet plasma membranes. Biochem. Pharmacol. **42:** 569-584.
42. HAXHIU, M. A., I. DRESHAJ, S. G. SCHÄFER & P. ERNSBERGER. 1994. Selective antihypertensive action of moxonidine is mediated mainly by I_1-imidazoline receptors in the rostral ventrolateral medulla. J. Cardiovasc. Pharmacol. **24** (Suppl. 1): S1-S8.
43. ERNSBERGER, P., H. L. ELLIOTT, H.-J. WEIMANN, A. RAAP, M. A. HAXHIU, E. HOFFERBER, A. LOW-KROGER, J. L. REID & H.-J. MEST. 1993. Moxonidine: A second-generation central antihypertensive agent. Cardiovasc. Drug Rev. **11:** 411-431.
44. GOMEZ, R. E., P. ERNSBERGER, G. FEINLAND & D. J. REIS. 1991. Rilmenidine lowers arterial pressure via imidazole receptors in brainstem C1 area. Eur. J. Pharmacol. **195:** 181-191.
45. NOSJEAN, A. & P. G. GUYENET. 1991. Role of ventrolateral medulla in sympatholytic effect of 8-OHDPAT in rats. Am. J. Physiol. Regul. Integr. Comp. Physiol. **260:** R600-R609.

46. Ernsberger, P., Y. R. Kou & N. R. Prabhakar. 1990. Imidazole binding sites in the ventrolateral medulla and carotid body are labeled by ^{125}I-p-iodoclonidine (^{125}I-PIC): Possible chemosensory role. FASEB J. **6:** A716.
47. Kou, Y. R., P. Ernsberger, P. A. Cragg, N. S. Cherniack & N. R. Prabhakar. 1991. Role of α_2-adrenergic receptors in the carotid body response to isocapnic hypoxia. Respir. Physiol. **83:** 353-364.
48. Quan, N., L. Xin, A. L. Ungar & C. M. Blatteis. 1992. Preoptic norepinephrine-induced hypothermia is mediated by α_2-adrenoceptors. Am. J. Physiol. Regul. Integr. Comp. Physiol. **262:** R407-R411.
49. Menon, M. K., R. L. Lloyd & L. J. Fitten. 1990. Antagonism of the hypothermic effect of clozapine in mice by centrally-active α_2-adrenergic antagonists and α_1-adrenergic agonists. Psychopharmacology **101:** 67-72.
50. Romanovsky, A. A., O. Shido, A. L. Ungar & C. M. Blatteis. 1993. Genesis of biphasic thermal response to intrapreoptically microinjected clonidine. Brain Res. Bull. **31:** 509-513.
51. Walker, B. R., J. A. M. Anderson & C. R. W. Edwards. 1992. Clonidine therapy for Shapiro's syndrome. Q. J. Med. **82:** 235-245.
52. Beleslin, D. B., A. H. Rezvani & R. D. Myers. 1986. Dissociation of locomotor impairment from mydriasis evoked by clonidine injected into cat's rostral hypothalamus. Brain Res. Bull. **17:** 379-385.
53. Timmermans, P. B. M. W. M., A. M. C. Schoop, H. Y. Kwa & P. A. Van zwieten. 1981. Characterization of α-adrenergic receptors participating in the central hypotensive and sedative effects of clonidine using yohimbine, rauwolscine and corynanthine. Eur. J. Pharmacol. **70:** 7-15.
54. Zornow, M. H. 1991. Ventilatory, hemodynamic and sedative effects of the α_2-adrenergic agonist, dexmedetomidine. Neuropharmacology **30:** 1065-1071.
55. Schmidt, U., H. Frerick, K. Kraft, N. Schenk & A. Low-Kroger. 1992. Hypertension: A possible risk in road traffic. J. Cardiovasc. Pharmacol. **20** (Suppl. 4): S50-S56.
56. De Sarro, G. B., C. Ascioti, F. Froio, V. Libri & G. Nistico. 1987. Evidence that locus coeruleus is the site where clonidine and drugs acting at α_1- and α_2-adrenoceptors affect sleep and arousal mechanisms. Br. J. Pharmacol. **90:** 675-685.
57. Tibiriça, E., J. Feldman, C. Mermet, L. Monassier, F. Gonon & P. Bousquet. 1991. Selectivity of rilmenidine for the nucleus reticularis lateralis, a ventrolateral medullary structure containing imidazoline-preferring receptors. Eur. J. Pharmacol. **209:** 213-221.
58. Powis, D. A. & P. F. Baker. 1986. α_2-Adrenoceptors do not regulate catecholamine secretion by bovine adrenal medullary cells: A study with clonidine. Mol. Pharmacol. **29:** 134-141.
59. Allan, D. R., S. B. Penner & D. D. Smyth. 1993. Renal imidazoline preferring sites and solute excretion in the rat. Br. J. Pharmacol. **108:** 870-875.
60. Regunathan, S., M. J. Evinger, M. P. Meeley & D. J. Reis. 1991. Effects of clonidine and other imidazole-receptor binding agents on second messenger systems and calcium influx in bovine adrenal chromaffin cells. Biochem. Pharmacol. **42:** 2011-2018.
61. Regunathan, S., M. P. Meeley & D. J. Reis. 1990. Effect of clonidine on second messenger systems in rat adrenal gland. Life Sci. **47:** 2127-2133.
62. Axelrod, J. 1990. Receptor-mediated activation of phospholipase A_2 and arachidonic acid release in signal transduction. Biochem. Soc. Trans. **18:** 503-508.
63. Ernsberger, P. 1992. I_1-Imidazoline binding sites in PC12 cells are up-regulated by nerve growth factor (NGF). Soc. Neurosci. Abstr. **18:** 622.
64. Siren, A-L. & H. Karppanen. 1980. Influence of analgesic antipyretics on the central cardiovascular effects of clonidine in rats. Prostaglandins **20:** 285-296.

Dynamic Properties of I_2-Receptors[a]

JARL E. S. WIKBERG

Department of Pharmaceutical Biosciences
Uppsala University
Uppsala, Sweden

An impressive number of studies have accumulated in support of the existence of "I-receptors"[1,2] that are responsive to imidazolines and related compounds. It has been suggested that two different classes of such receptors exist, termed I_1 and I_2. These receptor subtypes are distinguished by their selectivity for drugs, the I_1 type showing high affinity for clonidine and moxonidine, for example, and the I_2 type for idazoxan and guanabenz.[3,4] Although the I-receptors have mainly been identified with radioligand binding using ligands such as [^{3}H]clonidine and [^{3}H]idazoxan, mounting evidence also indicates that these sites are linked to the regulation of various physiological functions such as blood pressure,[5] neural activity in locus coeruleus neurons,[6] transmembrane ion fluxes,[7,8] prolactin[9,10] and insulin secretion,[11,12] specific mRNA expression,[13] the release of noradrenaline from sympathetic neurons,[14–16] as well as the regulation of the number of I-receptor sites themselves.[17,18] An endogenous compound that was previously known as clonidine-displacing substance (CDS) has been isolated and shown to be agmatine, the decarboxylated arginine.[19,20] However, although the pattern of a novel physiological system is beginning to emerge, a number of problems and contradictions still exist, and in many cases the link between imidazoline binding sites and physiological function is not clear. For example, ample evidence exists that I_2-binding sites are located on both the outer membrane of the mitochondria as well as the plasma membranal sites,[19,21–24] but the functional significance of these differentially located sites is at present completely unknown. Moreover, there are also many indications that imidazoline sites are even more heterogeneous than judged by their subdivision into I_1 and I_2. In the past, much interest was focused on amiloride because the compound seemed to delineate I_2-sites into two subtypes.[4] The reported differences were initially between species only, but recent data suggest that amiloride may also differentiate between different I_2-sites in the human as well.[23] We have also shown that I_2-sites in the guinea pig may be delineated further into two subclasses according to their selectivity for (+) and (–)medetomidine.[2] Inasmuch as guinea pig I_2-sites show all over low affinity for amiloride, these combined data seem to indicate that I_2-sites are a heterogeneous family of [^{3}H]idazoxan binding sites. However, our more recent studies provide evidence that at least some of the heterogeneity for I_2-sites may arise from the functional regulation of the I_2-binding sites.[25,26] Some of our previous results supporting the existence of the cooperative regulation of I_2-sites will be summarized. New data to support the existence of dynamic interconversion of I_2-sites will also be presented.

[a] This work was supported by the Swedish Medical Research Council (04X-05957).

EVIDENCE FOR THE EXISTENCE OF AFFINITY STATES OF I_2-RECEPTORS

The first indication for the existence of interconvertible [^{3}H]idazoxan binding sites came from studies on membranes prepared from the guinea pig kidney.[25] In the presence of 3 μM yohimbine to block α_2-adrenoceptors, [^{3}H]idazoxan appeared to bind to a single nonadrenergic site with a K_d of 11 nM and a B_{max} of 2.9 pmol/mg protein in the kidney membranes. A number of imidazoline and guanidinium compounds showed high affinity for the [^{3}H]idazoxan site, whereas catecholamines and many other classic receptor active compounds showed low affinity. Because the [^{3}H]idazoxan site had a much higher affinity for idazoxan itself and for drugs such as cirazoline and guanabenz than it did for clonidine, rilmenidine, and moxonidine, these sites are clearly of the I_2-type. However, of 29 compounds tested in competition with [^{3}H]idazoxan, 18 displayed strongly biphasic competition curves that could be resolved into two site fits. It thus appeared that the I_2-sites were heterogeneous and consisted of two different entities that were capable of binding [^{3}H]idazoxan with equal affinity but many other compounds with differing affinity. However, more careful analysis of the data indicated that the situation was more complicated, suggesting that the affinity of the I_2-sites was dynamically modulated by the binding of a drug to the I_2-sites themselves. In other words, it seemed as if the I_2-site was subject to cooperative regulation. Evidence for this contention is summarized as follows.

First, in the competition studies the apparent proportion of high and low affinity sites for the drugs that gave significant two-site fits varied considerably and depended on which drug was used (Figs. 1 and 2), a finding that is not expected for two independent binding sites. Many of the drugs tested displayed grossly differing affinity for the high and low affinity sites, allowing accurate determination of the proportion of sites. Yet the proportion of sites varied from ~$^{10}/_{90}$% to ~$^{50}/_{50}$%, depending on which drug was tested. Thus, for example, in the experiment presented in Figure 1D, (−)lofexidine showed a 316-fold difference for the apparent high and low affinity sites, 92% of the sites being of high and 8% of low affinity. Moreover, as shown in Figure 1C, histamine showed a 320-fold difference in the affinity for the high and low affinity sties, whereas the proportion of sites instead was 75% of high and 25% of low affinity. Furthermore, rilmenidine showed a 60-fold difference in affinity for the high and low affinity sites and 40% of the sites were of high and 60% of low affinity (Fig. 1A), and UK-14,304 showed a 108-fold difference in affinity for sites and a $^{72}/_{28}$% ratio for the proportion of sites (Fig. 1B). These varying proportions of sites were highly-reproducible between experiments, and analysis of variance showed high significance ($p < 0.0001$) due to the effect induced by the drug used in competition with [^{3}H]idazoxan. It should also be pointed out that the variation was not due to the use of different batches of kidney membranes, because all tests were performed on the same batch of membranes. Thus, these data suggest that the proportion of sites was dynamically modulated depending on which drug was present in the assay mixture.

The second indication that [^{3}H]idazoxan binding to the I_2-sites of the kidney membranes did not obey a simple law of massaction kinetics for two independent sites came from studies that included three different ligands in the assay mixture. We reasoned that if a compound that showed a grossly differing affinity for the two

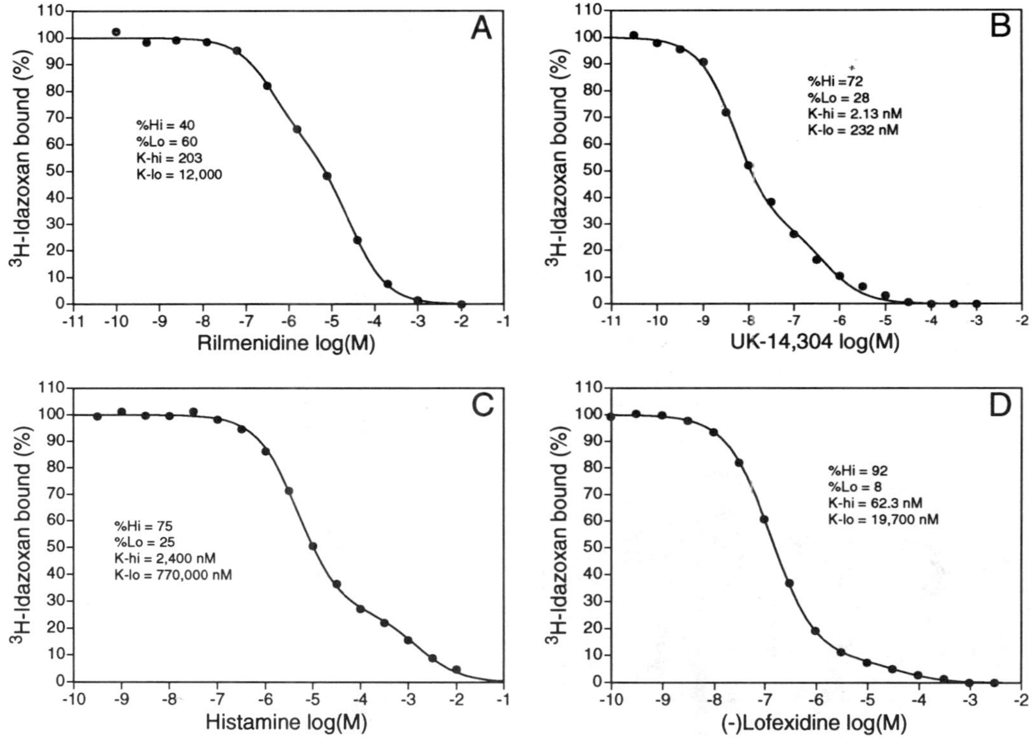

FIGURE 1. Competition curves of various drugs for guinea pig kidney I_2-sites using ~10 nM [^{3}H]idazoxan and 3 μM rauwolscine to prevent binding to α_2-adrenoceptors. The *solid curves* represent the computer drawn best fit, assuming that ligands bound to two independent sites according to the law of massaction. Shown also are the results for % sites and apparent K_ds of the competitors obtained from the two site fits. (Figure was adopted from Wikberg *et al.*[25] through the courtesy of *Pharmacology & Toxicology*.)

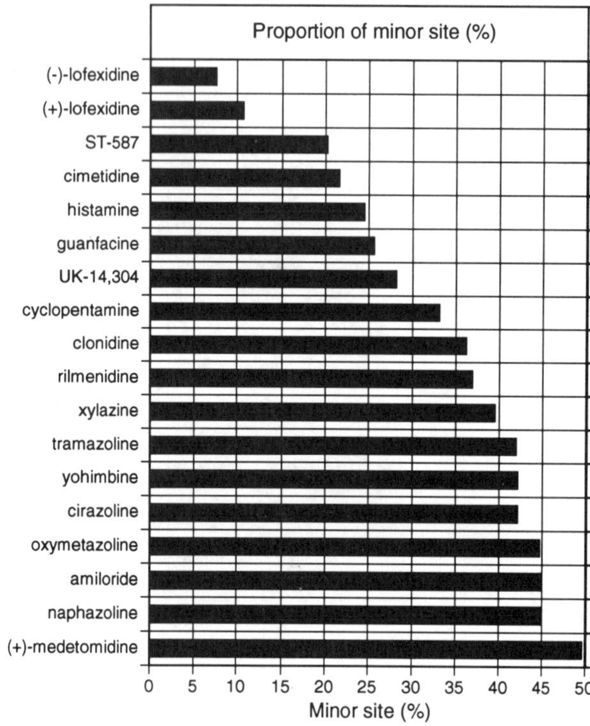

	K-d Hi (nM)	K-d Lo (nM)	Lo/Hi
(-)-lofexidine	61.5	14900	242
(+)-lofexidine	327	20000	61
ST-587	3.88	216	56
cimetidine	119000	2240000	19
histamine	3000	747000	249
guanfacine	607	7150	12
UK-14,304	2.93	261	89
cyclopentamine	3190	506000	159
clonidine	1170	38200	33
rilmenidine	237	10400	44
xylazine	9610	206000	21
tramazoline	24.8	500	20
yohimbine	10200	127000	12
cirazoline	0.363	5.47	15
oxymetazoline	1030	21600	21
amiloride	3350	97100	29
naphazoline	6.6	169	26
(+)-medetomidine	55.5	1170	21

FIGURE 2. Apparent proportions of the minor binding site for the competition curves of 18 drugs that resolved into two site fits using an assay on guinea pig kidney membranes with ~10 nM [^{3}H]idazoxan as radioligand. The two site model was based on the assumption that ligands bound to independent sites according to the law of massaction. Besides the data for the 18 compounds shown, 7 compounds, namely, idazoxan, (−)medetomidine, detomidine, guanabenz, phentolamine, clemastin, and prazosin, were not included in the figure because these compounds did not generate significant two site fits. Moreover, additionally four compounds were excluded because of their over-all low affinity for the kidney I_2 site. (Figure is based on data taken from Wikberg *et al.*[25] through the courtesy of *Pharmacology & Toxicology.*)

kidney I_2-sites was added in a fixed (masking) concentration that excluded the binding of [^{3}H]idazoxan to one of the sites but not to the other, a competition curve of a second drug that also bound to the two I_2-sites with differing affinities should reflect only the competition with [^{3}H]idazoxan at the site not being excluded by the first compound. In the presence of the masking drug the competition curve of the second drug is expected to become steeper, as the curve would reflect competition from only one remaining site. In these experiments we selected histamine as the masking drug because it displayed a large difference in the affinities for the two apparent I_2-sites. (Histamine showed a 3-μM K_d for the high affinity site and a 750-μM K_d for the low affinity I_2-site in the guinea pig kidney.) TABLE 1 summarizes the results of the competition of UK-14,304, rilmenidine, oxymetazoline, and clonidine with 10 nM [^{3}H]idazoxan in the presence and in the absence of masks of 50 and 200 μM

TABLE 1. Hill Coefficients (n_H) and IC_{50} Values of Competition Curves for UK-14,304, Rilmenidine, Oxymetazoline, and Clonidine on Guinea Pig Kidney I_2-Sites Obtained with and without Maks of 50 or 200 μM Histamine and using ~10 nM [^{3}H]Idazoxan as Radioligand[a]

	Experimental Data		Computer Simulated Data	
Drugs	n_H	IC_{50} (nM)	n_H	IC_{50} (nM)
UK-14,304	0.60	22.7	0.62	22.9
+50 μM histamine	0.54	198	0.91	323
+200 μM histamine	0.49	218	0.99	505
Rilmenidine	0.58	5,100	0.58	5,130
+50 μM histamine	0.50	14,100	0.77	510
+200 μM histamine	0.46	16,100	0.95	470
Oxymetazoline	0.69	10,300	0.69	10,400
+50 μM histamine	0.70	22,000	0.97	73,300
+200 μM histamine	0.71	19,300	1.00	100,000
Clonidine	0.59	16,400	0.60	16,500
+50 μM histamine	0.58	50,900	0.76	1,830
+200 μM histamine	0.62	52,000	0.95	1,610

[a] These results are for actual and simulated competition curves: the latter had been generated in the computer by assuming that ligands bound to two independent sites according to the law of massaction. Data were adapted from Wikberg *et al.*[25] through the courtesy of *Pharmacology & Toxicology*. For details regarding binding constants used in the simulation, refer to the original paper.

of histamine. For comparison a set of computer-simulated data was also generated by assuming that the ligands were binding to two independent sites according to the law of massaction. (In these simulations the dissociation constants of the drugs were assumed to be those obtained in the previous competition studies using the test compound in competition with [^{3}H]idazoxan as tabulated in FIGURE 2.) All data were then analyzed by fitting them to the four-parameter logistic function to obtain Hill coefficients (n_H) and IC_{50} values. As seen in TABLE 1 the simulated data gave Hill coefficients that were approaching unity with the increasing addition of histamine, as expected. By contrast to the simulated data, Hill coefficients remained well below unity in both the presence and absence of histamine for the experimental data. For UK-14,304 and rilmenidine Hill coefficients even decreased. Thus, these data indicate that although the masking drug histamine seems to bind with higher affinity to one of two apparent I_2-sites, at least two apparent binding sites still remain after the addition of histamine for the second competitor to bind to. Such behavior could not be explained without assuming that the second competitor was actually able to "recruit" binding sites from the sites that histamine had already occupied. Such behavior could only occur if the binding affinities of one I_2-site were actually able to modulate the binding affinity of an adjacent I_2-site. In other words, these data seemed to indicate that the I_2-binding sites were subjected to cooperative regulation.

If the I_2-site was subjected to cooperative regulation, most of the tested ligands must have been inducing negative cooperativity because their competition curves were shallow with Hill coefficients less than unity. However, one drug, namely detomidine, showed a Hill coefficient slightly above unity (1.29) in the guinea pig kidney. These steep competition curves could be reproduced in all tests performed ($n = 6$), and statistical analysis showed that the Hill coefficients were clearly significantly higher than unity. To shed more light on this phenomenon we obtained corresponding data using the guinea pig liver, as shown in FIGURE 3. In the presence of 3 μM rauwolscine to block α_2-adrenoceptors, [^{3}H]idazoxan bound to one apparent I_2-site in the liver with a K_d of 18 nM and a capacity of 0.67 pmol/mg protein (FIG. 3A). Six compounds that had shown n_Hs ranging from 0.57 to 1.29 in the kidney were tested in the liver. Also in the liver the Hill coefficients of these compounds varied widely, the n_Hs ranging from 0.43 to 2.02. As seen in FIGURE 3, the difference in the slopes of the shallowest curve in the liver, which was generated by UK-14,304, is very large compared to the slope of the steepest curve which was generated by detomidine. Moreover, as seen in FIGURE 3D, the n_Hs obtained in the liver and kidney were highly correlated, the order of n_Hs being UK-14,304 < rilmenidine = histamine < (+)medetomidine < (–)medetomidine < detomidine for both tissues. An interesting observation was also made that [^{3}H]idazoxan binding in the liver contained a minor component that required a higher concentration of detomidine for its displacement (FIG. 3B). By using a compound model consisting of two independent logistic functions, it was shown that in fact the Hill coefficient of the major very steep component was 3.24, whereas the n_H of the minor component was close to unity. Thus, the steep competition curves obtained for detomidine on liver I_2-sites suggest that detomidine binding to these sites is governed by a strong positive cooperative effect similar to that for the guinea pig kidney I_2-sites. The nature of the minor component of [^{3}H]idazoxan binding revealed by detomidine in the liver is currently not completely clear. Perhaps yet another idazoxan-binding site exists in the liver that is different from the major site, the latter being the one subjected to cooperative regulation.

KINETIC PROPERTIES OF GUINEA PIG I_2-RECEPTORS

Further evidence for a dynamic interconversion of I_2-receptors has come from kinetic studies on the binding of [^{3}H]idazoxan to guinea pig kidney I_2-sites. These studies were performed at 12°C rather than at the standard temperature of 25°C used for the other experiments just summarized in order to slow down the rate of association/dissociation of [^{3}H]idazoxan to allow more accurate characterization of its kinetics. Thus, in association experiments performed at 12°C [^{3}H]idazoxan was found to bind with three rate constants (FIG. 4A; TABLE 2). The capacity of the three apparent components amounted to 35, 27, and 38%, respectively, for the fast, intermediate, and slow sites. However, in the dissociation experiments [^{3}H]idazoxan appeared to dissociate from the kidney I_2-site only from two sites, a fast and an intermediate site with 81 and 19%, respectively, of total binding capacity (FIG. 4B; TABLE 2). It thus appeared as if during the association of [^{3}H]idazoxan to I_2-binding sites a substantial

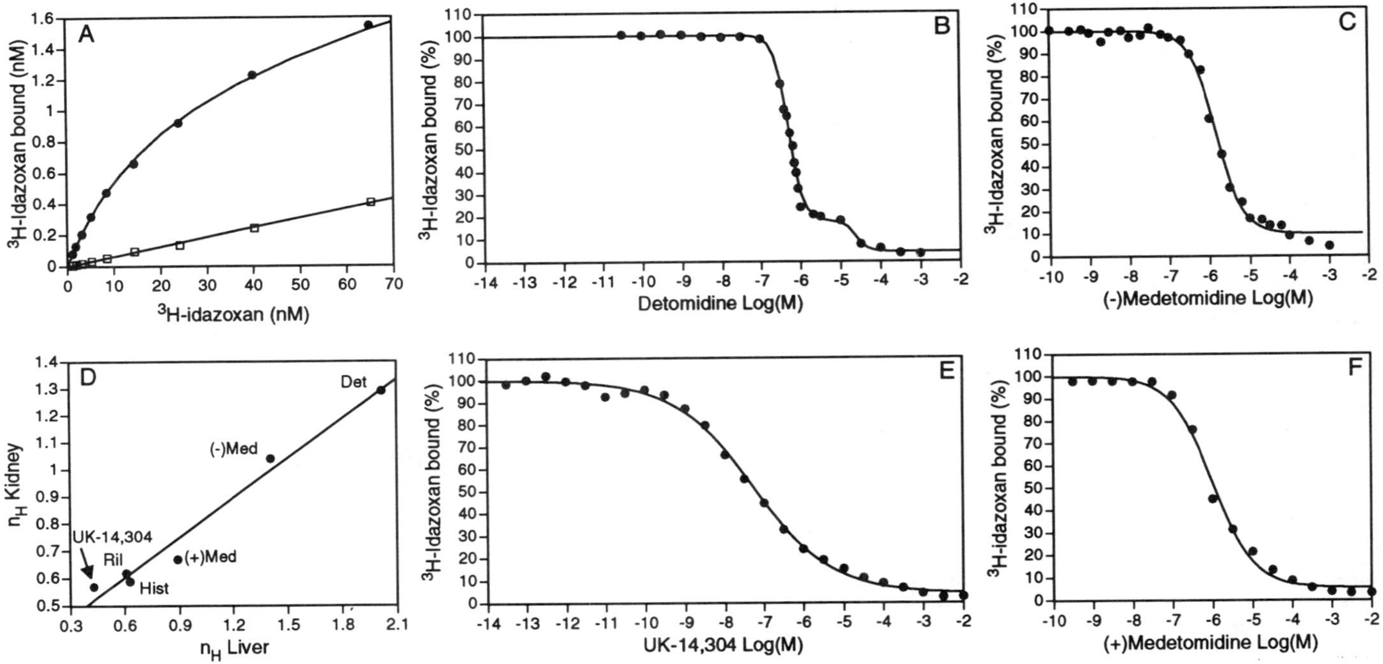

FIGURE 3. (**A**) Saturation study of [^{3}H]idazoxan in the guinea pig liver. Shown are total binding (J) and nonspecific binding (G), the latter determined in the presence of 3 μM nonlabeled idazoxan. (**B, C, E, & F**) Competition curves of, respectively, detomidine, (–)medetomidine, UK-14,304, and (+)medetomidine in the guinea pig liver using ~10 nM [^{3}H]idazoxan. (**D**) Correlation of Hill coefficient (n_H) of drugs obtained from fitting competition curves of the indicated drugs in competition with [^{3}H]idazoxan in guinea pig liver and kidney to the four-parameter logistic function. Correlation was highly significant ($r^2 = 0.98$; $p < 0.001$). For all assays 3 μM rauwolscine was included to prevent the binding of [^{3}H]idazoxan to α_2-adrenoceptors. (Figure was adopted from Rangel *et al.*[26] through the courtesy of *Pharmacology & Toxicology*.)

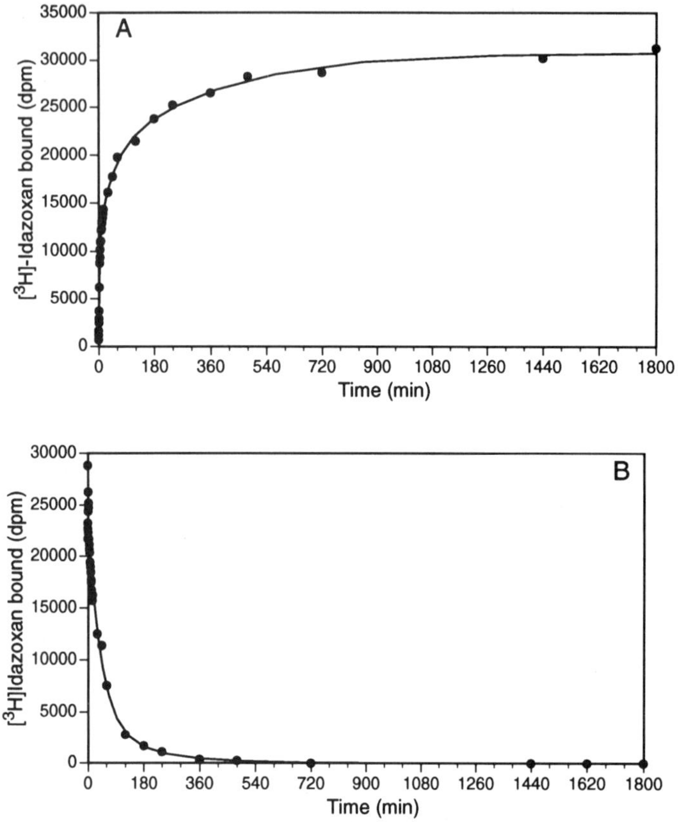

FIGURE 4. Kinetics of [^{3}H]idazoxan binding to guinea pig kidney I_2-sites at 12°C. (**A**) The association of 10 nM [^{3}H]idazoxan and (**B**) the dissociation of [^{3}H]idazoxan from the kidney I_2-sites are shown. In **B** the kidney membranes were first incubated with 10 nM [^{3}H]idazoxan for 48 hours at 12°C; thereafter, dissociation was achieved by diluting with unlabeled idazoxan (final concentration 3×10^{-5} M of unlabeled idazoxan). All time points represent specific [^{3}H]idazoxan binding which was obtained by subtracting binding in the presence of 3×10^{-6} M unlabeled idazoxan from binding in the absence of unlabeled idazoxan.

number of the slower sites has become converted to faster sites. The apparent dissociation constants for the fast and intermediate site were calculated to be 0.9 and 4.1 nM, respectively, from these data which corresponded to a somewhat higher apparent affinity for [^{3}H]idazoxan than that indicated by its equilibrium K_d determined at 25°C. However, in view of the overall complex behavior of the binding to I_2-sites indicated by both kinetic and equilibrium studies, these values should probably only be regarded as "apparent" and most likely reflect the combined action of processes of binding sites that are capable of undergoing complex interactions. Also the difference in temperature could have affected the calculated K_d values.

TABLE 2. Summary of Results from the Association and Dissociation of [^{3}H]Idazoxan to and from Guinea Pig Kidney I_2-Sites at 12°C[a]

Association		Dissociation	
k_1 ($min^{-1} \cdot nM^{-1}$)	R (%)	k_{-1} (min^{-1})	R (%)
2.78×10^{-2}	35	2.47×10^{-2}	81
1.60×10^{-3}	27	6.56×10^{-3}	19
2.28×10^{-4}	38	—	0

[a] R denotes the capacity of the site in % of the specific binding at steady state. For further details regarding the experiments, see the legend to FIGURE 4.

OTHER EVIDENCE FOR HETEROGENEITY OF GUINEA PIG I_2-BINDING SITES

From earlier studies we had obtained evidence that the imidazoline I_2-binding sites of the guinea pig cerebral cortex and ileum differed pharmacologically.[2] The most interesting difference involved the medetomidine stereoisomers, because in the cerebral cortex the (–) form of medetomidine was more than 10-fold less potent than was the (+) form, whereas in the ileum both isomers showed equally high affinity. Another marked difference was that for clonidine which was 15-fold more potent on the cortex I_2-sites than on the ileum I_2-sites. From these results it was of interest to compare the binding affinities of the cortex and ileum I_2-sites with those of the kidney and liver (FIG. 5; TABLE 3). As seen from the various correlations, the best correlation is for liver versus kidney I_2-sites ($r^2 = 0.92$). The second best correlation was for kidney versus ileum I_2-sites ($r^2 = 0.79$-0.83; TABLE 3). Data for the ileum and cerebral cortex had been based on the assumption that I_2-sites were homogeneous in these tissues and were not influenced by, for example, cooperative regulation.[2] Inasmuch as both the high and low affinity dissociation constants of two apparent I_2-sites in the liver, as well as the K_i values (based on the Cheng and Prusoff calculations) had been calculated, these three different values were used separately in correlations when the kidney was included. However, irrespective of which method was used in the calculations, there was a similar degree of correlation. The poorest correlation was between the kidney and cerebral cortex I_2-sites ($r^2 = 0.57$-0.63; TABLE 3). The more similar nature of the kidney and ileum (as well as the liver) I_2-sites is possibly indicated by the finding that all of these I_2-sites do not show stereoselectivity for (+) and (–)medetomidine, in contrast to the strong stereoselectivity of the medetomidine isomers for the cerebral cortex I_2-sites. It is also notable that all guinea pig I_2-sites show low affinities for amiloride (K_d of cerebral cortex = 40,500 nM, K_d of ileum = 292,000 nM, and K_i of kidney = 15,400 nM). Nevertheless, correlation analysis also indicates overall considerable heterogeneity between guinea pig [^{3}H]idazoxan binding sites. Whether this heterogeneity is due to a difference in the functional modulation of the binding affinities of sites or to structural differences remains to be established. It seems likely that the final answer will prove to be the combination of both structural and functional differences.

TABLE 3. Correlation (r^2) for Log(K_d) and Log(K_i) Values of 15 Drugs (Idazoxan, Cirazoline, Naphazoline, Clonidine, Oxymetazoline, (–)Lofexidine, (+)Lofexidine, UK-14,304, (–)Medetomidine, (+)Medetomidine, Detomidine, Guanabenz, Guanfacine, Amiloride, and Histamine) Determined in Competition with [^{3}H]Idazoxan in the Guinea Pig Cerebral Cortex, Ileum, and Kidney[a]

	Cerebral Cortex K_d	Ileum K_d	Kidney K_i	Kidney High Affinity K_d	Kidney Low Affinity K_d
Cerebral cortex	1.00	–	–	–	–
Ileum	0.77	1.00	–	–	–
Kidney K_I	0.57	0.83	1.00	–	–
Kidney high affinity K_d	0.63	0.79	0.92	1.00	–
Kidney low affinity K_d	0.63	0.81	0.79	0.74	1.00

[a] K_d values were determined using computer modeling, and K_i values were determined using the Cheng and Prusoff equation from the IC_{50} values of competition curves. Data were taken from Wikberg *et al.*[2,25] through the courtesy of *Pharmacology & Toxicology.*

CONCLUDING REMARKS

As just discussed, our data indicate that the idazoxan binding I_2-sites in the guinea pig represent complex entities that may be both heterogeneous and subjected to homeotrop allosteric regulation. Literature data also support the existence of heterogeneous I_2-binding sites. Thus, using a photoaffinity label, [^{125}I]AZIPI (a cirazoline analog), Lanier *et al.*[27] found evidence of three different imidazoline binding peptides in the rat. One of these showed an M_r of 61,000 and was sensitive to amiloride, and two others showed M_rs of 55,000 and 61,000 and were insensitive to amiloride. Photoaffinity labeling of all three peptides by [^{125}I]AZIPI was blocked by cirazoline, idazoxan, and guanabenz. Moreover, Limon *et al.*[22] purified a [^{3}H]idazoxan binding protein with an M_r of 60,000 from the rabbit kidney, whereas Wang *et al.*[28] purified another [^{3}H]idazoxan binding protein (M_r = 70,000) from the bovine adrenal chromaffin cells. In a recent study Bricca *et al.*[29] also provided evidence that I_2-sites in the human cerebral cortex were heterogeneous, because [^{3}H]idazoxan appeared to bind with two affinities to the cortex membranes. Moreover, the competition curves of some drugs were resolved into two-site fits, and the association and dissociation of [^{3}H]idazoxan to I-sites were stated to occur with a two rate constant. It is also

FIGURE 5. Correlation of log(K_d) and log(K_i) values for drugs determined in competition with [^{3}H]idazoxan obtained on membranes prepared from guinea pig cerebral cortex, ileum, kidney, and liver. Drugs used and K_i and K_d values were determined as detailed in the footnote to TABLE 3. For the liver the correlation is shown only for UK-14,304, rilmenidine, histamine, (+)medetomidine, (–)medetomidine, and detomidine. The data are taken from Rangel *et al.*[26] through the courtesy of *Pharmacology & Toxicology.*

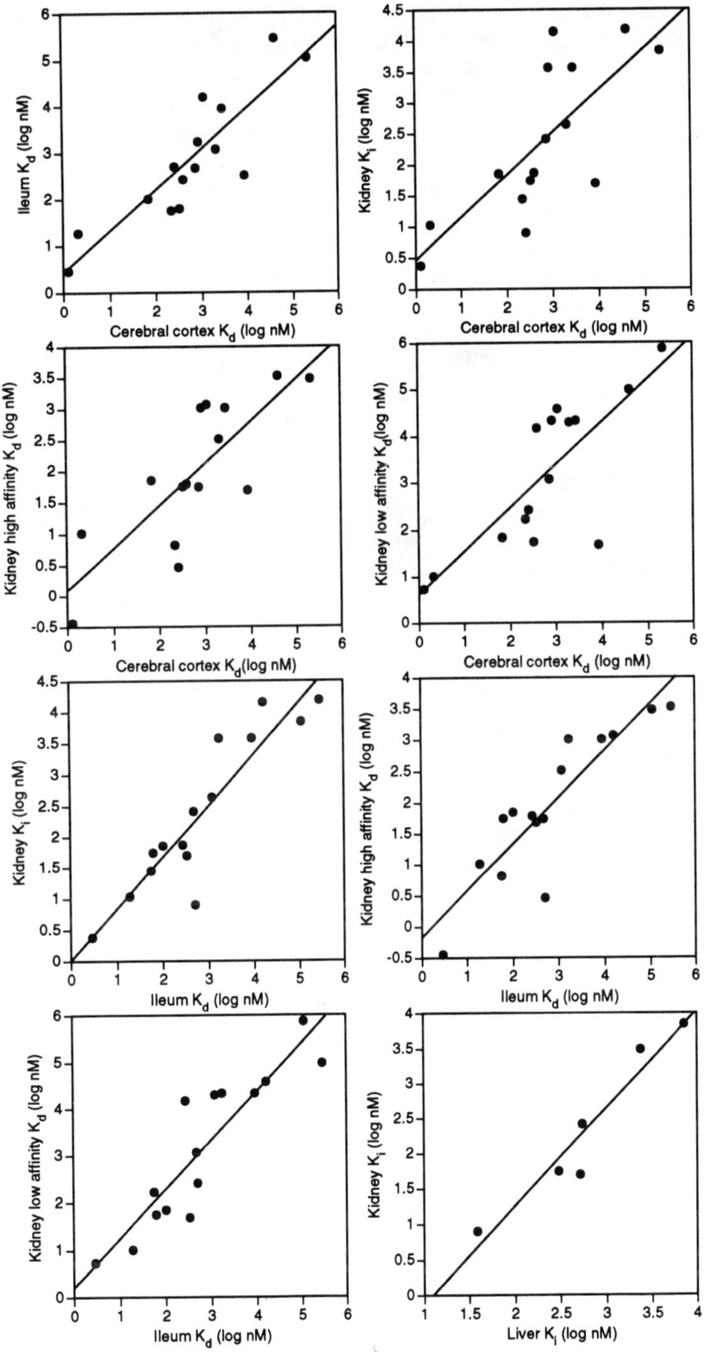
Ileum Kd (log nM)
Cerebral cortex Kd (log nM)
Kidney Ki (log nM)
Cerebral cortex Kd (log nM)
Kidney high affinity Kd (log nM)
Cerebral cortex Kd(log nM)
Kidney low affinity Kd(log nM)
Cerebral cortex Kd (log nM)
Kidney Ki (log nM)
Ileum Kd (log nM)
Kidney high affinity Kd (log nM)
Ileum Kd (log nM)
Kidney low affinity Kd (log nM)
Ileum Kd (log nM)
Kidney Ki (log nM)
Liver Ki (log nM)

FIGURE 5.

interesting that Bricca and coworkers stated that binding of [^{3}H]idazoxan was inhibited by clonidine in a "noncompetitive" manner, because their analysis of [^{3}H]idazoxan binding isotherms by the scatchard transform seemed to indicate a decrease in the apparent number of I_2-binding sites by clonidine. However, such an apparent decrease in the number of binding sites could possibly have been due to a more complex allosteric regulation, similar to the type that we described for guinea pig kidney I_2-sites, and their observation deserves further study. In this context it should be mentioned that others also found indications that [^{3}H]idazoxan binding to rabbit, human, and rat brain I_2-sites were heterogeneous in nature.[30,31] Another interesting observation was made recently by Li *et al.*[20] who showed that competition curves of the putative I-receptor signal substance agmatine for [^{3}H]idazoxan binding on I_2-receptors on bovine chromaffin cells are markedly shallow ($n_H = 0.5$). Moreover, agmatine induced the release of noradrenaline and adrenaline from bovine chromaffin cells. It is therefore tempting to speculate that the induction of negative cooperativity at the I_2-site is related to the functional properties of the compound. According to this interpretation, if a drug induces negative cooperativity ($n_H < 1$), it would be an agonist; if it was unable to induce any cooperativity ($n_H = 1$), it would be a competitive blocker; and if it induced positive cooperativity ($n_H > 1$), it would be an inverse agonist at the I_2-receptors. Admittedly this interpretation is currently only speculative and will have to be substantiated by further studies. Lastly, we have tested agmatine for its ability to bind to guinea pig liver I_2 sites. The Hill coefficient for agamatine was well below unity ($n_H = 0.74$), but surprisingly the substance showed a low affinity for guinea pig kidney I_2-sites ($K_i = 187{,}400$ nM). The apparent affinity of agmatine for kidney I_2-sites is therefore several 100-fold lower than its reported affinity for bovine adrenal chromaffin cell I_2-sites, something that again indicates the complex heterogeneity of the I-receptors.

SUMMARY

Evidence is presented that drug binding to I_2-sites in the guinea pig kidney and liver is subjected to dynamic homeotropic allosteric regulation. Drugs have been found that are capable of inducing both positive and negative cooperativity on both kidney and liver I_2-sites when studied in competition with [^{3}H]idazoxan. Moreover the kinetics of [^{3}H]idazoxan association with guinea pig kidney I_2-sites was complex and occurred with three different rate constants. However, on dissociation of [^{3}H]idazoxan from its binding sites in the kidney a substantial number of the slower sites had been converted to faster sites, thus supporting the idea of a dynamic interconversion of I_2 sites. The guinea pig liver and kidney I_2-sites appear to be similar in their pharmacological properties. They also bear some resemblance to the I_2-sites of the smooth muscle of the guinea pig ileum, but they appear to be more distant to the guinea pig cerebral cortex I_2-sites. Nevertheless, data indicate that considerable overall heterogeneity exists between the guinea pig kidney and liver I_2-sites and the guinea pig ileum and cerebral cortex I_2-sites.

REFERENCES

1. WIKBERG, J. E. S., S. UHLÉN & V. CHHAJLANI. 1991. Properties of membrane bound and solubilized idazoxan- (imidazoline) I-receptors and α_2-adrenoceptors. *In* Adrenoceptors: Structure, Mechanisms, Function.: 293-294. Brikhauser. Basel.
2. WIKBERG, J. E. S., S. UHLÉN & V. CHHAJLANI. 1991. Medetomidine stereoisomers delineate two closely related subtypes of idazoxan (imidazoline) I-receptors in the guinea pig. Eur. J. Pharmacol. **193:** 335-340.
3. MICHEL, M. C. & P. ERSNBERGER. 1992. Keeping an eye on the I site: Imidazoline-preferring receptors. TIPS **13:** 369-370.
4. MICHEL, M. C. & P. A. INSEL. 1989. Are there multiple imidazoline binding sites? TIPS **10:** 342-343.
5. BOUSQUET, P., G. BRICCA, M. DONTENWILL, J. FELDMAN, H. GRENEY, A. BELCOURT, J. STUTZMAN & E. TIBRICIA. 1992. From the α_2-adrenoceptors to the imidazoline preferring receptors. Fundam. Clin. Pharmacol. **6** (Suppl. 1): 15s-21s.
6. PINEDA, J., L. UGEDO & J. A. GARCIA-SEVILLA. 1993. Stimulatory effects of clonidine, cirazoline and rilmenidine on locus coeruleus noradrenergic neurones: Possible involvement of imidazoline-preferring receptors. Naunyn-Schmiedeberg's Arch. Pharmacol. **348:** 134-140.
7. REIS, D. J., S. REGUNATHAN, H. WANG, D. L. FEINSTEIN & M. P. MEELEY. 1992. Imidazoline receptors in the nervous system. Fundam. Clin. Pharmacol. **6:** 23S-29S.
8. IBBOTSON, T., G. EDWARDS & A. H. WESTON. 1993. Antagonism of levcromakalim by imidazoline- and guanidine-derivatives in rat portal vein: Involvement of the delayed rectifier. Br. J. Pharmacol. **110:** 1556-1564.
9. PREZIOSI, P., M. MARTIRE, P. NAVARRA, G. PISTRITTO & M. VACCA. 1989. Prolactin-lowering ability of ($\pm$)-idazoxan may be linked to a central noradrenergic-serotonergic interplay. J. Pharmacol. Exp. Ther. **249:** 256-263.
10. KRULICH, L., J. JURCOVICOVA & T. LE. 1989. Prolactin (PRL) release-inhibiting properties of the alpha 2 adrenergic receptor antagonist idazoxan: Comparison with yohimbine. Life Sci. **44:** 809-818.
11. SCHULTZ, A. & A. HASSELBLATT. 1989. An insulin-releasing property of imidazoline derivatives is not limited to compounds that block α-adrenoceptors. Naunyn-Schmiedeberg's Arch. Pharmacol. **340:** 321-327.
12. BROWN, C. A., S. L. CHAN, M. R. STILLINGS, S. A. SMITH & N. G. MORGAN. 1993. Antagonism of the stimulatory effects of efaroxan and glibenclamide in rat pancreatic islets by the imidazoline, RX801080. Br. J. Pharmacol. **110:** 1017-1022.
13. REGUNATHAN, S., D. L. FEINSTEIN & D. J. REIS. 1993. Expression of non-adrenergic imidazoline sites in rat cerebral cortical astrocytes. J. Neurosci. Res. **34:** 681-688.
14. GÖTHERT, M. & G. J. MOLDERINGS. 1991. Involvement of presynaptic imidazoline receptors in the alpha 2-adrenoceptor-independent inhibition of noradrenaline release by imidazoline derivatives. Naunyn-Schmiedeberg's Arch. Pharmacol. **343:** 271-282.
15. MOLDERINGS, G. J., F. HENTRICH & M. GÖTHERT. 1991. Pharmacological characterization of the imidazoline receptor which mediates inhibition of noradrenaline release in the rabbit pulmonary artery. Naunyn-Schmiedeberg's Arch. Pharmacol. **344:** 630-638.
16. FUDER, H. & P. SCHWARZ. 1993. Desensitization of inhibitory prejunctional alpha 2-adrenoceptors and putative imidazoline receptors on rabbit heart sympathetic nerves. Naunyn-Schmiedeberg's Arch. Pharmacol. **348:** 127-133.
17. OLMOS, G., A. MIRALLES, F. BARTUREN & J. A. GARCIA-SEVILLA. 1992. Characterization of brain imidazoline receptors in normotensive and hypertensive rats: Differential regulation by chronic imidazoline drug treatment. J. Pharmacol. Exp. Ther. **260:** 1000-1007.
18. HAMILTON, C. A., E. JARDINE & J. L. REID. 1993. Down-regulation of imidazoline sites in rabbit kidney. Eur. J. Pharmacol. **243:** 95-97.
19. TESSON, F., I. LIMON & A. PARINI. 1992. Tissue-specific localization of mitochondrial imidazoline-guanidinium receptive sites. Eur. J. Pharmacol. **219:** 335-338.

20. LI, G., S. REGUNATHAN, C. J. BARROW, J. ESHRAGHI, R. COOPER & D. J. REIS. 1994. Agmatine: An endogenous clonidine-displacing substance in the brain. Science **263:** 966-969.

21. TESSON, F. & A. PARINI. 1991. Identification of an imidazoline-guanidinium receptive site in mitochondria from rabbit cerebral cortex. Eur. J. Pharmacol. **208:** 81-83.

22. LIMON, I., I. COUPRY, S. M. LANIER & A. PARINI. 1992. Purification and characterization of mitochondrial imidazoline-guanidinium receptive site from rabbit kidney. J. Biol. Chem. **267:** 21645-21649.

23. DIAMANT, S., T. ELDAR-GEVA & D. ATLAS. 1992. Imidazoline binding sites in human placenta: Evidence for heterogeneity and a search for physiological function. Br. J. Pharmacol. **106:** 101-108.

24. LAFONTAN, M., D. LANGIN, M. PORTILLO & H. PARIS. 1992. Imidazoline binding sites in fat cells. Localization and pharmacologic differentiation from alpha 2-adrenergic receptors. Am. J. Hypertens. **5:** 72S-79S.

25. WIKBERG, J. E. S., S. UHLÉN & V. CHHAJLANI. 1992. Evidence that drug binding to non-adrenergic [^{3}H]idazoxan binding sites (I-receptors) occurs to interacting or interconvertible affinity forms of the receptor. Pharmacol. Toxicol. **70:** 208-219.

26. RANGEL, N., S. UHLÉN & J. E. S. WIKBERG. 1993. Detomidine binding to guinea pig liver imidazoline receptors (I-receptors) shows marked positive cooperativity. Pharmacol. Toxicol. **73:** 86-90.

27. LANIER, S. M., B. IVKOVIC, I. SINGH, J. L. NEUMEYER & V. BAKTHAVACHALAM. 1993. Visualization of multiple imidazoline/guanidinium-receptive sites. J. Biol. Chem. **268:** 16047-16051.

28. WANG, H., S. REGUNATHAN, M. P. MEELEY & D. J. REIS. 1992. Isolation and characterization of imidazoline receptor protein from bovine adrenal chromaffin cells. Mol. Pharmacol. **42:** 792-801.

29. BRICCA, G., M. GRENEY, H. G. M. DONTENWILL-KIEFFER, J. ZHANG, A. BELCOURT & P. BOUSQUET. 1993. Heterogeneity of the specific imidazoline binding of [^{3}H]idazoxan in the human cerebral cortex. Neurochem. Int. **22:** 153-163.

30. RENOUARD, A., P. S. WIDDOWSON & A. CORDI. 1993. [^{3}H]idazoxan binding to rabbit cerebral cortex recognises multiple imidazoline I_2-type receptors: Pharmacological characterization and relationship to monoamine oxidase. Br. J. Pharmacol. **109:** 625-631.

31. MIRALLES, A., G. OLMOS, M. SASTRE, F. BARTURNE, I. MARTIN & J. A. GARCIA-SEVILLA. 1993. Discrimination and pharmacological characterization of I_2-imidazoline sites with [^{3}H]idazoxan and alpha-2 adrenoceptors with [^{3}H]RX821002 (2-methoxy idazoxan) in the human and rat brains. J. Pharmacol. Exp. Ther. **264:** 1187-1197.

Imidazoline Receptors, Subclassification, and Drug-Induced Regulation

C. A. HAMILTON

Department of Medicine and Therapeutics
University of Glasgow
Western Infirmary
Glasgow G11 6NT, Scotland

It is now generally accepted that imidazoline receptors are distinct from α_2-adrenoceptors and that as α_2-adrenoceptors have been subclassified, subcategories of imidazoline receptors exist. Currently the division into I_1 and I_2 sites based on labeling with [^{3}H]clonidine and [^{3}H]idazoxan is widely used. However, this classification does not satisfy all the data. An additional division into I_{1A}, I_{1B}, I_{2A}, and I_{2B} based on affinities for cimetidine ($I_{1A} >> I_{1B}$) and amiloride ($I_{2A} >> I_{2B}$) has been suggested, but anomalies are still present.

CLASSIFICATION BASED ON RADIOLIGAND BINDING STUDIES

To date, classification has been based on radioligand binding characteristics. Here the choice of ligand, species, tissue, and methodology may all influence the result.

Choice of Ligand

As already mentioned, the binding of [^{3}H]clonidine and its analog [^{3}H]*p*-aminoclonidine (PAC) to nonadrenergic imidazoline sites is very different from that of [^{3}H]idazoxan and was reviewed extensively elsewhere.[1,2] Differences between the binding of [^{3}H]PAC and [^{3}H]clonidine may also exist. For example, cimetidine and histamine have been shown to displace [^{3}H]PAC binding to nonadrenergic sites in rat and bovine brain,[3,4] but they failed to displace [^{3}H]clonidine from rabbit or human brain.[5–7]

Other ligands have been used to label imidazoline sites. In 1989 Gould *et al.*[8] reported that [^{3}H]guanabenz labeled nonadrenergic sites in rat brain and kidney; however, neither clonidine nor phentolamine displaced binding at these sites. Rilmenidine was also shown to label nonadrenergic sites in rat cerebral cortex.[9] The affinities of a wide range of compounds for this site were examined and found not to be entirely compatible with rilmenidine labeling of either I_1 or I_2 sites, but corresponded most closely with binding at the I_{2B} site.

TABLE 1. Effect of Species on [^{3}H]Idazoxan Binding at I_2 Sites: Drug Affinities at Imidazoline Sites Labeled by [^{3}H]Idazoxan in Kidney

Species	Drug	K_I (nM)	Drug	K_I (nM)
Man	Guanabenz	9.6 ± 1.5[a]	Clonidine	2,575 ± 710[a]
Rabbit	—	1.8 ± 0.5[a]	—	303 ± 31[a]
Pig	—	18 ± 5[b]	—	10,000 ± 3,000[b]
Rat	—	87[c]	—	>10,000[c]

[a] Lachaud-Pettiti *et al.*, 1991.
[b] Vigne *et al.*, 1989 (IC_{50}).
[c] Michel *et al.*, 1989.

Special Variation

Marked variations in binding characteristics at imidazoline sites between species may also be observed as illustrated for [^{3}H]idazoxan binding in TABLE 1. Some of the anomalies previously discussed in [^{3}H]PAC and [^{3}H]clonidine binding at I_1 sites may in fact be species related.

Tissue Differences

In addition to species differences, differences in binding between tissues have been observed. This was most elegantly demonstrated by Wikberg and colleagues.[10] They reported that whereas cirazoline, guanabenz, and (+)medetomidine showed similar affinities for the [^{3}H]idazoxan binding site in guinea pig cortex and ileum, (−)medetomidine had a much higher affinity for the site in the ileum and clonidine for the site in the cortex (TABLE 2).

TABLE 2. Differences between Tissues in [^{3}H]Idazoxan Binding in Guinea Pig Tissues[a]

Drugs	K_I Cortex (nM)	K_I Ileum (nM)
Cirazoline	1.28	2.90
Guanabenz	68.4	102
(+)Medetomidine	728	473
(−)Medetomidine	8,550	330
Clonidine	1,110	16,200

[a] Reprinted, with permission, from Wikberg *et al.*[10]

TABLE 3. Conditions for Studying Binding to α_2, I_1, and I_2 Sites in Rabbit Kidney and Hindbrain Membranes[a]

Ligand	[^{3}H]Yohimbine	[^{3}H]Clonidine	[^{3}H]Idazoxan
Nonspecific binding	Phentolamine 10^{-5} M	Cirazoline 3×10^{-9} M	Phentolamine 10^{-5} M
Noradrenaline	—	2×10^{-5} M	2×10^{-5} M

[a] Noradrenaline was used to "mask" binding to α_2-adrenergic sites.

Multiple Binding Sites within Tissues

It is also possible that subtypes of I_1 and I_2 receptors coexist in the same tissue. In studies using unlabeled drugs to displace [^{3}H]clonidine and [^{3}H]idazoxan Hill coefficients have been reported to differ significantly from 1.[5,7] In some cases this could represent binding to I and α_2-adrenoceptor sites. However, in the studies cited, assay conditions were such that the radioligand bound exclusively to I sites. Under these conditions a Hill coefficient of less than 1 would suggest either negative cooperativity between sites or binding of the radioligand to more than one subtype of I site.

Methodological Considerations

The exact methodology is another factor that needs to be considered. The choice of ligand has already been discussed. The agent used to define nonspecific binding and the method used to distinguish between binding at adrenergic and nonadrenergic sites (masking adrenergic sites with unlabeled drug or fitting data to multisite models) are also important, as is the buffer composition. Both mono and divalent cations have been reported to modify binding and may be relevant if allosteric sites exist. It has been suggested that different interconvertible binding sites for imidazoline compounds might be included in the same receptor complex.[7] Different assay conditions might favor different conformations of the receptor, resulting in differing drug affinities.

RELATION BETWEEN BINDING AND FUNCTIONAL STUDIES

Functional studies provide an equally confusing picture. Some responses show a close correlation with binding at I_1 or I_2 sites, whereas others show little similarity. There appears to be a good correlation between vasodepressor potency and affinity for I_1 sites in specific brain regions such as the rostral ventrolateral medulla, as illustrated by Ernsberger and colleagues.[11]

In the kidney, inhibition of ^{22}Na uptake by imidazolines correlates closely with affinity for the I_2 site,[12] while effects on urine flow and solute excretion correlate better with binding at I_1 sites.[13]

In other functional studies it has been harder to correlate response with affinity for binding sites. Imidazolines appear to modify insulin secretion. A wide range of compounds with an imidazoline structure, including efaroxan, phentolamine, antazoline, and tolazoline, enhance insulin release from pancreatic β cells.[14] However, other compounds with a high affinity for imidazoline sites may be less effective, and potency in modulating insulin release does not correlate well with binding at either I_1 or I_2 sites.[15,16]

Imidazolines also regulate noradrenaline release from postganglionic sympathetic nerve endings in rabbit pulmonary artery,[17] rabbit heart,[18] and rat kidney.[19] Here there appear to be species differences between rabbit and rat, but in no case does a clear relationship exist between functional potency and I_1 or I_2 binding site affinity.

Thus, this brief survey demonstrates that no general consensus is found between binding and functional studies and that it is sometimes difficult to correlate the subtypes proposed from binding studies with those from functional studies. Moreover, within binding and functional studies different conclusions may be reached, depending on species, tissues studied, and methods. It is possible that the local receptor environment influences binding characteristics and functional responses and that differences exist between receptors in basal and activated states. Such factors could contribute to the apparent profusion of subtypes. Future studies into mechanisms of activation of the imidazoline receptors and their coupling to second messengers and at the level of the genome will clarify the picture.

RECEPTOR REGULATION

Studies into the way receptors are regulated may also provide information on subtype specificity. Homologous regulation would favor distinct entities for the populations concerned, whereas heterologous regulation would suggest close similarities or some complex interaction between subtypes. Regulation of imidazoline receptors may also be important when considering the pathophysiology and treatment of cardiovascular disease. One disadvantage of centrally acting antihypertensive agents is the frequency and severity of withdrawal symptoms on stopping treatment. This has been related to downregulation of the α_2-adrenoceptor on continued agonist exposure.[20] Preliminary reports suggest that withdrawal reactions are much less severe with drugs that are believed to act predominantly via imidazoline receptors.[21,22] In support of this hypothesis we have shown that α_2-adrenoceptor and imidazoline binding are differentially modified after chronic treatment with a range of compounds as discussed here.

REGULATION STUDIES IN KIDNEY AND HINDBRAIN FROM NEW ZEALAND WHITE RABBITS

Male New Zealand white rabbits were treated with a range of compounds with high affinity for α_2- and/or imidazoline receptors for 5-10 days. Drugs or vehicle was constantly infused into the femoral vein via osmotic minipumps, or if solubility was a problem, they were administered twice daily by intraperitoneal injection. Binding to sites labeled by [^{3}H]yohimbine, [^{3}H]idazoxan, and [^{3}H]clonidine in kidney

TABLE 4. Conditions for Chronic Drug Administration

Drug	Concentration (mg/kg/day)	Method of Administration	Durations of Treatment (days)
Noradrenaline	0.35	Osmotic minipump	10
Yohimbine	15	Intraperitoneal injection	5
Idazoxan	13.5	Osmotic minipump	6
Clonidine	1.4	Osmotic minipump	6
Guanabenz	4.8	Osmotic minipump	6
Rilmenidine	15.8	Osmotic minipump	6

and hindbrain was studied as detailed in TABLE 3. The following drugs (TABLE 4) were administered: noradrenaline, yohimbine, idazoxan, clonidine, guanabenz, and rilmenidine.

Small but significant pharmacological/functional effects were observed in most cases on treatment with these doses of the drugs. Briefly, both yohimbine and idazoxan at the doses selected antagonized the depressor response to intracisternally injected clonidine. Clonidine, guanabenz, and rilmenidine treatment resulted in significant bradycardia and decreases in blood pressure that just failed to reach significance. However, despite raising plasma noradrenaline levels approximately sixfold, no rise in blood pressure was observed in the catecholamine-treated group, although bradycardia was observed at 24 hours.

REGULATION IN HINDBRAIN

The effects of treatment on binding to hindbrain membranes are shown in TABLES 5 and 7. Treatment with clonidine and guanabenz, consistent with these drugs' agonist

TABLE 5. Effects of Chronic Drug Treatment on Binding in Hindbrain[a]

	E_{max} (fmol/mg protein)					
	[^{3}H]Yohimbine		[^{3}H]Clonidine		[^{3}H]Idazoxan	
Treatment	Vehicle	Treatment	Vehicle	Treatment	Vehicle	Treatment
Yohimbine	95 ± 12	120 ± 10	—	—	50 ± 5	55 ± 5
Idazoxan	126 ± 4	140 ± 1	25 ± 3	27 ± 4	50 ± 5	40 ± 5
Rilmenidine	83 ± 9	74 ± 7	25 ± 3	27 ± 4	55 ± 8	41 ± 5
Clonidine	83 ± 9	52 ± 5[b]	25 ± 3	34 ± 2	55 ± 8	56 ± 10
Guanabenz	108 ± 10	60 ± 11[b]	29 ± 3	29 ± 3	48 ± 4	45 ± 5

[a] Results expressed as mean ± SE; n = 6–8.
[b] Significantly different from vehicle, $p < 0.5$.

activity at α_2-adrenoceptors, decreased B_{max} for [^{3}H]yohimbine binding in hindbrain, whereas rilmenidine, which has a lower affinity for α_2-adrenoceptors, failed to cause a significant decrease in B_{max} for [^{3}H]yohimbine binding. The α_2-adrenoceptor antagonist idazoxan tended to increase B_{max}, but this did not reach significance. In contrast, no drug significantly altered B_{max} for either [^{3}H]clonidine or [^{3}H]idazoxan binding. No changes in K_D were observed with any drug or ligand. These results are consistent with the low incidence of CNS withdrawal symptoms reported for imidazoline receptor-preferring drugs. However, they contrast with a report by Olmos and coworkers[23] who showed that 7 days of treatment with cirazoline and idazoxan caused upregulation of [^{3}H]idazoxan binding to rat brain membranes. These differences could be species related, dose related, or methodological. Only crude hindbrain membranes, where overall receptor density is low, were examined in our study in the rabbit, whereas in the rat more discrete areas (cortex, medulla, and hypothalamus) were studied.

REGULATION IN KIDNEY

In kidney, noradrenaline treatment decreased and yohimbine treatment increased [^{3}H]yohimbine binding consistent with the respective agonist and antagonist activities of these compounds at α_2-adrenoceptors (TABLES 6 and 7). Neither idazoxan, clonidine, guanabenz, nor rilmenidine significantly modified [^{3}H]yohimbine binding. Guanabenz, idazoxan, and yohimbine significantly attenuated [^{3}H]idazoxan binding, whereas guanabenz and rilmenidine attenuated [^{3}H]clonidine binding (TABLE 6). Thus, in the kidney the pattern of desensitization is distinct for each ligand.

Looking at [^{3}H]clonidine binding in more detail, guanabenz and rilmenidine significantly decreased binding, while clonidine also tended to lower B_{max} and idazoxan to increase B_{max} consistent with their known agonist/antagonist activities at I_1 sites. More detailed comparison of the relative potencies of the compounds used cannot be made from a study of this type as only one dose of each drug was studied and pharmacokinetics were not investigated. Tissue distribution, metabolism, and

TABLE 6. Effects of Chronic Drug Treatment on Binding in Kidney[a]

	B_{max} (fmol/mg protein)					
	[^{3}H]Yohimbine		[^{3}H]Clonidine		[^{3}H]Idazoxan	
Treatment	Vehicle	Treatment	Vehicle	Treatment	Vehicle	Treatment
Noradrenaline	93 ± 16	30 ± 7[b]	–	–	220 ± 12	210 ± 30
Yohimbine	60 ± 5	115 ± 2[b]	–	–	220 ± 12	110 ± 5[b]
Idazoxan	60 ± 5	65 ± 3	152 ± 20	188 ± 22	188 ± 30	89 ± 8[b]
Rilmenidine	45 ± 6	44 ± 5	152 ± 20	76 ± 9[b]	188 ± 30	253 ± 50
Clonidine	45 ± 6	45 ± 7	152 ± 20	119 ± 20	188 ± 30	210 ± 46
Guanabenz	45 ± 4	45 ± 2	152 ± 20	77 ± 11[b]	188 ± 30	125 ± 15[b]

[a] Results expressed as mean ± SE; $n = 6$–8.
[b] Significantly different from vehicle, $p > 0.05$.

TABLE 7. Effects of Chronic Drug Treatment on Binding to α_2, I_1, and I_2 Sites in Kidney and Hindbrain

	[^{3}H]Yohimbine	[^{3}H]Clonidine	[^{3}H]Idazoxan
		Kidney	
Noradrenaline	↓	—	—
Yohimbine	↑	—	↓
Idazoxan	—	—	↓
Clonidine	—	—	—
Guanabenz	—	↓	↓
Rilmenidrine	—	↓	—
		Hindbrain	
Idazoxan	—	—	—
Clonidine	↓	—	—
Guanabenz	↓	—	—
Rilmenidine	—	—	—

— = no change in B_{max}.
↑ = B_{max} increased compared to that of vehicle-treated controls.
↓ = B_{max} decreased compared to that of vehicle-treated controls.

clearance will vary between drugs, and relative tissue concentrations may not reflect the amount of drug administered. For the same reasons comparisons between effects on binding in hindbrain and kidney must be treated with even greater caution.

Initially some of the results obtained for [^{3}H]idazoxan binding are surprising. Idazoxan acts as an antagonist at α_2 and I_1 sites, yet here chronic treatment resulted in a decrease in binding to I_2 sites consistent with agonist activity. Moreover, the effect appeared to be dose related, as we previously showed that treatment with a larger dose of idazoxan reduces binding even further.[24] Some functional studies also suggest that idazoxan could act as an agonist at this site. In proximal tubule cells from rabbit kidney idazoxan, like cirazoline, UK 14304, and rilmenidine, inhibited ^{22}Na uptake,[12] whereas clonidine, moxonidine, cirazoline, and idazoxan all inhibited noradrenaline release in rabbit aorta.[17] In contrast, in rat brain cirazoline and idazoxan both upregulated [^{3}H]idazoxan[23] consistent with antagonist activity. The decrease observed with yohimbine treatment was even more unexpected; however, the affinity of yohimbine for the [^{3}H]idazoxan-labeled site is in the micromolar range[5] and not negligible. Sufficient yohimbine may be present at the binding site for sufficient time to cause downregulation. However, clonidine and rilmenidine, which have approximately ninefold higher affinities for the site, did not modify binding. Fuder and Schwartz[18] similarly observed desensitization of both prejunctional α_2 and I receptors in rabbit heart on chronic treatment with oxymetazoline and aganodine. They were unable to distinguish between incomplete drug selectivity for I and α_2 receptors, structural similarity of the receptors, a functional interaction between the receptors, or true cross-desensitization. However, in our study, whereas yohimbine treatment caused upregulation of the α_2-adrenoceptor, it caused downregulation of the I_2 site; therefore, the results are not simply a consequence of structural similarities between receptors. It is possible that a functional interaction occurs between the α_2-adrenoceptor and the I_2-receptor in the kidney. Interaction between α_2-adrenoceptors and I_1-receptors in the CNS was previously proposed.[25]

COMPARISON OF REGULATION IN HINDBRAIN AND KIDNEY

As discussed previously, although caution must be exercised when comparing regulation in kidney and hindbrain, the different pattern observed in the two tissues could be a consequence of different subtypes of α_2 and I receptors predominating in kidney and hindbrain. The various subtypes could show different affinities for the drugs used as treatments or different sensitivities to modulation, giving rise to the distinct patterns of regulation observed in the experiments described here.

SUMMARY

Overall, as summarized in TABLE 6, a variety of responses to chronic drug treatment were observed depending on the drug, the tissue, and the ligand. Taken together these studies support the concept that the three ligands bind to distinct sites. In addition, they suggest that idazoxan and possibly yohimbine act as agonists at the I_2 site in kidney. Finally, the lack of regulation of the I_1 site in hindbrain is consistent with the low incidence of withdrawal symptoms reported with imidazoline-preferring drugs.

REFERENCES

1. MICHEL, M. C. & P. A. INSEL. 1989. Are there multiple imidazoline binding sites. TIPS **10:** 342-344.
2. HAMILTON, C. A. 1992. The role of imidazoline receptors in blood pressure regulation. Pharmacol. Ther. **54:** 231-248.
3. ERNSBERGER, P., M. P. MEELEY, J. J. MANN & D. J. REIS. 1987. Clonidine binds to imidazole binding sites as well as α_2 adrenoceptors in the ventrolateral medulla. Eur. J. Pharmacol. **134:** 1-13.
4. KAMISAKI, Y., T. ISHIKAWA, Y. TAKAO, H. OMODANI, N. KUNO & T. ITOH. 1990. Binding of [^{3}H]*p*-aminoclonidine to two sites, α_2 adrenoceptors and imidazoline binding sites: Distribution of imidazoline binding sites in rat brain. Brain Res. **514:** 15-21.
5. HAMILTON, C. A., M. A. YAKUBU, E. JARDINE & J. L. REID. 1991. Imidazole binding sites in rabbit kidney and forebrain membranes. J Auton. Pharmacol. **11:** 277-283.
6. BRICCA, G., J. ZHANG, H. GRENEY, H. DONTENWILL, A. STUTZMANN, A. BELCOURT & P. BOUSQUET. 1993. Relevance of the use of [^{3}H] clonidine to identify imidazoline receptors in the rabbit brainstem. Br. J. Pharmacol. **110:** 1537-1543.
7. BRICCA, G., H. GRENEY, J. ZHANG, M. DONTENWILL, J. STUTZMANN, A. BELCOURT & P. BOUSQUET. 1994. Human brain imidazoline receptors: Further characterisation with [^{3}H]clonidine. Eur. J. Pharmacol. **266:** 25-33.
8. GOULD, R. J., K. M. M. MURPHY, S. H. SNYDER, J. W. STRANDHOY & C. E. DUNLAP. 1989. Differential interaction of guanabenz with receptor binding sites in rat brain and kidney. Res. Commun. Chem. Pathol. Pharmacol. **66:** 3-19.

9. KING, P. R., A. L. GUNDLACH, B. JARROTT & W. J. LOUIS. 1992. α_2 Adrenoceptor and catecholamine-insensitive binding sites for [^{3}H] rilmenidine in membranes from rat cerebral cortex. Eur. J. Pharmacol. **218:** 101-108.
10. WIKBERG, J. E. S., S. UHLEIN & V. CHHAJLANI. 1991. Medetomidine stereoisomers delineate two closely related subtypes of idazoxan (imidazoline) 1-receptors in the guinea pig. Eur. J. Pharmacol. **193:** 335-340.
11. ERNSBERGER, P., R. GUILIANO, R. N. WILLETTE & D. J. REIS. 1990. Role of imidazole receptors in the vasodepressor response to clonidine analysis in the rostral ventrolateral medulla. J. Pharmacol. Exp. Ther. **25**(3): 408-418.
12. BIDET, M., P. POUJEOL & A. PARINI. 1990. Effect of imidazolines on Na^+ transport and intracellular pH in renal proximal tubule cells. Biochim. Biophys. Acta **1024:** 173-178.
13. ALLAN, D. R., S. S. PENNER & D. D. SMYTH. 1993. Renal imidazoline preferring sites and solute excretion in the rat. Br. J. Pharmacol. **108:** 870-875.
14. SCHULZ, A. & A. HASSELBLATT. 1989. An insulin-releasing property of imidazoline derivatives is not limited to compounds that block α-adrenoceptors. Naunyn-Schmiedeberg's Arch. Pharmacol. **340:** 321-327.
15. BROWN, C. A., A. C. LOWETH, S. A. SMITH & N. G. MORGAN. 1993. Stimulation of insulin secretion by imidazoline compounds is not duc to interaction with non-adrenoceptor idazoxan binding sites. Br. J. Pharmacol. **108:** 312-317.
16. CHAN, S. L. F., C. A. BROWN & N. G. MORGAN. 1993. Stimulation of insulin secretion by the imidazoline α_2-adrenoceptor antagonist efaroxan is mediated by a novel stereoselective binding site. Eur. J. Pharmacol. **230:** 375-378.
17. GÖTHERT, M. & G. J. MOLDERINGS. 1991. Involvement of presynaptic imidazoline receptors in the α_2-adrenoceptor-independent inhibition of noradrenaline release by imidazoline derivatives. Naunyn-Schmiedeberg's Arch. Pharmacol. **343:** 271-282.
18. FUDER, H. & P. SCHWARZ. 1993. Desensitisation of inhibitory prejunctional α_2-adrenoceptors and putative imidazoline receptors on rabbit heart sympathetic nerves. Naunyn-Schmiedeberg's Arch. Pharmacol. **348:** 127-133.
19. BOHMANN, C., P. SCHOLLMEYER & C. C. RUMP. 1994. Effects of imidazolines on noradrenaline release in rat isolated kidney. Naunyn-Schmiedeberg's Arch. Pharmacol. **349:** 118-124.
20. REID, J. L., B. C. CAMPBELL & C. A. HAMILTON. 1984. Withdrawal reactions following cessation of central and adrenergic receptor agonists. Hypertension **6** (Suppl. 11): 1171-1175.
21. LEWIS, S. J., B. JARROT & A. E. DOYLE. 1987. Rapid eye movement sleep and the associated cardiovascular changes during and after continuous 10 day infusion of clonidine in normotensive rats: Comparison with Lofexidine and S3341. J. Cardiovasc. Pharmacol. **9**(6): 711-718.
22. SANNAJUST, F., C. JULIEN, C. BARRES, C. CERUTTI, E. KOENIG-BERARD & J. SASSARD. 1989. Cardiovascular effects of rilmenidine, a new α_2-adrenoceptor agonist and clonidine in conscious spontaneously hypertensive rats. Clin. Exp. Pharmacol. Physiol. **16:** 837-848.
23. OLMOS, G., A. MIRALLES, F. BARTUREN & J. A. GARCIA-SEVILLA. 1992. Characterisation of brain imidazoline receptors in normotensive and hypertensive rats: Differential regulation by chronic imidazoline drug treatment. J. Pharmacol. Exp. Ther. **260:** 1000-1007.
24. YAKUBU, M. A., N. M. DEIGHTON, C. A. HAMILTON & J. L. REID. 1990. Differences in the regulation of [^{3}H]idazoxan and [^{3}H]yohimbine binding sites in the rabbit. Eur. J. Pharmacol. **176:** 305-311.
25. HEAD, G. A. & F. SANNAJUST. 1992. Importance of imidazoline receptors in the cardiovascular responses to clonidine and rilmenidine in conscious rabbits. Fundam. Clin. Pharmacol. **6** (Suppl. 1): 31S-34S.

Nonadrenergic Binding of [^{3}H]Atipamezole in Rat Lung

A Novel Imidazole Binding Site?[a]

BIRGITTA SJÖHOLM,[b-d] JUHA-MATTI SAVOLA,[e] AND MIKA SCHEININ[b]

[b]*Department of Pharmacology and*
[c]*Centre for Biotechnology*
University of Turku and
[e]*Orion-Farmos*
P.O. Box 425
FIN-20101 Turku, Finland

Both [^{3}H]idazoxan and [^{3}H]*para*-aminoclonidine were initially thought to be selective radioligands for α_2-adrenoceptors. However, they were subsequently also shown to bind extensively to nonadrenergic imidazoline binding sites in brain and other tissues.[1-5] Imidazoline-preferring binding sites or receptors were proposed to mediate the central antihypertensive[6] and insulin-increasing[7,8] actions of some imidazolines. These nonadrenergic binding sites may also mediate neuroprotective effects in an experimental stroke model.[9] Imidazoline receptors may also participate in the contraction of vascular and other smooth muscle.[8,10,11] Human blood platelets have nonadrenergic *para*-aminoclonidine binding sites that may be involved in platelet aggregation.[12] Their density may be increased in depression.[13]

Imidazoline-preferring receptors have been divided into several subclasses which differ from each other with respect to their relative affinities for ligands.[14] One, the I_1-receptor, preferentially binds clonidine and *para*-aminoclonidine, is localized to plasma membranes, and has limited tissue distribution. I_2-receptors preferentially bind idazoxan and are present in a number of tissues and cell types. I_2-receptors may be further subdivided into I_{2a} and I_{2b} types by the binding affinities of, for example, phentolamine and amiloride.[14]

Studies with subcellular fractions from human and rabbit liver have demonstrated that [^{3}H]idazoxan binding to I_{2a} sites is localized to mitochondrial fractions.[15] Blood platelets have I_{2b}-type binding sites in their internal membranes.[16] Therefore, in contrast to α_2-adrenoceptor binding, which is localized to the plasma membrane, [^{3}H]idazoxan binding to I_2 sites appears to be located intracellularly at least in liver cells and blood platelets.

[a]This study was supported by the Academy of Finland, the Technology Development Centre of Finland (TEKES), and Orion-Farmos.

[d]Address for correspondence: Birgitta Sjöholm, MSc, MediCity Research Laboratory, University of Turku, Tykistökatu 6A, FIN-20520 Turku, Finland.

Atipamezole has been described as a potent, specific, and selective α_2-adrenoceptor antagonist.[17,18] In receptor binding assays with [^{3}H]clonidine and [^{3}H]prazosin, the α_2/α_1 selectivity of atipamezole is substantially higher than that of yohimbine and idazoxan.[18] It also displays equal, high (K_i approximately 1 nM) affinity for the pharmacological and genetic α_{2A}-, α_{2B}-, and α_{2C}-adrenoceptor subtypes.[19–22] [^{3}H]Atipamezole was therefore of interest as a possible α_2-adrenergic radioligand. However, in neonatal rat lung membranes, [^{3}H]atipamezole identified an additional population of binding sites distinct from both classic α_2-adrenoceptors and idazoxan-defined imidazoline receptors.[23] The present study further characterizes the nonadrenergic binding of [^{3}H]atipamezole in rat lung.

MATERIALS AND METHODS

Membrane Preparations

All steps in membrane preparation were carried out at 4°C unless otherwise indicated. Protein concentrations were determined by the method of Bradford,[24] with bovine serum albumin as the reference standard.

Rat Lung

Lungs were removed from decapitated 1- to 3-day-old or adult rats (Sprague-Dawley of either sex) and stored at –20°C. The frozen tissue was homogenized using an Ultra-Turrax (2 × 15 s) in 20-25 volumes (compared to the original tissue wet weight) of ice-cold 50 mM Tris-HCl, 0.8 mM EDTA, pH 8.0 at 25°C. The homogenate was first centrifuged at 1,000 × *g* for 5 minutes to remove large particles. The supernatant was filtered through nylon mesh and then centrifuged at 48,000 × *g* for 15 minutes. The pellet was suspended in 20-25 volumes of the same buffer, and homogenization and centrifugation were repeated. The final pellet was resuspended in 10 volumes of 25 mM glycylglycine buffer, pH 7.4 at 25°C and stored at –70°C for binding assays.

Membrane Preparation for Solubilization

Frozen lungs were homogenized using an Ultra-Turrax in ice-cold 50 mM Tris-HCl, 5 mM EDTA, 0.1 mM phenylmethylsulfonyl fluoride, 10 μg/ml soybean trypsin inhibitor, and 200 μg/ml bacitracin, pH 7.5. Membranes were prepared as just described but with a third homogenization and centrifugation in Tris-HCl containing a lower concentration of EDTA (1.5 mM). The final pellets were diluted in 10 volumes of 25 mM glycylglycine buffer and stored at –70°C.

Solubilization

Membranes were pelleted at 48,000 × *g* for 10 minutes and resuspended in the same volume of ice-cold 25 mM glycylglycine buffer containing 6.5 mM 3-[(3-

cholamidopropyl)dimethylammonio]-1-propanesulfonate (CHAPS) and 1 mM dithiothreitol (DTT). The suspensions were gently sonicated for 1 minute and slowly agitated for 30 minutes. Sonication and agitation were repeated, and the suspensions were centrifuged at 120,000 × g for 90 minutes using a Sorvall Ultra Pro 80 ultracentrifuge and T 1270 rotor. The resulting supernatants were used directly for radioligand binding.

Binding Assays

Rat Lung Membranes

Receptor binding was measured by incubating 100 μl of diluted membrane suspension (5-125 μg protein per tube) with [^{3}H]atipamezole in a total volume of 1,000 μl of glycylglycine buffer. After 25 minutes in a 25°C water bath with shaking, incubations were terminated by rapid dilution with ice-cold Tris-buffer and rapid filtration through fiberglass filters (Whatman GF/B) using a Brandel M-48R (Gaithersburg, Maryland) cell harvester in a cold room (6°C). Filters were washed twice with 5 ml of the same buffer and placed into scintillation vials with HiSafe III (Wallac, Turku, Finland) for counting at 48% efficiency (Wallac 1410, Wallac). Specific binding was defined as the difference between total and nonspecific binding determined in parallel sets of tubes but in the presence of an excess of unlabeled atipamezole (1 μM). For saturation studies with adult rat lung membranes, the final concentrations of [^{3}H]atipamezole ranged from 0.08-10 nM.

Competition binding assays were performed at a radioligand concentration near the K_d of the radioligand. [^{3}H]Atipamezole and 12-16 concentrations of unlabeled drugs were incubated with rat lung membranes for 25 minutes at 25°C. Validation of these binding assays was reported earlier.[23]

Solubilized Receptors

The binding assays were carried out as just described, but with about 100 μg of protein per tube and with radioligand concentrations ranging from 0.15-25 nM. The CHAPS-solubilized receptors were incubated with [^{3}H]atipamezole for 3 hours at 5°C.

For kinetic studies of the association reaction, the solubilized receptors were incubated with [^{3}H]atipamezole at concentrations near the K_d at 5°C for periods ranging from 15 seconds to 240 minutes. The reaction was terminated by filtration. Reversibility of [^{3}H]atipamezole binding was assessed by preincubating the soluble preparations for 180 minutes in the presence of radioligand. Thereafter, an excess of competing drug (1 μM atipamezole) was added to displace bound [^{3}H]atipamezole, and specific binding was determined at time points between 0.5 and 16 hours.

Calculations

Receptor affinities (K_d) and receptor densities (B_{max}) were calculated from the results of saturation experiments using computer-assisted analysis of the data with the EBDA-LIGAND package (Elsevier Biosoft, Cambridge, UK), a nonlinear least squares computer program.[25] Association and dissociation data were analyzed with the KINETIC module, and 50% inhibitory concentrations of competing ligands (IC_{50}) were calculated with the EBDA-DRUG module of this program package. The reported proportions of specific binding of total bound radioactivity refer to graphically estimated values at radioligand concentrations equal to the calculated K_d value of each individual experiment.

The K_d values are presented as geometric means ± SE and the other results as arithmetic means ± SE.

Drugs and Chemicals

[^{3}H]Atipamezole (custom synthesized by Amersham, Buckinghamshire, UK; specific activity 83.9 Ci/mmol) was a gift from Orion-Farmos (Turku, Finland). Atipamezole [MPV-1248, 4-(2-ethyl-2,3-dihydro-1H-inden-2-yl)-1H-imidazole], detomidine, dexmedetomidine, idazoxan, *l*-medetomidine, MPV-207 [4-(2,6-dimethylbenzyl)imidazole], MPV-295 (4-[2-(2,6-dimethylphenyl)ethyl]imidazole), MPV-624 (1-[2-(2,6-dimethylphenyl)ethyl]imidazole), MPV-709B (2-[2-(2,6-dimethylphenyl)ethyl]imidazole), MPV-709D (2-[2-(2,6-dimethylphenyl)ethyl]imidazoline), and (*d,l*)-propranolol were synthesized by Orion-Farmos, Turku, Finland. (See FIGURE 1 for structures of these compounds.) *para*-Aminoclonidine, guanabenz, RX821002 [2-(2-methoxy-1,4-benzodioxan-2-yl)-2-imidazoline], and UK-14,304 [5-bromo-6-(2-imidazoline-2-yl-amino)-quinoxaline] were purchased from Research Biochemicals International (Natick, Massachusetts), phentolamine from Ciba-Geigy (Basel, Switzerland), and cirazoline from Synthelabo (Paris, France). The following compounds were obtained from Sigma (St. Louis, Missouri): agmatine, amiloride, bacitracin, 3-[(3-cholamidopropyl)dimethylammonio]-1-propanesulfonate (CHAPS), chlorpheniramine, cimetidine, clonidine, dopamine, dithiothreitol (DTT), (-)-noradrenaline, oxymetazoline, phenylmethylsulfonyl fluoride (PMSF), prazosin, serotonin, and soybean trypsin inhibitor. Other chemicals were of analytical or reagent grade and were purchased from commercial suppliers.

RESULTS

[^{3}H]Atipamezole Binding to Neonatal Rat Lung Membranes

A series of imidazoli(di)nes, imidazoles, and other substances used to characterize imidazoline receptors were tested for their ability to inhibit binding of [^{3}H]atipamezole (0.5 nM) to neonatal rat lung membranes. The specificity of binding was defined by 1 μM atipamezole. The proportion of specific binding was 67-86%. IC_{50} values for the inhibition of [^{3}H]atipamezole binding to neonatal rat lung membranes are shown in TABLE 1. Interestingly, the nonadrenergic binding of [^{3}H]atipamezole was potently displaced only by a number of α_2-adrenergic imidazoles, such as atipamezole (IC_{50} = 2.9 ± 1 nM), detomidine (IC_{50} = 7.3 ± 1 nM), *l*-medetomidine (IC_{50} = 7.8 ± 2 nM), dexmedetomidine (IC_{50} = 5.3 ± 1 nM), and MPV-207 (IC_{50} = 11.0 ± 1 nM), but other imidazoles such as the histamine H_2-antagonist cimetidine, histamine, or

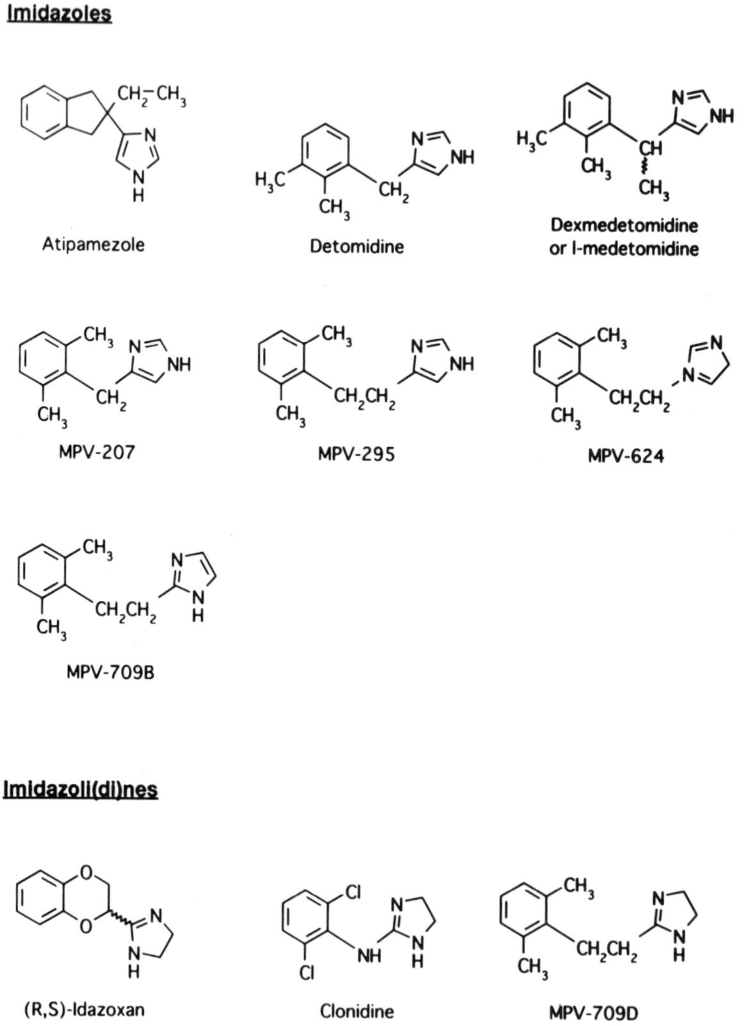

FIGURE 1. Structures of imidazole and imidazoline compounds used in competitive ligand binding assays. See text for chemical names.

imidazole-4-acetic acid did not recognize these binding sites. Furthermore, imidazoli(-di)nes such as clonidine, *para*-aminoclonidine, UK-14,304, phentolamine, idazoxan, and cirazoline were poorly effective in inhibiting the binding of [^{3}H]atipamezole. Neither guanabenz nor amiloride was able to inhibit the binding of [^{3}H]atipamezole. FIGURE 2 shows the inhibition of [^{3}H]atipamezole binding by *l*-medetomidine, dexmedetomidine, idazoxan, and *para*-aminoclonidine to neonatal rat lung membranes.

TABLE 1. Inhibition of [^{3}H]Atipamezole Binding to Membranes of Neonatal Rat Lung

Drug	IC_{50} (nM)	n_H	I_{max} (%)
Imidazoles			
Atipamezole	2.90	0.53	100
Dexmedetomidine	5.30	0.72	100
Detomidine	7.30	0.85	100
MPV-295	7.60	0.40	100
l-Medetomidine	7.80	0.90	100
MPV-207	11.0	0.62	100
MPV-624	230	0.54	100
MPV-709B	1,300	0.43	95
Chlorpheniramine	190,000	0.53	72
Cimetidine	200,000	0.17	58
Histamine	>1,000,000	–	38
Imidazoli(di)nes			
Idazoxan	24,000	0.54	69
MPV-709D	71,000	0.72	89
Cirazoline	80,000	0.19	64
Clonidine	560,000	0.24	56
Oxymetazoline	>500,000	–	34
para-Aminoclonidine	>1,000,000	–	42
UK-14,304	>1,000,000	–	43
Other compounds			
Guanabenz	140,000	0.89	88
(–)-Noradrenaline	>100,000	–	22
Amiloride	>1,000,000	–	35
Prazosin	>200,000	–	38
Serotonin	>1,000,000	–	42
Dopamine	>1,000,000	–	48
(*d,l*)-Propranolol	>1,000,000	–	42

NOTE: IC_{50} = concentration inhibiting 50% of specific [^{3}H]atipamezole binding; n_H = Hill number; I_{max} represents maximum inhibition of specific binding with highest concentration of compound tested.

[^{3}H]Atipamezole Binding to Solubilized Binding Sites from Neonatal Rat Lung

The association time course of [^{3}H]atipamezole binding to a CHAPS-solubilized preparation from neonatal rat lung membranes indicated that [^{3}H]atipamezole binding was biphasic. Approximately 40% of the binding equilibrated rather rapidly (k_{obs} = 0.31 ± 0.01 min^{-1}), and the remainder was bound more slowly (k_{obs} = 0.02 ± 0.01 min^{-1}) (n = 3). [^{3}H]Atipamezole binding was slowly reversed in a biphasic manner by the addition of 1 μM atipamezole. The calculated dissociation rate constants (k_{-1}) were 0.06 ± 0.02 min^{-1} (about 40%) and 0.004 ± 0.02 min^{-1} (about 60%) (n = 2).

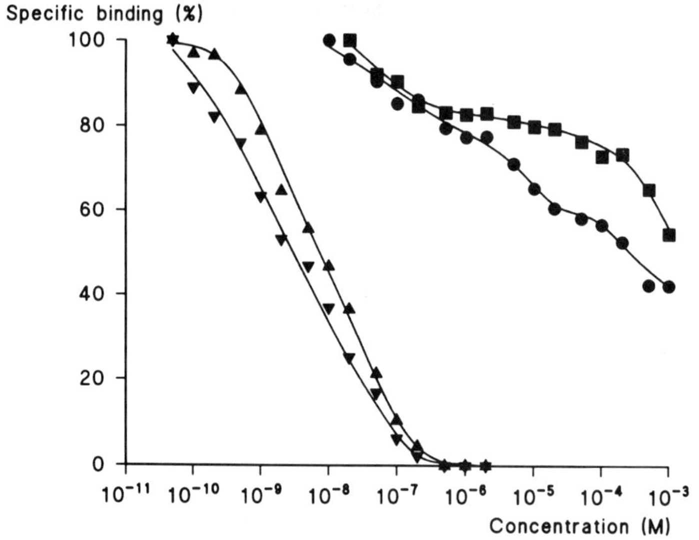

FIGURE 2. Inhibition of specific [^{3}H]atipamezole binding to neonatal rat lung membranes by *l*-medetomidine (▲), dexmedetomidine (▼), idazoxan (●), and *para*-aminoclonidine (■).

Saturation assays indicated that the solubilized neonatal rat lung preparation contained two populations of binding sites, but the employed radioligand concentrations were too low to characterize the low-affinity component accurately. The high-affinity component was characterized by a K_d of 0.46 ± 0.06 nM and a B_{max} of 150 ± 35 fmol/mg protein (n_H = 1.10 ± 0.05) (*n* = 8). The calculated IC_{50} values for inhibition of [^{3}H]atipamezole binding to solubilized receptors are shown in TABLE 2.

[^{3}H]Atipamezole Binding to Adult Rat Lung Membranes

In adult rat lung, [^{3}H]atipamezole labeled a single population of noncooperative sites, as indicated by linear Rosenthal plots (FIG. 3) and Hill coefficients not signifi-

TABLE 2. Inhibition of [^{3}H]Atipamezole Binding to CHAPS-Solubilized Receptors from Neonatal Rat Lung

Drug	IC_{50} (nM)	n_H	I_{max} (%)
Atipamezole	5.6	0.52	100
Dexmedetomidine	12.0	0.60	100
MPV-624	22.0	0.54	100
Oxymetazoline	47,000	0.26	55
(−)-Noradrenaline	>1,000,000	—	44

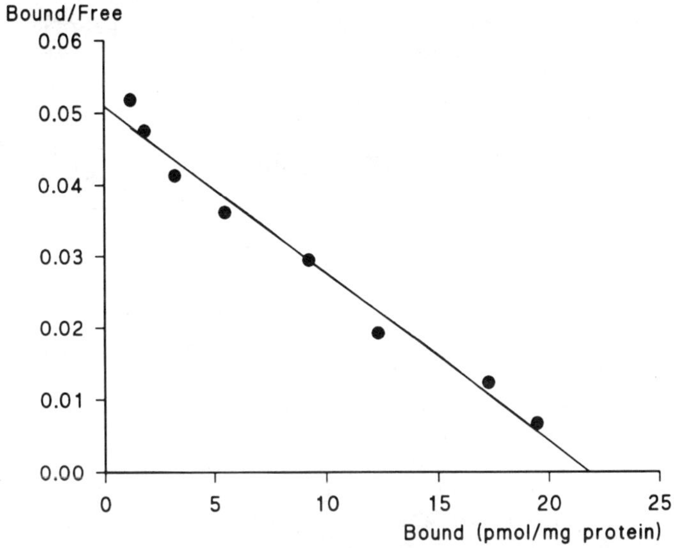

FIGURE 3. Representative Rosenthal plot of specific [^{3}H]atipamezole binding to adult rat lung membranes. K_d = 0.85 nM; B_{max} = 21.8 pmol/mg protein.

cantly different from unity. The total capacity of specific [^{3}H]atipamezole binding at equilibrium was 19.0 pmol/mg membrane protein with a K_d of 0.39 ± 0.11 nM (n = 6). The proportion of specific binding of total bound radioligand was 91 + 5%.

Competition binding assays indicated that the [^{3}H]atipamezole binding sites present in adult rat lung membranes had a similar pharmacological binding profile to that observed in neonatal rat lung membranes (TABLE 1). In contrast to that in neonatal lung membranes, no specific [^{3}H]rauwolscine binding could be detected in membranes from adult rat lung (results not shown). This difference was also reflected in the shapes of the displacement curves obtained with competing α_2-adrenergic ligands in these preparations, as the α_2-adrenergic component was missing from the experiments with adult lung membranes (Hill numbers ≃ 1). The binding of agmatine was tested only with adult rat lung membranes; it did not significantly inhibit [^{3}H]atipamezole binding even at a concentration of 1 mM.

DISCUSSION

In this study we demonstrated that [^{3}H]atipamezole identifies an atypical binding site in rat lung. Competitive binding assays were performed to determine the pharmacological specificity of nonadrenergic [^{3}H]atipamezole binding in neonatal and adult rat lung. Although [^{3}H]atipamezole was initially described as an α_2-adrenoceptor antagonist-type ^{3}H-ligand[23] this study indicates that in addition to α_2-adrenoceptors, [^{3}H]atipamezole also identifies a population of novel nonadrenergic binding sites in rat lung.

It has been suggested that there are three types of nonadrenergic imidazoline receptors.[14] Idazoxan has high affinity for all three types, none of which is recognized by catecholamines. The first, the I_1-receptor, is a [^{3}H]*para*-aminoclonidine binding site found in brain, kidney, and blood platelets. It is recognized by many imidazolines and imidazole compounds, including some that have very low affinity for α-adrenoceptors, such as the histamine H_2 antagonist cimetidine and imidazole-4-acetic acid.[1,26] These binding sites are not recognized by guanidino compounds (e.g., guanabenz) or benzazepine compounds (e.g., SK&F-86466). Another, the I_{2a}-receptor, or imidazoline-guanidinium receptive site, is a [^{3}H]idazoxan binding site found, for example, in rabbit kidney, rabbit adipocytes, rabbit urethra, and rat liver. It has high affinity for most imidazolines and the diuretic compound amiloride.[4,27–29] In contrast to the [^{3}H]*para*-aminoclonidine site just described, these sites have high affinity also for guanidino compounds. The third site, or I_{2b} receptor, is a [^{3}H]idazoxan binding site present, for example, in human kidney and guinea pig brain. It has high affinity for guanidino and benzazepine compounds and for some imidazolines (e.g., tolazoline and UK-14,304) but low affinity for other imidazolines (e.g., phentolamine and clonidine) and amiloride.[3,30] Agmatine, a putative endogenous ligand of imidazoline receptors, is bound to I_1 and I_{2a} sites with high affinity.[31] Dexmedetomidine and some related imidazole derivatives have relatively poor affinity for the previously described imidazoline binding sites.[16,32,33]

The nonadrenergic binding of [^{3}H]atipamezole does not fit into any of the three types of imidazoline receptors identified with [^{3}H]*para*-aminoclonidine and [^{3}H]idazoxan. First, the low affinities of phentolamine and clonidine indicate that this binding is not of the I_1 type of nonadrenergic binding described for [^{3}H]*para*-aminoclonidine. Second, the low affinities of idazoxan, guanabenz, and UK-14,304 exclude the I_{2a} and I_{2b} types of nonadrenergic imidazoline binding ([^{3}H]idazoxan binding sites).

The new [^{3}H]atipamezole binding site displays stringent structural requirements for competing ligands. As exemplified by the compounds MPV-709B and MPV-709D (see FIG. 1 for structures), substitution of an imidazole moiety for the imidazoline moiety has a dramatic effect on binding affinity. The remarkable sensitivity of this [^{3}H]atipamezole binding site for the imidazole-substitution pattern is clearly exemplified by comparison of MPV-295, a 4-imidazole substituted derivative, to MPV-709B, a 2-imidazole substituted derivative. The physico-chemical properties of these two compounds are obviously not equal; for example, their ionization constants (pKa) are different, which might cause this 200-fold difference in affinity in favor of the 4-imidazole substituted compound.

Other characteristics of the site labeled by [^{3}H]atipamezole are its high affinity for all tested imidazole derivatives (e.g., dexmedetomidine) and the insensitivity of this binding to the optical enantiomerism present in the bridge carbon of medetomidine. At α_2-adrenoceptors, the *dextro*-enantiomer dexmedetomidine has much higher affinity than the *levo*-enantiomer *l*-medetomidine.[34,35] Therefore, the [^{3}H]atipamezole site does not resemble any previously described adrenergic or imidazoline-preferring binding site or receptor.

No coupling mechanisms or second messenger pathways are known for the newly discovered [^{3}H]atipamezole binding site, and it is therefore still premature to call it a new receptor. Nevertheless, its high density in lung membranes offers hopes that it would render itself to enrichment, purification, and subsequent characterization by protein chemistry in order to better understand its biochemical properties. Binding assays may also prove useful in elucidating the functional characteristics of the binding site and its ligands; for example, coupling to guanine nucleotide regulatory proteins has been studied in many cases by manipulating the binding assay conditions prior to any knowledge of the molecular identity of the binding site. Finally, detection of [^{3}H]atipamezole binding sites in a transformed cell line (Sjöholm and Scheinin, unpublished data) offers hope that the possible coupling mechanisms of this putative receptor can be identified.

The physiological significance of the newly identified [^{3}H]atipamezole binding site in rat lung is still unknown. [^{3}H]Atipamezole is also bound to other rat tissues and cell lines with similar binding characteristics to those observed here for lung membranes, but with lower densities of binding sites[23] (Sjöholm and Scheinin, this meeting and unpublished results). The higher density of [^{3}H]atipamezole binding sites in adult compared to neonatal rat lungs may indicate developmental regulation of the cellular constituent(s) responsible for this binding. This binding site, at least to some extent, was solubilized from extensively washed neonatal rat lung membranes by the nonionic detergent CHAPS, suggesting that specific macromolecules in the plasma membrane account for this property. Rigorously performed subcellular fractionation experiments are still required to resolve the question of subcellular localization of the [^{3}H]atipamezole binding site. No endogenous ligand is known for the binding site characterized in this presentation. Agmatine was not bound to the [^{3}H]atipamezole binding site with high affinity, which tends to exclude this newly discovered endogenous ligand of at least some types of imidazoline receptors[31] from a similar role in the current class of imidazole binding sites.

REFERENCES

1. Ernsberger, P., M. P. Meeley, J. J. Mann & D. J. Reis. 1987. Clonidine binds to imidazole binding sites as well as α_2 adrenoceptors in the ventrolateral medulla. Eur. J. Pharmacol. **134:** 1-13.
2. Wikberg, J. E. S. 1989. High affinity binding of idazoxan to a non-catecholaminergic binding site in the central nervous system: Description of a putative idazoxan-receptor. Pharmacol. Toxicol. **64:** 152-155.
3. Michel, M. C., O.-E. Brodde, B. Schnepel, J. Behrendt, R. Tschada, H. J. Motulsky & P. A. Insel. 1989. [^{3}H]Idazoxan and some other α_2-adrenergic drugs also bind with high affinity to a nonadrenergic site. Mol. Pharmacol. **35:** 324-330.
4. Langin, D. & M. Lafontan. 1989. [^{3}H]Idazoxan binding at non-α_2-adrenoceptors in rabbit adipocyte membranes. Eur. J. Pharmacol. **159:** 199-203.
5. Michel, M. C., J. W. Regan, M. A. Gerhardt, R. R. Neubig, P. A. Insel & H. J. Motulsky. 1990. Nonadrenergic [^{3}H]idazoxan binding sites are physically distinct from α_2-adrenergic receptors. Mol. Pharmacol. **37:** 65-68.
6. Tibiriça, E., J. Feldman, C. Mermet, F. Gonon & P. Bousquet. 1991. An imidazoline-specific mechanism for the hypotensive effect of clonidine: A study with yohimbine and idazoxan. J. Pharmacol. Exp. Ther. **256:** 606-613.
7. Chan, S. L. F., C. A. Brown & N. G. Morgan. 1993. Stimulation of insulin secretion by the imidazoline α_2-adrenoceptor antagonist efaroxan is mediated by a novel, stereoselective, binding site. Eur. J. Pharmacol. **230:** 375-378.
8. Berdeu, D., R. Gross, G. Ribes, M.-M. Loubatières-Mariani & G. Bertrand. 1994. Effects of imidazolines and derivatives on insulin secretion and vascular resistance in perfused rat pancreas. Eur. J. Pharmacol. **254:** 119-125.

9. Maiese, K., L. Pek, S. B. Berger & D. J. Reis. 1992. Reduction in focal cerebral ischemia by agents acting at imidazole receptors. J. Cereb. Blood Flow Metab. **12:** 53-63.
10. Diamant, S. & D. Atlas. 1986. An endogenous brain substance, CDS (clonidine-displacing substance), inhibits the twitch response of rat vas deferens. Biochem. Biophys. Res. Commun. **134:** 184-190.
11. Felsen, D., P. Ernsberger, M. P. Meeley & D. J. Reis. 1987. Clonidine displacing substance is biologically active on smooth muscle. Eur. J. Pharmacol. **142:** 453-455.
12. Diamant, S., A. Eldor & D. Atlas. 1987. A low molecular weight brain substance interacts, similarly to clonidine, with alpha 2-adrenoceptors of human platelets. Eur. J. Pharmacol. **144:** 247-255.
13. Piletz, J. E., A. Halaris, A. Saran & M. R. Marler. 1991. Desipramine lowers 3H-para-aminoclonidine binding in platelets of depressed patients. Arch. Gen. Psychiatry **48:** 813-820.
14. Michel, M. C. & P. A. Insel. 1989. Are there multiple imidazoline binding sites? Trends Pharmacol. Sci. **10:** 342-344.
15. Tesson, F., C. Prip-Buus, A. Lemoine, J.-P. Pegorier & A. Parini. 1991. Subcellular distribution of imidazoline-guanidinium-receptive sites in human and rabbit liver. J. Biol. Chem. **266:** 155-160.
16. Piletz, J. E. & K. Sletten. 1993. Nonadrenergic imidazoline binding sites on human platelets. J. Pharmacol. Exp. Ther. **267:** 1493-1502.
17. Scheinin, H., E. MacDonald & M. Scheinin. 1988. Behavioural and neurochemical effects of atipamezole, a novel α_2-adrenoceptor antagonist. Eur. J. Pharmacol. **151:** 35-42.
18. Virtanen, R., J.-M. Savola & V. Saano. 1989. Highly selective and specific antagonism of central and peripheral α_2-adrenoceptors by atipamezole. Arch. Int. Pharmacodyn. Ther. **297:** 190-204.
19. Bylund, D. B., C. Ray-Prenger & T. J. Murphy. 1988. Alpha-2A and alpha-2B adrenergic receptor subtypes: Antagonist binding in tissues and cell lines containing only one subtype. J. Pharmacol. Exp. Ther. **245:** 600-607.
20. Blaxall, H. S., T. J. Murphy, J. C. Baker, C. Ray & D. B. Bylund. 1991. Characterization of the alpha-2C adrenergic receptor subtype in the opossum kidney and in the OK cell line. J. Pharmacol. Exp. Ther. **259:** 323-329.
21. Harrison, J. K., D. D. D'Angelo, D. Zeng & K. R. Lynch. 1991. Pharmacological characterization of rat α_2-adrenergic receptors. Mol. Pharmacol. **40:** 407-412.
22. Marjamäki, A., K. Luomala, S. Ala-Uotila & M. Scheinin. 1993. Use of recombinant human α_2-adrenoceptors to characterize subtype selectivity of antagonist binding. Eur. J. Pharmacol.-Mol. Pharmacol. Sect. **246:** 219-226.
23. Sjöholm, B., R. Voutilainen, K. Luomala, J.-M. Savola & M. Scheinin. 1992. Characterization of [^{3}H]atipamezole as a radioligand for α_2-adrenoceptors. Eur. J. Pharmacol. **215:** 109-117.
24. Bradford, M. M. 1976. A rapid sensitive method for the quantitation of protein utilizing the principle of protein dye binding. Anal. Biochem. **72:** 248-254.
25. McPherson, G. A. 1985. Analysis of radioligand binding experiments: A collection of computer programs for the IBM PC. J. Pharmacol. Methods **14:** 213-229.
26. Ernsberger, P., G. Feinland, M. P. Meeley & D. J. Reis. 1988. In rat renal cortex, clonidine binds to imidazole sites as well as α_2-adrenergic receptors. Am. J. Hypertens. **1:** 30A.
27. Parini, A., I. Coupry, R. M. Grahams, I. Uziella, D. Atlas & S. M. Lanier. 1989. Characterization of an imidazoline/guanidium receptive site distinct from the α_2-adrenergic receptor. J. Biol. Chem. **264:** 11874-11878.
28. Zonnenschein, R., S. Diamant & D. Atlas. 1990. Imidazoline receptors in rat liver cells: A novel receptor or a subtype of α_2-adrenoceptors? Eur. J. Pharmacol. **190:** 203-215.

29. YABLONSKY, F. & J. P. DAUSSE. 1991. Non-adrenergic binding sites for the 'α_2-antagonist' [^{3}H]idazoxan in the rabbit urethral smooth muscle. Biochem. Pharmacol. **41:** 701-707.

30. WIKBERG, J. E. S. & S. UHLÉN. 1990. Further characterization of the guinea pig cerebral cortex idazoxan receptor: Solubilization, distinction from the imidazole site, and demonstration of cirazoline as an idazoxan receptor-selective drug. J. Neurochem. **55:** 192-203.

31. LI, G., S. REGUNATHAN, C. J. BARROW, J. ESHRAGHI, R. COOPER & D. J. REIS. 1994. Agmatine: An endogenous clonidine-displacing substance in the brain. Science **263:** 966-968.

32. WIKBERG, J. E. S., S. UHLÉN & V. CHHAJLANI. 1991. Medetomidine stereoisomers delineate two closely related subtypes of idazoxan (imidazoline) I-receptors in the guinea pig. Eur. J. Pharmacol. **193:** 335-340.

33. RANGEL, N., S. UHLÉN & J. E. S. WIKBERG. 1993. Detomidine binding to guinea pig liver imidazoline receptors (I-receptors) shows marked positive cooperativity. Pharmacol. Toxicol. **73:** 86-90.

34. SAVOLA, J.-M. & R. VIRTANEN. 1991. Central alpha 2-adrenoceptors are highly stereoselective for dexmedetomidine, the dextro enantiomer of medetomidine. Eur. J. Pharmacol. **195:** 193-199.

35. JANSSON, C. C., A. MARJAMÄKI, K. LUOMALA, J.-M. SAVOLA, M. SCHEININ & K. E. O. ÅKERMAN. 1994. Coupling of human α_2-adrenoceptor subtypes to regulation of cAMP-production in transfected S115 cells. Eur. J. Pharmacol.-Mol. Pharmacol. Sect. **266:** 165-174.

Adrenergic and Imidazoline Receptor-Mediated Responses to UK-14,304-18 (Brimonidine) in Rabbits and Monkeys

A Species Difference

JAMES BURKE,[a] ALEX KHARLAMB,
TANVEER SHAN, EILEEN RUNDE, EDWIN PADILLO,
CYNTHIA MANLAPAZ, AND LARRY WHEELER

Department of Biological Sciences
Allergan Inc.
2525 Dupont Dr.
Irvine, California 92715

Glaucoma is a heterogenous group of ocular disorders in which there is progressive damage to the optic nerve, resulting in visual field loss and eventual blindness.[1] Elevated intraocular pressure (IOP) is an important risk factor for glaucoma.[2] Current treatment of primary open angle glaucoma, the most common form, is directed towards the reduction of IOP. Alpha-adrenergic receptor agonists reduce IOP[3] and have played a significant role in managing glaucoma.[4] Brimonidine (UK-14,304-18, AGN 190342), a relatively selective and potent alpha-2 adrenoceptor agonist,[5,6] lowers IOP in normotensive rabbits,[7–9] cats,[8] and monkeys[8,10,11] as well as glaucomatous or ocular hypertensive rabbits,[8] monkeys,[10,12] and humans.[12–14] Brimonidine is also highly effective in preventing elevated IOP following ocular laser surgery.[15–17] In addition to activating alpha-2 adrenoceptors, brimonidine binds to nonadrenergic "imidazoline" receptors[18,19] that can mediate biological effects such as hypotension[20] and potentially IOP. This study evaluates the pharmacology of the ocular and cardiovascular responses to topical administration of brimonidine in rabbits and monkeys by pretreatment with antagonists selective for alpha-adrenergic and imidazoline receptors and by radioligand binding displacement experiments using [^{3}H]brimonidine in membranes from rabbit ciliary body and monkey brain.

MATERIALS AND METHODS

Materials. The following compounds were used in these experiments: Brimonidine tartrate (UK-14,304-18), a gift from Pfizer, UK; rauwolscine HCl, prazosin HCl, and

[a]Correspondence and requests for reprints: James Burke, Department of Biological Sciences, Allergan Inc., 2525 Dupont Dr., Irvine, California 92715.

idazoxan HCl, purchased from Research Biochemical Inc., Natick, Massachusetts; and SKF 105854, a gift from SmithKline Beecham Labs., Philadelphia, Pennsylvania (formerly SmithKline Beckman). Drug concentrations in the *in vivo* experiments are expressed as a percentage (w/v) of the active ingredient, that is, base. The radioligand [^{3}H]brimonidine (UK-14,304; specific activity 63 Ci/mmol) was obtained from New England Nuclear, Boston, Massachusetts.

In Vivo. Experiments were performed in fully conscious drug-naive New Zealand White rabbits (albino) of either sex weighing 2-3 kg and female cynomolgus monkeys weighing 3-4 kg. Monkeys sat in specially designed chairs (Primate Products, San Francisco, California) throughout each experiment and occasionally were fed orange juice and fruit. Rabbits were put back into cages after each measurement. A 30R model Digilab pneumatonometer and a millimeter scale ruler were used to measure IOP and pupil diameter (PD), respectively. Pupil diameter was measured to the nearest 0.5 mm. Blood pressure and heart rate were measured noninvasively with a model BP100S automated sphygmomanometer from Nippon Colin, Japan. The pressure cuff of the sphygmomanometer was placed on the forearm of monkeys and the hindleg of rabbits. Drugs were formulated in distilled water and administered unilaterally as a single 50-μl eyedrop; the contralateral eyes received an equal volume of saline solution. Antagonists were administered 30 minutes prior to brimonidine instillation. Two baseline measurements were made before instillation of brimonidine, followed by periodic measurements up to 6 hours postinstillation.

Receptor Binding Assays

Tissue Preparation. Plasma membrane homogenate was prepared from rabbit iris/ciliary body, monkey cerebral cortex and brain stem, and rat cerebral cortex (as a control). Briefly, tissues (1 g) were homogenized in 25 ml of ice-cold buffer (5 mM Tris-HCl, pH 7.5) using a Polytron homogenizer for 30 seconds at setting #7 and centrifuged for 10-12 minutes at 300 × *g* at 4°C. The supernatant was filtered through two layers of gauze and diluted 1:2 with 50 mM Tris-HCl buffer, pH 7.4, then centrifuged at 49,000 × *g* for 15-20 minutes. The pellet fraction was washed three times, resuspended in Tris-HCl buffer, and centrifuged for 15-20 minutes at 49,000 × *g*. The pellet was then stored at –80°C until the binding assay.

Binding Studies. Frozen membrane pellet was resuspended in buffer containing 50 mM Tris-HCl, 0.1 mM $MnCl_2$, at pH 7.7. Rabbit iris/ciliary body was resuspended in a different buffer of the following composition: 50 mM Tris-HCl, 1 mM $MgCl_2$, and 2 mM EGTA, pH 7.5. Membrane protein homogenate (75-300 μg) was incubated with radioligand for 90 minutes at 22°C in a final volume of 250 or 500 μl. At the end of the incubation period, samples were filtered through glass fiber filters (Whatman GF/B) in a 24- or 96-well cell harvester and rapidly washed four times with 4 ml of iced-cold 50 mM Tris-HCl buffer, pH 7.4. The filters were then oven dried and transferred to scintillation vials containing 10 ml of Beckman's Ready Protein® scintillation cocktail for counting. Specific binding, defined by 10 μM phentolamine for competition studies, was as follows: 62% in rabbit iris/ciliary body using 2.4 nM [^{3}H]brimonidine and 90% in monkey and rat cerebral cortex and monkey brain stem using 1 nM [^{3}H]brimonidine. Protein concentrations were determined with a protein assay kit from Bio Rad (Hercules, California). Binding isotherms and equilibrium dissociation and affinity constants were analyzed and determined by the nonlinear least squares curve fitting program AccuFit Competition/Saturation from Beckman, Fullerton, California.

Data Presentation. Data from *in vivo* experiments are presented in graph form as the mean percentage of change from pre-drug baseline or absolute mean ± standard error versus time or in table form as mean percentage of change from baseline. Statistical analysis of the response to brimonidine or the antagonists was done by a paired Student's t test comparing data at each time point with pretreatment values (baseline) except for the rabbit IOP data. Intraocular pressure in rabbits tended to increase with time due to diurnal influences, so IOPs were compared between the drug-treated group and a separate vehicle-treated group and analyzed using an unpaired Student's t test. A p value of <0.05 was considered significant. To statistically evaluate the effect of antagonist pretreatment on the response to brimonidine, mean percentages of change from pre-drug levels in each experimental group were computed and compared with the mean percentages of change from the brimonidine only group at each time point using the unpaired Student's t test; $p < 0.05$ was considered significant.

RESULTS

Rabbits

The IOP responses to multiple concentrations (0.0001–0.1%) of unilateral ocular administration of brimonidine were determined in rabbits (FIG. 1). Starting IOPs ranged from 18–22 mm Hg. Brimonidine produced a concentration-dependent decrease in IOP. The 0.01% concentration was at the top of the concentration-response curve and produced a peak ocular hypotensive response of 23 ± 2% at 2 hours after drug administration. Higher concentrations (0.03 and 0.1%) increased the duration of action and produced a contralateral decrease in IOP. The 0.1% concentration produced ocular hypotension in the treated eye which lasted the duration of the 6-hour experiment. The IOP decrease in eyes treated with 0.01, 0.03, and 0.1% brimonidine was preceded by transient ocular hypertension. There were no pupillary responses to brimonidine in rabbits at the concentrations tested.

The effects of ocular administration of 0.03, 0.1, and 0.3% brimonidine on blood pressure and heart rate in rabbits are illustrated in FIGURE 2. Pre-drug blood pressure was about 100/50 mm Hg. Heart rate was about 230 beats/min. Brimonidine (0.03, 0.1%) did not produce a statistically significant change in blood pressure or heart rate. The 0.1% concentration reduced systolic blood pressure by 12 ± 8%. Systolic blood pressure was significantly reduced with the 0.3% concentration by 22 ± 6% at 1 hour following instillation of brimonidine. Heart rate was suppressed by 15 ± 3% at 1 hour. The hypotensive and bradycardic effects of brimonidine in rabbits were resolved by 2 hours after drug administration.

The effects of topical antagonist pretreatment on the IOP response to 0.01 and 0.03% brimonidine in rabbits are presented in TABLE 2. The antagonists rauwolscine, idazoxan, or SKF 105854 at the 1% concentration did not significantly alter IOP

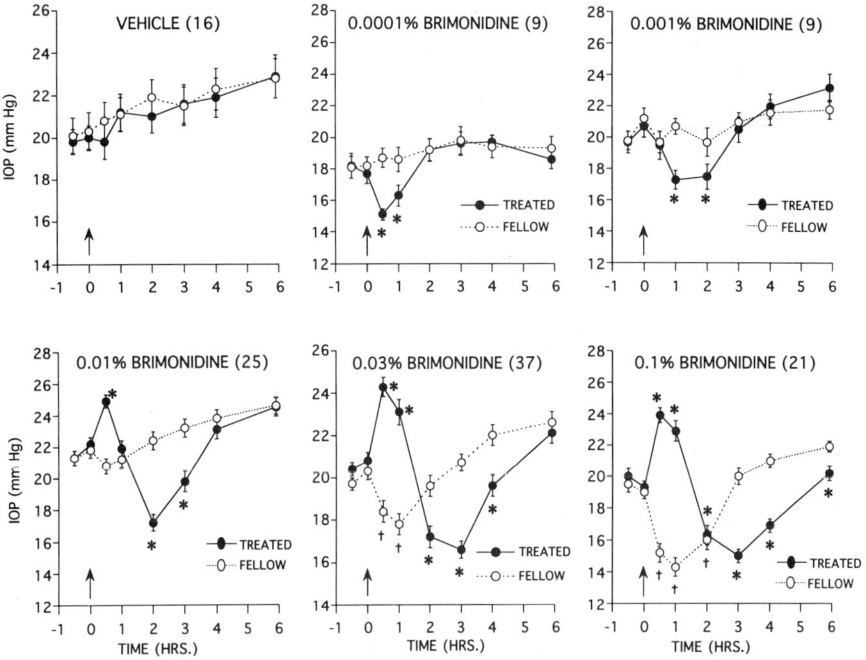

FIGURE 1. Intraocular pressure responses to unilateral topical administration of saline solution and 0.0001-0.1% brimonidine in conscious albino rabbits. Saline or brimonidine was administered as a single 50-μl drop. The treated eye is indicated by -●-, the untreated eye (fellow) by -○-. Values in parentheses represent the number of rabbits in each experiment. *Arrows* indicate time of drug administration. *Asterisks* indicate a statistically significant ($p < 0.05$) deviation from vehicle control for the treated eye; *daggers* indicate a statistically significant ($p < 0.05$) deviation from vehicle control for the untreated eye.

during the ocular hypotensive phase of brimonidine. Prazosin, 0.3%, lowered IOP by about 13%. The 0.01 and 0.03% concentrations of brimonidine decreased IOP by 23 and 20%, respectively. Rauwolscine inhibited the ocular hypotensive response to brimonidine by about 70%. Idazoxan inhibited the brimonidine-induced IOP decrease by about 37%. Neither SKF 105854 nor prazosin inhibited the IOP response to brimonidine in rabbits.

Monkeys

The effects of brimonidine (0.03-0.3%) on IOP, pupil diameter (PD), and cardiovascular variables in monkeys were evaluated. The results are shown in FIGURES 3 and 4. Pre-drug IOPs were 15-18 mm Hg; PD was about 4.5 mm. Brimonidine produced a concentration-related decrease in IOP and PD. The IOP responses were bilateral and symmetrical; the PD responses were unilateral. The maximum IOP response to 0.3% brimonidine was 43 ± 3% at 2 hours; the peak PD response was

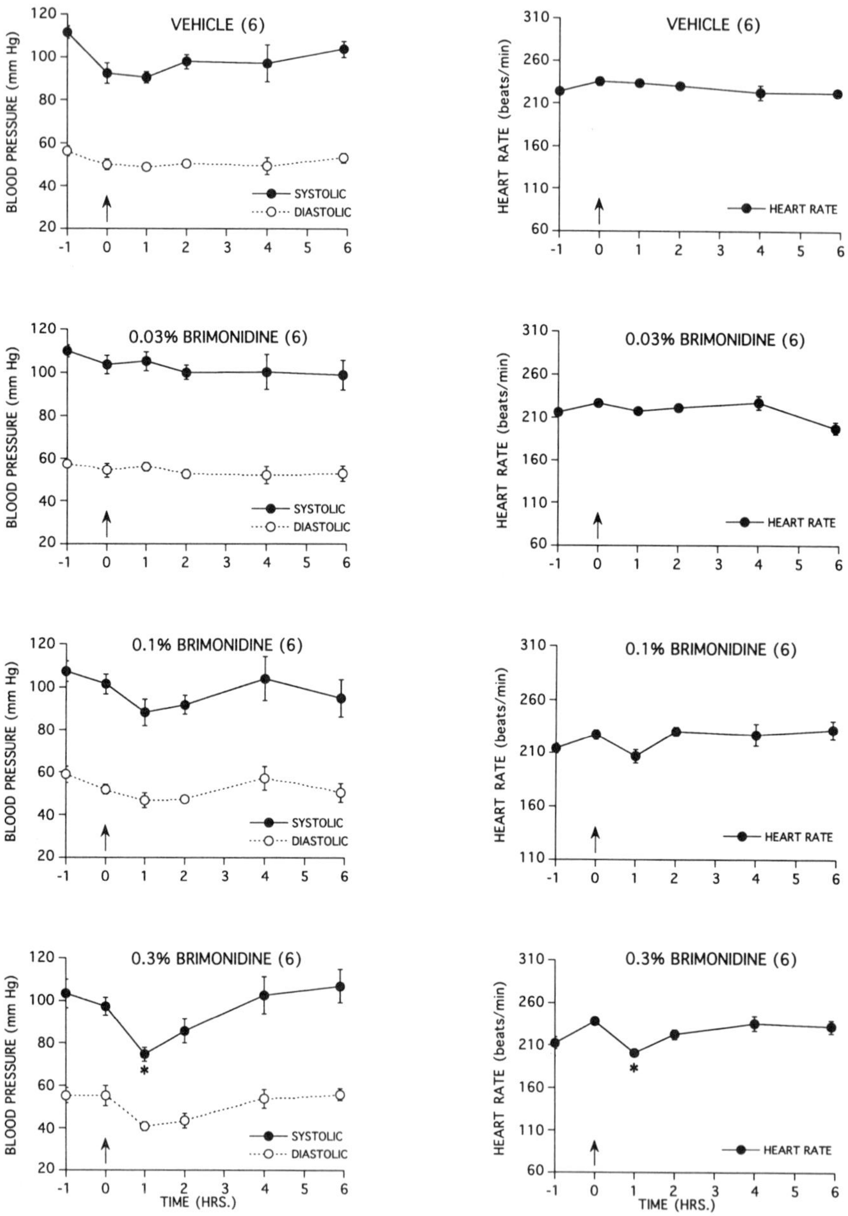

FIGURE 2. Blood pressure and heart rate responses to unilateral, ocular administration (50 μl) of a saline solution, 0.03%, 0.1%, and 0.3% brimonidine in conscious albino rabbits. Number in parentheses represents the *n* in each experiment. *Arrows* indicate time of drug administration. *Asterisks* represent a statistical difference from pretreatment values.

TABLE 1. *In Vitro* Pharmacology of Alpha-Adrenergic Receptor Antagonists

	Kb (nM)[a]		
Compound	Rabbit Aorta	Guinea Pig Atrium Prejunctional α_2	Dog Saphenous Vein Postjunctional α_2
Rauwolscine	1,300–1788[b,c]	3–5[d,e]	1–3[f,g]
Idazoxan	2,344[c]	13–18[e]	31–83[e,f]
SKF 105854	1,700[b]	>3,000[b]	39[b]
Prazosin	2–4[h]	>1,000[b,d]	501–1258[b,e,f]

[a] Receptor dissociation constant determined by schild analysis or by the formula Kb = [antagonist]/dose ratio –1 if only one antagonist concentration was used.

[b] Determined against (–) norepinephrine in the rabbit aorta and BHT-920 in the guinea pig atrium and dog saphenous vein.[13,22]

[c] Determined against (–) norepinephrine.[48]

[d] Determined against clonidine.[26]

[e] Determined against BHT-920.[17,21,24,36]

[f] Determined against BHT-933.[35]

[g] Determined against UK-14,304 (brimonidine).[2]

[h] Determined against (–) norepinephrine, α-methylnorepinephrine, phenylephrine, and BE 6143.[1]

TABLE 2. Effect of Antagonist Pretreatment on the Ocular Hypotensive Response to Topical Administration of Brimonidine in Rabbits

	% Change in Intraocular Pressure[a]		
Compound	+ Vehicle	+0.01% Brimonidine	+0.03% Brimonidine
Vehicle	0.6 ± 1.0 (15)	–22.5 ± 3.4 (6)	–19.7 ± 1.2 (37)
1% Rauwolscine	–1.2 ± 2.3 (9)	–5.0 ± 4.3 (6)[b]	–6.7 ± 2.7 (11)[b]
1% Idazoxan	–3.1 ± 1.3 (9)	–13.2 ± 2.2 (6)[c]	–13.0 ± 2.3 (7)[b]
1% SKF 105854	2.6 ± 3.7 (6)	–20.6 ± 3.9 (6)	–19.8 ± 3.4 (9)
0.3% Prazosin	–13.3 ± 3.3 (10)	nd	–22.3 ± 2.3 (15)

[a] Percent difference between the treated and contralateral eyes at 3 hours following administration of brimonidine. Mean ± SEM; values in parentheses represent the number of rabbits in each experiment.

[b] Significant difference between vehicle/brimonidine and antagonist/brimonidine data; $p < 0.05$, unpaired Student's t test.

[c] $p = 0.06$.

nd = not determined.

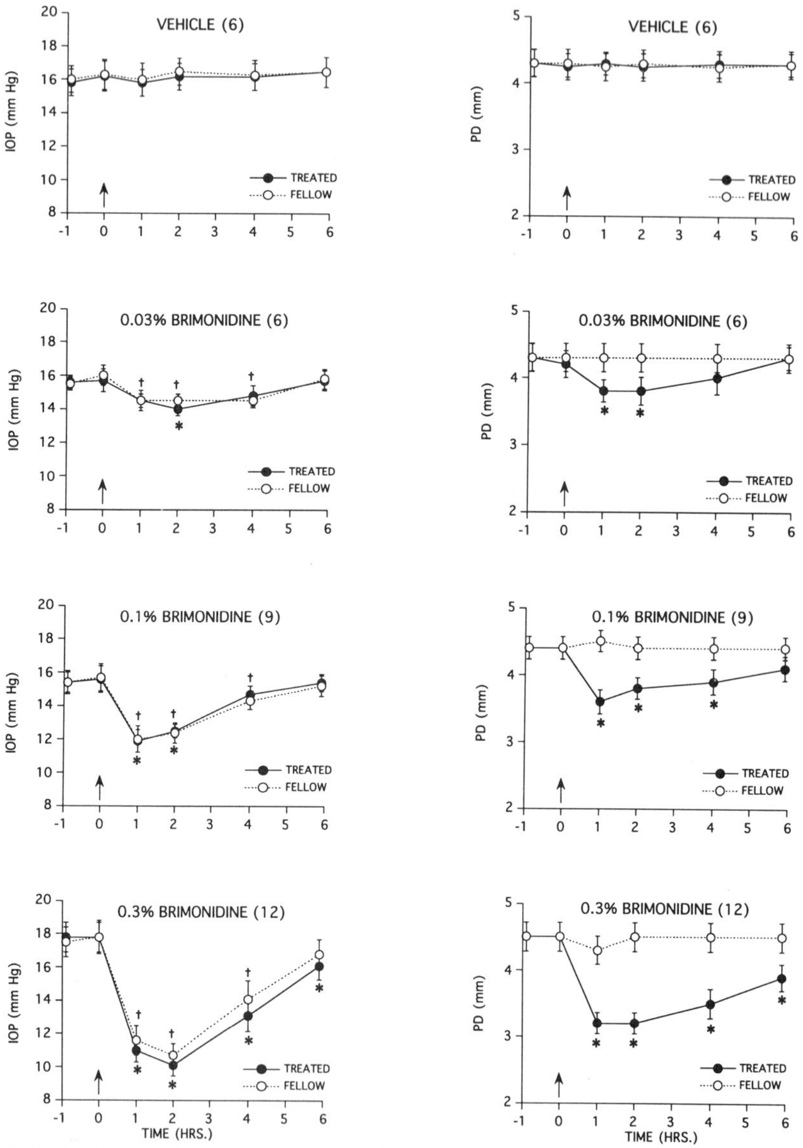

FIGURE 3. Intraocular pressure and pupil diameter responses to unilateral, ocular administration (50 μl) of saline solution, 0.03%, 0.1%, and 0.3% brimonidine in conscious normotensive cynomolgus monkeys. The treated eye is indicated by -●-, the untreated eye (fellow) by -○-. Number in parentheses represents the *n* in each experiment. *Arrows* indicate time of drug administration. *Asterisks* represent a statistical difference from pretreatment values; *daggers* represent a significant contralateral effect.

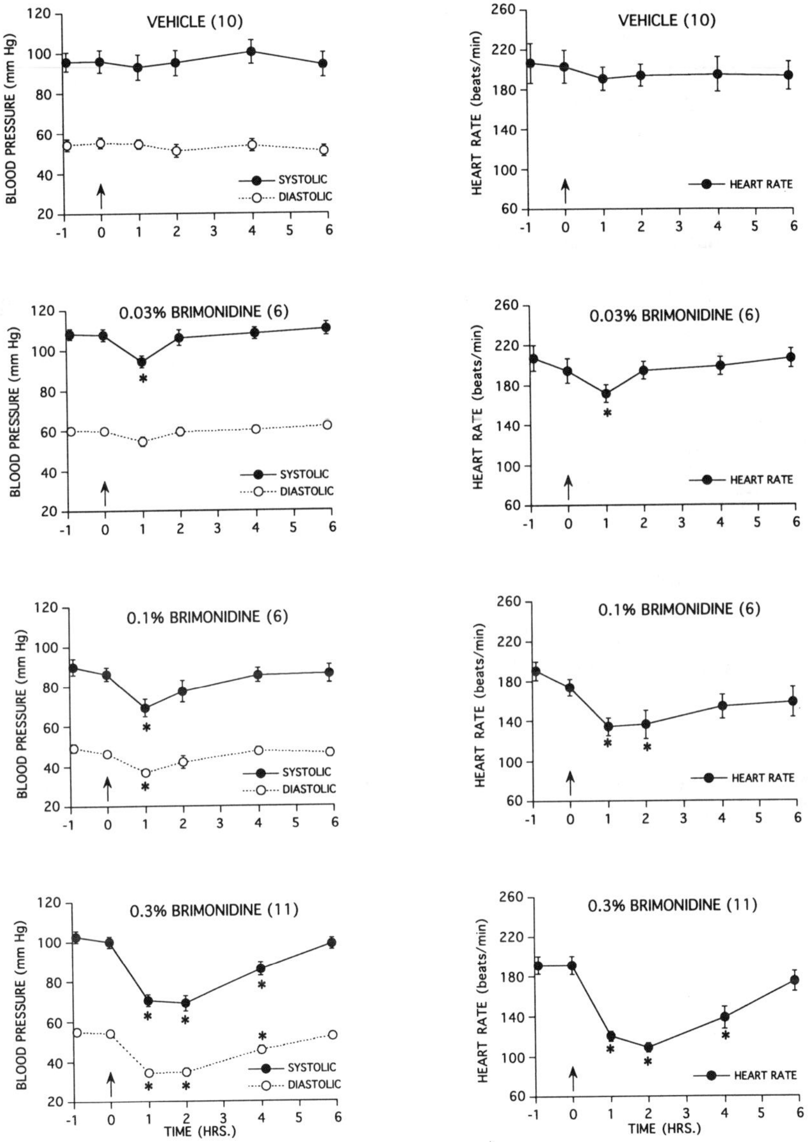

FIGURE 4. Blood pressure and heart rate responses to unilateral, ocular administration (50 μl) of saline solution, 0.03%, 0.1%, and 0.3% brimonidine in conscious cynomolgus monkeys. Number in parentheses represents the *n* in each experiment. *Arrows* indicate time of drug administration. *Asterisks* represent a statistical difference from pretreatment values.

23 ± 2% at 1 hour. Pre-drug blood pressure in monkeys was about 100/55 mm Hg; heart rate ranged from 180-200 beats/min. Brimonidine produced dose-related hypotension and bradycardia. The peak mean arterial blood pressure response to 0.3% brimonidine was 33.8 ± 2.9% at 2 hours and the maximum heart rate response was 42.0 ± 3.0% at 2 hours.

The IOP, PD, and cardiovascular responses to ocular application of the selective alpha-adrenoceptor antagonists rauwolscine, idazoxan, SKF 105854, and prazosin in monkeys are depicted in FIGURES 5 and 6. Rauwolscine and idazoxan transiently increased IOP and produced unilateral mydriasis. SKF 105854 did not alter IOP, PD, or heart rate but it decreased systolic blood pressure slightly. Prazosin did not change IOP but produced miosis and transient systolic hypotension and tachycardia.

The effects of the selective alpha antagonists on the IOP, PD, and cardiovascular responses to 0.3% brimonidine in monkeys are illustrated in FIGURE 7. Idazoxan significantly blocked the ocular hypotensive and cardiovascular responses to brimonidine. Inhibition was greater than 75%. Rauwolscine slightly reduced the hypotensive response to brimonidine at T = 1 hour only. Prazosin augmented the ocular hypotensive response to brimonidine. Both rauwolscine and idazoxan inhibited the miotic effect of brimonidine.

Radioligand Binding

The effects of selective alpha-adrenoceptor antagonists on [^{3}H]brimonidine binding in plasma membrane preparations from rabbit iris/ciliary body, and monkey and rat CNS tissues were determined. The results are shown in TABLE 3. Saturation studies in rabbit iris/ciliary body and monkey cerebral cortex show that [^{3}H]brimonidine bound to a single site with a K_d of 3.6 ± 0.7 and 0.48 ± 0.02 nM, respectively. Competition experiments show that rauwolscine was 10-fold more potent than idazoxan at displacing [^{3}H]brimonidine from rabbit ocular tissue. The ranked order of potency was rauwolscine > idazoxan > SKF 105854 > prazosin. In monkey and rat CNS tissue, idazoxan was more potent than rauwolscine at inhibiting [^{3}H]brimonidine binding. The ranked order of potency was idazoxan > rauwolscine > SKF 105854 > prazosin.

DISCUSSION

This investigation to determine the adrenoceptor subtype(s) that mediates the ocular hypotensive response to brimonidine (UK-14,304-18) was motivated by emerging evidence demonstrating the existence of multiple subtypes of the alpha-2 adrenoceptor and the imidazoline receptor. At least three subtypes of the alpha-2 adrenoceptor were identified by molecular cloning techniques.[21] Four subtypes were identified by radioligand binding methods.[22] Additionally, prejunctional and vascular postjunctional alpha-2 receptor subtypes were defined pharmacologically based on sensitivity to a series of 3-benzazepine analogs such as SKF 104078 and SKF 104856.[23–26] It was also shown that alpha-2 adrenoceptor agonists with an imidazoline moiety such as brimonidine can interact with a closely related but nonadrenergic receptor called the "imidazoline preferring receptor" or the "imidazoline" receptor.[27] This new

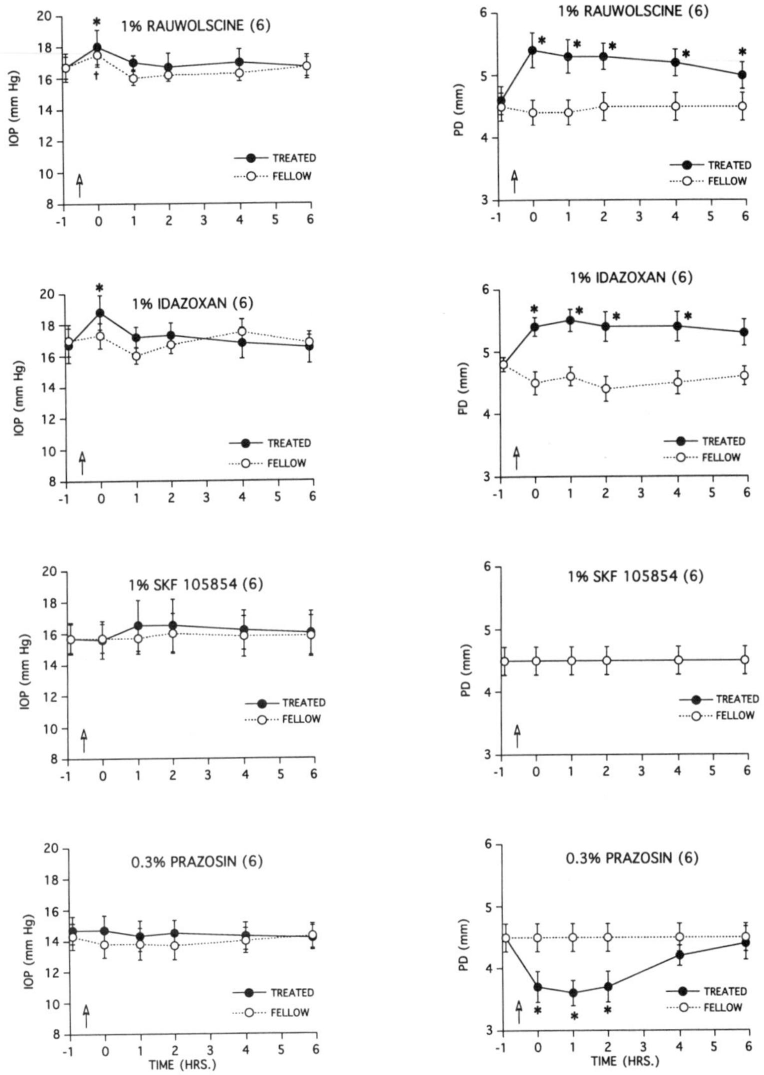

FIGURE 5. Intraocular pressure and pupil diameter responses to unilateral, ocular administration (50 μl) of the alpha-2 adrenoceptor antagonists rauwolscine, idazoxan, and SKF 105854, and the alpha-1 adrenoceptor antagonist prazosin in conscious cynomolgus monkeys. Number in parentheses represents the *n* in each experiment. The treated eye is indicated by -●-, the untreated eye (fellow) by -○-. *Arrows* indicate time of drug administration. *Asterisks* represent a statistical difference from vehicle control; *daggers* represent a significant contralateral effect.

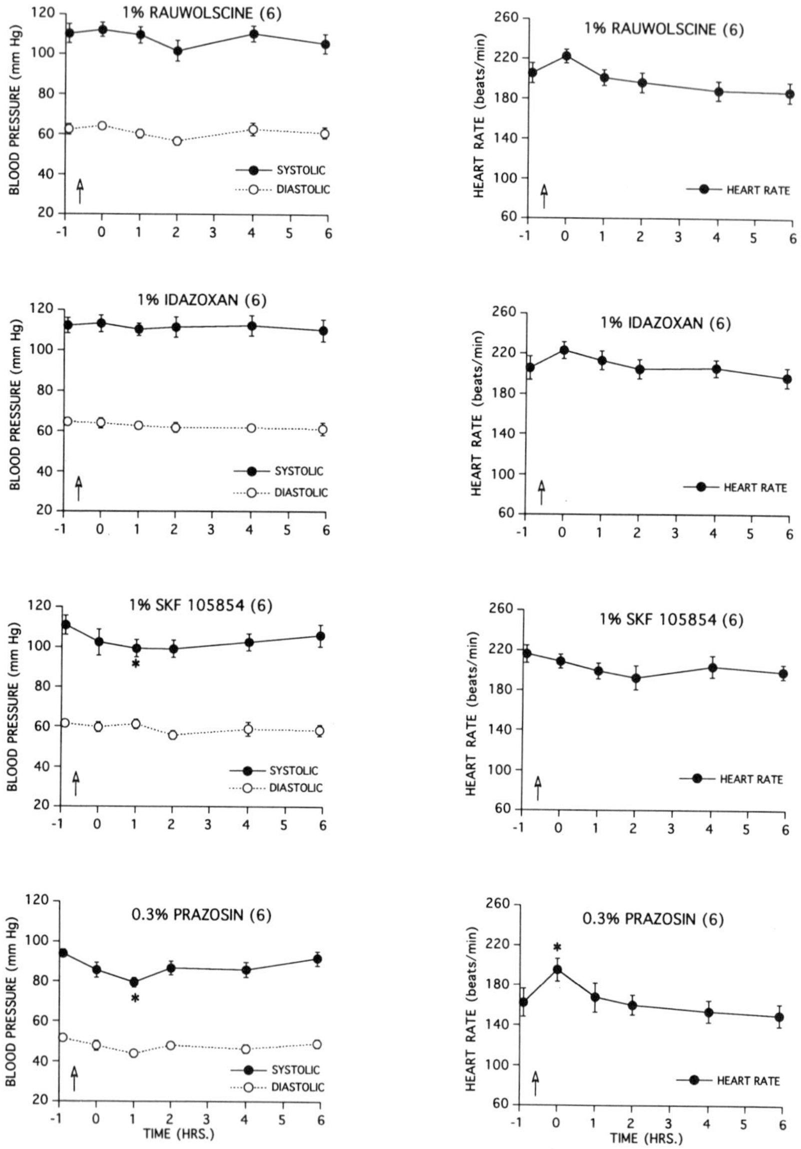

FIGURE 6. Blood pressure and heart rate responses to unilateral, ocular administration (50 μl) of the alpha-2 adrenoceptor antagonists rauwolscine, idazoxan, and SKF 105854 and the alpha-1 adrenoceptor antagonist prazosin in conscious cynomolgus monkeys. Number in parentheses represents the *n* in each experiment. *Arrows* indicate time of drug administration. *Asterisks* represent a statistical difference from pretreatment values.

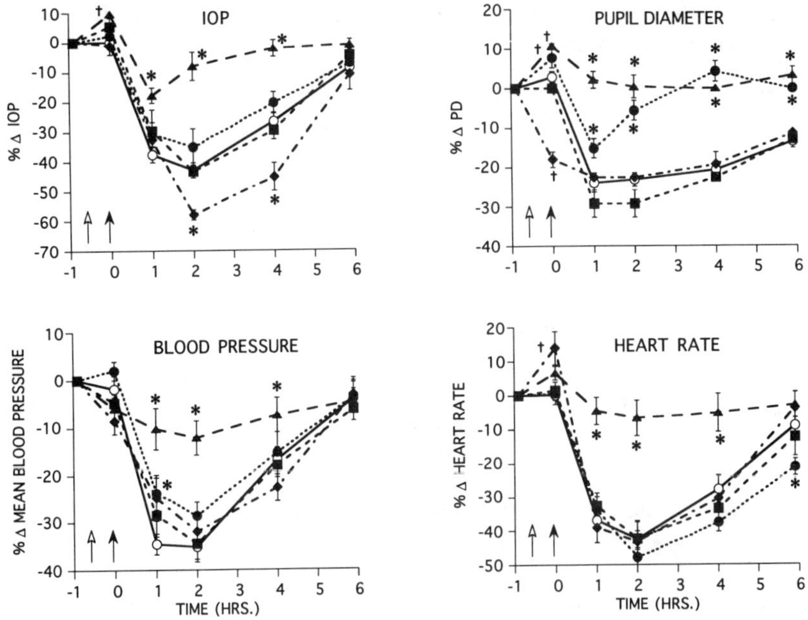

FIGURE 7. Effect of antagonist pretreatment (T = –0.5 hour) on the ocular and cardiovascular (mean arterial blood pressure, heart rate) responses to 0.3% brimonidine (T = 0 hour; n = 6–11) in conscious cynomolgus monkeys. Pretreatment = ○, saline (n = 12); ●, 1% rauwolscine (n = 6); ▲, 1% idazoxan (n = 11); ■, 1% SKF 105854 (n = 6); ◆, 0.3% prazosin (n = 6). Only the treated-eye responses are shown; the effects were bilateral. *Asterisks* represent a statistically significant difference ($p < 0.05$) between the effects of the antagonist/brimonidine and the saline/brimonidine groups. *Open* and *closed arrows* indicate application of antagonist and brimonidine, respectively.

imidazoline receptor has a low affinity for the natural adrenergic hormone norepinephrine and has been localized in the nucleus reticularis lateralis of the brain stem where it is implicated in the antihypertensive effect of clonidine and other substances.[28,29] The alpha-2 adrenoceptor antagonist idazoxan has been identified as a potent ligand for this putative receptor.[30]

This study evaluated the pharmacology of the IOP response to brimonidine in rabbits and monkeys by pretreatment with selective antagonists. The pharmacology of pupil size and cardiovascular responses to brimonidine in monkeys was also determined. The antagonists chosen for these experiments show selectivity for the following receptors: alpha-1 (prazosin), alpha-2 (rauwolscine), alpha-2/imidazoline (idazoxan), and vascular postjunctional alpha-2 (SKF 105854). These compounds were chosen because of their selectivity in functional assays as summarized in TABLE 1.

The results of this study showed that brimonidine lowered IOP in rabbits by stimulating an alpha-2 adrenoceptor. Unlike rauwolscine and idazoxan, the alpha-1 antagonist prazosin did not inhibit the ocular hypotensive response to brimonidine.

TABLE 3. Inhibition of [^{3}H]Brimonidine Binding in Plasma Membrane Homogenate from Rabbit Iris/Ciliary Body, Monkey Cerebral Cortex, Monkey Brain Stem, and Rat Cerebral Cortex by Selective Alpha-Adrenoceptor Antagonists

	K_i (nM)			
	Rabbit Iris/Ciliary Body[a]	Monkey Cerebral Cortex[b]	Monkey Brain Stem[c]	Rat Cerebral Cortex[d]
Rauwolscine	47 ± 11 (5)[e]	39 ± 9 (6)	45 ± 19 (2)	234 ± 35 (7)
Idazoxan	355 ± 76 (6)	14 ± 3 (6)	12 (1)	26 ± 4 (7)
SKF 105854	3,783 ± 1,038 (4)	345 ± 118 (5)	248 ± 142 (2)	708 ± 217 (6)
Prazosin	>10,000 (4)	>10,000 (1)	>10,000 (1)	6,912 ± 2,017 (4)

[a] $K_d = 3.6 \pm 0.7$ nM; $B_{max} = 205 \pm 13$ fmol/mg protein ($n = 4$).
[b] $K_d = 0.48 \pm 0.02$ nM; $B_{max} = 77 \pm 0.63$ fmol/mg protein ($n = 3$).
[c] Saturation experiment not done; K_d assumed to be 0.48 nM.
[d] $K_d = 1.4$ nM.[11]
[e] Mean ± SEM; number in parentheses = n.

Rauwolscine was more potent than idazoxan in inhibiting the IOP response to brimonidine, which agrees with the pharmacology of these compounds in *in vitro* functional assays. Reported data (TABLE 1) show that rauwolscine is more potent than idazoxan in inhibiting alpha-2 adrenoceptor-mediated functional responses to agonists. Rabbit IOP data also correlate with results from [^{3}H]brimonidine binding experiments in the rabbit iris/ciliary body which contains the machinery by which alpha-2 adrenoceptor agonists are presumed to lower IOP.[31,32] Binding data show the following ranked order of potency: rauwolscine > idazoxan > SKF 105854 > prazosin. The conclusion that brimonidine stimulates an alpha-2 adrenoceptor to lower IOP in the rabbit is also supported by other binding,[33,34] functional,[35,36] and autoradiographic[37] data which show the presence of alpha-2 adrenoceptors in rabbit iris/ciliary body. The vascular postjunctional alpha-2 adrenoceptor antagonist SKF 105854 did not inhibit the ocular hypotensive response to brimonidine in rabbits, suggesting that this receptor may not be involved in mediating the decrease in IOP.

In monkeys, brimonidine activated an imidazoline receptor to decrease IOP and cardiovascular parameters, as previously reported.[38–41] Brimonidine-induced ocular hypotension and bradycardia were augmented by prazosin and SKF 105854, respectively. Ocular and systemic hypotensive and bradycardic responses to brimonidine in monkeys were inhibited more by idazoxan than by rauwolscine. Rauwolscine only slightly attenuated the IOP response to brimonidine. This pharmacological profile is consistent with [^{3}H]brimonidine displacement experiments in monkey central nervous system (CNS) tissues which show that the ranked order of potency was idazoxan > rauwolscine > SKF 105854 > prazosin. These data are also supported by the observation that idazoxan is more potent than rauwolscine or its diastereoisomer yohimbine in blocking clonidine-induced hypotension[28,42] in rats and sedation[42] in mice.

Experiments with the selective antagonists also revealed that the miotic effect of brimonidine in monkeys is mediated by alpha-2 adrenoceptors, probably as a result of activating prejunctional autoreceptors to reduce sympathetic tone to the iris dilator muscle.[8] The vascular postjunctional alpha-2 adrenoceptor appears to be uninvolved in the miotic response to brimonidine in monkeys.

In addition to the species difference in the pharmacology of the IOP response to brimonidine, a difference in the site of action was also concluded from this study. In rabbits, unilateral administration of brimonidine produced unilateral IOP decreases at low concentrations and asymmetrical bilateral effects at higher concentrations, suggesting a local site of action. In the monkey, unilateral brimonidine produced bilateral IOP drops which were mostly symmetrical. The miotic effect was unilateral. In addition to symmetrical ocular hypotensive effects, brimonidine produced concomitant decreases in blood pressure and heart rate. The blood pressure response is predominantly CNS-mediated, because related compounds with poor CNS access are ineffective in producing this effect.[43] However, the bradycardic response includes a peripheral component.[44] Taken together, these data suggest that the IOP response to brimonidine in monkeys is predominantly CNS-mediated and that the response in rabbits is mediated by a local site of action.

Given a species difference in rabbits and monkeys with the pharmacology and site of action of the ocular hypotensive response to brimonidine, can the data infer the site of action and the pharmacology of brimonidine's response in humans? In man, unilateral application of brimonidine produces a unilateral IOP decrease without a corresponding decrease in cardiovascular variables.[45,46] This IOP and blood pressure profile is most similar to that of the rabbit in that the brimonidine IOP effect can be unilateral without concomitant hypotension. Taken together, the data allow the speculation that the IOP response to brimonidine in humans is mediated by peripheral receptor. There is some autoradiographic evidence[47] that [^{3}H]brimonidine binding to the human ciliary epithelium is displaced by phentolamine and not by prazosin which suggests that an alpha-2 adrenoceptor or an imidazoline receptor may be involved. The receptor (alpha-2 adrenergic or imidazoline) in the human eye remains to be identified.

SUMMARY

Brimonidine is a relatively selective alpha-2 adrenoceptor agonist that is being developed for the treatment of glaucoma. Because brimonidine is chemically related to clonidine and has affinity for the nonadrenergic imidazoline receptor, its ocular effects may be unrelated to alpha-2 receptor activation. The objective of this study was to determine the pharmacology of the intraocular pressure (IOP) response to brimonidine in rabbits and monkeys and the side effects (miosis, cardiovascular depression) in monkeys.

Conscious albino rabbits and cynomolgus monkeys were pretreated topically with the following receptor antagonists: rauwolscine (alpha-2), idazoxan (alpha-2, imidazoline), SKF 105854 (vascular postjunctional alpha-2), and prazosin (alpha-1). Intraocular pressure, pupil size, or blood pressure/heart rate was monitored noninvasively for 6 hours following dosing. Binding experiments were performed using [^{3}H]brimonidine in membrane preparations from rabbit iris/ciliary body and from monkey cerebral cortex and brain stem. In rabbits, the ocular hypotensive response to brimonidine was unilateral and was inhibited by rauwolscine > idazoxan >> SKF 105854 = prazosin; this ranked order of potency correlated with displacement of [^{3}H]brimonidine in the rabbit iris/ciliary body. In monkeys, brimonidine decreased IOP bilaterally and suppressed cardiovascular function suggesting a CNS site of action. Intraocular pressure and cardiovascular responses to brimonidine were inhibited by idazoxan >> rauwolscine > SKF 105854 = prazosin; a similar profile was obtained for displacement of [^{3}H]brimonidine in monkey brain tissue. Both rauwolscine and idazoxan inhibited the miotic response to brimonidine in monkeys.

Taken together, these results indicate that brimonidine stimulates an ocular alpha-2 adrenoceptor to decrease IOP in the rabbit and a CNS imidazoline receptor to decrease IOP, blood pressure, and heart rate in the cynomolgus monkey. The miotic response in the monkey is mediated by an alpha-2 adrenoceptor. The alpha-1 and vascular postjunctional alpha-2 adrenoceptors do not appear to play a role in mediating these responses.

REFERENCES

1. SHIELDS, M. B. 1987. Classifications of the glaucomas. *In* Textbook of Glaucoma.: 139-143. Williams & Wilkins. Baltimore.
2. SHIELDS, M. B. 1987. Primary open-angle glaucoma. *In* Textbook of Glaucoma.: 144-164. Williams & Wilkins. Baltimore.
3. INNEMEE, H. C., A. DE JONGE, J. C. A. VAN MEEL, P. B. M. W. M. TIMMERMANS & P. A. VAN ZWIETEN. 1981. The effect of selective $\alpha 1$ and $\alpha 2$-adrenoceptor stimulation on intraocular pressure in the conscious rabbit. Arch. Pharmacol. **316:** 294-298.
4. SHIELDS, M. B. 1987. Adrenergic stimulators. *In* Textbook of Glaucoma.: 387-401. Williams & Wilkins. Baltimore.
5. CAMBRIDGE, D. 1981. UK-14,304, a potent and selective alpha-2 agonist for the characterization of alpha-adrenoceptor subtype. Eur. J. Pharmacol. **72:** 413-415.
6. SUMMERS, R., B. JARROTT & W. LOUIS. 1980. Displacement of [^{3}H]clonidine binding by clonidine analogues in membranes from the rat cerebral cortex. Eur. J. Pharmacol. **66:** 233-241.
7. STEIDL, S., J. SERLE, S. PODOS & R.-F. WANG. 1987. Alpha adrenergic drugs and intraocular pressure (IOP). Invest. Ophthalmol. Vis. Sci. **28**(Suppl): 284.
8. BURKE, J. A. & D. E. POTTER. 1986. Ocular effects of a relatively selective alpha 2 agonist (UK-14,304-18) in cats, rabbits and monkeys. Curr. Eye Res. **5:** 665-676.
9. CROSSON, C. E., J. A. BURKE & D. E. POTTER. 1987. Ocular hypotensive activity of clonidine and related imidazolidines. Invest. Opthalmol. Vis. Sci. **28**(Suppl): 66.
10. SERLE, J. B., S. STEIDL, R. F. WANG, T. W. MITTAG & S. M. PODOS. 1991. Selective $\alpha 2$-adrenergic agonists B-HT 920 and UK14,304-18: Effects on aqueous humor dynamics in monkeys. Arch. Ophthalmol. **109:** 1158-1162.
11. STEIDL, S., J. SERLE, R.-F. WANG, S. PODOS & A. SIDDIQUE. 1988. Selective alpha adrenergic agents in cynomolgus monkeys. Invest. Ophthalmol. Vis. Sci. **29**(Suppl): 15.
12. SERLE, J. B., S. M. PODOS, G. P. ABUNDO, R. RITCH, A. L. COLEMAN, W. C. PANEK & C. E. MANTRAS. 1993. The effect of brimonidine tartrate in glaucoma patients on maximal medical therapy. Invest. Ophthalmol. Vis. Sci. **34**(Suppl): 929.
13. DERICK, R. J., T. R. WALTERS, A. L. ROBIN, H. S. BARNEBEY, N. T. CHOPLIN, E. P. KELLEY & J. F. STOECKER. 1993. Brimonidine tartrate: A one month dose response study. Invest. Ophthalmol. Vis. Sci. **34**(Suppl): 929.
14. WALTERS, T., R. REPASS, J. SARGENT, E. KELLEY, J. STOEKER, K. CHEN & D. HARPER. 1991. A pilot study of the efficacy and safety of AGN 190342-LF 0.02% and 0.08%

in patients with elevated intraocular pressure. Invest. Ophthalmol. Vis. Sci. **32**(Suppl): 988.

15. Barnebey, H. S., A. L. Robin. T. J. Zimmerman, J. C. Morrison, S. B. Hersh, R. A. Lewis, A. L. Coleman, D. J. Cinotti, J. Walt, K. S. Chen & E. Kelley. 1993. The efficacy of brimonidine in decreasing elevations in intraocular pressure after laser trabeculoplasty. Ophthalmology **100:** 1083-1088.
16. David, R., G. L. Spaeth, C. E. Clevenger, H. F. Perell, L. I. Siegel, J. C. Henry, M. C. Stiles, M. S. Passo, R. L. Stamper & J. G. Walt. 1993. Brimonidine in the prevention of intraocular pressure elevation following argon laser trabeculoplasty. Arch. Ophthalmol. **111:** 1387-1390.
17. Spaeth, G. L., R. David, C. E. Clevenger, H. F. Perell & L. I. Siegel. 1992. The effects of brimonidine tartrate on the incidence of intraocular pressure (IOP) spikes following argon laser trabeculoplasty (ALT). Invest. Ophthalmol. Vis. Sci. **33**(Suppl): 1159.
18. Lachaud, V., I. Coupry, R.-A. Podevin, J.-P. Dausse, E. Koenig & A. Parini. 1988. Interaction of clonidine and rilmenidine with imidazoline-preferring receptors. J. Hypertens. **6**(Suppl 4): S511-S513.
19. Ernsberger, P. R., K. L. Westbrooks, M. O. Christen & S. G. Schafer. 1992. A second generation of centrally acting antihypertensive agents act on putative I_1-imidazoline receptors. J. Cardiovasc. Pharmacol. **20**(Suppl. 4): S1-S10.
20. Bousquet, P., J. Feldman & J. Schwartz. 1984. Central cardiovascular effects of alpha adrenergic drugs: Differences between catecholamines and imidazolines. J. Pharmacol. Exp. Ther. **230:** 232-236.
21. Harrison, J. K., W. R. Pearson & K. R. Lynch. 1991. Molecular characterization of alpha 1- and alpha 2-adrenoceptors. Trends Pharmacol. Sci. **12:** 62-67.
22. Bylund, D. B. 1992. Subtypes of alpha 1- and alpha 2-adrenergic receptors. FASEB J. **6:** 832-839.
23. Daly, R. N., A. C. Sulpizio, B. Levitt, R. M. DeMarinis, J. W. Regan, R. R. Ruffolo & J. P. Hieble. 1988. Evidence for heterogeneity between pre- and postjunctional alpha-2 adrenoceptors using 9-substituted 3-benzazepines. J. Pharmacol. Exp. Ther. **247:** 127-128.
24. Ruffolo, R. R., A. C. Sulpizio, A. J. Nichols, R. M. DeMarinis & J. P. Hieble. 1987. Pharmacologic differentiation between pre- and postjunctional alpha 2-adrenoceptors by SK&F 104078. Naunyn-Schmiedeberg's Arch. Pharmacol. **336:** 415-418.
25. Hieble, J. P., A. C. Sulpizio, R. Edwards, H. Chapman, P. Young, S. P. Roberts, T. P. Blackburn, M. D. Wood, D. H. Shah, R. M. DeMarinis & R. R. Ruffolo, Jr. 1991. Additional evidence for functional subclassification of alpha-2 adrenoceptors based on a new selective antagonist, SK&F 104856. J. Pharmacol. Exp. Ther. **259:** 643-652.
26. Hieble, J. P., A. C. Sulpizio, A. J. Nichols, R. N. Willette & R. R. Ruffolo, Jr. 1988. Pharmacologic characterization of SK&F 104078, a novel alpha-2 adrenoceptor antagonist which discriminates between pre- and postjunctional alpha-2 adrenoceptors. J. Pharmacol. Exp. Ther. **247:** 645-652.
27. Lehmann, J., E. Koenig-Berard & P. Vitou. 1989. The imidazoline-preferring receptor. Life Sci. **45:** 1609-1615.
28. Tibirica, E., J. Feldman, C. Mermet, F. Gonon & P. Bousquet. 1991. An imidazoline-specific mechanism for the hypotensive effect of clonidine: A study with yohimbine and idazoxan. J. Pharmacol. Exp. Ther. **256:** 606-613.
29. Bousquet, P., G. Bricca, M. Dontenwill, J. Feldman, H. Greney, A. Belcourt, J. Stutzmann & E. Tibirica. 1992. From the alpha 2-adrenoceptors to the imidazoline preferring receptors. Fundam. Clin. Pharmacol. **6:** 15S-21S.
30. Wikberg, J. E. 1989. High affinity binding of idazoxan to a non-catecholaminergic binding site in the central nervous system: Description of a putative idazoxan-receptor. Pharmacol. Toxicol. **64:** 152-155.

31. Potter, D. E. 1981. Adrenergic pharmacology of aqueous humor dynamics. Pharmacol. Rev. **33:** 133-153.
32. Potter, D. E. & J. M. Rowland. 1981. Adrenergic drugs and intraocular pressure. Gen. Pharmacol. **12:** 1-13.
33. Jin, Y., A. Verstappen & T. Yorio. 1994. Characterization of alpha2-adrenoceptor binding sites in rabbit ciliary body membranes. Invest. Ophthalmol. Vis. Sci. **35:** 2500-2508.
34. Mittag, T. W. & A. Tormay. 1985. Adrenergic receptor subtypes in rabbit iris-ciliary body membranes: Classification by radioligand studies. Exp. Eye Res. **40:** 239-249.
35. Jumblatt, J. E., J. G. H. Liu & G. T. North. 1987. Alpha-2 adrenergic modulation of norepinephrine secretion in the perfused rabbit iris-ciliary body. Curr. Eye Res. **6:** 767-777.
36. Jin, Y., E. E. Elko, T. Tran & T. Yorio. 1989. Inhibition of adenylate cyclase in bovine ciliary process and rabbit iris ciliary body to alpha$_2$-adrenergic agonists. J. Ocul. Pharmacol. **5:** 189-197.
37. Elena, P. P., M. Kosina-Boix, P. Denis, G. Moulin & P. Lapalus. 1989. Alpha$_2$-adrenergic receptors in rat and rabbit eye: A tritium-sensitive film autoradiography. Ophthalmic Res. **21:** 309-314.
38. Burke, J., E. Padillo, T. Shan & A. Kharlamb. 1991. AGN190342 (UK-14304-18) stimulates an imidazoline receptor to lower intraocular pressure IOP in conscious monkeys. Invest. Ophthalmol. Vis. Sci. **32**(Suppl): 867.
39. Kharlamb, A. B., J. A. Burke & E. K. Runde. 1994. I1 imidazoline receptor subtype mediates ocular hypotensive and cardiovascular effects of brimonidine (UK 14,304) in cynomolgus monkeys. Invest. Ophthalmol. Vis. Sci. **35**(Suppl): 2048.
40. Burke, J., A. Kharlamb, T. Shan, E. Runde, E. Padillo & C. Manlapaz. 1992. The cynomolgus monkey has a CNS imidazoline receptor which mediated IOP and cardiovascular responses to UK-14,304 (AGN 190342). Fundam. Clin. Pharmacol. **6**(Suppl): 56s.
41. Kharlamb, A., T. Shan, E. Runde, E. Padillo, C. Machado & J. Burke. 1992. The cynomolgus monkey has a CNS imidazoline receptor which mediated IOP and cardiovascular responses to UK-14,304 (AGN 190342). Invest. Ophthalmol. Vis. Sci. **33**(Suppl): 1115.
42. Timmermans, P. B. M. W., J. Q. Qian, R. R. Ruffolo, Jr. & P. A. van Zwieten. 1984. A study of the selectivity and potency of rauwolscine, RX 781094 and RS 21361 as antagonists of alpha-1 and alpha-2 adrenoceptors. J. Pharmacol. Exp. Ther. **228:** 739-748.
43. Timmermans, P. B. M. W. M., W. Hoefke, H. Stahle & P. A. van Zwieten. 1980. SAR in clonidine-like drugs. Prog. Pharmacol. **3:** 1-104.
44. Hoefke, W., W. Kobinger & A. Walland. 1975. Relationship between activity and structure in derivatives of clonidine. Arzneim. Forsch. **25:** 786-793.
45. Harper, D. G., E. P. Kelley, J. G. Walt, J. Meinders & K. Chen. Allergan Clinical Research Report No: S342-108-8042 dated August 1990. The safety, comfort, onset and duration of action of AGN 190342-LF in normal subjects.
46. Harper, D., E. Kelley, T. Iacono, J. Meinders & K. Chen. Allergan Clinical Research Report No: S342-107-7831 dated February, 1990. The safety, comfort, onset and duration of action of AGN 190342-LF in normal subjects.
47. Matsuo, T. & M. S. Cynader. 1992. Localization of alpha-2 adrenergic receptors in the human eye. Ophthalmic. Res. **24:** 213-219.
48. Hieble, J. P. Personal Communication.
49. Hieble, J. P. & D. F. Woodward. 1984. Different characteristics of postjunctional alpha-adrenoceptors on arterial and venous smooth muscle. Naunyn-Schmiedeberg's Arch. Pharmacol. **328:** 44-50.

50. Hicks, P. E., M. Barras, G. Herman, P. Mauduit, J. M. Armstrong & B. Rossignol. 1991. Alpha-adrenoceptor subtypes in dog saphenous vein that mediate contraction and inositol phosphate production. Br. J. Pharmacol. **102:** 151-161.
51. Flavahan, N. A., T. J. Rimele, J. P. Cooke & P. M. Vanhoutte. 1984. Characterization of postjunctional alpha-1 and alpha-2 adrenoceptors activated by exogenous or nerve-released norepinephrine in the canine saphenous vein. J. Pharmacol. Exp. Ther. **230:** 699-705.
52. Hieble, J. P., A. C. Sulpizio, A. J. Nichols, R. M. DeMarinis, F. R. Pfeiffer, P. G. Lavanchy & R. R. Ruffolo, Jr. 1986. Pharmacological differentiation of pre- and post-junctional $alpha_2$-adrenoceptors. J. Hypertens. **4**(Suppl 6): S189-S192.
53. Rhodes, K. F. & J. F. Waterfall. 1987. The effects of some α-adrenoceptor antagonists on the responses of the canine saphenous vein to B-HT 933, UK-14,304 and methoxamine. Naunyn-Schmiedeberg's Arch. Pharmacol. **335:** 261-268.
54. Alabaster, V. A., R. F. Keir & C. J. Peters. 1985. Comparison of activity of alpha-adrenoceptor agonists and antagonists in dog and rabbit saphenous vein. Naunyn-Schmiedeberg's Arch. Pharmacol. **330:** 33-36.
55. Agrawal, D. K., C. R. Triggle & E. E. Daniel. 1984. Pharmacological characterization of the postsynaptic alpha adrenoceptors in vascular smooth muscle from canine and rat mesenteric vascular beds. J. Pharmacol. Exp. Ther. **229:** 831-838.
56. Cheung, Y.-D., D. Barnett & S. Nahorski. 1982. [^{3}H]Rauwolscine and [^{3}H]yohimbine binding to rat cerebral and human platelet membranes: Possible heterogeneity of α2-adrenoceptors. Eur. J. Pharmacol. **84:** 79-85.

2,6-Dimethyl Clonidine: α_2-Adrenoceptor or I_1-Imidazoline Receptor Agonist?

P. ERNSBERGER,[b] P. LI,[a] AND D. D. SMYTH[a]

[a]*Departments of Internal Medicine and Pharmacology & Therapeutics Faculty of Medicine University of Manitoba Winnipeg, Manitoba, Canada R3E 0W3*

[b]*Departments of Medicine, Pharmacology, and Neuroscience Case Western Reserve School of Medicine Cleveland, Ohio 44106*

Initially, studies in our laboratory had utilized clonidine and 2,6-dimethyl clonidine (DMC) as specific α_2-adrenoceptor agonists.[1] However, these two purported α_2-adrenoceptor agonists appeared to increase urine flow rate by two distinct mechanisms. Clonidine increased urine flow rate secondary to an increase in free water clearance,[2] and DMC increased urine flow rate secondary to an increase in osmolar clearance. Studies in rats with 1 kidney−1 clip hypertension also demonstrated a difference between clonidine and DMC.[3] In this model of acquired hypertension, the response to an intrarenal infusion of DMC, but not clonidine, was completely attenuated compared to that in 1 kidney sham control rats. This attenuation was also observed for selective I_1 imidazoline agonists.[4] Collectively, these studies were interpreted as clonidine and DMC acting at two unique sites and/or receptors, despite the close structural similarities of these compounds. We therefore determined the specificity of DMC for the I_1 binding site in isolated membranes *in vitro* and by the ability of relatively specific antagonists for the α_2 (rauwolscine) and imidazoline (idazoxan) receptors to block the renal actions of DMC.

METHODS

[^{3}H]Clonidine Binding Assays

[^{3}H]Clonidine was used for determination of specific binding to I_1 imidazoline sites and α_2-adrenoceptors in membranes from the fresh bovine ventrolateral medulla (VLM) as previously described.[5] On the day of the experiment, the previously prepared membranes were slowly thawed and resuspended at a concentration of 1 mg protein/ml in assay buffer, consisting of HEPES-Tris (5.0 mM, pH 7.7) containing 0.5 mM EGTA, 0.5 mM EDTA, and 0.5 mM $MgCl_2$. Assays were conducted in a total volume of 250 μl (125 μl membrane suspension, 25 μl radioligand, and 100 μl drug or vehicle). Incubations were carried out for 40 minutes at 25 °C. Nonspecific binding

TABLE 1. Renal Actions of 2,6-Dimethyl Clonidine (DMC) in the Presence and Absence of Idazoxan

	DMC (μg/kg/min)			DMC (1 μg/kg/min)	
Measurement	0	0.3	1	+ Idazoxan (0.3 mg/kg)	+ Rauwolscine (0.3 mg/kg)
Blood pressure (mm Hg)	116 ± 2	113 ± 4	127 ± 4[a]	118 ± 5[b]	124 ± 1[a]
Creatinine clearance (ml/min)	2.1 ± 0.1	1.7 ± 0.2	2.5 ± 0.2	1.8 ± 0.3[b]	2.0 ± 0.1[b]
Urine flow rate (μl/min)	26 ± 6	93 ± 21[a]	123 ± 14[a]	71 ± 8[a,b]	98 ± 8[a,b]
Sodium excretion (μEq/min)	5.0 ± 1.5	13.3 ± 2.9[a]	16.2 ± 2.8[a]	11.4 ± 1.7[b]	17.0 ± 2.8[a]
Osmolar clearance (μl/min)	98 ± 14	154 ± 18[a]	190 ± 19[a]	120 ± 16[b]	175 ± 15[a]

[a] $p < 0.05$ vs saline vehicle (DMC 0 μg/kg/min).
[b] $p < 0.05$ vs DMC 1 μg/kg/min.

was defined in the presence of piperoxan (0.1 mM). Specific α_2-adrenoceptor binding was defined by epinephrine (0.1 mM). Incubations were terminated by vacuum filtration over Reeves-Angel glass fiber filters using a cell harvester (Brandel). Filters were washed four times with 5 ml ice-cold assay buffer, placed in scintillation vials, covered with 4 ml of scintillation cocktail (BioSafe II, Research Products International), and counted at 50% efficiency (Beckman LS5801). Data were analyzed using the LIGAND program for nonlinear curve-fitting.

In Vivo *Renal Function*

Male Sprague-Dawley rats (200-225 g) were prepared as previously described.[1] Renal function was isolated in anesthetized (pentobarbitone) rats to allow direct intrarenal infusion of the study compounds. Following stabilization (45 minutes), a 15-minute control urine collection was obtained. This was followed by the continuous infusion of saline solution (vehicle control) or DMC (0.3 and 1 μg/kg/min) at a rate of 20 μl/min into the renal artery. Only one infusion rate was investigated in each rat. Four consecutive 15-minute urine samples were collected during this infusion. The effects of α_2-adrenoceptor or imidazoline receptor blockade on DMC (1 μg/kg/min) were determined by administering as a bolus dose (0.2 ml over 3 minutes) either rauwolscine (0.3 mg/kg) or idazoxan (0.3 mg/kg) 30 minutes following the start of the stabilization period. Data are presented as the mean ± the least squares standard error of the mean.

RESULTS AND DISCUSSION

Binding of [^{3}H]clonidine to VLM brain stem membranes was high-affinity, specific, and saturable. The saturation binding kinetics at I_1-imidazoline and α_2-adrenoceptor binding sites were determined separately with epinephrine (0.1 mM) to selectively mask α_2-adrenoceptors. The B_{max} and K_d for I_1-imidazoline sites were 191 ± 63 fmol/mg protein and 8.3 ± 3.3 nM, whereas values for α_2-adrenoceptor sites were 112 ± 12 fmol/mg protein and 1.1 ± 0.2 nM. This represented a sevenfold higher affinity of [^{3}H]clonidine for α_2-adrenoceptors over I_1-imidazoline sites. Of the specific sites labeled by [^{3}H]clonidine, rauwolscine bound with high affinity to 58 ± 2% (K_i = 19 ± 2 nM). A much lower affinity was observed for the remaining sites (K_i = 48 ± 24 μM). The clonidine analog DMC dose-dependently and completely inhibited specific [^{3}H]clonidine binding in these VLM membranes. Nonlinear curve-fitting to a two-site model revealed that DMC bound with high affinity to 42 ± 3% of the total specific [^{3}H]clonidine binding sites (K_i = 3.9 ± 0.5 nM), but it showed nearly fivefold lower affinity for the remaining 58% of the sites (K_i = 18.9 ± 1.9 nM). These data are consistent with at least a fivefold selectivity of DMC for I_1-imidazoline over α_2-adrenoceptor binding sites.

DMC (0, 0.3, and 1 μg/kg/min) produced a dose-related increase in urine flow rate. This increase was secondary to an increase in osmolar clearance. At the highest infusion rate, DMC produced a slight but significant increase in blood pressure and creatinine clearance. In another group of animals receiving DMC (1 μg/kg/min), rauwolscine (0.3 mg/kg) partially decreased the increase in urine flow rate, but it failed to alter the increase in sodium excretion and osmolar clearance. Idazoxan (0.3 mg/kg) significantly decreased the increase in urine flow rate, sodium excretion, and osmolar clearance. In a previous study in our laboratory,[6] we found that the same dose of rauwolscine but not idazoxan decreased the response to clonidine, indicating the relative selectivity of idazoxan over rauwolscine in antagonism of the renal actions of DMC.

In summary, the previously purported α_2-adrenoceptor agonist DMC possesses a fivefold greater affinity for the I_1-imidazoline receptor over the α_2-adrenoceptor. Idazoxan preferentially antagonized the renal effects of DMC, as compared to clonidine, whereas the reverse was true for rauwolscine. These data coupled with previous studies that have identified similarities between other I_1-imidazoline agonists and DMC (increased solute excretion, attenuated renal response in 1 kidney–1 clip hypertension) indicate that previously documented effects of DMC were most conceivably mediated by stimulation of the imidazoline receptor.

REFERENCES

1. SMYTH, D. D., P. LI, D. E. BLANDFORD & S. B. PENNER. 1992. J. Pharmacol. Exp. Ther. **261:** 1080-1086.
2. BLANDFORD, D. E. & D. D. SMYTH. 1988. J. Pharmacol. Exp. Ther. **247:** 1181-1186.
3. LI, P. & D. D. SMYTH. 1993. J. Pharmacol. Exp. Ther. **267:** 1395-1400.
4. LI, P., S. B. PENNER & D. D. SMYTH. 1994. Br. J. Pharmacol. **112:** 200-206.
5. ERNSBERGER, P., H. D. DAMON, L. M. GRAFF, S. G. SCHAFER & M. O. CHRISTEN. 1993. J. Pharmacol. Exp. Ther. **264:** 172-182.
6. ALLAN, D. R., S. B. PENNER & D. D. SMYTH. 1993. Br. J. Pharmacol. **108:** 870-875.

Imidazoline Binding Sites

Pharmacological and Molecular Characteristics

ANGELO PARINI

INSERM Unit 388
Institut Louis Bugnard
CHU Rangueil
1, av. J. Poulhès
31054 Toulouse Cédex, France

The concept of imidazoline receptors originated in the mid-1980s from the experiences of Bousquet and co-workers[1] who showed that the imidazoline alpha-adrenergic agonists ST587, cirazoline, and clonidine induced a centrally mediated decrease in blood pressure by a mechanism independent of the stimulation of adrenergic receptors. Thereafter, the existence of putative imidazoline receptors was further supported by radioligand binding studies. In 1987 two reports identified binding sites displaying pharmacological properties compatible with those of potential imidazoline receptors. Indeed, one of these studies showed that in the ventrolateral medulla of the bovine brain, 30% of the binding of the α_2-adrenergic agonist [^{3}H]*p*-aminoclonidine was insensitive to noradrenaline and solely recognized by imidazole derivatives.[2] At the same time, Coupry *et al.*,[3] using the α_2-adrenergic antagonist [^{3}H]idazoxan, identified a binding site in rabbit renal cortex displaying a high affinity for imidazoline derivatives and not recognized by catecholamines. The high affinity for imidazoline derivatives and the insensitivity to catecholamines allowed definition of both of these central and peripheral binding sites as "imidazoline receptors." However, marked differences in ligand recognition properties suggested that binding sites labeled by [^{3}H]*p*-aminoclonidine and [^{3}H]idazoxan could represent two subtypes of the same class of membrane proteins. This hypothesis was confirmed by several binding studies that allowed the characterization of imidazoline binding sites (IBS) in different tissues and species.[4–10] From these studies, two major subtypes of IBS have been identified based on their high (I_1BS) or low (I_2BS) affinity for clonidine, rilmenidine, and moxonidine.[11,12] In addition, I_2BS has been further subclassified into amiloride-sensitive (I_{2A}BS) and amiloride-insensitive (I_{2B}BS) binding sites.[8] Imidazoline binding sites have been characterized mostly using [^{3}H]*p*-aminoclonidine, [^{3}H]clonidine, and [^{125}I]iodoclonidine. In contrast, [^{3}H]idazoxan has long been used as a selective ligand for I_2BS. Recently, the selectivity of [^{3}H]idazoxan for I_2BS has been questioned. Indeed, we showed that in isolated cells from rabbit renal proximal tubule, [^{3}H]idazoxan recognizes I_1BS and I_2BS with identical affinity.[13]

The molecular characterization of I_1BS has largely been hindered by the low density of these binding sites and the high nonspecific binding of their selective radioligands. Some studies reported that [^{3}H]clonidine binding to membrane preparations from bovine ventrolateral medulla[14] and adrenal chromaffin cells[15] was inhibited by guanylnucleotides, suggesting that I_1BS could belong to the family of G-protein coupled receptors. However, the fact that these results have not been confirmed in human brain[16] or with [^{3}H]*p*-aminoclonidine as radioligand[17] indicates that coupling to G-protein could be limited to one specific subtype of I_1BS. To date, additional data on the structural properties of I_1BS are not available. Recently, two novel tools for the study of IBS were developed: the photoaffinity probe 2-[3-azido-4-[^{125}I]iodophenoxy]methyl imidazoline ([^{125}I]AZIPI),[18] a potential ligand for I_1BS which was successfully used for the molecular characterization of I_2BS, and an antiidiotypic antibody capable of recognizing both I_1BS and I_2BS.[19] It is conceivable that in the next few years, the use of these tools will greatly improve our knowledges in I_1BS structure. The molecular properties of I_2BS have been greatly elucidated. Although the structural and functional relationships of I_1BS with α_2-adrenergic receptors remain unclear, I_2BS and α_2-adrenergic receptors were demonstrated to be distinct entities in biochemical and molecular biology studies. Indeed, we showed that I_2BS and α_2-adrenergic receptors from rabbit renal cortex can be physically separated by using two different chromatography procedures.[20] These findings were then confirmed by Michel *et al.*[21] who showed that transfection of COS-7 cells with recombinant cDNA encoding for two subtypes of α_2-adrenergic receptors led to the expression of these receptors but not of I_2BS. The demonstration that I_2BS is a mitochondrial protein[22] while α_2-adrenergic receptors are located in the plasma membranes[23] definitely showed that these two binding sites belong to distinct families of membrane proteins and differ in their signal-transduction pathways as well as their functional activities.

To date, it has been shown that in different tissues, I_2BS is mainly localized in the mitochondria.[24] Some of the structural properties of mitochondrial I_2BS have been determined by protein purification procedures or photoaffinity labeling with [^{125}I]AZIPI.[18] Starting from bovine adrenal chromaffin cells, Wang and coworkers[25] purified a 70-kD protein labeled by [^{3}H]idazoxan and displaying ligand recognition properties characteristic of the IBS. It is noteworthy that this 70-kD protein is recognized with high affinity by clonidine and amiloride, two selective ligands for I_1BS and I_2BS, respectively. It is currently still unclear if this pharmacological profile identifies a particular subclass of IBS.

Using [^{125}I]AZIPI, Lanier *et al.*[18] identified two peptides with apparent molecular weights of 55 and 61 kD. The photolabeling of these two species was not prevented either by the alpha-adrenergic antagonists rauwolscine and prazosin or by epinephrine. In contrast, the photoincorporation of [^{125}I]AZIPI was inhibited by a series of imidazoline and guanidinium derivatives and poorly affected by clonidine, indicating that the 55- and 61-kD proteins both belong to the I_2BS subtype. The fact that amiloride exhibits higher affinity for the 61-kD protein than the 55-kD protein suggests that these two peptides could identify I_{2A}BS and I_{2B}BS, respectively. This very interesting finding supports a possible relationship between pharmacological and structural heterogeneity of I_2BS and shows that I_{2A}BS and I_{2B}BS can be coexpressed in the same tissue. Interestingly, photoaffinity labeling in membrane preparation from PC12 cells revealed an additional 55-kD peptide which, even if recognized by a series of I_2BS ligands, was insensitive to idazoxan. Taken together, the results indicate that I_2BS represents a heterogeneous family of membrane proteins displaying different pharmacological and structural characteristics.

In our laboratory, we focus our efforts on the molecular and functional characterization of mitochondrial I_2BS from rabbit renal cortex. Using two chromatography steps, we purified to apparent homogeneity a 60-kD protein displaying a pharmacological profile typical of I_2BS.[26] Additional experiments showed that this I_2BS is an acidic (pI 5.5), not glycosylated protein whose binding activity is regulated by K^+[23] and which is highly sensitive to H^+ at a physiological pH.[26] To identify the amino acid residues directly involved in ligand interaction with I_2BS and in regulation of ligand binding by H^+ and K^+, we studied the effects of chemical agents modifying different amino acid residues on [^{3}H]idazoxan binding to solubilized I_2BS. The fact that modifying agents for carboxyl and sulfhydryl groups and tyrosine residues did not affect [^{3}H]idazoxan binding even at very high concentrations shows that glutamate, aspartate, cysteine, and tyrosine residues are neither associated with I_2BS ligand binding domain nor involved in allosteric regulation of binding activity.[27] By contrast, [^{3}H]idazoxan binding to I_2BS was completely abolished in a dose-dependent manner by diethylpyrocarbonate (DEPC) and bromophenacyl bromide, two histidine-modifying agents. The fact that in protection experiments, the effect of DEPC on [^{3}H]idazoxan binding was not prevented either by saturating concentrations of amiloride and guanabenz, two I_2BS high affinity ligands, or by 150 mM KCl, indicates that histidine residues critical for ligand binding to I_2BS are not located within ligand and K^+ binding sites. By contrast, these histidine residues seem to be associated with the I_2BS domain involved in ligand-binding regulation by H^+. Indeed, the pH regulation of [^{3}H]idazoxan binding was completely prevented by treatment of solubilized preparations with DEPC. These studies allowed the identification of three different I_2BS domains, a ligand binding site and two distinct regulatory regions involved in the regulation of ligand binding by K^+ and H^+.

Our results point to close similarities between I_2BS and monoamine oxidases (MAOs), enzymes catalyzing the oxidative deamination of endogenous (i.e., catecholamines and serotonin) and exogenous amines.[28] Indeed, I_2BS and MAO share identical submitochondrial localization, similar apparent molecular weight,[26] and the same regulation by K^+ and modification of histidine residues *in vitro*.[23,27,29,30] The association of I_2BS with MAOs and their involvement in the regulation of enzyme activity are further strengthened by several findings showing that I_2BS and MAO are colocalized in different areas of rabbit[31] and human brain,[32] their density increases simultaneously in human brain in the elderly,[32] and in Alzheimer's disease,[33] and I_2BS density is downregulated by rat chronic treatment with the MAO inhibitors clorgyline and pargyline.[34] Recently, using biochemical and molecular biology approaches, we showed that I_2BS is a site located in an MAO domain distinct from the catalytic site. The functional relevance of these binding sites is supported by the demonstration that imidazoline derivatives inhibit MAO activity *in vitro*[35] and display centrally mediated pharmacological activities that could be mediated by a modification of MAO activity.[36,37]

In conclusion, by showing that I_2BS and MAO are a single protein we definitely identified a member of the IBS family with a specific membrane protein. Further studies will be necessary to define the structural and functional relationship of the other IBS with MAO.

REFERENCES

1. BOUSQUET, P., J. FELDMAN & J. SCHWARTZ. 1984. Central cardiovascular effects of alpha adrenergic drugs: Differences between catecholamines and imidazolines. J. Pharmacol. Exp. Ther. **230:** 232-236.
2. ERNSBERGER, P., M. P. MEELEY, J. J. MANN & D. J. REIS. 1987. Clonidine binds to imidazole binding sites as well as α_2-adrenoceptors in the ventrolateral medulla. Eur. J. Pharmacol. **134:** 1-13.
3. COUPRY, I., R. A. PODEVIN, J. P. DAUSSE & A. PARINI. 1987. Evidence for imidazoline binding sites in basolateral membranes from rabbit kidney. Biochem. Biophys. Res. Commun. **147:** 1055-1060.
4. BRICCA, G., H. GRENEY, M. DONTENWILL-KIEFFER, J. ZHANG, A. BELCOURT & P. BOUSQUET. 1993. Heterogeneity of the specific imidazoline binding of [^{3}H]idazoxan in the human cerebral cortex. Neurochem. Int. **22:** 153-163.
5. LACHAUD-PETTITI, V., R. A. PODEVIN, Y. CHRETIEN & A. PARINI. 1991. Imidazoline-guanidinium receptive sites in basolateral membranes from human kidney. Eur. J. Pharmacol. **206:** 23-31.
6. LACOMBE, C., V. VIALLARD & H. PARIS. 1993. Identification of α_2-adrenoceptors and of non-adrenergic idazoxan binding sites in pancreatic islets from young and adult hamsters. Int. J. Biochem. **25:** 1077-1083.
7. DIAMANT, S., T. ELDAR-GEVA & D. ATLAS. 1992. Imidazoline binding sites in human placenta: Evidence for heterogeneity and a search for physiological function. Br. J. Pharmacol. **106:** 101-108.
8. ZONNENSCHEIN, R., S. DIAMANT & D. ATLAS. 1990. Imidazoline receptors in rat liver cells: A novel receptor or a subtype of α_2-adrenoceptors? Eur. J. Pharmacol. **190:** 203-215.
9. LANGIN, D. & M. LAFONTAN. 1989. [^{3}H]idazoxan binding at non-α_2-adrenoceptors in rabbit adipocyte membranes. Eur. J. Pharmacol. **159:** 199-203.
10. TESSON, F. & A. PARINI. 1991. Identification of an imidazoline-guanidinium receptive site in mitochondria from rabbit cerebral cortex. Eur. J. Pharmacol. **208:** 81-83.
11. MICHEL, M. C. & P. A. INSEL. 1989. Are there multiple imidazoline binding sites? Trends Pharmacol. Sci. **10:** 342-344.
12. LANIER, S. & A. PARINI. 1994. *In* The RBI Handbook of Receptor Classification. J. W. Kebabian & J. L. Neumeyer, Eds.: 40-41. Natick.
13. GARGALIDIS-MOUDANOS, C. & A. PARINI. 1994. Identification of I_1 and I_2 imidazoline binding sites in rabbit renal proximal tubule cells using [^{3}H]idazoxan. II International Symposium on Imidazoline Receptors. Abstract book S28.
14. ERNSBERGER, P., K. WESTBROOKS & L. GRAFF. 1991. Imidazole binding sites labeled by clonidine analogs (I_1 sites): Putative G-protein-linked receptors. FASEB J. **5:** A1066.
15. MOLDERINGS, G. J., D. MOURA, K. FINK, H. BÖNISCH & M. GÖTHERT. 1993. Binding of [^{3}H]clonidine to I_1-imidazoline sites in bovine adrenal medullary membranes. Naunyn-Schmiedeberg's Arch. Pharmacol. **348:** 70-76.
16. BRICCA, G., H. GRENEY, J. ZHANG, M. DONTENWILL, J. STUTZMANN, A. BELCOURT & P. BOUSQUET. 1994. Human brain imidazoline receptors: Further characterization with [^{3}H]clonidine. Eur. J. Pharmacol. **266:** 25-33.
17. MACKINNON, A. C., M. STEWART, H. J. OLVERMAN, M. SPEDDING & C. M. BROWN. 1993. [^{3}H]*p*-Aminoclonidine and [^{3}H]idazoxan label different populations of imidazoline sites on rat kidney. Eur. J. Pharmacol. **232:** 79-87.
18. LANIER, S. M., B. IVKOVIC, I. SINGH, J. L. NEUMEYER & V. BAKTHAVACHALAM. 1993. Visualization of multiple imidazoline/guanidinium-receptive sites. J. Biol. Chem. **268:** 16047-16051.
19. BENNAÏ, F., H. GRENEY, J. STUTZMANN, P. BOUSQUET & M. DONTENWILL. 1994. II International Symposium on Imidazoline Receptors. Abstract book, S11.

20. PARINI, A., I. COUPRY, R. M. GRAHAM, I. UZIELLIA, D. ATLAS & S. M. LANIER. 1989. Characterization of an imidazoline-guanidinium receptive site distinct from the α_2-adrenergic receptor. J. Biol. Chem. **264:** 11874-11878.
21. MICHEL, M. C., J. W. REGAN, M. A. GERHARDT, R. R. NEUBIG, P. A. INSEL & J. J. MOTULSKY. 1990. Nonadrenergic [^{3}H]idazoxan binding sites are physically distinct from α_2-adrenergic receptors. Mol. Pharmacol. **37:** 65-68.
22. TESSON, F., C. PRIP-BUUS, A. LEMOINE, J. P. PEGORIER & A. PARINI. 1991. Subcellular distribution of imidazoline-guanidinium receptive sites in human and rabbit liver. J. Biol. Chem. **266:** 155-160.
23. COUPRY, I., D. ATLAS, R. A. PODEVIN, I. UZIELLI & A. PARINI. 1990. Imidazoline-guanidinium receptive site in renal proximal tubule: Asymmetric distribution, regulation by cations and interaction with an endogenous clonidine displacing substance. J. Pharmacol. Exp. Ther. **252:** 293-299.
24. TESSON, F., I. LIMON & A. PARINI. 1992. Tissue-specific localization of mitochondrial imidazoline-guanidinium receptive sites. Eur. J. Pharmacol. **219:** 335-338.
25. WANG, H., S. REGUNATHAN, M. P. MEELEY & D. J. REIS. 1992. Isolation and characterization of imidazoline receptor protein from bovine adrenal chromaffin cells. Mol. Pharmacol. **42:** 792-801.
26. LIMON, I., I. COUPRY, S. M. LANIER & A. PARINI. 1992. Purification and characterization of mitochondrial imidazoline-guanidinium receptive site from rabbit kidney. J. Biol. Chem. **267:** 21645-21649.
27. PARINI, A., I, COUPRY, F. TESSON, I. LIMON & C. GARGALIDIS-MOUDANOS. 1992. Imidazoline-guanidinium receptive site (IGRS): Mitochondrial distribution and regulation properties. Fundam. Clin. Pharmacol. **6:** 35s-40s.
28. SHIH, J. C., J. GRIMSBY & K. CHEN. 1990. The expression of human MAO-A and B genes. J. Neural Transm. **32:** 41-47.
29. HIRAMATSU, A., S. TSURUSHIIN & K. T. YASUNOBU. 1975. Evidence for essential histidine residues in bovine-liver mitochondrial monoamine oxidase. Eur. J. Biochem. **57:** 587-593.
30. LIMON, I. & A. PARINI. Unpublished results.
31. RENOUARD, A., P. S. WIDDOWSON & A. CORDI. 1993. [^{3}H]idazoxan binding to rabbit cerebral cortex recognizes multiple imidazoline I_2-type receptors: Pharmacological characterization and relationship to monoamine oxidase. Br. J. Pharmacol. **109:** 625-631.
32. SASTRE, M. & J. A. GARCIA-SEVILLA. 1993. Opposite age-dependant changes of α_{2A}-adrenoceptors and nonadrenoceptors [^{3}H]idazoxan binding sites (I_2-imidazoline sites) in the human brain: Strong correlation of I_2 with monoamine oxidase-B sites. J. Neurochem. **61:** 881-889.
33. RUIZ, J., I. MARTIN, L. F. CALLADO, J. J. MEANA, F. BARTUREN & J. A. SEVILLA. 1993. Non-adrenoceptor [^{3}H]idazoxan binding sites (I_2-imidazoline sites) are increased in postmortem brain from patients with Alzheimer's disease. Neurosci. Lett. **160:** 109-112.
34. OLMOS, G., A. M. GABILONDO, A. MIRALLES, P. V. ESCRIBA & J. A. GARCIA-SEVILLA. 1993. Chronic treatment with the monoamine oxidase inhibitors clorgyline and pargyline down-regulates non-adrenoceptor [^{3}H]idazoxan binding sites in the rat brain. Br. J. Pharmacol. **108:** 597-603.
35. TESSON, F., I. LIMON & A. PARINI. 1994. Inhibition of monoamine oxydase by imidazoline and guanidinium derivatives. II International Symposium on Imidazoline Receptors. Abstracts book, S22.

36. Jackson, H. C., S. L. Dickinson & D. J. Nutt. 1991. Exploring the pharmacology of the pro-convulsant effects of α_2-adrenoceptor antagonists in mice. Psychopharmacology **105:** 558-562.
37. Venault, P., F. Jacquot, E. Save, S. Sara & G. Chapouthier. 1993. Anxiogenic-like effects of yohimbine and idazoxan in two behavioral situations in mice. Life Sci. **52:** 639-645.

Use of High Affinity, Radioiodinated Probes for Identification of Imidazoline/Guanidinium Receptive Sites[a]

STEPHEN M. LANIER,[b] BILJANA LANIER,
VENKATESALU BAKTHAVACHALAM,[c]
CHRISTOPHER R. McGRATH,[c] AND
JOHN L. NEUMEYER[c]

Department of Pharmacology
Medical University of South Carolina
Charleston, South Carolina 29425

[c]*Research Biochemicals International*
Natick, Massachusetts 01760

Proteins that bind imidazoline and guanidinium compounds are of particular interest given the diverse effects elicited by these compounds both centrally and in peripheral tissues. Although such compounds (i.e., cirazoline, guanabenz, clonidine, and amiloride) interact with various neurotransmitter receptors or ion transporters/exchangers, they also bind with high affinity to unidentified membrane proteins termed "imidazoline binding sites" or "imidazoline/guanidinium receptive sites" (IGRS).[1-3] The isolation and partial characterization of endogenous substances that also recognize these proteins led to the suggestion that the components of this system constitute a hormonal/neurotransmitter receptor system. However, the role of IGRS and related proteins in signal transfer and cell response is unclear. To address this issue, we developed a series of functionalized derivatives of cirazoline, a ligand that exhibits high affinity for such entities, that could be used as probes for functional/structural characterization of this family of imidazoline binding proteins. Four molecules are described: AMPI (2-[3-aminophenoxy]methyl imidazoline), ^{125}I-AMIPI (2-[3-amino-4-{^{125}I}iodophenoxy]methyl imidazoline), ^{125}I-AZIPI (2-[3-azido-4-{^{125}I}iodophenoxy]methyl imidazoline), and AMPI-FLU (2-[3-fluoresceinylthiouredophenoxy]-methyl imidazoline).

[a]This work was supported by National Institutes of Health grants R3-GM46605 (J.L.N. and S.M.L.), RO1-NS24821 (S.M.L.), The Council for Tobacco Research grant 2235 (S.M.L.), and an institutional research award from the Medical University of South Carolina (S.M.L.)

[b]Address for correspondence: Stephen M. Lanier, Ph.D., Department of Pharmacology, Medical University of South Carolina, 171 Ashley Avenue, Charleston, South Carolina 29425.

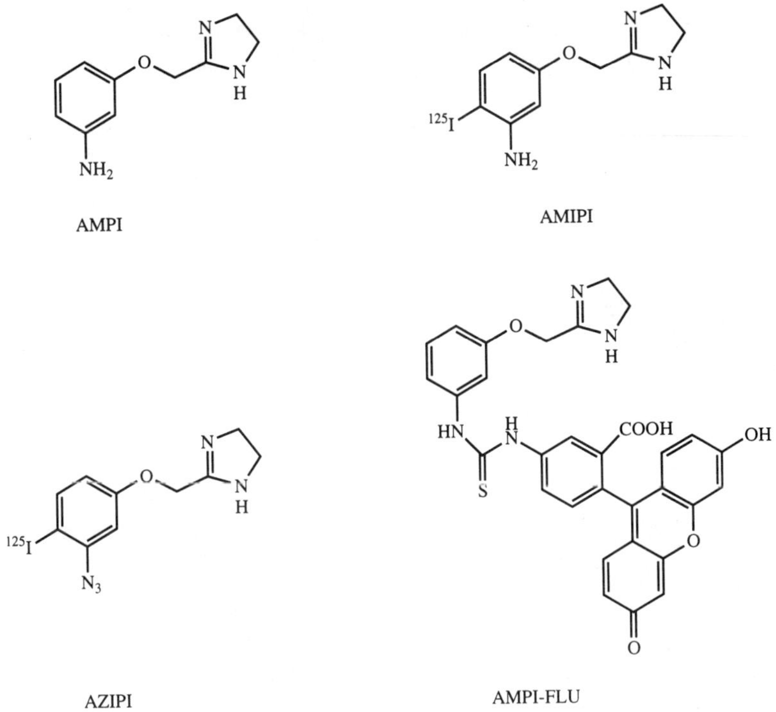

FIGURE 1. Structure of functionalized molecules for the study of the imidazoline/guanidinium receptive site and related imidazoline binding proteins.

RESULTS AND DISCUSSION

Each molecule in FIGURE 1 provides a useful tool in approaching different aspects of the structure/function of IGRS and related proteins. Because of its high affinity and selectivity for IGRS, the imidazoline cirazoline was selected for structural modification. As a preliminary structure/activity relationship study indicated that the cyclopropyl moiety is not required for ligand recognition by IGRS, we synthesized a series of 2-(2-aryloxy)-methylimidazolines. The desired imidazoline derivatives were synthesized by the reaction of the respective phenols with 2-chloromethylimidazoline in the presence of sodium ethoxide in ethanol. The key intermediate for the preparation of functionalized ligands was AMPI, which was isolated as a byproduct from the reaction of 3-acetamidophenol with chloromethylimidazoline. Iodination of AMPI with ICl generated a mixture of mono- and diiodonated derivatives which were readily separated by flash chromatography. ^{125}I-AMIPI and ^{125}I-AZIPI were prepared as described.[4,5] The fluoresceinated probe was obtained by condensation of AMPI with fluorescein isothiocyanate in pH 8 buffer.

AMPI retains high affinity and selectivity for IGRS in rabbit kidney and rat liver (K_i = 2-5 nM). By virtue of the phenyl-amino moiety, three approaches to structural analysis of IGRS are feasible. A ligand-affinity matrix can be generated by coupling of the molecule via the amino- group to 1-1′-carbonyldilmidazole-activated Trisacryl GF-2000 (Pierce, Rockford, Illinois). Wang *et al.*[6] used this approach to couple *p*-aminoclonidine to a similar matrix that proved useful in protein purification. An AMPI-Trisacryl matrix may be more efficient because of the higher affinity of AMPI and the potential recognition of multiple imidazoline binding proteins.

Secondly, the phenyl-amino moiety allows radioiodination of AMPI to generate ^{125}I-AMIPI, a high affinity, high specific activity radioligand for identification and intratissue localization of IGRS and related proteins.[5] ^{125}I-AMIPI exhibits high affinity for IGRS in rabbit kidney and liver (K_D = 2.0 ± 0.9 nM, B_{max} = 554 ± 201 fmol/mg [n = 5], rabbit kidney; K_D = 3.75 ± 0.25 nM, B_{max} = 73 ± 10 fmol/mg, rat liver). ^{125}I-AMIPI exhibits low nonspecific binding, and various imidazoline/guanidinium analogs inhibit binding to a similar degree (FIG. 2). ^{125}I-AMIPI binding is inhibited by a series of compounds with a rank order of potency consistent with its identification as an imidazoline/guanidinium receptive site.[1,2] Because of these characteristics, ^{125}I-AMIPI is useful for intratissue or intracellular localization of binding proteins.[5]

The third approach to structure/function analysis of IGRS is provided by the photoaffinity adduct ^{125}I-AZIPI. The phenyl-amino substitution of ^{125}I-AMIPI is converted to the arylazide by diazotization and reaction with sodium azide to generate a photosensitive ligand. ^{125}I-AZIPI was used to identify the ligand binding subunit of IGRS in various tissues. The autoradiograph in FIGURE 3 indicates that ^{125}I-AZIPI labels two peptide species with M_r of ~ 55,000 and 61,000. Photoincorporation of ^{125}I-AZIPI is inhibited by the imidazolines cirazoline and idazoxan and the guanidinium guanabenz (FIG. 3, unpublished observations).

To further aid in the study of the family of imidazoline binding proteins, we synthesized two fluorescent AMPI derivatives (FIG. 1). However, substitution of the fluorescein moiety at either the 3 or the 4 position on the phenyl ring led to a 50- to 100-fold decrease in affinity as determined in competition binding studies using [^{3}H]idazoxan in rabbit kidney membranes.

AMPI and its radiolabeled analogs are derivatives of cirazoline, a molecule that exhibits high affinity for IGRS and related imidazoline binding proteins and is functionally active at sites in the central nervous system that are involved in blood pressure regulation. Thus, the molecules described in this report should be useful in the study of both I1 and I2 imidazoline binding proteins. Indeed, photolabeling the ligand binding subunit of IGRS in various tissues and species indicates a family of binding proteins[4] that differ in either their M_r and/or their ligand recognition properties.

SUMMARY

Various pharmacologically active compounds with an imidazoline or guanidinium moiety are recognized by membrane bound proteins that appear structurally and functionally distinct from known hormone receptors. Such entities are termed imidazoline binding sites, I receptors, or imidazoline/guanidinium receptive sites (IGRS). To facilitate the identification and structural analysis of IGRS, we developed functionalized molecular probes exhibiting high affinity and selectivity for IGRS. The molecu-

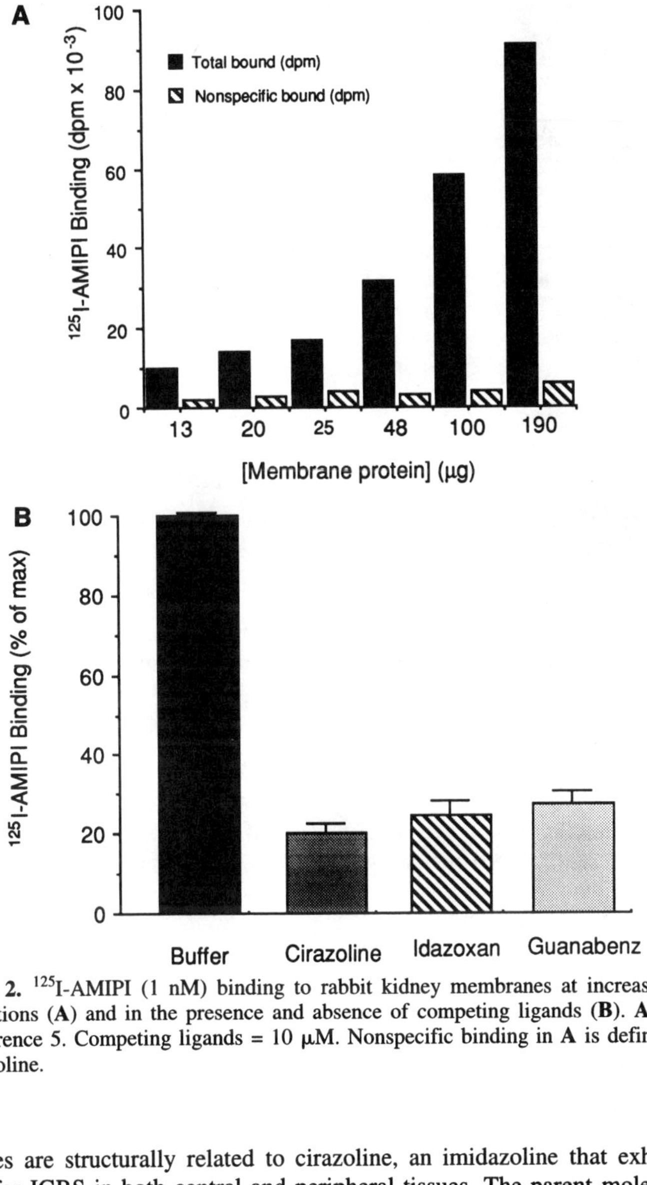

FIGURE 2. ^{125}I-AMIPI (1 nM) binding to rabbit kidney membranes at increasing protein concentrations (**A**) and in the presence and absence of competing ligands (**B**). **A** is adapted from reference 5. Competing ligands = 10 μM. Nonspecific binding in **A** is defined with 10 μM cirazoline.

lar probes are structurally related to cirazoline, an imidazoline that exhibits high affinity for IGRS in both central and peripheral tissues. The parent molecule 2-[3-aminophenoxy]methyl imidazoline was radioiodinated to yield 2-[3-amino-4-{^{125}I} iodophenoxy]methyl imidazoline (^{125}I-AMIPI), which was used to identify IGRS in brain and peripheral tissues. ^{125}I-AMIPI was converted to the photosensitive arylazide derivative (^{125}I-AMIPI) and used to identify the M_r of the ligand binding subunit of IGRS in various tissues including brain, pancreas, kidney, and liver. The results of these studies indicate that there are multiple binding proteins for these molecules

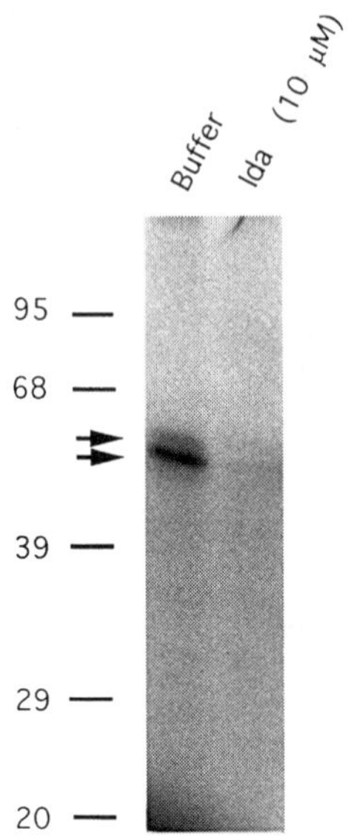

FIGURE 3. Photoaffinity labeling of rat brain membranes using ^{125}I-AZIPI. Brain membrane protein 100 μg was incubated with 2 nM ^{125}I-AZIPI and photolabeled as described.[5]

that differ in their apparent molecular weight, tissue distribution, intratissue location, and ligand recognition properties.

ACKNOWLEDGMENTS

The authors thank Dr. Angelo Parini for continued discussion. Because of space considerations, the citation list is limited; a complete reference list is found in the manuscripts cited.

REFERENCES

1. Parini, A., I. Coupry, R. M. Graham, I. Uzielli, D. Atlas & S. M. Lanier. 1989. Characterization of an imidazoline/guanidinium receptive site distinct from the alpha2-adrenergic receptor. J. Biol. Chem. **264:** 11874-11878.
2. Coupry, I., D. Atlas, R. A. Podevin, I. Uzielli & A. Parini. 1989. Imidazoline-guanidinium receptive site in renal proximal tubule: Asymmetric distribution, regulation by cations and interaction with an endogenous clonidine displacing substance. J. Pharmacol. Exp. Ther. **252:** 293-299.
3. Limon, I., I. Coupry, S. M. Lanier & A. Parini. 1992. Purification and characterization of mitochondrial imidazoline-guanidinium receptive site from rabbit kidney. J. Biol. Chem. **267:** 21645.
4. Lanier, S. M., B. Ivkovic, I. Singh, J. L. Neumeyer & V. Bakthavachalam. 1993. Visualization of multiple imidazoline/guanidinium receptive sites. J. Biol. Chem. **268:** 16047-16051.
5. Ivkovic, B., J. L. Neumeyer W. Zhang, V. Bakthavachalam & S. Lanier. 1994. Synthesis and characterization of 2-[3-amino-4{^{125}I} iodophenoxy]methyl imidazoline:

A functionalized ligand exhibiting high affinity for imidazoline/guanidinium receptive sites. Mol. Pharmacol. **46:** 15–23.

6. WANG, H. S., S. REGUNATHAN, M. P. MEELEY & D. J. REIS. 1992. Isolation and characterization of imidazoline receptor protein from bovine adrenal chromaffin cells. Mol. Pharmacol. **42:** 792–801.

[^{3}H]Cirazoline as a Tool for the Characterization of Imidazoline Sites

I. ANGEL,[a] M. LE ROUZIC, C. PIMOULE, D. GRAHAM, AND S. ARBILLA

Synthélabo Recherche
10, rue des Carrières
BP 248
92504 Rueil-Malmaison Cedex, France

Compounds chemically related to the imidazoline structure such as idazoxan, clonidine, or bromoxidine were reported to possess high affinity for α_2-adrenoceptors. Recent studies have demonstrated that such compounds also bind with high affinity to nonadrenergic sites. These studies were carried out in a variety of tissues and by using [^{3}H]idazoxan, [^{3}H]clonidine, or [^{3}H]*p*-aminoclonidine ([^{3}H]PAC) as radioligands.[1,2]

The heterogeneity of α_2-adrenergic receptors (α_2-AR) was confirmed following the identification of three structurally distinct human α_2-AR subtypes, so-called α_2C_2-, α_2C_4-, and α_2C_{10}-AR for their chromosomal localization.[3] From the different pharmacological behavior of guanidinium-imidazoline derivatives on α_2-AR compared to imidazoline sites and from differences in their distribution and biochemical properties it was suggested that these sites represent distinct physical and pharmacological entities.[4-6]

Wikberg and Uhlén[7] demonstrated that cirazoline, an α_1-adrenoceptor agonist, is among the ligands with the highest affinity for [^{3}H]idazoxan sites, its affinity for nonadrenergic sites being 500-fold higher than that of α_1-adrenoceptors and between 50- and 200-fold higher than that of [^{3}H]-adrenoceptors. In a previous preliminary study, we identified [^{3}H]cirazoline as a selective ligand that labels imidazoline sites in rat and kidney membranes.[8]

In the present study we further characterize the binding of [^{3}H]cirazoline to rat brain and kidney as well as to beta cell HIT T15 membrane homogenates. Furthermore, to confirm the differences between the α_2-AR and the imidazoline recognition site on a molecular level, we compared the specific binding of [^{3}H]cirazoline to these tissues to α_2-AR binding, using [^{3}H]rauwolscine in chinese hamster ovary (CHO) cell lines stably transfected with either genomic or cDNA encoding the human α_2-AR.

[a] To whom correspondence should be addressed.

MATERIALS AND METHODS

Materials

Amiloride, benextramine, clonidine, (-)adrenaline, phentolamine, prazosin, propranolol, and yohimbine were purchased from Sigma (Paris, France) and rauwolscine from Laboratoires Sarget (Mérignac, France). [^{3}H]rauwolscine (75-80 Ci.mmol^{-1}) was from NEN, France. Bromoxidine, idazoxan, cirazoline, and [^{3}H]cirazoline (25 Ci.mmol^{-1}) were synthesized in the chemistry department of Synthélabo Recherche (Bagneux, France).

Membrane Homogenate Preparations

Chinese hamster ovary (courtesy of Drs. Lefkowitz and Caron, Duke University, Durham, North Carolina) and HIT T15 (ATCC) cells were grown in HAM-F12 (L-glutamine) and RPMI 1640 (L-glutamine) medium, respectively, with fetal calf serum 10%, penicillin (50 units/ml), streptomycin (50 μg/ml), 95% air, and 5% CO_2 at 37°C. Dexamethasone (0.1 μM) was added 18-24 hours before CHO-cell confluency. Membrane preparations from cell lines or whole brain and kidney (from Sprague-Dawley rats, C. River, France, 200-250 g) were performed and stored at –80°C, respectively, in a hypotonic 2 mM Tris/HCl and isotonic 50 mM Tris/HCl buffer (in mM:EDTA 3, $MgCl_2$ 3.5, pH 7.4, 4°C). Membranes were obtained using a polytron homogenizer and rinsed twice at 46,500 *g* (20 min, 4°C). Brain and kidney pellets were resuspended in isotonic buffer (incubation medium) and incubated 30 minutes at 25°C to remove endogenous inhibitors. Before use, thawed CHO and HIT T15 cell membranes were washed and rehomogenized in incubation buffer. Proteins were determined using the BIORAD method.

Brain or kidney membranes (500 μg/ml) and HIT T15 or CHO homogenates (400 μg/ml) were incubated in the presence of [^{3}H]cirazoline in 250 μl of incubation medium (30 min, 25°C). Chinese hamster ovary cell membranes (40 μg/ml) were incubated in the presence of [^{3}H]rauwolscine in 1 ml of incubation buffer (60 min, 25°C). [^{3}H]Rauwolscine and [^{3}H]cirazoline binding assays were stopped by vacuum filtration over Whatman GF/B and GF/C filters, respectively.

Filters were washed with 3 × 5 ml of isotonic buffer (4°C). Nonspecific binding of [^{3}H]rauwolscine was defined by 10 μM phentolamine and of [^{3}H]cirazoline by 10 μM of unlabeled cirazoline or 1 μM of bromoxidine.

Competition studies were carried out under standard incubation conditions with 5 nM of [^{3}H]cirazoline and a range of concentrations of different drugs (1 pM to 1 mM). At 5 nM [^{3}H]cirazoline the nonspecific binding did not exceed 25 and 30% of total binding for kidney and brain homogenates, respectively. Specific binding represented approximately 0.09% of the total radioligand concentration (n = 20).

Association studies were performed by incubating the membranes (500 μg Prot/ml) with 1 nM [^{3}H]cirazoline at 25°C for periods ranging from 10 seconds to 120 minutes. Reversibility was assessed by adding an excess of bromoxidine (10 μM) after equilibrium was reached.

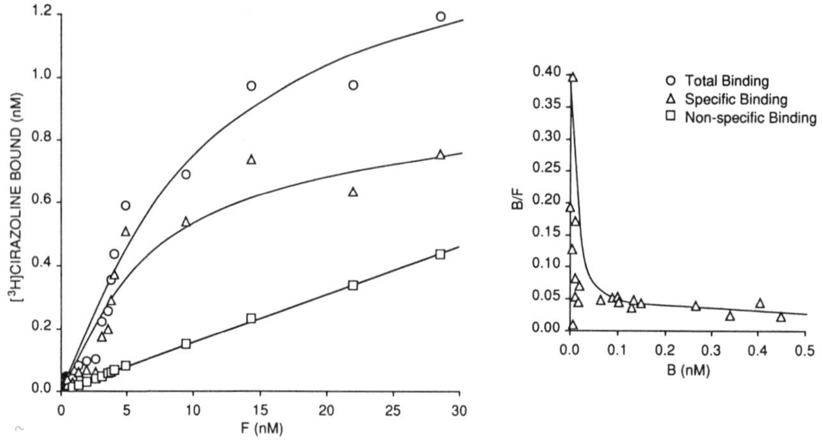

FIGURE 1. (Left) Saturation analysis of [^{3}H]cirazoline binding (1 pM to 50 nM) to brain membrane homogenates: (○) total [^{3}H]cirazoline binding, (△) specific binding, and (□) nonspecific binding (in the presence of 10 μM of unlabeled bromoxidine). **(Right)** Scatchard analysis plot of [^{3}H]cirazoline specific binding, indicating two classes of sites.

Data Analysis

Results are expressed as the mean ± SEM of *n* experiments performed in triplicate. B_{max}, K_d, and K_i parameters were derived from nonlinear regression analysis using the LIGAND program.[9] Pseudo Hill coefficients (n_H) were deduced from the slope of the Hill plot of inhibition data ($r > 0.92$). Appropriate single or two-site model fitting was obtained using the differential F test.

RESULTS

Characterization of [^{3}H]Cirazoline Binding in Brain and Kidney Homogenates

Specific binding studies have shown that [^{3}H]cirazoline, in a range of concentrations from 1 pM to 50 nM, binds to brain and kidney membrane homogenates in a reversible manner (to be discussed). Nonspecific binding of [^{3}H]cirazoline was a linear function of the radioligand concentration (FIGS. 1 and 2). Saturation analysis revealed two classes of high affinity binding sites for each homogenate preparation ($p < 0.001$; $n = 4$). Binding parameters of rat brain and kidney homogenate membranes were not significantly different. For higher affinity binding sites, the dissociation constant, K_d, of [^{3}H]cirazoline on brain membranes was 0.38 ± 0.05 nM, and the maximal binding density, B_{max}, was 46.15 ± 1.69 fmol/mg protein. Likewise, using kidney homogenates, the K_d was 0.39 ± 0.06 nM and the B_{max} was 30.53 ± 1.94 fmol/mg protein. The maximal binding capacity of the low affinity binding sites was 457 ± 30 fmol/mg protein and 426 ± 29 fmol/mg protein using rat brain and kidney

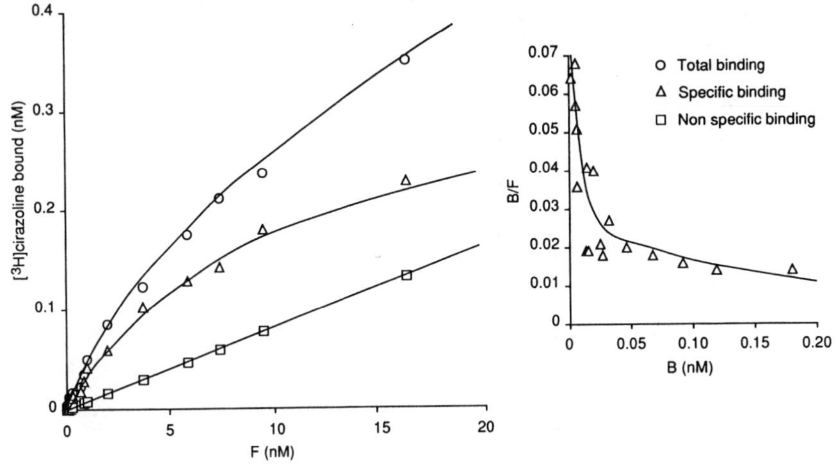

FIGURE 2. **(Left)** Saturation analysis of [^{3}H]cirazoline binding (1 pM to 50 nM) to kidney membrane homogenates: (○) total [^{3}H]cirazoline binding, (△) specific binding, and (□) nonspecific binding (in the presence of 10 μM of unlabeled bromoxidine). (**Right**) Scatchard analysis plot of [^{3}H]cirazoline specific binding, indicating two classes of sites to kidney membrane.

membranes, respectively. The K_d of [^{3}H]cirazoline was 5.61 ± 0.67 nM using brain homogenates and 9.14 ± 0.79 nM using kidney membranes (TABLE 1).

To further characterize the binding parameters of [^{3}H]cirazoline, kinetic studies of association and dissociation were carried out in rat brain and kidney membrane homogenates. In both tissues, association of [^{3}H]cirazoline was rapid, with equilibrium reached after 5-6 minutes (FIGS. 3 and 4). Analysis of association constants reveals that the association is biphasic. Similar constants of association rates were obtained for brain and kidney membranes (TABLE 3). Upon the addition of bromoxidine (10 μM), a rapid, biphasic dissociation of bound [^{3}H]cirazoline was observed (FIGS. 3 and 4, TABLE 3).

Competition studies of the lower affinity site of [^{3}H]cirazoline binding to kidney and brain homogenates were performed with alkaloid, phenylethylamine, imidazoline, and guanidinium derivatives (TABLE 2). These studies were carried out at 5 nM of

TABLE 1. Binding Parameters of [^{3}H]Cirazoline in Rat Brain and Kidney Homogenates

	Brain		Kidney	
Binding Parameter	Higher Affinity	Lower Affinity	Higher Affinity	Lower Affinity
B_{max} (fmol/mg protein)	46.15 ± 1.69	457 ± 30	30.53 ± 1.94	426 ± 29
K_d (nM)	0.38 ± 0.05	5.61 ± 0.67	0.39 ± 0.06	9.14 ± 0.79

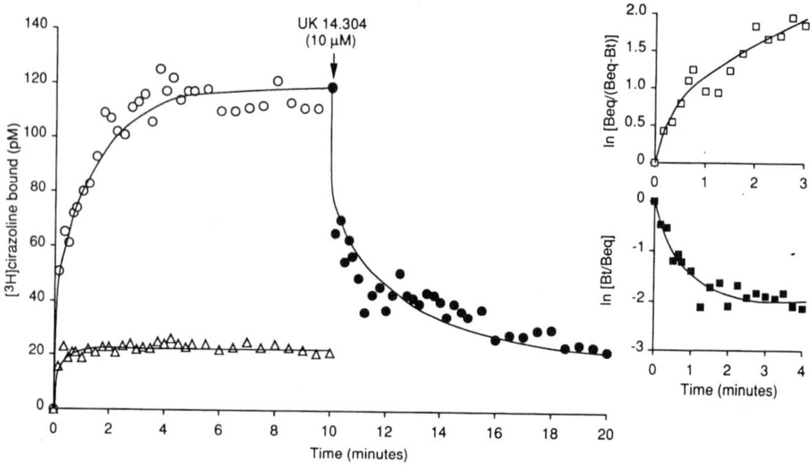

FIGURE 3. Time course of association and dissociation of [^{3}H]cirazoline binding to rat brain homogenates. Association of [^{3}H]cirazoline (1 nM) was performed at 25°C at the indicated time points. Dissociation at equilibrium was initiated with 10 μM of bromoxidine (UK 14304). (○, □) association of total binding, (△, ■) association of nonspecific binding, and (●) dissociation of specific binding.

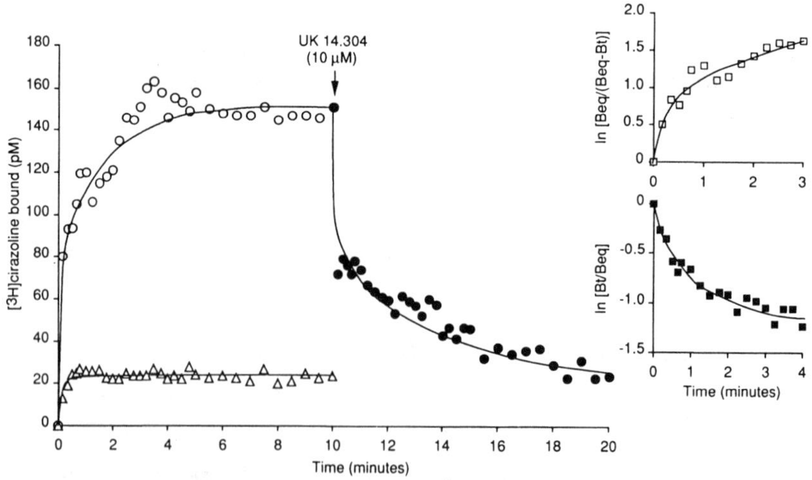

FIGURE 4. Time course of association and dissociation of [^{3}H]cirazoline binding to rat kidney homogenates. Association of [^{3}H]cirazoline (1 nM) was performed at 25°C at the indicated time points. Dissociation at equilibrium was initiated with 10 μM of bromoxidine (UK 14304). (○, □) association of total binding, (△, ■) association of nonspecific binding, and (●) dissociation of specific binding to rat kidney homogenates. Experimental condition as in FIGURE 3.

TABLE 2. Inhibition of [^{3}H]Cirazoline Binding by Drugs[a]

Ligands	K_i (μM)	
	Brain	Kidney
Cirazoline	0.004 ± 0.001	0.002 ± 0.004
Clonidine	0.050 ± 0.006	0.012 ± 0.004
Bromoxidine	0.011 ± 0.002	0.012 ± 0.001
Idazoxan	0.158 ± 0.044	0.302 ± 0.004
Amiloride	0.214 ± 0.024	0.105 ± 0.031
Naphazoline	0.193 ± 0.071	0.44 ± 0.10
Guanabenz	1.97 ± 0.45	1.97 ± 0.88
SL 84.0418	1.97 ± 0.71	2.84 ± 1.46
SL 86.0715	1.66 ± 0.76	6.92 ± 0.98
Tolazoline	4.70 ± 1.10	14.21 ± 1.98
SL 86.0714	34.70 ± 6.52	13.18 ± 7.78
Oxymetazoline	22.06 ± 12.18	30.58 ± 4.79
PAC	19.25 ± 2.69	33.10 ± 12.3
Rauwolscine	32.50 ± 10.50	26.71 ± 12.30
Phentolamine	13.98 ± 5.36	14.76 ± 7.06
Benextramine	17.10 ± 3.53	46.30 ± 12.20
BDF 8933	35.16 ± 18.95	39.68 ± 2.81
Efaroxan	219 ± 55	201 ± 72
Yohimbine	232 ± 46	293 ± 99
Propanolol	395 ± 160	418 ± 176
(–)Epinephrine	635 ± 213	230 ± 92
Prazosine	1772 ± 242	1514 ± 384
Cimetidine	>1 mM	>1 mM
Digitoxine	>1 mM	>1 mM
Digoxine	>1 mM	>1 mM
Ouabagenine	>1 mM	>1 mM
Ouabaïne	>1 mM	>1 mM

[a] Inhibition studies of [^{3}H]cirazoline (5 nM) binding to brain and kidney membranes were performed with imidazoline, guanidinium, phenylethylamine, and alkaloid compounds. Nonspecific binding was determined using 10 μM of unlabeled bromoxidine. The dissociation constant K_i of each drug is expressed in μM. Data represent the mean ± SEM of four individual experiments. (See Materials and Methods for details.)

[^{3}H]cirazoline, conditions in which the higher affinity site is saturated, and therefore they represent the pharmacological profile of the low affinity binding sites. Catecholamines (adrenaline) and non-imidazoline adrenoceptor ligands such as benextramine, prazosin, propranolol, rauwolscine, or yohimbine did not compete with [^{3}H]cirazoline binding (K_i > 10 μM). Under our experimental conditions, only guanidinium and imidazoline derivatives except phentolamine (K_i > 10 μM) had high affinity for [^{3}H]cirazoline binding sites. Unlabeled cirazoline, clonidine, and bromoxidine had the highest affinity for [^{3}H]cirazoline binding (K_i < 0.05 μM) (TABLE 2). None of the compounds tested showed significant differences in their affinity for [^{3}H]cirazoline binding sites in either brain or kidney homogenates.

TABLE 3. Kinetic Parameters of [^{3}H]Cirazoline Binding to Rat Brain and Kidney Membrane Homogenates[a]

	Brain		Kidney	
Kinetic Parameter	Higher Affinity	Lower Affinity	Higher Affinity	Lower Affinity
Kobs (min)	0.69 ± 0.11	1.05 ± 0.05	0.91 ± 0.01	1.30 ± 0.04
K_{-1} (min^{-1}, nM^{-1})	0.28 ± 0.06	0.10 ± 0.02	0.37 ± 0.02	0.16 ± 0.03
$t_{1/2}$ association (min)	2.65 ± 0.50	8.83 ± 2.99	1.85 ± 0.10	4.57 ± 0.88
K_1 (min^{-1})	0.12 ± 0.01	0.85 ± 0.01	0.20 ± 0.04	0.99 ± 0.07
$t_{1/2}$ dissociation (min)	5.86 ± 0.62	0.81 ± 0.01	3.83 ± 0.98	0.70 ± 0.05
K_D apparent (nM)	0.46 ± 0.09	10.8 ± 3.63	0.56 ± 0.16	6.74 ± 1.82

[a] Values obtained as described in FIGURE 3.

Characterization of [^{3}H]Cirazoline Binding to HIT T15 Cells

In pancreatic beta cell line HIT T15 membranes, saturation analysis of [^{3}H]cirazoline binding revealed two classes of high affinity binding sites (FIG. 5). For the higher affinity site, the dissociation constant, K_d, was 0.22 ± 0.05 nM and the maximal binding density, B_{max}, 56.5 ± 12.7 fmol/mg protein. The K_d for the lower affinity site was 11.12 ± 1.29 nM and the B_{max} 207.5 ± 18.12 (TABLE 4).

Competition studies were carried out at 1 nM of [^{3}H]cirazoline using imidazoline, phenylethylamines, and guanidinium derivatives. Except cirazoline, which revealed a biphasic inhibition curve, all other compounds displayed monophasic inhibition curves (FIG. 6). As could be seen, this site was not recognized (K_i > 10 μM) by non-imidazoline alpha- or beta-adrenoceptor ligands such as rauwolscine, prazosin, epinephrine, or propranolol.

Furthermore, a good correlation ($r = 0.96$, $p < 0.01$) was obtained between drug affinities towards [^{3}H]cirazoline binding in HIT T15 cells and in rat kidney (FIG. 7) or brain homogenates ($r = 0.98$, $p < 0.01$) (data not shown).

Comparison between [^{3}H]Cirazoline Binding and α_2-Adrenoceptors

Specific binding of 1 nM [^{3}H]cirazoline represented 75.8 ± 1.5%, 74.2 ± 2.5%, and 65.4 ± 3.1% of total binding using brain, kidney, and HIT T15 membrane homogenates, respectively (FIG. 8). Specific binding to α_2C_2-, α_2C_4-, and α_2C_{10}-AR CHO cell line preparations, respectively, represented 1.5 ± 1.0%, 3.7 ± 0.6%, and 2.5 ± 1.5% of total binding using [^{3}H]cirazoline. However, using 1 nM [^{3}H]rauwolscine it represented 86.7 ± 0.3%, 88.0 ± 1.5%, and 83.2 ± 0.7% (FIG. 8).

These data indicate that the transfected cells contain large proportions of their respective cloned α_2-adrenoceptor subtype, but they do not possess measurable imidazoline binding sites.

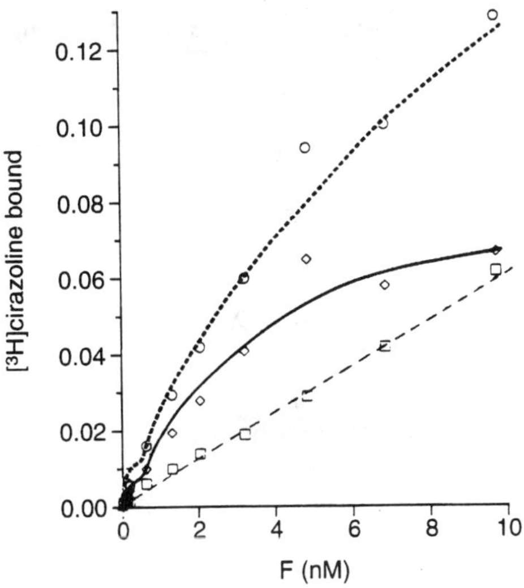

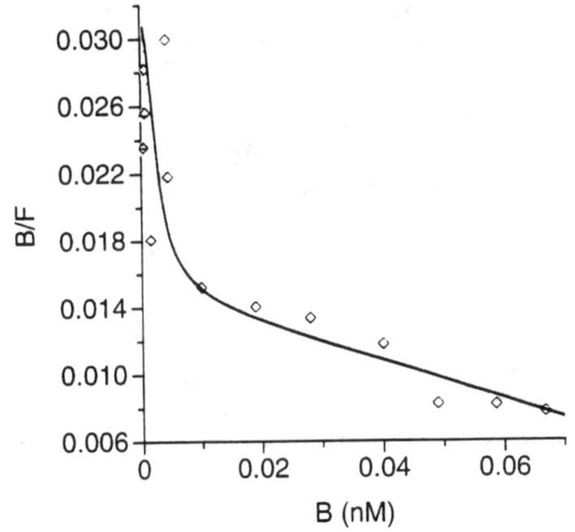

FIGURE 5. (**Top**) Saturation analysis of [^{3}H]cirazoline binding (1 pM to 50 nM) to HIT T15 cell membrane homogenates: (○) total [^{3}H]cirazoline binding, (◇) specific binding, and (□) nonspecific binding (in the presence of 10 μM of unlabeled bromoxidine). (**Bottom**) Scatchard analysis plot of [^{3}H]cirazoline specific binding, indicating two classes of sites to HIT T15 membranes.

DISCUSSION

Previous studies have demonstrated that cirazoline has the highest affinity for the nonadrenergic imidazoline recognition site, with an inhibition constant of 0.5–1 nM, compared to other imidazoline and guanidinium derivatives.[2,7] Studies to characterize the nonadrenergic imidazoline binding site are usually performed with radioligands such as [^{3}H]clonidine, [^{3}H]PAC, [^{3}H]bromoxidine, or [^{3}H]idazoxan

TABLE 4. Binding Parameters of [^{3}H]Cirazoline in HIT T15 Cell Homogenates

Binding Parameter	Higher Affinity	Lower Affinity
B_{max} (fmol/mg protein)	56.5 ± 12.7	207.5 ± 18.12
K_d (nM)	0.22 ± 0.05	11.12 ± 1.29

which are less specific for imidazoline binding sites than is cirazoline.[1,2,7,10] Therefore, such studies were carried out in the presence of adrenaline, noradrenaline, or phentolamine in order to occupy α_2-adrenoceptor sites. In most of these binding studies, relative affinity for α_2-adrenoceptors and imidazoline density of the sites depend on the radioligand used.[1,2,9,1] In the present study, with the highly selective ligand [^{3}H]cirazoline, we characterized nonadrenergic imidazoline binding sites in rat brain and kidney and in HIT T15 cell membrane homogenates in the absence of any additional drugs in the incubation medium to occupy the α_2-adrenoceptors.

Using [^{3}H]cirazoline, an α_1-AR agonist and highly selective drug on imidazoline sites,[7] we characterized two classes of high affinity binding sites on rat brain and kidney membranes and on HIT T15 cell homogenates. We previously described that the pharmacological profile of [^{3}H]cirazoline binding sites using rat brain and kidney membranes corresponds to imidazoline sites.[8] Furthermore, because idazoxan and clonidine have similar affinities to the [^{3}H]cirazoline binding site, it was suggested that this site does not discriminate between the so-called I_1 and I_2 imidazoline sites. In the present study, no [^{3}H]cirazoline binding was detected using the α_2-AR transfected CHO cell lines, indicating that they are devoid of imidazoline sites. However, our studies show high levels of specific binding of [^{3}H]rauwolscine, a selective α_2-AR antagonist, to the three CHO cell membranes.

These data confirm that imidazoline sites, as labeled by [^{3}H]cirazoline, are neither a subunit nor a subtype of α_2-adrenoceptors. They thereby confirm and extend previous observations in α_2-adrenoceptor transfected COS-7 cells using [^{3}H]idazoxan.

Saturation experiments, performed with a range of [^{3}H]cirazoline concentrations from 1 pM to 50 nM did not reveal significant differences between brain and kidney preparations or between these tissues and HIT T15 cells. In all tissues, the dissociation constants of the high and low affinity binding sites are about 0.2-0.4 nM and 5-12 nM, respectively. The receptor density of the high affinity sites is ~50 fmol/mg protein, whereas the B_{max} of the low affinity sites is ~470 fmol/mg protein.

The two categories of [^{3}H]cirazoline binding sites just characterized have similar affinities to the two classes of imidazoline-preferring binding sites described by Greney and coworkers[12] using [^{3}H]idazoxan in the human nucleus reticularis lateralis. In addition, the dissociation constants of the low affinity [^{3}H]cirazoline binding sites are in the same range as are the K_d values observed using [^{3}H]clonidine, [^{3}H]PAC, or [^{3}H]idazoxan.[1,2,10] However, in both tissues the receptor density of [^{3}H]cirazoline binding sites is different from the results reported by other authors.[1,2,10–12] These differences may be due to the radioligand used to characterize imidazoline-preferring sites and to the different experimental conditions.

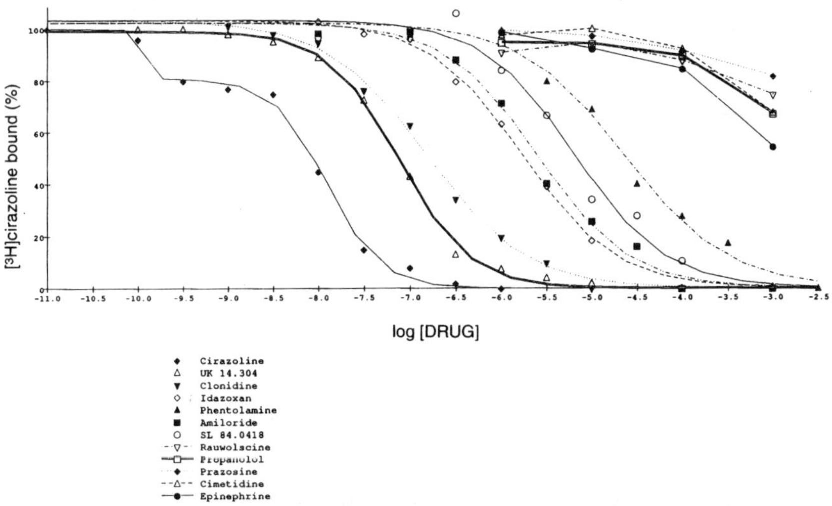

FIGURE 6. Inhibition curves of [^{3}H]cirazoline binding to HIT T15 cell homogenates. The indicated concentrations of competing drugs were added to the incubation medium (250 μl final volume) containing homogenate preparation (400 μg/ml) and 1 nM of [^{3}H]cirazoline for 30 minutes at 25°C.

Competition experiments were performed using 5 nM of [^{3}H]cirazoline, a concentration at which near saturation of the high affinity binding site is obtained. Under these conditions two groups of compounds could be distinguished. The alkaloid and phenylethylamine derivatives show very low affinity (>10 μM). This included adrenoceptor ligands of different chemical classes and pharmacological activities such as (-)adrenaline (α- and β-agonist), prazosin (α_1-antagonist), benextramine, rauwolscine, and yohimbine (α_2-antagonists) as well as propranolol, a β-adrenoceptor antagonist (TABLE 1). The second group included the guanidinium and imidazoline derivatives. Phentolamine did not compete with [^{3}H]cirazoline on these binding sites despite its imidazoline moiety (K_i > 10 μM).

Similar low affinity of phentolamine for imidazoline sites was previously observed by other authors using [^{3}H]bromoxidine or [^{3}H]idazoxan as radioligands.[2,13] Idazoxan and amiloride possess similar intermediate affinity (K_i ~ 0.15 μM), and the highest affinities were observed with bromoxidine, clonidine, and cirazoline (K_i ~ 10-50 nM). The decreasing order of inhibitor potencies in brain membranes was cirazoline > bromoxidine > clonidine > idazoxan > amiloride, whereas in membranes from kidney preparations it was cirazoline > clonidine ≥ bromoxidine > amiloride > idazoxan. It is now evident that imidazoline-preferring sites can be classified into at least two groups.[14] The I_1 types, so-called clonidine-preferring sites, bind [^{3}H]PAC and [^{3}H]clonidine, preferentially showing low affinity for amiloride. The I_2 types or idazoxan-preferring sites are labeled by [^{3}H]idazoxan and can be subdivided into two subtypes such as I_{2a} and I_{2b}, possessing respectively high and low affinity for amiloride. The high potencies of clonidine and bromoxidine to inhibit [^{3}H]cirazoline

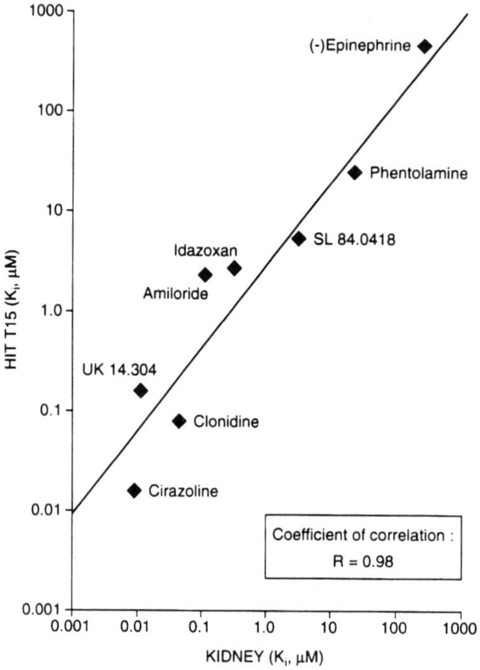

FIGURE 7. Correlation between drug affinities to [^{3}H]cirazoline binding in HIT T15 cells and rat kidney homogenates. Inhibition constants for HIT T15 cells and brain homogenates (K_i, µM) were used as described in FIGURE 6 and TABLE 1, respectively.

binding would support the terminology of I_1 sites. However, the [^{3}H]cirazoline binding sites are well recognized by guanidinium compounds such as amiloride, and therefore they appear to be different from the sites labeled by [^{3}H]PAC.[15] On the other hand, the high affinity of unlabeled cirazoline and the potencies of idazoxan and amiloride rather indicate that we are dealing with I_2 sites. These results suggest that [^{3}H]cirazoline may not distinguish between the I_1 and the I_2 sites or that it represents another, as yet uncharacterized subtype of imidazoline-preferring site.

In summary, [^{3}H]cirazoline is a novel high affinity radioligand that specifically labels imidazoline-guanidinium recognition sites without significant α- or β-adrenoceptor binding on rat brain and kidney membranes and HIT T15 cells. These special properties make [^{3}H]cirazoline a very useful ligand for further studies on imidazoline recognition sites. In addition, cirazoline is a useful tool for *in vitro* and *in vivo* characterization of the function associated with imidazoline recognition sites.

SUMMARY

Recent studies have shown that cirazoline, an α_1-adrenoceptor agonist, has greater affinity than do other imidazoline or guanidinium compounds at imidazoline recogni-

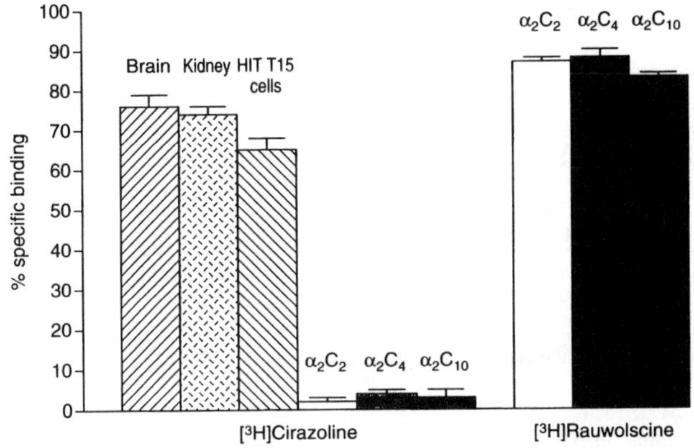

FIGURE 8. Specific binding of [^{3}H]cirazoline and [^{3}H]rauwolscine on membrane homogenates. Specific bindings are expressed as percentage of total binding and results as the mean ± SEM of *n* experiments performed in triplicate (*n* = 4 for [^{3}H]cirazoline binding; *n* = 3–7 for [^{3}H]rauwolscine binding).

tion sites. In this report we used [^{3}H]cirazoline as a probe to characterize imidazoline recognition sites present in membrane homogenates of rat brain and kidney as well as pancreatic beta HIT T15 cells.

Specific binding of [^{3}H]cirazoline to these various homogenates was saturable and reversible and was resolved into two classes of high affinity binding sites. Competition inhibition studies of [^{3}H]cirazoline binding to these different membrane preparations were performed with alkaloid, phenylethylamine, imidazoline, and guanidinium compounds. Catecholamines and non-imidazoline adrenoceptor ligands such as epinephrine, benextramine, prazosin, propranolol, rauwolscine, or adrenoceptor ligands such as epinephrine, benextramine, prazosin, propranolol, rauwolscine, or yohimbine did not compete with [^{3}H]cirazoline (K_i > 10 μM). Under our experimental conditions, only guanidinium and imidazoline derivatives had high affinities for [^{3}H]cirazoline binding sites. Unlabeled cirazoline, clonidine, bromoxidine, idazoxan, and amiloride had the highest affinities with this respective rank order. These results suggest that [^{3}H]cirazoline is a novel high affinity radioligand that specifically labels nonadrenergic imidazoline-guanidinium sites in the brain, kidney, and beta cells. Furthermore, the obtained rank order of inhibition suggests that [^{3}H]cirazoline binding does not distinguish between I_1 and I_2 sites. In addition, we compared the specific binding of [^{3}H]cirazoline with that of the α_2-adrenoceptor antagonist [^{3}H]rauwolscine in chinese hamster ovary (CHO) cell lines stably expressing human α_2C_2-, α_2C_4-, and α_2C_{10}-adrenoceptor subtypes. Using [^{3}H]rauwolscine as a probe, each of these transfected cell lines expressed high levels for the three different α_2-adrenoceptor subtypes (B_{max} values were between 2 and 7 pmol.mg-1 protein). In contrast, none of these cell lines displayed measurable imidazoline recognition sites.

In summary, [^{3}H]cirazoline is a novel high affinity radioligand that specifically labels imidazoline recognition sites without significant α- or β-adrenoceptor binding.

Furthermore, our results using α_2-adrenoceptor transfected cells confirm that the imidazoline recognition sites and each of the cloned α_2-adrenoceptor subtypes represent distinct macromolecular entities.

REFERENCES

1. ERNSBERGER, P., M. P. MEELEY, J. MANN & D. J. REIS. 1987. Clonidine binds to imidazole binding sites as well as α_2-adrenoceptors in the ventrolateral medulla. Eur. J. Pharmacol. **134:** 1.
2. HAMILTON, C. A., M. A. YAKUBU, E. JARDINE & J. L. REID. 1991. Imidazole binding sites in rabbit kidney and forebrain membranes. J. Auton. Pharmacol. **11:** 277.
3. LOMASNEY, J. W., W. LORENZ, L. F. ALLEN, K. KING, J. W. REGAN, T. L. YANG-FENG, M. G. CARON & R. J. LEFKOWITZ. 1990. Expansion of the α_2-adrenergic receptor family: Cloning and characterization of a human α_2-adrenergic receptor subtype, the gene for which is located on chromosome 2. Proc. Natl. Acad. Sci. USA **87:** 5094.
4. MICHEL, M. D., J. W. REGAN, M. A. GERHARDT, R. R. NENBIG, P. INSEL & H. J. MOTULSKY. 1989. Nonadrenergic [^{3}H]idazoxan binding sites are physically distinct from α_2-adrenergic receptors. Mol. Pharmacol. **35:** 324-330.
5. DE VOS, H., A. COUVENTS, J. DEKEYSER, J. P. DEBAKER, I. J. B. VAN MEGEN, G. EBINGER & G. VAUQUELIN. 1991. Autoradiographic distribution of α_2-adrenoceptors, NAIBS and 5-HT$_{1A}$ receptors in human brain using [^{3}H]idazoxan and [^{3}H]rauwolscine. Brain Res. **566:** 13.
6. MIRALLES, A., G. OLMOS, M. SASTRE, F. BARUREN, I. MARTIN & J. A. GARCIA-SEVILLA. 1993. Discrimination and pharmacological characterization of I$_2$-imidazoline sites with [^{3}H]idazoxan and α_2-adrenoceptor with [^{3}H]RX 821002 (2-methoxy idazoxan) in the human and rat brain. J. Pharmacol. Exp. Ther. **264:** 1187.
7. WIKBERG, J. E. S. & S. UHLÉN. 1990. Further characterization of the guinea pig cerebral cortex idazoxan receptor: Solubilization, distinction from imidazoline site, and demonstration of cirazoline as an idazoxan receptor selective drug. J. Neurochem. **55:** 192.
8. ANGEL, I., H. SCHOEMAKER, J. ALLEN, M. LE ROUZIC, C. PIMOULE, S. ARBILLA & S. Z. LANGER. 1992. The use of [^{3}H]cirazoline, a novel, selective and high affinity ligand to label the imidazoline recognition site. Fundam. Clin. Pharmacol. **6**(suppl. 1): P15, 54S.
9. MUNSON, P. J. & D. RODBARD. 1980. LIGAND: A versatile computerised approach for characterization of ligand binding systems. Anal. Biochem. **107:** 220.
10. COUPRY, I., R. A. PODEVIN, J. P. DAUSSE & A. PARINI. 1987. Evidence for imidazoline binding site in basolateral membranes from rabbit kidney. Biochem. Biophys. Res. Commun. **147:** 1055.
11. ZONNENSCHEIN, R., S. DIAMANT & D. ATLAS. 1990. Imidazoline receptor in rat liver cell: A novel receptor or a subtype of α_2-adrenoceptor? Eur. J. Pharmacol. **190:** 203.
12. GRENEY, H., G. BRICCA, M. DONTENWILL, A. BELCOURT & P. BOUSQUET. 1992. [^{3}H]idazoxan and [^{3}H]clonidine binding in the human Nucleus Reticularis Lateralis (NRL). Fundam. Clin. Pharmacol. **6**(suppl. 1): 51S.
13. LANGIN, D. & M. LAFONTAN. 1989. [^{3}H]idazoxan binding at non-$\alpha_2$adrenoceptors in rabbit adipocyte membranes. Eur. J. Pharmacol. **159:** 199.
14. REIS, D. J., S. REGUNATHAN, H. WANG, D. L. FEINSTEIN & M. P. MELLEY. 1992. Imidazoline receptors in the nervous system. Fundam. Clin. Pharmacol. **6** (suppl. 1): 23S.
15. MICHEL, M. C. & P. A. INSEL. 1989. Are there multiple imidazoline binding sites? Trends Pharmacol. Sci. **10:** 342.

Functional Studies of Specific Imidazoline-2 Receptor Ligands

D. J. NUTT,[a,b] N. FRENCH,[c] S. HANDLEY,[d]
A. HUDSON,[a] S. HUSBANDS,[e] H. JACKSON,[a]
S. JORDAN,[d] M. D. LALIES,[a] J. LEWIS,[e] L. LIONE,[a]
N. MALLARD,[a] AND J. PRATT [c]

[a]*Psychopharmacology Unit*
University of Bristol
School of Medical Sciences
University Walk
Bristol, BS8 1TD, UK

[c]*Department of Physiology & Pharmacology*
University of Strathclyde
Glasgow, G1 1XW, UK

[d]*Department of Pharmacology*
Aston University
Aston Triangle
Birmingham, B4 7ET, UK

[e]*Department of Organic Chemistry*
University of Bristol
Cantocks Close
Bristol, BS8 1TS, UK

Although the concept of imidazoline receptors has been in the literature for nearly a decade, their function remains elusive. In part, this reflects the lack of availability of selective ligands. We addressed this problem using two classical pharmaceutical approaches; the screening of existing compounds and the synthesis of new ones. This approach led us to the discovery of a series of I_2 selective drugs which are now available for the functional assessment of I_2 receptors. It is now clear that the pharmacology of imidazoline receptors is distinct from that of α_2-adrenoceptors and that they can be subdivided into I_1 and I_2 receptors.[1] Although selective compounds for I_1 receptors, such as moxonidine and rilmenidine, have become available, compounds selective for I_2 receptors are only now being reported. The selectivity of several

[b] Address for correspondence: D. J. Nutt, Psychopharmacology Unit, University of Bristol, School of Medical Sciences, University Walk, Bristol BS8 1TD, UK.

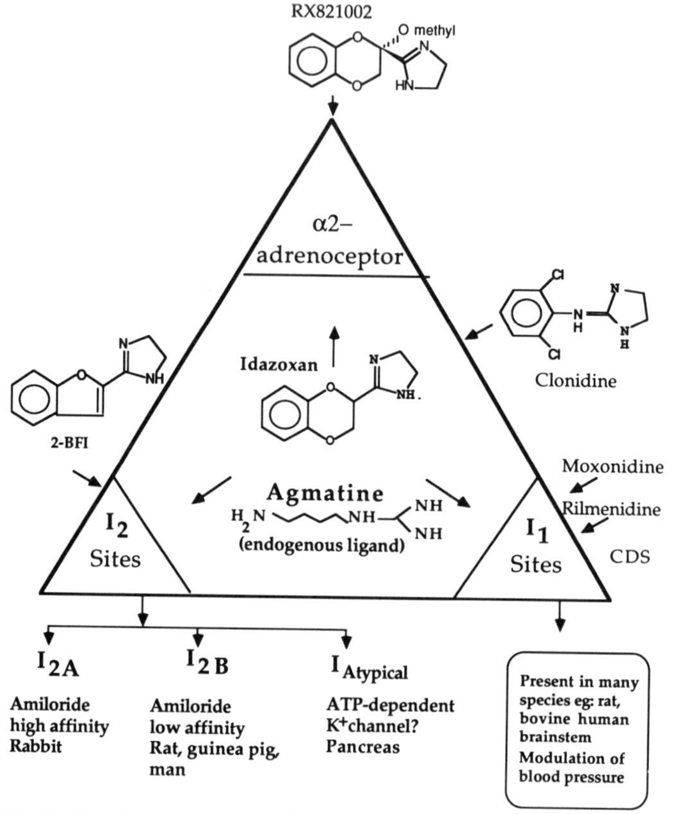

FIGURE 1. A triangular perspective of the pharmacological relationships between α_2-adrenoceptors, I_1- and I_2-receptors. Agmatine and idazoxan appear as global ligands for all three receptors; in contrast, RX821002, moxonidine, and 2-BFI (2-(2-benzofuranyl)-2-imidazoline) are specific for α_2-adrenoceptors, I_1- and I_2-receptors, respectively. Agmatine was recently (Li *et al.,* 1994) determined as the active constituent of clonidine displacing substance (CDS) and is the proposed endogenous ligand for imidazoline receptors.

drugs in relation to imidazoline-1, imidazoline-2, and α_2-adrenoceptors is illustrated in FIGURE 1.

RECEPTOR BINDING STUDIES

A key finding in the evolution of the concept of imidazoline receptors was that the α_2-adrenoceptor antagonist idazoxan bound to more sites than could be displaced by adrenaline. (For a review see ref. 2.) The "non-adrenoceptor" or I_2 component of [^{3}H]idazoxan binding was reported in central and peripheral tissues of several species; however, linking a function to these novel sites has been hindered by the lack of specific drugs. Recognizing the need for such compounds with which to dissect possible physiological and pharmacological functions mediated by I_2 receptors, we began to explore a series of idazoxan homologues that might display relative selectivity for imidazoline over α_2-adrenoceptor sites. Initially this procedure was facilitated by the known structure activity relationships in the series (FIG. 1).

Notably, the imidazoline ring of idazoxan is attached at the 2 position of the benzodioxan moiety and results in this drug having similar affinity at both α_2-adrenoceptors and imidazoline receptors. Additional substitution in the 2 position, as in 2-methoxy-idazoxan (RX821002), dramatically reduces affinity for imidazoline binding sites, whereas that for α_2-adrenoceptors is increased (TABLE 1). Indeed, labeled methoxy-idazoxan ([^{3}H]RX821002) has proved to be a superior ligand for radioligand binding and the autoradiography of α_2-adrenoceptors.[3,4]

By contrast, we found that six and/or seven aromatic substituted analogs of idazoxan show selectivity for I_2 sites, in that affinity for α_2-adrenoceptors is reduced whereas that for the former sites is maintained.[5] Examples of these semiselective I_2 site ligands include 6- or 7-fluoro-idazoxan (RX801023) and 6,7-dichloro-idazoxan, the latter compound being some 40-fold selective (TABLE 1). Much more selective for I_2 sites is the 1,3 benzodioxan isomer of idazoxan, RX821029 (2-(1,3-benzodioxanyl)-2-imidazoline), which has a fourfold higher affinity for I_2 sites than does idazoxan, and low affinity for α_2-adrenoceptors, giving an overall selectivity of about 160-fold (TABLE 1). In 1992, RX821029 represented one of the best available ligands, with its binding selectivity in a range of species reported at the "First Symposium on Imidazoline Preferring Receptors."[6] Subsequently we explored its pharmacology *in vivo* in several paradigms (to be described). However, because the 1,3 dioxan nucleus of RX821029 is rather susceptible to hydrolysis, a more stable analog was sought. This came in the form of RX801077 (2-(2-benzofuranyl)-2-imidazoline, 2-BFI), which demonstrated a low nanomolar affinity at imidazoline receptors and a remarkable 2,800-fold selectivity vis-à-vis α_2-adrenoceptors (TABLE 1). With support from the National Institutes of Health we were able to check its specificity by Novascreen. It is remarkably clean, at 10 μM producing less than 20% inhibition of binding to all the receptors screened including adenosine, glycine, kainate, MK801, NMDA, PCP, quisqualate, sigma, benzodiazepine, GABA, dopamine, 5HT, adrenergic, and interleukin; calcium, chloride, and potassium channels; and peptide receptors and second messengers. Relatively little affinity for I_1 receptors is predicted for these series of compounds, because clonidine has low affinity in displacing the parent compound (idazoxan) from the imidazoline sites it labels.

SYNTHESIS OF NEW LIGANDS FOR I_2 RECEPTORS

From the structure-activity lessons learned from the screening approach we were able to direct new synthesis with imidazoline-receptor ligands as the specific goal. Two *N*-containing derivatives were made where the oxygen atoms in the benzodioxan ring of idazoxan were replaced with two atoms of nitrogen (BU239, 2-(4,5-dihydroimidaz-2-yl)-quinoxaline) or one nitrogen atom and one carbon atom (BU224, 2-(4,5-dihydroimidaz-2-yl)-quinoline). Subsequent binding studies have shown that these new ligands show excellent selectivity for I_2 receptors over adrenoceptors,[7] particu-

TABLE 1. Binding Affinities (K_i) of Idazoxan and Related Compounds for I_2 Sites and α_2-Adrenoceptors in Rabbit Whole Brain Membranes[a]

Compound	K_i (nM)		
	I_2 Sites	α_2-Adrenoceptors	α_2/I_2 Sites
Idazoxan 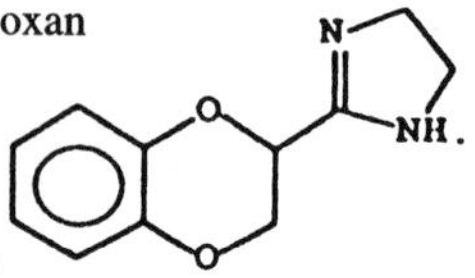	10.6 ± 1.4	55.4 ± 4.9	5.2
RX821002 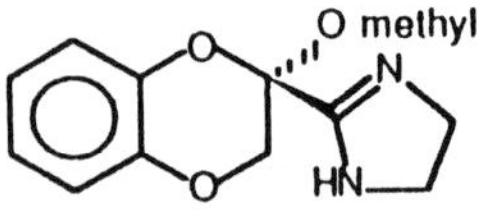	>30,000	6.9 ± 0.2	>0.0002
6- or 7-fluoro-idazoxan 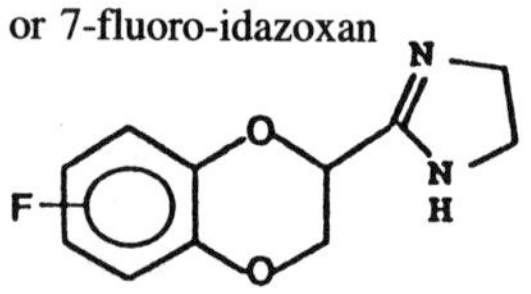	4.9 ± 0.5	142.8 ± 8.7	29.1
6,7-dichloro-idazoxan 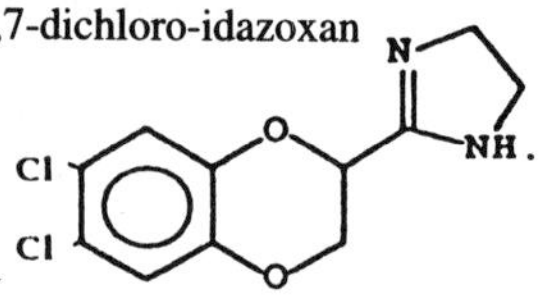	8.7 ± 0.7	359.0 ± 16.4	41.3
RX821029 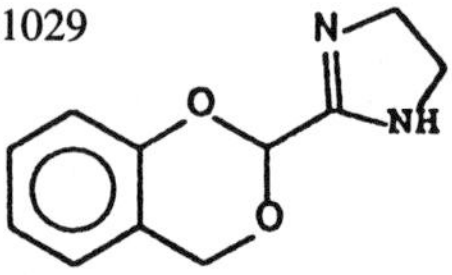	2.7 ± 0.5	436.2 ± 26.0	161.5
2-BFI 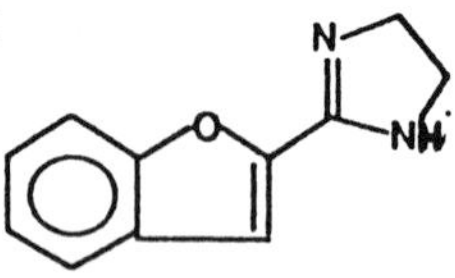	1.3 ± 0.5	3736.0 ± 147.0	2,873.8
BU224, BU224* 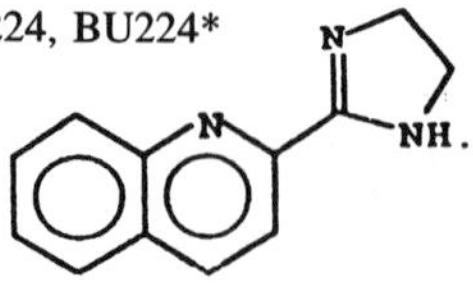	3.28 ± 0.29 0.43 ± 0.80, 36.9 ± 2.49	12,100 ± 640 (2-site fit at I_2 sites)	3,175

TABLE 1. Binding Affinities (K_i) of Idazoxan and Related Compounds for I_2 Sites and α_2-Adrenoceptors in Rabbit Whole Brain Membranes[a] (*Cont'd*)

Compound	K_i (nM)		
	I_2 Sites	α_2-Adrenoceptors	α_2/I_2 Sites
BU239 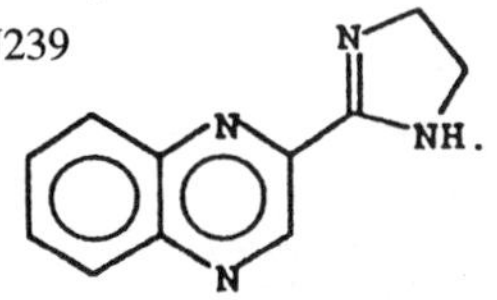	2.40 ± 0.31	7,296 ± 203	3,040

[a] Data represent the mean K_i values ± SE of the mean for data from four experiments performed in triplicate. I_2-site was determined by labeling with 2–3 nM [^{3}H]idazoxan in the presence of rauwolscine (5 μM) to preclude labeling of α_2-adrenoceptors. α_2-Adrenoceptors were labeled with 1–2 nM [^{3}H]RX821002. The specific components of binding were determined with RX801023 (10 μM) and rauwolscine (10 μM), respectively. Data (*) for BU224 binding to I_2 sites best fit a two-site model of radioligand binding, as compared to a one-site model using the differential F value defined by the equation:

$$F = \frac{(SS_1 - SS_2)/(df_1 - df_2)}{SS_2/(df_2)}$$

where SS_1 is the sum of the squares error for the single site fit, SS_2 is the sum of the squares error for the two-site fit, and df_1 and df_2 are the degrees of freedom for one- and two-site models, respectively. The two-site fit was assumed to describe a displacement curve better than a one-site fit as the determined F value had a $p < 0.05$.

larly in rabbit whole brain membranes (TABLE 1, FIG. 2), where BU224 appears comparable in selectivity to 2-BFI. The potential utility of these new compounds awaits further testing, but it is possible that they will show selectivity for potential subtypes of the I_2 receptor as BU224 displaces [^{3}H]idazoxan binding to I_2 receptors in a complex manner (FIG. 2).

[^{3}H]2-BFI, A NEW RADIOLIGAND FOR THE STUDY OF I_2 RECEPTORS

Although [^{3}H]idazoxan has proved invaluable in research into imidazoline receptors, its high affinity for α_2-adrenoceptors can be confounding and a radiolabeled selective I_2 ligand would represent a significant achievement. To produce one we synthesized a dibrominated derivative of 2-BFI for subsequent tritium exchange. Amersham International has taken this dibromo variant to yield [^{3}H]2-BFI with a specific activity of 58 Ci/mmol. Initial saturation studies show it to have potential as a ligand (FIG. 3). In rabbit whole brain membranes, [^{3}H]2-BFI binds to an apparent single population of sites with high affinity (K_D, 2.65 nM) and in a saturable manner (B_{max}, 451 fmol/mg protein). At concentrations approaching its K_D value, the nonspecific component of [^{3}H]2-BFI binding accounted for less than 9% of the total radioligand bound. In displacement studies, [^{3}H]2-BFI binding (1 nM) was inhibited by 2-

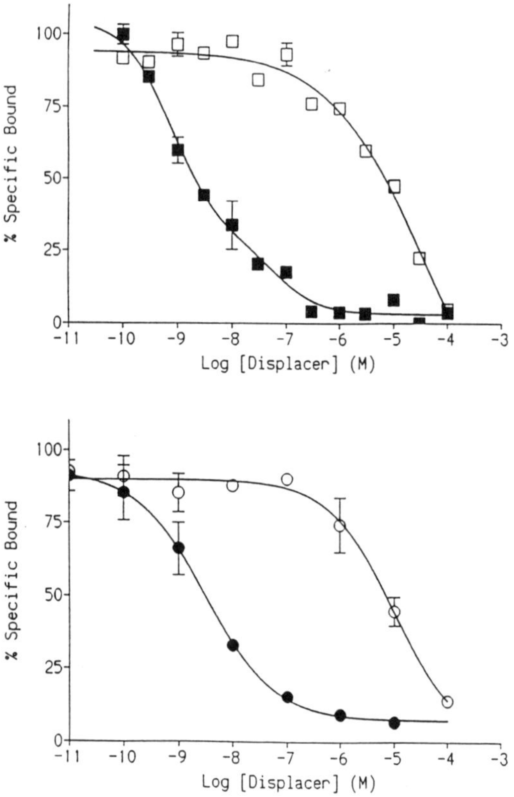

FIGURE 2. Concentration-dependent displacement of [^{3}H]-idazoxan (*closed symbols,* 2 nM in the presence of 5 μM rauwolscine to preclude binding to α_2-adrenoceptors) and [^{3}H]-RX821002 (*open symbols,* 2 nM) from I_2-receptors and α_2-adrenoceptors, respectively, in rabbit whole brain membranes. Washed membranes and radioligand were incubated for 30 minutes (50 mM Tris-HCl + 1 mM Mg^{2+}, pH 7.4, room temperature) with displacing drugs (BU224, *upper panel;* BU239, *lower panel*). Specific binding was determined by 10 μM fluoro-idazoxan (I_2 sites) or 10 μM rauwolscine (α_2-adrenoceptors) and bound ligand separated by filtration. Data represent the mean ± SE of the mean (*vertical bars* from four experiments performed in triplicate.

BFI, idazoxan, and BU224 in a concentration-dependent manner, with the latter two compounds displacing from two apparent sites with Hill slopes less than 1 (FIG. 3). Corresponding K_i values for these drugs were consistent with the displacement of bound [^{3}H]2-BFI from I_2 receptors (2-BFI, K_i = 4.0 nM; idazoxan, K_{iL} = 36 nM and K_{iH} = 0.7 nM; BU224, K_{iL} = 4.1 nM and K_{iH} = 0.004 nM); however, the nature of these biphasic displacement curves is currently being investigated. Interestingly, the I_1 site-selective drug moxonidine was of very low affinity (K_i = 251 μM) for the sites labeled by this new radioligand, as was the α_2-adrenoceptor antagonist rauwolscine (FIG. 3). [^{3}H]2-BFI is presently being evaluated as a ligand for I_2 receptors in a range of species, including its suitability for autoradiographical studies, but these initial studies in rabbit indicate that [^{3}H]2-BFI will be an excellent radioligand for the study of I_2 receptors.

NEUROCHEMICAL EFFECTS OF IMIDAZOLINE-2 RECEPTOR LIGANDS

Imidazoline receptors have been postulated to be involved in a number of neuropsychiatric conditions, particularly depressive states. For example, it has been reported

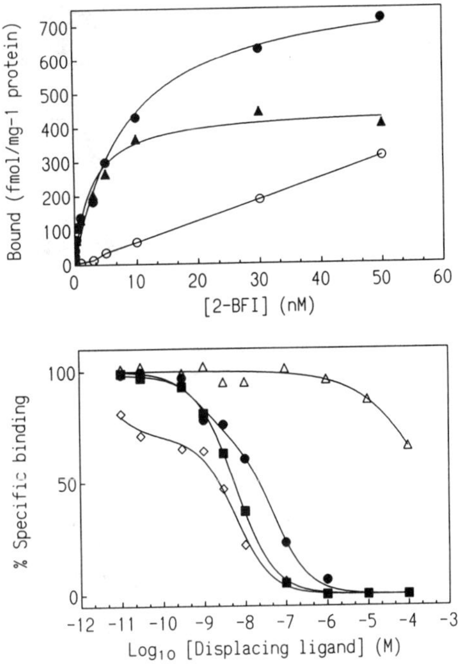

FIGURE 3. Saturation (**upper panel**) and displacement (**lower panel**) studies with [^{3}H]-2-BFI in rabbit whole brain membranes, a potential new ligand for the study of I_2-receptors. Specific binding was determined with fluoro-idazoxan under similar experimental conditions outlined in FIGURE 2. (**Upper panel**) Specific binding (▲) was determined by deducting nonspecific (○) from the total (●) ligand bound. (**Lower panel**) Displacement of specific [^{3}H]-2-BFI binding (1 nM) by BU224 (◇), 2-BFI (■), idazoxan (●), and moxonidine (△).

that I_1 receptors are elevated in platelets of some depressed humans,[8] and increased I_2 receptor binding also has been found in brains taken from suicide victims.[9] Moreover, I_2 receptors in rat brain cortex can be downregulated by long-term treatment of rats with antidepressant monoamine oxidase inhibitors clorgyline and pargyline.[10] Inasmuch as depressive states involve changes in brain function, which is reflected by neurochemical changes, we have begun to explore the effects of selective I_2 ligands on brain monoamines using both *in vitro* and *in vivo* methods.

In an *in vitro* model (measuring release of pre-loaded radiolabeled noradrenaline from superfused rat hippocampal slices), it was found that 2-BFI increased potassium-evoked (^{3}H)noradrenaline release. It was further noted that this action was the opposite of that produced by the selective noradrenaline uptake blocker desipramine (DMI) TABLE 2). Furthermore, preliminary studies on the uptake of [^{3}H]NA into the slices revealed 2-BFI to be at least 100-fold less active than DMI in blocking the uptake (data not shown). Thus, these initial studies showed that this selective I_2 receptor ligand increases noradrenaline availability by a mechanism other than inhibition of presynaptic reuptake.

Because we have also been testing the effects of these compounds in animal behavioral models (to be described), our work has focused mainly on neurochemical studies in the whole animal, with intact neuronal networks, rather than in isolated synaptic preparations. Using the technique of *in vivo* microdialysis (under chloral hydrate anesthesia) it was found that RX821029 (10 mg/kg ip) significantly reduced extracellular levels of the dopamine metabolites DOPAC and HVA as well as the

TABLE 2. Effects of 2-BFI and Desipramine on Potassium-Evoked [^{3}H]Noradrenaline Release from Rat Hippocampal Slices[a]

Drug (μM)	Mean S2/S1 Ratio	
	2-BFI	DMI
0	0.79	0.79
0.5	0.81	—
1.0	0.86	0.78
5.0	0.90	0.71
10.0	0.99	0.65

[a] Cross-chopped hippocampal slices were pre-loaded with [^{3}H]NA, then superfused in chambers, and subjected to two pulses of 25 mM K^+ (S1 and S2) with the drug present in the superfusion medium for S2 only. Hence, S2/S1 ratios were calculated and compared in each case.

serotonin metabolite 5HIAA, while causing a transient increase in extracellular dopamine in rat striatum.[11] Of the metabolites, DOPAC and 5HIAA were the more significantly affected, beginning to decrease almost immediately, while the effect on HVA levels began about 45 minutes later. These effects also occurred and were more pronounced, particularly on the metabolites, following administration of 2-BFI (20 mg/kg ip). The data for dopamine and its primary metabolite DOPAC are shown in FIGURE 4a and b, respectively. The effects of these two imidazoline ligands could be explained by a decrease in turnover of the monoamines, which in turn could arise form inhibition of monoamine oxidase. Indeed, similar profiles of effect have been observed in microdialysis studies of the MAO_A inhibitors clorgyline[12] and moclobemide.[13] This similarity is intriguing, particularly in light of the previously mentioned studies of the interactions of MAOIs with I_2 receptors.[10] Furthermore, it was recently shown that both RX821029 and 2-BFI decrease MAO activity in rat liver homogenates.[14]

We also began to investigate the effects of these two compounds on noradrenaline release in the rat frontal cortex and hippocampus. Following their administration (at 5 mg/kg ip) a transient increase in extracellular noradrenaline was observed in both brain areas, with regional differences being apparent (FIG. 5). The shorter duration of effect in the hippocampus may be due to more rapid clearance of noradrenaline by the process of neuronal uptake, which is the primary mechanism controlling noradrenaline levels in the synaptic cleft in this region.[15] This would further suggest that RX821029 and 2-BFI do not inhibit reuptake. Again, these effects may be attributed to decreased monoamine turnover via inhibition of MAO; however, without information on metabolite levels, facilitation of presynaptic release (or even a combination of both mechanisms) should not be discounted.

Another approach that can be used to explore the role of imidazoline receptors in the brain is to examine the effect of selective ligands on cerebral metabolism using the autoradiographic 2-deoxyglucose procedure.[16,17] Because the rate of metabolic activity in any particular brain region is directly proportional to functional activity,

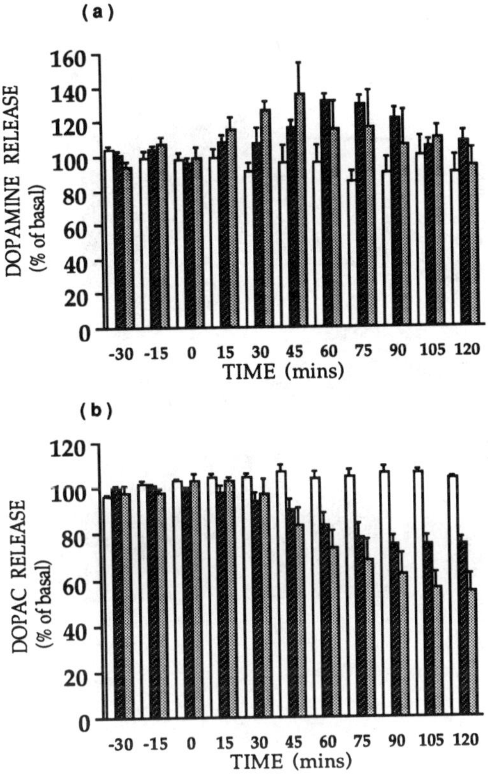

FIGURE 4. The *in vivo* effects of saline solution (□), RX821029 (10 mg/kg ip, ■), and 2-BFI (20 mg/kg ip, ▩) on extracellular levels of (**a**) dopamine and (**b**) dopac in the rat striatum.

application of this procedure therefore allows the determination of the regions (and by extrapolation, neural circuits) functionally important in the actions of the ligand(s) under investigation.

Experiments performed using 2-BFI, administered at doses of 1, 3, and 10 mg kg^{-1} iv, have shown that changes in the rate of metabolic activity are localized to a small number of brain nuclei[18] (TABLE 3). The changes observed in the arcuate nucleus and ventromedial thalamus are of particular interest, because both structures appear to be involved in feeding behavior[19,20] and Jackson *et al.*[21] have proposed that the ability of idazoxan to increase food intake in rats may involve its interaction at I_2 receptors. It is also noteworthy that although the pattern of changes observed following 2-BFI administration is different from that obtained for the selective α_2-adrenoceptor antagonist RX811059, it overlaps slightly with that obtained for idazoxan.[18] Thus, in the motor cortex, somatosensory cortex, and the molecular layer of the CA2 field of the hippocampus, both 2-BFI and idazoxan evoked reductions in metabolism that were of similar magnitude. These data suggest that there may be some overlap in the psychopharmacological profile of idazoxan and I_2 receptor ligands.

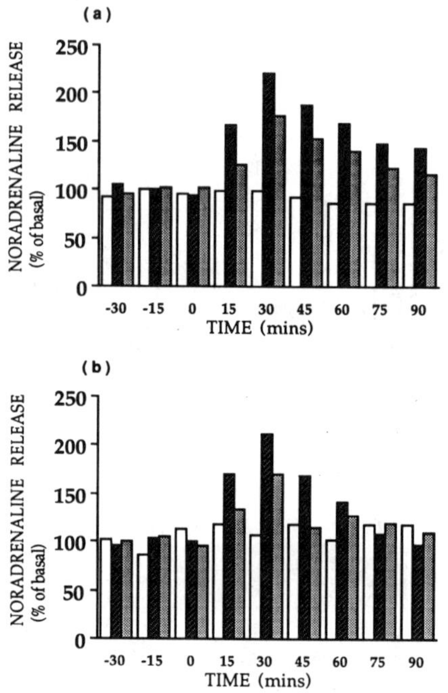

FIGURE 5. The *in vivo* effects of saline solution (□), RX821029 (5 mg/kg ip, ■), and 2-BFI (5 mg/kg ip, ▩) on extracellular levels of noradrenaline in (**a**) frontal cortex and (**b**) hippocampus.

TABLE 3. Summary of Brain Regions in Which LCGU Was Significantly Altered by 2-BFI

Structure	Control (5)	2-BFI ($mgkg^{-1}$ iv) 1 (5)	3 (5)	10 (5)
Inferior colliculus	126 ± 8	126 ± 11	159 ± 6*	131 ± 6
Superior olive	93 ± 4	99 ± 4	111 ± 6*	90 ± 3
CA2 stratum	82 ± 3	76 ± 3	77 ± 4	68 ± 1*
Subiculum	76 ± 3	75 ± 3	77 ± 4	96 ± 3*
Fornix	60 ± 2	61 ± 4	62 ± 2	80 ± 4*
Somatosensory	103 ± 3	96 ± 4	93 ± 2*	89 ± 2*
Motor cortex	101 ± 3	95 ± 4	94 ± 3	89 ± 1*
Mediodorsal	72 ± 2	64 ± 2	69 ± 3	81 ± 3*
Ventromedial	88 ± 2	92 ± 4	92 ± 3	103 ± 2*
Arcuate nucleus	46 ± 2	59 ± 2*	55 ± 1*	—

[a] Data are presented as mean glucose use (μmol/100 g/min) ± SEM, with group numbers in parentheses. Statistical significance was analyzed using one-way ANOVA followed by Newman-Keuls multiple range test. *$p < 0.05$ compared with control.

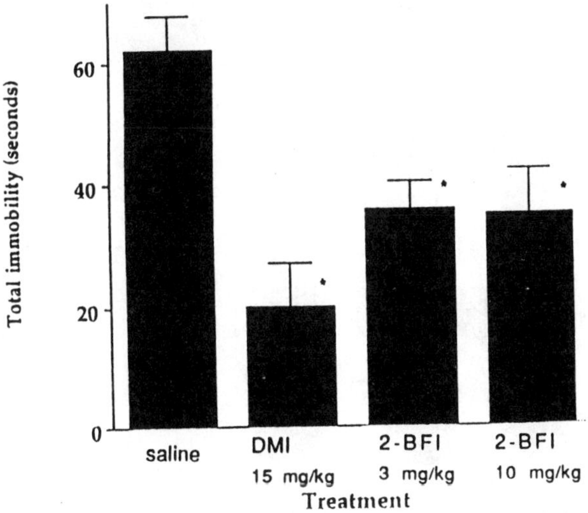

FIGURE 6. The effect of subchronic treatments (t = 0.5, 19, and 23 hours after first swim) of DMI and 2-BFI upon immobility (0–5 minutes) in a second 8-minute forced swim (t) = 24 hours) in male Wistar rats (n = 8 per group).

BEHAVIORAL STUDIES

Where does one start in exploring the behavioral actions of a new class of compounds that have not been derived with a specific therapeutic goal in mind? One way is to extrapolate from the clinical findings with nonselective imidazoline receptor compounds such as the antidepressant idazoxan.[22] Another is to make inferences from the emerging knowledge of their pharmacology. One obvious area where both overlap is depression, particularly in light of the aforementioned studies reporting changes in these receptors during depressive states and their interactions with MAO.

The best validated animal model of depression is the "Forced Swim" behavioral despair test.[23] In this model, 2-BFI decreased the immobility time of the rats (FIG. 6). As demonstrated by the positive control antidepressant desipramine, this is indicative of antidepressant activity. Thus, 2-BFI has a similar antidepressant effect as does DMI in this model, but through a mechanism other than uptake blockade, presumably increasing monamine availability. We have put other imidazoline-selective ligands through the same test with some, though not all, showing significant antidepressant activity.

A related behavior that may reflect imidazoline receptor function is feeding. We showed a few years ago that idazoxan, but not selective α_2-adrenoceptor antagonists, would increase food intake in nonfood-deprived rats.[21] To test the theory that imidazoline receptors were involved, the experiments were repeated with some of the new I_2 selective compounds. Both RX821029 and 2-BFI increased eating as predicted (FIG. 7). It is interesting that a series of I_2 compounds were recently patented for the stimulation of appetite (UK Patent Application, GB, 2 262 739 A., 1993). Perhaps

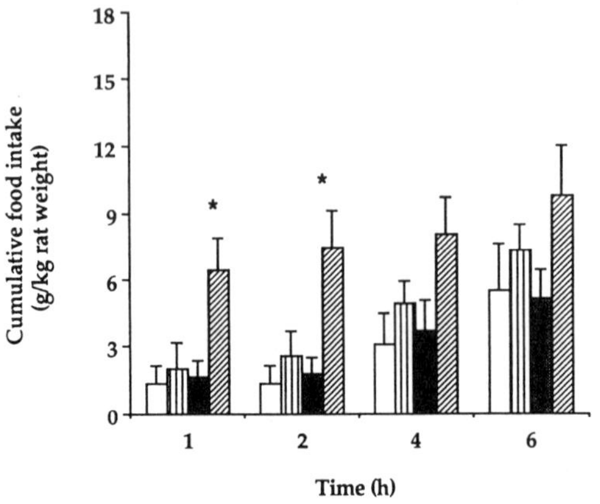

FIGURE 7. Effects of vehicle (□) and 2-BFI at 1 (▥), 3 (■), and 10 (▨) mg/kg ip on food intake over 6 hours in non-food-deprived rats.

these compounds will have some value in the treatment of anorexia, a life-threatening condition for which drug therapy has so far been sadly ineffective. If we assume that the current batch of I_2 ligands are agonists (on the grounds that they mimic the action of the presumed endogenous transmitter agmatine[24]), then it is possible that antagonists could be appetite suppressants useful in obesity.

DRUG DISCRIMINATION

This is one of the newer techniques in psychopharmacology that uses the ability of animals (usually rats, pigeons, or humans) to discriminate between their effects to probe the comparative pharmacology of drugs. For example, rats can be trained to discriminate an intraperitoneal injection of clonidine from saline solution. When presented with other α_2-adrenoceptor agonists, they perceive these drugs as being similar in effect and so respond on the clonidine cue.[25] The clonidine cue is centrally mediated as it is blocked by RX821002 but not by the peripheral α_2-adrenoceptor antagonist L659,284. When rats trained on clonidine (0.02 mg/kg ip) are presented with I_2 selective drugs, they show partial generalization. This can be interpreted as the I_2 drugs sharing some but not all the subjective effects of clonidine, perhaps sedation, dry mouth, and the like. We have also trained rats to discriminate the α_2-adrenoceptor antagonist RX811059. As with clonidine, 2-BFI also showed partial generalization to this cue (72% substitution at 7 mg/kg). One explanation would be the increase in brain noradrenaline that both classes of drug produce. Conversely, it is possible to train rats to discriminate an I_2 ligand. Using 2-BFI (7 mg/kg ip), we were able to train rats to fully discriminate this from saline solution. Of great interest was the finding that the 2-BFI cue was mimicked not only by other I_2 ligands, but also by the selective α_2-adrenoceptor antagonists RX811059 (3 mg/kg) and fluparoxan (3 mg/kg), by idazoxan (10 mg/kg), and by the MAOIs moclobemide (30 mg/kg) and pargyline (30 mg/kg). In contrast, neither clonidine (0.2 mg/kg) nor guanabenz

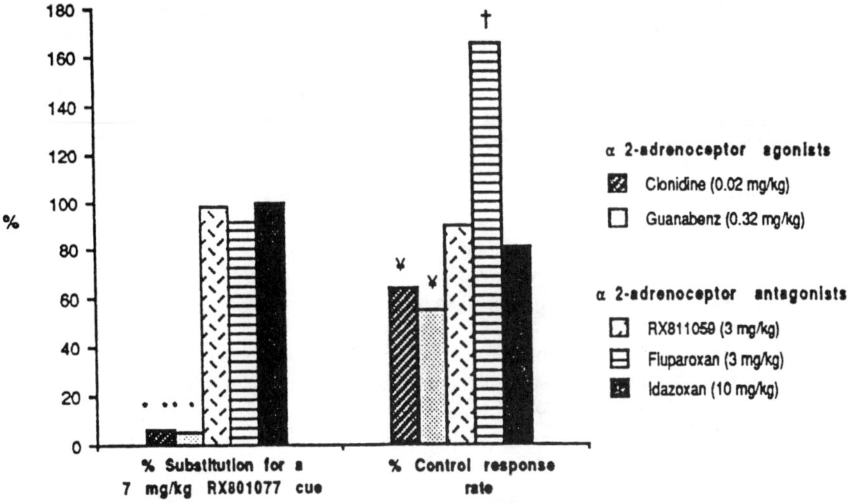

FIGURE 8. Ability of a variety of α_2-adrenoceptor agents to substitute for an RX801077 (7 mg/kg ip)-induced cue. $^{**}p < 0.01$ versus previous 7 mg/kg RX801077 training session (Student's paired *t* test), i.e., inability to substitute for an RX801077 cue. ¥p 0.05, $^{\dagger}p < 0.01$ versus response rate in previous 0.9% saline control training session (Student's paired *t* test).

(0.32 mg/kg) substituted at all (FIG. 8). These findings are consistent with the possibility that the I_2 cue is mediated by central noradrenaline release. Further studies, such as antagonism by α_1- and β-adrenoceptor antagonists, would be warranted.

CONCLUSIONS

The role of the imidazoline-2 receptor is becoming clearer thanks to the development of selective ligands and their use in functional models. The neurochemical effects of drugs such as 2-BFI suggest a new way of manipulating monoamine availability. I_2 selective drugs may offer new ways of treating psychiatric and neurologic disorders such as depression, eating disorders, and Parkinson's disease.

ACKNOWLEDGMENTS

We thank Dr. P. Towers of Amersham International (Cardiff, UK) for the kind gift of [^{3}H]2-BFI and Nancy Nutt for her help in compiling the manuscript.

REFERENCES

1. ERNSBERGER, P. 1992. Heterogeneity of imidazoline binding sites: Proposed I_1 and I_2 subtypes. Fundam. Clin. Pharmacol. **6** (suppl. 1): 55s, P18.

2. Michel, M. C. & P. A. Insel. 1989. Are there multiple imidazoline binding sites? Trends Pharmacol. Sci. **10:** 342-344.
3. Hudson, A. L., N. J. Mallard, R. Tyacke & D. J. Nutt. 1992. (^{3}H)-RX821002: A highly selective ligand for the identification of α_2-adrenoceptors in the rat brain. Mol. Neuropharmacol. **1:** 219-229.
4. Langin, D., M. Lafontan, M. R. Stillings & H. Paris. 1989. (^{3}H)-RX821002: A new tool for the identification of α2A-adrenoceptors. Eur. J. Pharmacol. **167:** 95-104.
5. Hudson, A. L., N. J. Mallard, J. Wilson, R. Tyacke & D. J. Nutt. 1992. Affinity and selectivity of 6 and/or 7 aromatic substituted analogues of idazoxan for rabbit brain non-adrenoceptor idazoxan binding sites. Br. J. Pharmacol. **105:** 158P.
6. Hudson, A. L., N. J. Mallard, D. Nutt & C. Chapleo. 1992. RX821029: A selective ligand for non-adrenoceptor idazoxan binding sites in mammalian brain. Fundam. Clin. Pharmacol. **6** (suppl. 1): 45s, I4.
7. Hudson, A. L., S. Husbands, J. W. Lewis & D. J. Nutt. 1994. Affinity and selectivity of BU224 and BU239 for rabbit brain non-adrenoceptor idazoxan binding sites (I_2-sites). Br. J. Pharmacol. **112:** 320P.
8. Piletz, J. E., K. R. Slatten & A. E. Halaris. 1992. Imidazoline-preferring binding sites in human platelets: Plasma membrane subtype I_1 is elevated in mood disorders. Fundam. Clin. Pharmacol. **6** (suppl. 1): III 19.
9. Barturen, F., J. J. Meanna, I. Martin & J. A. Garcia-Sevilla. 1992. Increased non-adrenoceptor (^{3}H)-idazoxan binding sites (NAIBS) in postmortem brains of depressed suicide victims. Fundam. Clin. Pharmacol. **6** (suppl. 1): 57s, P27.
10. Olmos, G., A. M. Gabilondo, A. Miralles, P. V. Escriba & J. A. Garcia-Sevilla. 1993. Chronic treatment with the monoamine oxidase inhibitors clorgyline and pargyline down-regulates non-adrenoceptor (^{3}H)-idazoxan binding sites in the rat brian. Br. J. Pharmacol. **108:** 597-603.
11. Lalies, M. D. & D. J. Nutt. 1993. A microdialysis study of the neurochemical effects in rat striatum of RX821029, a selective ligand for non-adrenoceptor binding sites. Br. J. Pharmacol. **109:** 85P.
12. Butcher, S. P., I. S. Fairbrother, J. S. Kelly & G. W. Arbuthnott. 1990. Effects of selective monamine oxidase inhibitors on the *in vivo* release and metabolism of dopamine in the rat striatum. J. Neurochem. **55:** 981-988.
13. Colzi, A., F. d'Agostini, R. Kettler, E. Borroni & M. Da Prada. 1990. Effect of selective and reversible MAO inhibitors on dopamine outflow in rat striatum: A microdialysis study. J. Neural Transm. **32:** (suppl.): 79-84.
14. Carpéné, C., P. Collon, A. Remaury, A. Cordi, A. Hudson, D. J. Nutt & M. Lafontan. 1995. Inhibition of amine oxidase activity by derivatives recognising non-adrenergic imidazoline binding sites (NAIBS). J. Pharmacol. Exp. Ther. In press.
15. Thomas, D. N. & R. B. Holman. 1991. A microdialysis study of the regulation of endogenous noradrenaline release in the rat hippocampus. J. Neurochem. **56:** 1741-1746.
16. Sokoloff, L., M. Reivich, C. Kennedy, M. H. Des Rosiers, C. S. Patlak, K. D. Pettigrew, O. Sakurada & M. Shinohara. 1977. The (^{14}C)-deoxyglucose method for the measurement of local cerebral glucose utilization: Theory, procedure, and normal values in the conscious and anaesthetized albino rat. J. Neurochem. **28:** 897-916.
17. McCulloch, J. 1982. Mapping functional alterations in the CNS with (^{14}C)deoxyglucose. *In* Handbook of Psychopharmacology, Vol. 15. L. L. Iversen, S. D. Iversen & S. H. Snyder, Eds.: 381-442. Plenum Press. New York.
18. French, N. 1994. Functional evaluation of α_2-adrenoceptors and imidazoline-preferring I_2 sites in the central nervous system. PhD Thesis. University of Strathclyde.
19. Braun, J. J., P. S. Lasiter & S. W. Kiefer. 1982. The gustatory neocortex of the rat. Physiol. Psychol. **10:** 13-45.

20. Beck, B., M. Jhanwaruniyal, A. Burlet, M. Chapleurchateau, S. F. Leibowitz & C. Burlet. 1990. Rapid and localized alterations of neuropeptide Y in discrete hypothalamic nuclei with feeding status. Brain Res. **528:** 245-249.
21. Jackson, H. C., I. J. Griffin & D. J. Nutt. 1991. The effects of idazoxan and other α_2-adrenoceptor antagonists on food and water intake in the rat. Br. J. Pharmacol. **104:** 258-262.
22. Osman, O. T., M. V. Rudorfer & W. Z. Potter. 1989. Idazoxan: A selective α_2-adrenoceptor antagonist and effective sustained antidepressant in two bipolar depressed patients. Arch. Gen. Psychiatry **46:** 958-959.
23. Porsolt, R. D., G. Anton, N. Blavet & M. Jalfre. 1978. Behavioural despair in rats: A new model sensitive to antidepressant treatments. Eur. J. Pharmacol. **47:** 379-391.
24. Li, G., S. Regunathan, C. J. Barrow, J. Eshraghi, R. Cooper & D. J. Reis. 1994. Agmatine: An endogenous clonidine-displacing substance in the brain. Science **263:** 966-969.
25. Jordan, S., H. C. Jackson, D. J. Nutt & S. L. Handley. 1993. Central α_2-adrenoceptors are responsible for the clonidine-induced cue in rat drug discrimination studies. Psychopharmacology **110:** 209-212.

Antiidiotypic Antibodies as Tools to Study Imidazoline Receptors

FATIMA BENNAI, HUGUES GRENEY, ANNICK MOLINES, PASCAL BOUSQUET, AND MONIQUE DONTENWILL [a]

Laboratoire de Pharmacologie Cardiovasculaire et Rénale
CNRS URA 589
Université Louis Pasteur
Faculté de Médecine
11 rue Humann
67000 Strasbourg, France

Clonidine is a well known centrally acting hypotensive substance that has been described as an α_2-adrenoceptor agonist. However, pharmacological investigations into the mechanism of the hypotensive action of this drug have led to the discovery of a new class of receptors, the imidazoline receptors (IR). Binding studies using [^{3}H]clonidine confirmed the existence of binding sites insensitive to catecholamines and sensitive to imidazolines in the brain of different species.[1,2] Idazoxan, an imidazoline with a benzodioxan moiety, proved to be a potent and selective antagonist of the effects of clonidine.[3] Therefore, [^{3}H]idazoxan was used to label nonadrenergic, imidazoline-specific binding sites in the brain as well as in other tissues in various species. Heterogeneity in imidazoline binding sites was already proposed in light of differences in the affinities of drugs as revealed by binding studies. For example, amiloride was able to displace [^{3}H]idazoxan-specific binding in the nanomolar range in rabbit brain, kidney, and liver,[4,5] whereas micromolar affinities were found in the human brain and kidney.[6,7] [^{3}H]Idazoxan imidazoline-specific binding sites sensitive to amiloride were already purified from peripheral tissues such as rabbit kidney[8] and bovine chromaffin cells.[9] These proteins exhibited molecular weights of 60 and 70 kD, respectively. Using an imidazoline binding protein-specific photoaffinity probe, AZIPI, Lanier *et al.*[10] labeled two major proteins with molecular weights of 55 and 60 kD in various rat and rabbit tissues including kidney. Both proteins were sensitive to idazoxan and cirazoline, but only the latter was sensitive to amiloride. Thus, heterogeneity of imidazoline binding sites seems to be related to the existence of different binding proteins as suggested by these different approaches.

IMIDAZOLINE BINDING SITES IN THE HUMAN BRAIN

In the human brain, at least in the nucleus reticularis lateralis/rostral ventro lateral medulla (NRL/RVLM) region, [^{3}H]idazoxan and [^{3}H]clonidine imidazoline-specific

[a] Corresponding author.

TABLE 1. Comparison of Selectivity Spectra of Anti-Idazoxan Antibodies and [^{3}H]Idazoxan Imidazoline-Specific Binding Sites in Human RVLM Membranes[a]

	Selectivity (IC_{50} nM)	
	Anti-Idazoxan Antibodies (K_d [nM], 2.2 ± 1.5)	[^{3}H]Idazoxan Binding Sites (K_d [nM], 6.6 ± 0.1
Amino-Idazoxan	2.3 ± 0.6	–
D,L Idazoxan	1.1 ± 0.4	5.7 ± 3
I Idazoxan	3.4 ± 0.5	147 ± 37
Cirazoline	1,800 ± 1,300	4.9 ± 2.3
Clonidine	2,400 ± 400	2,400 ± 500
Noradrenaline	>1 mM	>10 μM
Adrenaline	>10 μM	>10 μM
Amiloride	>1 mM	>10 μM

[a] Affinities of various compounds for antibodies are taken from ref. 20 and for binding sites from ref. 11.

TABLE 2. Comparison of Selectivity Spectra of Anti-Clonidine Antibodies (mFE7) and [^{3}H]Clonidine Imidazoline-Specific Binding Sites in Human RVLM Membranes[a]

	Selectivity (IC_{50} nM)	
	Anti-Clonidine Antibodies (K_d [nM], 490 ± 90)[b] (K_d [nM], 180 ± 25)[c]	[^{3}H]Clonidine Binding Sites (K_d [nM], 71 ± 13)
Cirazoline	8,100 ± 1,300	4.5 ± 0.6
(±) Idazoxan	56,300 ± 11,700	152 ± 46
Clonidine	430 ± 100	412 ± 97
p-Aminoclonidine	147 ± 7	4,900 ± 400
Noradrenaline	>1 mM	>20 μM
Histamine	>1 mM	>20 μM

[a] Affinities of various compounds for antibodies are taken from ref. 21 and for binding sites from ref. 12.
[b] Refers to [^{3}H]clonidine binding.
[c] Refers to [^{3}H]*p*-aminoclonidine binding.

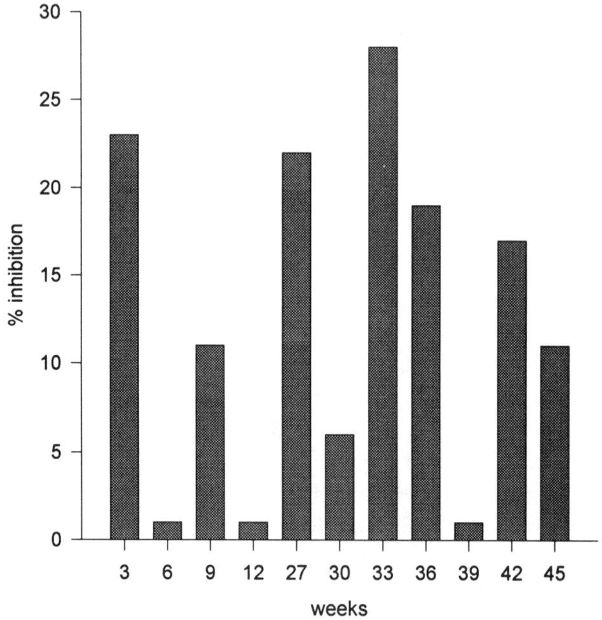

FIGURE 1. Inhibition of [^{3}H]idazoxan binding to polyclonal anti-idazoxan antibodies by AbId1. Immunization of rabbit 1 was performed with purified anti-idazoxan antibodies. Different serum samples were tested by radioimmunoassay, as described elsewhere,[20] during the time course of the immunization protocol.

catecholamine-insensitive binding sites coexist.[6,11,12] On the other hand, no [^{3}H]*p*-aminoclonidine binding sites insensitive to catecholamines could be detected in the same region of the human brain. Thus, in the human brain contrary to the bovine brain,[13] [^{3}H]clonidine and [^{3}H]*p*-aminoclonidine do not bind to the same imidazoline-specific binding sites.

As just mentioned, amiloride sensitivity has been used to classify imidazoline binding sites. In the human brain, they are insensitive to amiloride whichever radioligand is used, [^{3}H]idazoxan or [^{3}H]clonidine.[14] Moreover, we have shown that these binding sites are also insensitive to 2-(3-aminophenoxy)methyl imidazoline (AMPI), the precursor of AZIPI.[14] In parallel experiments rabbit imidazoline binding sites (brain and kidney) were highly sensitive to these two drugs. Thus, amiloride and AMPI were able to discriminate between human and rabbit imidazoline binding sites but not between [^{3}H]idazoxan and [^{3}H]clonidine imidazoline binding sites in the human brain. In fact, a strong correlation in the affinities of imidazoline derivatives for idazoxan binding sites versus clonidine binding sites was found as far as affinities of clonidine and idazoxan themselves were put out of the correlation.[15] As a matter of fact, the identity between clonidine binding site and idazoxan binding sites remains to be established, as noncompetitive interaction between them has been described.[12] To answer the questions about the relationships between clonidine and idazoxan

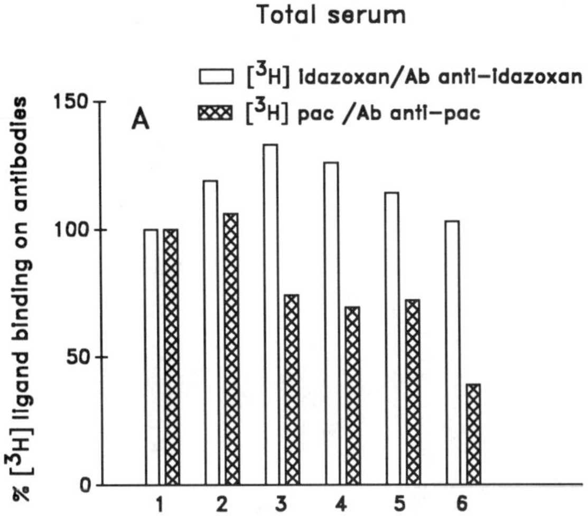

FIGURE 2. Inhibitory activity of AbId3 on anti-ligand antibodies. Rabbit 3 was immunized with purified anti-idazoxan antibodies. Immune serum samples were tested on either [^{3}H]idazoxan binding to anti-idazoxan antibodies (*empty bars*) or [^{3}H]*p*-aminoclonidine binding to monoclonal anti-clonidine antibodies (*filled bars*).

binding sites in the human brain and about the characteristics of the protein(s) bearing these sites, purification was undertaken and new specific tools were developed.

ANTIRECEPTOR ANTIBODIES

Specific antibodies against imidazoline receptors not only can help in the purification strategy but also are expected to be useful in characterizing binding proteins and in cloning the gene of these proteins by expression cloning strategy. One way to obtain antibodies directed against membrane receptors is to inject in animals the purified receptor as the antigen. In a recent approach, anti-ligand antibodies used as the template of the receptor are first prepared and, in turn, injected into animals to obtain anti-idiotypic antibodies. The latter may contain a subpopulation of anti-receptor antibodies. We chose this strategy because it was used successfully for other receptors.[16–18] The main advantage of this procedure is that anti-receptor antibodies can be raised when purified receptor is unavailable or in limited amount. Furthermore, because such antibodies are directed against the binding site of the receptor, agonistic or antagonistic activities on the receptor are often observed.[18]

Anti-Ligand Antibodies

As clonidine and idazoxan were the ligands most widely used to study imidazoline binding sites and as idazoxan acts as an antagonist towards the effects of clonidine on

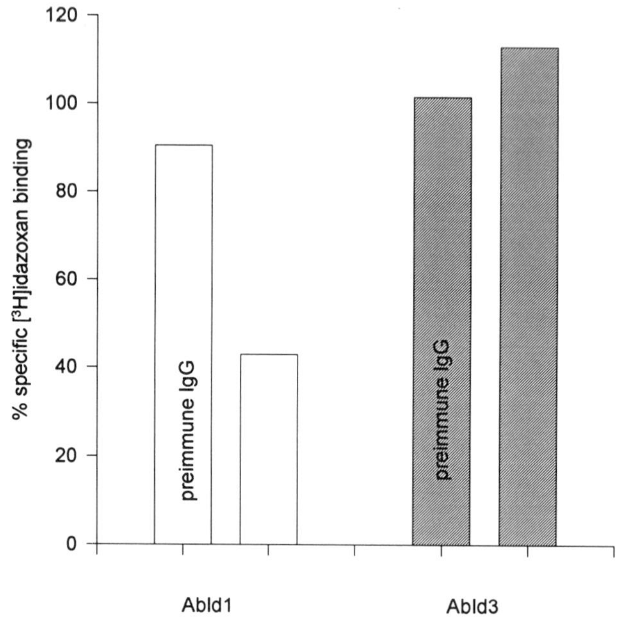

FIGURE 3. Inhibition of binding of [^{3}H]idazoxan to human brain NRL/RVLM membranes with purified anti-idiotypic antibodies. Total specific binding of [^{3}H]idazoxan was determined in the presence of 10 μM (-)-norepinephrine to mask α_2-adrenoceptors, and nonspecific binding was defined by 10 μM cirazoline. Either AbId1 (*empty bars*) or AbId3 (*filled bars*) 20 μg was allowed to react with human brain RVLM membranes. A typical experiment is shown.

imidazoline receptors, we used both drugs as antigens to obtain anti-ligand antibodies. Amino-derivatives of clonidine and idazoxan were coupled to hemocyanin to raise specific antibodies and their presence in the serum of immunized rabbits or mice was assessed by radioimmunoassay as described elsewhere.[19,20] We thus obtained specific anti-imidazoline antibodies: polyclonal anti-idazoxan and anti-clonidine antibodies[19,20] as well as monoclonal anti-clonidine antibodies.[21]

To increase the probability of obtaining anti-receptor antibodies, binding properties similar to those of the receptor are required for anti-ligand antibodies. In TABLES 1 and 2, we compare the properties of anti-idazoxan and anti-clonidine antibodies with [^{3}H]idazoxan and [^{3}H]clonidine imidazoline-specific binding sites, respectively, in the human NRL/RVLM region.[11,12] Our antibodies are highly selective for their antigens; as the imidazoline receptors themselves, they were able to recognize imidazolines but not catecholamines. We first plan to use the anti-idazoxan antibodies as antigen in rabbits in the anti-idiotypic strategy, as their selectivity looked more like that of the receptor (TABLE 1).

Anti-Idiotypic Antibodies

Effect on Anti-Ligand Antibodies. The anti-idazoxan antibodies were purified from rabbit serum by affinity chromatography on idazoxan-sepharose columns. They

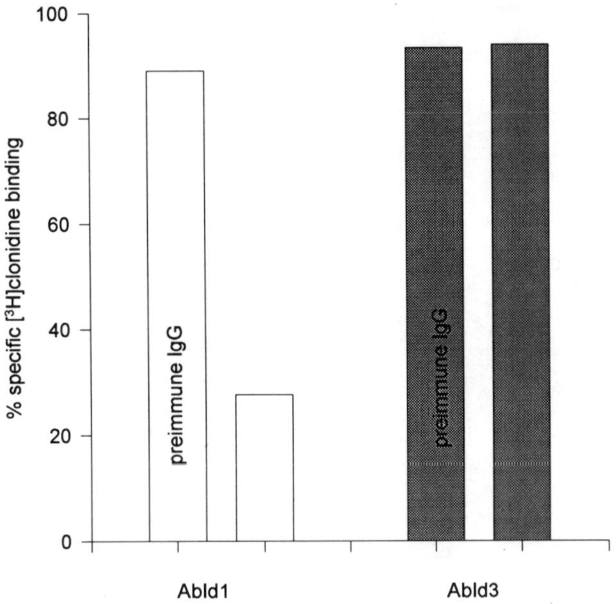

FIGURE 4. Inhibition of [^{3}H]clonidine binding to human RVLM membranes by purified anti-idiotypic antibodies. Total specific binding of [^{3}H]clonidine was determined in the presence of 10 μM (-)-norepinephrine to mask α_2-adrenoceptors, and 10 μM cirazoline defined nonspecific binding. Inhibitory activity of 20 μg AbId1 (*empty bars*) or AbId3 (*filled bars*) and a typical experiment are shown.

were then injected monthly in different rabbits, and the presence of anti-idiotypic antibodies were assessed by the ability to inhibit [^{3}H]imidazoline binding on the anti-ligand antibodies. The appearance of anti-idiotypic antibodies were cyclical (FIG. 1), and the specificity of the responses towards the anti-ligand antibodies seemed to differ in each immunized rabbit (FIG. 2). In fact, in rabbit 1 anti-idiotypic antibodies (AbId1) able to inhibit [^{3}H]idazoxan binding on anti-idazoxan antibodies were obtained (FIG. 1), whereas in rabbit 3, antibodies (AbId3) able to inhibit [^{3}H]*p*-aminoclonidine on anti-clonidine antibodies were only raised (FIG. 2). Thus, injection of anti-idazoxan antibodies as immunogens led to anti-idiotypic antibodies with very variable characteristics.

Effect on Imidazoline Receptors. Anti-idiotypic antibodies from the two rabbits were purified on either anti-idazoxan antibodies or anti-clonidine antibodies and their activity was tested on noncatecholamine [^{3}H]imidazoline-specific binding in human NRL/RVLM membranes. AbId1 was able to inhibit [^{3}H]idazoxan imidazoline-specific binding in these preparations, 20 μg purified antibodies achieving greater than 50% inhibition, whereas the same amount of either AbId3 or rabbit preimmune immunoglobulins failed to have any significant effect (FIG. 3).

Experiments were also conduced on [^{3}H]clonidine imidazoline-specific binding in human NRL membranes. Results shown on FIGURE 4 indicate that AbId1 recognize

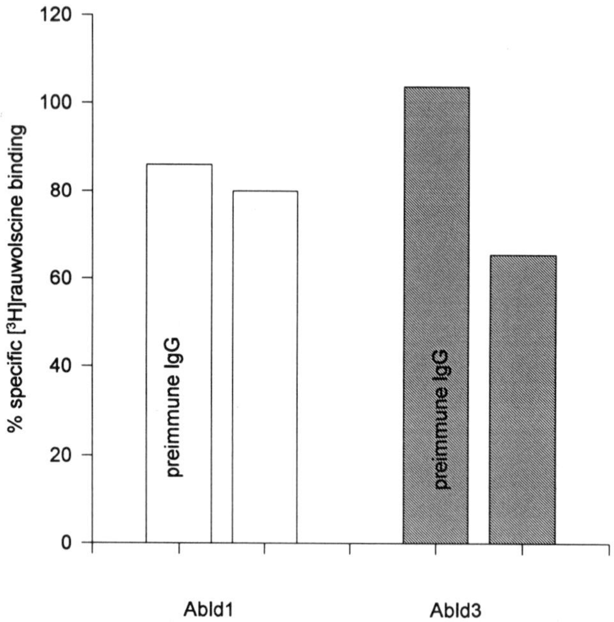

FIGURE 5. Inhibition of binding of [^{3}H]rauwolscine to rat cerebral cortex membranes by anti-idiotypic antibodies. Nonspecific binding was defined by 10 μM phentolamine. Inhibition by 20 μg AbId1 (*empty bars*) or AbId3 (*full bars*) and a typical experiment are shown.

these clonidine binding sites, as 20 μg of anti-idiotypic antibodies inhibit about 70% of the binding. As was the case when [^{3}H]idazoxan was used as radioligand, neither AbId3 nor preimmune immunoglobulins were able to inhibit [^{3}H]clonidine binding.

The effects of anti-idiotypic antibodies on α_2-adrenoceptors were checked in rat cortex membranes. [^{3}H]Rauwolscine or [^{3}H]clonidine was used to label the adrenoceptors, and in both cases AbId1 proved unable to significantly inhibit specific binding on these receptors. However, 35% inhibition of [^{3}H]clonidine binding on α_2-adrenoceptors was obtained with 20 μg AbId3, whereas nonimmune immunoglobulins failed to show any effect (Fig. 5).

The results just depicted led us to further study the specific interaction of AbId1 with imidazoline binding sites in the human brain. Solubilized imidazoline binding sites from the human NRL region maintained their binding characteristics.[6] Immunoprecipitation experiments of imidazoline binding sites were performed in the presence of purified AbId1 or nonimmune immunoglobulins. AbId1 20 μg was able to deplete 56.5 ± 13.0% of [^{3}H]idazoxan binding sites from solubilisate when nonimmune IgG had no effect.

Western blot analyses were performed with purified AbId1. When human NRL membranes were subjected to SDS-PAGE and immunoblotting, several bands ranging from 60-30 kD were revealed. However, when partially purified imidazoline binding proteins from the same human brain region were subjected to the same procedure, one 43-kD protein was stained with the antibodies. Nonimmune serum failed to label any band. This 43-kD protein appeared to be related to imidazoline binding sites in the human brain.

CONCLUSIONS

The anti-idiotypic strategy produced new tools to study imidazoline binding sites. Injection of anti-idazoxan antibodies into rabbits led to the production of anti-idiotypic antibodies with either imidazoline binding sites or α_2-adrenoceptor recognition potencies. These results confirm the finding that idazoxan itself behaves as an α_2-adrenoceptor ligand as well as an imidazoline receptor-specific ligand. Subpopulations of polyclonal anti-idazoxan antibodies may exist with different fine specificities[20] leading to the production of different anti-idiotypic antibodies.

Some interesting tools in the field of imidazoline receptors have been described as specific antibodies directed against bovine adrenal gland imidazoline binding sites[22] and photoaffinity labeling ligands[10] which allowed the vizualization of binding proteins exhibiting different molecular weights (70-, 55-, and 60-kD proteins). The use of anti-idiotypic antibodies, in turn, confirmed the heterogeneity of imidazoline binding sites already supported by binding data, as they were able to reveal a yet unknown 43-kD protein, at least in the human NRL/RVLM region. At the time, no differences between [^{3}H]clonidine and [^{3}H]idazoxan binding proteins in the human brain were noted, because the antibodies recognized both binding sites and also because the fraction containing the 43-kD protein exhibited imidazoline-specific binding with both radioligands.[23] Further characterization of this protein is in progress. Comparison of this protein with the others described in peripheral tissues as well as in other species will therefore be possible.

REFERENCES

1. Bricca, G., M. Dontenwill, A. Molines, J. Feldman, A. Belcourt & P. Bousquet. 1989. The imidazoline preferring receptors: Binding in bovine, rat and human brainstem. Eur. J. Pharmacol. **62:** 1-9.
2. Hamilton, C. A., M. A. Yakubu, E. Jardine & J. L. Reid. 1991. Imidazole binding sites in rabbit kidney and forebrain membranes. J. Auton. Pharmacol. **11:** 277.
3. Tibirica, E, J. Feldman, C. Mermet, F. Gonon & P. Bousquet. 1991. An imidazoline specific mechanism for the hypotensive effect of clonidine: A study with yohimbine and idazoxan. J. Pharmacol. Exp. Ther. **256:** 606-613.
4. Lachaud-Petiti, V., R. A. Podevin, Y. Chretien & A. Parini. 1991. Imidazoline-guanidinium and α_2 adrenergic binding sites in basolateral membranes from human kidney. Eur. J. Pharmacol. Mol. Pharmacol. **206:** 23.
5. Tesson, F., C. Prip-Buus, A. Lemoine, J. P. Pregorier & A. Parini. 1991. Subcellular distribution of imidazoline-guanidinium receptive sites in human and rabbit liver: Major localization to the mitochondrial outer membrane. J. Biol. Chem. **266:** 155-160.
6. Greney, H., G. Bricca, M. Dontenwill, J. Stutzmann, P. Bousquet & A. Belcourt. 1994. Characterization of imidazoline binding protein(s) solubilized from human brainstem: Studies with [^{3}H]idazoxan and [^{3}H]clonidine. Neurochem. Int. **25:** 183-191.
7. Michel, M. & P. Insel. 1989. Are there multiple imidazoline binding sites? Trends Pharmacol. Sci. **10:** 342-343.

8. LIMON, I., I. COUPRY, S. M. LANIER & A. PARINI. 1992. Purification and characterization of mitochondrial imidazoline-guanidinium receptive site from rabbit kidney. J. Biol. Chem. **267:** 21645-21649.
9. WANG, H., S. REGUNATHAN, M. P. MEELEY & D. J. REIS. 1992. Isolation and characterization of imidazoline receptor protein from bovine adrenal chromaffin cells. Mol. Pharmacol. **42:** 792-801.
10. LANIER, S. M., B. IVKOVIC, I. SINGH, J. L. NEUMEYER & V. BAKTHAVACHALAM. 1993. Visualization of multiple imidazoline/guanidinium-receptive sites. J. Biol. Chem. **268:** 16047-16051.
11. BRICCA, G., H. GRENEY, M. DONTENWILL-KIEFFER, J. ZHANG, A. BELCOURT & P. BOUSQUET. 1993. Heterogeneity of the specific imidazoline binding of idazoxan in the human cerebral cortex. Neurochem. Int. **22:** 153-163.
12. BRICCA, G., H. GRENEY, J. ZHANG, M. DONTENWILL, J. STUTZMAN, A. BELCOURT & P. BOUSQUET. 1994. Human brain imidazoline receptors: Further characterization with clonidine. Eur. J. Pharmacol. Mol. Pharmacol. **266:** 25.
13. ERNSBERGER, P., T. H. DAMON, L. GRAFF, S. G. SCHÄFFER & M. O. CHRISTEN. 1993. Moxonidine, a centrally acting antihypertensive agent, is a selective ligand for I1 imidazoline sites J. Pharmacol. Exp. Ther. **264:** 172-182.
14. GRENEY, H., A. MOLINES, P. BOUSQUET & M. DONTENWILL. 1994. Heterogeneity of imidazoline binding sites revealed by a cirazoline derivative. Eur. J. Pharmacol. **271:** 533-536.
15. DE VOS, H., G. BRICCA, J. DE KEYSER, J. P. DE BAKER, P. BOUSQUET & G. VAUQUELIN. 1994. Imidazoline receptors, non-adrenergic idazoxan binding sites and α_2-adrenoceptors in the human central nervous system. Neuroscience **59:** 589-598.
16. GUILLET, J. G., S. V. KAVERI, O. DURIEU, C. DELAVIER, J. HOEBEKE & A. D. STROSBERG. 1985. β-Adrenergic agonist activity of a monoclonal anti-idiotypic antibody. Proc. Natl. Acad. Sci. USA **82:** 1781-1784.
17. ELAZAR, Z., H. KANETY, M. SCHREIBER & S. FUCHS. 1988. Anti-idiotypes against a monoclonal anti-haloperidol antibody bind to dopamine receptors. Life Sci. **42:** 1987-1993.
18. WAI, M. K. & H. H. TAI. 1993. Monoclonal anti-idiotypic antibody to a potent thromboxane A_2 receptor antagonist and its interaction with thromboxane A_2 receptor. J. Biol. Chem. **268:** 6364-6370.
19. DONTENWILL, M., G. BRICCA, A. MOLINES, P. BOUSQUET & A. BELCOURT. 1988. Production and characterization of anti-clonidine antibodies not cross reacting with catecholamines. Eur. J. Pharmacol. **149:** 249-255.
20. BENNAI, F., H. GRENEY, G. BRICCA, A. BELCOURT, P. BOUSQUET & M. DONTENWILL. 1993. Polyclonal anti-idazoxan antibodies: Characterization and purification. Eur. J. Pharmacol. (Mol. Pharmacol.) **246:** 45-51.
21. DONTENWILL, M., A. MOLINES, G. BRICCA, J. STUTZMANN, J. KEMPF, A. BELCOURT & P. BOUSQUET. 1992. Production and characterization of an iminoimidazolidine specific monoclonal antibody using para-aminoclonidine as antigen. Life Sci. **50:** 1859-1868.
22. WANG, H., S. REGUNATHAN, D. A. RUGGIERO & D. J. REIS. 1993. Production and characterization of antibodies specific for the imidazoline receptor protein. Mol. Pharmacol. **48:** 509-515.
23. GRENEY, H., F. BENNAI, A. MOLINES, M. DONTENWILL, A. BELCOURT & P. BOUSQUET. 1994. Isolation of a human cerebral imidazoline specific binding protein. Eur. J. Pharmacol. **265:** R1-R2.

Structure-Affinity Relationship of Cirazoline Derivatives for Idazoxan Imidazoline-Specific Sites in the Human Cerebral Cortex and the Rabbit Kidney

L. BRASILI,[a] M. PIGINI,[b] J. D. EHRHARDT,[c]
A. MOLINES,[c] H. GRENEY,[c] P. BOUSQUET,[c] AND
M. DONTENWILL [c]

[a]*Dipartimento di Scienze Farmaceutiche*
Università di Modena
Via Campi 183
41100, Modena, Italy

[b]*Dipartimento di Scienze Chimiche*
Università di Camerino
Via S. Agostino 1
62032 Camerino (MC), Italy

[c]*Laboratoire de Pharmacologie Cardiovasculaire et Rénale*
CNRS URA 589
Faculté de Médecine
Université Louis Pasteur
11 rue Humann
67000-Strasbourg, France

Idazoxan, an α_2-adrenoceptor antagonist belonging to the imidazoline class,[1] binds to a second site in a wide range of tissues, with an affinity comparable to that determined at α_2-adrenoreceptors. This site has been named the nonadrenergic idazoxan binding site (NAIBS). Furthermore, in light of recent affinity values derived from ligand-binding studies, the existence of more than one nonadrenoreceptor binding site has been suggested. In fact, [^{3}H]clonidine labeled sites that seem to be distinct from NAIBS.[2] In contrast to idazoxan, which potently inhibited the binding of [^{3}H]clonidine to imidazoline binding sites, clonidine barely displaced [^{3}H]idazoxan binding to NAIBS and did so in a noncompetitive fashion.[3] The different affinities reported in ligand-binding studies for the K^+-sparing diuretic amiloride point to the heterogeneity of NAIBS, which has been described as amiloride-sensitive in rabbit tissues but not in others.[4–9]

To gain further insight into NAIBS heterogeneity, we compared the affinities of a series of cirazoline derivatives for [^{3}H]idazoxan binding sites in the rabbit kidney and human cerebral cortex membranes.

TABLE 1. Inhibition Constants (K_i) of Cirazoline **1**, Cirazoline Derivatives **2-9**, Idazoxan, and Amiloride at [^{3}H]Idazoxan Binding Sites of Rabbit Kidney and Human Cortex

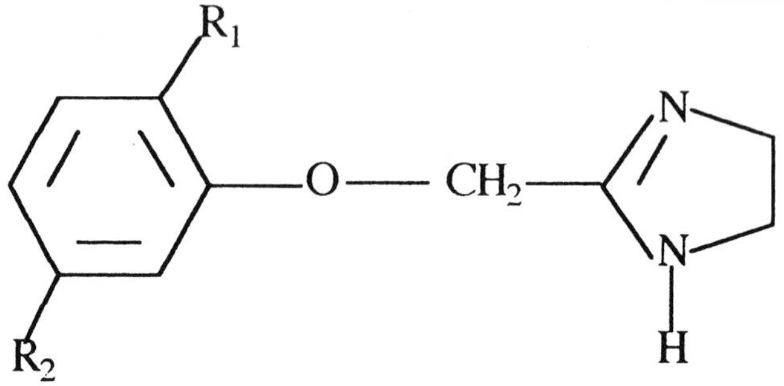

	R_1	R_2	Rabbit Kidney (K_i nM)	Human Cortex (K_i nM)
1	Cyclopropyl	H	0.5 ± 0.2	0.95 ± 0.25
2	H	H	0.28 ± 0.2	2.4 ± 0.5
3	CH_3	H	1.7 ± 0.8	3.0 ± 1.1
4	C_2H_5	H	1.8 ± 0.7	6.5 ± 1.4
5	nC_3H_7	H	8.6 ± 3.9	42.1 ± 11.6
6	iC_3H_7	H	11.4 ± 2.2	11.1 ± 3.3
7	CH_2-CH=CH_2	H	1.9 ± 0.5	5.4 ± 1.3
8	H	NH_2	10.6 ± 1.8	2,900 ± 180
9	CH_3	NH_2	1.4 ± 0.4	41.4 ± 14.5
	Idazoxan		4.3 ± 1.6	19.5 ± 8.2
	Amiloride		140 ± 28	2,500 ± 960

MATERIALS AND METHODS

[^{3}H]Idazoxan (43 Ci/mmol) was from Amersham, cirazoline and idazoxan were gifts from LERS Synthelabo and Servier, respectively, (–)norepinephrine and amiloride were purchased from Sigma, AMPI was provided by Research Biochemicals Internationals as part of the Chemical Synthesis Program of the U.S. National Institute of Mental Health, contract NOIMH00007, and cirazoline derivatives were synthesized in our laboratories.

Rabbit kidney and human cerebral cortex membranes were prepared as previously described.[10,11] Binding tests were run in triplicate at 25°C for 45 minutes, as already reported.[12] Competition experiments were performed in a final volume of 0.4 ml containing 0.2-0.5 mg protein, radioligands, and competitors (0.1 nM to 1 mM final concentration). Nonspecific binding was determined in the presence of 10 μM cirazoline. All binding studies were performed in the presence of 10 μM (–)norepinephrine with 0.005% ascorbic acid in order to mask α_2-adrenoceptors. [^{3}H]Idazoxan was used at 4 nM in rabbit kidney and at 8 nM in human cortex.

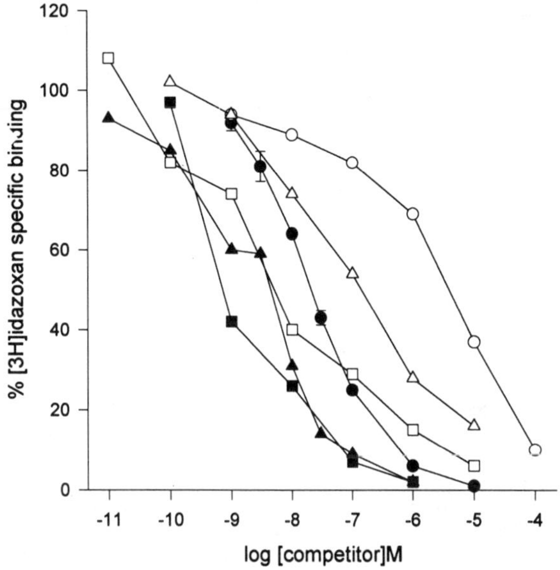

FIGURE 1. Inhibition of [^{3}H]idazoxan binding to rabbit kidney (■, ▲, ●) and human cortex (□, △, ○) by **2** (□, ■), **8** (△, ▲), and **9** (○, ●).

RESULTS AND DISCUSSION

The results of binding studies of cirazoline derivatives, together with those obtained with cirazoline, idazoxan, and amiloride for comparison, are reported in TABLE 1.

Cirazoline and idazoxan bind with high affinities to the two tissues; however, idazoxan is about 8-20 times less effective than cirazoline in displacing [^{3}H]idazoxan binding. Amiloride binds with moderate (K_i = 140 nM) and low (K_i = 2,500 nM) affinity to rabbit kidney and human cortex, respectively, with an affinity ratio of about 8.

As for cirazoline derivatives, removal of the cyclopropyl group from cirazoline **1** to give compound **2** maintains high affinity, indicating that the cyclopropyl moiety is not an essential feature for binding to NAIBS. However, the small opposite effects on the two preparations make compound 2 10-fold selective for NAIBS on rabbit kidney.

Replacement of cyclopropyl with other alkyl substituents (**3**-**6**) also maintains affinity in the nanomolar range, but a slight decrease is observed as the substituent size increases. Of the alkyl groups studied, the branched ones afford the more pronounced negative effect: compounds **5** and **6** cause a 40- and 20-fold decrease in affinity in human cortex and rabbit kidney, respectively.

Introduction of the NH_2 group in meta position on the phenyl ring (**8**, AMPI) causes a 40- and 1,200-fold decrease in affinity in rabbit kidney and human cortex, respectively. Thus, AMPI discriminates the two binding sites with an affinity 300 times higher in rabbit kidney. When NH_2 is substituted on compound **3**, no change in affinity is observed in rabbit kidney, whereas in human cortex there is a 14-fold decrease. The resulting compound **9** (methyl-AMPI) thus maintains an appreciable (30-fold) selectivity for rabbit kidney. These results indicate that the NH_2 group affects ligand binding to NAIBS of the two preparations differently, the human cortex being more sensitive to it.

In conclusion, this work shows that modification of the cirazoline structure generates high affinity ligands. Some of them (**2**, **9** and particularly **8**) display significant differences in affinity values between rabbit kidney and human cortex, further supporting the existence of NAIBS heterogeneity.

REFERENCES

1. Chapleo, B. C., J. C. Doxey, P. L. Meyers & A. G. Roach. 1981. Br. J. Pharmacol. **74:** 842P.
2. De Vos, H., G. Bricca, J. De Keyser, J. P. De Baker, P. Bousquet & G. Vauquelin. 1994. Neurosciences **59:** 589-598.
3. Bricca, G., H. Greney, M. Dontenwill-Kieffer, J. Zhang, A. Belcourt & P. Bousquet. 1993. Neurochem. Int. **22:** 153-163.
4. Yablonsky, F. & J. P. Dausse. 1989. Eur. J. Pharmacol. **164:** 167-170.
5. Lauchaud-Pettiti, V., R. A. Podevin, Y. Chretien & A. Parini. 1991. Eur. J. Pharmacol. Mol. Pharmacol. **206:** 23-31.
6. Brown, C. M., A. C. MacKinnon, J. C. MacGrath, M. Spedding & A. T. Kilpatrick. 1990. Br. J. Pharmacol. **99:** 803-809.
7. Zonnenchein, R., S. Diamant & D. Atlas. 1990. Eur. J. Pharmacol. **190:** 203-215.
8. Langin, D. & M. Lafontan. 1989. Eur. J. Pharmacol. **159:** 199-203.
9. Coupry, I., D. Atlas, I. Uzilli, S. M. Lanier & A. Parini. 1989. FASEB J. **3:** A118.
10. Hamilton, C. A., M. A. Yakubu, E. Jardine & J. L. Reid. 1991. J. Auton. Pharmacol. **11:** 277-283.
11. Bricca, G., H. Greney, J. Zhang, M. Dontenwill, J. Stutzman, A. Belcourt & P. Bousquet. 1994. Eur. J. Pharmacol. Mol. Pharmacol. **266:** 25-33.
12. Bricca, G., M. Dontenwill, A. Molines, J. Feldman, A. Belcourt & P. Bousquet. 1988. Eur. J. Pharmacol. **180:** 401-402.

Pancreatic β-Cells Express an Imidazoline Binding Site That Is Distinct from I_1 and I_2 Sites[a]

SUSAN L. F. CHAN, COLIN A. BROWN, KAY E. SCARPELLO, AND NOEL G. MORGAN

Cellular Pharmacology Group
Department of Biological Sciences
Keele University
Keele, Staffs., ST5 5BG, UK

Recent studies both *in vitro* and *in vivo* have documented the augmentation of insulin secretion by imidazoline derivatives such as efaroxan and phentolamine.[1–3] The mechanism of action of these imidazoline insulin secretagogues involves blockade of ATP-regulated potassium channels present in the plasma membrane of pancreatic β-cells.[4–6] This results in depolarization of the membrane with the subsequent opening of voltage-dependent calcium channels, initiating the process of secretory granule exocytosis. The putative imidazoline binding site responsible for control of K^+_{ATP} channel activity was recently shown to exhibit stereoselectivity.[7] Moreover, the secretory response is subject to downregulation in the presence of agonists (imidazoline secretagogues).[7]

Imidazoline binding sites were previously described in insulin-secreting cells: I_2-receptors were labeled with [^{3}H]idazoxan in pancreatic islets[8,9] and in RINm5F insulinoma cells[10]; [^{125}I]*p*-iodoclonidine binding in the HIT-T15 B-cell line was potently displaced by the I_2-selective ligand idazoxan, but only weakly displaced by the I_1-selective ligand efaroxan, thus suggesting a relative lack of I_1-imidazoline binding sites in insulin-secreting β-cells (Brown and Morgan, unpublished observations). However, the emerging pharmacology of both defined I_1 and I_2 sites does not correlate well with the observed responses in islets. In the present report, we describe observations that support the concept that the islet imidazoline binding site may represent a novel subtype of imidazoline receptor.

EXPERIMENTAL PROCEDURES

Insulin Secretion Experiments

Islets of Langerhans were isolated from the pancreas of male Wistar rats by collagenase digestion.[11] Static incubations and desensitization experiments were performed as described by Chan *et al.*[7]

[a] This work was supported by the Wellcome Trust, the MRC, the British Diabetic Association, and the North Staffs. Hospital Trust for their financial support.

TABLE 1. Effects of Culture Conditions on the Ability of Imidazoline Insulin Secretagogues to Antagonize the Inhibition of Insulin Secretion Induced by Diazoxide[a]

Incubation Conditions	Culture Conditions			
	Efaroxan	Phentolamine	UK14,304	Idazoxan
Glucose 20 mM	2.50 ± 0.29	1.79 ± 0.35	1.30 ± 0.08	1.52 ± 0.28
+ Diazoxide	1.08 ± 0.16	0.43 ± 0.08	0.46 ± 0.13	0.63 ± 0.12
+ Diazoxide + Efaroxan	1.11 ± 0.37	0.37 ± 0.10	1.17 ± 0.24[b]	1.32 ± 0.29[c]
+ Diazoxide + Phentolamine	1.42 ± 0.15	0.20 ± 0.10	1.34 ± 0.34[b]	1.19 ± 0.13[c]

[a] Groups of isolated rat islets were cultured for 18 hours in medium supplemented with 100 μM efaroxan, phentolamine, idazoxan, or UK14,304. Islets were washed and incubated in the presence of 20 mM of glucose and diazoxide, 250 μM; noradrenaline, 10 μM; efaroxan, 100 μM; and phentolamine, 100 μM, as indicated. Insulin secretion (ng/islet/h) was measured after 60 minutes of incubation at 37°C. Results represent mean values ± SEM for 7-8 observations.

[b] $p < 0.001$ and [c] $p < 0.01$ relative to incubation with diazoxide in the absence of imidazoline.

Radioligand Binding Experiments

Membranes prepared from RINm5F cells (50-100 μg of protein), [^{3}H]RX821002, and competing drugs were incubated in a final volume of 100 μl of TEM buffer for 60 minutes at 25°C. The reaction was stopped by adding 3 ml of ice-cold TEM buffer. Bound and free radioligand was separated by filtration through Whatman GF/B filters under vacuum. Filters were washed and their radioactivity was measured after the addition of scintillant. For saturation studies, [^{3}H]RX821002 was used over a range of 1-100 nM, and nonspecific binding was defined in the presence of 100 μM of phentolamine. In experiments designed to specifically study binding at the "imidazoline" site, incubations were performed in the presence of 10 μM of adrenaline to block α_2-adrenoceptors.

RESULTS AND DISCUSSION

Desensitization of Imidazoline-Mediated Stimulation of Insulin Release

Imidazoline insulin secretagogues can reverse the inhibitory actions of diazoxide (K^+ channel opener) on glucose-induced insulin release and thus provide a convenient assay system to study the functional interaction of imidazolines with islet cells. When isolated rat islets were cultured in the presence of either efaroxan or phentolamine (100 μM) for 18 hours, they were refractory to the effects of these compounds on subsequent reexposure (TABLE 1). However, culture of islets with either idazoxan or UK14,304 (non-secretagogue I_2 ligands) did not lead to any loss of such response.

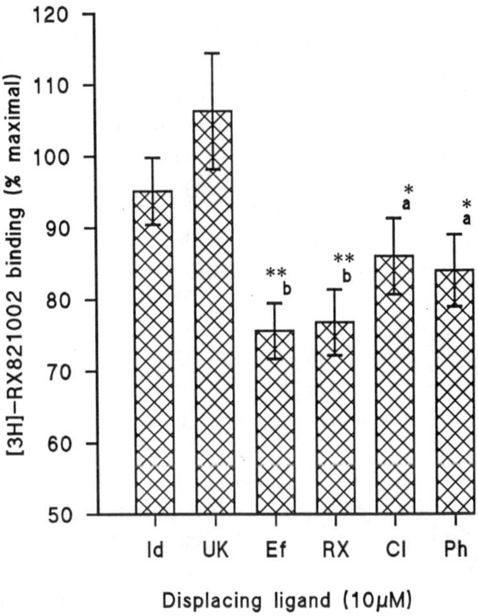

FIGURE 1. Effects of α-adrenergic compounds on the binding of [^{3}H]RX821002 to the low affinity binding site in RINm5F membranes. Competition binding studies were performed on RINm5F membranes with a single concentration (50 nM) of [^{3}H]RX821002. Competing agents present at a concentration of 10 μM were idazoxan (Id), UK14,304 (UK), efaroxan (Ef), RX821002 (RX), clonidine (Cl), and phentolamine (Ph). Adrenaline (10 μM) was included in all tubes to block α_2-adrenoceptors. Data are expressed as percentages of control binding measured in the absence of a competitor. Each point represents the mean ± SEM for 12-18 observations. *$p < 0.01$ and **$p < 0.001$ versus the control (100%).

By contrast, culture with UK14,304 did lead to loss of sensitivity to noradrenaline, consistent with downregulation of α_2-adrenoceptors. These results indicate that both efaroxan and phentolamine act as agonists at the islet imidazoline receptor and that this site is pharmacologically distinct from previously characterized α_2, I_1, and I_2 sites.

Identification of RX821002 Binding Sites in RINm5F Cell Membranes

[^{3}H]RX821002 was employed in labeling and characterizing the binding site responsible for mediating the secretagogue activity of imidazoline compounds. This compound, which does not interact significantly with imidazoline sites in other tissues,[12] act agonistically to stimulate insulin release and also to antagonize the inhibitory effect of diazoxide. Because of the limited availability of islet tissue, radioligand binding experiments were performed with the insulin-secreting cell line RINm5F (Fig. 1).

The binding of [^{3}H]RX821002 to RINm5F membranes was rapid, with steady state being reached within 20 minutes at 25°C (data not shown). When employed at concentrations between 1 and 10 nM, [^{3}H]RX821002 labeled a high affinity site that was saturable. However, raising the concentration of [^{3}H]RX821002 revealed the presence of a second binding site of much lower affinity, which remained less than fully saturated even at a radioligand concentration of 100 nM. The high affinity site bound [^{3}H]RX821002 with a K_D of 2.01 ± 0.7 nM, consistent with interaction with α_2-adrenoceptors. By contrast, the second site exhibited an estimated K_D of 145 ± 22 nM, but it bound [^{3}H]RX821002 with much higher capacity (B_{max}: 364 ± 22 fmol/mg protein for four separate observations) than did the α_2-adrenoceptor site (B_{max}: 34 ± 10.4 mol/mg protein).

Competition studies employing imidazoline compounds of varying secretagogue activity revealed that the pharmacological profile of the low affinity site correlates well with that observed in secretion experiments. Thus, efaroxan, RX821002, phentolamine, and clonidine, all of which can augment insulin release, inhibited the specific binding of [^{3}H]RX821002 to the low affinity "imidazoline" binding site, whereas UK14,304 and idazoxan had little effect on binding of the radioligand to this site.

The results described in this paper strongly suggest that the low affinity [^{3}H]RX821002 binding site may represent the functional receptor responsible for the secretagogue activity of imidazoline compounds in the endocrine pancreas and that it has a pharmacological profile distinct from those of I_1 and I_2 sites.

REFERENCES

1. CHAN, S. L. F. 1993. Clin. Sci. **85:** 671-677.
2. CHAN, S. L. F., R. D. HURST & N. G. MORGAN. 1988. Biochem. Soc. Trans. **16:** 1005-1006.
3. SCHULZ, A. & A. HASSELBLATT. 1988. Naunyn-Schmiedeberg's Arch. Pharmacol. **337:** 637-643.
4. CHAN, S. L. F., M. J. DUNNE, M. R. STILLINGS & N. G. MORGAN. 1991. Eur. J. Pharmacol. **204:** 41-48.
5. DUNNE, M. J. 1991. Br. J. Pharmacol. **103:** 1847-1850.
6. JONAS, J.-C., T. D. PLANT & J. C. HENQUIN. 1992. Br. J. Pharmacol. **107:** 8-14.
7. CHAN, S. L. F., C. A. BROWN & N. G. MORGAN. 1993. Eur. J. Pharmacol. **230:** 375-378.
8. BROWN, C. A., A. C. LOWETH, S. A. SMITH & N. G. MORGAN. 1993. Br. J. Pharmacol. **108:** 312-317.
9. LACOMBE, C., V. VIALLARD & H. PARIS. 1993. Int. J. Biochem. **25:** 1077-1083.
10. REMAURY, A. & H. PARIS. 1992. Pharmacol. Exp. Ther. **260:** 417-426.
11. MONTAGUE, W. & K. W. TAYLOR. 1968. Biochem. J. **109:** 333-339.
12. LANGIN, D., M. LAFONTAN, M. R. STILLINGS & H. PARIS. 1989. Eur. J. Pharmacol. **167:** 95-104.

Imidazoline Binding Sites on PC12 Cells and Bovine Chromaffin Cells

G. STEFFEN, A. DENDORFER, AND P. DOMINIAK[a]

Institute of Pharmacology
Medical University of Lübeck
Ratzeburger Allee 160
D-23538 Lübeck, Germany

Several investigators have demonstrated that imidazolines such as clonidine and moxonidine bind not only to α_2-adrenoceptors but also to imidazoline binding sites.[1] Binding studies in the ventrolateral medulla in combination with functional studies clearly showed that almost exclusively I_1-receptors, and not α_2-adrenoceptors, were responsible for mediating blood pressure reduction.[2] However, imidazoline binding sites have also been described in peripheral organs such as the kidney, pancreatic islets, and adrenal medulla.[3] Whereas I-binding sites of the kidney mediate sodium and water diuresis, those of the adrenal medulla induce epinephrine release.[3] Although a receptor protein with a molecular size of 70 kD was isolated from chromaffin cells of the adrenal medulla which shows specific binding sites for imidazolines but not α_2-adrenoceptor ligands, almost nothing was known about second messenger systems until now.[4] The comparative ability of the imidazolines to displace either tritiated clonidine or tritiated idazoxan binding can be used to distinguish I_1 from I_2 type binding sites,[3] but only a small amount of data exist on the physiologic and/or pharmacologic effects of those compounds on I_2-receptors. The present study characterizes I_1- and I_2-binding sites on a PC12 cell line and on bovine chromaffin cells and measures the intracellular calcium signal to establish a cellular model for investigating second messenger systems of putative imidazoline receptors.

MATERIALS AND METHODS

Binding assays were performed according to Regunathan *et al.*[5] Briefly, the medullas of bovine adrenal glands were dissected from the cortex, minced, and homogenized in ice-cold 10 mM HEPES buffer containing 330 mM sucrose at pH 7.4. The supernatant of a 600-*g* centrifugation was centrifuged again for 20 minutes by 30,000 *g* at 4°C to achieve the crude membrane fraction. Separation of mitochondrial membranes from plasma membranes was performed by differential centrifugation of 6,500 *g* (15 min) for mitochondrial membranes and 30,000 *g* (20 min) for plasma membranes. Pellets were washed three times and resuspended in 50 mM Tris HC1 containing 5 mM EDTA buffer (pH 7.4). Crude membrane fractions were also prepared from PC12 cells cultured in Dulbecco's Modified Eagle's Medium consisting

[a] To whom correspondence should be addressed.

of 10% horse serum, 5% fetal calf serum, 1% glutamine, and the antibiotics penicillin and streptomycin.

Saturation and competition experiments were performed on mitochondrial and plasma membrane fractions of adrenal medullary tissue and crude membrane fractions of PC12 cells. I_1-binding sites have been characterized by [^{3}H]clonidine using 100 μM naphazoline or phentolamine to define nonspecific binding. I_2-binding sites have been characterized by [^{3}H]idazoxan; nonspecific binding was eliminated by 10 μM idazoxan. α_2-Adrenoceptors have been masked by 0.1 mM epinephrine. After filtration through a glass fiber filter, radioactivity was determined in a liquid scintillation counter.

Intracellular calcium concentrations were measured using the fluorescent calcium indicator Fura-2 with dual wave length excitation of 340 and 380 nm. Emission at 510 nm was measured with a fluorescence microscope. To block α_2-adrenoceptor-mediated calcium currents the α_2-adrenoceptor antagonist rauwolscine (10 μM) was added to each sample. Stimulation tests were performed using the imidazolines clonidine, moxonidine, rilmenidine, idazoxan, and the guanidine agmatine.

RESULTS

On crude membrane fractions of the adrenal medulla, binding with [^{3}H]clonidine yielded a B_{max} of 110 fmol/mg protein and a K_d of 51 nM for the high affinity state (not depicted), with a specific binding of 51%. To define I_2-binding, idazoxan was used in purified mitochondrial and plasma membrane fractions. Binding site number (B_{max} = 218 fmol/mg protein) and the affinity (K_d = 18 nM) of mitochondrial membranes (specific binding 71%) were higher than those of plasma membranes (B_{max} = 123 fmol/mg protein; K_d = 30 nM; FIG. 1a and b), with a specific binding of 63%.

Competition experiments on mitochondrial membranes with both imidazolines clonidine and idazoxan demonstrated full displacement of idazoxan. The inhibition constant (K_i) for idazoxan was 200 nM. Clonidine displaces idazoxan at a 100-fold higher concentration (K_i = 20 μM; FIG. 2). Clonidine and idazoxan bind to PC12 cell membranes. However, binding was not saturable in concentrations of 1–20 nM (not depicted).

A small number of the chromaffin cells investigated responded to stimulation by the imidazolines clonidine, moxonidine, rilmenidine, idazoxan, and the guanidine agmatine with an increase in intracellular free calcium concentration. At high concentrations of each substance used (10–100 μM), a calcium signal could be observed in 20% of the cells tested (not depicted).

DISCUSSION

Our results with [^{3}H]clonidine on crude membrane fractions of the adrenal medulla demonstrated the existence of I_1-binding sites on those tissues and confirm previous reports.[1,5,6] Although the I_1-receptor density of our binding assay showed comparable data, the K_d value (51 nM) is about 10-fold higher than that reported in refs. 1, 5,

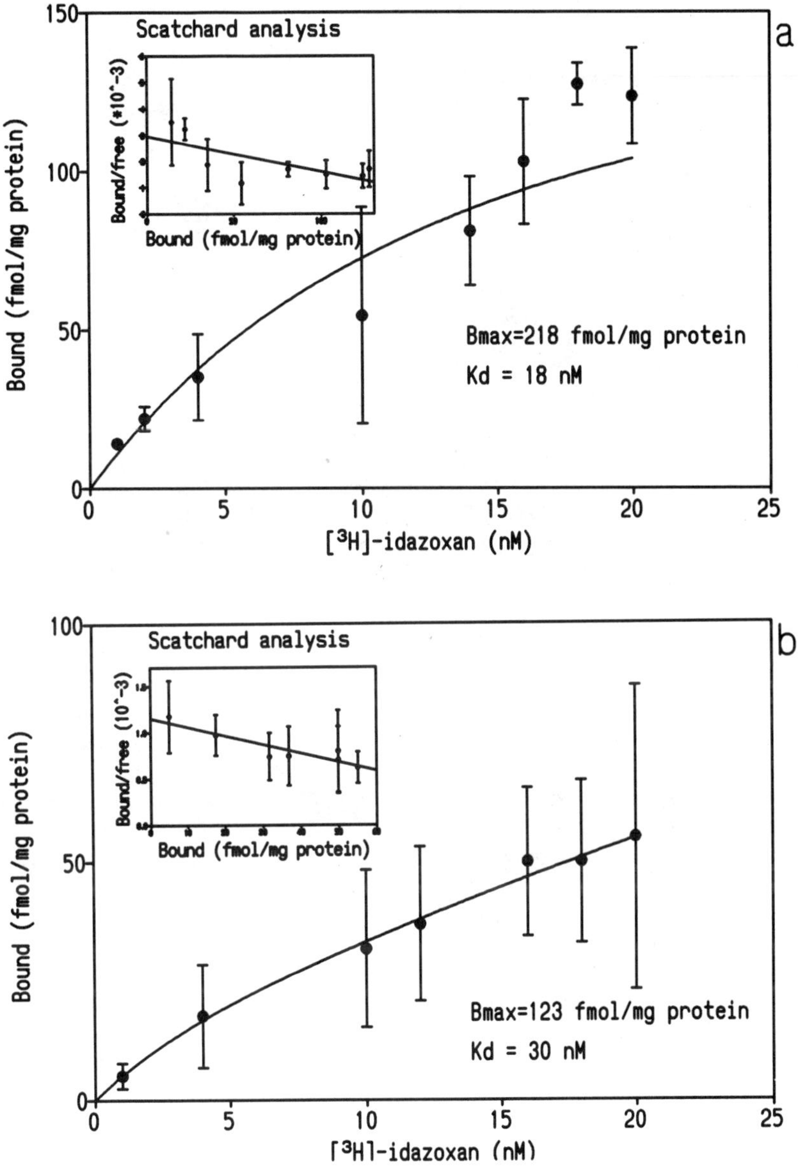

FIGURE 1. (**a**) Specific [^{3}H]idazoxan binding to mitochondrial membranes of bovine adrenal medulla. Idazoxan 10 μM was used to define nonspecific binding and epinephrine 100 μM to mask α_2-adrenoceptors. (**b**) Specific [^{3}H]idazoxan binding to plasma membranes of bovine adrenal medulla. Idazoxan and epinephrine were added at the same concentrations as those described for FIGURE **1a.**

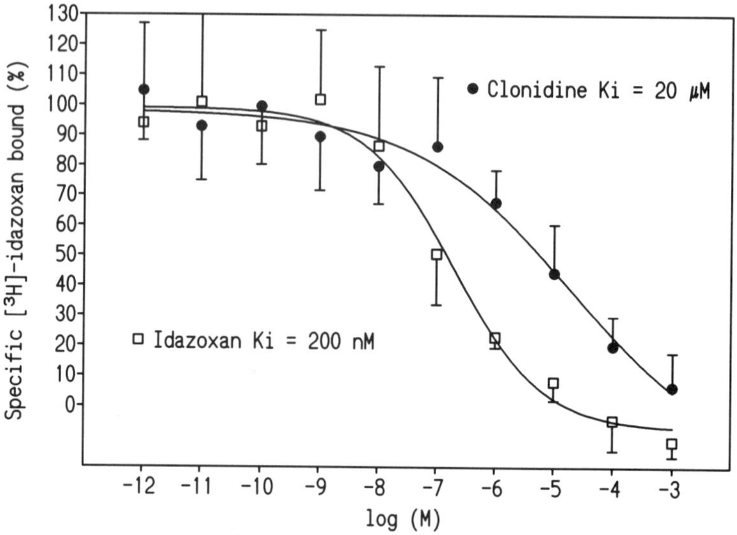

FIGURE 2. Displacement of [^{3}H]idazoxan binding by increasing concentrations of idazoxan and clonidine on mitochondrial membranes of bovine adrenal medulla. Idazoxan 10 μM was used to define nonspecific binding and epinephrine 100 μM to mask α_2-adrenoceptors.

and 6. However, experiments with various imidazolines in pithed spontaneously hypertensive rats (SHR) performed to induce stimulation-dependent catecholamine release demonstrated a similar range of imidazoline concentrations for stimulating epinephrine release.[7]

Binding experiments with [^{3}H]idazoxan on purified plasma membranes of the adrenal medulla yielded an I_2-receptor density and affinity almost identical to that of the I_1-receptor. The I_2-binding sites of purified mitochondrial membranes (B_{max} = 218 fmol/mg protein), however, were double and the affinity was half when compared to the plasma membrane fraction. If one evaluates the literature data, both types of imidazoline receptors have been described.[5,6] Whereas[5] reported almost I_1-binding sites on bovine adrenal medullary membranes,[6] was able to demonstrate imidazoline binding sites which preferred I_2-binding rather than I_1. Interestingly, the latter also observed I_2-binding sites predominantly localized on adrenal medullary mitochondrial membranes.[6] The functional role of these binding sites, however, remains unclear.

Clonidine displaced idazoxan from I_2-binding sites of mitochondrial membranes (FIG. 2). Although the affinity for clonidine was 100-fold less than that of idazoxan, it also possesses an affinity to that binding site. The fact that 20 μM of clonidine displaces 5 nM of idazoxan from 50% of the binding sites leads to the conclusion that the I_2-binding number for clonidine is 109 fmol/mg protein. If one substitutes the affinity constant of that displacement to a saturation curve for clonidine at a concentration of 18 nM, the result may be a small binding number of 0.2 fmol/mg protein, but the true binding number is 28 fmol/mg protein at a concentration of 18 nM. The difference may be due to specific clonidine preferring I_1-binding sites to which idazoxan binds with low affinity only. From these data a small number of I_1-binding sites on adrenal medullary mitochondrial membranes is concluded.

For stimulation of the intracellular calcium signal, various imidazoline receptor modulators on I_1- and I_2-binding sites were tested. Because all of them increased the Fura-2 signal, it is supposed that the imidazoline receptor function of both subtypes is coupled to an intracellular increase in calcium.[8] Whether the putative imidazoline receptor-induced epinephrine release from the adrenal medulla[7,9] is dependent on intracellular calcium increase or not has to be elucidated.

The binding experiments on PC12 cells showed some specific binding for I_2-sites, but saturation was not obtained. Therefore, they cannot serve as a cellular model for investigation of signal transduction. In conclusion, chromaffin cells possess both subtypes for imidazoline binding sites on either purified plasma membrane or mitochondrial membrane fractions. Binding of both subtypes with imidazoline receptor modulators resulted in an increase in intracellular calcium signal as a parameter for signal transduction. Therefore, bovine chromaffin cells are a suitable model for studying the second messenger system of putative imidazoline receptors.

SUMMARY

Stimulating actions of imidazolines on the adrenal medulla were demonstrated by several laboratories. As only a few data about signal transduction exist, the aim of the present study is to establish a cellular model on which both subtypes of imidazoline receptors are present. Binding studies using [^{3}H]clonidine for I_1-sites and [^{3}H]idazoxan for I_2-sites were performed on bovine chromaffin cells and a PC12 cell line. The intracellular calcium signal was determined by a Fura-2 signal using a fluorescence microscope. Both subtypes of imidazoline binding sites are present on either crude membrane fractions, purified plasma membranes, and mitochondrial membranes of the adrenal medulla. Although the density of I_1- and I_2-sites on the plasma membrane fraction is almost equal, on the mitochondrial membrane fraction B_{max} of I_2-binding sites was double that of the I_1-binding number. An increase in intracellular calcium signal could be obtained during stimulation of chromaffin cells by various I_1- and I_2-receptor modulators. Because a saturation of I_2-binding could not be obtained in PC12 cells, adrenal medullary chromaffin cells may be a more suitable model for investigating imidazoline receptor signal transduction.

REFERENCES

1. Ernsberger, P. R., K. L. Westbrooks, M. O. Christen & S. G. Schäfer. 1992. A second generation of centrally acting antihypertensive agents act on putative I_1-imidazoline receptors. J. Cardiovasc. Pharmacol. **20** (Suppl. 4): S1–S10.
2. Ernsberger, P. R., P. Giuliano, R. N. Willette & D. J. Reis. 1990. Role of imidazoline receptors in the vasodepressor response to clonidine analogues in the rostral ventrolateral medulla. J. Pharmacol. Exp. Ther. **253:** 408–418.
3. Hamilton, C. A. 1992. The role of imidazoline receptors in blood pressure regulation. Pharmacol. Ther. **54:** 231–248.
4. Wang, H., S. Regunathan, M. P. Meeley & D. J. Reis. 1992. Isolation and characterization of imidazoline receptor protein from bovine adrenal chromaffin cells. Mol. Pharmacol. **42:** 792–801.

5. Regunathan, S., M. P. Meeley & D. J. Reis. 1993. Expression of non-adrenergic imidazoline sites in chromaffin cells and mitochondrial membranes of bovine adrenal medulla. Biochem. Pharmacol. **45:** 1667-1675.
6. Molderings, G. J., D. Moura, K. Fink, H. Bönisch & M. Göthert. 1993. Binding of [^{3}H]clonidine to I_1-imidazoline sites in bovine adrenal medullary membranes. Naunyn-Schmiedeberg's Arch. Pharmacol. **348:** 70-76.
7. Häuser, W., J. Gütting, Th. Nguyen & P. Dominiak. 1995. Influence of imidazolines on catecholamine release in pithed spontaneously hypertensive rats (SHR). This volume.
8. Regunathan, S., M. J. Evinger, M. P. Meeley & D. J. Reis. 1991. Effects of clonidine and other imidazole-receptor binding agents on second messenger systems and calcium influx in bovine adrenal chromaffin cells. Biochem. Pharmacol. **42:** 2011-2018.
9. Li, G., S. Regunathan, C. J. Barrow, J. Eshragi, R. Cooper & D. J. Reis. 1994. Agmatine: An endogenous clonidine-displacing substance in the brain. Science **263:** 966-969.

Optimization of Radioligand Binding Assays for I_1-Imidazoline Sites[a]

PAUL ERNSBERGER,[b] JOHN E. PILETZ,[c]
LYNETTE M. GRAFF, AND MARILYN E. GRAVES

Departments of Medicine, Pharmacology, and Neuroscience
Case Western Reserve School of Medicine
10900 Euclid Ave.
Cleveland, Ohio 44106–4982

[c]*Departments of Psychiatry and Pharmacology*
University of Mississippi Medical Center
2500 N. State St
Jackson, Mississippi 39216–4505

In cell membranes from the rostral ventrolateral medulla oblongata (RVLM) of bovine brainstem, [^{3}H]clonidine binds with nearly equal nanomolar affinities to both α_2-adrenergic and I_1-imidazoline sites.[1–5] Human platelet I_1-imidazoline sites have also been characterized and are nearly identical to the bovine RVLM sites, because correlation coefficients for pK_i values are over 0.9.[6] Even as evidence accumulates for cell-specific expression of I_1-imidazoline binding sites and their probable bioactivity, several groups have had difficulty labeling I_1-sites with radiolabeled clonidine analogs. For example, bovine adrenomedullary chromaffin cells were reported to lack specific binding sites for [^{3}H]clonidine,[7] even though abundant I_1-imidazoline sites had been found on identically prepared cells.[6] Others report low-affinity (K_d > 20 nM) high-capacity binding sites for [^{3}H]clonidine which are clearly distinct from either the I_1- or I_2-imidazoline subtypes in terms of overall affinity and order of potency of the imidazolines and are tentatively termed "I_3-imidazoline binding sites."[6] In this series of experiments, we optimized conditions for selectively labeling high-affinity I_1-imidazoline sites.

METHODS AND TECHNICAL NOTES

Membrane Isolation

Bovine brains and adrenal glands were processed as previously described.[3–5] Outdated human platelets were used to prepare purified plasma membranes by pre-

[a] The work was supported by the National Institutes of Health (HL44514 to P.E.) and the National Institute of Mental Health (MH49248 to J.E.P. and P.E.). L. M. Graff was partially supported by the Cornell Tradition undergraduate work-study program.

[b] Address all correspondence to: Paul Ernsberger, Division of Hypertension, Case Western Reserve University, Cleveland, OH 44106-4982.

viously reported methods.[6,9–11] The features of these previously reported procedures which are essential for the success of the assay are given below.

Tissue may be fresh or quick-frozen. Homogenization should be in isotonic buffer containing protease inhibitors. Our experience suggests that brain I_1-imidazoline binding sites can be degraded by serine and calpain-like proteases and metallopeptidases. Failure to inhibit proteases can lead to a selective loss of I_1-sites even when α_2- and I_2-sites are preserved. Gentle homogenization with teflon or glass pestles or nitrogen cavitation (for isolated cells) give the highest yields, followed by polytronic disruption on a moderate setting. High intensity sonication can lead to selective loss of I_1-sites. For organs such as kidney and prostate, collagen should be removed by straining or collagenase treatment, because clonidine binds nonspecifically to collagen.

Membranes must be prepared with low-speed centrifugation to remove nuclei and cell debris. Resuspension with chelator (5.0 mM EDTA) is required for detection of I_1-sites. Isolation of plasma membranes by sucrose gradient is helpful especially for platelets,[6,9] but also for brain and chromaffin-like cells.[12] To remove endogenous interferences, preincubation (i.e., 30 minutes at 25°C) and several washes in hypotonic buffer are required. Dry pellets flash-frozen in liquid nitrogen can be stored at –70°C under nitrogen gas for up to 6 months. Final protein levels should be minimized to reduce nonspecific binding. We use up to 100 μg/well for assays of bovine RVLM using [^{3}H]clonidine, but with purified plasma membranes from PC12 cells as little as 2 μg/well is sufficient when [^{125}]PIC is used.

[^{3}H]Clonidine and [^{125}I]PIC Binding Assays

Binding assays of I_1-sites with [^{3}H]clonidine or [^{125}I]PIC were performed as described.[3–6,9] Essential features of the assay are as follows.

Radioligand should be stored under nitrogen at –70°C in ethanol and diluted in water or assay buffer immediately before use. [^{3}H]*p*-Aminoclonidine and [^{3}H]moxonidine are chemically unstable. Low concentrations ($<K_D$) of radioligand minimize labeling of low-affinity high-capacity I_3-sites or nonspecific sites.

Blank drug concentrations should be high enough to inhibit >90% of I_1-sites but low enough to avoid uncovering I_3-sites. We use 10 μM of BDF-6143 or moxonidine or 100 μM of oxymetazoline.

Assay buffer optimum pH is 7.7-8.5. Tris or HEPES are preferable to glycylglycine. Osmotic strength should be minimized. Divalent cations inhibit binding; we add 0.5 mM of both EGTA and EDTA with only 0.5 mM of $MgCl_2$. K^+ is a potent inhibitor; we avoid pH electrodes containing KCl and never adjust pH with KOH. Atmospheric dust is a significant source of K^+ and other ions. We rinse all glassware immediately prior to use. Failure of the I_1-binding assay in one lab was traced to a defective deionization resin in the water purification system.

Incubation is best carried out at 21°-22°C, keeping all components on ice prior to incubation and mixing thoroughly by vortexing. Temperatures of 25°C and 37°C reduce I_1 binding, and at 4°C labeling of I_1-sites does not occur. Association is rapid at I_1-sites,[9] so 30 minutes of incubation is sufficient. If α_2-adrenergic sites are present, 40-60 minutes may be required.

Filtration is critical for the I_1-imidazoline binding assay because of extremely rapid dissociation.[9] The harvester must be preequilibrated with 0°C wash buffer, and the vacuum must be strong with a high flow rate. We use an Edwards EM8 (½ hp) with a 12-place harvester. Teflon tubing speeds filtration and eliminates clonidine binding to tygon tubing. Pretreatment of filters 4-16 hours at 4°C with 0.3% polyethylimine is helpful. GF/C type filters bind less radioligand and can accept up to about 50 μg protein/well. For higher protein, GF/B type filters are acceptable. After sample aspiration, filters should be washed with 4 pulses of about 4 ml buffer in less than 10 seconds before breaking vacuum. Interruption of the vacuum or buffer source during filtration may lead to loss of I_1 binding.

RESULTS

The importance of an appropriate definition of nonspecific binding is illustrated in FIGURE 1A. Clonidine and its close analog lofexidine inhibit [^{125}I]PIC binding to bovine RVLM membranes in a biphasic manner. Inhibition curves were better fit by a 2-site rather than a 1-site model as demonstrated by an F-test on the residual variance ($p < 0.01$ in each case). One might suppose that the high affinity component is the α_2-site, whereas the low affinity component is the I_1-site. In fact, clonidine and lofexidine have nearly identical affinity for α_2- and I_1-binding sites in RVLM, which together comprise the high affinity component. The low affinity component of binding appears to be an I_3-imidazoline site, given the requirement for micromolar concentrations of imidazolines for inhibition. This low affinity should normally be included in nonspecific binding, because it is usually not inhibited by 10 μM of BDF-6143 or moxonidine. Therefore, the use of 10^{-4} M of clonidine to define nonspecific binding would lead to inclusion of low affinity, high capacity I_3-sites.

The importance of assay buffer composition is illustrated in FIGURES 1B and 2. FIGURE 1B shows the effect of ionic additives on [^{3}H]clonidine binding to bovine adrenomedullary cell membranes, which lack α_2-adrenergic receptors.[4,5] In the control condition, 0.5 mM EGTA and 0.5 mM $MgCl_2$ are present. Specific [^{3}H]clonidine binding to I_1-imidazoline sites was potently inhibited by Mg^{2+}. Even 1.0 mM inhibited by nearly 20%. The addition of low concentrations of EDTA · Na_4 slightly increased binding, probably by chelating residual endogenous divalent cations not removed by treatment with 5.0 mM EDTA during membrane preparation. Also shown in FIGURE 1B is the marked sensitivity of the I_1-imidazoline binding assay to K^+. Concentrations as low as 1 mM produced significant inhibition of binding, and 30 mM K^+ abolished specific [^{3}H]clonidine binding.

Optimization of buffer molarity and composition in a [^{125}I]PIC binding assay in human platelet plasma membranes is shown in FIGURE 2. The upper left quadrant shows that the apparent density of α_2-sites decreased with increasing buffer molarity from 5-170 mM. This inhibition of α_2 binding was entirely due to the effect of Na^+ added as NaOH to maintain pH, because the effect could be duplicated by NaCl, as shown in the upper right. This effect on B_{max} is probably due to a decrease in the affinity of a portion of the α_2-sites below the detection limit of the assay. Increasing buffer molarity also decreased the apparent density of I_1 binding, as shown in the lower left of FIGURE 2. In contrast to α_2 binding, I_1 binding was only slightly decreased

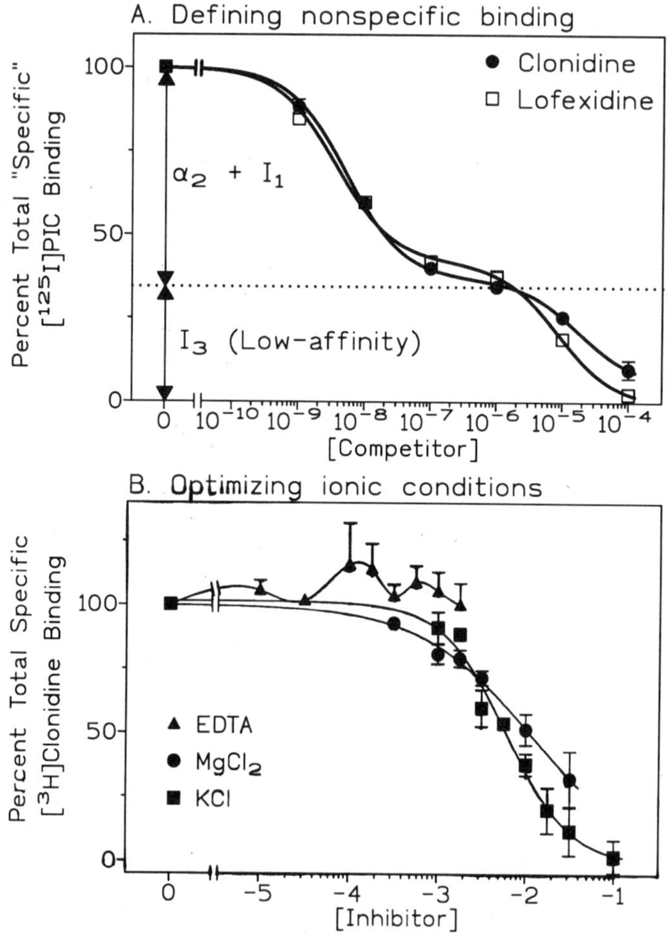

FIGURE 1. Pitfalls of the I_1-imidazoline binding assay: nonspecific binding and ionic conditions. (**A**) Dose-dependent inhibition of [^{125}I]*p*-iodoclonidine ([^{125}I]PIC; 0.2 nM) binding to RVLM membranes (30 μg protein) by increasing concentrations of clonidine and lofexidine. Data represent the means ± standard error of four experiments in triplicate. "Nonspecific" binding was defined by 1.0 mM clonidine. The *dotted line* indicates the level of nonspecific binding as defined by 1.0 μM clonidine or 10 μM BDF-6143. (**B**) Inhibition or facilitation of [^{3}H]clonidine (2 nM) binding to adrenomedullary cell membranes (20 μg protein) by cations or a chelator. Data represent the means ± standard error of five to eight experiments in triplicate. Nonspecific binding was defined by 0.1 mM naphazoline.

by $NaCl_2$. Thus, there appeared to be a direct effect of buffer molarity on I_1 binding. This effect was not specific to HEPES, because similar inhibition was obtained with Tris-HCl or Tris-HEPES buffers (not shown). Adding 5.0 mM of $MgCl_2$ increased the density of α_2 binding, as expected (FIG. 2, upper left). By contrast, Mg^{2+} decreased the density of I_1 binding sites by as much as two thirds, consistent with results in

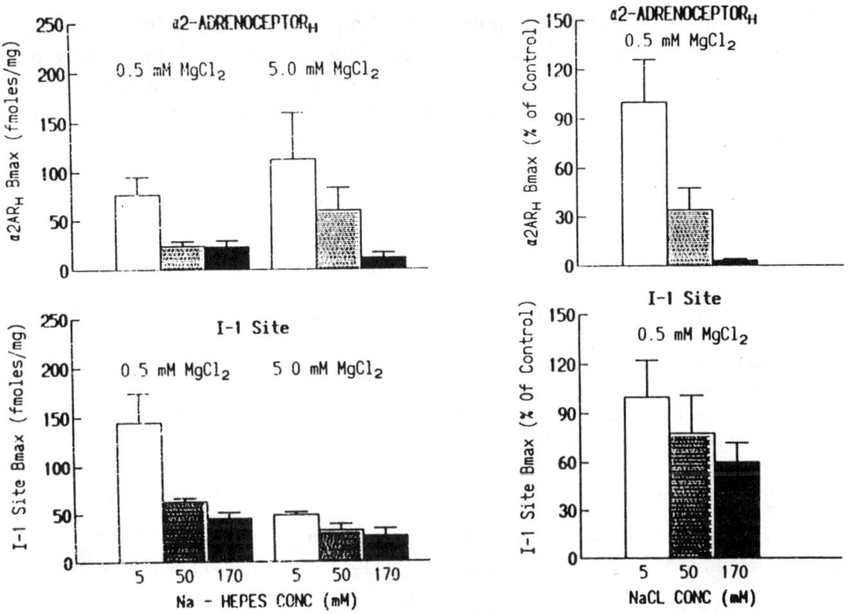

FIGURE 2. Optimization of the I_1-imidazoline binding assay buffer. [^{125}I]PIC saturation binding experiments (7 concentrations of radioligand for each site) were performed on human platelet plasma membranes in the presence of a 10-μM norepinephrine mask for I_1 binding. Nonspecific binding was defined by 10 μM of norepinephrine for α_2 and 0.1 mM moxonidine for I_1. Reactions were buffered at pH 7.4 at 21°C and contained 0.5 mM EGTA and 0.1 mM ascorbic acid, but varied as indicated in the concentration of HEPES-Na^+ buffer, $MgCl_2$, and NaCl. The notation "α_2AR_H" designates that under these assay conditions, radioligand binding appeared to be the high affinity state of the α_2-adrenergic receptor. Under all conditions, we observed only minor effects on radioligand affinity: K_D values at α_2 sites ranged from 0.7–3.5 nM, and K_D values at I_1-sites ranged from 5.1–8.5 nM.

adrenomedullary cells (FIG. 1B). Increasing buffer molarity in the presence of 5.0 mM Mg^{2+} further decreased I_1 binding to almost undetectable levels (FIG. 2). Therefore, in all subsequent experiments in both of our laboratories, we used 5.0 mM HEPES with 0.5 mM $MgCl_2$ to optimize I_1 binding.

DISCUSSION

We reported a number of technical "tips" on the I_1-imidazoline radioligand binding assay using either [^{3}H]clonidine, [^{3}H]*p*-aminoclonidine, or [^{125}I]PIC. We also demonstrate some of the steps that went into the development of an optimum assay by our two laboratories. The assay is relatively sensitive to minor changes in conditions compared to typical assays with highly potent antagonist ligands. Furthermore, conditions that are optimum for α_2 binding, such as mM Mg^{2+}, may obscure I_1 binding. We suggest that any alterations in these established protocols should be tested prior to adoption.

The RVLM assay has practical predictive validity, because the affinity of various imidazolines for the bovine RVLM I_1-imidazoline site, as determined under the conditions described herein, predicts their potency in lowering blood pressure in rats[2,3] and in humans.[4] Similarly, levels of human platelet I_1-imidazoline sites are associated with mood disorders.[10,11]

REFERENCES

1. Ernsberger, P., M. P. Meeley, J. J. Mann & D. J. Reis. 1987. Eur. J. Pharmacol. **134:** 1-13.
2. Ernsberger, P., R. Giuliano, R. N. Willette & D. J. Reis. 1990. J. Pharmacol. Exp. Ther. **253:** 408-418.
3. Ernsberger, P. R., K. L. Westbrooks, M. O. Christen & S. G. Schäfer. 1992. J. Cardiovasc. Pharmacol. **20** (Suppl. 4): S1-S10.
4. Ernsberger, P., T. H. Damon, L. M. Graff, M. O. Christen & S. G. Schäfer. 1993. J. Pharmacol. Exp. Ther. **264:** 172-182.
5. Ernsberger, P., M. A. Haxhiu, L. M. Graff, L. A. Collins, I. Dreshaj, D. L. Grove, M. E. Graves, S. G. Schäfer & M. O. Christen. 1994. Cardiovasc. Drugs Ther. **8:** 27-41.
6. Piletz, J. E. & K. Sletten. 1993. J. Pharmacol. Exp. Ther. **267:** 1493-1502.
7. Regunathan, S., M. P. Meeley & D. J. Reis. 1993. Biochem. Pharmacol. **45:** 1667-1675.
8. Ernsberger, P., G. Feinland, M. J. Evinger, M. P. Meeley & D. J. Reis. 1989. Am J. Hypertens. **2:** 93A.
9. Piletz, J. E., A. C. Andorn, J. R. Unnerstall & A. Halaris. 1991. Biochem. Pharmacol. **42:** 569-584.
10. Piletz, J. E., A. Halaris, A. Saran & M. R. Marler. 1991. Arch. Gen. Psychiatry **48:** 813-820.
11. Halbreich, U., J. E. Piletz, S. Carson, A. Halaris & N. Rojansky. 1993. Biol. Psychiatry **34:** 676-686.
12. Ernsberger, P., M. E. Graves & D. L. Grove. 1994. FASEB J. **8:** A355.

Pharmacological and Immunological Characterization of Solubilized 130–140- and 66-kD Imidazoline Receptors in the Rat Brain[a]

PABLO V. ESCRIBÁ, ANDRÉS OZAITA, ANTONIO MIRALLES,[b] DONALD J. REIS, AND JESÚS A. GARCÍA-SEVILLA

Laboratory of Neuropharmacology
Department of Fundamental Biology and Health Sciences
University of the Balearic Islands
Palma de Mallorca, Spain

[b]*Division of Neurobiology*
Department of Neurology and Neurosciences
Cornell University Medical College
New York, New York 10021

Several compounds with imidazoli(di)ne/guanidine structure, such as clonidine, idazoxan, and their derivatives, were currently used to label α_2-adrenoceptors. However, several studies demonstrated that they also bind with high affinity to other proteins, termed imidazoline receptors (IRs). The pharmacological and functional differences between IRs and α_2-adrenoceptors affirm not only that they are different molecular entities, but also that IRs are not a new α_2-adrenoceptor subtype. They have been found in the CNS and in other tissues and organs of various species and have been associated with regulation of blood pressure, control of catecholamine release, reduction of focal cerebral ischemia, and control of the expression of glial fibrillary acidic protein, among other effects.[1] The rank order of potency for various ligands has been used to classify them as I_1-IRs, or clonidine-preferring IRs, and I_2-IRs, or idazoxan-preferring IRs. The putative endogenous ligand and neurotransmitter for these receptors has been purified from the clonidine-displacing substance (CDS) mixture and determined to be agmatine.[2]

Despite all the biochemical and physiological evidence of the existence of IRs, little is known about the proteins that bind and mediate the functional effects promoted by imidazoline drugs. In the present study we sought the molecular identification and characterization of the multiple IR proteins that have been found in rat brain,

[a]This work was supported by grants CDTI-LASA 92-0178, FIS 93/0641, and DGICYT PM91-0069 to J.A.G.-S. A.O. was supported by a fellowship from the Basque Government.

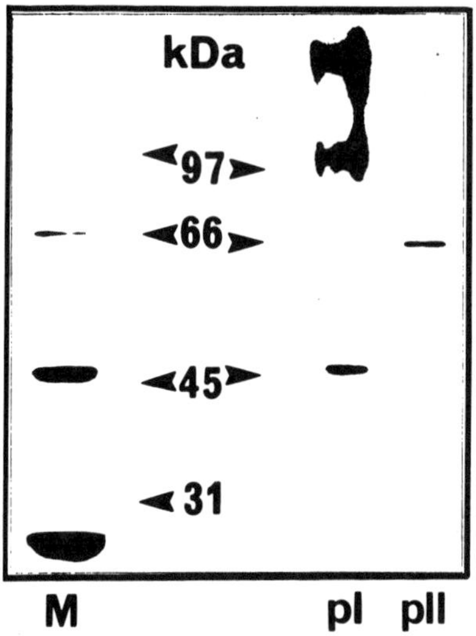

FIGURE 1. Immunoblotting, using an anti-imidazoline receptor (IR) antibody, of rat brain cortical (crude) membranes (M) before solubilization, and after solubilization with CHAPS and gel filtration chromatography (pI = IR protein eluted in the first peak of chromatography; pII = IR protein eluted in the second peak).

namely, 29/30-, 45-, and 66-kD IR proteins[3] (FIG. 1). For this purpose rat brain membranes were solubilized with CHAPS (50 mM Tris-HCl, pH 7.5, 0.5% CHAPS) and the solubilized IRs were submitted to gel filtration chromatography (2.5 × 90 cm, Sephadex G-100 column; eluting with 50 mM Tris-HCl, pH 7.5, 1 mM CHAPS).

Four peaks of [^{3}H]idazoxan specific binding were discriminated, but only the first two were further characterized by pharmacological and immunological means because of their relative abundance (both peaks defined about 95% of the total [^{3}H]idazoxan binding to solubilized IRs). These peaks (peak 1 [pI] and peak 2 [pII]) appeared to have very different structural features. The pI-IR had an M_r of 130,000-140,000, as determined by gel filtration chromatography, but immunoblot analysis using specific anti-bovine adrenal medulla chromaffin cell imidazoline receptor antiserum, showed the presence of a 45-kD peptide (FIG. 1). In any case, neither the 130-140-kD complex nor the 45-kD peptide had an M_r similar to those previously described, suggesting that this IR has structural dissimilarities to formerly visualized IR proteins[4,5] and that this 45-kD IR protein can be associated or coupled with other proteins forming an oligomeric complex. Experiments of displacement of [^{3}H]idazoxan binding by cirazoline showed that the presence of the nonhydrolyable GTP analog GppNHp did not alter the binding parameters of the competition experiments. This suggests that pI-IR is not coupled to G proteins, but this depends on the unknown agonist/antagonist effects of idazoxan and cirazoline at pI-IR. Conversely, pII-IR appeared to have a similar M_r (approximately 66,000) as measured by gel filtration chromatography or electrophoresis/immunoblot (FIG. 1). The M_r of pII-IR was similar to those previously described in rat liver and bovine adrenal medulla,[4,5] suggesting possible structural similarities among these IRs.

Displacement experiments of [^{3}H]clonidine binding by various imidazoli(di)ne/guanidine drugs (cirazoline, guanabenz, amiloride, efaroxan, and agmatine), in the presence of 10^{-6} M epinephrine to prevent binding of radioligand to α_2-adrenoceptors, yielded K_i values in the high micromolar range for pI- and pII-IR, indicating that none of them are clonidine (I_1)-preferring IRs. On the other hand, although displacement of [^{3}H]idazoxan binding by the same drugs showed slight differences between pI- and pII-IR, they had a similar rank order of potency for displacement of this radioligand: cirazoline (6.6 nM) > guanabenz (87 nM) > amiloride (272 nM) >> efaroxan (83 μM) ≈ agmatine (96 μM) for pI-IR, and cirazoline (4.8 nM) > guanabenz (18 nM) >> efaroxan (210 μM) for pII-IR. These pharmacological profiles are similar to those found in native (nonsolubilized) rat brain cortical membranes.[6] Therefore, these IRs belong to the idazoxan (I_2)-preferring type, suggesting that at least this receptor family is formed by more than one member. The relative low K_i value for amiloride indicates that this IR can be classified as I_{2B}-IR. The affinity of pI-IR for agmatine was similar to that obtained for efaroxan, and the K_i value for the former was higher than that found for bovine adrenal chromaffin cell I_2-IRs.[2]

The present study evidences the heterogeneous molecular nature of the I_2-IR, showing that these receptors can form a subfamily of IRs made up of molecular entities of different structure and suggests the need for receptor reclassification.

REFERENCES

1. Ernsberger, P., M. P. Meeley, J. J. Mann & D. J. Reis. 1987. Clonidine binds to imidazole binding sites as well as α_2-adrenoceptors in the ventrolateral medulla. Eur. J. Pharmacol. **134:** 1-13.
2. Li, G., S. Regunathan, C. J. Barrow, J. Eshraghi, R. Cooper & D. J. Reis. 1994. Agmatine: An endogenous clonidine-displacing substance in the brain. Science **263:** 966-969.
3. Escribá, P. V., M. Sastre, H. Wang, S. Regunathan, D. J. Reis & J. A. Garcia-Sevilla. 1994. Immunodetection of putative imidazoline receptor proteins in the human and rat brain and other tissues. Neurosci. Lett. **178:** 81-84.
4. Lanier, S. M., B. Ivkovic, I. Singh, J. L. Neumeyer & V. Bakthavachalem. 1993. Visualization of multiple imidazoline/guanidinium-receptive sites. J. Biol. Chem. **268:** 16047-16051.
5. Wang, H., S. Regunathan, M. P. Meeley & D. J. Reis. 1992. Isolation and characterization of imidazoline receptor protein from bovine adrenal chromaffin cells. Mol. Pharmacol. **42:** 792-801.
6. Miralles, A., G. Olmos, M. Sastre, F. Barturen, I. Martin & J. A. Garcia-Sevilla. 1993. Discrimination and pharmacological characterization of I_2-imidazoline sites with [^{3}H]idazoxan and alpha-2 adrenoceptors with [^{3}H]RX821002 (2-methoxy idazoxan) in the human and rat brains. J. Pharmacol. Exp. Ther. **264:** 1187-1197.

[^{3}H]Idazoxan Binds to Mitochondrial I_2 Imidazoline Binding Sites in Isolated Cells from Rabbit Kidney Proximal Tubule[a]

C. GARGALIDIS-MOUDANOS AND A. PARINI[b]

INSERM U388
CHU Rangueil
31054 Toulouse, France

In previous studies we demonstrated that in many tissues, including rabbit kidney, I_2 imidazoline binding sites (I_2BS) are mainly located in the outer membrane of mitochondia.[1,2] Recent studies performed in our laboratory have shown a strong relation between these binding sites and monoamine oxidases (MAO), two mitochondrial enzymes. In effect, I_2BS seemed to be associated with MAO in mitochondrial membrane preparations, and imidazoline and guanidinium derivatives inhibited MAO activity. To investigate whether imidazoline compounds can cross the plasma membrane and interact with mitochondrial I_2BS, we performed binding studies in isolated cells from rabbit kidney proximal tubule.

Isolated proximal cells were prepared as previously described by Poujeol and Vandewalle.[3] Binding experiments were carried out at 4°C to minimize the possible internalization of the ligand-binding site complex. During these studies, cells were incubated with [^{3}H]idazoxan, considered as I_2BS selective ligand and in the presence of 10 μM rauwolscine to mask α_2-adrenergic receptors.

[^{3}H]Idazoxan binding was time-dependent, reversible, saturable, and of high affinity. Scatchard analysis of binding isotherms revealed only one class of binding sites. The equilibrium dissociation constant (K_d) was 4.7 ± 1.8 nM and the maximal number of sites (B_{max}) 848 ± 193 fmol/mg of protein (n = 5).

In competition experiments, imidazoline and guanidinium ligands produced concentration-dependent inhibition of specific [^{3}H]idazoxan binding. Competition curves of [^{3}H]idazoxan binding by cirazoline and guanabenz were monophasic with inhibition constants corresponding to these previously reported for mitochondrial I_2BS.[4,5] By contrast, radioligand displacement by clonidine and rilmenidine was biphasic with both high and low affinity components, corresponding to I_1 and to I_2 binding sites, respectively[5] (TABLE 1).

[a] This research was supported by grants from the Foundation "Alexander S. Onassis" and the Institut de Recherches Internationales Servier.

[b] Address for correspondence: Dr. Angelo Parini, INSERM U388—Institut Louis Bugnard—CHU Rangueil—Bât. L3, 31054 Toulouse Cedex, France.

TABLE 1. Inhibition Constants of Various Imidazoline and Guanidinium Compounds with [^{3}H]Idazoxan Specific Binding in Rabbit Kidney Isolated Cells

Competitors	K_{IH} (nM)	K_{IL} (nM)
Cirazoline	19.1 ± 8.7	–
Guanabenz	6.6 ± 2.8	–
Clonidine	58.2 ± 27[a]	4,179 ± 2,663
Rilmenidine	7 ± 3.5[a]	5,189 ± 1,816

[a] 30% of specific binding.

To verify whether imidazoline ligands can enter cells and bind to intracellular sites, we permeabilized isolated cells with digitonin. For cell permeabilization we used digitonin concentrations which did not induce I_2BS solubilization or binding inhibition.

Cell treatment with increasing digitonin concentrations led to a dose-dependent release of lactate dehydrogenase, a cytosolic enzyme, and decreased [^{3}H]idazoxan binding by 60%, suggesting that cell permeabilization induced a loss of ligand bound to intact cells. In addition, when we compared binding data in cells treated with digitonin or sonicated prior to binding experiments, we observed the same maximal I_2BS density, indicating that 60% of the ligand was free in the cytosol.

Taken together, these data suggest that imidazoline ligands can penetrate the cells and interact with mitochondrial sites and that idazoxan recognizes with the same affinity to both I_1 and I_2 binding sites. Thus, it is conceivable that *in vivo,* imidazoline derivatives elicit these pharmacological activities through interaction with mitochondrial I_2BS.

REFERENCES

1. TESSON, F., C. PRIP-BUUS, A. LEMOINE, J. P. PEGORIER & A. PARINI. 1991. J. Biol. Chem. **266:** 155-160.
2. TESSON, F., I. LIMON & A. PARINI. 1992. Eur. J. Pharmacol. **219:** 335-338.
3. POUJEOL, P. & A. VANDEWALLE. 1985. Am. J. Physiol. **249:** F461-F471.
4. PARINI, A., I. COUPRY, R. M. GRAHAM, I. UZIELLI, D. ATLAS & S. M. LANIER. 1989. J. Biol. Chem. **264:** 11874-11878.
5. LACHAUD, V., I. COUPRY, R. A. PODEVIN, J. P. DAUSSE, E. KOENIG & A. PARINI. 1988. J. Hypertens. **6** (suppl 4): S511-S513.

Castration of Rats Upregulates the Imidazoline Receptor in Prostate

Y. NASSIR,[a] S. REGUNATHAN,[b] D. J. REIS,[b]
E. D. VAUGHAN, JR.,[a] AND D. FELSEN[a,c]

Cornell University Medical College
[a]Department of Urology
and
[b]Division of Neurobiology
New York, New York 10021

The prostate gland is an accessory reproductive gland found in all male animals.[1] Its secretions form a major component of the seminal fluid in the ejaculate. Prostate growth is androgen dependent and castration causes involution of the prostate accompanied by decreases in serum testosterone.[2] In response to castration, upregulation of prostatic benzodiazepine receptors and downregulation of muscarinic cholinergic receptors occurs.[3,4] Recently, we demonstrated the presence of imidazoline I_1 receptors in canine prostate.[5] With autoradiography, these receptors were localized to the epithelial portion of the prostate. However, their function is unknown. In the present experiments, we sought to determine if imidazoline receptors are present in rat prostate as well. Furthermore, we were interested in the effects of castration on the binding characteristics of these receptors.

MATERIALS AND METHODS

Animals and Surgical Procedures. Bilateral orchiectomy was performed on 42 male Sprague-Dawley rats under metofane anesthesia; sham operation was conducted on 15 animals. Three (n = 20) and 7 days (n = 10) after castration animals were sacrificed by CO_2 gas, and ventral prostates were immediately placed in ice-cold sucrose (0.32 M)-HEPES (5 mM) buffer (pH 7.4) containing 0.1 mM EDTA.

Ligand Binding. Ligand binding assays were carried out using [^{3}H]idazoxan with membranes prepared from ventral prostate.

RESULTS

FIGURE 1 shows that [^{3}H]idazoxan bound with high affinity (K_d 15.8 nM) to rat prostate membranes in a specific and saturable manner (B_{max} 895 fmol/mg protein). Binding was inhibited by drugs in the following order: cirazoline > clonidine > UK 14304 > guanabenz, indicating an imidazoline I_2 type of receptor. Binding was also

[c]To whom correspondence should be addressed.

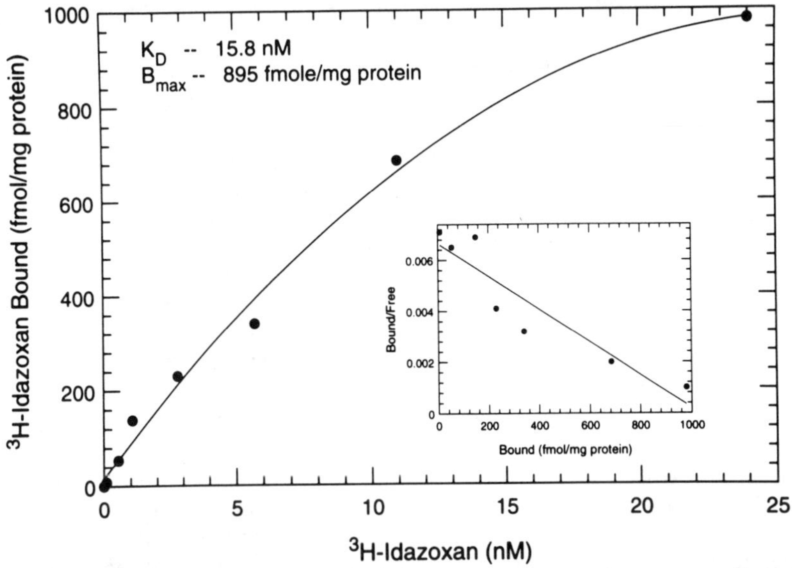

FIGURE 1. [^{3}H]Idazoxan binds with high affinity to membranes of rat prostate. Saturation binding of [^{3}H]idazoxan to membranes of rat ventral prostate. Membranes (about 100 μg protein) were incubated with various concentrations of [^{3}H]idazoxan in the presence of 10 μM of epinephrine; nonspecific binding was defined by 10 μM of idazoxan. Values are from one representative experiment repeated with three different cell membrane preparations. Scatchard analysis of specific binding is given in the *inset*. The kinetic constants are from all three experiments analyzed by LIGAND.

inhibited by agmatine,[6] an endogenous ligand for the imidazoline receptor (data not shown). Castration decreased ventral prostate weight by 50% in 3 days and by over 80% in 7 days. Three days after castration, the K_d was unaffected, whereas the B_{max} was almost doubled from 986 ± 124 to 1,911 ± 418 fmol/mg protein (TABLE 1). Although only one measurement was made using 7-day castrates, the B_{max} continued to increase over that time.

TABLE 1. Effect of Castration on Imidazoline Binding in Rat Prostate

	K_d Prostate	B_{max} Prostate
Sham	25.0 ± 2.3	986.3 ± 124.4
Castration, 3rd day	30.6 ± 8.2	1,911.5 ± 418.4
Castration, 7th day	42	2,928

DISCUSSION

Results of the present study demonstrate for the first time the presence of imidazoline I_2 receptors in the prostate. This was documented by (1) specific, saturable binding of [^{3}H]idazoxan in membranes, (2) specific displacement by agmatine, and (3) displacement with other imidazoline I_2-preferring compounds. Binding to the imidazoline I_2 receptor was upregulated by castration.

With [^{3}H]*para*-aminoclonidine as a ligand, no significant nonadrenergic binding was noted, indicating a lack of imidazoline I_1 receptors in the rat prostate. This finding differs from that in canine prostate[5] and human prostate (Felsen, unpublished data). Although the reason for this is not known, it is of interest that humans and dogs are the only two species to exhibit naturally occurring age-dependent benign prostatic hyperplasia.[1] Abundant imidazoline I_2 receptors were found when [^{3}H]idazoxan was used as a ligand.

In the present study, castration led to an increase in the B_{max} for imidazoline I_2 receptors, with no change in the K_d at 3 days. At 7 days, there was a further increase in the B_{max}, and the K_d appeared to increase, but the small amount of available tissue did not allow for repetition of these experiments. These results are similar to those found for the peripheral benzodiazepine receptor in rabbit prostate. After 3 weeks of castration the mean density of the peripheral benzodiazepine receptor was almost doubled, as was the K_d.[3] The peripheral benzodiazepine receptor is found in highest concentration in the mitochondrial outer membrane,[7] a finding similar to that of imidazoline I_2 receptors in adrenochromaffin cells.[8] Localization of the imidazoline I_2 receptor in prostate has not yet been described. The muscarinic receptor in prostate, by contrast, is downregulated by castration. The K_d is unchanged, but the B_{max} decreased by 73%. These changes were restored by testosterone treatment.[4] Changes in neurotransmitter receptors following castration may be related to interactions of these receptors with cell growth or differentiation (peripheral benzodiazepine receptor) or with secretion (muscarinic receptor). Whether the imidazoline receptor is involved in these areas is not known at this time. However, the effects of various neurotransmitters on secretion were examined by Wang *et al.*,[9] who demonstrated that exogenous clonidine given to rats had no effect on prostate secretion, whereas other neurotransmitters, including norepinephrine, phenylephrine, and carbachol, were active. As clonidine and agmatine may not have identical biological activities, this does not rule out the possibility of involvement of imidazoline I_2 receptors in prostatic secretion. In summary, this is the first demonstration of imidazoline receptors in rat prostate. That imidazoline binding can be affected by castration suggests an interplay between the prostate androgen milieu and imidazoline receptors. These relationships need to be further explored.

REFERENCES

1. Lepor, H. & R. K. Lawson. (Eds.) 1993. Prostate Diseases. W. B. Saunders. Philadelphia.
2. Kerr, J. F. R. & J. Searle. 1973. Virchows Arch. Abt. B Zellpathol. **13:** 87-102.
3. Batra, S. & J. Alenfall. 1992. Life Sci. **51:** 1211-1215.
4. Shapiro, E., A. R. Miller & H. Lepor. 1985. J. Urol. **134:** 179-182.
5. Felsen, D., P. Ernsberger, P. M. Sutaria, R. J. Nejat, P. Nguyen, M. May, D. Breslin, D. N. Marion & E. D. Vaughan, Jr. 1994. J. Pharmacol. Exp. Ther. **268:** 1063-1071.

6. Li, G., S. Regunathan, C. J. Barrow, J. Eshraghi, R. Cooper & D. J. Reis. 1994. Science **263:** 969-971.
7. Anholt, R. R. H., P. L. Pedersen, E. B. De Souza & S. H. Snyder. 1986. J. Biol. Chem. **261:** 576-583.
8. Regunathan, S., M. P. Meeley & D. J. Reis. 1993. Biochem. Pharmacol. **45:** 1667-1675.
9. Wang, J. M., K. E. McKenna & C. Lee. 1991. The Prostate **18:** 289-301.

I_2-Imidazoline Receptors in the Healthy and Pathologic Human Brain[a]

JESÚS A. GARCÍA-SEVILLA,[b] ANTONIO MIRALLES,[b] MAGDALENA SASTRE,[b] PABLO V. ESCRIBÁ,[b] GABRIEL OLMOS,[b] AND J. JAVIER MEANA[c]

[b]*Laboratory of Neuropharmacology*
Department of Fundamental Biology and Health Sciences
University of the Balearic Islands
E-07071 Palma de Mallorca, Spain

[c]*Department of Pharmacology*
University of the Basque Country
E-48940 Leioa, Spain

During the last few years, various α_2-adrenoceptor radioligands with an imidazoline structure, such as [^{3}H]idazoxan and [^{3}H]clonidine/*p*-aminoclonidine, have been shown to also recognize with high affinity nonadrenoceptor binding sites in various tissues, including the brain.[1–3] It is now believed that these sites exist in two main types. The clonidine-preferring type, the I_1-imidazoline receptor, has been localized in the lower brain stem, which is involved in the regulation of blood pressure,[4,5] as well as in other brain regions of the human brain,[6] and also in the adrenal medulla.[7] The idazoxan-preferring subtype, the I_2-imidazoline receptor, is expressed in several tissues,[8–10] including the brain,[6,11] and at the subcellular level is located on the mitochondrial outer membrane.[12–14] Various possible functional roles have been assigned to this new putative I_2-imidazoline receptor identified by [^{3}H]idazoxan. At present, however, the nature and physiological function of the brain I_2-imidazoline receptor remain unknown. This paper reviews briefly recent findings on the characterization, modulation, and clinical implications of I_2-imidazoline receptors in the human brain.

DISCRIMINATION BETWEEN I_2-IMIDAZOLINE RECEPTORS AND α_2-ADRENOCEPTORS IN THE HUMAN BRAIN USING [^{3}H]IDAZOXAN

Competition curves for (-)-epinephrine against [^{3}H]idazoxan binding to human brain cortical membranes revealed heterogeneous interaction with two binding sites

[a]The studies performed in the authors' laboratories were supported by DGICYT grant PM 91-0069, Gobierno Vasco grant PVG 91-18, CDTI-LASA Laboratorios grant 92-0178, FIS grant 93-0641, and CICYT grant SAF 93-0178.

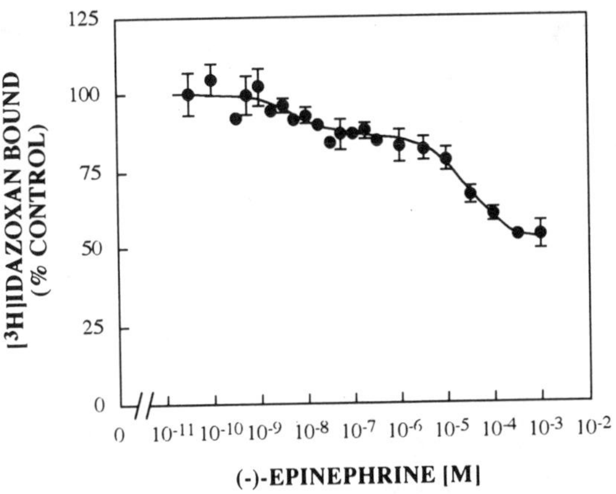

FIGURE 1. Inhibition of binding of [^{3}H]idazoxan to human cortical α_2-adrenoceptors and I_2-imidazoline receptors by (-)-epinephrine.

for the radioligand, one with very low affinity for the catecholamine. Computer-assisted nonlinear analysis indicated that (-)-epinephrine inhibited the binding of [^{3}H]idazoxan with high (K_{iH} = 4 nM) and very low (K_{iL} = 16,000 nM or greater) potency, the proportion of sites being about 30 and 70%, respectively. In the human brain, a relevant portion of total [^{3}H]idazoxan binding (about 50%) remained as not displaceable by 10^{-3} M (-)-epinephrine[15] (FIG. 1). It is recognized that the binding of [^{3}H]idazoxan inhibited by high concentrations of (-)-epinephrine (> 10^{-6} M) is not to an adrenoceptor, because it lacks the pattern of stereoselectivity for epinephrine enantiomers.[16–18] Therefore, the portion of [^{3}H]idazoxan binding inhibited by low concentrations of (-)-epinephrine (up to 10^{-6} M) is assumed to be the stereoselective α_2-adrenoceptor population, because the competition curve displayed a plateau at this concentration, whereas that not displaceable or displaced only by high concentrations of the catecholamine (range 10^{-6}-10^{-3} M) is assumed to correspond to the population of imidazoline receptors.[15]

DIFFERENTIAL PHARMACOLOGICAL CHARACTERIZATION OF I_2-IMIDAZOLINE RECEPTORS AND α_{2A}-ADRENOCEPTORS IN THE HUMAN BRAIN

Competition studies with various drugs were performed in human cortical membranes (Brodmann's area 9) against [^{3}H]idazoxan (in the presence of 10^{-6} M epinephrine to prevent binding to α_2-adrenoceptors) or [^{3}H]RX821002 (2-methoxy idazoxan), a specific antagonist radioligand for α_2-adrenoceptors.[19] Competition curves for guanoxan, cirazoline, naphazoline, clonidine, and bromoxidine (UK 14304) against [^{3}H]idazoxan were shallow and biphasic (FIG. 2A). Computer curve fitting revealed that these compounds inhibited [^{3}H]idazoxan from two distinct binding sites, which suggested the existence of high and low affinity states/forms for the I_2-imidazoline

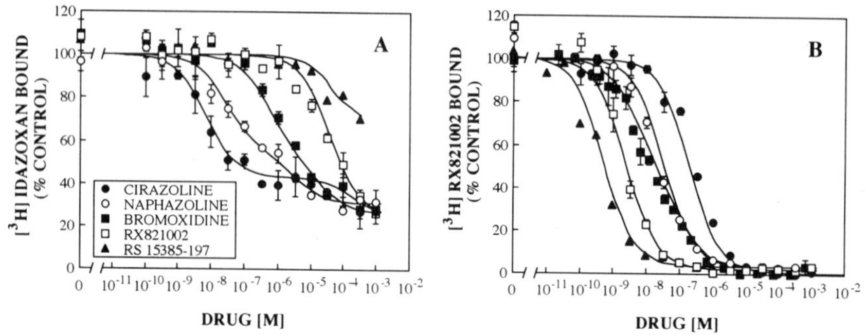

FIGURE 2. (**A**) Inhibition of binding of [^{3}H]idazoxan to I_2-imidazoline receptors by various drugs in the human brain. (**B**) Inhibition of binding of [^{3}H]RX821002 (2-methoxy idazoxan) to α_2-adrenoceptors by various drugs in the human brain. From Miralles *et al.*[15]

receptors (TABLE 1). In contrast, competition of the same drugs for [^{3}H]RX821002 binding sites resulted in steep curves except for the α_2-adrenoceptor agonists clonidine and bromoxidine (UK 14304) (TABLE 1, FIG. 2B). In marked contrast with the high potency displayed by the selective α_2-adrenoceptor antagonists RS15385-197, RX821002, and yohimbine against the binding of [^{3}H]RX821002 (K_i values in the low nanomolar range), these compounds were very weak at inhibiting [^{3}H]idazoxan

TABLE 1. Binding Parameters of Various Drugs to Imidazoline Receptors and α_{2A}-Adrenoceptors in the Human Brain[a]

	I_2-Imidazoline Receptors			α_{2A}-Adrenoceptors		
Drug	K_{iH} (nM)	K_{iL} (nM)	%R_H	K_{iH} (nM)	K_{iL} (nM)	%R_H
Guanoxan	1.3	231	55	82		100
Cirazoline	2.4	1,900	88	76		100
Idazoxan	18		100	12		100
Naphazoline	22	2,500	65	11		100
Clonidine	435	56,500	40	2.9	45	53
Bromoxidine	532	14,000	69	1.8	31	61
Rilmenidine	2,100		100	1,700		100
Amiloride	2,500		100	>100,000		100
Yohimbine	14,000		100	1.5		100
RX821002	32,000		100	1.0		100
RS 15385-197	>100,000			0.2		100
Agmatine	>100,000			10,600		100
1-Arginine	>100,000			>100,000		

[a] Data taken from Miralles *et al.*[15]

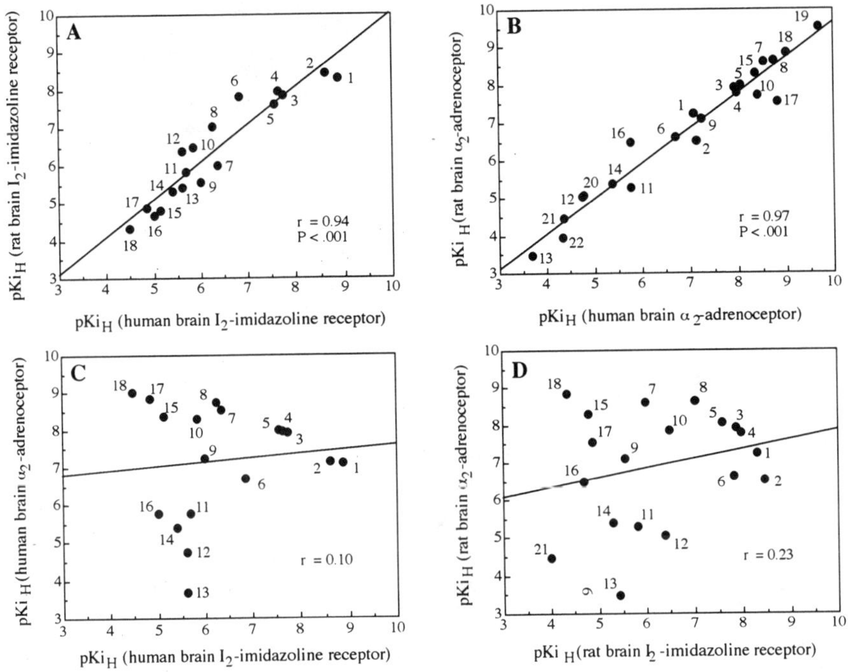

FIGURE 3. Correlations between drug affinities (expressed as pK_{iH} values, that is, the negative logarithm of the inhibition constant for the high affinity site) for human and rat brain I_2-imidazoline receptors (**A**), human and rat brain α_2-adrenoceptors (**B**), human brain I_2-imidazoline receptors and α_2-adrenoceptors (**C**), and rat brain I_2-imidazoline receptors and α_2-adrenoceptors (**D**). The lines represent regressions of the correlations. (1) Guanoxan, (2) cirazoline, (3) idazoxan, (4) naphazoline, (5) guanabenz, (6) tolazoline, (7) clonidine, (8) bromoxidine (UK 14304), (9) medetomidine, (10) oxymetazoline, (11) rilmenidine, (12) phenylbiguanide, (13) amiloride, (14) guanethidine, (15) phentolamine, (16) xylazine, (17) yohimbine, (18) RX821002 (2-methoxy idazoxan), (19) RS15385-197, (20) prazosin, (21) dimaprit, and (22) cimetidine. From Miralles *et al.*[15]

from the I_2-imidazoline receptor (K_i values in the high micromolar range) (TABLE 1, FIG. 2).

These data, along with the correlations of drug affinities (FIG. 3), confirmed the nonadrenergic character of these imidazoline sites and indicated that α_{2A}-adrenoceptors and I_2-imidazoline receptors are different molecular entities. The I_2-imidazoline receptor of the human brain appears to belong to the I_{2b}-subtype according to the low affinity displayed by amiloride (TABLE 1).[20] Recently, one component of the so-called clonidine-displacing substance (CDS) has been purified from bovine brain and determined to be agmatine (decarboxylated arginine). Agmatine has been proposed to be the common neurotransmitter for imidazoline receptors, because of its ability to bind to I_1- and I_2-imidazoline receptors as well as to α_2-adrenoceptors.[21] In the human brain, however, agmatine and its precursor arginine displayed very low affinities for I_2-imidazoline receptors and α_{2A}-adrenoceptors (TABLE 1).

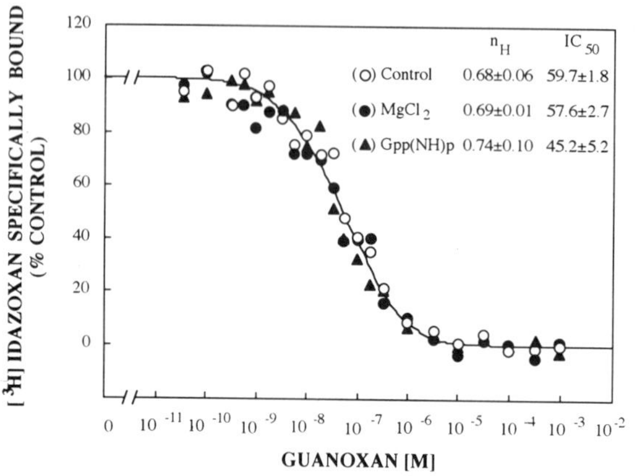

FIGURE 4. Inhibition of specific binding of [^{3}H]idazoxan to rat cortical I_2-imidazoline receptors by guanoxan in the absence (○) or presence of 5 mM $MgCl_2$ (●) or 5 mM $MgCl_2$ plus 0.1 mM Gpp(NH)p (▲). From Miralles *et al.*[15]

The possible existence of different affinity states for the I_2-imidazoline receptor was tested by studying the effect of Gpp(NH)p, a nonhydrolyzable analog of GTP, on [^{3}H]idazoxan binding, but no change in affinity for any of the competing drugs (e.g., guanoxan, clonidine, bromoxidine, and guanabenz) was observed Fig. 4),[10,22–25] suggesting that these putative receptors are not coupled to "classical" G-proteins. Other studies have found that GTP interacts with [^{3}H]idazoxan binding sites.[26] However, the good correlation between the inhibition constants (K_i) of several drugs for the high and low affinity populations of I_2-imidazoline receptors ($r = 0.89$ in the human brain and $r = 0.92$ in the rat brain)[15] further supports the concept of heterogeneity of this novel receptor with a possible interdependence of the two affinity sites.[25]

IMMUNODETECTION OF PUTATIVE IMIDAZOLINE RECEPTOR PROTEINS IN THE HUMAN BRAIN

Recently, proteins of 55–70 kD, identified as idazoxan-preferring imidazoline receptors, have been purified from mitochondrial rabbit kidney[27] and bovine adrenal chromaffin cells[28] or visualized by photoaffinity labeling in the rat liver and PC-12 cells.[29] To further characterize the nature of imidazoline receptors in the human brain, the immunological detection of imidazoline receptor proteins was assessed by means of a specific antiserum raised against the imidazoline receptor purified from bovine adrenal chromaffin cells.[28,30] Solubilized imidazoline receptors from human and rat brain tissues were separated by gel electrophoresis (SDS-PAGE), transferred to nitrocellulose membranes (western blotting), labeled with the specific antiserum, and detected by enhanced chemiluminescence.[31]

Various human and rat brain regions (cerebral cortex, caudate nucleus, hypothalamus, hippocampus, medulla oblongata, and cerebellum) expressed a double band of

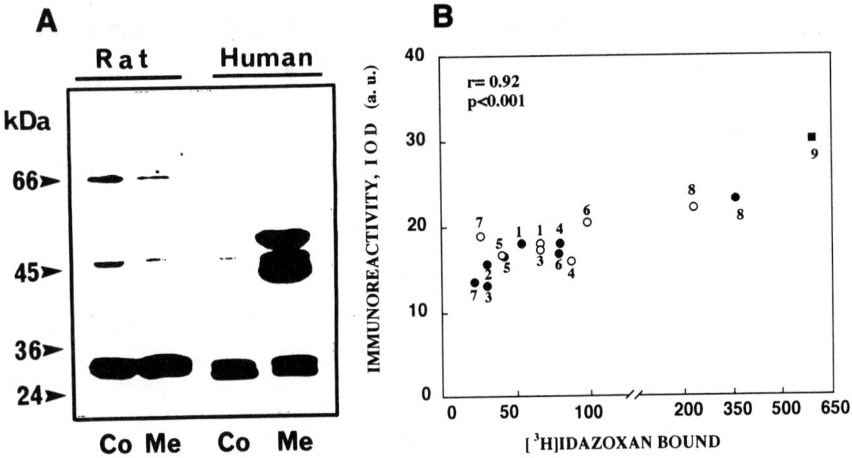

FIGURE 5. (**A**) Immunological detection of imidazoline receptor proteins in the human and rat brain cortex (Co) and medulla oblongata (Me). (**B**) Correlation between the immunoreactivity of imidazoline receptor proteins and the density of I_2-imidazoline receptors in various human (○) and rat (●) tissues. (1) Frontal cortex, (2) parietooccipital cortex, (3) caudate nucleus, (4) hypothalamus, (5) hippocampus, (6) medulla oblongata, (7) cerebellum, (8) liver, and (9) bovine adrenal medulla. Modified from Escribá *et al.*[32]

29-30 kD. Less intense bands of 47 and 66 kD were also present in most regions of the rat brain. In the human brain, however, only the 47-kD band was detected in some regions. The human medulla oblongata also expressed bands of 44 and 50 kD (FIG. 5A).[32] The strong correlation (r = 0.92) between the density of I_2-imidazoline receptors and the integrated optical density of all bands in various human and rat brain regions and tissues (FIG. 5B) suggests that the proteins detected by immunoblot analysis correspond to the pharmacological characterized I_2-imidazoline receptors. By contrast, the lack of correlation ($r = 0.03$) between the density of α_2-adrenoceptors and the immunoreactivity of the bands in the same tissues clearly indicated that the molecular entities detected by immunoblotting are not α_2-adrenoceptors.[32]

Recent observations indicate that treatment of rats with idazoxan, cirazoline, and other drugs results in parallel changes in the density of brain I_2-imidazoline receptors (B_{max} for [^{3}H]idazoxan binding) and in the immunoreactivity of the various bands, further suggesting a correspondence between the pharmacological and immunological defined receptors.

EFFECT OF AGING ON I_2-IMIDAZOLINE RECEPTORS AND α_{2A}-ADRENOCEPTORS IN THE HUMAN BRAIN

The density of most neurotransmitter receptors declines with aging as a consequence of the degenerative process that occurs in the mammalian brain. As observed for agonist radioligands,[33,34] the density of α_{2A}-adrenoceptors labeled by the antagonist [^{3}H]RX821002 in the frontal cortex showed a significant negative correlation with

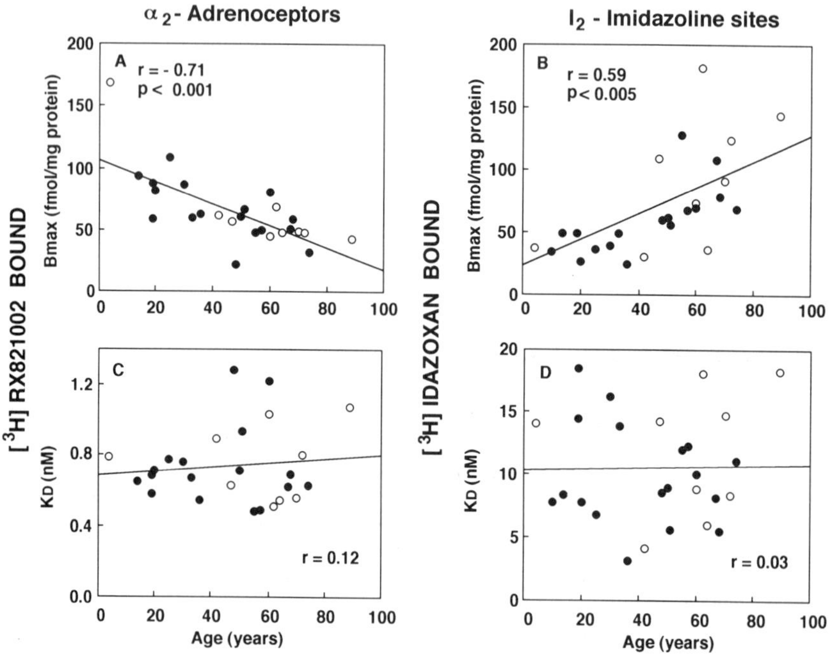

FIGURE 6. Age-related decrease in [^{3}H]RX821002 binding sites (α_{2A}-adrenoceptors) (**A**) and age-related increase in [^{3}H]idazoxan binding sites (I_2-imidazoline sites) (**B**) in the frontal cortex of 27 control subjects (range 4-89 years; ●, man; ○, woman). Binding affinities of the radioligands were not age-related (**C,D**). From Sastre and García-Sevilla.[35]

the age of the subject at death (FIG. 6A). In marked contrast, the density of I_2-imidazoline receptors labeled by [^{3}H]idazoxan in the same subjects clearly showed a positive correlation with age (FIG. 6B). Binding affinities expressed as K_d values for [^{3}H]RX821002 and [^{3}H]idazoxan were not age-related (FIG. 6C and D). For the α_{2A}-adrenoceptor the average decrease per decade was 8.8 fmol/mg of protein (9% mean decline per decade from 20-70 years) and for the I_2-imidazoline receptor the average increase per decade was 10.5 fmol/mg of protein (16% mean rise per decade from 20-70 years) (FIG. 6A and B).[35]

These results demonstrated that the density of brain α_2-adrenoceptors and I_2-imidazoline receptors shows opposite age-related changes, which further supports the concept that these receptors are independent molecular entities.

HUMAN BRAIN MONOAMINE OXIDASE: EFFECT OF AGING AND RELATION OF MONOAMINE OXIDASE-B TO I_2-IMIDAZOLINE RECEPTORS

Age-related increases in monoamine oxidase (MAO) activity have been reported in the human brain.[36,37] Because of the age-related increase in the density of brain

I_2-imidazoline receptors[35] and the association of this site with mitochondrial outer membranes,[13] two features shared by the enzyme MAO, the possible relationship between I_2-imidazoline receptors and MAO was investigated. In the frontal cortex of an age-selected group of subjects, the density of MAO-B sites labeled by [^{3}H]Ro 19-6327 (lazabemide) showed a significant positive correlation with the age of the subject at death (FIG. 7A). In contrast, no significant correlation was found between the density of MAO-A sites labeled by [^{3}H]Ro 41-1049 in the same subjects and age. For the density of MAO-B sites labeled by [^{3}H]Ro 19-6327 the average increase per decade was 45 fmol/mg of protein (50% mean rise per decade from 20–70 years) (FIG. 7A). This rise in brain MAO-B levels per decade was more than four times greater in absolute values than the increase per decade observed for the I_2-imidazoline receptor in the same subjects.[35]

Because the density of [^{3}H]Ro 19-6327 sites, which reflects the concentration of MAO-B molecules, and I_2-imidazoline receptors correlated with the age of the same subjects at death, the density of both binding sites also showed the expected positive correlation, that is, the greater the density of I_2-imidazoline receptors, the greater the concentration of MAO-B (FIG. 7B). In contrast, no correlation was found between the concentration of MAO-A and the density of I_2-imidazoline receptors in the same subjects. A multiple regression analysis between age and the densities of I_2-imidazoline and MAO-B sites in the same subjects also showed a marked positive correlation ($r = 0.90$).[35]

In the frontal cortex, the density of MAO-B sites ([^{3}H]Ro 19-6327, B_{max} = 2,500 fmol/mg of protein) was much greater than the density of I_2-imidazoline receptors ([^{3}H]idazoxan, B_{max} = 72 fmol/mg of protein), and the B_{max} ratio in the same subjects was about 35. The ratio of density of sites (MAO-B/I_2) did not correlate with the age of the subject at death (FIG. 7C), suggesting that the MAO-B/I_2 ratio is rather constant during the process of aging.[35]

Because of the correlation between MAO-B and I_2-imidazoline receptors (FIG. 7B) and the strong association of these two variables with the age of the subjects, the possibility of a direct interaction of [^{3}H]idazoxan with the MAO-B was investigated. The compound Ro 16-6491, a selective MAO-B inhibitor, displayed high affinity (K_i = 42 nM) and idazoxan very low affinity (K_i = 89 μM) against [^{3}H]Ro 19-6327 binding to human brian MAO-B.[35] Similarly, idazoxan also displayed very low affinity (K_i = 40 μM) against the binding of [^{3}H]Ro 41-1049 to brain MAO-A.[38]

These results excluded a direct interaction of [^{3}H]idazoxan with the active center of both MAO isoenzymes in the human brain. However, I_2-imidazoline receptors and MAO-B sites show a clear coexpression not only in the human frontal cortex during the process of aging,[35] but also in various regions of the human brain (predominance of MAO-B) (FIG. 8A) as well as the rat brain (predominance of MAO-A) (FIG. 8B).[35] On the other hand, chronic treatment with various irreversible MAO inhibitors, such as clorgyline and phenelzine, resulted in marked decreases (30–50%) in the density of I_2-imidazoline receptors in the rat brain[38] (Alemany *et al.*, this volume). These novel findings suggest an indirect mechanism for the *in vivo* effects of irreversible MAO inhibitors on I_2-imidazoline receptors that could be related to their clinical antidepressant properties.

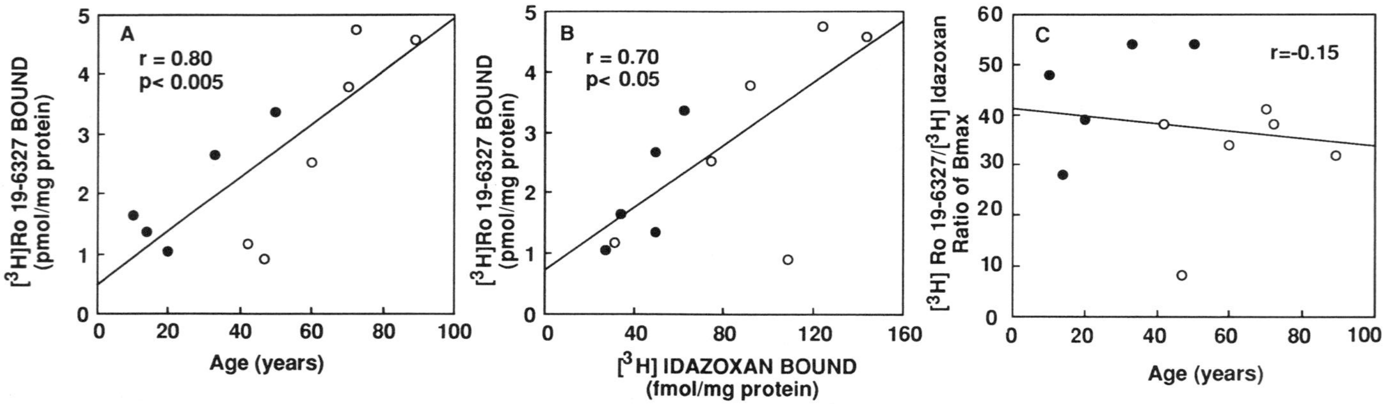

FIGURE 7. (**A**) Age-related increase in [^{3}H]Ro 19-6327 (lazabemide) binding sites (MAO-B sites) in the frontal cortex of 11 control subjects (range 10–89 years, ●, man; ○, woman). (**B**) Correlation between the density of sites (B_{max} values) for the binding of [^{3}H]idazoxan (I_2-imidazoline sites) and [^{3}H]Ro 19-6327 (MAO-B sites) in the frontal cortex of 11 control subjects. (**C**) Lack of correlation between the ratio of density of sites (MAO-B/I_2-imidazoline sites) and the age of the subjects (range 10–89 years). From Sastre and García-Sevilla.[35]

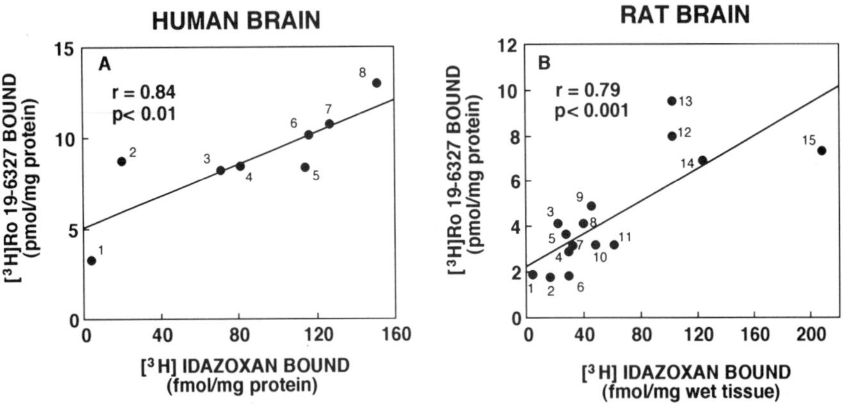

FIGURE 8. (**A**) Correlation between the autoradiographic density of sites for the binding of [^{3}H]idazoxan (I_2-imidazoline sites; data taken from De Vos *et al.*, Brain Res. **566:** 13–20, 1991) and [^{3}H]Ro 19-6327 (MAO-B sites; data taken from Saura *et al.* 1992. J. Neurosci. **12:** 1977–1999) in various regions of the human brain. (1) Cerebellum, (2) formatio reticularis, (3) parahippocampal cortex, (4) pons, (5) hippocampus, (6) putamen, (7) caudate nucleus, and (8) substantia nigra. (**B**) Correlation between the autoradiographic density of sites for the binding of [^{3}H]idazoxan (I_2-imidazoline sites; data taken from Mallard *et al.* 1992. Br. J. Pharmacol. **106:** 1019–1027) and [^{3}H]Ro 19-6327 (MAO-B sites; data taken from Saura *et al.*, 1992) in various regions of the rat brain. (1) Corpus callosum, (2) inferior colliculus, (3) nucleus tractus solitarii, (4) frontal cortex, (5) central gray, (6) caudate/putamen, (7) cingulate cortex, (8) ventromedial hypothalamus, (9) medial habenular nucleus, (10) superior colliculus, (11) hippocampus, CA_1 field, (12) ependymal layer, (13) arcuate nucleus, (14) interpeduncular nucleus, and (15) area postrema. From Sastre and García-Sevilla.[35]

BRAIN I_2-IMIDAZOLINE RECEPTORS AND REGULATION OF THE GLIAL FIBRILLARY ACIDIC PROTEIN

At present the nature and physiological function of the brain I_2-imidazoline receptor remain unknown. In the monkey and human brain the more abundant isoenzyme MAO-B is located in serotenergic neurons and glial cells.[39,40] Moreover, the well-known age-related rise in brain MAO-B is entirely due to an increase in the extraneuronal component of the enzyme,[41] which suggests some form of reactive proliferation of glial cells during aging, specially of astrocytes that are enriched in MAO-B.[42,43] Recently, age-related increases in levels of the astrocyte glial fibrillary acidic protein (GFAP) or its specific mRNA were demonstrated throughout the rodent brain, suggesting that the astrocyte hypertrophy observed with aging involves an accumulation of glial filaments.[44,45] In this context it is important to note that rat cerebral cortical astrocytes were shown to possess functional I_2-imidazoline receptors ([^{3}H]idazoxan binding) and that idazoxan induces a marked increase in GFAP mRNA levels in these glial cells.[46] Therefore, it was postulated that age-related increases in the density of I_2-imidazoline receptors in the human brain reflected the astrocyte hypertrophy associated with aging.[35]

To test this hypothesis, it was of interest to assess whether treatment with imidazoline drugs could modulate in parallel the density of I_2-imidazoline receptors and

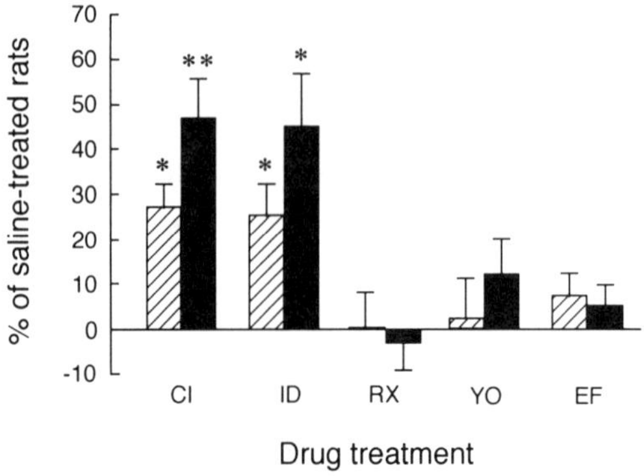

FIGURE 9. Effect of chronic treatment (7 days) with cirazoline (CI) (1 mg kg^{-1}), idazoxan (ID) (10 mg kg^{-1}), RX821002 (2-methoxy idazoxan) (RX) (10 mg kg^{-1}), yohimbine (YO) (10 mg kg^{-1}), or efaroxan (EF) (10 mg kg^{-1}) on the density (B_{max}) of I_2-imidazoline receptors (*hatched bars*) and GFAP immunoreactivity (*solid bars*) in the rat cerebral cortex. $*p < 0.05$ and $**p < 0.01$ as compared with saline-treated group (ANOVA followed by Scheffé's test). A strong correlation ($r = 0.97$; $p < 0.01$) exists between changes in the density of I_2-imidazoline receptors and GFAP immunoreactivity after the various treatments. From Olmos *et al.*[47]

GFAP levels, quantitated by immunoblotting, in the rat brain. Chronic treatment with the imidazoline drugs cirazoline and idazoxan but not with the structurally related α_2-adrenoceptor antagonists RX821002 (2-methoxy idazoxan) and efaroxan markedly increased (45%) GFAP immunoreactivity in rat cerebral cortex. Chronic treatment with yohimbine, a non-imidazoline α_2-adrenoceptor antagonist, did not significantly alter GFAP immunoreactivity in rat brain (FIG. 9).[47] Moreover, similar chronic treatments with cirazoline and idazoxan, but not with RX821002, efaroxan, or yohimbine also increased (25%) the density of I_2-imidazoline receptors in the rat brain. A strong correlation ($r = 0.97$) was found when the mean percentage changes in GFAP immunoreactivity were related to the mean percentage changes in I_2-imidazoline receptors after various drug treatments (FIG. 9).[47]

These studies suggested a direct physiological function of glial I_2-imidazoline receptors in the regulation of GFAP levels. Because these I_2-imidazoline receptors are mainly located on mitochondrial membranes[12–14] and are associated with calcium mobilization,[48] it is possible that upregulation of I_2-imidazoline receptors could be related to increased mobilization of mitochondrial calcium stores, and this in turn may influence the pattern of GFAP gene expression.

BRAIN I_2-IMIDAZOLINE RECEPTORS IN ALZHEIMER'S DISEASE

The loss of α_2-adrenoceptors, among other neurotransmitter receptors, was demonstrated in Alzheimer's disease as a consequence of the degenerative process that occurs in the brain.[34,49] Moreover, relevant increases in brain MAO-B activity were also demonstrated in Alzheimer's and other neurodegenerative disorders.[50,51] This

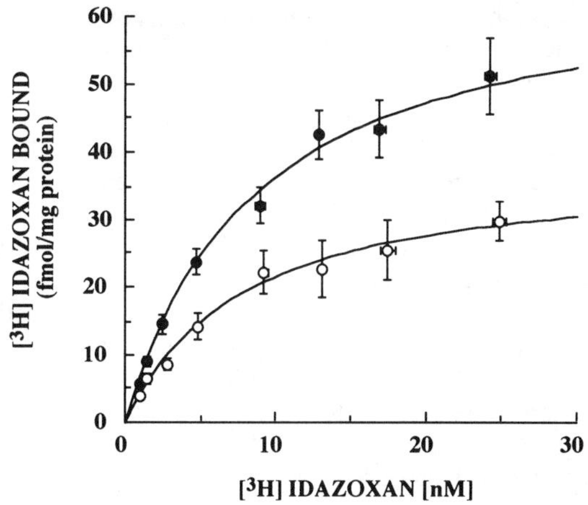

FIGURE 10. Specific binding of [^{3}H]idazoxan to I_2-imidazoline receptors in postmortem human brain membranes from the prefrontal cortex of control subjects (○) and patients with Alzheimer's disease (●). The hyperbolic plots yielded a K_d value of 7.7 ± 2.6 nM with a B_{max} value of 37 ± 5 fmol/mg protein for controls (○) and K_d value of 8.3 ± 1.7 nM with a B_{max} value of 66 ± 6 fmol/mg protein for patients with Alzheimer's disease (●), displaying binding isotherms clearly different from each other. Modified from Ruiz *et al.*[54]

finding has been associated with the glial presence of the enzyme[51] and could be related to the well-known astrocytic hypertrophy (higher GFAP and mRNA for GFAP levels) associated with Alzheimer's disease.[52,53] Therefore, it was postulated that the density of I_2-imidazoline receptors might be increased in Alzheimer's disease, in which a decrease in α_2-adrenoceptor density had been demonstrated.

As expected, the density of I_2-imidazoline receptors was significantly increased (63%) in the prefrontal cortex (Brodmann's area 9) of a well-defined group of postmortem brains from patients with a clinical diagnosis and neuropathological confirmation of Alzheimer's disease than in age-matched controls (FIG. 10).[54] This finding stresses the potential important role of this novel receptor and indicates that the quantitation of I_2-imidazoline receptors could be useful as a new glia (astrocyte) marker in the postmortem human brain for the study of normal aging and neurodegenerative disorders such as Alzheimer's disease.

BRAIN I_2-IMIDAZOLINE RECEPTORS IN MAJOR DEPRESSION AND SUICIDE

Endogenous depression appears to be associated with increased brain α_2-adrenoceptors, and downregulation of these inhibitory receptors has been involved in the mechanism of action of various antidepressant drugs.[55,56] In recent pilot studies, the density of imidazoline receptors also was reported to be increased in platelets (I_1-type)[57] and postmortem brain (I_2-type)[58] of depressed patients. In this latter study, competition curves for norepinephrine against the binding of [^{3}H]idazoxan in the frontal cortex of control subjects and depressed suicides were biphasic with a plateau at similar concentrations of the catecholamine (10^{-6} M). However, the proportion of [^{3}H]idazoxan binding sites displaced by high concentrations of norepinephrine (10^{-6} M to 10^{-3} M; i.e., I_2-imidazoline receptors) was significantly different between the suicide (64%, 21 fmol/mg protein) and control (46%, 9 fmol/mg protein) groups.[58] This finding was the first experimental evidence of a possible abnormal higher density of I_2-imidazoline receptors in the brain of depressed suicide victims. In the peripheral nervous system, imidazoline-preferring receptors mediate presynaptic inhibition of norepinephrine release.[59,60] Because reduced noradrenergic function in the central nervous system was postulated in the etiology of depression, enhanced imidazoline receptor (and α_2-adrenoceptor) activity and with it decreased norepinephrine release were suggested in the pathogenesis of endogenous depression.

In recent studies (Sastre *et al.*, this volume), however, the density of I_2-imidazoline receptors quantified in saturation studies with [^{3}H]idazoxan was reduced (B_{max} decreased by 40%) in the frontal cortex of a heterogeneous group of undiagnosed suicide victims. In this brain region, immunoreactivity of the 29-30-kD imidazoline receptor protein was also significantly reduced (18%) in suicide victims. By contrast, immunoreactivity of the 45-kD imidazoline receptor protein was significantly increased (47%) in the same suicide victims (Sastre *et al.*, this volume). Further experimental work is needed to assess the possible relevance of imidazoline receptors in the pathogenesis of depression.

REFERENCES

1. MICHEL, M. C. & P. A. INSEL. 1989. Are there multiple imidazoline binding sites? Trends Pharmacol. Sci. **10:** 342-344.
2. ATLAS, D. 1991. Clonidine-displacing substance (CDS) and its putative imidazoline receptor. New leads for further divergence of α_2-adrenergic receptor activity. Biochem. Pharmacol. **41:** 1541-1549.
3. MICHEL, M. C. & P. ERNSBERGER. 1992. Keeping an eye on the I site: Imidazoline-preferring receptors. Trends Pharmacol. Sci. **13:** 369-370.
4. BOUSQUET, P., J. FELDMAN & J. SCHWARTZ. 1984. Central cardiovascular effects of alpha adrenergic drugs: Differences between catecholamines and imidazolines. J. Pharmacol. Exp. Ther. **230:** 232-236.
5. ERNSBERGER, P. R., T. H. DAMON, L. M. GRAFF, S. G. SCHÄFFER & M. ODILE CHRISTEN. 1993. Moxonidine, a centrally antihypertensive agent, is a selective ligand for I_1-imidazoline sites. J. Pharmacol. Exp. Ther. **264:** 172-182.
6. DE VOS, H., G. BRICCA, J. DE KEYSER, J. P. DE BACKER, P. BOUSQUET & G. VAUQUELIN. 1994. Imidazoline receptors, non-adrenergic idazoxan binding sites and α_2-adrenoceptors in the human central nervous system. Neuroscience **59:** 589-598.
7. MOLDERINGS, G. J., D. MOURA, K. FINK, H. BÖNISH & M. GÖTHERT. 1993. Binding of [^{3}H]clonidine to I_1-imidazoline sites in bovine adrenal medullary membranes. Naunyn-Schmiedeberg's Arch. Pharmacol. **348:** 70-76.
8. LANGIN, D. & M. LAFONTAN. 1989. [^{3}H]Idazoxan binding at non-α_2-adrenoceptors in rabbit adipocyte membranes. Eur. J. Pharmacol. **159:** 199-203.
9. COUPRY, I., D. ATLAS, R.-A. PODEVIN, I. UZIELLI & A. PARINI. 1989. Imidazoline-guanidinium receptive site in renal proximal tubule: Asymmetric distribution, regulation by cations and interaction with an endogenous clonidine displacing substance. J. Pharmacol. Exp. Ther. **252:** 293-299.
10. ZONNENSCHEIN, R., S. DIAMANT & D. ATLAS. 1990. Imidazoline receptors in rat liver cells: A novel receptor or a subtype of α_2-adrenoceptor? Eur. J. Pharmacol. **190:** 203-215.
11. MALLARD, N. J., A. L. HUDSON & D. J. NUTT. 1992. Characterization and autoradiographical localization of non-adrenoceptor binding sites in the rat brain. Br. J. Pharmacol. **106:** 1019-1027.
12. TESSON, F. & A. PARINI. 1991. Identification of an imidazoline-guanidinium receptive site in mitochondria from rabbit cerebral cortex. Eur. J. Pharmacol. **208:** 81-83.

13. Tesson, F., C. Prip-Buus, A. Lemoine, J.-P. Pegorier & A. Parini. 1991. Subcellular distribution of imidazoline-guanidinium-receptive sites in human and rabbit liver. J. Biol. Chem. **266:**155-160.
14. Regunathan, S., M. P. Meeley & D. J. Reis. 1993. Expression of non-adrenergic imidazoline sites in chromaffin cells and mitochondrial membranes of bovine adrenal medulla. Biochem. Pharmacol. **45:** 1667-1675.
15. Miralles, A., G. Olmos, M. Sastre, F. Barturen, I. Martin & J. A. García-Sevilla. 1993. Discrimination and pharmacological characterization of I_2-imidazoline sites with [^{3}H]idazoxan and alpha-2 adrenoceptors with [^{3}H]RX821002 (2-methoxy idazoxan) in the human and rat brains. J. Pharmacol. Exp. Ther. **264:** 1187-1197.
16. Convents, A., D. Convents, J.-P. De Backer, J. De Keyser & G. Vauquelin. 1989. High affinity binding of ^{3}H-rauwolscine and ^{3}H-RX781094 to α_2 adrenergic receptors and non-stereoselective sites in human and rabbit brain cortex membranes. Biochem. Pharmacol. **38:** 455-463.
17. Wikberg, J. E. S. 1989. High affinity binding of idazoxan to a non-catecholaminergic binding site in the central nervous system: Description of a putative idazoxan-receptor. Pharmacol. Toxicol. **64:** 152-155.
18. Wikberg, J. E. S. & S. Uhlén. 1990. Further characterization of the guinea pig cerebral cortex idazoxan receptor: Solubilization, distribution from the imidazole site, and demonstration of cirazoline as an idazoxan receptor selective drug. J. Neurochem. **55:** 192-203.
19. Sastre, M. & J. A. García-Sevilla. 1994. α_2-Adrenoceptor subtypes identified by [^{3}H]RX821002 binding in the human brain: The agonist guanoxabenz does not discriminate different forms of the predominant α_{2A}-subtype. J. Neurochem. **63:** 1077-1085.
20. Diamant, S., T. Eldar-Geva & D. Atlas. 1992. Imidazoline binding sites in human placenta: Evidence for heterogeneity and a search for physiological function. Br. J. Pharmacol. **106:** 101-108.
21. Li, G., S. Regunathan, C. J. Barrow, J. Eshraghi, R. Cooper & D. J. Reis. 1994. Agmatine: An endogenous clonidine-displacing substance in the brain. Science **263:** 966-969.
22. Michel, M. C., O.-E. Brodde, B. Schnepel, J. Behrendt, R. Tschada, H. J. Motulsky & P. A. Insel. 1989. [^{3}H]Idazoxan and some other α_2-adrenergic drugs also bind with high affinity to a nonadrenergic site. Mol. Pharmacol. **35:** 324-330.
23. Lachaud-Pettiti, V., P. R.-A., Y. Chrétien & A. Parini. 1991. Imidazoline-guanidinium and α_2-adrenergic binding sites in basolateral membranes from human kidney. Eur. J. Pharmacol. **206:** 23-31.
24. Yablonsky, F. & J. P. Dausse. 1991. Non-adrenergic binding sites for the ''α_2-antagonist'' [^{3}H]idazoxan in the rabbit urethral smooth muscle. Pharmacological and biochemical characterization. Biochem. Pharmacol. **41:** 701-707.
25. Wikberg, J. E. S., S. Uhlén & V. Chhajlani. 1992. Evidence that drug binding to non-adrenergic [^{3}H]-idazoxan binding sites (I-receptors) occurs to interacting or interconvertible affinity forms of the receptor. Pharmacol. Toxicol. **70:** 208-219.
26. Hussain, J. F., D. A. Kendall & V. G. Wilson. 1993. Species-selective binding of [^{3}H]-idazoxan to α_2-adrenoceptors and non-adrenoceptor, imidazoline binding sites in the central nervous system. Br. J. Pharmacol. **109:** 831-837.
27. Limon, I., I. Coupry, S. M. Lanier & A. Parini. 1992. Purification and characterization of mitochondrial imidazoline-guanidinium receptive sites from rabbit kidney. J. Biol. Chem. **267:** 21645-21649.
28. Wang, H., S. Regunathan, M. P. Meeley & D. J. Reis. 1992. Isolation and characterization of imidazoline receptor protein from bovine adrenal chromaffin cells. Mol. Pharmacol. **42:** 792-801.
29. Lanier, S., B. Ivkovic, I. Singh, J. L. Neumeyer & V. Bakthavachalam. 1993. Visualization of multiple imidazoline/guanidinium-receptive sites. J. Biol. Chem. **268:** 16047-16051.

30. WANG, H., S. REGUNATHAN, D. A. RUGGIERO & D. J. REIS. 1993. Production and characterization of antibodies specific for the imidazoline receptor protein. Mol. Pharmacol. **43:** 509-515.
31. ESCRIBÁ, P. V., M. SASTRE & J. A. GARCÍA-SEVILLA. 1994. Increased density of guanine nucleotide-binding proteins in the postmortem brains of heroin addicts. Arch. Gen. Psychiatry **51:** 494-501.
32. ESCRIBÁ, P. V., M. SASTRE, H. WANG, S. REGUNATHAN, D. J. REIS & J. A. GARCÍA-SEVILLA. 1994. Immunodetection of putative imidazoline receptor proteins in the human and rat brain and other tissues. Neurosci. Lett. **178:** 81-84.
33. PASCUAL, J., C. DEL ARCO, A. M. GONZÁLEZ, E. DEL OLMO & A. PAZOS. 1991. Regionally specific age-dependent decline in α_2-adrenoceptors: An autoradiographic study in human brain. Neurosci. Lett. **133:** 279-283.
34. MEANA, J. J., F. BARTUREN, M. A. GARRO, J. A. GARCÍA-SEVILLA, A. FONTÁN & J. J. ZARRANZ. 1992. Decreased density of presynaptic α_2-adrenoceptors in postmortem brains of patients with Alzheimer's disease. J. Neurochem. **58:** 1896-1904.
35. SASTRE, M. & J. A. GARCÍA-SEVILLA. 1993. Opposite age-dependent changes of α_{2A}-adrenoceptors and non-adrenoceptor [^{3}H]idazoxan binding sites (I_2-imidazoline sites) in the human brain: strong correlation of I_2 with MAO-B sites. J. Neurochem. **61:** 881-889.
36. FOWLER, C. J., A. WIBERG, L. ORELASND, J. MARCUSSON & B. WINBLAD. 1980. The effect of age on the activity and molecular properties of human brain monoamine oxidase. J. Neural Transm. **49:** 1-20.
37. MANN, J. J. & M. STANLEY. 1984. Postmortem monoamine oxidase enzyme kinetics in the frontal cortex of suicide victims and controls. Acta Psychiatr. Scand. **69:** 135-139.
38. OLMOS, G., A. M. GABILONDO, A. MIRALLES, P. V. ESCRIBA & J. A. GARCÍA-SEVILLA. 1993. Chronic treatment with the monoamine oxidase inhibitors clorgyline and pargyline down-regulates non-adrenoceptor [^{3}H]-idazoxan binding sites in the rat brain. Br. J. Pharmacol. **108:** 597-603.
39. WESTLUND, K. L., R. M. DENNEY, C. M. KOCHERSPERGER, R. M. ROSE & C. W. ABELL. 1985. Distinct monoamine oxidase A and B populations in primate brain. Science **230:** 181-183.
40. WESTLUND, K. L., R. M. DENNEY, R. M. ROSE & C. W. ABELL. 1988. Localization of distinct monoamine oxidase A and monoamine oxidase B populations in human brainstem. Neuroscience **25:** 439-456.
41. ARAI, Y., A. SENSTRÖM & L. ORELAND. 1985. The effect of age on extra- and intraneuronal MAO-A and B activities in rat brain. Biogenic Amines **2:** 65-71.
42. LEVITT, P., J. E. PINTER & X. O. BREAKEFIELD. 1982. Immunocytochemical demonstration of monoamine oxidase B in brain astrocytes and serotonergic neurons. Proc. Natl. Acad. Sci. USA **79:** 6385-6389.
43. NAKAMURA, S., T. KAWAMATA, I. AKIGUCHI, M. KAMEYAMA, N. NAKAMURA & J. KIMURA. 1990. Expression of monoamine oxidase B activity in astrocytes of senile plaques. Acta Neuropathol. **80:** 419-425.
44. GOSS, J. R., C. E. FINCH & D. G. MORGAN. 1991. Age-related changes in glial fibrillary acidic protein mRNA in the mouse brain. Neurobiol. Aging **12:** 165-170.
45. O'CALLAGHAN, J. P. & D. B. MILLER. 1991. The concentration of glial fibrillary acidic protein increases with age in the mouse and rat brain. Neurobiol. Aging **12:** 171-174.
46. REGUNATHAN, S., D. L. FEINSTEIN & D. J. REIS. 1993. Expression of non-adrenergic imidazoline sites in rat cerebral cortical astrocytes. J. Neurosci. Res. **34:** 681-688.
47. OLMOS, G., R. ALEMANY, P. V. ESCRIBÁ & J. A. GARCÍA-SEVILLA. 1994. The effects of chronic imidazoline drug treatment on glial fibrillary acidic protein concentrations in rat brain. Br. J. Pharmacol. **111:** 997-1002.
48. REGUNATHAN, S., M. J. EVINGER, M. P. MEELEY & D. J. REIS. 1991. Effects of clonidine and other imidazole-receptor binding agents on second messenger systems and calcium influx in bovine adrenal chromaffin cells. biochem. Pharmacol. **42:** 2011-2018.

49. PASCUAL, J., B. GRIJALBA, J. A. GARCÍA-SEVILLA, J. J. ZARRANZ & A. PAZOS. 1992. Loss of high affinity α_2-adrenoceptors in Alzheimer's disease: An autoradiographic study in frontal cortex and hippocampus. Neurosci. Lett. **142:** 36-40.
50. JOSSAN, S. S., P. G. GILLBERG, C. G. GOTTFRIES, I. KARLSSON & L. ORELAND. 1991. Monoamine oxidase B in brains from patients with Alzheimer's disease: A biochemical and autoradiographical study. Neuroscience **45:** 1-12.
51. SHERIF, F., C. G. GOTTFRIES, I. ALAFUZOFF & L. ORELAND. 1992. Brain gamma-aminobutyrate aminotransferase (GABA-T) and monoamine oxidase (MAO) in patients with Alzheimer's disease. J. Neural Transm. (P-D Sect.) **4:** 227-240.
52. DUFFY, P. E., M. RAPPORT & L. GRAF. 1980. Glial fibrillary acid protein and Alzheimer-like senile dementia. Neurology **30:** 778-782.
53. LE PRINCE, G., P. DELAERE, C. FAGES, C. DUYCKAERTS, J.-J. HAUW & M. TARDY. 1993. Alterations of glial fibrillary acidic protein mRNA level in the aging brain and in senile dementia of the Alzheimer type. Neurosci. Lett. **151:** 71-73.
54. RUIZ, J., I. MARTÍN, L. F. CALLADO, J. J. MEANA, F. BARTUREN & J. A. GARCÍA-SEVILLA. 1993. Non-adrenoceptor [^{3}H]idazoxan binding sites (I_2-imidazoline sites) are increased in postmortem brain from patients with Alzheimer's disease. Neurosci. Lett. **160:** 109-112.
55. MEANA, J. J., F. BARTUREN & J. A. GARCÍA-SEVILLA. 1992. α_2-Adrenoceptors in the brain of suicide victims: Increased receptor density associated with major depression. Biol. Psychiatry **31:** 471-490.
56. GONZÁLEZ, A. M., J. PASCUAL, J. J. MEANA, F. BARTUREN, C. DEL ARCO, A. PAZOS & J. A. GARCÍA-SEVILLA. 1994. Autoradiographic demonstration of increased α_2-adrenoceptor agonist binding sites in the hippocampus and frontal cortex of depressed suicide victims. J. Neurochem. **63:** 256-265.
57. PILETZ, J. E., A. HALARIS, A. SARAN & M. R. MARLER. 1991. Desipramine lowers tritiated para-aminoclonidine binding in platelets of depressed patients. Arch. Gen. Psychiatry **48:** 813-820.
58. MEANA, J. J., F. BARTUREN, I. MARTÍN & J. A. GARCÍA-SEVILLA. 1993. Evidence of increased non-adrenoceptor [^{3}H]idazoxan binding sites in the frontal cortex of depressed suicide victims. Biol. Psychiatry **34:** 498-501.
59. GÖTHERT, M. & G. J. MOLDERINGS. 1991. Involvement of presynaptic imidazoline receptors in the α_2-adrenoceptor-independent inhibition of noradrenaline release by imidazoline derivatives. Naunyn-Schmiedeberg's Arch. Pharmacol. **343:** 271-282.
60. MOLDERINGS, G. J., F. HENTRICH & M. GÖTHERT. 1991. Pharmacological characterization of the imidazoline receptor which mediates inhibition of noradrenaline release in the rabbit pulmonary artery. Naunyn-Schmiedeberg's Arch. Pharmacol. **344:** 630-638.

Distribution of Nonadrenergic $[^3H]$Rilmenidine Binding in Rat Brain and Kidney[a]

P. R. KING, S. SUZUKI,[b] W. J. LOUIS, AND A. L. GUNDLACH[c]

University of Melbourne
Clinical Pharmacology and Therapeutics Unit
Department of Medicine
Austin and Heidelberg Repatriation Hospitals
Heidelberg, Victoria, 3084, Australia

Over the last 5 years, imidazoline receptor (I-receptor) binding sites in both brain and peripheral tissues have been extensively studied using various labeled imidazolines including $[^3H]$*para*-aminoclonidine (PAC), $[^3H]$clonidine (CLON), and $[^3H]$idazoxan (IDAZ).[1–7] Previous studies in our laboratory have shown that in the rat cerebral cortex the oxazoline $[^3H]$rilmenidine (RIL) labeled a population of α_{2A}-adrenoceptors (i.e., those with a high affinity for oxymetazoline and a low affinity for prazosin and ARC-239) and a population of I-receptor binding sites (i.e., catecholamine-insensitive and sensitive to imidazolines, oxazolines, and guanidinium compounds).[8,9] Differential sensitivity for PAC, CLON, IDAZ, and RIL was one of the criteria for the recent classification of I_1- and I_2-subtypes of I-receptor.[7,10,11] Further subclassification of the I_2-subtype has also been described: amiloride-sensitivity (nM affinity) denoting the I_{2A}-receptor and insensitivity to amiloride (nM-μM affinity) the I_{2B}-receptor.[10,11] However, the finding that oxazoline compounds label an I-receptor binding site that is inhibited with higher affinity by oxazolines than imidazolines suggests the existence of additional subtypes.

In recent studies described herein, we used autoradiography to examine the distribution and relative density of both α_{2A}-adrenoceptors and nonadrenergic I-receptors in the rat CNS labeled by $[^3H]$RIL and to compare the distribution of binding sites labeled by $[^3H]$RIL with that of the α_2-agonist and putative I_1-receptor ligand $[^3H]$PAC and the α_2-antagonist and putative I_2-receptor ligand $[^3H]$IDAZ in sections of rat kidney.

[a] This work was supported by the National Health and Medical Research Council of Australia and Servier Laboratories, Paris, France.

[b] Present address: Department of Pharmacology, Aichi Medical University, Nagakute-Cho, Aichi Prefecture, Japan.

[c] Address correspondence to: Dr. A. L. Gundlach, Clinical Pharmacology & Therapeutics Unit, Department of Medicine, Austin Hospital, Heidelberg, Victoria, 3084, Australia.

MATERIALS AND METHODS

Tissue Preparation

Male Wistar-Kyoto (WKY) rats (250 g) were stunned and decapitated. Brains, spinal cords, and/or kidneys were rapidly excised, dissected free of surrounding fat and connective tissue (kidney), and then frozen over liquid nitrogen and stored at −70°C. Tissues were mounted onto cryostat chucks, and 10 μm horizontal, coronal, or sagittal (brain) or longitudinal sections (kidney) were cut at −15°C on a cryostat (Reichert-Jung Cryocut 1800, Leica Instruments, Nusslock, Germany) and thaw-mounted onto cold gelatin/chrome alum-coated slides. Between 2 and 4 sections were mounted on each slide, air dried at room temperature, and stored with silica gel as desiccant at −70°C for up to 2 weeks before use.

[^{3}H]Rilmenidine Autoradiography

Frozen slide-mounted sections were equilibrated to room temperature and then preincubated for 20 minutes in 50 mM Tris-HCl, pH 7.6, to help remove endogenous inhibitors. Slides were then dried at room temperature. Sections were covered with 50 mM Tris-HCl, pH 7.6 (+ 10 μM pargyline, 25 μM EDTA), containing 40 nM [^{3}H]RIL (38 Ci/mmol, Servier Laboratories, Paris, France; 200 μl/section) in either the absence or the presence of 1 μM adrenaline and incubated for 60 minutes at room temperature to assess total binding and I-receptor binding, respectively. Nonspecific binding was determined by the inclusion of 10 μM BAY a6781 in the incubation buffer. To assess the pharmacological sensitivity of [^{3}H]RIL binding to kidney sections some incubations contained fixed concentrations of putative I-receptor and α_2-adrenoceptor compounds (see Results). Incubations were terminated by washing sections in three changes of buffer for a total of 4 minutes. Sections were then dried by a stream of cool air and desiccated overnight.

[^{3}H]Para-Aminoclonidine and [^{3}H]Idazoxan Autoradiography

To label α_2-adrenoceptors and/or I_1-receptors, kidney sections were preincubated for 30 minutes at room temperature in Mg^{2+}-free Krebs buffer followed by 10 minutes in 170 mM Tris-HCl, pH 7.6 (+ 10 mM $MgCl_2$), to remove endogenous catecholamines and then dried. Sections were then incubated with 4-5 nM [^{3}H]PAC (56 Ci/mmol, Du Pont-NEN, Boston, Massachusetts; concentration ~K_d [^{3}H]PAC for binding in kidney[5]) in the absence or the presence of adrenaline (1 μM) or methoxyidazoxan (10 μM) to block α_2-adrenoceptors, IAA (100 μM) to block I_1-receptors, and amiloride (100 μM) to block I_{2A}-receptors. Nonspecific binding was determined in the presence of 10 μM phentolamine. Unbound ligand was removed by washing for 2 × 5 minutes in ice-cold buffer. To label α_2-adrenoceptors and/or I_2-receptors, sections were preincubated for 15 minutes in Tris-HCl, pH 7.6 (+ 10 mM $MgCl_2$, 1 mM EDTA, 10 μM pargyline), and dried at room temperature. Sections were incubated with 4-5 nM [^{3}H]IDAZ (42 Ci/mmol, Amersham International plc., Amersham, BK, UK; concentration ~K_d in kidney[12]) in the absence or the presence of displacing drugs (see above) for 30 minutes at room temperature in 50 mM Tris-HCl, pH 7.6 (+ 10 mM $MgCl_2$), followed by a 2-minute wash in ice-cold buffer. Nonspecific binding was determined in the presence of 100 μM cirazoline. All sections were air-dried and desiccated overnight.

Autoradiograms of the [^{3}H]RIL-labeled slide-mounted sections were prepared by apposition of labeled sections to tritium-sensitive film (Hyperfilm-^{3}H, Amersham International plc., Amersham, BK, UK) in standard X-ray cassettes at 4°C for 7-10 weeks for brain, 15-17 weeks for spinal cord, and 20-30 days for kidney. [^{3}H]PAC- and [^{3}H]IDAZ-labeled kidney sections were apposed to film for 16-17 weeks. Tritium standards (Amersham) were co-exposed with each film. Films were developed and autoradiograms were analyzed using computer-aided densitometry on an MCID system (Imaging Research Inc., St. Catharines, Ontario, Canada).

RESULTS

Localization of [^{3}H]Rilmenidine-Labeled α_2-Adrenoceptor Binding Sites in Rat Central Nervous System

[^{3}H]Rilmenidine binding to both α_2-adrenoceptor and I-receptor sites was broadly distributed throughout the CNS (FIG. 1; TABLE 1). High levels of [^{3}H]RIL binding to α_{2A}-adrenoceptors were observed in the entorhinal, piriform, and perirhinal cortices and throughout all subregions of the anterior olfactory nucleus (TABLE 1). In the septal region, the bed nucleus stria terminalis and lateral septum (dorsal, intermediate, and ventral) also contained high densities of α_{2A}-adrenoceptors. Low densities of [^{3}H]RIL binding to α_{2A}-adrenoceptors were detected in the basal ganglia (except for the accumbens shell which had high levels), in the CA1-CA3 regions of the hippocampus, and in the dentate gyrus. High to very high densities of α_{2A}-adrenoceptor binding were found in the reuniens, paraventricular and centromedial thalamic nuclei, and throughout the ventral amygdala. In the hypothalamus, very high levels were present in the arcuate nucleus, whereas the medial mammillary nucleus and dorsomedial and posterior hypothalamic areas contained high levels of binding (TABLE 1).

Medium to high densities of specific α_{2A}-adrenoceptor binding were observed in discrete regions of the pons, such as the locus coeruleus and lateral parabrachial nucleus, and in the medulla oblongata including the nucleus of the solitary tract, dorsal motor nucleus of the vagus, and area postrema. Labeling of other medullary areas such as the facial and ambiguus nuclei and rostral ventrolateral medulla was low, as was binding to all spinal cord regions. The pineal gland contained a medium density of [^{3}H]RIL binding to α_{2A}-adrenoceptor binding sites (TABLE 1).

Localization of [^{3}H]Rilmenidine-Labeled I-Receptor Binding Sites in Rat Central Nervous System

I-receptor sites were broadly distributed throughout the CNS (FIG. 1; TABLE 1). Piriform, entorhinal, and perirhinal cortices contained medium densities, as did the basal ganglia, most septal areas, and limbic regions including the hippocampus and amygdala. In the thalamus, medium to high levels of [^{3}H]RIL I-receptor binding were present throughout, with the highest levels in paraventricular, centromedial, and reuniens nuclei. The hypothalamus contained high densities of I-receptor binding sites with very high levels in the arcuate nucleus, dorsomedial nucleus, and posterior hypothalamic area. In the midbrain, medium to high levels of I-receptor binding were found in the oculomotor nucleus, the central grey, and the reticular part of the substantia nigra. In the pons, high levels of I-receptor binding sites were present in the dorsal raphe and the parabrachial and motor trigeminal nuclei. The locus coeruleus had a very high density of I-receptor-associated [^{3}H]RIL binding. In the medulla, high to very high densities of I-receptor binding sites were present in the nucleus of the solitary tract, the facial, hypoglossal, and ambiguus nuclei, dorsal motor nucleus of the vagus, and area postrema. All other structures of the medulla and spinal cord exhibited low to medium densities of I-receptor binding sites. The pineal gland possessed the highest density of [^{3}H]RIL binding to I-receptor binding sites.

Localization of [^{3}H]Rilmenidine, [^{3}H]Idazoxan, and [^{3}H]Para-Aminoclonidine Binding in Rat Kidney

In sections of rat kidney, specific [^{3}H]RIL, [^{3}H]IDAZ, and [^{3}H]PAC binding was heterogeneously and differentially distributed (FIG. 2; TABLE 2). High densities of specific [^{3}H]RIL binding were present in the cortex and in the outer stripe of the outer medulla, with lower densities in remaining regions (FIG. 2; TABLE 2). In the cortex and outer stripe, specific [^{3}H]RIL binding was virtually insensitive to adrenergic drugs, whereas in the inner medulla these agents were more effective (% inhibition): 1 μM adrenaline and 10 μM methoxyidazoxan, 15-30%; 500 nM oxymetazoline (α_{2A}-adrenoceptor selective), 15%; but 500 nM ARC239 (α_{2B}-adrenoceptor selective), 0%, suggesting the presence of a population of α_{2A}-adrenoceptors (10-20 fmol/mg) in the inner medullary areas. Guanfacine (100 nM), an α_{2A}-adrenoceptor selective drug that also has relatively high affinity for [^{3}H]RIL-labeled nonadrenergic sites in brain,[9] displaced 30-40% of specific [^{3}H]RIL binding in both cortex/outer stripe and medulla (FIG. 2; TABLE 2).

In the rat kidney, [^{3}H]PAC and [^{3}H]IDAZ (at concentrations ~K_d for α_{2A}-adrenoceptor and I-receptor binding to kidney)[5,12] specifically labeled a high density of sites in the inner medulla (FIG. 2; TABLE 2). Specific binding of [^{3}H]PAC and [^{3}H]IDAZ was fully displaced by 1 μM adrenaline or 10 μM methoxyidazoxan. Thus, in these experiments, little or no specific binding to I-receptors was detected. Under identical conditions, I-receptor binding sites for [^{3}H]PAC have been reported in rat kidney based on their inhibition by a putative I_1-selective drug, imidazole-4-acetic acid (IAA),[5] but in the present study, binding of [^{3}H]PAC, [^{3}H]IDAZ, and [^{3}H]RIL was unaffected by 100 μM IAA. High concentrations (100 μM) of amiloride, a putative I_{2A}-R ligand, inhibited 65-85% of specific [^{3}H]IDAZ binding. Consistent with these results, however, is the finding that amiloride allosterically inhibited [^{3}H]IDAZ binding to α_2-adrenoceptors.[15]

TABLE 1. Relative Density of α_2-Adrenoceptor and I-Receptor Binding Sites Labeled by [^{3}H]RIL in Selected Regions of the Rat Central Nervous System[a]

Region		α_2-Adrenoceptor (fmol/mg, mean ± SD)	I-Receptor (fmol/mg, mean ± SD)	% I-Receptor
Cortex	Piriform	27 ± 9	22 ± 11	45
	Entorhinal	39 ± 22	20 ± 13	34
	Perirhinal	24 ± 9	20 ± 12	45
Olfactory	Anterior olfactory n.	27 ± 8	20 ± 9	43
Basal ganglia	Caudate putamen	7 ± 1	19 ± 8	73
	Accumbens n.	12 ± 6	23 ± 11	66
	Accumbens shell	26 ± 9	22 ± 7	46
Septal area	Lateral septum	28 ± 8	21 ± 9	43
	Bed n. stria terminalis	37 ± 18	29 ± 5	44
Hippocampus	Pyramidal cell layer	10 ± 5	36 ± 12	78
	Dentate gyrus (overall)	5 ± 1	23 ± 8	82
Amygdala	Overall	49 ± 17	25 ± 12	34
Thalamus	Paraventricular n.			
	Anterior	29 ± 8	34 ± 3	54
	Posterior	31 ± 13	27 ± 7	47
	Central medial n.	29 ± 13	29 ± 3	50
	Anterolateral thalamic area	23 ± 11	25 ± 1	52
	Reuniens n.	45 ± 17	29 ± 5	39
	Geniculate nuclei	16 ± 7	15 ± 5	48
Hypothalamus	Arcuate n.	42 ± 19	35 ± 6	45
	Medial mammillary n. medial	21 ± 19	27 ± 6	56
	Dorsomedial n.*	33 ± 20	43 ± 2	57
	Posterior hypothalamic area*	34 ± 2	35 ± 7	52
Midbrain	Central grey	16 ± 8	23 ± 11	59
	Interpeduncular n.*	15 ± 8	25 ± 13	63
	Oculomotor n.	5 ± 3	25 ± 14	83
	Substantia nigra, reticular	7 ± 5	23 ± 11	77
	Superior colliculus, overall	8 ± 4	18 ± 9	69
Pons	Dorsal raphe n.	13 ± 4	27 ± 12	68
	Locus coeruleus	25 ± 15	37 ± 11	60
	Lateral parabrachial n.	23 ± 10	27 ± 5	54
	Motor trigeminal n.	6 ± 2	31 ± 7	84
	Pontine n.	8 ± 2	17 ± 7	68
Medulla	Facial n.	4 ± 4	33 ± 6	89
	Hypoglossal n.	16 ± 2	36 ± 12	69
	N. solitary tract	32 ± 8	26 ± 7	45
	Dorsal motor n. vagus*	20 ± 14	35 ± 21	64
	Area postrema	24 ± 3	36 ± 12	60
	Ambiguus n.	15 ± 4	35 ± 14	70
	Rostral ventrolateral medulla	7 ± 4	19 ± 5	73
Cerebellum	Overall	5 ± 3	25 ± 7	83

TABLE 1. Relative Density of α_2-Adrenoceptor and I-Receptor Binding Sites Labeled by [^{3}H]RIL in Selected Regions of the Rat Central Nervous System[a] (*Continued*)

Region		α_2-Adrenoceptor (fmol/mg, mean ± SD)	I-Receptor (fmol/mg, mean ± SD)	% I-Receptor
Spinal Cord	Layer 1-3	10 ± 4	20 ± 6	67
	Layer 8	3 ± 4	22 ± 9	88
	Layer 9	5 ± 3	23 ± 10	82
	Intermediolateral cell column	11 ± 4	18 ± 8	62
Other	Pineal gland	22 ± 15	94 ± 21	81

[a] Coronal sections (10 μm) were labeled as described in Methods with 40 nM [^{3}H]rilmenidine in the absence and presence of 1 μM of adrenaline. Nonspecific binding was measured in the presence of 10 μM of BAY a6781. Tissues were apposed to Hyperfilm-^{3}H for 8 weeks, and images were analyzed using computer-assisted densitometry. Optical densities were converted to fmol/mg tissue equivalent by reference to tritium impregnated polymer microscale standards. Readings for each brain structure were derived from multiple autoradiogram images from 3-5 animals, except those marked with an *asterisk,* which were from two animals. Brain regions were identified by reference to the atlas of Paxinos and Watson.[13] n = nucleus.

DISCUSSION

Central Nervous System

Semiquantitative mapping of α_2-adrenoceptor and nonadrenergic (I-receptor) binding sites in rat brain and spinal cord labeled by [^{3}H]RIL has established the widespread distribution of both receptor types. This and other autoradiographic studies (e.g., Mallard *et al.*[16]) allow a comparison between the distribution of I-receptor binding sites labeled by [^{3}H]RIL and [^{3}H]IDAZ, the putative I_2-receptor ligand, in the rat CNS. In some brain regions there are close similarities in the distribution and relative density of I-receptor sites labeled by these two ligands. Most notably, both ligands bind to high to very high relative densities of sites in the area postrema, arcuate nucleus, interpeduncular nucleus, and pineal gland[16] (FIG 1). Apparent differences between nonadrenergic sites labeled by [^{3}H]IDAZ and [^{3}H]RIL occur in other brain regions. In the hippocampus, high levels of total [^{3}H]IDAZ binding are present only over the stratum lacunosum moleculare of CA1, and 40-60% of this binding is reported to be I-receptor binding.[16,17] In contrast, [^{3}H]RIL appears to label the pyramidal cell layer of CA1-CA3 of the hippocampus. The pattern of binding obtained with [^{3}H]IDAZ and [^{3}H]RIL is also different in the cranial motor nuclei including the motor trigeminal, facial, and hypoglossal nuclei, dorsal motor nucleus of the vagus, nucleus ambiguus, and over spinal cord motor neurons, where [^{3}H]RIL binding appears punctate and in relatively high density compared to that in adjacent areas. In contrast, [^{3}H]IDAZ binding was more evenly distributed and present in a lower relative density in these areas.[16] In the cerebellum, [^{3}H]IDAZ preferentially labels

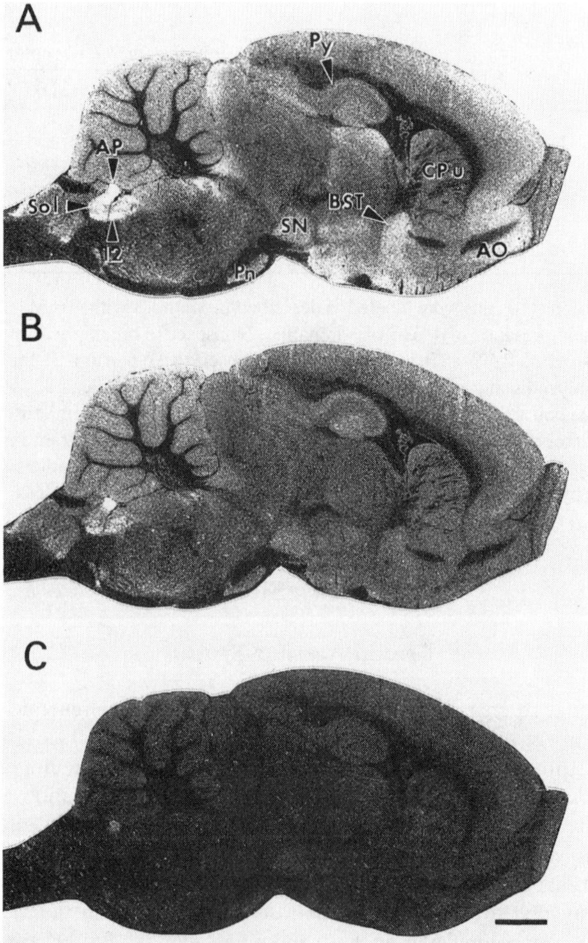

FIGURE 1. Autoradiograms of the anatomical distribution of (**A**) total, (**B**) adrenaline-insensitive I-receptor, and (**C**) nonspecific [^{3}H]rilmenidine (RIL) binding to 10 μm sagittal sections of rat brain. Brain sections were incubated with 40 nM [^{3}H]RIL in the absence or presence of 1 μM of adrenaline or 10 μM of BAYa6781 for 60 minutes at room temperature. Adrenaline blocked α_{2A}-adrenergic binding of [^{3}H]RIL, allowing visualization of I-receptor binding sites. Sections were exposed to [^{3}H]-sensitive Hyperfilm for 8–10 weeks at 4°C. Autoradiograms were directly printed from the film. White to black represents highest to lowest densities of binding. Structures were identified by reference to the atlas of Paxinos and Watson.[13] Bar = 2.5 mm.

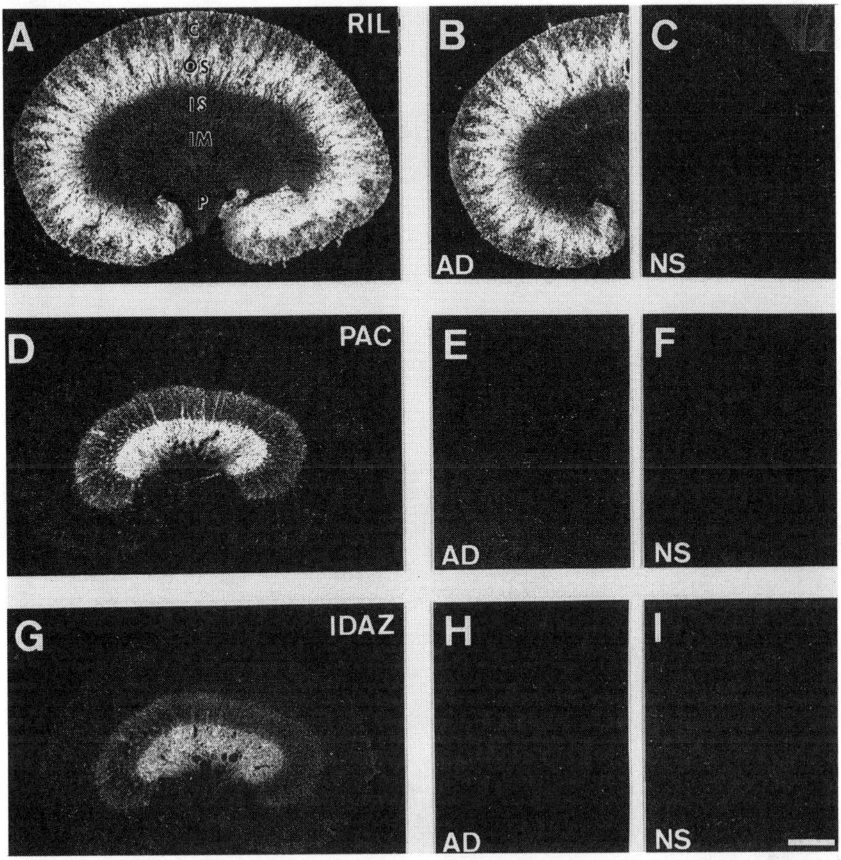

FIGURE 2. Autoradiograms of the distribution of [^{3}H]rilmenidine (RIL) **(top row)**, [^{3}H]*para*-aminoclonidine (PAC) **(middle row)**, and [^{3}H]idazoxan (IDAZ) **(bottom row)** to longitudinal sections of rat kidney. Adrenaline (1 μM) was added to eliminate α_2-adrenoceptor binding **(B,E,H)** of each radioligand. Nonspecific binding was estimated by inclusion of 10 μM of BAYa6781 (**C**), 10 μM phentolamine (**F**), and 100 μM cirazoline (**I**) for [^{3}H]RIL, [^{3}H]PAC, and [^{3}H]IDAZ, respectively. Kidney structures were identified by reference to Kriz and Bankir.[14] Bar = 2.5 mm.

the granule cell layer over the molecular layer (a ratio of 2 : 1),[16] while [^{3}H]RIL does not delineate the two major cerebellar layers.

The coexistence of [^{3}H]IDAZ-labeled I_2-receptors and the monoamine oxidase enzyme MAO_B labeled with [^{3}H]RO 19-6327 (lazabemide) was recently reported in rat and human brain.[18–21] In addition, the density of I_2-receptors in rat brain can be decreased by chronic treatment of rats with selective MAO_B inhibitors.[19] However, the I_2-receptor appears not to be the catalytic site of MAO_B, as selective ligands for the two sites do not cross-react and the density of binding sites for the MAO_B ligand

TABLE 2. Regional Distribution and Pharmacology of [^{3}H]Rilmenidine (RIL), [^{3}H]*para*-Aminoclonidine (PAC), and [^{3}H]Idazoxan (IDAZ) Binding in Sections of Rat Kidney[a]

		Region			
Radioligand	Conditions/Drug Added	Cortex	Outer Stripe	Inner Stripe	Inner Medulla
[^{3}H]Rilmenidine	Total binding (fmol/mg)	126 ± 8.2	262 ± 28	58 ± 2.7	57 ± 3.3
	Residual specific binding (%) with:				
	Adrenaline 1 μM	89 ± 3	93 ± 2	81 ± 4	73 ± 4
	Methoxyidazoxan 10 μM	90 ± 4	94 ± 3	85 ± 4	75 ± 2
	ARC-239 500 nM	113 ± 5	108 ± 3	122 ± 2	112 ± 4
	Oxymetazoline 500 nM	102 ± 3	103 ± 2	97 ± 4	85 ± 3
	IAA 100 μM	102 ± 9	102 ± 6	105 ± 17	107 ± 19
	Guanfacine 100 nM	64 ± 2	59 ± 2	65 ± 4	59 ± 2
[^{3}H]*Para*-Aminoclonidine	Total binding (fmol/mg)	8.9 ± 1.1	6.7 ± 0.4	23 ± 1.5	95 ± 7.6
	Residual specific binding (%) with:				
	Adrenaline 1 μM	10 ± 3	34 ± 5	8 ± 1	2 ± 1
	Methoxyidazoxan 10 μM	6 ± 2	11 ± 5	4 ± 1	1 ± 0
	IAA 100 μM	93 ± 17	86 ± 12	100 ± 6	96 ± 6
	Amiloride 100 μM	45 ± 4	58 ± 8	74 ± 3	84 ± 7
[^{3}H]Idazoxan	Total binding (fmol/mg)	13 ± 1.0	11 ± 1.0	24 ± 3.0	77 ± 9.0
	Residual specific binding (%) with:				
	Adrenaline 1 μM	28 ± 6	25 ± 3	8 ± 1	3 ± 1
	Methoxyidazoxan 10 μM	25 ± 2	23 ± 2	6 ± 0	3 ± 1
	IAA 100 μM	100 ± 7	99 ± 13	97 ± 12	99 ± 12
	Amiloride 100 μM	21 ± 2	25 ± 4	32 ± 4	37 ± 5

[a] Rat kidney sections were incubated with 40 nM [^{3}H]RIL, 4-5 nM [^{3}H]PAC, and [^{3}H]IDAZ in the presence or absence of drugs, washed, dried, and apposed to Hyperfilm-^{3}H for 20-30 days for [^{3}H]RIL and 16-17 weeks for [^{3}H]PAC and [^{3}H]IDAZ, respectively. Film images were analyzed by computer-aided densitometry. Resulting mean optical density values were transformed to fmol/mg tissue equivalent by reference to [^{3}H]-microscale standards (Amersham). Specific binding (fmol/mg) was derived by subtraction of mean nonspecific from total binding. Values are mean ± SEM of multiple readings from duplicate, serial sections from four animals. IAA = imidazole-4-acetic acid.

is 30-40 times higher than that of I_2-receptors detected with [^{3}H]IDAZ. Nonetheless, this suggests that the I_2-receptor may be associated with the outer mitochondrial membrane in neurons and other cells such as glia that express the MAO_B enzyme.

The exact relation between oxazoline and imidazoline binding site distributions and putative functional effects of drugs at I-receptors is not clear at present. However, medullary and supramedullary regions involved with blood pressure regulation such as the nucleus of the solitary tract, area postrema, locus coeruleus, and some cortical regions have been shown, using [^{3}H]RIL, to contain both α_{2A}-adrenoceptor and I-receptor binding sites. These findings are consistent with the effects of α_2-adrenoceptor/I-receptor drugs on blood pressure following microinjection into the ventrolateral medulla region of the rat[2,22] and rabbit.[23]

The high levels of [^{3}H]RIL binding seen in several hypothalamic areas in the present study is consistent with studies of [^{3}H]IDAZ binding in brain of normotensive and genetically hypertensive rats which revealed that the hypothalamus possessed the highest density of I_2-receptor binding sites.[24] These sites may have functional significance, as idazoxan treatment modulates the level of I_2-receptor binding sites seen in normotensive and hypertensive strains, suggesting involvement of I_2-receptors in hypothalamic and other brain regions in blood pressure control. Actions of idazoxan at I-receptors in hypothalamus have also been implicated in the control of feeding behavior,[25] a finding consistent with the presence of I-receptors labeled by both [^{3}H]IDAZ[16] and [^{3}H]RIL in the arcuate and dorsomedial hypothalamic nuclei and other parts of the hypothalamus involved in feeding mechanisms.[26,27]

Kidney

Comparative autoradiographic studies have demonstrated the differential localization of [^{3}H]RIL, [^{3}H]PAC, and [^{3}H]IDAZ binding sites in the rat kidney. Under the experimental conditions employed, the nanomolar concentrations of ligand used (approximately equivalent to the K_d for each drug to label I_1- and I_2-receptor binding sites)[5,12] appear to label only α_2-adrenoceptor sites (i.e., binding is fully inhibited by adrenaline and methoxyidazoxan). These results are consistent with the reported moderate affinity (K_d –120 nM) of [^{3}H]PAC for I_1-receptors in rat kidney,[12] but are in contrast to the reported enrichment of I_1-receptor binding sites labeled by [^{3}H]PAC in the inner stripe of the renal outer medulla.[5] [^{3}H]Idazoxan has been used to label α_2-adrenoceptors and putative I_2-receptor binding sites in rat kidney homogenates.[7,28] However, in the present study we saw no clear evidence of any high affinity I_2-receptor binding in kidney sections. This may be due to the differential affinities (K_ds) of the various α_2-adrenoceptor and I_2-receptor populations present and the resultant differential occupancy, and although both α_2-adrenoceptors and I_2-receptors reportedly have nanomolar affinity for [^{3}H]IDAZ,[28,29] it is possible that the use of higher concentrations of [^{3}H]IDAZ (and [^{3}H]PAC) and reduced washing time of sections may reveal other low affinity I-receptor sites.

The pattern of labeling in kidney with [^{3}H]IDAZ was identical to that seen with [^{3}H]PAC and dissimilar to that described using [^{3}H]rauwolscine.[30,31] Autoradiographic studies with [^{3}H]rauwolscine show significantly greater binding in rat kidney to cortical tubules, particularly the proximal tubules[30–32] and lower levels of binding in glomeruli, medullary tubules, and blood vessels.[30,32–34] The unique distribution of [^{3}H]rauwolscine binding compared with [^{3}H]PAC and [^{3}H]IDAZ binding described here may, in part, be due to the α_{2B}-adrenoceptor-selective nature of rauwolscine (and yohimbine) in rat kidney[35–37] and the α_{2A}-adrenoceptor selective characteristics of both [^{3}H]PAC and [^{3}H]IDAZ.[17,29] Indeed, the pharmacology of [^{3}H]rauwolscine binding to rat proximal tubule membranes displays an α_{2B}-adrenoceptor profile.[38] Hence, both distribution patterns represent a partial view of overall α_2-adrenoceptor localization in the rat kidney. These results are consistent with the enrichment of α_{2A}-adrenoceptor mRNA associated with collecting ducts in the inner stripe of the outer and inner medulla and the high expression of α_{2B}-adrenoceptor mRNA by proximal tubules in the outer stripe of the outer medulla,[39,40] a region enriched in [^{3}H]rauwolscine binding.[30]

[^{3}H]Rilmenidine was previously reported to label a high affinity α_{2A}-adrenoceptor and a lower affinity $I_{2\text{-like}}$ receptor binding site in rat brain homogenates and frozen sections.[9,41] In rat kidney, which contains both α_{2A}- and α_{2B}-adrenoceptors, [^{3}H]RIL detected a small population of α_{2A}-adrenoceptors in the inner medulla. In contrast, it labeled a very high density of nonadrenergic [^{3}H]RIL binding sites over proximal straight tubules in the outer stripe of the outer medulla and in the cortex (medullary rays) which were inhibited by another oxazoline and guanfacine, but not appreciably by cirazoline or idazoxan (data not shown). Rilmenidine has been described as being a high affinity I_1-receptor ligand, similar to clonidine and moxonidine,[7,42] but current results indicate otherwise. Instead, these findings suggest that a unique, high affinity, nonadrenergic binding site population may exist for oxazolines in various mammalian tissues including brain and kidney[41,43] that have tissue-specific differences to I-receptor binding sites labeled by [^{3}H]PAC and [^{3}H]IDAZ. Oxazoline radioligands will thus be useful in further investigating the differential characteristics of these renal sites and the nature and functional importance of these putative drug receptors in all types of tissues.

ACKNOWLEDGMENTS

We wish to thank Marina Skiba for assistance and Kate Ashley and Allison Quinn for the preparation of this manuscript.

REFERENCES

1. Ernsberger, P., M. P. Meeley, J. J. Mann & D. J. Reis. 1987. Clonidine binds to imidazole binding sites as well as α_2-adrenoceptors in the ventrolateral medulla. Eur. J. Pharmacol. **134:** 1–13.
2. Ernsberger, P., R. Giuliano, R. N. Willette & D. J. Reis. 1990. Role of imidazole receptors in the vasodepressor response to clonidine analogs in the rostral ventrolateral medulla. J. Pharmacol. Exp. Ther. **253:** 408–418.
3. Wikberg, J. E. S. 1989. High affinity binding of idazoxan to a non-catecholaminergic binding site in the central nervous system: Description of a putative idazoxan-receptor. Pharmacol. Toxicol. **64:** 152–155.
4. Wikberg, J. E. S. & S. Uhlén. 1990. Further characterization of the guinea pig cerebral cortex idazoxan receptor: Solubilization, distinction from the imidazole site, and demon-

stration of cirazoline as an idazoxan receptor-selective drug. J. Neurochem. **55:** 192-203.

5. ERNSBERGER, P., G. FEINLAND, M. P. MEELEY & D. J. REIS. 1990. Characterization and visualization of clonidine-sensitive imidazole sites in rat kidney which recognize clonidine-displacing substance. Am. J. Hypertens. **3:** 90-97.
6. ZONNENSCHEIN, R., S. DIAMANT & D. ATLAS. 1990. Imidazoline receptors in rat liver cells: A novel receptor or a subtype of α_2-adrenoceptors? Eur. J. Pharmacol. **190:** 203-215.
7. MICHEL, M. & P. ERNSBERGER. 1992. Keeping an eye on the I-site: Imidazoline-preferring receptors. Trends Pharmacol. Sci. **13:** 369-370.
8. KING, P. R., A. L. GUNDLACH, B. JARROTT & W. J. LOUIS. 1992. Nature of [^{3}H]rilmenidine binding to membranes of rat cerebral cortex. Am. J. Hypertens. **5:** 64S-68S.
9. KING, P. R., A. L. GUNDLACH, B. JARROTT & W. J. LOUIS. 1992. α_2-Adrenoceptor and catecholamine-insensitive binding sites for [^{3}H]rilmenidine in membranes from rat cerebral cortex. Eur. J. Pharmacol. **218:** 101-108.
10. MICHEL, M. C. & P. A. INSEL. 1989. Are there multiple imidazoline binding sites? Trends Pharmacol. Sci. **10:**342-344.
11. REIS, D. J., S. REGUNATHAN, H. WANG, D. L. FEINSTEIN & M. P. MEELEY. 1992. Imidazoline receptors in the nervous system. Fund. Clin. Pharmacol. **6** (Suppl. 1): 23s-29s.
12. MACKINNON, A. C., M. STEWART, H. J. OLVERMAN, M. SPEDDING & C. M. BROWN. 1993. [^{3}H]p-aminoclonidine and [^{3}H]idazoxan label different populations of imidazoline sites on rat kidney. Eur. J. Pharmacol. **232:** 79-87.
13. PAXINOS, G. & C. WATSON. 1986. The Rat Brain in Stereotaxic Coordinates, 2nd Ed. Academic Press. Sydney.
14. KRIZ, W. & L. BANKIR. 1988. A standard nomenclature for structures of the kidney. Kidney Int. **33:** 1-7.
15. HOWARD, M. J., R. J. HUGHES, H. J. MOTULSKY, M. D. MULLEN & P. A. INSEL. 1987. Interactions of amiloride with α- and β-adrenergic receptors: Amiloride reveals an allosteric site on α_2-adrenergic receptors. Mol. Pharmacol. **32:** 53-58.
16. MALLARD, N. J., A. L. HUDSON & D. J. NUTT. 1992. Characterization and autoradiographical localization of non-adrenoceptor idazoxan binding sites in the rat brain. Br. J. Pharmacol. **106:** 1019-1027.
17. RAYMON, H. K., T. D. SMITH & F. M. LESLIE. 1992. Further pharmacological characterization of [^{3}H]idazoxan binding sites in rat brain: Evidence for predominant labeling of α_2-adrenergic receptors. Brain Res. **582:** 261-267.
18. MIRALLES, A., G. OLMOS, M. SASTRE, F. BARTUREN, I. MARTIN & J. A. GARCÍA-SEVILLA. 1993. Discrimination and pharmacological characterization of I_2-imidazoline sites with [^{3}H]idazoxan and *alpha*-2 adrenoceptors with [^{3}H]RX821002 (2-methoxy idazoxan) in the human and rat brains. J. Pharmacol. Exp. Ther. **264:** 1187-1197.
19. OLMOS, G., A. M. GABILONDO, A. MIRALLES, P. V. ESCRIBÁ & J. A. GARCÍA-SEVILLA. 1993. Chronic treatment with the monoamine oxidase inhibitors clorgyline and pargyline down-regulates non-adrenoceptor [^{3}H]-idazoxan binding sites in the rat brain. Br. J. Pharmacol. **108:** 597-603.
20. SASTRE, M. & J. A. GARCÍA-SEVILLA. 1993. Opposite age-dependent changes of α_{2A}-adrenoceptors and nonadrenoceptor [^{3}H]idazoxan binding sites (I_2-imidazoline sites) in the human brain: Strong correlation of I_2 with monoamine oxidase-B sites. J. Neurochem. **61:** 881-889.
21. SAURA, J., R. KETTLER, M. DA PRADA & J. G. RICHARDS. 1992. Quantitative enzyme radioautography with ^{3}H-Ro 41-1049 and ^{3}H-Ro 19-6327 *in vitro*: Localization and abundance of MAO-A and MAO-B in rat CNS, peripheral organs, and human brain. J. Neurosci. **12:** 1977-1999.
22. GOMEZ, R. E., P. ERNSBERGER, G. FEINLAND & D. J. REIS. 1991. Rilmenidine lowers arterial pressure via imidazole receptors in brainstem C1 area. Eur. J. Pharmacol. **195:** 181-191.

23. FELDMAN, J., E. TIBIRICA, G. BRICCA, M. DONTENWILL, A. BELCOURT & P. BOUSQUET. 1990. Evidence for the involvement of imidazoline receptors in the central hypotensive effect of rilmenidine in the rabbit. Br. J. Pharmacol. **100:** 600-604.
24. OLMOS, G., A. MIRALLES, F. BARTUREN & J. A. GARCÍA-SEVILLA. 1991. Repeated idazoxan increases brain imidazoline receptors in normotensive (WKY) but not in hypertensive (SHR) rats. J. Neurochem. **57:** 1811-1813.
25. JACKSON, H. C., I. J. GRIFFIN & D. J. NUTT. 1991. The effects of idazoxan and other α_2-adrenoceptor antagonists on food and water intake in the rat. Br. J. Pharmacol. **104:** 258-262.
26. LEIBOWITZ, S. F. 1980. Neurochemical systems of the hypothalamus. Control of feeding and drinking behavior and water-electrolyte excretion. *In* Handbook of the Hypothalamus, Vol 3. P. J. Morgane & J. Panksepp, Eds.: 299-437. Marcel Dekker. New York.
27. LUITEN, P. G., G. J. TER HORST & A. B. STEFFENS. 1987. The hypothalamus, intrinsic connections and outflow pathways to the endocrine system in relation to the control of feeding and metabolism. Prog. Neurobiol. **28:** 1-54.
28. MICHEL, M. C., O.-E. BRODDE, B. SCHNEPEL, J. BEHRENDT, R. TSCHADA, H. J. MOTULSKY & P. A. INSEL. 1988. [^{3}H]Idazoxan and some other α_2-adrenergic drugs also bind with high affinity to a nonadrenergic site. Mol. Pharmacol. **35:** 324-330.
29. BROWN, C. M., A. C. MACKINNON, J. C. MCGRATH, M. SPEDDING & A. T. KILPATRICK. 1990. α_2-Adrenoceptor subtypes and imidazoline-like binding sites in the rat brain. Br. J. Pharmacol. **99:** 803-809.
30. STEPHENSON, J. A. & R. J. SUMMERS. 1985. Light microscopic autoradiography of the distribution of [^{3}H]rauwolscine binding to α_2-adrenoceptors in rat kidney. Eur. J. Pharmacol. **116:** 271-278.
31. CALIANOS, T., II, & K. H. MUNTZ. 1990. Autoradiographic quantification of adrenergic receptors in rat kidney. Kidney Int. **38:** 39-46.
32. SUMMERS, R. J., J. A. STEPHENSON, S. LIPE & C. B. NEYLON. 1985. α_2-Adrenoceptors in dog kidney: Autoradiographic localization and putative functions. Clin. Sci. **68** (Suppl. 10): 105s-109s.
33. INSEL, P. A., M. D. SNAVELY, D. P. HEALY, P. A. MUNZEL, C. L. POTENZA & E. P. NORD. 1985. Radioligand binding and functional assays demonstrate postsynaptic alpha$_2$-receptors on proximal tubules of rat and rabbit kidney. J. Cardiovasc. Pharmacol. **7** (Suppl. 8): S9-S17.
34. MUNTZ, K. H., L. MEYER, S. GADOL & T. A. CALIANOS. 1986. Alpha-2 adrenergic receptor localization in the rat heart and kidney using autoradiography and tritiated rauwolscine. J. Pharmacol. Exp. Ther. **236:** 542-547.
35. UHLÉN, S. & J. E. S. WIKBERG. 1991. Rat spinal cord α_2-adrenoceptors are of the α_{2A}-subtype: Comparison with α_{2A}- and α_{2B}-adrenoceptors in rat spleen, cerebral cortex and kidney using ^{3}H-RX821002 ligand binding. Pharmacol. Toxicol. **69:** 341-350.
36. UHLÉN, S. & J. E. S. WIKBERG. 1991. Delineation of rat kidney α_{2A}- and α_{2B}-adrenoceptors with [^{3}H]RX821002 radioligand binding: Computer modelling reveals that guanfacine is an α_{2A}-selective compound. Eur. J. Pharmacol. **202:** 235-243.
37. XIA, Y., S. UHLÉN, V. CHHAJLANI, E. J. LIEN & J. E. S. WIKBERG. 1993. Further evidence for the existence of two forms of α_{2B}-adrenoceptors in rat. Pharmacol. Toxicol. **72:** 40-49.
38. STANKO, C. K., M. I. VANDEL, R. BOSE & D. D. SMYTH. 1990. Characterization of α_2-adrenoceptors in the rat: Proximal tubule, renal membrane and whole kidney studies. Eur. J. Pharmacol. **175:** 13-20.
39. NICHOLAS, A. P., V. PIERIBONE & T. HÖKFELT. 1993. Distributions of mRNAs for alpha-2 adrenergic receptor subtypes in rat brain: An *in situ* hybridization study. J. Comp. Neurol. **328:** 575-594.
40. MEISTER, B., Å. DAGERLIND, A. P. NICHOLAS & T. HÖKFELT. 1994. Patterns of messenger RNA expression for adrenergic receptor subtypes in the rat kidney. J. Pharmacol. Exp. Ther. **268:** 1605-1611.

41. KING, P. R., A. L. GUNDLACH & W. J. LOUIS. 1995. Quantitative autoradiographic localization in rat brain of α_2-adrenergic and non-adrenergic I-receptor binding sites labelled by [^{3}H]rilmenidine. Brain Res. In press.
42. ERNSBERGER, P., T. H. DAMON, L. M. GRAFF, S. G. SCHAFER & M. O. CHRISTEN. 1993. Moxonidine, a centrally acting antihypertensive agent, is a selective ligand for I_1-imidazoline sites. J. Pharmacol. Exp. Ther. **264:** 172–182.
43. KING, P. R., S. SUZUKI, T. NERO, A. L. GUNDLACH & W. J. LOUIS. 1993. Molecular studies of alpha-2-adrenoceptor and catecholamine-insensitive imidazoline-receptor binding sites using [^{3}H]-rilmenidine in rat brain and kidney. J. Hypertens. **11** (Suppl. 5): S160–S161.

Distribution of Imidazoline Receptor Binding Protein in the Central Nervous System

DAVID A. RUGGIERO,
SOUNDARARAJAN REGUNATHAN, HONG WANG,
TERESA A. MILNER, AND DONALD J. REIS

Division of Neurobiology
Department of Neurology and Neuroscience
Cornell University Medical College
New York, New York 10021

Knowledge of the distribution of imidazoline (I-) receptors within the central nervous system (CNS) is incomplete. The difficulty to define their regional localization by receptor autoradiography, the only available method, is complicated by the fact that all ligands that bind to I-receptors also bind to α_2-adrenergic receptors.[1,2] Thus, identification of I-receptors has been indirect and deduced either by: (1) cross-comparison, on adjacent autoradiographs, of the distribution of ligands binding to I- and α_2-adrenergic receptors (e.g., [^{3}H]clonidine or [^{3}H]idazoxan) with agents selective for α_2-adrenergic receptors (e.g., [^{3}H]rauwolscine)[3]; or (2) mapping the distribution of [^{3}H]idazoxan on sections coincubated with saturating concentrations of selective α_2-adrenergic receptor agonists or antagonists to mask the α_2-adrenergic component.[4-6] Such studies have suggested that I-receptors may be selectively expressed in several regions of the brain[4-6] and in some nonneuronal cells.[6,7]

We recently isolated a protein from adrenal chromaffin cells that meets pharmacological criteria as an I-receptor binding protein (IRBP)[8] and appears to represent part of or the entire I-receptor. The protein, which has a molecular mass of 70 kD, is biochemically and pharmacologically distinct from α_2-adrenergic receptors and has been used to generate polyclonal antibodies. These IRBP antibodies can bind the IRBP, immunoprecipitate the receptor, and block binding of ligands to I- but not α_2-adrenergic receptors.[9] Moreover, the antibody has been used to localize, by immunocytochemistry, I-receptors in adrenal chromaffin cells,[9] glomus cells of the carotid body,[10] and vascular smooth muscle and endothelium.[11] Because of its selectivity for I-receptors and its lack of cross-reactivity with the α_2-adrenergic receptor, we have used this antibody to map the distribution of I-receptors in the rat CNS by light and electron microscopic immunocytochemistry.[12]

METHODS

We partially purified an idazoxan binding protein from detergent extracts of bovine chromaffin cell membranes,[8] cells that do not express α_2-adrenergic receptors.[13]

Imidazoline binding proteins were solubilized from membranes and isolated by a ligand affinity resin Idazoxan-PharmaLink agarose (IDA-agarose) by single step affinity chromatography. CHAPS-solubilized membrane proteins were partially purified as assessed by SDS-PAGE analysis and [^{3}H]idazoxan binding assays. The purified proteins have a major protein component of 70 kD and demonstrated a drug-selective profile similar to that of the receptors in membrane preparations, that is, sensitivity to imidazolines (cirazoline > clonidine) and insensitivity to catecholamines (rauwolscine ≈ epinephrine). The purified imidazoline binding proteins from IDA-Agarose and PAC-ReactiGel have similar affinity for [^{3}H]idazoxan (K_D = 3.7 nM and 4.9 nm, respectively). Polyclonal antibodies in rabbits were raised against purified, native imidazoline-binding proteins. These antibodies, labeled a ≈70 kD protein in Western blots of bovine adrenal chromaffin cell membranes, produced a maximal 40% specific immunoinhibition of [^{3}H]idazoxan binding to the chromaffin cell membrane I-receptor and specifically immunoextracted 75% of all imidazoline-binding activity associated with solubilized chromaffin cell membrane proteins. Immunocytochemical studies using these antibodies revealed specific labeling of cultured chromaffin cells.[9]

Light and Electron Microscopic Immunocytochemistry. Sprague-Dawley rats (300–350 g) were deeply anesthetized with pentobarbital intraperitoneally and perfused transcardially with heparinized normal saline solution followed by 4% paraformaldehyde (LM) or 3.75% acrolein and 2% paraformaldehyde (EM) in phosphate-buffered saline solution (PBS, pH 7.4). Brain and cervicothoracic spinal cord segments were blocked, postfixed, cryoprotected, and sectioned on a sledge microtome or vibratome for light and electron microscopy, respectively. After blocking nonspecific binding sites, tissues were incubated in the primary antiserum (diluted 1 : 2000 or 1 : 10,000) at 4°C for 48 to 72 hours and processed immunocytochemically with the avidin-biotin-peroxidase complex method (Vectastain Elite kit, Vector Labs) as described in our previous publications.[9,12,14] Alternate tissues were incubated in preimmune rabbit serum as the control.

RESULTS

Regional Distribution of IRBP-Immunoreactivity (I-Receptors) in the CNS

In the CNS, the receptor-binding protein (IRBP) was detected in neurons and glia by light and electron microscopy. IRBP-like immunoreactivity (IRBP-LI) was seen in processes and neuropil, surrounding predominantly unlabeled profiles of neuronal perikarya and intraparenchymal blood vessels. Ependyma were unstained. Immunoreactive elements in the CNS were often intermingled, selectively localized to related nuclei, and characterized by marked regional heterogeneity. Dorsal root ganglion cells were also labeled, which was consistent with evidence of processes in the spinal dorsal horn. These findings provided suggestive evidence that the receptor might be localized to systems of neural crest origin.[15] Other tissues derived from neural crest, including a subpopulation of chromaffin cells in the adrenal medulla and myenteric plexus, were also labeled. Control tissues incubated in preimmune serum were devoid of IRBP-LI. The schematic drawing in FIGURE 1 illustrates regions in the CNS labeled with our antibody.

Spinal Cord. Immunostained fibers were detected in superficial laminae of the dorsal horn where they were heavily concentrated in laminae I and II, a terminal field of primary nociceptive and visceral afferents.[16,17] Another heavily labeled field was seen in the lateral cervical and lateral spinal nuclei embedded in the lateral funiculus. A prominent plexus was detected in the lateral horn where processes surrounded preganglionic neurons in the thoracic intermediolateral cell column. IRBP-LI was contained in fibrous astrocytes radially organized along septa and characterized by long unbranched glial processes and also in proximity to the central canal.

Medulla Oblongata. The neuropil was labeled in a superficial lamina of the spinal trigeminal nucleus caudalis (STNc) and directly contiguous with a field in the upper cervical dorsal horn. Processes in the STN were concentrated in a region that relays first order noxious and thermal afferents to the thalamus.[18] Immunoreactive elements were heavily concentrated in the nucleus tractus solitarii (NTS) and dorsal motor nucleus (DMX), recipients of primary visceral afferents of cranial nerves VII (nervus intermedius), IX, and X.[cf19] A weak band of immunoreactivity was discerned in the neuropil of the lateral tegmental field. This field in medullary reticular formation formed a diagonal arc extending between the solitary nuclear complex (NTS-X) and the ventrolateral medulla.[20]

In the NTS-X, densities of immunolabeled processes in the sensory, NTS, and parasympathetic motor, DMX compartments were distinguished from background levels in the surrounding area, such as the hypoglossal nucleus (FIG. 2A and B). Processes were heavily labeled in the subnucleus centralis, a column in the medial division of NTS targeted by primary esophageal afferents and which selectively projects to esophageo-motoneurons in the compact division of the nucleus ambiguus (NA).[cf19]

The compact and semicompact divisions of the NA were also characterized by high concentrations of immunolabeled processes that yielded evidence of receptor localization to structurally interconnected and, perhaps, functionally integrated systems. In the NA, immunolabeled punctate processes resembled terminal boutons and formed close appositions with large unlabeled visceral motoneurons (FIG. 2C).

In the medial NTS, a dense plexus was present in the subpostremal division and was composed of glial processes admixed with smaller punctate terminal-type processes. This region is transitional between a subnucleus of NTS receiving input from the gastrointestinal tract[21] and the area postrema, a circumventricular organ.[22] The neuropil of the area postrema, was diffusely labeled. Immunoreactivity in the area postrema was light although clearly differentiated from profiles of unlabeled neuronal cell bodies and the plexus of fenestrated capillaries that distinguishes the chemoreceptor organ. In the DMX, immunoreactive processes surrounded unlabeled preganglionic perikarya. Immunoreactive processes were also observed in the inferior olive and concentrated in accessory subnuclei connected with the fastigial nucleus and cerebellar vermis.

Pons/Midbrain. In the pons, densities of immunoreactive processes invested neuronal soma in several subnuclei of the lateral parabrachial nucleus. Perhaps significant is axonal transport evidence that this field is a principal recipient of ascending projections from sensory relay nuclei also enriched in the receptor-binding protein,

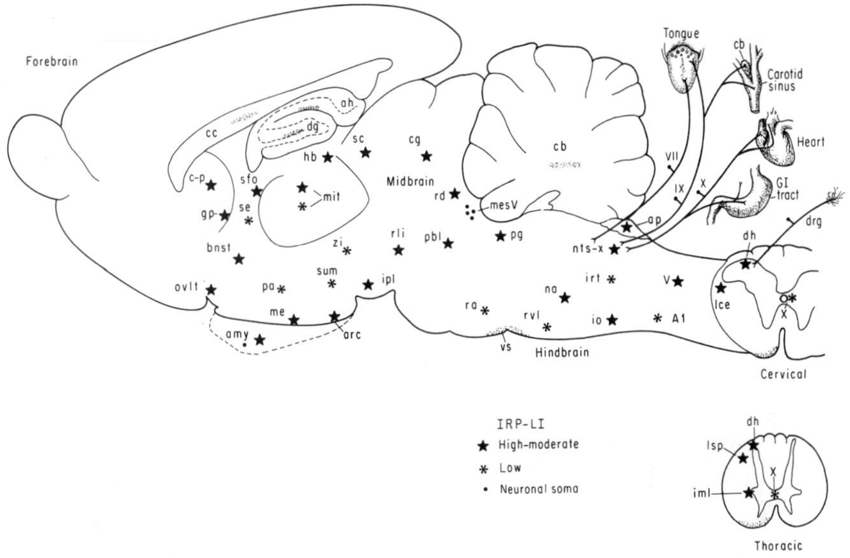

FIGURE 1. Schematic diagram illustrates in sagittal section locations of presumptive imidazoline receptor sites in the central nervous system. The distribution of receptor-binding protein immunoreactivity suggests that the I-receptor may be selectively localized to nuclei engaged in sensory processing, visceral reflex regulation, and behavior. Sites demonstrating high or low densities of receptor-binding protein immunoreactivity are indicated. Dorsal root ganglia, carotid body, and adrenal medulla were also labeled. *Stippled areas* indicate immunoreactivity confined to elements with a morphological appearance of glia. Abbreviations: A1 = noradrenergic nucleus in caudal ventrolateral medulla; ah = hippocampal area; amy = amygdala; ap = area postrema; arc = arcuate nucleus of mediobasal hypothalamus; bnst = bed nucleus of stria terminalis; cb = cerebellum or carotid body; cc = corpus callosum; ce = central nucleus of the amygdala; cg = central (periaqueductal) gray; c-p = caudate-putamen; dh = dorsal horn; dg = dentate gyrus; dmv = dorsal motor nucleus; drg = dorsal root ganglion; fs = fundus striati; gp = globus pallidus; hb = habenular (medial subnucleus); iml = intermediolateral cell column; io = inferior olive; ipl = interpeduncular nucleus; irt = intermediate reticular zone; lcn = lateral cervical nucleus; lsp = lateral spinal nucleus; me = median eminence; mesV = mesencephalic trigeminal nucleus; mit = midline-intralaminar thalamus; na = nucleus ambiguus; nh = nucleus hypoglossi; nts-x = solitary nuclear complex (principal recipient of first order visceral afferents); ovlt = organum vasculosum lamina terminalis; pbl = lateral (dorsal) parabrachial nucleus (bidirectional relay of visceral and nociceptive afferents to forebrain and lower brain stem); pa = paraventricular nucleus of hypothalamus; pg = periventricular gray; ra = raphe; rd = nucleus raphe dorsalis; rli = nucleus rostral-linearis; rvl = rostral ventrolateral reticular nucleus (C1 adrenergic vasopressor area); sc = subcommissural organ; scp = superior cerebellar peduncle; se = septum; sfo = subfornical organ; sum = supramammillary nucleus; tr = solitary tract; vhc = ventral hippocampal commissure; vs = ventral medullary surface; zi = zona incerta; V = trigeminal nucleus (caudalis); VII = facial nerve (nervus intermedius); IX = glossopharyngeal nerve; X = vagus nerve or lamina X; XII = hypoglossal nerve.

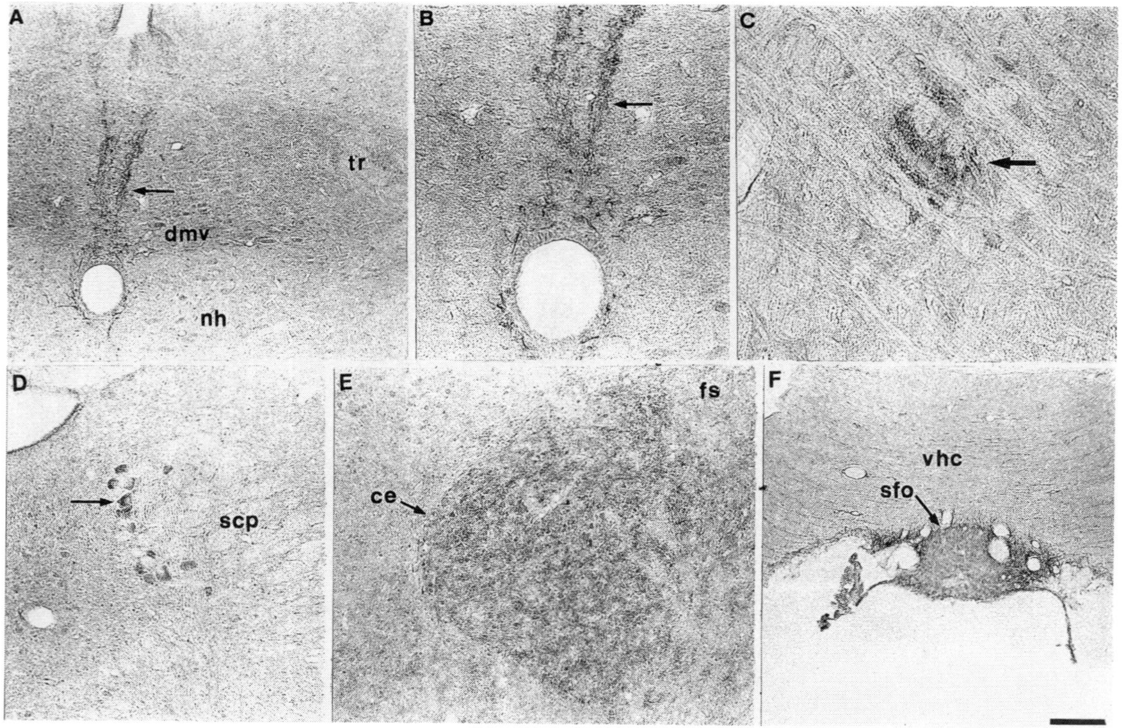

FIGURE 2. Photomicrographs of transverse sections of rat brain demonstrate circumscribed fields of imidazoline-receptor binding protein immunoreactivity. (**A**) Nucleus tractus solitarii (NTS) and dorsal vagal motor nucleus (dmv) (**B**); *arrows* in **A** and **B** point to a midline plexus in NTS formed by immunoreactive glia and neuronal processes; (**C**) nucleus ambiguus, semicompact division (sc); (**D**) mesencephalic trigeminal neurons (*arrow*) labeled intracellularly (superior cerebellar peduncle, scp); (**E**) central nucleus of the amygdala (ce) (fundus striati, fs); (**F**) subfornical organ (sfo) (ventral hippocampal commissure, vhc). Bar = 120 μm in **A, D, F** and 50 μ in **B, C, E**.

for example, the spinal dorsal horn, trigeminal nucleus caudalis, and NTS-AP.[23] Intracellular labeling was detected in neurons in the mesencephalic trigeminal nucleus, an intraparenchymal homolog of the dorsal root ganglion[24] (FIG. 2D). Light to moderate densities were distinguished in the neuropil of the central gray and mesopontine raphe.

Mesodiencephalic Junction. In the epithalamus, processes surrounded cell bodies in the medial habenular (Hb) nucleus. Another labeled field was identified in the interpeduncular nucleus, a mesencephalic target of the Hb-interpeduncular tract and an outlet to the brain stem raphe and central gray.[25,26]

Forebrain. High densities were selectively localized to the central nucleus of the amygdala (FIG. 2E) and along efferent (amygdalofugal) projection fields. The most heavily labeled field in the thalamus was detected in a periventricular division, the nucleus paraventricularis thalami (n.Pvt).[27,28] Other members of the midline-intralaminar thalamus were labeled, though far less intensely.

Another prominent field was the arcuate nucleus of the mediobasal hypothalamus. Immunoreactivity also was present in glial-type processes in the lamina interna and a subependymal zone of the median eminence. Concentrations equivalent to those seen in the central nucleus of the amygdala were detected along its efferent pathway in the bed nucleus of the stria terminalis. Lower densities in the medial amygdaloid nucleus and amygdalo-striatal zone (fundus striati) were distinct from background levels seen in the claustrum and cerebral cortex. A prominent plexus was also observed in the medial segment of the globus pallidus and caudate putamen.

The subfornical organ and organum vasculosum lamina terminalis[22] were characterized by a diffuse granular immunoreaction product similar to that seen in other circumventricular organs (FIG. 2F). IRBP-LI was also present in the median and periventricular preoptic nuclei, septum, and hippocampus. Lower densities characterized other sites illustrated in FIGURE 1.

Ultrastructural Analysis. Electron microscopic studies were performed to ultrastructurally localize the receptor binding protein. Samples were taken from the dense subpostremal plexus in NTS transitional with the area postrema. Most immunoreaction product was identified in astroglia that were marked by immunostained glial filaments and labeled mitochondria. Immunoreactivity was also detected in unmyelinated axons containing microtubules and axon terminals which formed predominantly asymmetric synapses with small unlabeled dendrites. Immunoreactive terminals were characterized by labeling of small clear vesicles associated with synaptic specializations and larger dense core vesicles associated with plasmalemmal membranes apposed to glia (FIG. 3).

DISCUSSION

Mapping of I-Receptors in the Central Nervous System with an Antibody to the IRBP

The distribution of I-receptors was mapped within the CNS of rat with an antibody to a 70-kD protein isolated by ligand affinity chromatography from bovine adrenal chromaffin cells.[8,9] This protein, the IRBP, has all of the ligand binding properties

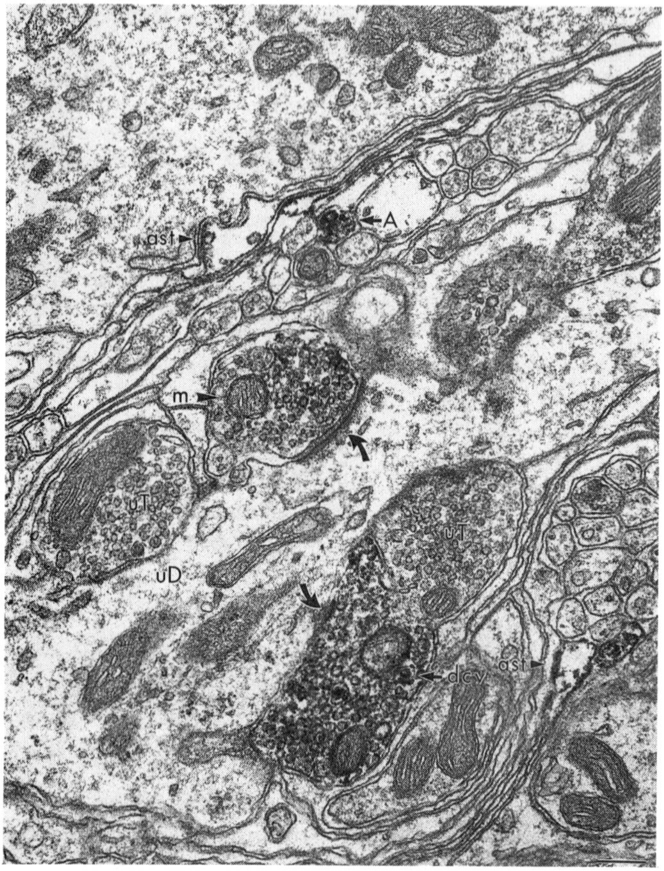

FIGURE 3. Ultrastructural localization of IRBP-LI within the nucleus tractus solitarii. Two immunolabeled terminals contact (*curved arrows*) the shaft of the same small unlabeled dendrite (uD). One IRBP-containing terminal (*top*) forms an asymmetric synapse with the dendrite. Both terminals contain mitochondria (m) rimmed with IRBP-LI. The dendrite is also contacted by two unlabeled terminals (uT). Most of the large dense-core vesicles (dcvs) within the immunolabeled terminal are located away from the dendrite in apposition to the plasmalemma of an astrocytic process. IRBP-LI is also found within an axon (A) and astrocytic processes (ast). Bar = 0.25 μm.

of I- but not α_2-adrenergic receptors, inhibits binding of I- but not adrenergic ligands to their receptors, precipitates I-receptors in tissue homogenates,[8] and can localize the I- but not α_2-adrenergic receptors in other tissues by immunocytochemistry.[9–11] Whether the IRBP represents the entire I-receptor or is a subunit thereof is not known, because the protein has not been cloned. It is also not clear if the antibody recognizes a protein common to all I-receptors or is selective for I-receptors of the I_1 or I_2 subclasses. However, the fact that a similar protein was isolated from idazoxan- or *para*-aminoclonidine-affinity columns and that the antibody labels neurons in the brain stem and glia, presumably representing tissues expressing receptors of different subtypes, suggests that the antibody recognizes both.

Our immunocytochemical method for localization of I-receptors is far more specific than the only other method used to map I-receptor distribution in the CNS, receptor autoradiography.[3–6] Because all radioligands employed label both I- and α_2-adrenergic receptors, identification of I-receptor sites could only be deduced by subtraction of areas labeled by ligands selective for α_2-adrenergic receptors[3] or by incubation of radioligands with masking concentrations of adrenergic agonists or antagonists.[4–6] It is of interest, however, that the regional distributions of IRBP-LI that we demonstrated substantially overlap nonadrenergic binding sites (NAIBS), presumably an I-receptor, defined autoradiographically on sections coincubated with [^{3}H]idazoxan and saturating concentrations of α_2-adrenoceptor ligands.[4–6] Sites containing IRBP-LI and the NAIBS were localized to circumventricular organs, the arcuate hypothalamic nucleus, midbrain central gray, and the interpeduncular nucleus. We also discovered I-receptors in regions interpreted as sparsely populated or devoid of NAIBS. Such differences could reflect the higher resolution afforded by immunocytochemistry. Alternatively, our antibody to the IRBP may detect I-receptors of all subclasses, whereas [^{3}H]idazoxan may label only a subset.

I-Receptors in the CNS: Functional Implications

Cytologically, IRBP-LI was localized to neurons and glia. In neurons the protein appeared primarily in axon terminals as suggested by light microscopy and proved ultrastructurally. Indeed, the only neuronal perikarya noted in the brain were those labeled in the mesencephalic nucleus of the Vth nerve. The relatively selective labeling of terminals supports pharmacological evidence that I-receptors may control transmitter release presynaptically.[29] The absence of detectable levels of IRBP-LI in neuronal soma may reflect differences in receptor density. The presence of I-receptors in astrocytes was confirmed ultrastructurally and is consistent with observations *in vitro*[30] that I-receptors of the I_2 subclass are expressed in cultured astrocytes.

I-receptors were expressed along the entire extent of the CNS from spinal cord to hippocampal cortex. Moreover, their distribution reflected a grouping into specific functional units, suggesting a role of I-receptors in specific neuronal functions, particularly processing of primary somatic and visceral afferents, autonomic and neuroendocrine functions, chemoreception, and the integration of emotional behaviors.

I-Receptors and Pain

I-receptors were heavily represented at many stations in the CNS involved in the perception of and responses to pain. I-receptors were contained in processes of small diameter with the morphological characteristics of primary nociceptive afferents, within superficial spinal laminae[16,17] as well as sites in the brain stem that depress nociceptive reflexes[31] or relay pain signals to the forebrain including the trigeminal nucleus caudalis,[18] lateral medullary reticular formation, lateral parabrachial nucleus, and a nucleus in midbrain raphe associated with the Edinger-Westphal complex and responding selectively to intraperitoneal inflammation.[16] Especially striking were densities of IRBP-LI in areas of the midbrain central gray and forebrain (e.g., amygdala) subserving the affective significance of and emotional responses evoked by pain.[32,33] The identified systems may be responsible for mediating some of the potent analgesic responses to clonidine.[34]

I-Receptors and the Central Autonomic Nervous System

I-receptors were also concentrated in areas of the CNS subserving autonomic regulation. They were heavily concentrated in the NTS, the site of termination of primary visceral afferents,[cf19] in internuncial autonomic networks in lateral tegmental field[20,35] and parabrachial complex[36] of the lower brain stem, and in visceral motor nuclei including parasympathetic motor neurons of the dorsal vagal complex and the dense plexus in the spinal sympathetic nucleus, IML. I-receptors were also expressed in suprapontine regions known to represent essential links in the central autonomic core including the central gray of the midbrain, and forebrain nuclei (e.g., hypothalamus and amygdala) with well-established roles in regulation of autonomic function.

While it is generally believed that I-receptors mediated the hypotensive actions of clonidine and rilmenidine, largely within the rostral ventrolateral reticular nucleus (n.RVL),[37–39] the density of I-receptors in this nucleus was relatively low. Higher concentrations were localized to other intramedullary centers mediating the baroreceptor reflex including the caudal ventrolateral medulla and the NTS.[35] The relative paucity of receptors in the n.RVL, however, might reflect the small number of sympathoexcitatory reticulospinal vasomotor neurons in the n.RVL[40] or that the I-receptor density is low. The paucity of I-receptors in the n.RVL contrasts with prominent labeling of α_2-adrenoceptor mRNA[41] and [^{3}H]PAC,[42] evidence that α_2-adrenergic receptors may constitute the predominant receptor site.

I-Receptors and the Neuroendocrine System

A role of the I-receptor in neuroendocrine control is suggested by the presence of glial IRBP-LI in periventricular regions of hypothalamus that regulate release of pituitary hormones. I-receptors in the arcuate nucleus and median eminence, a circumventricular organ complex, may subserve regulation of adenohypophyseal hormones, such as the hypophysiotropic stress hormone CRF or prolactin, by modulating transmission or release into the portal plexus. These findings may relate to earlier reports on the actions of clonidine in releasing arginine vasopressin (AVP) or growth hormone.[43,44] The effects were attributed to actions on α_2-adrenergic receptors because the responses were blocked by idazoxan. However, because idazoxan also binds to I-receptors, it is conceivable that some of the neuroendocrine actions are mediated by this receptor.

I-receptors may also modulate the action of circulating agents at circumventricular (neurohaemal) organs (CVOs). Thus, a prominent plexus was identified in the area postrema and NTS that receive input from primary visceral afferents and that concentrate receptors for vasoactive neuropeptides such as arginine vasopressin and angiotensin II (AII).[cf19,45,46] Receptors were also detected in glial membranes that separate the neuronal compartment from ependyma lining the ventricular space, that is, the stratum gliosum subependymale. I-receptors also were localized to the subfornical organ and its neural networks subserving AII-mediated thirst and release of vasopressin,[47,48] and aforementioned sites in the area postrema-subpostremal complex that are coactivated in response to hypovolemia,[46] osmotic stimulation of the gastric mucosae,[49] or toxic insult.[50]

I-Receptors in Nuclei Subserving Behavior

The functional significance of I-receptors in the habenular-interpeduncular system may relate to its role in behavioral regulation of autonomic and neuroendocrine function.[cf25] Sites enriched in the receptor transmit via the medial habenular and interpeduncular nuclei signals converging from limbic (septal and preoptic) and autonomic motor (superior cervical ganglia) neurons to serotonergic nuclei in the midbrain raphe[25,26] that influence the activity of widespread regions of the CNS.

The I-receptors were localized to sites in the lateral parabrachial nucleus, central gray, and mesopontine raphe nuclei that transmit visceral and nociceptive signals to the forebrain. The role of these receptor sites may relate to their extensive bidirectional mesopontine pathways, such as collaterals to the midline thalamus (n.Pvt) and NTS (DMX) implicated in coordinating thalamocortical and visceral reflex components of organized behavior.[27]

I-receptors were also localized in the forebrain to neural networks subserving affective and expressive components of emotional behavior. I-receptors in the amygdala, fundus striati, and central gray may relate to their roles as outlets for the autonomic, endocrine, and somatomotor components of emotional expression.[cf32,33] Signals processed by these nuclei are polymodal and derive from multiple sensory relay nuclei, the intralaminar thalamus and areas of supramodal cerebral cortex endowed with complex integrative and storage capacity. Another presumptive receptor site in the midline thalamus (n.Pvt) receives input from identical regions including the central nucleus of the amygdala[27] that transmit signals implicated as emotionally charged.[32] The functional significance of I-receptors in the n.Pvt may relate to the hypothesized role of this periventricular nucleus in processing visceral and nociceptive-stimuli,[27] subserving detection of agents circulating in the cerebrospinal fluid[28] and the adaptive response to stress.[51,52] Signals processed by the n.Pvt are transmitted to the prefrontal cortex, amygdala, and a site in ventral striatum, the n. accumbens involved in motivated behaviors.[28,53] The midline thalamus, when stimulated, provokes increases in food and water intake behavior[54] and mediates autonomic responses elicited during learning tasks.[55] Recently demonstrated were increases in functional activity of the n.Pvt associated with conditioned drug intake behavior[56] and stress.[51,52] Perhaps the effects of idazoxan on food and water intake[57] are mediated, in part, by I-receptors. In conclusion, the I-receptors may regulate integrative behaviors related to stress and the preservation of homeostasis.

Relationship of I-Receptors to α_2-Adrenoceptors

Our findings demonstrate that the distribution of I-receptors and α_2-adrenoceptors, while partially overlapping in large part, differ substantially. Alpha$_2$-adrenoceptors are localized to many regions not expressing detectable levels of IRBP-LI. Throughout the nervous system, nuclei recognized as endowed with high densities of α_2-adrenoceptors[3–6,42] or mRNAs for α_2-adrenoceptor subtypes[41] either did not express detectable levels of immunoreactivity or were labeled sparsely or inconsistently.

The cerebral cortex and olfactory tubercle, although devoid of IRBP-LI, were enriched in α_2-adrenoceptor mRNA[41] or sites binding [^{3}H]PAC and [^{3}H]idazoxan.[2–4,6,42] Similar populations of α_2-adrenoceptors were radiolabeled with [^{3}H]rauwolscine in the human brain and interpreted as associated with the adrenergic binding component of [^{3}H]idazoxan.[5] Subcortical densities of [^{3}H]PAC or rauwolscine binding sites[3,6,42] in regions of rat forebrain characterized by imperceptible IRBP-LI include the nucleus accumbens, basolateral amygdala, thalamus, notably the reticular nucleus and anterior nuclear group, areas of hypothalamus, and substantia nigra. Similarly, in the lower brain stem and spinal cord sites binding [^{3}H]PAC[42] yet devoid of detectable IRBP-LI include the tectum, locus ceruleus, cranial nerve motor nuclei, and ventral horn. These sites are thus interpreted as predominantly or exclusively adrenergic.

SUMMARY

I-receptors can be localized immunocytochemically in rat nervous system with polyclonal antibodies to an IRBP. I-receptors are cytoplasmic and detected in neuronal perikarya, processes, and glia. Labeled neuronal perikarya in the CNS are uncommon and localized to the mesencephalic trigeminal nucleus. I-receptors are heavily represented in *primary sensory systems* including: somatosensory systems (spinal and trigeminal) and visceral afferent systems (NTS), in central networks subserving *autonomic* regulation, *neuroendocrine* control and *emotional behaviors,* in *circumventricular (neurohaemal) organs* and in *nonneuronal cells* including astrocytes with regional densities paralleling neuronal innervation. The distributions of I-receptors and α_2-adrenergic receptors partially differ. I-receptors in the CNS appear to relate broadly to the visceral brain and its afferent inputs, particularly pain. Its functions may relate to regulation of integrative behaviors related to stress.

REFERENCES

1. Ernsberger, P., M. P. Meeley, J. J. Mann & D. J. Reis. 1987. Clonidine binds to imidazole binding sites as well as α_2-adrenoceptors in the ventrolateral medulla. Eur. J. Pharmacol. **134:** 1-14.
2. Kamisaki, Y., T. Ishikawa, Y. Takao, H. Omodani, N. Kuno & T. Itoh. 1990. Binding of [^{3}H]*p*-aminoclonidine to two sites, α_2-adrenoceptors and imidazoline binding sites: Distribution of imidazoline binding sites in rat brain. Brain Res. **514:** 15-21.
3. Boyajian, C. L., S. E. Loughlin & F. M. Leslie. 1987. Anatomical evidence for alpha-2 adrenoceptor heterogeneity: Differential autoradiographic distributions of [^{3}H]rauwolscine and [^{3}H]idazoxan in rat brain. J. Pharmacol. Exp. Ther. **241:** 1079-1091.
4. Bruning, G., P. Kaulen & H. G. Baumgarten. 1987. Quantitative autoradiographic localization of α_2-antagonist binding sites in rat brain using [^{3}H]idazoxan. Neurosci. Lett. **83:** 333-337.
5. De Vos, H., A. Convents, J. De Keyser, J.-P. De Backer, I. J. B. Van Megen, G. Ebinger & G. Vauquelin. 1991. Autoradiographic distribution of α_2 adrenoceptors,

NAIBS, and 5-HT_{1A} receptors in human brain using [^{3}H]idazoxan and [^{3}H]rauwolscine. Brain Res. **566:** 13-20.
6. Mallard, N. J., A. L. Hudson & D. J. Nutt. 1992. Characterization and autoradiographical localization of non-adrenoceptor idazoxan binding sites in rat brain. Br. J. Pharmacol. **106:** 1019-1027.
7. Ernsberger, P., G. Feinland, M. P. Meeley & D. J. Reis. 1990. Characterization and visualization of clonidine-sensitive imidazole sites in rat kidney which recognize clonidine-displacing substance. Am. J. Hypertens. **3:** 90-97.
8. Wang, H., S. Regunathan, M. P. Meeley & D. J. Reis. 1992. Isolation and characterization of imidazoline receptor proteins from bovine adrenal chromaffin cells. Mol. Pharmacol. **42:** 792-801.
9. Wang, H., S. Regunathan, D. A. Ruggiero & D. J. Reis. 1993. Production and characterization of antibodies specific for the imidazoline receptor protein. Mol. Pharmacol. **43:** 509-515.
10. Youngson, C., S. Regunathan, H. Wang, G. Li & D. J. Reis. 1995. Co-expression of imidazoline receptors and agmatine in rat carotid body. Proceedings of the II International Symposium on Imidazoline Receptors. Ann. N.Y. Acad. Sci. This volume.
11. Regunathan, S., C. Youngson, H. Wang & D. J. Reis. 1995. Imidazoline receptors in vascular smooth muscle and endothelial cells. Proceedings of the II International Symposium on Imidazoline Receptors. Ann. N.Y. Acad. Sci. This volume.
12. Ruggiero, D. A., T. A. Milner, H. Wang, S. Regunathan & D. J. Reis. 1992. Imidazoline receptor-associated binding protein in central nervous system. Soc. Neurosci. Abstr. **18:** 541.
13. Regunathan, S., M. P. Meeley & D. J. Reis. 1993. Expression of nonadrenergic imidazoline sites in chromaffin cells and mitochondrial membranes of bovine adrenal medulla. Biochem. Pharmacol. **45:** 1667-1675.
14. Milner, T. A., D. J. Reis, V. M. Pickel, S. A. Aicher & R. Giuliano. 1993. Ultrastructural localization and afferent sources of corticotropin-releasing factor in the rat rostral ventrolateral medulla: Implications for central cardiovascular regulation. J. Comp. Neurol. **333:** 151-167.
15. Anderson, D. J. 1993. Molecular control of cell fate in the neural crest: The sympathoadrenal lineage. Ann. Rev. Neurosci. **16:** 129-158.
16. Lantéri-Minet, M., P. Isnardon, J. De Pommery & D. Menétrey. 1993. Spinal and hindbrain structures involved in visceroception and visceronociception as revealed by the expression of Fos, Jun and Krox-24 proteins. Neuroscience **55:** 737-753.
17. Menétrey, D., A. Gannon, J. D. Levine & A. I. Basbaum. 1989. Expression of c-fos protein in interneurons and projection neurons of the rat spinal cord in response to noxious somatic, articular, and visceral stimulation. J. Comp. Neurol. **285:** 177-195.
18. Price, D. D., R. Dubner & J. W. Hu. 1976. Trigeminothalamic neurons in nucleus caudalis responsive to tactile, thermal, and nociceptive stimulation of monkey's face. J. Neurophysiol. **39:** 936-953.
19. Ruggiero, D. A., V. M. Pickel, T. A. Milner, M. Anwar, K. Otake, E. P. Mtui & D. Park. 1994. Viscerosensory processing in nucleus tractus solitarii: Structural and neurochemical substrates. *In* Nucleus of the Solitary Tract. R. A. Barraco, Ed.: 3-34. CRC Press. Boca Raton, FL.
20. Ruggiero, D. A., S. Cravo, V. Arango & D. J. Reis. 1989. Central control of the circulation by the rostral ventrolateral reticular nucleus: Anatomical substrates. *In* Progress in Brain Research 81. J. Ciriello, C. Polosa & M. Caverson, Eds.: 49-79. Elsevier. Amsterdam.
21. Norgren, R. & G. P. Smith. 1988. Central distribution of subdiaphragmatic vagal branches in the rat. J. Comp. Neurol. **273:** 207-223.
22. McKinley, M. J. & B. J. Oldfield. 1990. Circumventricular organs. *In* The Human Nervous System. G. Paxinos, Ed.: 415-438. Academic Press. San Diego, CA.

23. Cechetto, D. F., D. G. Standaert & C. B. Saper. 1985. Spinal and trigeminal dorsal horn projections to the parabrachial nucleus in the rat. J. Comp. Neurol. **240:** 153-160.
24. Ruggiero, D. A., C. A. Ross, M. Kumada & D. J. Reis. 1982. Reevaluation of projections from the mesencephalic trigeminal nucleus to the medulla and spinal cord: New projections. A combined retrograde and anterograde horseradish peroxidase study. J. Comp. Neurol. **206:** 278-292.
25. Jones, E. G. 1985. The Thalamus. Plenum Press. New York.
26. Groenewegen, H. J., S. Ahlenius, S. N. Haber, N. W. Kowall & W. J. H. Nauta. 1986. Cytoarchitecture, fiber connections, and some histochemical aspects of the interpeduncular nucleus in the rat. J. Comp. Neurol. **249:** 65-102.
27. Otake, K., D. J. Reis & D. A. Ruggiero. 1994. Afferents to the midline thalamus issue collaterals to the nucleus tractus solitarii: An anatomical basis for thalamic and visceral reflex integration. J. Neurosci. **14:** 5694-5707.
28. Su, H.-S. & M. Bentivoglio. 1990. Thalamic midline cell populations projecting to the nucleus accumbens, amygdala and hippocampus in the rat. J. Comp. Neurol. **297:** 592-593.
29. Gothert, M. & G. J. Molderings. 1991. Involvement of presynaptic imidazoline receptors in the alpha-2-adrenoceptor-independent inhibition of noradrenaline release by imidazoline derivatives. Naunyn-Schmiedeberg's Arch. Pharmacol. **343:** 271-282.
30. Regunathan, S., D. L. Feinstein & D. J. Reis. 1993. Expression of non-adrenergic imidazoline sites in rat cerebral cortical astrocytes. J. Neurosci. Res. **34:** 681-688.
31. Gebhart, G. F. & M. Ossipov. 1986. Characterization of inhibition of the spinal tail-flick reflex in the rat from the medullary lateral reticular nucleus. J. Neurosci. **6:** 701-713.
32. LeDoux, J. E. 1994. Emotion, memory and the brain. Sci. Am. **270:** 50-57.
33. Keay, K. A. & R. Bandler. 1992. Deep and superficial noxious stimulation increases Fos-like immunoreactivity in different regions of the midbrain periaqueductal grey of the rat. Neurosci. Lett. **154:** 23-26.
34. Fielding, S., J. Wilker, M. Hynes, M. Szewczak, W. J. Novick, Jr. & H. Lal. 1978. A comparison of clonidine with morphine for antinociceptive and antiwithdrawal actions. J. Pharmacol. Exp. Ther. **207:** 899-905.
35. Reis, D. J., E. V. Golanov, D. A. Ruggiero & M.-K. Sun. 1994. Sympatho-excitatory neurons of the rostral medulla are oxygen sensors and essential elements in the tonic and reflex control of the systemic and cerebral circulations. J. Hypertens. (Suppl.) **12:** S159-S180.
36. Fulwiler, C. E. & C. B. Saper. 1984. Subnuclear organization of the efferent connections of the parabrachial nucleus in the rat. Brain Res. Rev. **7:** 229-259.
37. Bousquet, P., J. Feldman & J. Schwartz. 1984. Central cardiovascular effects of alpha adrenergic drugs: Differences between catecholamines and imidazolines. J. Pharmacol. Exp. Ther. **230:** 232-236.
38. Ernsberger, P., R. Giuliano, R. N. Willette & D. J. Reis. 1990. Role of imidazole receptors in the vasodepressor response to clonidine analogs in the rostral ventrolateral medulla. J. Pharmacol. Exp. Ther. **253:** 408-418.
39. Gomez, R. E., P. Ernsberger, G. Feinland & D. J. Reis. 1991. Rilmenidine lowers arterial pressure via imidazole receptors in brainstem C1 area. Eur. J. Pharmacol. **195:** 181-191.
40. Ruggiero, D. A., S. L. Cravo, E. Golanov, R. Gomez, M. Anwar & D. J. Reis. 1994. Adrenergic and non-adrenergic spinal projections of a cardiovascular-active pressor area of medulla oblongata: Quantitative topographic analysis. Brain Res. **663:** 107-120.
41. Nicholas, A. P., V. Pieribone & T. Hökfelt. 1993. Distributions of mRNAs for α_2-adrenergic receptor subtypes in rat brain: An *in situ* hybridization study. J. Comp. Neurol. **328:** 575-594.

42. UNNERSTALL, J. R., T. KOPAJTIC & M. J. KUHAR. 1984. Distribution of α_2-agonist binding sites in the rat and human central nervous system: Analysis of functional, anatomic correlates of the pharmacologic effects of clonidine and related adrenergic agents. Brain Res. Rev. **7:** 69-101.
43. CHAPMAN, I. M., R. KAPOOR & J. O. WILLOUGHBY. 1993. Interactions between the effects of opioid, serotonin and alpha-2-adrenergic receptor agonists on growth hormone release in the male rat. Intrahypothalamic administration. Neuroendocrinology **57:** 921-927.
44. BAILEY, A. R., G. CLARKE & J. B. WAKERLEY. 1994. Inhibition of supraoptic vasopressin neurones following systemic clonidine. Neuropharmacology **33:** 211-214.
45. BONHAM, A. C. & E. M. HASSER. 1993. Area postrema and aortic or vagal afferents converge to excite cells in nucleus tractus solitarius. Am. J. Physiol. **264:** H1674-H1685.
46. CHAN, R. K. W. & P. E. SAWCHENKO. 1994. Spatially and temporally differentiated patterns of *c-fos* expression in brainstem catecholaminergic cell groups induced by cardiovascular challenges in the rat. J. Comp. Neurol. **348:** 433-460.
47. MISELIS, R. R. 1981. The efferent projections of the subfornical organ of the rat: A circumventricular organ within a neural network subserving water balance. Brain Res. **230:** 1-23.
48. SIMPSON, J. B., A. N. EPSTEIN & S. S. CAMARDO. 1978. Localization of receptors for thc dipsogenic action of angiotensin II in the subfornical organ of the rat. J. Comp. Physiol. Psychol. **92:** 2676-2682.
49. KOBASHI, M., H. ICHIKAWA, T. SUGIMOTO & A. ADACHI. 1993. Response of neurons in the solitary tract nucleus, area postrema and lateral parabrachial nucleus to gastric load of hypertonic saline. Neurosci. Lett. **158:** 47-50.
50. MILLER, A. D. & D. A. RUGGIERO. 1994. Emetic reflex arc revealed by expression of the immediate-early gene c-*fos* in the cat. J. Neurosci. **14:** 871-888.
51. CHASTRETTE, N., D. W. PFAFF & R. B. GIBBS. 1991. Effects of daytime and nighttime stress on Fos-like immunoreactivity in the paraventricular nucleus of the hypothalamus, the habenula, and the posterior paraventricular nucleus of the thalamus. Brain Res. **563:** 339-344.
52. IMAKI, T., T. SHIBASAKI, M. HOTTA & H. DEMURA. 1993. Intracerebroventricular administration of corticotropin-releasing factor induces c-*fos* mRNA expression in brain regions related to stress responses: Comparison with pattern of c-*fos* mRNA induction after stress. Brain Res. **616:** 114-125.
53. BERENDSE, H. W. & H. J. GROENEWEGEN. 1991. Restricted termination fields of the midline and intralaminar thalamic nuclei in the rat. Neuroscience **42:** 73-102.
54. ROBINSON, B. W. & M. MISHKIN. 1968. Alimentary responses to forebrain stimulation in monkeys. Exp. Brain Res. **4:** 330-366.
55. BUCHANAN, S. L., R. H. THOMPSON & D. A. POWELL. 1989. Midline thalamic lesions enhance conditioned bradycardia and the cardiac orienting reflex. Psychobiology **17:** 300-306.
56. BROWN, E. E., G. S. ROBERTSON & H. C. FIBIGER. 1992. Evidence for conditional neuronal activation following exposure to a cocaine-paired environment: Role of forebrain limbic structures. J. Neurosci. **12:** 4112-4121.
57. JACKSON, H. C., I. J. GRIFFIN & D. J. NUTT. 1991. The effects of idazoxan and other α_2-adrenoceptor antagonists on food and water intake in the rat. Br. J. Pharmacol. **104:** 258-262.

In Situ Hybridization

A Complementary Method to Radioligand-Mediated Autoradiography for Localizing Adrenergic, Alpha-2 Receptor-Producing Cells[a]

ANTHONY P. NICHOLAS,[b] VINCENT PIERIBONE,
ÅKE DAGERLIND, BJÖRN MEISTER, ROBERT ELDE,
AND TOMAS HÖKFELT

Department of Histology and Neurobiology
Karolinska Institute
Stockholm, 10401, Sweden

After Ahlquist[1] first pharmacologically differentiated adrenoceptors into alpha and beta subtypes, alpha receptors were further separated into alpha-1 and alpha-2 subgroups based on many different criteria. It was first believed that the difference between the alpha receptor subgroups was based on anatomical location,[2–4] causing Langer[2] to propose that postsynaptic adrenergic receptors be named "alpha-1" and presynaptic adrenergic receptors be named "alpha-2." It was later discovered that these two receptors differed in their physiological actions,[5,6] secondary messengers,[7–9] and pharmacological affinities for adrenergic agonists and antagonists.[10–12] The present nomenclature relies heavily on the latter method to differentiate between subsets of adrenoceptors.

From many binding studies, it was suspected that the alpha-2 receptor subset could be further subdivided, because different "specific" alpha-2 agonists and antagonists often showed different binding properties in rat tissues.[10,12,13–28] Some of these drugs were also used as radioligands in various autoradiographic studies to localize alpha-2 receptors in the rat central nervous system[16,28–38] and kidney.[11,39–49] However, different labeling patterns were sometimes obtained depending on the radioligand used.[20,31,32,34] It was therefore suggested that many of these drugs clearly did not bind only to one alpha-2 receptor subtype.[19,27,28] In some cases, "specific" alpha-2 receptor radioligands were shown to bind to other monoamine receptors,[19,22,50,51] as well as to nonadrenergic receptors,[51,52] including imidazoline receptors,[19,24,51,53–61] calling into question the true localization of central alpha-2 adrenoceptors as defined by the quantitative autoradiographic technique.[62]

[a] Publication of this manuscript was made possible thanks to Drs. J. Whitaker, R. Powers, and L. Harrell and financial support from National Institutes of Health grants P01AG10163 and F32HL08494-01A1, Swedish MRC grant 2887, and the Department of Neurology, UAB, Birmingham, Alabama.

[b] Present address: Department of Neurology, University of Alabama at Birmingham, UAB Station, 12th Floor, Jefferson Tower, Birmingham, AL 35294-0007.

The present alpha-2 receptor nomenclature was originally based on competitive binding of radioligands in specific cell lines that supposedly possess only one of the alpha-2 receptor subtypes. The alpha-2A receptor was first shown in human platelets and demonstrated a rank order of affinity for the following radioligands: rauwolscine > phentolamine > oxymetazoline > prazosin.[21,63] With molecular cloning techniques, the gene for this receptor was found on human chromosome 10 and was thus designated as C10.[63]

Around the same time, another alpha-2 receptor was discovered in a neonatal rat lung cell line with a different rank order of affinity for radioligands compared to that shown with the alpha-2A receptor.[21,23,64,65] This new receptor, which bound rauwolscine > phentolamine > prazosin > oxymetazoline, was called the alpha-2B receptor. Molecular cloning techniques using a rat kidney cDNA library produced a sequence-dubbed RNG that produced adrenergic receptors consistent with the alpha-2B binding profile.[65] The human homologue for RNG[64,66] was later found on chromosome 2,[67] therefore named C2, and an 82% amino acid homology was shown between these two alpha-2B receptors.[68]

Controversy ensued when another alpha-2 receptor subtype was discovered in the OKY opossum kidney cell line.[69,70] This receptor, pharmacologically named alpha-2C, had the same rank potency as the alpha-2B receptor listed above, but it exhibited a much higher affinity for rauwolscine than did either the alpha-2A or 2B subtypes. In all of the previously mentioned experiments involving the binding properties of alpha-2A, 2B, and 2C receptors, quantitative differences between the K_i values for the competing ligands were considered difficult to interpret because these experiments were performed in different laboratories under different incubation conditions. However, a human sequence found on chromosome 4, named C4,[71] and its rat homologue, named RG10,[72–74] produced alpha-2 receptors that were consistent with the affinity profile for the alpha-2C receptor. The human and rat alpha-2C receptors showed a 90% amino acid homology.[68]

A search for the rat homologue for the human alpha-2A receptor (C10) produced a receptor from a rat salivary gland cell line[24,75] with 89% amino acid homology with the human alpha-2A receptor,[68] and its sequence was named RG20.[73] Unfortunately, the rank order of affinity for this receptor was phentolamine > rauwolscine = oxymetazoline > prazosin, which was different from the alpha-2A profile expressed in the human from the C10 sequence.[73] Therefore, it was argued that RG20 represented a unique receptor and was pharmacologically designated as the alpha-2D receptor.[73]

Using molecular biological techniques, sequences for the human alpha-2 receptor subtypes (C10, C2, and C4)[63,64,67,71] and their rat homologues (RG20, RNG, and RG10)[65,72–74] were transfected into cell lines that ensured only one receptor subtype be expressed, and affinity binding experiments were performed (TABLE 1).[63–65,67,71–74,76] Again, individual K_i values differed, most likely because of differences in the cell lines used and in the methodologies of the various laboratories, but other than a few exceptions (bolded values, TABLE 1), a definite trend became apparent (TABLE 2). Thus, the human C10 sequence correlated pharmacologically with the alpha-2A receptor, with high affinity for yohimbine and phentolamine, as do the alpha-2B and alpha-2C receptors, but unlike the latter two subtypes, the alpha-2A receptor had low affinity for prazosin. The alpha-2B and 2C receptors differed primarily in their affinity for noradrenaline, oxymetazoline, and adrenaline. Thus, the alpha-2B receptor, represented by the C2 and RNG sequences, bound noradrenaline > oxymetazoline > adrenaline, whereas the alpha-2C receptor, represented by the C4 and RG10 sequences, bound oxymetazoline >> adrenaline > noradrenaline. Pharmacologically, however, the alpha-2D receptor was totally unlike all of the other alpha-2 receptor subsets. This receptor, represented by the RG20 sequence, bound phentolamine > oxymetazoline > yohimbine >> prazosin. Unfortunately, the rank order of noradrenaline (and adrenaline) binding for this receptor remains to be clarified.

It is interesting that imidazoles such as idazoxan, oxymetazoline, and *para*-aminoclonidine bind, sometimes very strongly, to a majority of these receptors (TABLE 1). Perhaps the binding profiles of the various alpha-2 receptors can be studied further in future experiments using only one cell line (such as COS1 or COS7) and various alpha-1, alpha-2, and beta adrenoceptor ligands as well as other imidazoline-like ligands such as rilmenidine, cirazoline, and guanabenz.

When the RNG, RG10, and RG20 sequences were used in Northern blot analysis of homogenized regions of rat brain tissue, the general locations of these three receptor mRNA subtypes were identified,[26,68,72,73,77] but at that time data on the specific neuronal populations that expressed this genetic material were nonexistent. That prompted our laboratory to design specific oligonucleotide probes to the three known alpha-2 receptor sequences in order to identify cell groups in the rat that expressed alpha-2 receptor subtype-specific mRNA labeling using the *in situ* hybridization technique.[78–81]

METHODOLOGICAL CONSIDERATIONS

As the alpha-2 adrenergic receptors are members of a family of receptors with regions of high homology, especially in the highly conserved seven transmembrane-spanning areas,[82] it was imperative to choose regions of each sequence that were unique to each specific receptor. Using the MacVector software program, we compared a receptor sequence of interest to all of the other known adrenergic receptor sequences and found areas of low versus high homology (FIG. 1). Regions of interest were then examined for the optimum percentage of guanine plus cytosine (60–70% in our laboratory for 48 mer length probes) (FIG. 2), and the selected probe sequences were checked against GenBank to assure that the probe would not additionally bind to some other known rodent sequence. It was interesting that a majority of probes made were from the region to be translated into the large cytoplasmic loop (FIG. 3), believed to be the most unique and functionally essential part of the receptor.[82,83] After the sequences were selected, they were synthesized commercially (Scandinavian Gene Synthesis AB, Köping, Sweden) and used in *in situ* hybridization experiments as previously described.[78–81,84,85]

Interpretation of control experiments for *in situ* hybridization is fraught with peril, and no method alone can absolutely assure specificity (see ref. 86). As previously stated, designing the probe with other partially homologous sequences in mind and comparing the probe to known sequences on GenBank are paramount. Other control experiments attempted in our various studies of adrenergic receptor subtypes included the use of Northern blots,[81,85] sense probes,[80,86] a partially homologous hybrid antisense probe,[80] competitive binding with excess nonradioactive probes,[78–81,85,86] and production of identical labeling patterns obtained from two probes designed from different regions of the same sequence.[85]

Also considered essential was the practice of using unfixed frozen tissue, because fixation may produce false-positive labeling and a poor signal-to-noise ratio.[87] Other details concerning facets of the *in situ* hybridization protocol were published previously.[86,88,89]

CELLULAR LOCALIZATION OF mRNAs FOR ALPHA-2 SUBSET SPECIFIC RECEPTORS USING *IN SITU* HYBRIDIZATION

Using unique oligonucleotide probes to RNG, RG10, and RG20, we demonstrated specific labeling patterns in various rat tissues.[78–81] These are in general agreement with the results of past Northern blot analysis of homogenized kidney and regions of rat brain tissue[26,68,72,73,77]; however, use of *in situ* hybridization provides visualization of the actual cell groups expressing these mRNAs.

In our *in situ* hybridization experiments examining the rat brain, RNG labeling was confined to the thalamus.[80] However, RG10 labeling was demonstrated in many brain regions including the islands of Calleja, olfactory bulb (FIG. 4a), cerebellar cortex (FIG. 4b), cerebral cortex, hippocampal formation (FIG. 4c), striatum[80] (FIG. 4d), and dorsal root ganglia[78,80] (FIG. 6c). RG20 labeling was seen primarily in layer six of the cerebral cortex (FIG. 5a), locus coeruleus (LC; FIG. 5b), amygdala, hypothalamic paraventricular nucleus (FIG. 5c), pontine nuclei, nucleus tractus solitarii, ventrolateral medullary reticular formation (RVL; FIG. 5d), and the dorsal horn and intermediolateral cell column (IML) of the thoracic spinal cord (FIG. 6a).[80] Additionally RG20 nuclear labeling of unknown significance was also apparent in the reticular thalamic nucleus, vestibular nuclei, trapezoid nuclei, and deep cerebellar nuclei.[80] Recent *in situ* hybridization studies from other laboratories using oligonucleotide probes[90] and riboprobes[91] demonstrated similar labeling patterns; however, these studies did not demonstrate our unusual nuclear labeling patterns as in the thalamic reticular nucleus. This suggests that our labeling of these structures may be artifactual, possibly produced by cross-binding to an as yet undiscovered nuclear sequence. An alternative explanation is that this may represent additional binding demonstrated only by our unique oligoprobe to a truncated version of the RG20 sequence that is seen in or near the nucleus which may possibly later be translated or remain functionally inactive. In contrast, other studies showed mRNA labeling where we saw none. For example, labeling was demonstrated in the cerebellum using an RG20 riboprobe and in the RVL and dorsal raphe using an RG10 riboprobe.[91] This may represent the superior sensitivity of the riboprobe technique. However, this may also represent artifact, because in the latter study, tissue slices were postfixed with paraformaldehyde which has been shown to sometimes produce false-positive labeling with oligoprobes in such cell-dense regions as the hippocampus and cerebellar cortex.[87] On the other hand, we demonstrated RG10 labeling in the thalamus and in regions of the hippocampus that were seen not with the RG10 riboprobe[91] but with other RG10 oligoprobes.[90]

In the rat kidney, RNG labeling was very intense (FIG. 7),[78–81] especially in the outer stripe of the outer medulla in the region of segment S3 of the proximal tubules,

TABLE 1. Competitive Affinities for Adrenergic Receptor Ligands (K_i, nM)

Ref./Cell Line												
α_2A **C10**	YOH	RAU	Idazox	Phent	OXY	PAC	WB	SKF	Cory	–AD	Praz	–NA
Lanier *et al.*,[73]/COS1	7	9	10	11	15			101		1,434	2,790	3,456
Lomasney *et al.*,[67]/ COS7		7.1		**6.2**	13.2	13	47	97	1,188	1,671	2,237	3,677
Kobilka *et al.*,[63]/XO	4.4	**3.5**	18	27		140			**2,200**	1,900	**880**	7,500
Regan *et al.*,[71]/COS7	1.6		10	10	11	74	**7.8**	86	710	1,000	1,800	2,400
α_2B **C2**	YOH	Chlor	Phent	RAU	SKF	PAC	WB	Praz	Cory	–NA	OXY	–AD
Lomasney *et al.*, [67]/COS7			9.2	11	105	120	132	293	1,002	1,265	1,506	1,851
Weinshenk *et al.* [64]/Ltk⁻	1.2 ± 0.1			**0.32 ± 0.5**		4.5 ± 1.9	17 ± 6	23 ± 4	77 ± 9		213 ± 13255 ± 68	
α_2B **RNG**												
Zeng *et al.* [65]/COS1	5	7						27		374	613	
Harrison *et al.*, [76]/COS1	6.9 ± 1.4		12.4 ± 2		72 ± 26			**36.3 ± 8.3**	273 ± 21209 ± 34430 ± 87			
α_2C **C4**	RAU	YOH	WB	Idazox	Phent	Praz	SKF	OXY	Cory	BHT	–AD	–NA
Lomasney *et al.*,[67]/ COS7	2.1				14.4	67.7	**41**	125			318	606
Regan *et al.*,[71]/COS7		0.93	0.94	17	33	41	**33**	62	73	140	170	240

TABLE 1. Competitive Affinities for Adrenergic Receptor Ligands (K_i, nM) (*Continued*)

Ref./Cell Line α_2C **RG10**	RAU	YOH	WB	Idazox	Phent	Praz	SKF	OXY	Cory	BHT	–AD	–NA
Lanier *et al.*,[73]/COS1	1	4		8	14	85	89	290			1,071	2,143
Flordellis *et al.*,[72]/ COS7	0.76	1.49				50		148				
Voigt *et al.*,[74]/COS7		1.5	1.6			20		34	81			
Harrison *et al.*,[76]/ COS1		2.6 ± 0.6		13.4 ± 2	22.7 ± 6.1	80 ± 13.9	87 ± 29	128 ± 35	287 ± 41	328 ± 22		388 ± 22
α_2D **RG20**	Phent	Idazox	OXY	RAU	YOH	SKF	Praz	–AD	–NA			
Lanier *et al.*,[73]/COS1	6	7	43	51	61	531	1,586	2,100	3,576			
Harrison *et al.*, [76]/COS1	3.2 ± 1	7 ±1.6	12.3 ± 2.5		42.6 ± 7.2	162 ± 26	2,530 ± 328		**633 ± 122**			

ABBREVIATIONS: AD = (–)adrenaline; BHT = BHT-920; Chlor = chlorpromazine; Cory = corynanthine; Idazox = idazoxan; Ltk^- = Mouse fibroblast Ltk– cells; –NA = (–)noradrenaline; OXY– oxymetazoline; PAC = *para*-aminoclonidine; Phent = phentolamine; Praz = prazosin; RAU = rauwolscine; SKF = SKF-104078; WB = WB-4101; XO = *Xenopus laevis* oocytes; YOH = yohimbine. Results with ± standard error of the mean are shown when available.

TABLE 2. Alpha-2 Receptor Subtype Nomenclature

Subtype	Ligand Affinity Profile	Sequences
α_2A	YOH > PHENT > OXY > AD > PRAZ > NA	C10
α_2B	YOH > PHENT > PRAZ > NA > OXY > AD	C2, RNG
α_2C	YOH > PHENT > PRAZ > OXY > AD > NA	C4, RG10
α_2D	PHENT > OXY > YOH > PRAZ (> AD > NA)	RG20

ABBREVIATIONS: AD = adrenaline; NA = noradrenaline; OXY = oxymetazoline; PHENT = phentolamine; PRAZ = prazosin; YOH = yohimbine.

with rays radiating into the cortex.[81] RG10 labeling in the rat kidney was demonstrated weakly in tubular rays of the outer and inner medulla (FIG. 7), whereas strong RG20 labeling was seen in the ureter, the inner stripe of the outer medulla, and the entire inner medulla (FIG. 7), the latter representing labeling of collecting tubules.[81]

In other peripheral tissues, RG10 labeling was seen[78] in presumed noradrenergic ganglia-type cells of the adrenal medulla[92] and in the sympathetic superior cervical

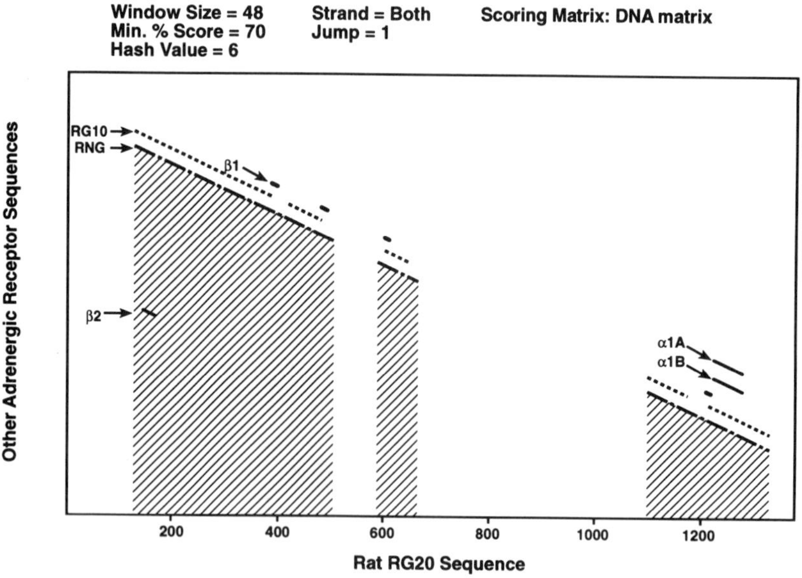

FIGURE 1. Comparison of genetic sequences within the family of adrenergic receptors. To design a unique probe to the RG20 sequence (alpha-2D adrenergic receptor), the RG20 nucleotide base sequence (X axis) was compared to the DNA sequences of the other known rat adrenergic receptor sequences (Y axis). *Shaded areas* represent areas of high homology between the other adrenergic receptors (identified parallel lines) and the RG20 sequence. These regions should be avoided when attempting to design an RG20 oligoprobe.

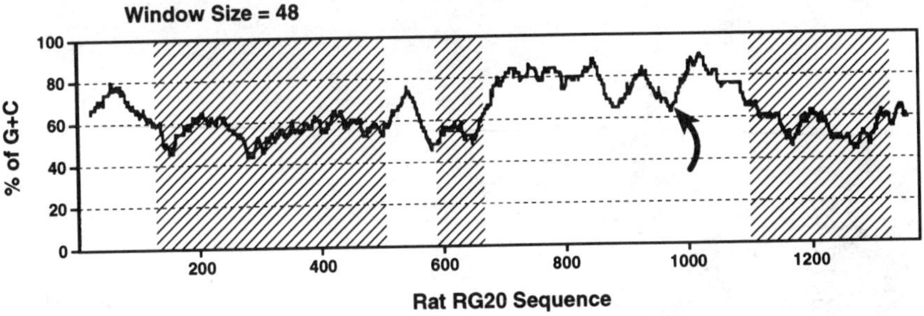

FIGURE 2. Guanine plus cytosine percentage (G + C %) of RG20 sequence with areas of high adrenergic receptor sequence homology (*shaded regions*). *Curved arrow* represents an area of low homology within the proper G + C % window (60–70%) from which the RG20 oligoprobe was made.

ganglia[78]; however, a recent *in situ* hybridization study contends that RG20 labeling is predominant in this latter structure.[93]

HYBRIDIZATION STUDIES VERSUS RADIOLIGAND-MEDIATED AUTORADIOGRAPHY

A comparison between *in situ* hybridization results and earlier autoradiographic ligand binding studies reveals an added specificity provided by molecular biological techniques. Even though McCune and collaborators[90] suggest that [^{3}H]rauwolscine binding correlates highly with the location of RG10-positive cells (alpha-2C receptors) and [^{3}H]idazoxan labeling correlates well with the location of RG20-positive cells (alpha-2D receptors), there is strong evidence that in addition to binding to alpha-2 receptors, [^{3}H]idazoxan also binds to imidazoline receptors[19,51,53,56,57,60,61] and

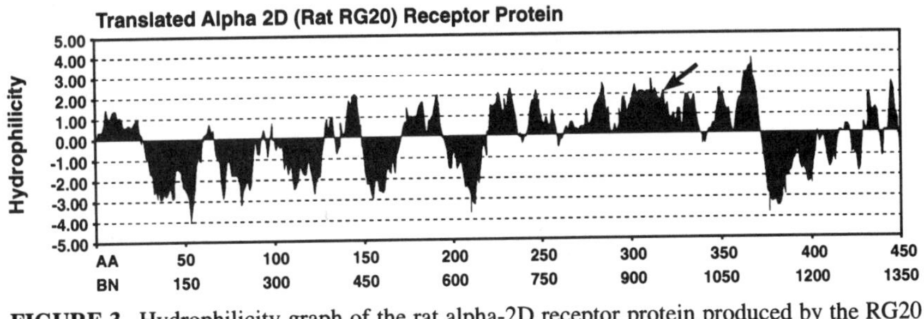

FIGURE 3. Hydrophilicity graph of the rat alpha-2D receptor protein produced by the RG20 sequence, with amino acid (AA) and base numbers (BN) shown on the X axis. Hydrophilic (above 0 line; positive values) and hydrophobic regions (below 0 line; negative values) demonstrate a typical adrenergic receptor with seven transmembrane areas (hydrophobic) and the large intracellular loop (hydrophilic) between the fifth and sixth transmembrane areas containing the region from which the RG20 oligoprobe was made (*arrow*).

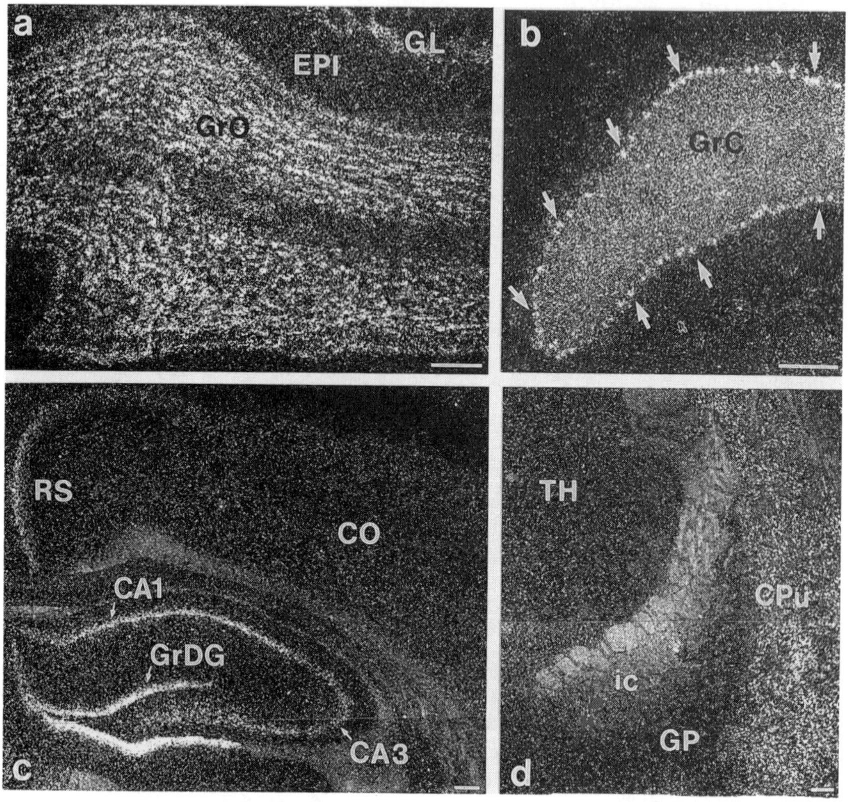

FIGURE 4. Examples of RG10 labeling in emulsion-dipped coronal sections of rat brain. (**a**) In the olfactory bulb, RG10 labeling is strong in the granular layer (GrO) and moderate in the glomerular layer (GL). (**b**) In the cerebellar cortex, RG10 labeling is seen in the granular cell layer (GrC) as well as in the purkinje cell layer (*arrows*). (**c**) At the level of the diencephalon, RG10 labeling is seen specifically in the retrosplenal region (RS) of the cerebral cortex (CO) and in the hippocampal formation. (**d**) RG10 labeling is strong in the caudate/putamen (CPu), but minimal to nonexistent in the thalamus (TH) and globus pallidus (GP), respectively. EPI = external plexiform layer; CA1, CA3 = regions of the hippocampus; GrDG = granule layer of the dentate gyrus; ic = internal capsule. Bars = 200 μm. Used with permission from Wiley-Liss, Inc. and *The Journal of Comparative Neurology.*

[^{3}H]rauwolscine also binds to serotonin-1A receptors.[19,50,51] In addition, lack of specificity of the alpha-2 receptor radioligands is exemplified by the fact that [^{3}H]idazoxan,[20] [^{3}H]rauwolscine,[20] [^{3}H]*para*-aminoclonidine,[33] and [^{125}I]iodoclonidine[38] all label the locus coeruleus, but only RG20 labeling is seen in this structure.[80,90,91] Therefore, using *in situ* hybridization and sequence-specific probes, a major postsynaptic alpha-2 receptor subtype can be clarified in a certain brain region autoradiographically labeled by one or more of the less specific alpha-2 receptor radioligands.

However, simply comparing autoradiographic data to *in situ* results may not differentiate all of the alpha-2 receptor subtypes involved, because there is an inherent

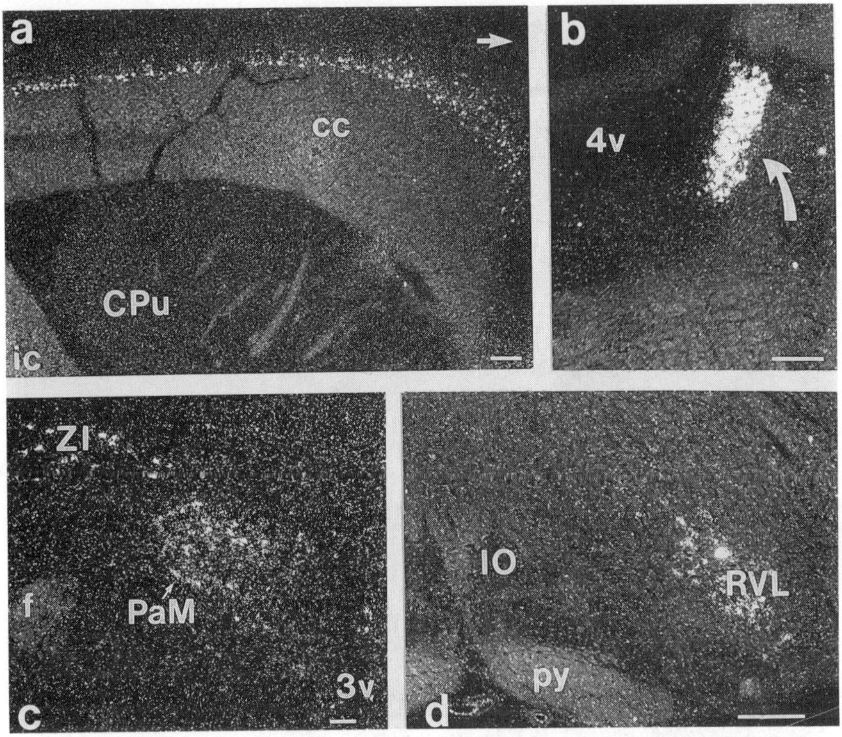

FIGURE 5. Examples of RG20 labeling in emulsion-dipped longitudinal (**a**) and coronal (**b**-**d**) sections of rat brain. (**a**) RG20 labeling is seen in layer six of the cerebral cortex above the corpus callosum (cc; *arrow* points rostrally). (**b**) In the pons, strong RG20 labeling is seen in the locus coeruleus (*curved arrow*). (**c**) In the hypothalamus, RG20 labeling is prominent in the magnocellular paraventricular nucleus (PaM) and in the zona incerta (ZI). (**d**) In the medulla oblongata, RG20 labeling is seen in the rostral ventrolateral reticular formation (RVL). cc = corpus callosum; CPu = caudate/putamen; ic = internal capsule; 4v = fourth ventricle; f = fornix; 3v = third ventricle; IO = inferior olivary nucleus; py = pyramidal tract. Bars = 200 μm. Used with permission from Wiley-Liss, Inc. and *The Journal of Comparative Neurology.*

difference between the manner in which structures are labeled with the two methods. For example, only cell bodies that express mRNA for a certain protein are labeled with *in situ* hybridization techniques,[89] whereas with autoradiographic techniques, radioligands bound to pre- or postsynaptic receptors localized on axonal and dendritic processes also are visualized.[62] Therefore, even though alpha-2 ligand binding and one alpha-2 mRNA subtype may be present in certain brain structure, the possibility of presynaptic receptors in this region cannot be ruled out (to be discussed).

In the kidney, previous autoradiographic studies using various alpha-2 ligands have indicated the presence of alpha receptors in rat renal cortex,[39,45,94,95] with the highest amount of labeling at the corticomedullary junction.[43] Some studies have further defined the localization of alpha-2 receptors to proximal tubules,[42,44,48] primar-

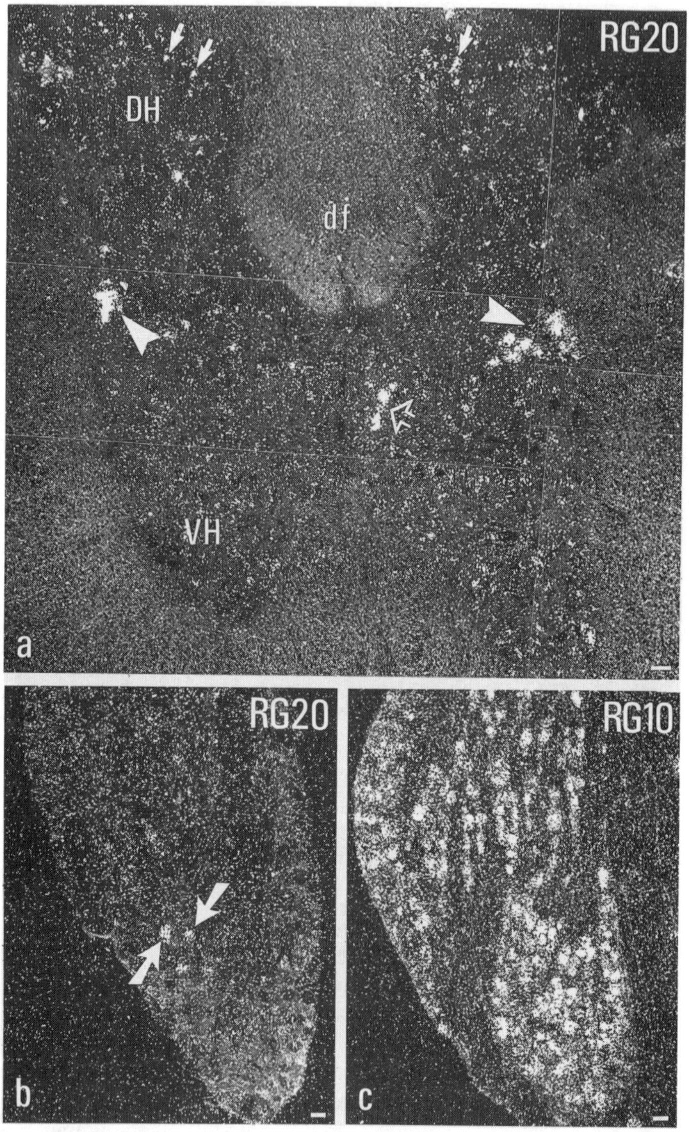

FIGURE 6. Dark field autoradiographs of the spinal cord (**a**) and dorsal root ganglion (**b,c**) after hybridization with probes to mRNA for the RG20 (**a,b**) and RG10 (**c**) alpha-2 receptors. (**a**) The strongest RG20 labeling in the spinal cord is found in the intermediolateral cell column (*arrowheads*), but numerous cells are also seen in the dorsal horn (*arrows*) and near the central canal, with some strongly labeled cells in the border area between laminae 7 and 10 (*open arrow*). (**b,c**) Most dorsal root ganglion cells are RG10 positive (**c**), but few exhibit an RG20 signal (*arrows* in **b**). Bars = 50 μm. Used with permission from Wiley-Liss, Inc. and *The Journal of Comparative Neurology.*

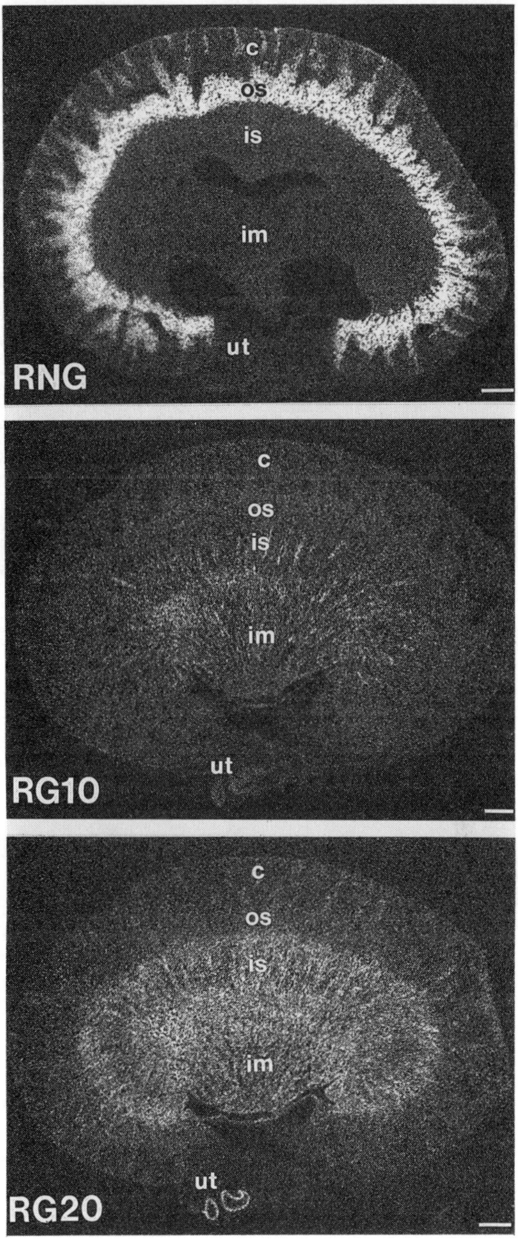

FIGURE 7. Film autoradiograms of sections of rat kidney after *in situ* hybridization with oligonucleotide probes to the alpha-2B (RNG), alpha-2C (RG10), and alpha-2D (RG20) adrenoceptor mRNAs. Very strong hybridization signal with the RNG probe is observed in the outer stripe (os) of the outer medulla extending into the cortex (c), whereas the inner medullary stripe (is) and the inner medulla (im) are devoid of labeling. Only a weak hybridization signal is seen in the inner medulla using the RG10 probe, whereas the ureter (ut), inner medullary stripe, and inner medulla are labeled strongly using the RG20 probe. Bars = 1 mm. Used with permission from *The Journal of Pharmacology and Experimental Therapeutics.*

ily in the proximal convoluted tubules,[96] and a higher concentration associated with the basolateral membrane.[97] In our studies of the kidney,[78,79,81] predominant labeling with the alpha-2B (RNG) and also the alpha-2C (RG10) probe was observed in segment S3, which represents the straight portion of proximal tubules, suggesting the presence of these receptors in this part of the nephron.[81]

In addition to proximal and collecting tubules, however, previous autoradiographic studies have also shown alpha-2 radioligand labeling associated with glomeruli[40,49] and renal vascular structures such as arterioles and the vasa recti,[43,45,49] whereas these structure were not labeled with any of the alpha-2 probes. This apparent discrepancy has many possible explanations. First, *in situ* hybridization using oligonucleotide probes may not be a sensitive enough technique to detect possible very low levels of alpha-2 mRNA in these structures. As recently demonstrated in the rat brain,[91] perhaps the use of full-length riboprobes in future kidney experiments will have greater sensitivity. Again, another possible explanation may lie in the inherent differences of structures labeled with *in situ* hybridization versus radioligand-mediated autoradiography. For example, using the latter technique, pre- and postsynaptic receptors are both labeled, whereas using *in situ* hybridization, only the cell bodies that make mRNA for these receptors are labeled. Therefore, inasmuch as it was previously demonstrated that there is a rich sympathetic innervation to structures such as glomeruli and renal vessels[98] as well as to tubule cells,[46,99] perhaps alpha-2 radioligand binding of these structures represents presynaptic alpha-2 receptors. This may correlate with *in situ* hybridization studies showing intense RG10[78] and RG20[93] labeling of neuronal cell bodies in sympathetic ganglia.

Pharmacological studies using [^{3}H]-RX821002 (an alpha-2A antagonist) in conjunction with guanoxabenz (an alpha-2 agonist) and ARC 239 (an alpha-2B antagonist) in competition binding experiments have indicated three subtypes of alpha-2 receptors in the kidney.[27] This is in agreement with our studies showing hybridization signal with probes to the three alpha-2 receptor subtypes in the kidney.[81]

HYBRIDIZATION STUDIES AS AN ADJUNCT TO RADIOLIGAND-MEDIATED AUTORADIOGRAPHY

Alpha-2 Receptors and Bulbospinal Antinociception Pathways

One interesting result of our *in situ* hybridization experiments was the finding of a majority of primary sensory neurons containing RG10 mRNA. This suggests that a percentage of radioligand binding seen in the dorsal horn[16,28,33,35] may represent presynaptic alpha-2C receptors made in dorsal root ganglion cells. Thus, presynaptic alpha-2 receptors made in non-catecholaminergic cells, such as primary sensory neurons, may be used to regulate the release of other neurotransmitters from these cells. In this case, possible catecholamine sources include noradrenaline from the locus coeruleus[100] or, as shown in the primate, adrenaline from the rostral ventrolateral reticular formation (RVL).[101] It can be theorized that these catecholamines may interact primarily with alpha-2C receptors on primary afferent terminals, thus influencing the release of excitatory neurotransmitters such as glutamate and substance P. This idea is supported by several lines of evidence. Thus, using different methodologies,

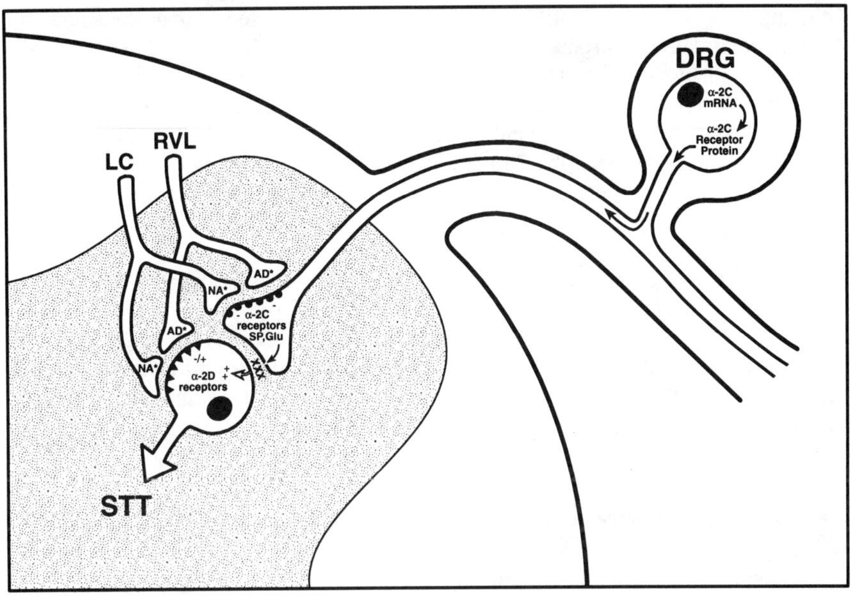

FIGURE 8. In primary sensory neurons of dorsal root ganglia (DRG), alpha-2C mRNA (RG10) is translated into alpha-2C receptor protein. This is transported into the dorsal horn and expressed on presynaptic terminals that normally release excitatory neurotransmitters such as substance P (SP) and glutamate (Glu). Adrenaline (AD) produced in the C1 cells of the rostral ventrolateral medulla (RVL, in primates) and noradrenaline (NA) produced in the locus coeruleus (LC) may be released onto presynaptic alpha-2C receptors which then shut off (XXX) the release of Glu and SP. NA and AD theoretically may also be released directly onto alpha-2D receptors found postsynaptically on spinothalamic tract (STT) cells (or other dorsal horn interneurons) and/or presynaptically (*asterisks*) on terminals from neurons in the RVL or LC, because RG20 labeling is seen in neurons of the RVL, LC, and dorsal horn.

noradrenaline and alpha-adrenergic agonists were shown to exert analgesic effects, presumably via alpha-2 adrenoceptors.[102–108] Furthermore, a previous electron microscopic study showed noradrenergic terminals synapsing on non-noradrenergic terminals in the dorsal horn.[109] Finally, an *in vivo* study demonstrated that substance P release in the dorsal horn produced by peripheral noxious stimuli can be blocked by local application of noradrenaline to the dorsal horn and that this reaction can be reversed by the alpha-2 antagonist yohimbine.[110]

Therefore, perhaps bulbospinal pathways from the adrenergic C1 cell group[111] in the RVL and/or the noradrenergic locus coeruleus neurons release their catecholamine neurotransmitters primarily onto presynaptic alpha-2C receptors found on terminals of primary afferents, thus antagonizing the release of nociceptive neurotransmitters such as glutamate or substance P onto spinothalamic tract cells in the dorsal horn (FIG. 8). Theoretically these catecholamines may also influence dorsal horn neurons directly, because our studies have shown a number of RG20-positive (alpha-2D receptor-producing) cells in this area. This is supported by an exhaustive competitive binding study demonstrating that the major postsynaptic receptor in the spinal cord is consistent with that produced by the RG20 sequence.[28] In the cat, [^{3}H]rauwolscine binding in the dorsal horn decreased by 20% after dorsal root ganglionectomy[112] and negligibly after cervical hemisection,[113] suggesting that a vast majority of adrenoceptors labeled with this radioligand are intrinsic to the dorsal horn. Since, as stated previously, many of the alpha-2 radioligands cross-react with other receptors,[19,22,50–61] the contribution of alpha-2 receptor subtypes from bulbospinal and primary sensory versus intrinsic dorsal horn neurons, especially in the rat and human, remain to be determined.

Furthermore, both putative catecholaminergic bulbospinal pathways from the locus coeruleus and the C1 area may be autoregulated presynaptically via alpha-2D receptors, because we have shown that cells in both of these areas of the brain stem contained RG20 labeling.[80] This is partially supported by experiments with a dorsal horn slice preparation showing inhibition of electrically stimulated intraspinal [^{3}H]noradrenaline release by the alpha-2 agonist clonidine.[114]

Alpha-2 Receptors and Bulbospinal Cardiovascular Pathways

Even though a wealth of evidence suggests that imidazoline receptors may be more important than alpha-2 receptors in regulating cardiovascular responses within the medulla oblongata,[58,115–121] we performed a number of experiments utilizing tract tracing, immunocytochemical, *in situ* hybridization, and physiological techniques to understand the cardiovascular regulatory functioning of catecholamines and alpha receptors in the IML of the thoracic spinal cord. Using tricolor immunocytochemistry, we showed that adrenergic C1 neurons[111] in the RVL may produce glutamate and substance P in addition to a catecholamine.[122] Because neurons in this pressor area of the brain stem[123] have been shown with anterograde[124–128] and retrograde tract-tracing techniques[124,125,129–132] to project to the IML second synapse directly on sympathetic preganglionic neurons,[133] we propose a temporal theory for cardiovascular regulation of this bulbospinal system (Fig. 9). We suggest that minute-to-minute excitatory drive to sympathetic preganglionic neurons is mediated by the amino acid glutamate, which has been shown to briefly excite these neurons.[134] Thus, C1 neurons and/or adjacent noncatecholamine, glutaminergic, bulbospinal pacemaker cells[135] may tonically release this excitatory neurotransmitter directly into the IML.[126,136] In times of increased sympathetic need, the peptide substance P may then be produced and released from bulbospinal C1 neurons into the IML, resulting in a more long-lasting excitatory response.[137,138] When the pressor response requires termination, it can be theorized that adrenaline may then be released from C1 terminals and produce an inhibitory effect via alpha-2D autoreceptors shutting off the release of substance P and glutamate presynaptically and/or postsynaptically by alpha-2D receptors. This is consistent with our results showing intense RG20 labeling in the C1 area and in the IML using *in situ* hybridization techniques.[80]

Because noradrenaline produced primarily from the locus coeruleus is also released in the IML, debate raged concerning whether noradrenaline and/or adrenaline inhibited or excited sympathetic preganglionic neurons.[139–149] However, during the last few years it was shown that catecholamines may produce their effects on different

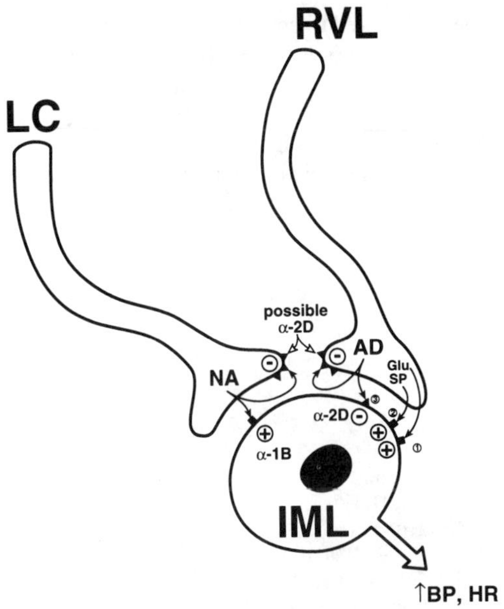

FIGURE 9. When excited, sympathetic preganglionic neurons in the intermediolateral cell column (IML) normally produce increases in blood pressure (BP) and heart rate (HR). We propose that C1 neurons in the rostral ventrolateral medulla (RVL) first release glutamate (Glu) then substance P (SP) to excite these neurons in times of stress. Sympathoinhibition is later obtained by the release of adrenaline (AD) which preferentially binds to postsynaptic alpha-2D receptors. Noradrenaline (NA) produced in the locus coeruleus (LC) is also released into the IML producing excitation of these cells via postsynaptic alpha-1B receptors. Autoregulation via presynaptic alpha-2D receptors is also possible, as RG20 labeling is seen in the LC and RVL.

adrenergic receptors depending on the concentration released. For example, microinjections of 1-2 pmol (presumed physiologic) quantities of noradrenaline in the IML produced tachycardia that was blocked only by alpha-1 receptors, whereas higher doses produced bradycardia that was blocked only by alpha-2 receptors.[150] This is supported by our own experiments[151] utilizing an intrathecal drug administration technique.[152] We demonstrated that low doses (1 mg) of intrathecally injected noradrenaline produced pressor responses and tachycardia blocked by the alpha-1 receptor antagonist prazosin and the same dose of adrenaline produced depressor responses and bradycardia blocked by the alpha-2 receptor inhibitor yohimbine.[150] When doses were increased, opposite effects were produced, and it may be proposed that these catecholamines probably saturated their preferred receptors and started to bind to less preferred adrenergic receptors at the higher doses. This also implies that if large amounts of catecholamines are released in the IML (or in the dorsal horn), they may exert their influences not only at the most immediate synaptic site, but also at more distant sites, as predicted by the "volume transmission" theory (see ref. 153). Autoradiography demonstrates binding of alpha-1[154] and alpha-2 radioligands[16,33] in the IML, but our *in situ* hybridization experiments specifically identify these receptors as the alpha-1B[85] and alpha-2D (RG20)[80] receptors, respectively. Finally, noradrenaline release from the LC may also be autoregulated by presynaptic alpha-2D receptors, as we showed intense RG20 labeling in this structure.[80]

CONCLUSION

Because it is now known that adrenergic drugs previously used as radioligands cannot differentiate between subsets of alpha-2 receptors, *in situ* hybridization has been employed to complement radioligand-mediated autoradiography to better localize cells that produce specific alpha-2 receptors. Inasmuch as the hybridization signal can be quantified, specific receptor mRNAs also can be compared in tissue from control animals versus that from diseased or pharmacologically manipulated animals.[79,87-89] Radioligand-mediated autoradiography and *in situ* hybridization combined with knowledge obtained with tract tracing and immunohistochemistry may help neuroscientists better understand the neuroanatomy of catecholamine pathways. With this knowledge, pharmacologists may better design new medicines that are specific for subsets of adrenergic receptors in specific tissues and brain areas, which may produce more efficacious results and fewer side effects.

REFERENCES

1. Ahlquist, R. P. 1948. Am. J. Physiol. **153:** 586-600.
2. Langer, S. Z. 1974. Biochem. Pharmacol. **23:** 1793-1800.
3. Starke, K., T. Endo & H. D. Taube. 1975. Naunyn-Schmiedeberg's Arch. Pharmacol. **291:** 55-78.
4. Taube, H. D., K. Starke & E. Borowski. 1977. Naunyn-Schmiedeberg's Arch. Pharmacol. **299:** 123-141.
5. Greenberg, D. A., D. C. U'Prichard & S. H. Snyder. 1976. Life Sci. **19:** 69-76.
6. Berthelsen, S. & W. A. Pettinger. 1977. Life Sci. **21:** 595-606.
7. Wikberg, J. E. S. 1979. Acta Physiol. Scand. (Suppl.) **468:** 1-99.
8. Fain, J. N. & J. A. García-Sáinz. 1980. Life Sci. **26:** 1183-1194.
9. Hoffman, B. B., D. Mullikin-Kilpatrick & R. J. Lefkowitz. 1980. J. Biol. Chem. **255:** 4645-4652.
10. U'Prichard, D. C., W. D. Bechtel, B. M. Rouot & S. H. Snyder. 1979. Mol. Pharmacol. **16:** 47-60.
11. U'Prichard, D. C. & S. H. Snyder. 1979. Life Sci. **24:** 79-88.
12. Bylund, D. B. & D. C. U'Prichard. 1983. Int. Rev. Neurobiol. **24:** 343-431.
13. Bylund, D. B. 1981. Pharmacologist **23:** 215.
14. Perry, B. D. & D. C. U'Prichard. 1981. Eur. J. Pharmacol. **76:** 461-464.
15. Bylund, D. B. & C. Ray. 1983. Trans. Am. Soc. Neurochem. **14:** 160.
16. Seybold, V. S. & R. P. Elde. 1984. J. Neurosci. **4:** 2533-2542.
17. Bylund, D. B. 1985. Pharmacol. Biochem. Behav. **22:** 835-843.
18. Maura, G., A. Gemignani & M. Raiteri. 1985. Eur. J. Pharmacol. **116:** 335-339.
19. Boyajian, C. L. & F. M. Leslie. 1987. J. Pharmacol. Exp. Ther. **241:** 1092-1098.
20. Boyaijan, C. L., S. E. Loughlin & F. M. Leslie. 1987. J. Pharmacol. Exp. Ther. **241:** 1079-1091.
21. Broadhurst, A. M., B. S. Alexander & M. D. Wood. 1988. Life Sci. **43:** 83-92.
22. Bylund, D. B., C. Ray-Prenger & T. J. Murphy. 1988. J. Pharmacol. Exp. Ther. **245:** 600-607.
23. Lanier, S. M., C. J. Homey, C. Patenaude & R. M. Graham. 1988. J. Biol. Chem. **263:** 14491-14496.
24. Michel, A. D., D. N. Loury & R. L. Whiting. 1989. Br. J. Pharmacol. **98:** 890-897.
25. Brown, C. M., A. C. MacKinnon, J. C. McGrath, M. Spedding & A. T. Kilpatrick. 1990. Br. J. Pharmacol. **99:** 481-486.
26. Lorenz, W., J. W. Lomasney, S. Collins, J. W. Regan, M. G. Caron & R. J. Lefkowitz. 1990. Mol. Pharmacol. **38:** 599-603.
27. Uhlén, S. & J. E. S. Wikberg. 1991. Br. J. Pharmacol. **104:** 657-664.
28. Uhlén, S. & J. E. S. Wikberg. 1991. Pharmacol. Toxicol. **69:** 341-350.
29. Young, W. S., III, & M. J. Kuhar. 1979. Eur. J. Pharmacol. **59:** 317-319.
30. Young, W. S., III, & M. J. Kuhar. 1980. Proc. Natl. Acad. Sci. USA **77:** 1696-1700.

31. CHEUNG, Y.-D., D. B. BARNETT & S. R. NAHORSKI. 1982. Eur. J. Pharmacol. **84:** 79-85.
32. ROUOT, B., M. C. QUENNEDEY & J. SCHWARTZ. 1982. Naunyn-Schmiedeberg's Arch. Pharmacol. **321:** 253-259.
33. UNNERSTALL, J. R., T. A. KOPAJTIC & M. J. KUHAR. 1984. Brain Res. Rev. **7:** 69-101.
34. PETRASH, A. C. & D. B. BYLUND. 1986. Life Sci. **38:** 2129-2137.
35. SULLIVAN, A. F., M. R. DASHWOOD & A. H. DICKENSON. 1987. Eur. J. Pharmacol. **138:** 169-177.
36. CUMMINGS, S. & V. SEYBOLD. 1988. Neuroendocrinology **47:** 523-532.
37. CHAMBA, G., D. WEISSMANN, C. ROUSSET, B. RENAUD & J. F. PUJOL. 1991. Brain Res. Bull. **26:** 185-193.
38. LUQUE, J. M., A. GUILLAMON & B. H. HWANG. 1991. Neurosci. Lett. **131:** 163-166.
39. MCPHERSON, G. A. & R. J. SUMMERS. 1981. J. Pharm. Pharmacol. **33:** 189-191.
40. MCPHERSON, G. A. & R. J. SUMMERS. 1983. Eur. J. Pharmacol. **90:** 333-341.
41. SUMMERS, R. J. 1984. Fed. Proc. **43:** 2917-2922.
42. INSEL, P. A., M. D. SNAVELY, D. P. HEALEY, P. A. MÜNZEL, C. L. POTENZA & E. P. NORD. 1985. J. Cardiovasc. Pharmacol. **7** (Suppl. 8): S9-S17.
43. SRIPANIDKULCHAI, B. & J. M. WYSS. 1985. J. Cell. Biol. **101:** 133a.
44. STEPHENSON, J. A. & R. J. SUMMERS. 1985. Eur. J. Pharmacol. **116:** 271-278.
45. MUNTZ, K. H., L. MEYER, S. GADOL & T. A. CALIANOS. 1986. J. Pharmacol. Exp. Ther. **236:** 542-547.
46. SRIPANIDKULCHAI, B. & J. M. WYSS. 1987. Brain Res. **400:** 91-100.
47. SRIPANIDKULCHAI, B., R. DAWSON, S. OPARIL & J. M. WYSS. 1987. Am. J. Physiol. **252:** F283-F290.
48. SUNDARESAN, P. R., T. L. FORTIN & S. L. KELVIE. 1987. Am. J. Physiol. **253:** F848-F856.
49. CALIANOS II, T. & K. H. MUNTZ. 1990. Kidney Int. **38:** 39-46.
50. CONVENTS, A., D. CONVENTS, J.-P. DE BACKER, J. DE KEYSER & G. VAUQUELIN. 1989. Biochem. Pharmacol. **38:** 455-463.
51. DE VOS, H., A. CONVENTS, J. DE KEYSER, J.-P. DE BACKER, I. J. B. VAN MEGEN, G. EBINGER & G. VANQUELIN. 1991. Brain Res. **566:** 13-20.
52. WIKBERG, J. E. S. 1989. Pharmacol. Toxicol. **64:** 152-155.
53. ERNSBERGER, P., M. P. MEELEY, J. J. MANN & D. J. REIS. 1987. Eur. J. Pharmacol. **134:** 1-13.
54. BRICCA, G., M. DONTENWILL, A. MOLINES, J. FELDMAN, A. BELCOURT & P. BOUSQUET. 1988. Eur. J. Pharmacol. **150:** 401-402.
55. BRICCA, G., M. DONTENWILL, A. MOLINES, J. FELDMAN, A. BELCOURT & P. BOUSQUET. 1989. Eur. J. Pharmacol. **162:** 1-9.
56. MICHEL, M. C., O.-E. BRODDE, B. SCHNEPEL, R. TSCHADA, H. J. MOTULSKY & P. A. INSEL. 1989. Mol. Pharmacol. **35:** 324-330.
57. PARINI, A., I. COUPRY, R. M. GRAHAM, I. UZIRLLI, D. ATLAS & S. M. LANIER. 1989. J. Biol. Chem. **264:** 11874-11878.
58. ERNSBERGER, P., R. GIULIANO, R. N. WILLETTE & D. J. REIS. 1990. J. Pharmacol. Exp. Ther. **253:** 408-418.
59. KAMISAKI, Y., T. ISHIKAWA, Y. TAKAO, H. OMODANI, N. KUNO & T. ITHO. 1990. Brain Res. **514:** 15-21.
60. MICHEL, M. C., J. W. REGAN, M. A. GERHARDT, R. R. NEUBIG, P. A. INSEL & H. J. MOTULSKY. 1990. Mol. Pharmacol. **37:** 65-68.
61. MALLARD, N. J., A. L. HUDSON & D. J. NUTT. 1992. Br. J. Pharmacol. **106:** 1019-1027.
62. YOUNG, W. S., III & M. J. KUHAR. 1979. Brain Res. **179:** 255-270.
63. KOBILKA, B. K., H. MATSUI, T. S. KOBILKA, T. L. YANG-FENG, U. FRANCKE, M. G. CARON, R. J. LEFKOWITZ & J. W. REGAN. 1987. Science **238:** 650-656.
64. WEINSHANK, R. L., J. M. ZGOMBICK, M. MACCHI, N. ADHAM, H. LITCHBLAU, T. A. BRANCHEK & P. R. HARTIG. 1990. Mol. Pharmacol. **38:** 681-688.

65. Zeng, D., J. K. Harrison, D. D'Angelo, C. M. Barber, A. L. Tucker, Z. Lu & K. R. Lynch. 1990. Proc. Natl. Acad. Sci. USA **87:** 3102-3106.
66. Lomasney, J. W., S. Coteccchia, R. J. Lefkowitz & M. G. Caron. 1991. Bioch. Biophys. Acta **1095:** 127-139.
67. Lomasney, J. W., W. Lorenz, L. F. Allen, K. King, J. W. Regan, T. L. Yang-Feng, M. G. Caron & R. J. Lefkowitz. 1990. Proc. Natl. Acad. Sci. USA **87:** 5094-5098.
68. Zeng, D. & K. R. Lynch. 1991. Mol. Brain Res. **10:** 219-225.
69. Bylund, D. B. 1988. Trends Pharmacol. Sci. **9:** 356-361.
70. Murphy, T. J. & D. B. Bylund. 1988. J. Pharmacol. Exp. Ther. **244:** 571-578.
71. Regan, J. W., T. S. Kobilka, T. L. Yang-Feng, M. G. Caron, R. J. Lefkowitz & B. K. Kobilka. 1988. Proc. Natl. Acad. Sci. USA **85:** 6301-6305.
72. Flordellis, C. S., D. E. Handy, M. R. Bresnahan, V. I. Zannis & H. Gavras. 1991. Proc. Natl. Acad. Sci. USA **88:** 1019-1023.
73. Lanier, S. M., S. Downing, E. Duzic & C. J. Homcy. 1991. J. Biol. Chem. **266:** 10470-10478.
74. Voigt, M. M., S. K. McCune, R. Y. Kanterman & C. C. Felder. 1991. FEBS Lett. **278:** 45-50.
75. Feller, D. J. & D. B. Bylund. 1984. J. Pharmacol. Exp. Ther. **228:** 275-282.
76. Harrison, J. K., D. D. D'Angelo, D. Zeng & K. R. Lynch. 1991. Mol. Pharmacol. **40:** 407-412.
77. McCune, S. K. & M. M. Voigt. 1991. J. Mol. Neurosci. **3:** 29-37.
78. Nicholas, A. P., V. A. Pieribone, R. Elde & T. Hökfelt. 1991. Mol. Cell. Neurosci. **2:** 344-350.
79. Meister, B., Å. Dagerlind, A. P. Nicholas & T. Hökfelt. 1993. Eur. J. Pharmacol. **247:** 229-232.
80. Nicholas, A. P., V. Pieribone & T. Hökfelt. 1993. J. Comp. Neurol. **328:** 575-594.
81. Meister, B., Å. Dagerlind, A. P. Nicholas & T. Hökfelt. 1994. J. Pharmacol. Exp. Ther. **268:** 1605-1611.
82. Lefkowitz, R. J., S. Cotecchia, M. A. Kjelsberg, J. Pitcher, W. J. Koch, J. Inglese & M. G. Caron. 1993. Adrenergic receptors: Recent insights into their mechanism of activation and desensitization. *In* Advances in Second Messenger and Phosphoprotein Research. B. Brown & P. R. M. Dobson, Eds. Vol. **28:** 1-9. Raven Press. New York.
83. Ren, Q., H. Kurose, R. J. Lefkowitz & S. Cotecchia. 1993. J. Biol. Chem. **268:** 16483-16487.
84. Nicholas, A. P., V. A. Pieribone & T. Hökfelt. 1993. Neuroscience **56:** 1023-1039.
85. Pieribone, V. A., A. P. Nicholas, Å. Dagerlind & T. Hökfelt. 1994. J. Neurosci. **14:** 4252-4268.
86. Dagerlind, Å. 1994. The Adrenal Medulla. In Situ Hybridization and Immunohistochemical Studies on Neuronal and Hormonal Regulation of Catecholamine Synthesizing Enzymes, Peptides and Immediate-Early Genes. Repro Print, AB. Stockholm, Sweden.
87. Dagerlind, Å., K. Friberg, A. J. Bean & T. Hökfelt. 1992. Histochemistry **98:** 39-49.
88. Schalling, M. 1990. In Situ Hybridization Studies on Regulatory Molecules in Neural and Endocrine Tissues with Special Reference to Expression of Coexisting Peptides. ISBN. Stockholm, Sweden.
89. Young III, W. S. 1990. In situ hybridization histochemistry. *In* Handbook of Chemical Neuroanatomy: Analysis of Neuronal Microcircuits and Synaptic Interactions. A. Björklund, T. Hökfelt, F. G. Wouterlood & A. N. van den Pol, Eds. Vol. **8:** 481-512. Elsevier. Amsterdam, The Netherlands.
90. McCune, S. K., M. M. Voigt & J. M. Hill. 1993. Neuroscience **57:** 143-151.
91. Scheinin, M., J. W. Lomasney, D. M. Hayden-Hixson, U. B. Schambra, M. G. Caron, R. J. Lefkowitz & R. T. Fremeau. 1994. Mol. Brain Res. **21:** 133-149.
92. Dagerlind, Å., M. Goldstein & T. Hökfelt. 1990. NeuroReport **1:** 137-140.
93. Vidovic, M., D. Cohen & C. E. Hill. 1994. Mol. Brain Res. **22:** 49-56.

94. Snavely, M. D. & P. A. Insel. 1982. Mol. Pharmacol. **22:** 532-546.
95. Woodcock, E. A. & C. I. Johnston. 1982. Mol. Pharmacol. **22:** 589-594.
96. Umemura, S., D. Marver, D. D. Smyth & W. A. Pettinger. 1985. Am. J. Physiol. **249:** F28-F33.
97. Matsushima, Y., S. Akabane & K. Ito. 1986. Biochem. Pharmacol. **35:** 2593-2600.
98. DiBona, G. F. 1982. Rev. Physiol. Biochem. Pharmacol. **94:** 75-181.
99. Barajas, L., K. Powers & P. Wang. 1984. Am. J. Physiol. **247:** F50-F60.
100. Nygren, L.-G. & L. Olson. 1977. Brain Res. **132:** 85-93.
101. Carlton, S. M., C. N. Honda, W. S. Willcockson, M. Lacrampe, D. Zhang, L. Denoroy, J. M. Chung & W. D. Willis. 1991. Brain Res. **543:** 77-90.
102. Engberg, I. & R. W. Ryall. 1966. J. Physiol. **185:** 298-322.
103. Headley, P. M., A. W. Duggan & B. T. Griersmith. 1978. Brain Res. **145:** 185-189.
104. Reddy, S. V. R., J. L. Maderdrut & T. L. Yaksh. 1980. J. Pharmacol. Exp. Ther. **213:** 525-533.
105. Yaksh, T. L. & S.V. R. Reddy. 1981. Anesthesiology **54:**451-467.
106. Barbaro, N. M., D. L. Hammond & H. L. Fields. 1985. Brain Res. **343:** 223-229.
107. Jeftinija, S., K. Semba & M. Randić. 1981. Brain Res. **219:** 456-463.
108. Fleetwood-Walker, S. M., R. Mitchell, P. J. Hope, V. Molony & A. Iggo. 1985. Brain Res. **334:** 243-254.
109. Satoh, K., A. Kashiba, H. Kimura & T. Maeda. 1982. Cell Tissue Res. **222:** 359-378.
110. Kuraishi, Y., N. Hirota, Y. Sato, S. Kaneko, M. Satoh & H. Takagi. 1985. Brain Res. **359:** 177-182.
111. Hökfelt, T., K. Fuxe, M. Goldstein & O. Johansson. 1974. Brain Res. **66:** 235-251.
112. Howe, J. R., T. L. Yaksh & V. L. W. Go. 1987. Neuroscience **21:** 385-394.
113. Howe, J. R., T. L. Yaksh & G. M. Tyce. 1987. Neuroscience **21:** 377-384.
114. Reimann, W. & F. Schneider. 1989. Eur. J. Pharmacol. **167:** 161-166.
115. Meeley, M. P., P. R. Ernsberger, A. R. Granata & D. J. Reis. 1986. Life Sci. **38:** 1119-1126.
116. Bousquet, P., J. Feldman & J. Schwartz. 1984. J. Pharmacol. Exp. Ther. **230:** 232-236.
117. Bousquet, P., J. Feldman & D. Atlas. 1986. Eur. J. Pharmacol. **124:** 167-170.
118. Atlas, D. 1991. Biochem. Pharmacol. **41:** 1541-1549.
119. Gomez, R. E., P. Ernsberger, G. Feinland & D. J. Reis. 1991. Eur. J. Pharmacol. **195:**181-191.
120. Hamilton, C. A. 1992. Pharmacol. Ther. **54:** 231-248.
121. Reis, D. J., S. Regunathan & M. P. Meeley. 1992. Am. J. Hypertens. **5:** 51S-57S.
122. Nicholas, A. P., A. C. Cuello, M. Goldstein & T. Hökfelt. 1990. NeuroReport **1:** 235-238.
123. Ross, C. A., D. A. Ruggiero, D. H. Park, T. H. Joh, A. F. Sved, J. Fernandez-Pardal, J. M. Saavedra & D. J. Reis. 1984. J. Neurosci. **4:** 474-494.
124. Loewy, A. D., J. H. Wallach & S. McKellar. 1981. Brain Res. Rev. **3:** 63-80.
125. Ross, C. A., D. A. Ruggiero, T. H. Joh, D. H. Park & D. J. Reis. 1984. J. Comp. Neurol. **228:** 168-185.
126. Morrison, S. F., J. Callaway, T. A. Milner & D. J. Reis. 1991. Brain Res. **562:** 126-135.
127. Nicholas, A. P. & M. B. Hancock. 1991. Neurosci. Lett. **122:** 91-95.
128. Zagon, A. & S. J. Bacon. 1991. Eur. J. Neurosci. **3:** 55-65.
129. Hancock, M. B. & A. P. Nicholas. 1987. Brain Res. Bull. **18:** 213-219.
130. Tucker, D. C., C. B. Saper, D. A. Ruggiero & D. J. Reis. 1987. J. Comp. Neurol. **259:** 591-603.
131. Nicholas, A. P. & M. B. Hancock. 1988. Synapse **2:** 569-576.
132. Minson, J., I. Llewellyn-Smith, A. Neville, P. Somogyi & J. Chalmers. 1990. Auton. Nerv. Syst. **30:** 209-220.

133. MILNER, T. A., S. F. MORRISON, C. ABATE & D. J. REIS. 1988. Brain Res. **448:** 205-222.
134. BACKMAN, S. B. & J. L. HENRY. 1984. Can. J. Physiol. Pharmacol. **62:** 248-251.
135. SUN, M.-K., B. S. YOUNG, J. T. HACKETT & P. G. GUYENET. 1988. Brain Res. **451:** 345-349.
136. KAPOOR, V., J. MINSON & J. CHALMERS. 1992. NeuroReport **3:** 55-58.
137. BACKMAN, S. B. & J. L. HENRY. 1983. Brain Res. **277:** 370-374.
138. GILBEY, M. P., K. E. MCKENNA & L. P. SCHRAMM. 1983. Neurosci. Lett. **41:** 157-159.
139. FUXE, K., P. BOLME, L. F. AGNATI, G. JONSSON, K. ANDERSON, C. KÖHLER & T. HÖKFELT. 1980. On the role of central adrenaline neurons in central cardiovascular regulation. *In* Central Adrenaline Neurons: Basic Aspects and Their Role in Cardiovascular Functions. K. Fuxe, M. Goldstein, B. Hökfelt, & T. Hökfelt, Eds.: 161-182. Pergamon Press. New York.
140. COOTE, J. H., V. H. MACLEOD, S. FLEETWOOD-WALKER & M. P. GILBEY. 1981. Brain Res. **215:** 135-145.
141. GUYENET, P. G. & J. B. J. CABOT. 1981. Neuroscience **1:** 908-917.
142. FRANZ, D. N., P. W. MADSEN, R. G. PETERSON & C. SANGDEE. 1982. Clin. Exp. Hypertens. **A4:** 543-562.
143. GUYENET, P. G. & R. L. STORNETTA. 1982. Brain Res. **235:** 271-283.
144. KADZIELAWA, K. 1983. Neuropharmacology **22:** 3-17.
145. SANGDEE, C. & D. N. FRANZ. 1983. Neurosci. Lett. **37:** 167-173.
146. REIS, D. J., A. R. GRANATA, T. H. JOH, C. A. ROSS, D. A. RUGGIERO & D. H. PARK. 1984. Hypertension **6** (Suppl. II): 7-15.
147. NISHI, S., M. YOSHIMURA & C. POLOSA. 1987. Synaptic potentials and putative transmitter actions in sympathetic preganglionic neurons. *In* Organization of the Autonomic Nervous System: Central and Peripheral Mechanisms. J. Ciriello, F. R. Calaresu, L. P. Renaud & C. Polosa, Eds.: 15-26. Alan R. Liss, Inc. New York.
148. COOTE, J. H. 1988. Rev. Physiol. Biochem. Pharmacol. **110:** 147-285.
149. MCCALL, R. B. 1988. Ann. Rev. Physiol. **50:** 553-564.
150. SUNDARAM, K., J. MURUGAIAN & H. SAPRU. 1991. Brain Res. **544:** 227-234.
151. GRADIN, K., A. P. NICHOLAS, P. HJEMDAHL, T. SVENSSON & T. HÖKFELT. 1992. J. Cardivasc. Pharmacol. **20:** 367-374.
152. YAKSH, T. L. & T. A. RUDY. 1976. Physiol. Behav. **17:** 1031-1036.
153. FUXE, K. & L. F. AGNATI. 1991. Volume Transmission in the Brain. Raven Press. New York.
154. MARKS, S. A., R. D. STEIN, M. R. DASHWOOD & M. P. GILBEY. 1990. Brain Res. **530:** 321-324.

Potassium Channels, Imidazolines, and Insulin-Secreting Cells[a]

MARK J. DUNNE,[b,c] ELIZABETH A. HARDING,[b] JONATHAN H. JAGGAR,[b] PAUL E. SQUIRES,[b] RIXIN LIANG,[b] CHARLOTTE KANE,[b] ROGER F. L. JAMES,[d] AND NICHOLAS J. M. LONDON[d]

[b]*Cell Biology Research Group*
Department of Biomedical Science
University of Sheffield
Sheffield S10 2TN, UK

[d]*Department of Surgery*
Leicester University
Clinical Sciences Building
Royal Leicester Infirmary
Leicester LE2 7LX, UK

It has been recognized for many years that certain imidazoline-derived α-adrenergic receptor antagonists will promote insulin secretion from the β-cells of the pancreatic islets of Langerhans. These effects occur both *in vivo* and *in vitro* and are mediated through interaction with the β-cell membrane at a site dissimilar to the α-adrenoceptor. Through the combined approaches of radiolabeled tracer studies and electrophysiology, there is good evidence that many of these compounds will inhibit potassium channels. Most electrophysiological studies have focused on a particular type of K^+ channel, the adenosine-triphosphate (ATP)-regulated K^+ (K^+_{ATP}) channel. These channels are inhibited by imidazolines, and because they also play a major role in governing changes in β-cell membrane potential, channel block accounts for the fact that such compounds will depolarize the membrane, promote the generation of Ca^{2+} action potentials, elevate intracellular Ca^{2+}, and elicit insulin secretion. In this short review we provide some background on the effects of imidazolines on insulin-secreting cells, and we present recent data from our own laboratory which show that imidazolines block other types of K^+ channels in the β-cell (including human tissue), that they elevate intracellular Ca^{2+}, and finally that the site of interaction with the K^+

[a]Human pancreatic β-cell work in our laboratories is funded by the British Diabetic Association (M.D., R.F.L.J., N.J.M.L.) and the Wellcome Trust (M.D.). Other experimental work was supported by the Medical Research Council, The Yorkshire Cancer Research Campaign, The Royal Society, and the University of Sheffield Research Stimulation Fund.

[c]Address for correspondence: Dr. M. J. Dunne, Department of Biomedical Science, University of Sheffield, Sheffield S10 2TN, UK.

channel involves a complicated interaction with intracellular adenosine-diphosphate (ADP).

Insulin is the major hormone associated with plasma glucose homeostasis, and therefore regulation of its secretion from β-cells of the pancreatic islets of Langerhans is a highly complex, integrated process. The most predominant physiological stimulus that elicits secretion is increased plasma glucose concentration. However, against the background of changes in the nutrient status of the blood, many hormones, neuropeptides, and neurotransmitters will also modulate secretion. In the first place, the insulin-secreting β-cells are one of several types of cells found in the islets, and although they dominate the islet cell population, insulin secretion is inhibited by the paracrine influences of the δ-cells which release somatostatin and is augmented by the hormone glucagon secreted from the islet α-cells. Secondly, the islets have a rich blood supply, and therefore secretion can be regulated through circulatory hormones, particularly gastrointestinal hormones, and finally they are also richly innervated by autonomic nerve fibers which terminate within the pancreatic islet. Although less obvious than the effects of glucose on the β-cell, the regulation of insulin secretion by hormones and neurotransmitters is also of major physiological importance. The cephalic phase of insulin release beginning prior to glucose absorption, is thought to be mediated by acetylcholine released from parasympathetic nerve terminals, whilst stimulation of the sympathetic nervous system results in noradrenaline release and the inhibition of secretion.[1] The consequence of this is that during exercise or stress, plasma glucose levels will be maintained for functional utilization of carbohydrate.

The precise subcellular mechanism(s) by which adrenergic inhibition of secretion occurs has yet to be evaluated, but it is likely to involve more than one regulatory process, including activation of K^+ channels.[1,2] It has even been suggested that exaggerated sympathetic innervation of the islets could account for the impaired secretory responses to glucose observed in certain Type II or non-insulin-dependent diabetics.[3] Indeed, administration of α-adrenoceptor antagonists has been favorably argued to be one possible mechanism with which to effectively treat such individuals, because of the ability of these compounds to overcome the pathological inhibition of secretion by noradrenaline. To support this there is convincing evidence that α-adrenergic receptor antagonists elevate plasma insulin levels *in vivo;* however, the mechanisms that account for the effect are **not** solely attributable to blockage of the α-adrenergic receptor. The most obvious demonstration of this is that *in vitro* α-adrenergic receptor antagonists promote insulin secretion from β-cells in the absence of agonists to the α-adrenoceptor.[4] This was first shown for phentolamine, a nonselective α_1- and α_2-adrenergic receptor antagonist.[4,5] Phentolamine promotes secretion both *in vivo* and *in vitro* and will also reverse the effects of another compound, diazoxide, a benzothiadiazine derivative and a potent inhibitor of insulin secretion from the β-cell.[6] Similar effects have now also been reported for other imidazoline and non-imidazoline α-adrenergic antagonists. In addition to their effects on secretion, compounds such as phentolamine, clonidine, and efaroxan have also been shown to alleviate the diazoxide-induced enhanced rate of $^{86}Rb^+$ efflux from pre-loaded islets of Langerhans.[7–9] These findings provided the first insight to at least one of the subcellular sites of action of such molecules, and this involves the modulation of potassium channels. With the technique of patch-clamp electrophysiology to record single ion channel electrical events in the mid-1980s, diazoxide was shown to be a

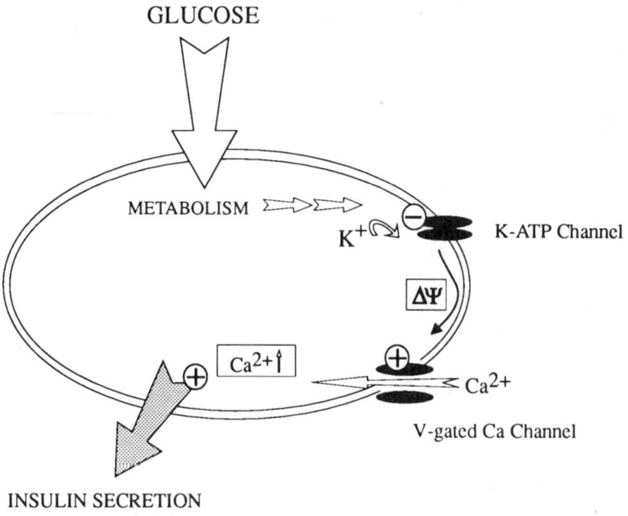

FIGURE 1. Schematic representation of the role of ATP-regulated K^+ channels (K-ATP) and voltage-gated Ca^{2+} channels in glucose-induced insulin secretion from pancreatic β-cells. Glucose metabolism induces K^+_{ATP} channel closure, leading to depolarization of the membrane, Ca^{2+} influx, and an increase in $[Ca^{2+}]_i$.

potent activator of a type of K^+ channel regulated by intracellular ATP, the ATP-sensitive potassium (K^+_{ATP}) channel.[10,11] These channels are now recognized to be inhibited by many imidazoline-based compounds.

Ion channels have many varied and important roles in governing the mechanisms of insulin secretion from the pancreatic β-cell. For glucose-evoked secretion, signal transduction events depend on a rise in the β-cell intracellular calcium on concentration ($[Ca^{2+}]_i$). This is associated with voltage-dependent calcium influx across the plasma membrane and not mobilization of calcium from intracellular calcium stores[12] (FIG. 1). To adjust the rate of insulin release in response to fluctuations in the availability of glucose, the β-cell has adopted several discrete control mechanisms to regulate $[Ca^{2+}]_i$. The most dominant of these is a complex pattern of electrophysiological events that involve changes in the β-cell membrane potential and the remodeling of ionic fluxes across the plasma membrane and intracellular membrane systems of the cell.[13] In brief, when the concentration of glucose is below that required to elicit insulin secretion, the membrane potential is silent and generally between approximately –60 and –70 mV. This is determined by a high resting potassium permeability, maintained by the Na^+K^+-ATPase and open K^+ channels.[14–16] Raising the glucose concentration above the threshold for secretion (>7 mM) promotes slow depolarization of the membrane sufficient to bring the β-cell to a critical threshold potential at which electrical activity is initiated. The pattern of electrical activity consists of cycles or *slow waves* of depolarization and hyperpolization of the membrane culminating in the generation of *plateau potentials* with superimposed *voltage-dependent* Ca^{2+} *spikes*. These action potential-like spikes result from Ca^{2+} influx across the membrane.[17,18] This occurs through voltage-gated Ca^{2+} channels, the activities of which are governed by the change in cell membrane potential. Open Ca^{2+} channels are therefore responsible for the rise in $[Ca^{2+}]_i$, and in this manner changes in electrical activity of the β-cell provide a mechanism for linking glucose metabolism to an increase in $[Ca^{2+}]_i$, and the subsequent release of insulin from the cells by exocytosis (FIG. 1).

By regulating the duration and periodicity of the plateau potentials, $[Ca^{2+}]_i$ levels can be critically controlled. However, whilst Ca^{2+} channels play a prominent role in Ca^{2+} signaling mechanisms, their activity is mainly regulated by changes in the cell membrane potential, which in turn is governed by the opening and closing of K^+ channels. Indeed, K^+ channels are the most widely studied of all the ion channels in the pancreatic β-cells. They are also the most varied group of ion channels which reflects upon the number of different roles that K^+ channels have to fulfill: (1) they control the resting cell membrane potential, (2) they regulate action potentials (duration and periodicity), and (3) they themselves are the target site for a number of hormones and neurotransmitters that are able to promote or inhibit secretion by mechanisms that are in part dependent upon changes in the cell membrane potential.[2,13,18]

There are at least five and possibly six clearly defined types of K^+ selective channels described in β-cells.[18] The functional role(s) and regulatory properties of two of these channels are highly relevant to our discussion of the effects of imidazolines on insulin-secreting cells: the **K^+_{ATP} channel** and the **K^+_{Ca} channel.** In the resting cell, open events from K^+_{ATP} channels are responsible for establishing the cell membrane potential.[16,19] When glucose is present at stimulatory concentrations, openings from these channels in intact β-cells cease, and it is this event that initiates a depolarization of the membrane.[16] K^+_{ATP} channels are in many ways equivalent to a "metabolism sensor," effectively coupling metabolic changes in the nutrient status of the cell to plasma membrane electrophysiological events (FIG. 1). As glucose will initiate changes in the intracellular nucleotide concentrations of ATP and ADP, and because both of these nucleotides have complex effects on the K^+ channels, the link between metabolism and channel activity is thought to occur following an increase in the ratio of ATP to ADP.[14,15,20] In contrast to K^+_{ATP} channels, the large Ca^{2+}- and voltage-gated (K^+_{Ca} or K^+_{maxi}) K^+ channels **do not** operate in intact resting cells because the $[Ca^{2+}]_i$ is not high enough or the membrane potential favorable enough to support openings from these channels.[21,22] However, the channels do play a very important role in rapidly repolarizing the Ca-spike potential that occurs following Ca^{2+} influx. Evidence for this is both circumstantial, a combination of the rapid depolarization and influx of Ca^{2+} being favorable for channel openings, and direct, because selective blockers of these channels increase the duration of the spike potentials.[14]

Two experimental observations initially prompted us to study the actions of imidazoline-based compounds on K^+_{ATP} channels in insulin-secreting cells by patch-clamp techniques. Firstly, phentolamine and efaroxan were shown to prevent diazoxide-induced inhibition of insulin secretion from β-cells, and secondly, both compounds significantly reduced the increased rate of $^{86}Rb^+$ efflux by diazoxide from pre-loaded islets of Langerhans.[6,9,22,23] Using excised cell-free patches of membrane and open cells, it was successfully demonstrated that the imidazolines phentolamine (FIG. 2A), efaroxan (FIG. 2B), antazoline, tolazoline, clonidine, and to a lesser extent

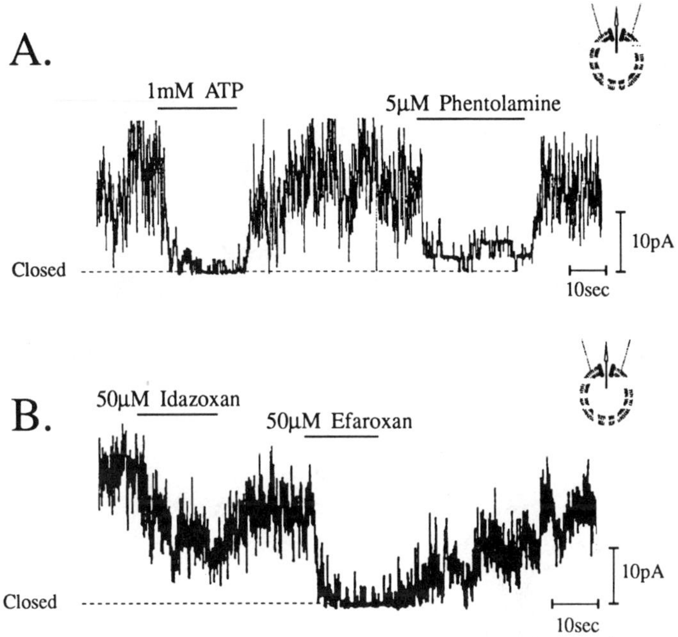

FIGURE 2. Imidazoline-induced block of K^+_{ATP} channels in RINm5F insulin-secreting cells. Both records are from separate cells. Single channel current data were obtained using the "open-cell" configuration (see cartoon). In these as in all subsequent records shown in this paper, openings from K^+ channels were recorded at 0 mV voltage-clamp and are shown as upward deflections from baseline ("closed"). Ionic gradients were quasi-physiological with a 140-mM NaCl-rich solution in the pipette (outside face of the membrane) and a 140-mM KCl-rich solution in the bath (inside face of the membrane). (**A**) Inhibition of K^+_{ATP} channels by 1 mM ATP and 5 μM phentolamine; (**B**) effects of 50 μM idazoxan and 50 μM efaroxan.

RX811059A (FIG. 3) and idazoxan (FIGS. 2B and 4) will all block ATP-regulated K^+ channels.[24–27] Several of these compounds were also found to inhibit whole-cell K^+_{ATP} currents in rodent tissue.[8,26,27] In our studies using the insulin-secreting cell line RINm5F we found that the relative potency of these molecules was as follows: phentolamine (K_I [50% blockade of channels] = 0.7 μM) > efaroxan (K_I = ~12 μM) > antazoline (K_I = <25 μM) > tolazoline (K_I = ~25 μM) > idazoxan (K_I = > 100 μM). Furthermore, we have also shown that blockade of channels is also associated with a significant change in cell electrical activity; phentolamine, efaroxan, and the imidazoline RX811059A will all depolarize the β-cell membrane in single isolated cells and evoke the generation of Ca^{2+}-dependent action potentials, effects that can be effectively reversed by the K^+_{ATP} channel agonist diazoxide (FIG. 5). Recent studies from our laboratory have concentrated more on the mechanisms of channel blockade, particularly interactions between phentolamine, efaroxan, and idazoxan with K^+_{ATP} channels. We find that each compound will inhibit channels from either side of the plasma membrane, that is, when added to the inside face of the membrane in inside-

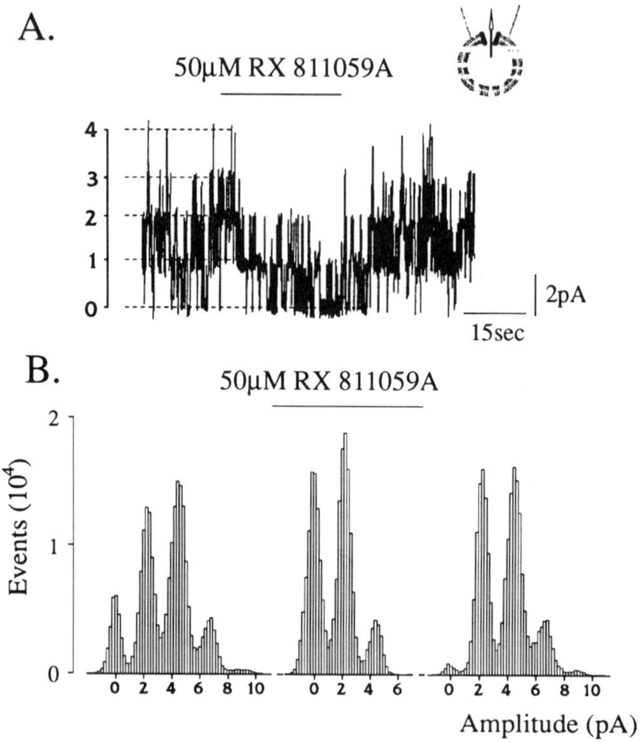

FIGURE 3. Block of K^+_{ATP} channels in RINm5F cells by the imidazoline RX811059A. **Panel A** shows single channel current data, and **panel B** effects of 50 μM RX811059A on K^+_{ATP} channel amplitude. Data in each panel come from separate cells.

out or open-cell patch-clamp experiments (FIG. 2) or the outside face of the membrane in outside-out patch-clamp or whole-cell experiments (FIG. 6). These observations, although somewhat conceptually simple, provide us with an insight as to the precise interaction of the compounds with the channels. As both agents are readily effective from either side of the membrane (and more importantly do not appear to be more efficient at blocking channels from either one side or the other), the data suggests that the "receptor" or "binding site" for the imidazolines ("I-receptor") is either equally accessible from the cytosolic and the extracellular face of the membrane or, alternatively, that it is located within the lipid bilayer of the plasma membrane. As the imidazolines are effective in blocking channels from the inside of the membrane, the idea that the compounds bind to an extracellular facing receptor binding site which then interacts with the channel cannot possibly hold. Perhaps the most obvious implication of these data is that the imidazolines interact directly with the channel (see below). There have been suggestions that the K^+_{ATP} channel is the endogenous binding site for the imidazolines, but the fact that efaroxan, idazoxan, and phentolamine will all inhibit other types of K^+ channels (FIG. 9) in both human and rodent

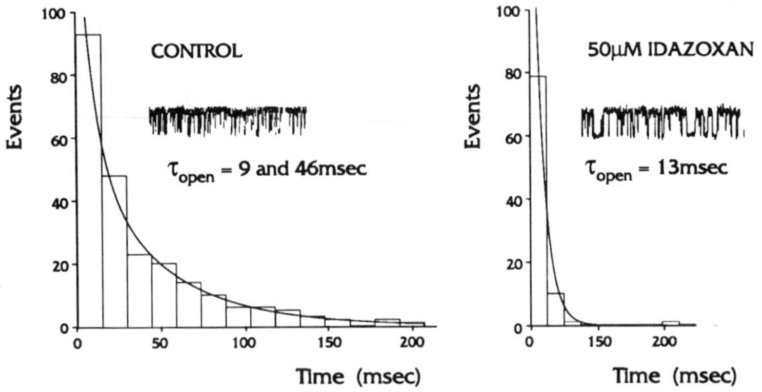

FIGURE 4. Inhibition of K^+_{ATP} channels in RINm5F cells by the imidazoline idazoxan. Data show the action of idazoxan on the opening rate constants of K^+_{ATP} channels. Only one channel was present in this patch, and kinetic data indicate that idazoxan reduced the opening rate constants associated with this channel from τ_{open} = 9 and 46 ms (control) to τ_{closed} = 13 ms.

β-cells clearly disagrees with this hypothesis. In RINm5F cells, analysis of single-channel current data at different concentrations of efaroxan and phentolamine has yielded Hill coefficients of approximately 1 for both compounds, suggesting that each compound interacts with a single class of binding site associated with the K^+_{ATP} channel.[24,25] Through close examination and quantitative analysis of single-channel current events, far more information relating to the precise mechanism of interaction can be gathered. For instance, by plotting amplitude histogram profiles and analyzing changes in channel opening probability, we know that phentolamine, efaroxan, idazoxan (FIG. 4), and RX811059A (FIG. 3) all inhibit channels through a mechanism associated with a decrease in the opening frequency of the channel and not by a decrease in the amplitude of the single-channel current event (see also FIGS. 2 and 6). For idazoxan we have examined the details of channel block through kinetic analysis of channel behavior. From these studies in membrane patches from RINm5F cells that contained just a single K^+_{ATP} channel, we found that the predominant action of idazoxan was to change the rate constant governing the duration of channel openings, τ_{open}. In the absence of idazoxan the sum of K^+_{ATP} channel openings was best fitted by two opening rate constants, τ_1 = 9 ms and τ_2 = 46 ms, whilst in the presence of the compound the longer opening rate constant was significantly shorter, τ_2 = 13 ms (FIG. 4). There were no significant effects of idazoxan on the closing rate constants, τ_{closed}, suggesting that the imidazoline is able to inhibit channel gating not by preventing the channel from opening, but by dramatically increasing the time the channel spends in the closed state. We also recently reported similar effects for polymyxin B on channel gating in insulin-secreting cells.[28]

At the single channel current level it is also possible to study interactions between imidazolines and K^+ channel activators. By these approaches we have been able to demonstrate that efaroxan-inhibited channels can be activated by diazoxide,[24] thereby explaining how diazoxide can alleviate the effects of efaroxan on $^{86}Rb^+$ efflux, the cell membrane potential (FIG. 5), and secretion, but it is not possible to say with

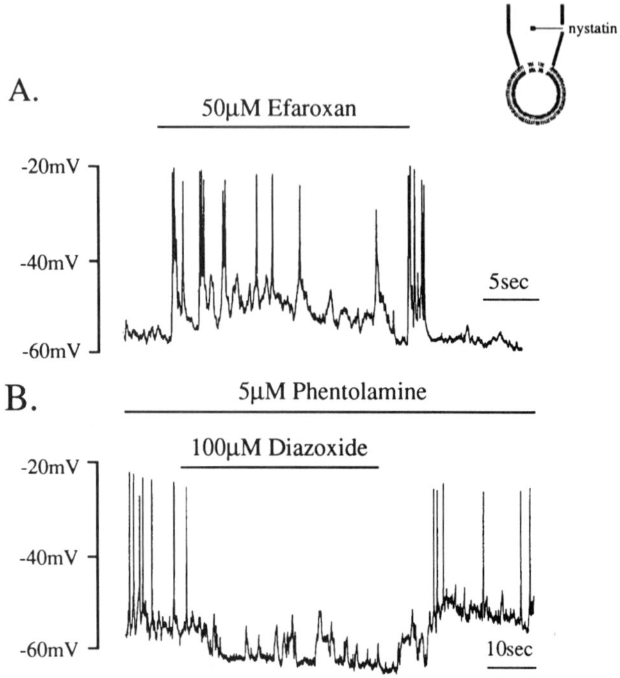

FIGURE 5. Effects of 50 μM efaroxan and 5 μM phentolamine on cell membrane potential of isolated RINm5F cells. Both records come from separate cells and data obtained using the nystatin perforated patch technique (200 μg/ml nystatin current-clamp configuration). **(A)** Efaroxan causes depolarization of the membrane and generation of Ca^{2+} action potentials; **(B)** phentolamine-induced depolarization is alleviated by diazoxide (100 μM). Diazoxide causes a repolarization of the cell, which reversibly blocks the generation of action potentials.

certainty from these types of experiments if both compounds interact with the same site on the channel. However, it is interesting to note that intracellular ADP has both a pronounced effect on the ability of diazoxide to open K^+_{ATP} channels[29] **and** a major influence on the gating of K^+_{ATP} channels in the presence of either efaroxan, phentolamine, or idazoxan, data that may suggest a common interaction(s) with an ADP binding site on the channel?

LIGAND BINDING TO K⁺ CHANNELS OR RECEPTOR-MEDIATED EVENTS?

Although there is currently a well-founded interest in the modulation of K^+_{ATP} channels by the imidazolines, the pharmacology of these particular ion channels has mainly been described in the literature through the pharmacology of "K^+ channel activator compounds,"[15] and more completely in connection with the potent antidiabetic sulfonylurea drugs.[30] These compounds have pronounced effects on insulin

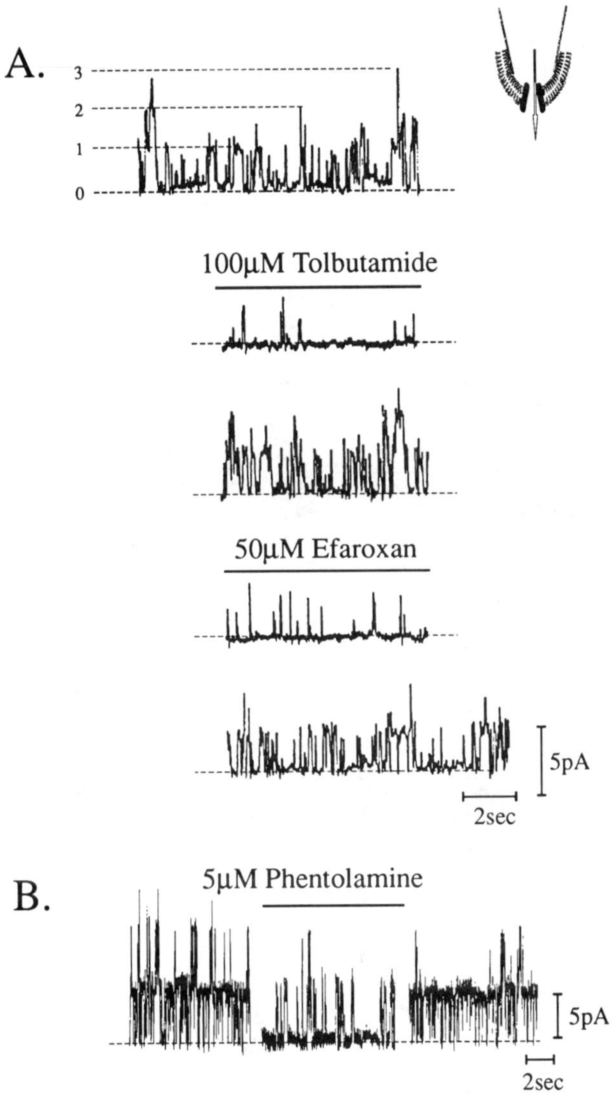

FIGURE 6. Blockade of K^+_{ATP} channels by imidazolines from the external face of the cell membrane. Data in **A and B** come from two separate RINm5F cells obtained using the outside-out patch technique. **A** shows that 50 μM of efaroxan is as effective at blocking channels as 100 μM tolbutamide (all data from the same cell). **B** shows phentolamine (5 μM) induced inhibition of K^+_{ATP} channels.[25] Note that the amplitude of single channel current events is not significantly influenced by the compounds.

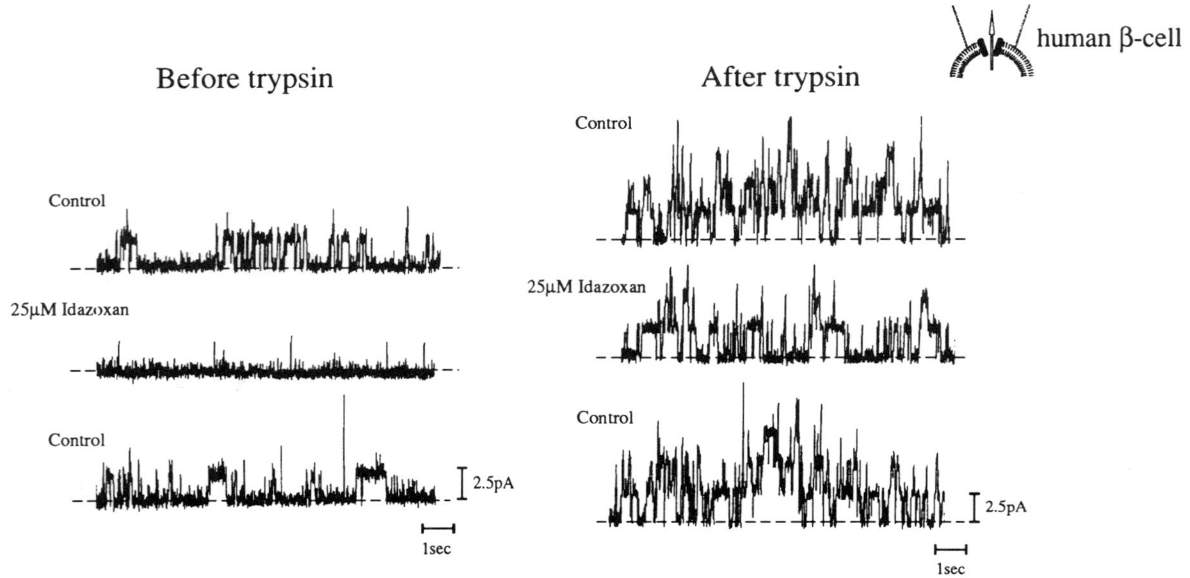

FIGURE 7. Inhibition K^+_{ATP} channels by 25 μM idazoxan in human pancreatic β-cells is not affected by exposure of the patch of membrane to 1 mg/ml of trypsin (not shown). The most marked effect of trypsin on K^+_{ATP} channels is an increase in channel activity by reversing channel rundown/inactivation; compare the number of channel openings in the right panel with those in the left one. Human β-cells were prepared using the methods described by Squires *et al.*[17]

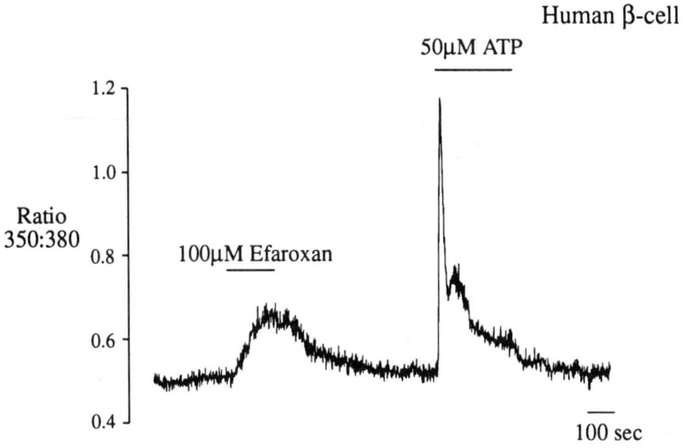

FIGURE 8. Elevation of intracellular Ca^{2+} ($[Ca^{2+}]_i$) by efaroxan (100 μM) and ATP (50 μM) in a single isolated human pancreatic β-cell. Increases in $[Ca^{2+}]_i$ were examined using the Ca^{2+} indicator fura-2. Changes in the fluorescence of the excitation spectra are expressed as a ratio of 350/380 and represent changes in $[Ca^{2+}]_i$. Note how efaroxan causes a smaller rise in cytosolic Ca^{2+} than does the purinergic receptor agonist ATP. The rate of rise in $[Ca^{2+}]_i$ is also different with each compound: efaroxan causes a slow rise in $[Ca^{2+}]_i$, whilst ATP promotes a very rapid increase. The ATP-evoked Ca^{2+} signal is characteristically biphasic, consisting of a large transient increase in $[Ca^{2+}]_i$ (Ca^{2+} mobilization) followed by a small, more sustained rise in $[Ca^{2+}]_i$ (Ca^{2+} influx).[17]

secretion, acting as highly potent secretagogues that mediate their actions by blocking K^+_{ATP} channels.[30,31] Channel block leads to depolarization of the membrane and an increase in $[Ca^{2+}]_i$.[30] As with the imidazolines and their interactions with the imidazoline receptor, important questions have been raised concerning the identity of the "sulfonylurea receptor" and its relationship to the K^+_{ATP} channel. Ligand-binding studies have indicated that in intact cells the sulfonylurea glibenclamide has a remarkably high affinity for K^+_{ATP} channels in pancreatic β-cells.[30] However, the compounds are much less effective in excised patches, and as demonstrated for K^+_{ATP} channels in neurons,[32] this might suggest that the sulfonylurea receptor is not a functional component of the channel but rather somehow "associated" with it. Thus, when patches of membrane are excised from an intact cell, displacement of the sulfonylurea receptor from the channel might be expected. It has also been strongly suggested that the sulfonylurea receptor is part of the K^+_{ATP} channel, and whilst this is still widely debated, experimental evidence from patch-clamp studies suggests that the receptor site and the channel are "interactive" rather than being a single protein: (1) as just discussed, the effects of the sulfonylureas are "lost" with time in excised patches of membrane (Harding and Dunne, unpublished data), and (2) the sulfonylurea block of K^+_{ATP} channels can be "removed" by treatment of the cytosolic face of the plasma membrane with the proteolytic enzyme trypsin, and (3) sulfonylureas require the presence of Mg^{2+} to close the channels.[33] Indeed, the trypsin experiments are particularly noteworthy. Several papers recently described the actions of trypsin

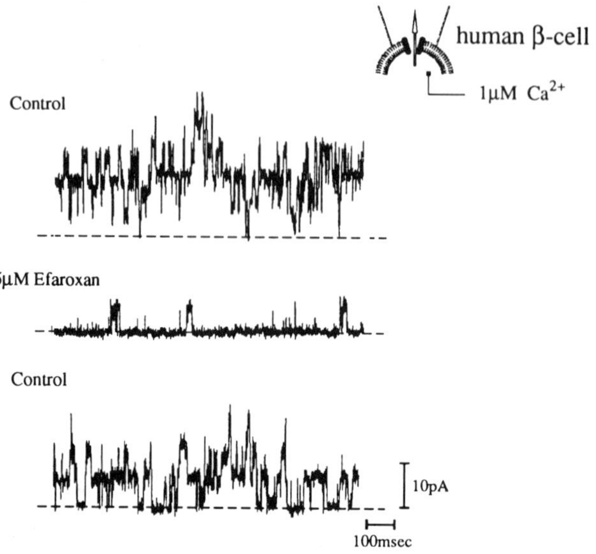

FIGURE 9. Block of the large Ca^{2+}- and voltage-activated K^+ (K_{Ca}) channel in human pancreatic β-cells by 25 μM idazoxan and efaroxan. All records were obtained from the same patch (inside out). In each experiment K_{Ca} channels were activated by an increase in the intracellular Ca^{2+} concentration to 1 μM as indicated.

on K^+_{ATP} channels in both cardiac myocytes and insulin-secreting cells.[33–35] Exposing the cytosolic face of the plasma membrane to the proteolytic enzyme has several dramatic effects on channel gating. Most significantly this procedure reduces (eliminates?) the rate of channel run-down,[33,35] but perhaps more relevant to this paper is that trypsinization of the membrane removes the block of K^+_{ATP} channels by the sulfonylureas.[34,35] This has been taken as evidence that the sulfonylurea receptor is not an integral part of the K^+_{ATP} channel protein, an idea supported by the fact that following trypsinization, K^+_{ATP} channels can still be inhibited by intracellular ATP and ADP.[33–35] With these points in mind, it is therefore very important to note that inhibition of K^+_{ATP} channels by efaroxan, phentolamine, and idazoxan was **not** significantly affected by treatment of the inside of the plasma membrane with trypsin (FIG. 7), and neither were the effects of the compounds significantly diminished over the time course of an experiment in cell-free patches (which can on some occasions last for many minutes). Both these pieces of data suggest, firstly, that the interaction of the imidazolines with K^+_{ATP} channels is dissimilar to the mechanism(s) that underlie the actions of the sulfonylureas on the channels, and, secondly, that the imidazoline derivatives directly bind to the K^+_{ATP} channel. FIGURE 7 clearly shows the block of K^+_{ATP} channels in a human β-cell by idazoxan, and these effects are not influenced by exposure of the inside face of the membrane to trypsin. Note also how trypsin has increased the channel open-state probability through a reverse of run-down.

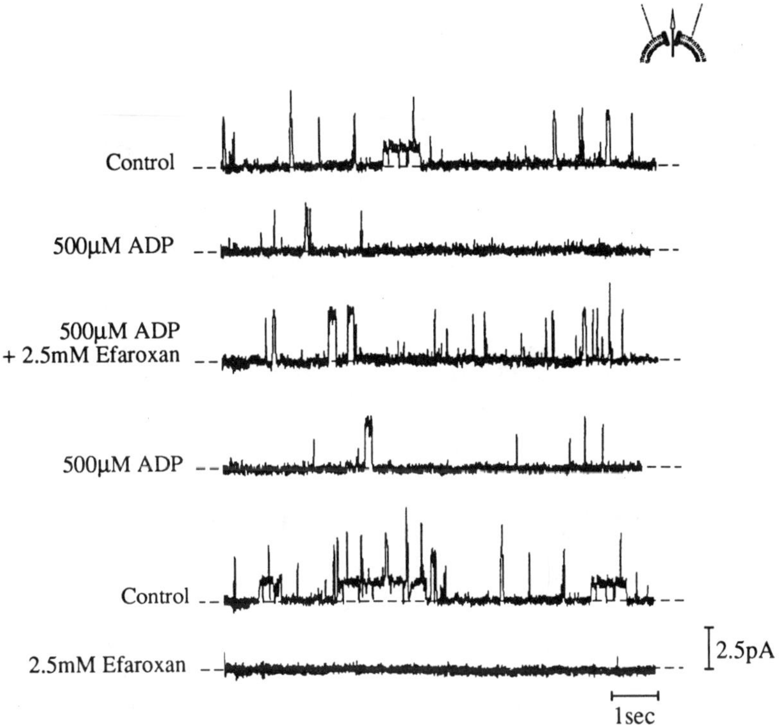

FIGURE 10. Effects of efaroxan on ADP-inhibited K^+_{ATP} channels in a rat pancreatic β-cell. Efaroxan increases channel open-state probability in the presence of ADP which by itself also closes K^+_{ATP} channels. Also note how openings from a smaller channel, the non-selective cation channel (?), are present in the "control" solutions, but are not seen when either ADP, efaroxan, or ADP + efaroxan is present.

K⁺ CHANNELS, INTRACELLULAR ADP, AND HUMAN PANCREATIC β-CELLS

Most investigations of the effects of imidazolines on insulin-secreting cells have been carried out using either rodent β-cells or insulin-secreting cell lines. In our laboratory we recently examined the actions of a number of α_2-adrenergic antagonists on isolated β-cells from transplantable human islets of Langerhans. These studies are important particularly because few papers have been published on the pharmacological characterization of ion channels in human tissue, and it is also imperative that a much greater understanding of ion channels in these types of β-cells be forwarded. Overall, the effects of the imidazolines efaroxan and idazoxan on ion channels and intracellular calcium signals are very similar to their actions reported in rodent β-cells and insulin-secreting cell lines (FIGS. 7 and 8). We did, however, find in both electrophysiological studies and single-cell microfluorimetry experiments that phentolamine was much less readily able to influence ion channels in human cells

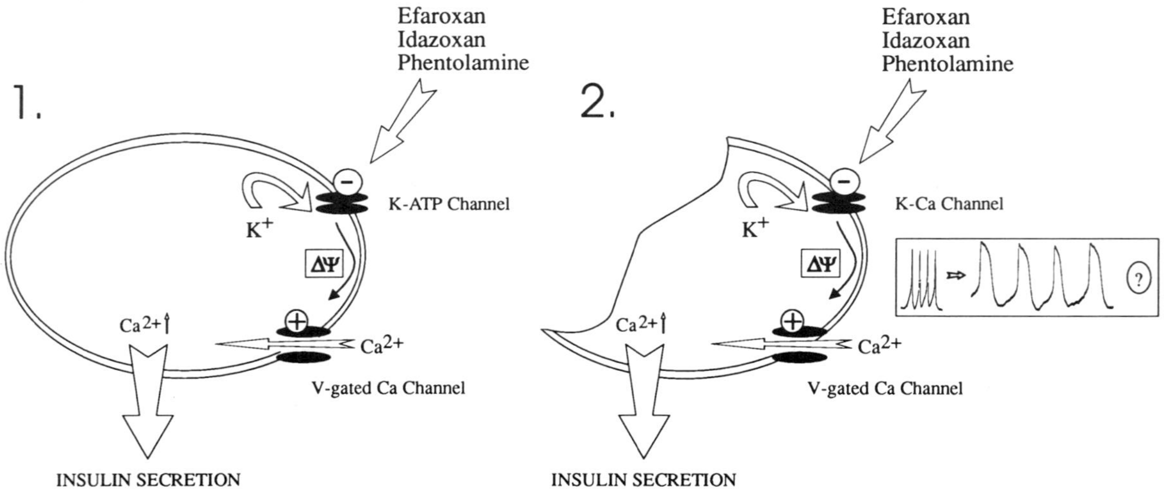

FIGURE 11. Schematic diagrams that summarize the effects of K^+ channel modulation by imidazolines in pancreatic β-cells. **Panel 1** shows K^+_{ATP} channel block, whilst **panel 2** shows K_{Ca} channel block. Inhibiting K^+_{Ca} channels, we suggest, will likely increase the duration of the voltage-gated Ca^{2+} spike potentials.

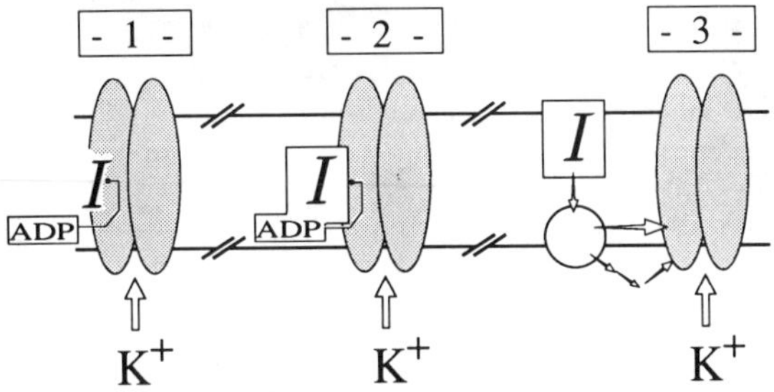

FIGURE 12. Schematic representation to model imidazoline "I"-induced block of K^+_{ATP} channels in β-cells. As our recent data suggest a strong involvement of ADP, "1" proposes direct ligand binding with the channel, "2" suggests receptor interaction with the channel, whilst "3" suggests that receptor activation could involve more distal events—G-protein, protein kinase, etc. On balance, our data more readily support hypotheses 1 and 2.

than in studies with model tissues. In human β-cells we also studied the actions of efaroxan, idazoxan, and phentolamine on the large Ca^{2+} and voltage-gated K^+ channel and again found that each of the compounds was able to inhibit these channels (FIG. 9). These data suggest that their actions in human tissue were not solely restricted to blockade of K^+_{ATP} channels. In these types of experiments (FIG. 9) K_{Ca} channels do not spontaneously gate in excised patches and must be activated by either a change in cell membrane potential or an increase in $[Ca^{2+}]_i$; under these conditions each of the compounds consistently closed the channels through an effect on the channel open-state probability. Certain imidazolines have also been reported to inhibit whole-cell voltage-gated K^+ currents in mouse β-cells.[27]

Using both human and rodent β-cells we also recently examined interactions between K^+ channels, intracellular ADP, and imidazolines. Intracellular nucleotides are important to the gating mechanisms of both the K^+_{ATP} channel and the K^+_{Ca} channel. In β-cells K^+_{ATP} channels are not significantly influenced by changes in cell membrane potential or intracellular Ca^{2+} concentration. The actions of internal ATP dominate in the intact cell, because ATP is a blocker of these channels and is present at cellular concentrations in the millimolar range.[19] However, intracellular ADP also plays an important role in governing channel activity; ADP will promote a concentration-dependent decrease in channel openings at >200 μM, whilst concentrations <200 μM tend to activate K^+_{ATP} channels.[15] Despite this, ADP will always activate K^+_{ATP} channels in the presence of ATP, and this effect is seen even for concentrations of ADP which by themselves act as blockers of K^+_{ATP} channels.[36] When one considers that the approximate intracellular ADP concentration has been estimated to be between 500 μM and 2 mM, then the predominant effects of ADP *in vivo* will be to open K^+_{ATP} channels that are tonically inhibited by cytosolic ATP levels.[36] This by itself would explain why K^+_{ATP} channels are open in the intact cell despite the fact that (1) the K_I value equivalent to 50% inhibition of the channels is in the sub-100 μM concentration range, and (2) it has been estimated that approximately 5 mM ATP are present in the intact cell.[14,15,19] (In addition, we also have evidence that the cytosolic nucleotides NADP, NADH, NADP, and NADPH significantly modulate the gating of K^+_{ATP} channels in both human and clonal insulin-secreting cells.[37]) Besides providing one explanation as to how K^+_{ATP} channels are operational in the intact cell, the ADP data also provides the most attractive hypothesis for explaining how changes in the nutrient status of the β-cell can be coupled to changes in cellular electrophysiology, that is, through an increase in the ATP-ADP ratio.[14,15,20] Recent experiments from our laboratory have also shown that ATP and ADP affect the gating of K^+_{Ca} channels in human β-cells.[38] Both nucleotides will close K^+_{Ca} channels by mechanisms that we believe are mediated through the action(s) of protein kinases associated with the channel. Since intracellular ADP has such a significant modulatory role on the regulation of K^+ channels in the β-cell, it is not surprising then that interactions of ADP with pharmacological modulators of these channels have also provided important insights into the molecular mechanism of actions of such compounds.[29,30] Thus, whilst the blocking actions of the sulfonylureas on K^+_{ATP} channels are potentiated by intracellular ADP,[30] the nucleotide more readily augments the activation of K^+_{ATP} channels by diazoxide and cromakalim than ATP.[29] In this context we recently obtained unexpected results when we examined interactions between ADP and the imidazolines efaroxan, idazoxan, and phentolamine on K^+ channels using cell-free inside-out patches from both human and rodent tissue (FIG. 10); we found that in sharp contrast to enhancing the degree of channel blockade by the compounds, adding intracellular ADP effectively removed the inhibition of the channels by the imidazolines. As the individual concentrations of efaroxan, idazoxan, phentolamine, and ADP used in our experiments caused channel closure, it is hard to reconcile a suitable mechanism of action that explains how the simultaneous addition of both compounds will cause such a marked change in channel activity. From our data it is clear that ADP does not stimulate K^+_{ATP} channels in the presence of an imidazoline, nor can imidazolines ''activate'' ADP-inhibited channels *per se* (FIG. 10). We came to this conclusion based on the fact that the peak K^+_{ATP} current seen in the presence of both types of compounds was lower than that seen in the immediate control solutions. Although not shown, we also have evidence that ''relief'' of a channel block by ADP is more significant with the more prominent inhibition of K^+_{ATP} channels, that is, phentolamine > efaroxan > idazoxan. ADP will also remove the inhibition of K^+_{Ca} channels by imidazolines. This may support the hypothesis that the imidazoline binding site is also an ADP binding protein. The precise identity of the site is not obvious, but trypsinization *does not* influence the ADP-induced modulation of imidazoline-inhibited channels. As nucleotide diphosphates are important molecules for the regulation of K^+_{ATP} channels in the β-cell, we have also studied the interactions between imidazolines and GDP. These experiments show that GDP does not prevent phentolamine-induced block of K^+_{ATP} channels (Liang and Dunne, unpublished data). This would indicate that the imidazoline site on the channel is not located at the nucleotide-diphosphate binding site.[29]

Finally, a putative endogenous ligand for the imidazoline receptor was recently described in the CNS.[39] This compound, agmatine, also promotes insulin secretion from the pancreas[40] which may indicate that agmatine also acts as a peripheral neurotransmitter (see N. G. Morgan *et al.*, this volume). Its effects on K^+ channels have not been previously investigated, but preliminary data from our laboratory indicate that agmatine will inhibit ATP-sensitive K^+ channels in a concentration-related manner, but it has little direct effect below 1 mM.

In summary, recent data from our laboratory have shown that imidazoline-derived α-adrenergic receptor antagonists, which for many years have been recognized as being able to promote insulin secretion from β-cells, mediate their effects through blockade of ATP-regulated K^+ channels. Inhibiting K^+_{ATP} channels facilitates depolarization of the β-cell membrane potential, promoting the opening of voltage-gated Ca^{2+} channels, Ca^{2+} influx, and an increase in $[Ca^{2+}]_i$ (FIG. 11(1)). We also find that efaroxan, idazoxan, and phentolamine will each inhibit K^+_{Ca} channels in addition to their effect on K^+_{ATP} channels. As these channels are involved in the repolarization of the pancreatic β-cell Ca^{2+} action potential, channel inhibition would tend to increase the duration of the Ca^{2+} spike, thereby augmenting Ca^{2+} influx (FIG. 11(2)). Details of the precise molecular interactions of these ligand binding sites with K^+ channels are slowly emerging, and we believe that we now have very good data to support the hypothesis that ADP and imidazoline-evoked channel closure share similar binding domains on K^+ channels in pancreatic β-cells (FIG. 12).

REFERENCES

1. MORGAN, N. G., R. D. HURST & S. L. F. CHAN. 1989. Control of insulin secretion by α_2-receptor activation. Proc. 5th Jap. α-Receptor Conference (Excerpta Medica, Tokyo).: 79-91.
2. RORSMAN, P., K. BOKVIST, C. ÄMMÄLÄ, P. ARKHAMMAR, P.-O. BERGREN, O. LARSSON & K. WAHLANDER. 1991. Activation by adrenaline of a low conductance G protein-dependent K^+ channel in mouse pancreatic β-cells. Nature **349:** 77-79.
3. ROBERTSON, R. P. & D. PORT. 1973. Adrenergic modulation of basal insulin secretion in man. Diabetes **22:** 1-8.
4. EFENDIC, S., E. CERASI & R. LUFT. 1975. Effect of phentolamine and preperfusion with glucose on insulin release from the isolated perfused pancreas from fasted and fed rats. Diabetologia **11:** 407-410.
5. SCHULZ, A. & A. HASSELBLATT. 1988. Phentolamine, a deceptive tool to investigate sympathetic nervous control of insulin release. Naunyn-Schmiedeberg's Arch. Pharmacol. **337:** 637-643.
6. BURR, I. M., E. B. MARLISS, W. STAUFFACHER & A. E. RENOLD. 1971. Diazoxide effects on biphasic insulin release: 'Adrenergic' suppression and enhancement in the perfused rat pancreas. J. Clin. Invest. **50:** 1444-1450.
7. SCHULZ, A. & A. HASSELBLATT. 1989. An insulin-releasing property of imidazoline derivatives is not limited to compounds that block α-adrenoreceptors. Naunyn-Schmiedeberg's Arch. Pharmacol. **340:** 321-327.
8. PLANT, T. & J. C. HENQUIN. 1990. Phentolamine and yohimbine inhibit ATP-sensitive K^+ channels in mouse β-cells. Br. J. Pharmacol. **101:** 115-120.
9. CHAN, S. L. F. & N. G. MORGAN. 1990. Stimulation of insulin secretion by efaroxan may involve interaction with potassium channels. Eur. J. Pharmacol. **176:** 97-101.
10. TRUBE, G., P. RORSMAN & T. OHNO-SHOSAKU. 1986. Opposite effects of tolbutamide and diazoxide on the ATP-dependent K^+ channel in mouse pancreatic β-cells. Pflügers Arch. **407:** 493-499.
11. DUNNE, M. J., M. C. ILLOT & O. H. PETERSEN. 1987. Interaction of diazoxide, tolbutamide and ATP^{4-} on nucleotide-dependent K^+ channels in an insulin-secreting cell line. J. Membr. Biol. **99:** 215-224.
12. WOLLHEIM, C. B. & G. W. G. SHARP. 1981. Regulation of insulin release by calcium. Physiol. Rev. **61:** 914-973.
13. HENQUIN, J. C. 1990. Glucose-induced electrical activity in β-cells. Feedback control of ATP-sensitive K^+ channels by Ca^{2+}. Diabetes **39:** 457-460.

14. PETERSEN, O. H. & I. FINDLAY. 1987. Electrophysiology of the pancreas. Physiol. Rev. **67:** 1054-1116.
15. DUNNE, M. J. & O. H. PETERSEN. 1991. Potassium selective ion channels in insulin-secreting cells: Physiology, pharmacology and their role in stimulus-secreting coupling. Biochim. Biophys. Acta **1071:** 67-85.
16. ASHCROFT, F. M., S. J. H. ASHCROFT & D. E. HARRISON. 1984. Glucose induces closure of single potassium channels in isolated rat pancreatic β-cells. Nature **312:** 446-448.
17. SQUIRES, P. E., R. L. F. JAMES, N. J. M. LONDON & M. J. DUNNE. 1994. ATP-induced intracellular Ca^{2+}-signals in isolated human insulin-secreting cells. Pflügers Arch. **427:** 181-183.
18. DUNNE, M. J., E. A. HARDING, J. H. JAGGAR & P. E. SQUIRES. 1994. Ion channels and the molecular control of insulin secretion. Biochem. Soc. Trans. **22:** 6-12.
19. COOK, D. L. & C. N. HALES. 1984. Intracellular ATP directly blocks K^+ channels in pancreatic β-cells. Nature **311:** 271-273.
20. DUNNE, M. J. & O. H. PETERSEN. 1986. Intracellular ADP activates K^+ channels that are inhibited by ATP in an insulin-secreting cell line. FEBS Lett. **208:** 59-62.
21. COOK, D. L., M. IKEUCHI & W. Y. FUJIMOTO. 1984. Lowering of pHi inhibits Ca^{2+}-activated K^+ channels in pancreatic β-cells. Nature **311:** 269-271.
22. FINDLAY, I., M. J. DUNNE & O. H. PETERSEN. 1985. High conductance potassium channels in pancreatic islet cells are activated and inactivated by internal calcium. J. Membr. Biol. **83:** 169-175.
23. HENQUIN, J. C., S. CHARLES, M. NENQUIN, F. MATHOT & T. TAMAGAWA. 1982. Diazoxide and D600 inhibition of insulin release. Different mechanisms explain the specificity for different stimulus. Diabetes **31:** 776-783.
24. CHAN, S. L. F., M. J. DUNNE, M. R. STILLINGS & N. G. MORGAN. 1991. The α_2 adrenoceptor antagonist efaroxan modulates K^+_{ATP} channels in insulin-secreting cells. Eur. J. Pharmacol. **204:** 41-48.
25. DUNNE, M. J. 1991. Block of ATP-regulated potassium channels by phentolamine and other α-adrenoceptor antagonists. Br. J. Pharmacol. **103:** 1847-1850.
26. PLANT, T. D., J. C. JONAS & J. C. HENQUIN. 1991. Clonidine inhibits ATP sensitive potassium channels in mouse pancreatic β-cells. Br. J. Pharmacol. **104:** 385-390.
27. JONAS, J. C., T. D. PLANT & J. C. HENQUIN. 1992. Imidazoline antagonists of α_2-adrenoceptors increase insulin release *in vitro* by inhibiting ATP-sensitive potassium channels in pancreatic β-cells. Br. J. Pharmacol. **107:** 8-14.
28. DUNNE, M. J., J. H. JAGGAR, E. A. HARDING & P. E. SQUIRES. 1994. Protein kinase C and the modulation of ion channel activity on insulin-secreting cells. *In* Insulin Secretion and Pancreatic β-cell Research. Chapt. 36. P. R. Flatt & S. Lenzen, Eds. Smith Gordon Press. London, UK. In press.
29. JAGGAR, J. H., E. A. HARDING, B. J. AYTON & M. J. DUNNE. 1993. Interaction of diazoxide and cromakalim with ATP-regulated K^+ channels in rodent and clonal insulin-secreting cells. J. Mol. Endocrinol. **10:** 59-70.
30. ASHCROFT, S. J. H. & F. M. ASHCROFT. 1992. The sulphonylurea receptor. Biochim. Biophys. Acta **1175:** 45-59.
31. STURGESS, N., M. L. J. ASHFORD, D. L. COOK & C. N. HALES. 1985. The sulphonylurea receptor may be an ATP-sensitive potassium channel. Lancet **8453:** 474-475.
32. ASHFORD, M. L. J., P. R. BODEN & J. M. TREHERNE. 1990. Tolbutamide excites rat glucoreceptive ventromedial hypothalamic neurones by indirect inhibition of ATP-K^+ channels. Br. J. Pharmacol. **101:** 531.
33. LEE, K., S. E. OZANNE, I. C. M. ROWE, C. N. HALES & M. L. J. ASHFORD. 1994. The effects of trypsin on ATP-sensitive potassium channel properties and sulphonylurea receptor in the CR1-91 insulin-secreting cell line. Mol. Pharmacol. **46:** 176-185.
34. NICHOLS, C. N. & A. LOPATIN. 1993. Trypsin and α-chymotrypsin treatment abolished glibenclamide sensitivity of K^+_{ATP} channels in rat ventricular myocytes. Pflügers Arch. **422:** 617-619.

35. PROKS, P. & F. M. ASHCROFT. 1993. Modification of potassium-ATP channels in pancreatic β-cells by trypsin. Pflügers Arch. **424:** 63-72.
36. DUNNE, M. J., J. WEST-JORDAN, R. J. ABRAHAM, R. T. H. EDWARDS & O. H. PETERSEN. 1988. The gating of nucleotide dependent K^+ channels in insulin-secreting cells can be modulated by changes in the ratio ATP^{4-}/ADP^{3-} and by non-hydrolysable analogues of ATP and ADP. J. Membr. Biol. **104:** 165-177.
37. DUNNE, M. J., I. FINDLAY & O. H. PETERSEN. 1988. The effects of pyridine nucleotides on the gating of ATP-sensitive K^+ channels in insulin-secreting cells. J. Membr. Biol. **102:** 205-216.
38. JAGGAR, J. H., E. A. HARDING, R. F. L. JAMES, K. HINCHLIFFE & M. J. DUNNE. 1994. Modulation of the calcium- and voltage activated K^+ (K^+_{Ca}) channel by intracellular nucleotides in isolated human insulin-secreting β-cells. J. Physiol. **477:** 84P.
39. LI, G., G. REGUATHAN, C. J. BARROW, J. ESHRAGI, A. R. COOPER & D. J. RIES. 1994. Agmatine: An endogenous clonidine displacing substance in the brain. Science **263:** 966-969.
40. SENER, A., P. LEBRUN, F. BLACHIER & W. J. MALAISSE. 1989. Stimulus-secretion coupling of arginine-induced insulin release. Biochem. Pharmacol. **38:** 327-330.

A New Therapeutic Approach for Sickle Cell Disease

Blockade of the Red Cell Ca^{2+}-Activated K^+ Channel by Clotrimazole[a]

CARLO BRUGNARA,[b,g] LUCIA DE FRANCESCHI,[d] CARRIE C. ARMSBY,[b] NACERA SAADANE,[e] MARIE TRUDEL,[e] YVES BEUZARD,[f] ANN RITTENHOUSE,[c] NADER RIFAI,[b] ORAH PLATT,[b] AND SETH L. ALPER [c]

[b]*Departments of Pathology, Laboratory Medicine and Medicine*
The Children's Hospital

[c]*Molecular Medicine and Renal Units*
Beth Israel Hospital
and
[c]*Department of Cell Biology*
Harvard Medical School
Boston, Massachusetts 02115

[d]*University of Verona*
Verona, Italy

[e]*Institut de Recherches Cliniques de Montreal*
Montreal, Canada

[f]*INSERM unit 91*
Creteil, France

When the intracellular free Ca^{2+} concentration increases in human erythrocytes, a large amount of K^+, Cl^- and water is lost because of the opening of a Ca^{2+}-activated K^+ channel, first described by Gardos.[1] The Gardos channel of human red cells belongs to a large family of Ca^{2+}-activated K^+ channels present in numerous cell types.[2] The electrophysiological properties of the Gardos channel and its regulation by Ca^{2+} and metabolism have been studied in detail.[3–7] Binding studies using ^{125}I-charybdotoxin (ChTX), a 37-aa peptide that inhibits K^+ flux through the channel, indicate the presence of 126 ± 24 binding sites per erythrocyte.[8] Recently, the cDNA of a high conductance Ca^{2+}-activated K^+ channel (maxi-K^+ channel) of mouse brain

[a]This work was supported by National Heart, Lung, and Blood Institute grant HL-15157 and by the Medical Research Council of Canada.

[g]Address correspondence and reprint requests to: Carlo Brugnara, MD, Department of Laboratory Medicine, The Children's Hospital, 300 Longwood Avenue, Bader 760, Boston, MA 02115.

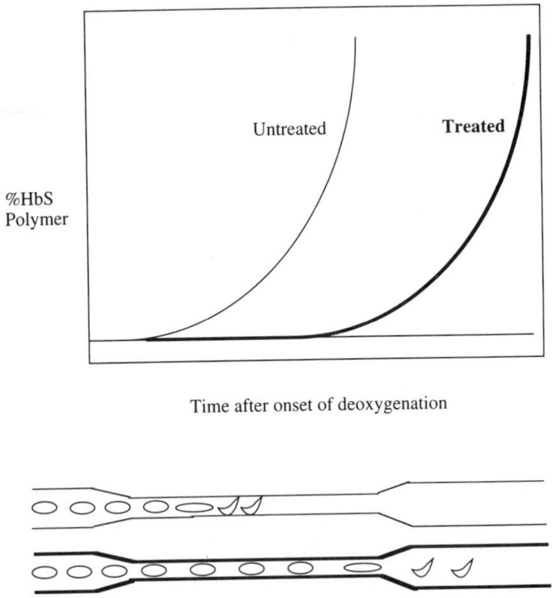

FIGURE 1. Expected change in delay time for Hb S polymerization induced by a hypothetical treatment that reduces Hb S concentration and its expected effect on capillary sickling and occlusion. (**Top**) A graphic simulation of the delay time for polymer formation in a normal cell and a cell with reduced Hb S concentration. (**Bottom**) The formation of sickle cells and occlusion of a capillary in untreated SS blood and (*in bold*) postcapillary sickling with no occlusion in cells with reduced Hb S concentration and prolonged delay time.

and skeletal muscle has been cloned. It encodes 6–8 hydrophobic, putative transmembrane spans of the Shaker-type K^+ channel and four additional hydrophobic segments, homologous to those of the Drosophila *slo* gene.[9] The molecular identity of the intermediate conductance Gardos channel of erythrocytes and the degree of its structural relationship to the *slo* gene family remain unknown.

The physiological function of the Gardos channel is unknown. The loss of cell K^+ via this channel may play a role in the maturation of normal reticulocytes[10] and in the prevention of colloido-osmotic lysis during transient complement activation on the red cell membrane.[11] The Gardos channel, either alone or in conjunction with K-Cl cotransport, plays a major role in the K^+ depletion observed in sickle erythrocytes.[12–16] Sickle cell dehydration is central to the pathophysiology of sickle cell (SS) disease. Because of the extreme dependency of the delay time for Hb S polymerization on Hb S concentration,[17] small increases in cell Hb S concentration dramatically enhance Hb S polymerization and cell sickling. For this reason, decreasing cellular Hb S concentration is a potential therapeutic strategy for sickle cell disease. Figure 1 schematizes the expected changes in delay time and morphologic sickling induced by a reduction in cell hemoglobin concentration.

Induction of hyponatremia and cell swelling by desmopressin (DDAVP) has been shown to lead to decreased red cell mean corpuscular hemoglobin concentration

(MCHC) and sickling,[18] but another study failed to confirm these findings,[19] probably because the levels of clinical hyponatremia required to induce significant cell swelling are very difficult to maintain. Other current therapeutic strategies in clinical testing are aimed at decreasing Hb S concentration by increasing Hb F concentration with either hydroxyurea (alone or with erythropoietin) or butyric acid derivatives.[20–23]

A different therapeutic approach to preventing sickle cell dehydration involves specific inhibition of K^+ loss in sickle erythrocytes. This approach is based on the possibility of obtaining significant inhibition of the Gardos channel *in vivo* by using a specific blocker for the channel. Various inhibitors of the red cell Gardos channel have been described: quinine (IC_{50} = 5 μM in K-free medium, 100 μM with 5 mM K[24]), carbocyanine (IC_{50} = 20–50 nM[25]), nifedipine (IC_{50} = 4 μM[26]), and nitrendipine (IC_{50} = 130 nM[27]). None of them has been used in sickle cell anemia because of toxicity or side effects.

Peptide toxins derived from scorpion venom have been shown to be specific ligands of the red cell Ca^{2+}-activated K^+ channel.[8,14] Charybdotoxin (ChTX), a protein isolated from the venom of the scorpion *Leiurus quinquestriatus,*[28] is a 4.3-kD polypeptide that inhibits Ca^{2+}-activated and voltage-gated K^+ channels.[28,29] The three-dimensional structure of the peptide has been solved by nuclear magnetic resonance imaging,[30] and the recombinant gene product has been overexpressed in *Escherichia coli.*[31] A radioiodinated derivate of ChTX has been used to study toxin binding in different cell types that express toxin-sensitive channel activity.[8,29,32] However, since ChTX interacts not only with Ca^{2+}-activated K^+ channels, but also with voltage-gated K^+ channels as well as with other potential targets, binding data should be interpreted with caution in cells possessing both types of K^+ channels. Human and rabbit red cells offer the advantage of having only Ca^{2+}-activated K^+ channels (Gardos channels). Thus, ChTX in these cells can be considered a specific inhibitor of the Gardos channel.

Iberiotoxin (IbTX), a 37-aa peptide toxin isolated from the venom of the scorpion *Buthus tamulus,* is a selective inhibitor both of K^+ transport mediated by and of ^{125}I-ChTX binding to Ca^{2+}-gated K^+ channels in bovine aortic smooth muscle preparations.[33,34] The sequence homology between IbTX and ChTX is 68%. Iberiotoxin is less positively charged than ChTX and, as true for ChTX, its three-dimensional structure has also been solved with nuclear magnetic resonance studies.[35] We demonstrated that IbTX does not inhibit either Ca^{2+}-activated K^+ transport or ^{125}I-ChTX binding in human red cells, whereas it does inhibit both K^+ transport and ligand binding in rabbit red cells, although with lower potency than does ChTX.[8]

A third member of this family, kaliotoxin (KTX), was recently described as an inhibitor of high conductance Ca^{2+}-activated K^+ channels in nerve cells from the mollusc *Helix pomatia* (65 pS) and in the rabbit celiac ganglia sympathetic neurons (150 pS; K_d of 10–20 nM[36]). Kaliotoxin has been purified from the venom of the scorpion *Androctonus mauretanicus mauretanicus.* It is a single 4-kD polypeptide chain, 38 aa in length.[37] Kaliotoxin displays sequence homology with other scorpion venom-derived inhibitors of Ca^{2+}-activated or voltage-gated K^+ channels: KTX shares 44% amino acid sequence identity with ChTX and 44% with IbTX.[36,37]

Kaliotoxin-induced channel blockade is voltage-independent, suggesting mechanistic differences from the channel blockade induced by ChTX. Our data indicate that KTX specifically blocks Ca^{2+}-activated K^+ transport in human red cells, with IC_{50} values of 2.2. ± 0.9 nM compared with values of 14 ± 2 pM for ChTX.

None of these peptide toxin inhibitors or the previously mentioned drug inhibitors has been used clinically because of the intrinsic toxicity of the former or the low specificity of the latter, requiring high dosages with associated serious side effects.

IMIDAZOLE ANTIMYCOTICS AND THE HUMAN RED CELL GARDOS CHANNEL

A recent development in the study of the red cell Gardos channel has been the description of the inhibitory effect of clotrimazole (CLT) and other imidazole antimycotics.[38] These imidazole derivatives are potent antimycotic agents based on their inhibitory effect on cytochrome P-450 and they affect K^+ as well as Ca^{2+} movement across cell membranes. Clotrimazole is the most potent inhibitor of the Ca^{2+}-activated K^+ channel, whereas econazole and miconazole are the most potent inhibitors of plasma membrane Ca^{2+}-channels.[39] Inhibition of ChTX-sensitive, Ca^{2+}-gated K^+ channels of other cell types (rat thymocytes, Ehrlich ascites tumor cells) was also demonstrated by Alvarez *et al.*[38] The effect of CLT and other imidazoles on the Gardos channel was interpreted as indicative of the involvement of cytochrome P-450-related hemoprotein in the regulation of this channel, as had been proposed for the effect of imidazoles on plasma membrane Ca^{2+} channels.[39]

We have shown that CLT specifically inhibits the Gardos channel of sickle erythrocytes by measuring the inhibition of Ca^{2+}-activated 86Rb$^+$ influx (IC_{50} = 51 ± 15 nM[16]). In our report we provided evidence for a direct interaction of CLT with the external portion of the Gardos channel by showing that CLT specifically displaces ^{125}I-ChTX bound to human red cells with an IC_{50} (12 ± 4 nM) similar to that for K^+ transport inhibition. Thus, CLT acts more like a pore blocker, because it displaces ChTX which is known to bind close to the mouth of the channel rather than via a hemeprotein-based intracellular regulator. *In vitro* pharmacological studies suggest that neither type I nor type II imidazoline receptors are detectable on human red cells, and so the action of CLT and other imidazole inhibitors on Gardos channel activity cannot be ascribed to interaction with such receptors (Coupry *et al.*, this volume). In additional experiments, we demonstrated that the use of CLT prevented dehydration of sickle cell induced in the presence of Ca^{2+} by either exposure to A 23187 or cyclic oxygenation-deoxygenation.[16] These *in vitro* studies indicated that CLT could specifically block K^+ loss and dehydration mediated by the Gardos channel in sickle cells.

In vitro studies by others have also indicated that when combined with a compound that stabilizes the oxyconformation of Hb S, CLT induces an additive reduction in the clogging rate of a micropore filter and may attenuate formation of irreversibly sickled cells.[40]

TEST OF ORAL CLOTRIMAZOLE ADMINISTRATION IN A TRANSGENIC MOUSE MODEL FOR SICKLE CELL DISEASE

A necessary step before using CLT in normal volunteers and patients with sickle cell disease is to demonstrate its efficacy in an animal model for sickle cell disease.

There are several transgenic mouse models for sickle cell disease. We have used a recently developed transgenic mouse model, the SAD mouse, to study the effect of oral administration of clotrimazole. The SAD mouse model is based on the transgenic expression of the triply mutated Hb SAD,[41] which combines the β 6 Glu-Val mutation with two additional mutations known to promote Hb S polymerization, Hb Antilles (β ^{23}Ile) and Hb D-Punjab (β^{121}Gln). Hb SAD is capable of polymerization at concentrations much lower than those necessary to induce polymerization of Hb S. SAD-1 mouse red cells contain 19% Hb SAD, and β thal/SAD 1 mouse red cells contain 26% Hb SAD. β thal/SAD 1 mice are characterized by: increased mortality during early fetal development; anemia during late fetal development; increased cell turnover with reticulocytosis; and microvascular occlusions and organ damage similar to human SS disease. Ion transport *in vitro* and the effect of oral CLT have been studied in SAD mice, and the results are summarized in the next paragraphs.[42]

The presence of Hb SAD is associated with significant K loss and dehydration compared with the thalassemic parental strains. Cell Na content is unchanged, but cell K content is reduced from 490 to 288 mmol/kg Hb. Mean corpuscular hemoglobin concentration is increased from 33 to 41 g/dL when Hb SAD is expressed. There are, however, no significant differences in the magnitude of Rb flux mediated by the Gardos pathway or its sensitivity to CLT *in vitro.* Clotrimazole markedly inhibited Ca^{2+}-activated Rb influx, with an IC_{50} of 75 nM, similar to that reported for sickle erythrocytes.[16] Experiments were carried out to determine if CLT could also inhibit *in vitro* dehydration of SAD 1 erythrocytes. Cell density was estimated by the phthalate density profiles. β thal/SAD 1 red cells have increased density compared with those of two other mouse strains. The addition of ionophore in the presence of external Ca induces substantial dehydration and an increase in cell density. The addition of CLT almost completely prevents this Ca-dependent dehydration.

The effect of oral administration of CLT was studied in short-term (1 week) and long-term (1 month) experiments. A liquid suspension of CLT was given orally twice a day with gavage. Oral administration of CLT (160 mg/kg per day) induced, after 24 hours, significant inhibition of Ca^{2+}-activated Rb^+ transport, measured in whole mouse blood, which persisted during the 7 days of treatment and disappeared after withdrawal of the drug. Similar inhibition of transport by CLT was observed in control animals, in β thal/SAD 1 animals, and when miconazole (120 mg/kg day) was given to the β thal/SAD 1 mice. Clotrimazole administration had no effect on the cell K^+ content of controls, whereas a significant increase in K^+ content was noted when either CLT or miconazole, another imidazole antimycotic, was given to β thal/SAD 1 mice. The increase in cell K content disappeared when CLT was withdrawn. Thus, only in the β thal/SAD 1 mouse was inhibition of Ca^{2+}-activated K^+ transport induced by CLT administration associated with increased erythrocyte K^+ content, decreased MCHC and red cell density, and increased hematocrit. All these changes were reversible when CLT was withdrawn. No side effects or toxicity was observed with the use of CLT.

However, any ideal treatment for sickle cell disease should be effective and well tolerated over the long term. Therefore, the effect of 1 month of oral CLT administration at a lower dose (80 mg/kg/day) was tested in control and β thal/SAD 1 mice and compared with that in animals treated with suspension medium alone. As was true for short-term treatment, oral administration of CLT for as long as 1 month induced significant and continuing inhibition of Ca^{2+}-activated K^+ channels measured as Ca^{2+}-activated $^{86}Rb^+$ influx in whole blood as compared with baseline values and untreated control animals. Inhibition of transport by CLT was observed in both SAD 1 and control animals. Inhibition of Ca^{2+}-activated K^+ transport was associated with increased cell K^+ content only in the β thal/SAD 1 mice treated with CLT, but not in the treated control animals. Clotrimazole administration had no effect on the cell K^+ content of nontransgenic mice. Over 28 days of treatment, neither β thal nor β thal/SAD 1 mice had any ill effects or weight loss compared with untreated animals. Thus, only in the β thal/SAD 1 mice was CLT administration associated with a significant decrease in MCHC, which led to increased hematocrit at constant Hb values. Red cell density, measured with the phthalate density profile, was also decreased by CLT administration.

The following conclusions can be drawn from studies in SAD mice: (1) CLT inhibits K^+ movement via the Ca^{2+}-activated K^+ channel of mouse erythrocytes *in vitro* and *in vivo;* (2) oral administration of CLT prevents erythrocyte dehydration in β thal/SAD 1 mice. These results suggest that CLT could be used to prevent sickle cell dehydration *in vivo.* However, the effect of oral CLT administration on the human red cell Gardos channel and the relationship between blood levels of CLT and inhibition of red cell K^+ transport needed to be characterized before the onset of clinical testing of CLT in patients with sickle cell disease.

ORAL CLOTRIMAZOLE ADMINISTRATION IN NORMAL VOLUNTEERS

Studies were carried out in normal volunteers to determine (1) whether oral administration of CLT is associated with inhibition of the human red cell Gardos channel and (2) the relationship between CLT plasma levels and % inhibition of the Gardos channel. Four normal healthy volunteers, 3 males and 1 female, aged 39 (A), 38 (B), 44 (C), and 45 years (D), respectively, ingested 1 g of CLT (dose range 13-18 mg/kg body weight; Mycelex, 500 mg tablets; Miles Pharmaceuticals, West Haven, Connecticut). Venous blood was sampled at 2, 4, 6, 8, 10, 24, and 36 hours following the ingestion of CLT. A newly developed HPLC assay was used to determine plasma and whole blood CLT levels.[43] In all subjects, measurable levels of CLT were achieved following ingestion of 1 gr of CLT, with peak values varying from 3.3 μM (subject D, at 4 hours) to 0.3 μM (subject B, at 2 hours). This is in agreement with previous studies on the pharmacokinetics of CLT, which have shown great variability in plasma levels among normal subjects.[44–48] Two additional peaks, most likely representing CLT metabolites, were also present in plasma following CLT administration. Levels of metabolite B peaked at 4-6 hours with nominal maximum concentrations ranging from 1.9 μM (subject C, at 6 hours) to 1.6 μM (A and D at 6 hours, B at 4 hours). Metabolite A levels also peaked at 6 hours with maximum nominal values ranging from 2.6 μM (subject D, at 6 hours) to 0.8 μM (subject B, at 4 hours).

Oral administration of CLT was associated with measurable inhibition of Ca^{2+}-activated K^+ flux measured in whole blood as Ca^{2+}-activated $^{86}Rb^+$ influx. Inhibition ranging from a low of 45% (subject B, at 8 hours) to a high of 92% (subject C, at 4 hours) was observed. An apparent IC_{50} value of 0.88 ± 0.3 μM was determined

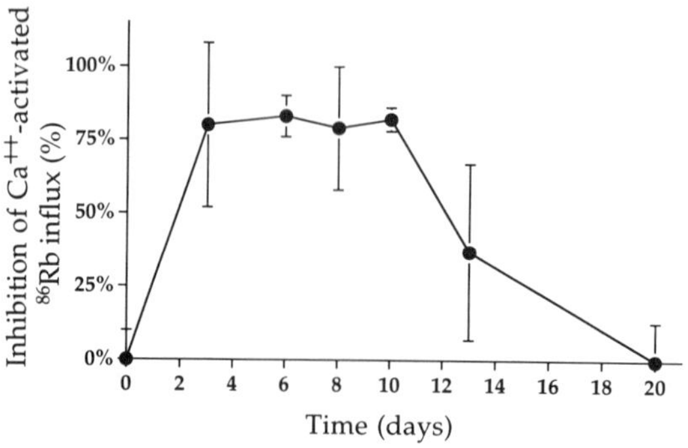

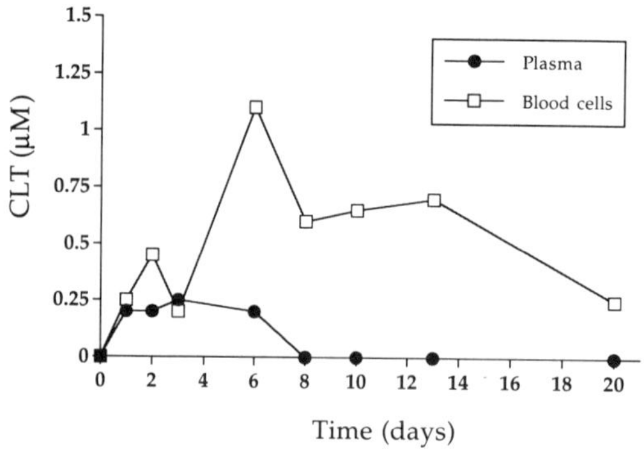

FIGURE 2. **(Top)** Inhibition of Ca^{2+}-activated ^{86}Rb influx by oral administration of clotrimazole (CLT) as a function of time following administration of 500 mg of CLT twice a day for 6 days in one normal subject. **(Bottom)** Plasma and red cell levels of CLT as a function of time following administration of 500 mg of CLT twice a day for 6 days in the same subject.

for the measured inhibition of red cell Ca^{2+}-activated K^+ influx by the summed levels of CLT and the two major CLT metabolites, which is close to the IC_{50} value for the *in vitro* addition of CLT to whole blood (1.7 ± 0.7 μM).

In a separate experiment, two subjects, one male (A, 72 kg) and one female (D, 56 kg), ingested one 500-mg CLT tablet every 12 hours for 6 days, corresponding to a daily CLT dose of 14 and 18 mg/kg, respectively. Blood was collected 6 hours after the AM dose of CLT on days 3 and 6 for CBC, reticulocyte count, red cell and reticulocyte cellular indices (cell volume, hemoglobin concentration, and content), and for CLT blood levels. Similar assays were repeated at the indicated intervals after discontinuation of drug therapy on day 6. As shown in FIGURE 2, there was significant inhibition of Ca^{2+}-activated K^+ influx during CLT administration, which persisted for at least 7 days after the drug was discontinued.

Despite significant and continued inhibition of Ca^{2+}-activated K^+ movement in whole blood during and after 6 days of oral treatment with CLT, the two subjects displayed no significant changes in reticulocyte count, mean reticulocyte volume

(MCVr), and mean reticulocyte hemoglobin concentration (CHCMr). Moreover, no changes were observed in the distribution of cell volume and hemoglobin concentration in mature red cells of normal subjects. Thus, a direct role of the Gardos pathway in determining the mean volume of normal human reticulocytes could not demonstrated in these studies. However, this lack of effect on normal red cell hydration state is similar to that in mouse thalassemic controls for the thal/SAD 1 transgenic mice.

Oral administration of CLT was not associated with significant side effects in any of the subjects studied. In particular no nausea, vertigo, vomiting, or diarrhea was observed or reported. No changes were observed in liver function tests, plasma creatinine, or blood urea nitrogen.

From these *in vitro* and *in vivo* data, it seems reasonable to conclude that oral CLT might represent a new therapeutic approach to the treatment of sickle cell disease. All the data collected so far in sickle cells *in vitro,* in a transgenic mouse model of human sickle cell disease, and in normal healthy volunteers indicate that oral administration of CLT should induce a significant and persistent blockade of the Gardos channel in patients with sickle cell disease. It remains to be determined if this blockade will lead to therapeutically meaningful changes in erythrocyte density and cell K^+ content and if the absence of significant toxicity or side effects observed in normal subjects can be replicated in patients with sickle cell disease. A pilot clinical trial designed to test this hypothesis is underway at Children's Hospital in Boston in patients with sickle cell disease.

REFERENCES

1. Gardos, G. 1958. The function of calcium in the potassium permeability of human erythrocytes. Biochim. Biophys. Acta **30:** 653-654.
2. Latorre, R., A. Oberhauser, P. Labarca & O. Alvarez. 1989. Varieties of calcium-activated potassium channels. Ann. Rev. Physiol. **51:** 385-399.
3. Lew, V. L. & H. G. Ferreira. 1976. Variable Ca sensitivity of a K-selective channel in intact red-cell membranes. Nature **263:** 336-338.
4. Romero, P. J., C. E. Ortiz & C. Melitto. 1990. Metabolic control of the K^+ channel of human red cells. J. Membr. Biol. **116:** 19-29.
5. Christopherson, P. 1991. Ca-activated K channel from human erythrocyte membranes: Single channel rectification and selectivity. J. Membr. Biol. **119:** 75-83.
6. Grygorczyk, R., W. Schwarz & H. Passow. 1984. Ca-activated K channels in human red cells. Biophys. J. **45:** 693-698.
7. Hamill, O. P. 1983. Potassium and chloride channels in red blood cells. *In* B. Sackmann & E. Neher, Eds.: 451-471. Single Channel Recording. Plenum Press. New York.
8. Brugnara, C., L. De Franceschi & S. L. Alper. 1993. Ca^{2+}-activated K^+ transport of human and rabbit erythrocytes: Comparison of binding and transport inhibition by scorpion toxins. J. Biol. Chem. **268:** 8760-8768.
9. Butler, A., S. Tsunoda, D. P. McCobb, A. Wei & L. Salkoff. 1993. *mSlo,* a complex mouse gene encoding "maxi" calcium-activated potassium channels. Science **261:** 221-224.
10. Lew, V. L., C. J. Freeman, O. E. Ortiz & R. M. Bookchin. 1991. A mathematical model of the volume, pH and ion content regulation in reticulocytes. J. Clin. Invest. **87:** 100-112.
11. Halperin, J. A., C. Brugnara & A. Nicholson-Weller. 1989. Ca^{++} activated K^+ efflux limits complement mediated lysis of human erythrocytes. J. Clin. Invest. **83:** 1466-1471.

12. GLADER, B. E. & D. G. NATHAN. 1978. Cation permeability alterations during sickling: Relationship to cation composition and cellular hydration of irreversibly sickled cells. Blood **51:** 983-989.
13. OHNISHI, S. T., H. KATAGI & C. KATAGI. 1989. Inhibition of the *in vitro* formation of dense cells and of irreversibly sickled cells by charybdotoxin, a specific inhibitor of calcium-activated potassium efflux. Biochim. Biophys. Acta **1010:** 199-203.
14. WOLFF, D., X. CECCHI, A. SPALVINS & M. CANESSA. 1988. Charybdotoxin blocks with high affinity the Ca-activated K^+ channel of Hb A and Hb S red cells: Individual differences in the number of channels. J. Membr. Biol. **106:** 243-252.
15. BOOKCHIN, R. M., O. E. ORTIZ & V. L. LEW. 1991. Evidence for a direct reticulocyte origin of dense red cells in sickle cell anemia. J. Clin. Invest. **87:** 113-124.
16. BRUGNARA, C., L. DE FRANCESCHI & S. L. ALPER 1993. Inhibition of Ca^{2+}-dependent K^+ transport and cell dehydration in sickle erythrocytes by CLT and other imidazole derivatives. J. Clin. Invest. **92:** 520-526.
17. EATON, W. A. & J. HOFRICHTER. 1987. Hemoglobin S gelation and sickle cell disease. Blood **70:** 1245-1266.
18. ROSA, R. M., B. E. BIERER, R. THOMAS, J. S. STOFF, M. KRUSKALL, S. ROBINSON, H. F. BUNN & F. H. EPSTEIN. 1980. A study of induced hyponatremia in the prevention and treatment of sickle cell crisis. N. Engl. J. Med. **303:** 1138-1143.
19. CHARACHE, S. & W. G. WALKER. 1981. Failure of desmopressin to lower serum sodium or prevent crisis in patients with sickle cell anemia. Blood **58:** 892-896.
20. PLATT, O. S., S. H. ORKIN, G. DOVER, G. P. BEARDSLEY, B. MILLER & D. G. NATHAN. 1984. Hydroxyurea enhances fetal hemoglobin production in sickle cell anemia. J. Clin. Invest. **74:** 652-656.
21. GOLDBERG, M. A., C. BRUGNARA, G. J. DOVER, L. SHAPIRA, S. CHARACHE & H. F. BUNN. 1990. Treatment of sickle cell anemia with hydroxyurea and erythropoietin. N. Engl. J. Med. **323:** 366-372.
22. RODGERS, G., G. J. DOVER, N. UYESAKA, C. T. NOGUCHI, A. N. SCHECHTER & A. W. NIENHUIS. 1993. Augmentation by erythropoietin of the fetal-hemoglobin response to hydroxyurea in sickle cell disease. N. Engl. J. Med. **328:** 73-80.
23. PERRINE, S., G. GINDER, D. V. FALLER, G. H. DOVER, T. IKUTA, H. E. WITKOWSKA, S. CAI, E. P. VICHINSKY & N. OLIVIERI. 1993. A short-term trial of butyrate to stimulate fetal-globin-gene expression in the β-globin disorders. N. Engl. J. Med. **328:** 81-86.
24. REICHSTEIN, E. & A. ROTHSTEIN. 1981. Effects of quinine on Ca^{++}-induced K^+ efflux from human red blood cells. J. Membr. Biol. **59:** 57-63.
25. SIMONS, T. J. 1976. Carbocyanine dyes inhibit Ca-dependent K efflux from human red cell ghosts. Nature **264:** 467-469.
26. KAJI, D. 1990. Nifedipine inhibits calcium-activated K transport in human erythrocytes. Am. J. Physiol. **259:** (Cell Physiol. 28): C332-C339.
27. ELLORY, J. C., K. KIRK, S. J. CULLIFORD, G. B. NASH & J. STUART. 1992. Nitrendipine is a potent inhibitor of the Ca^{2+}-activated K^+ channel of human erythrocytes. FEBS Lett. **296:** 219-221.
28. MILLER, C., E. MOCKZYDLOWSKI, R. LATORRE & M. PHILLIPS. 1985. Charybdotoxin, a protein inhibitor of single Ca^{2+}-activated K^+ channels from mammalian skeletal muscle. Nature **313:** 316-318.
29. DEUTSCH, C., M. PRICE, S. LEE, V. F. KING & M. L. GARCIA. 1991. Characterization of high affinity binding sites for charybdotoxin in human T lymphocytes. J. Biol. Chem. **266:** 3668-3674.
30. BONTEMS, F., C. ROUMESTAND, P. BOYOT, B. GILQUIN, Y. DOLJANSKY, A. MENEZ & F. TOMA. 1991. Three dimensional structure of natural charybdotoxin in aqueus solution by 1H-NMR. Eur. J. Biochem. **196:** 19-28.
31. PARK, C. S., S. F. HAUDOORFF & C. MILLER. 1991. Design, synthetic and functional expression of a gene for charybdotoxin, a peptide blocker of K^+ channels. Proc. Natl. Acad. Sci. USA **88:** 2046-2050.

32. VASQUEZ, J., P. FEIGENBAUM, V. F. KATZ, J. P. REUBEN, L. ROY-CONTACIN, R. S. SLAUGHTER, G. J. KACZOROWSKI & M. L. GARCIA. 1989. Characterization of high affinity binding sites for charybdotoxin in sarcolemmal membranes from bovine aortic smooth muscle. J. Biol. Chem. **264:** 20902-20909.
33. GALVEZ, A., G. GIMENEZ-GALLEGO, J. P. REUBEN, L. ROY-CONTANCIN, P. FEIGENBAUM, G. J. KACZOROWSKI & M. L. GARCIA. 1990. Purification and characterization of a unique, potent, peptidyl probe for the high conductance calcium-activated potassium channel from venom of the scorpio *Buthus tamulus*. J. Biol. Chem. **265:** 11083-11090.
34. CANDIA, S., M. L. GARCIA & R. LATORRE. 1992. Mode of action of iberiotoxin, a potent blocker of the large conductance Ca^{2+}-activated K^+ channel. Biophys. J. **63:** 583-590.
35. JOHNSON, B. A. & E. E. SUGG. 1992. Determination of the three-dimensional structure of iberiotoxin in solution by 1H nuclear magnetic resonance spectroscopy. Biochemistry **31:** 8151-8159.
36. CREST, M., G. JACQET, M. GOLA, H. ZERROUK, A. BENSLIMANE, H. ROCHAT, P. MANSUELLE & M. F. MARTIN-EACLAIRE. 1992. Kaliotoxin, a novel peptidyl inhibitor of neuronal BK-Type Ca^{2+}-activated K^+ channels characterized from *Androctonus mauretanicus mauretanicus* Venum. J. Biol. Chem. **267:** 1640-1647.
37. ROMI, R., M. CREST, M. GOLA, F. SAMPIERI, G. JACQUET, H. ZERROUK, P. MANSUELLE, O. SOROKINE, A. V. DORSSELAER, H. ROCHAT, M. F. MARTIN-EAUCLAIRE & J. V. RIETSCHOTEN. 1993. Synthesis and characterization of kaliotoxin. Is the 26-32 sequence essential for potassium channel recognition? J. Biol. Chem. **268:** 26302-26309.
38. ALVAREZ, J., M. MONTERO & J. GARCIA-SANCHO. 1992. High affinity inhibition of Ca^{2+}-dependent K^+ channels by cytochrome P-450 inhibitors. J. Biol. Chem. **267:** 11789-11793.
39. MONTERO, M., J. ALVAREZ & J. GARCIA-SANCHO. 1991. Agonist-induced Ca^{2+}-influx in human neutrophils is secondary to the emptying of intracellular calcium stores. Biochem. J. **277:** 73-79.
40. STUART, J., B. O. MOJIMINIYI, P. C. STONE, S. J. CULLIFORD & J. C. ELLORY. 1994. Additive *in vitro* effects of anti-sickling drugs. Br. J. Haematol. **86:** 820-823.
41. TRUDEL, M., N. SAADANE, M. C. GAREL, J. BARDAKDJIAN-MICHAU, Y. BLOUQUIT, J. L. GUERQUIN-KERN, P. ROUYER-FESSARD, D. VIDAUD, A. PACHNIS, P. H. ROMEO, Y. BEUZARD & F. COSTANTINI. 1991. Towards a transgenic mouse model of sickle cell disease: hemoglobin sad. EMBO J. **11:** 3157-3165.
42. DE FRANCESCHI, L., N. SAADANE, M. TRUDEL, S. L. ALPER, C. BRUGNARA & Y. BEUZARD. 1994. Treatment with oral clotrimazole blocks Ca^{2+}-activated K^+ transport and reverses erythrocyte dehydration in transgenic SAD mice. A model for therapy of sickle cell disease. J. Clin. Invest. **93:** 1670-1676.
43. SAKAMOTO, M., C. BRUGNARA & N. RIFAI. 1994. Development of a reverse phase HPLC assay for the determination of clotrimazole. Clin. Chem. **40:** 1078a.
44. WEUTA, H. 1974. Clinical studies with oral clotrimazole. Postgrad. Med. J. **50** (Suppl 1): 45-49.
45. GOODMAN, A. G., T. W. RALL, A. S. NIES & P. TAYLOR. 1990. Goodman and Gilman's. The Pharmacological Basis of Therapeutics. Pergamon Press. New York.
46. SAWYER, P. R., R. N. BROGDEN, R. M. PINDER, T. M. SPEIGHT & G. S. AVERY. 1975. Clotrimazole: A review of its antifungal activity and therapeutic efficacy. Drugs **9:** 424-447.
47. SEO, M., H. IIDA & Y. MIURA. 1977. Basic experiments with clotrimazole administered orally. Curr. Med. Res. Op. **5:** 169-178.
48. DUHM, B., W. MAUL, H. MEDENWALD, K. PATZSCHKE, J. PUETTER & L. A. WEGNER. 1974. The pharmacokinetics of clotrimazole 14C. Postgrad. Med. J. **50** (Suppl. 1): 13-16.

Clonidine and Cirazoline Inhibit Activation of Nicotinic Channels in PC-12 Cells[a]

I. F. MUSGRAVE,[b] D. KRAUTWURST,
J. HESCHELER, AND G. SCHULTZ

Institut für Pharmakologie
Freie Universität Berlin
Thielallee 69-73
D-14195 Berlin, Germany

Alpha$_2$-adrenergic drugs such as clonidine and cirazoline have repeatedly been shown to bind with nanomolar affinity to distinct nonadrenergic sites. (For reviews see references 1-4.) There is consensus that these sites can be divided into two distinct groups, the imidazoline I_1 sites and the imidazoline I_2 sites.[5,6] Amongst other characteristics, I_1 sites have a higher affinity for clonidine and a lower affinity for idazoxan than do I_2 sites. I_1 sites have also been linked to signal-transducing G-proteins,[5,7] while I_2 sites do not appear to be linked to G-proteins.[8]

Although great progress has been made in defining the binding properties and tissue and cellular localization of these sites (see, for example, refs. 8-11), far less is known about the functional roles of these imidazoline sites and their mechanisms of signal transduction. The best evidence for a functional role of these sites is the association between the I_1 sites in the brain stem and the hypotension produced by various adrenergic drugs with imidazoline structures[12–14] (see also reviews just cited). The mechanism underlying this effect is so far unknown, but it may involve inhibition of neurotransmitter secretion.[12] To determine if I_1 sites could be acting by modulating Ca^{2+} entry into neurons, we examined the effects of clonidine and cirazoline on Ca^{2+} entry in PC-12 cells, a neuronal cell line that has been reported to possess the I_1 site and to have no α_2-adrenoceptors.[5] We report here that clonidine and cirazoline inhibit nicotine-induced Ca^{2+} entry, apparently by directly blocking the intrinsic ion channel of the nicotinic acetylcholine receptor. This effect may underlie some reported effects putatively involving action at imidazoline binding sites.

METHODS

Rat pheochromocytoma (PC-12) cells (American Type Culture Collection, Rockville, Maryland) were grown in Dulbecco's modified Eagle's medium supplemented

[a] This research was supported by grants from the Deutsche Forschungsgemeinschaft and the Fonds der Chemischen Industrie.

[b] I.F.M. is an Australian NH&MRC fellow.

[c] Present address: Prince Henry's Institute of Medical Research, P.O. Box 5152, Clayton, Victoria 3168, Australia.

with 4.5 mg/ml glucose, 10% (v/v) fetal calf serum, 5% (v/v) horse serum, 1% (v/v) nonessential amino acids, 4 mM glutamine, 100 U/ml penicillin, and 100 μg/ml streptomycin in a humidified atmosphere with 5% CO_2. Cells were grown to confluence before harvesting. For electrophysiological experiments, cells were grown on glass coverslips in the aforementioned medium. Cells were differentiated by the addition of nerve growth factor (NGF; final concentration 100 ng/ml) to the medium for 7 days. In some experiments pertussis toxin (100 ng/ml) or vehicle was added to the medium 24 hours before the experiments.

Intracellular Calcium

Undifferentiated PC-12 cells were detached from their flasks by gentle agitation and harvested by centrifugation for 10 minutes at 250 *g*. Cells were washed once and resuspended to 1×10^7 cells/ml in a buffer consisting of 138 mM NaCl, 6 mM KCl, 1 mM $MgSO_4$, 1 mM $CaCl_2$, 1 mM Na_2HPO_4, 5 mM $NaHCO_3$, 5.5 mM glucose, and 20 mM HEPES supplemented with 1% (w/v) bovine serum albumin and adjusted to pH 7.4 with NaOH (loading buffer). PC-12 cells were loaded with fura-2 acetoxymethyl ester (fura-2/AM) by a modification of the method used by Musgrave *et al.*[15] Briefly, cells were incubated with fura-2/AM and pluronic acid F-127 at final concentrations of 4 μM and 0.5% w/v, respectively, in loading buffer and incubated with shaking for 60 minutes at 37 °C. After this the cells were centrifuged and washed once with loading buffer, then resuspended in buffer containing 138 mM NaCl, 6 mM KCl, 1 mM $MgCl_2$, 5.5 mM glucose, and 20 mM HEPES, pH 7.4 (measurement buffer) at 2×10^6 cells/ml. Fluorescence was measured ratiometrically using a Perkin Elmer LS 50 B spectrofluorimeter (Buckinghamshire, UK) with excitation wavelengths of 340 and 380 nm and an emission wavelength of 510 nm.

An aliquot of cells was suspended in acryl fluorescence cuvettes (Sarstedt, Nümbrecht, Germany) in measurement buffer to a final concentration of 1×10^6 cells/ml. This suspension was supplemented with Ca^{2+} for a final concentration of 1 mM. Drugs or vehicle was also added as described in Results. The cuvettes were preincubated for 2 minutes at 37 °C before measurements began. Basal calcium levels were monitored for a further minute before adding stimuli. Fluorescent signals were calibrated after lysis of the cells with 0.1% (w/v) Triton X-100 (maximal fluorescence) and the subsequent addition of 20 mM EGTA (minimal fluorescence). $[Ca^{2+}]_i$ values were calculated according to equation 5 by Grynkiewicz *et al.*[16]

Electrophysiology

Glass coverslips with adherent differentiated PC-12 cells were transferred into a perfusion chamber (0.2 ml) mounted on an inverted microscope and perfused at a rate of 4 ml/min. Patch pipettes were prepared from glass capillaries (Jencons, Leight Buzzard, UK) according to the method of Hamill *et al.*[17] The average resistance of the electrode was 4-6 MΩ. Cells were recorded in a whole cell configuration. Pipettes were filled with a solution containing 125 mM CsCl, 1 mM $MgCl_2$, 3 mM Mg-ATP, 10 mM EGTA, and 10 mM HEPES adjusted to pH 7.4 at 37 °C with CsOH. Electrical access to the cells was gained by suction. Membrane potentials and whole-cell currents were measured according to the method of Hamill *et al.*,[17] using a List LM/EPC7 patch-clamp amplifier (List Electronics, Darmstadt, Germany). The capacitance was between 5 and 15 pF. PC-12 cells were superfused externally with standard Tyrode's solution (140 mM NaCl, 5.4 mM KCl, 1 mM $MgCl_2$, 1.8 mM $CaCl_2$, 10 mM glucose, and 10 mM HEPES, pH 7.4) containing drugs or vehicle as described in Results. To avoid desensitization to nicotine, application of nicotine was brief (about 5 seconds) and each repeat application was separated by 3 minutes. Cells were preincubated with adrenergic drugs during the 1 minute before application of nicotine.

Binding

Membranes (a kind gift of Dr. F. Hucho, Department of Biochemistry, Freie Universität Berlin) were obtained from *Torpedo marmorata* electroplaques by the method of Schiebler and Hucho.[18] Membrane protein represented between 80 and 90% nicotinic acetylcholine receptors as determined by α-bungarotoxin binding.[19] Binding was performed by a modification of the method of Min and Weiland.[20] An aliquot of membranes (final concentration 0.02 mg/ml) was incubated in polypropylene tubes in 100 μl of binding buffer with a final concentration of NaCl 150 mM, HEPES 50 mM, EDTA 0.5 mM, and BSA 1% w/v, pH 7.4. Tubes were preincubated for 3 minutes at 25 °C and binding was initiated by the addition of [^{3}H]phencyclidine (100 nM). After 30 minutes the reaction was terminated by quickly diluting the aliquots with 2 ml of ice-cold wash buffer (NaCl 150 mM, HEPES 50 mM, and EDTA 0.5 mM, pH 7.4) and rapid filtration on Whatman GF/B glass filters presoaked with 1% (w/v) polyethylene imine (Fluka, Buchs, Switzerland) followed by two washes. Binding was estimated by liquid scintillation counting. Nonspecific binding was determined with triphenylmethylphosphonium (100 μM).

Materials

[^{3}H]Phencyclidine (42 Ci/mmol) was obtained from New England Nuclear (Dreieich, Germany). Adrenaline HCl, yohimbine HCl, clonidine HCl, isobutyl methylxanthine (IBMX), triphenylmethylphosphonium (TPP), and nicotine were obtained from Sigma Chemie (Deisenhofen, Germany). Cirazoline was a kind gift of Syntelabo (Bagneaux, France). 5-Allyl-2-amino-5,6,7,8-tetrahydro-4*H*-thiazolo-[4,5-*d*]-azepin-dihydrochloride (BHT 920) was a kind gift of Boehringer Ingelheim (Ingelheim, Germany). NGF (7S fraction) was obtained from Boehringer Mannheim (Mannheim, Germany). Pertussis toxin was obtained from List Biological Laboratories (Campbell, USA). Fura-2 acetoxymethyl ester was obtained from Calbiochem (Frankfurt/Main, Germany). Pluronic F-127 was purchased from Molecular Probes (Oregon, USA). Stock solutions were made by dissolving drugs in doubly deionized water to 10 mM with the exception of adrenaline, which was dissolved in 0.1 mM HCl. Fura-2 acetoxymethyl ester was dissolved in dry DMSO to give a stock solution of 2 mM. Pluronic F-127 was dissolved in dry DMSO to give a 25% (w/v) solution. All other chemicals were of the highest purity commercially available.

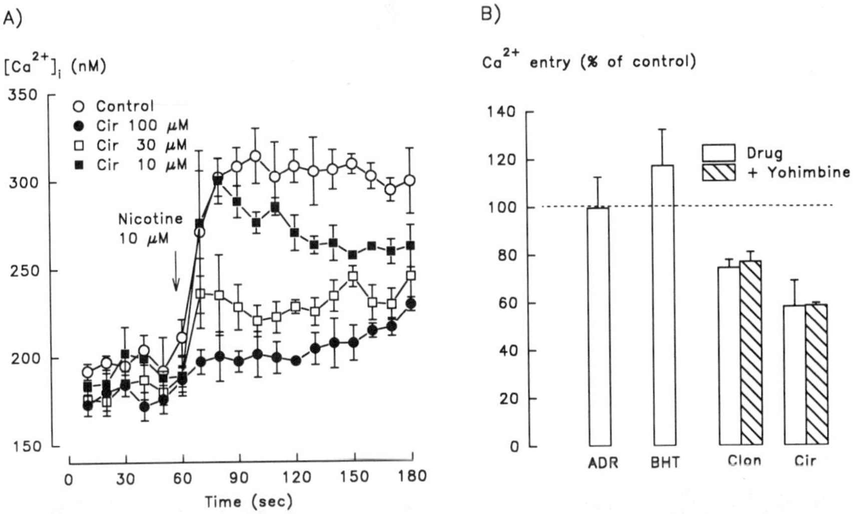

FIGURE 1. Modulation of nicotine-induced increases in intracellular Ca^{2+} in undifferentiated PC-12 cells by adrenergic drugs. Cells were loaded with Fura-2 as described in Methods. **(A)** Original results from an experiment representative of four separate experiments performed in duplicate. *Arrow* indicates the addition of nicotine (10 μM). Drugs or vehicle was added 3 minutes before nicotine. Cirazoline (Cir; 10–100 μM) inhibited the nicotine-induced increase in $[Ca^{2+}]_i$ in a concentration-dependent manner. Similar results were seen with clonidine. **(B)** Effects of adrenaline (ADR; 100 μM), BHT 920 (BHT; 100 μM), clonidine (Clon; 10 μM), and cirazoline (Cir; 30 μM) on nicotine-induced Ca^{2+} entry. Ca^{2+} was calculated by subtracting $[Ca^{2+}]_i$ in experiments performed in the absence of external Ca^{2+} from $[Ca^{2+}]_i$ in experiments performed in the presence of $[Ca^{2+}]_i$. The substantial inhibition of Ca^{2+} entry produced by clonidine and cirazoline was unaffected by preincubation with yohimbine (100 μM). Results are the mean ± SD of 2-3 separate experiments performed in duplicate.

RESULTS

The intracellular calcium concentration ($[Ca^{2+}]_i$) of undifferentiated PC-12 cells was significantly increased by the application of nicotine (10 μM; FIG. 1). Clonidine and cirazoline had no effect on $[Ca^{2+}]_i$ by themselves (data not shown). However, both clonidine and cirazoline inhibited the nicotine-induced increase in $[Ca^{2+}]_i$ in a concentration-dependent manner (FIGS. 1A and 2A). Adrenaline (100 μM) and BHT 920 (100 μM) had no effect on either basal $[Ca^{2+}]_i$ (data not shown) or the nicotine-induced increase in $[Ca^{2+}]_i$ (FIG. 1B). Inhibition of nicotine-induced increases in $[Ca^{2+}]_i$ by clonidine (10 μM) or cirazoline (30 μM) was not reversed by yohimbine (100 μM; FIG. 1B).

Pretreatment of cells with pertussis toxin for 24 hours did not affect the inhibition of the nicotine-induced increase in $[Ca^{2+}]_i$ by clonidine or cirazoline (FIG. 2A). Pretreatment of PC-12 cells with IBMX (100 μM) also did not affect the inhibitory effect of cirazoline (30 μM, FIG 2b).

Nicotine (10 μM) stimulated a whole-cell inward current in differentiated PC-12 cells voltage clamped at a holding potential of –60 mV (FIG. 3). The nicotine-

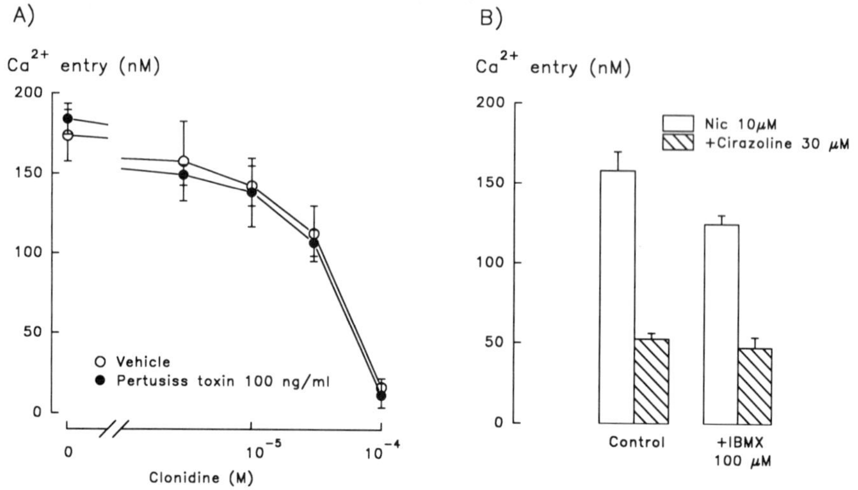

FIGURE 2. Effect of pertussis toxin and isobutylmethyxanthine (IBMX) on Ca^{2+} entry in undifferentiated PC-12 cells. PC-12 cells were loaded with Fura-2 as described in Methods. (**A**) The results from a single experiment performed in duplicate. Cells were pretreated for 24 hours with vehicle (*open symbols*) or pertussis toxin (100 ng/ml; *closed symbols*). Pretreatment with pertussis toxin had no effect on the inhibition of nicotine-induced Ca^{2+} entry by clonidine. Similar results were obtained with cirazoline. (**B**) Preincubation of cells with IBMX for 10 minutes had a small effect on the nicotine-induced Ca^{2+} entry (*open bars*) but not on its inhibition by cirazoline (30 μM; *hatched bars*). Values shown are the mean ± SD (n = 2).

stimulated inward current was inhibited by clonidine and cirazoline (Fig. 3) but not by adrenaline (100 μM). This inhibition was not blocked by yohimbine (100 μM; Fig. 3A). Pretreatment of cells with pertussis toxin (100 ng/ml for 24 hours) or perfusion of cells with GDP-β-S (2 mM) via the patch pipette had no effect on either the nicotine-induced increase in current density (control, 46 ± 17 pA/pF; pertussis toxin, 50 ± 18 pA/pF; GDP-β-S, 40 ± 20 pA/pF; n = 3) or its inhibition by cirazoline (control, 24 ± 12 pA/pF; pertussis toxin, 15 ± 9 pA/pF; GDP-β-S, 23 ± 11 pA/pF; n = 3).

[^{3}H]Phencyclidine bound specifically and saturably to *Torpedo* membranes and was increased approximately 20-fold in the presence of nicotine (10 μM, data not shown). Although cirazoline displaces specific [^{3}H]phencyclidine binding in the absence of nicotine (data not shown) due to better sensitivity of the assay in the presence of nicotine, all displacement experiments reported here were performed with nicotine. Specific [^{3}H]phencyclidine binding in the presence of nicotine was concentration-dependently displaced by clonidine and cirazoline, but not by adrenaline (Fig. 4).

DISCUSSION

PC-12 cells have been used as a model of sympathetic neurons.[21] Both undifferentiated and differentiated PC-12 cells are reported to have imidazoline I_1 sites without

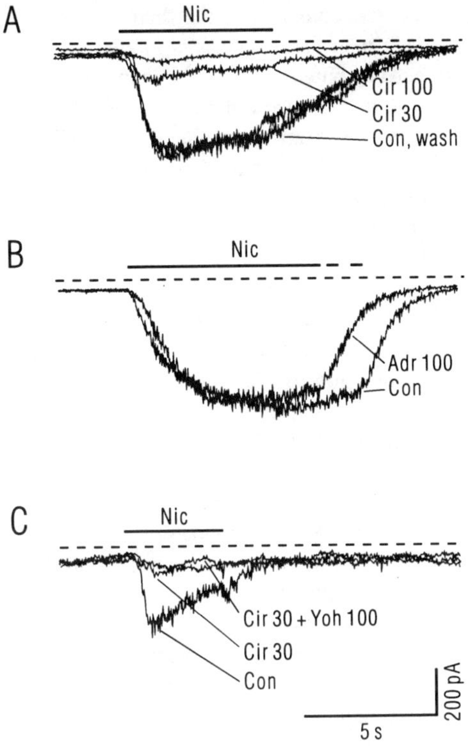

FIGURE 3. Effects of various adrenergic agents on nicotine-induced whole-cell current in NGF-differentiated PC-12 cells. Whole-cell currents were continuously recorded at a holding potential of −60 mV. Nicotine (10 μM) was present during the time indicated by the bars. Cells were superfused with adrenergic agents (numbers are concentrations in μM) during 1 minute prior to application of nicotine. *Broken lines* indicate zero current level. (**A**) Original current traces recorded from one cell without preincubation (Con, wash) and after 1 minute of preincubation with cirazoline (30 or 100 μM). (**B**) Nicotine-induced currents recorded from one cell before (Con) and after preincubation with 100 μM adrenaline (Adr 100). (**C**) Nicotine-induced currents recorded from one cell before (Con) and after preincubation with 30 μM cirazoline (Cir 30) or 30 mM cirazoline plus 100 μM yohimbine (Cir 30 + Yoh 100). **A** and **B** have the same scaling as does **C.** Traces are representative of 3-4 similar experiments.

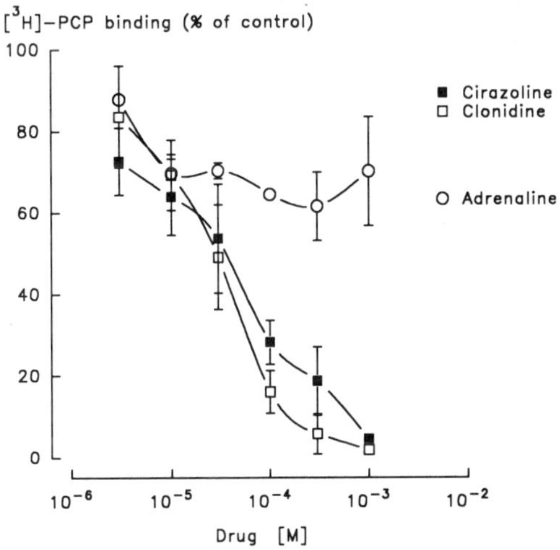

FIGURE 4. Effect of adrenergic drugs on binding of [^{3}H]phencyclidine to membranes from *Torpedo* electroplaques. [^{3}H]Phencyclidine (100 nM; [^{3}H]PCP) binding to *Torpedo* membranes was performed as described in Methods. Nonspecific binding was defined with TPP and usually was less than 10% of total binding. Clonidine and cirazoline (*open and closed squares,* respectively) concentration-dependently displaced specific [^{3}H]phencyclidine binding (expressed as % of control values). Adrenaline, in contrast, produced a small concentration-independent displacement of [^{3}H]PCP. Results are the mean ± SD of two separate experiments performed in triplicate.

α_2-adrenoceptors,[5] making them a potential model to study the role of imidazoline sites in neurons. A neuronal model is especially important, as the best characterized effect attributed to I_1 sites, production of hypotension, is presumably mediated by action on brain stem neurons.[12] Furthermore, stimulation-induced secretion of catecholamines from sympathetic neurons[22,23] and adrenal medullary cells[24,25] is also inhibited by imidazolines independently of α_2-adrenoceptors, but the role of I_1 sites in these effects is not clear. Given the pivotal importance of Ca^{2+} entry in secretion of neurotransmitters, we studied whether imidazoline compounds could inhibit stimulus-induced Ca^{2+} entry in PC-12 cells.

Nicotine increased $[Ca^{2+}]_i$ in undifferentiated PC-12 cells loaded with fura-2 in the presence of extracellular Ca^{2+}, but not in its absence. This suggests that nicotine is stimulating the influx of Ca^{2+}. In our clone of PC-12 cells this appears to be due to stimulation of nicotinic acetylcholine receptors (nicotinic AchR), which contain an intrinsic nonselective cation channel that is permeable to Ca^{2+}. This nicotine-stimulated influx of Ca^{2+} was inhibited by two drugs with an imidazoline structure, clonidine, an α_2-adrenoceptor agonist, and cirazoline, an α_2-adrenoceptor antagonist, in a concentration dependent manner (10–100 μM). In contrast, adrenaline, a phenylethylamine α_2-adrenoceptor agonist, and BHT 920, an azepine α_2-adrenoceptor agonist, had no effect on nicotine-induced Ca^{2+} influx. This suggests that the action of the imidazolines clonidine and cirazoline is not mediated by α_2-adrenoceptors. Further support for this idea is the failure of the non-imidazoline α_2-adrenoceptor antagonist yohimbine to reverse the effects of clonidine and cirazoline. These effects are similar to the α_2-adrenoceptor-independent inhibition of catecholamine release from sympathetic nerves[22,23] and adrenomedullary cells.[24,25]

In NGF-differentiated PC-12 cells with a neuronal phenotype, nicotine induced an inward current which was blocked by both cirazoline and clonidine, in agreement with Ca^{2+} entry data. As for Ca^{2+} entry, this inhibition of nicotine-induced current was not mimicked by adrenaline or blocked by yohimbine. Taken together with the reported absence of α_2-adrenoceptors on PC-12 cells,[5] these results suggest that clonidine and cirazoline are inhibiting activation of the intrinsic cation channel of the nicotinic AchR via a mechanism independent of α_2-adrenoceptors. As the most obvious feature shared by clonidine and cirazoline is their high affinity for imidazoline I_1 sites,[7] an action via imidazoline I_1 sites is an attractive possibility.

Binding of imidazolines to I_1 sites was reported to be inhibited by stable GTP analogs, although more weakly than inhibition of binding to α_2-adrenoceptors, prompting speculation that I_1 sites are coupled to signal-transducing G-proteins.[5,7] We studied the possible involvement of G-proteins in the inhibition of nicotinic effects by clonidine and cirazoline by pretreatment of PC-12 cells with pertussis toxin, to inactivate G-proteins of the G_i/G_o family,[26] and perfusion of PC-12 cells with GDP-β-S, which inhibits activation of all G-proteins.[27] Pertussis toxin had no effect on the inhibition of nicotine-stimulated ion entry by clonidine or cirazoline in either differentiated or undifferentiated PC-12 cells. Furthermore, perfusion of PC-12 cells via the patch pipette with GDP-β-S also had no effect on the inhibition of nicotine-stimulated ion entry by cirazoline. Together, these results suggest that G-proteins are not involved in the effects of cirazoline and clonidine on nicotinic AchR.

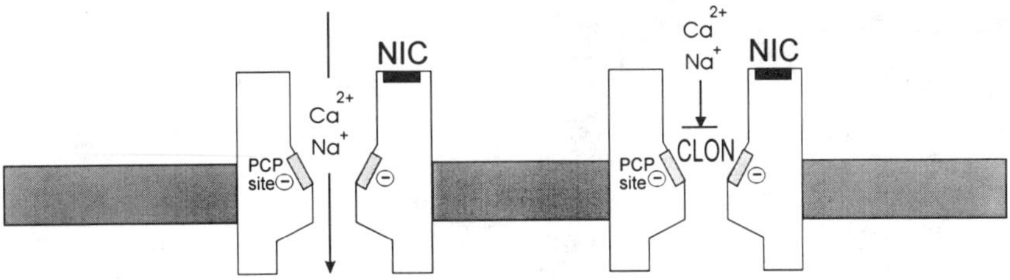

FIGURE 5. Schematic representation of a nicotinic acetylcholine receptor, showing the presumed mode of action of clonidine and cirazoline. The acetylcholine receptor has an intrinsic ion channel, permeable to cations such as Na^+ and Ca^{2+}. The inner part of the pore contains a ring of charged amino acids to which compounds such as phencyclidine and TPP bind (PCP site). We propose that clonidine and cirazoline block nicotine-induced Na^+ and Ca^{2+} entry into PC-12 cells by binding to this site.

Modulation of cyclic nucleotide levels by these drugs appears to be ruled out, as pretreatment with IBMX (100 μM) had no effect on the inhibition of nicotine-induced Ca^{2+} entry by cirazoline.

We then investigated the possibility that cirazoline and clonidine were acting directly on the nicotinic AchR. Indeed, Powis and Baker[25] suggest that the effect of clonidine on nicotine-induced catecholamine secretion from bovine adrenal medullary cells is due to blocking of the nicotinic AchR. Several cationic drugs, such as phencyclidine and TPP, block the nicotinic AchR channel by binding to negatively charged groups within the pore itself,[28,29] so we investigated whether clonidine or cirazoline, also cationic drugs, displaced [^{3}H]phencyclidine from nicotinic AchR. For these experiments we used purified membranes from *Torpedo* electroplaques. Although the nicotinic AchR in these membranes have a different subunit composition to that of neuronal nicotinic AchR,[28,29] they have the advantage of being a highly purified and well characterized source of nicotinic AchR.

[^{3}H]Phencyclidine bound saturably to *Torpedo* membranes, and this binding was increased approximately 20-fold by the addition of nicotine, in agreement with previous reports.[20] Both clonidine and cirazoline displaced [^{3}H]phencyclidine in the presence of nicotine in a concentration-dependent manner. The concentration range over which displacement occurred paralled the concentration range over which these drugs inhibited nicotine-induced ion entry (10-100 μM). In contrast, adrenaline produced only a small, concentration-independent displacement of [^{3}H]phencyclidine. This effect of clonidine and cirazoline could be due to displacement of nicotine. However, clonidine is not considered to bind to the nicotine binding site,[24,25] and in preliminary experiments nicotine did not displace [^{3}H]clonidine binding to *Torpedo* membranes (data not shown).

Thus, cirazoline and clonidine appear to exert their effects on nicotine-induced ion entry by binding to a cation binding site(s) in the channel of the nicotinic AchR, as do phencyclidine and TPP (FIG. 5). This action presumably underlies the reported inhibition of nicotinic stimulation-induced catecholamine secretion from adrenal medullary cells.[24,25] The similarity between concentration response curves for the α_2-adrenoceptor-independent inhibition of noradrenaline release from sympathetic nerves by clonidine and cirazoline and the blockade of nicotinic AchR by clonidine and cirazoline is intriguing. Although there is no evidence that sympathetic transmission is supported by nicotinic AchR, phencyclidine binding sites are present in other ligand-gated ion channels.[29] Therefore, an effect on sympathetic transmission via a facilitatory ion channel is a distinct possibility.

Could the phencyclidine binding site on the nicotinic acetylcholine receptor actually be the I_1 receptor? At first sight it would seem not, as the apparent affinities of cirazoline and clonidine for the nicotinic AchR are several orders of magnitude lower than the affinity of these drugs for I_1 sites (see above). However, I_1 sites also have a low affinity site,[7] which has micromolar affinity for imidazolines, like the $[^3H]$phencyclidine binding site. Furthermore, binding of imidazolines to the high affinity I_1 receptor is strongly inhibited by cations such as Na^+ and Ca^{2+},[7] and the $[^3H]$phencyclidine binding assay used here is performed in the presence of 150 mM NaCl. Naturally, a connection between nicotinic AchR and the imidazoline I_1 site is purely speculative at this stage and requires demonstration of high affinity binding of $[^3H]$clonidine to purified nicotinic AchR, preferably of the neuronal type, to establish any possible link.

In conclusion, we have demonstrated that the imidazoline α_2-adrenoceptor drugs cirazoline and clonidine block the intrinsic ion channel of the nicotinic AchR from *Torpedo* membranes. This action probably underlies the inhibition of nicotinic-induced secretion of catecholamines from adrenal medullary cells by clonidine and other imidazolines. An action at nicotinic AchR, or other members of the ligand-gated ion channel super family, may explain other reports of nonadrenoceptor imidazoline effects.

SUMMARY

Clonidine and cirazoline bind with high affinity to a nonadrenergic site in the brain stem, the so-called imidazoline I_1 receptor. Our aim was to determine the mechanism by which these receptors act and their possible linkage to signal-transducing heterotrimeric G-proteins. We examined the effects of clonidine and cirazoline on PC-12 cells, a neuronal cell line that is reported to possess the I_1 site and have no α_2-adrenoceptors. In undifferentiated PC-12 cells loaded with the Ca^{2+} indicator dye fura-2, clonidine and cirazoline (10–100 μM) inhibited the increase in $[Ca^{2+}]_i$ produced by nicotine (10 μM). This inhibition was not reversed by yohimbine (100 μM), and adrenaline and BHT 920 were ineffective at 100 μM. This effect was not inhibited by pretreatment with pertussis toxin (24 hours, 100 ng/ml) and not modulated by pretreatment with IBMX (100 μM). The nicotine-induced increase in $[Ca^{2+}]_i$ is apparently due to Ca^{2+} entering via the intrinsic ion channel of the nicotinic acetylcholine receptor. Clonidine and cirazoline inhibited the inward current produced by nicotine (10 μM) as measured by the whole cell patch-clamp technique in differentiated PC-12 cells, recorded at a holding potential of –60 mV. In agreement with the results found with fura-2, inhibition of inward current was concentration dependent and not blocked by yohimbine (100 μM) or mimicked by adrenaline (100 μM). Pretreatment of PC-12 cells with pertussis toxin or infusion of GDP-β-S (2 mM) via the patch pipette did not alter the inhibition of the nicotine-induced inward current by clonidine or cirazoline. Clonidine and cirazoline, but not adrenaline, displaced [^{3}H]phencyclidine from *Torpedo* electroplaque membranes enriched in nicotinic acetylcholine receptors in a concentration-dependent manner (10-100 μM). Taken together, these results suggest that clonidine and cirazoline inhibit Na^+ and Ca^{2+} entry through the nicotinic acetylcholine receptor via a nonadrenergic mechanism that is independent of G-proteins and cyclic nucleotides, presumably by direct blockade of the intrinsic ion channel of the nicotinic acetylcholine receptor.

REFERENCES

1. MICHEL, M. C. & P. A. INSEL. 1989. Are there multiple imidazoline binding sites? Trends Pharmacol. Sci. **10:** 342-344.
2. ATLAS, D. 1991. Clonidine displacing substance (CDS) and its putative imidazoline receptor. Biochem. Pharmacol. **41:** 1541-1549.
3. BOUSQUET, P., G. BRICCA, M. DONTENWILL, J. FELDMAN, H. GRENEY, A. BELCOURT, J. STUTZMANN & E. TIBIRICA. 1992. From the α_2-adrenoceptors to the imidazoline preferring receptors. Fundam. Clin. Pharmacol. **6** (Suppl 1): 15S-21S.
4. KILPATRICK, A. T., C. C. BROWN & A. C. MACKINNON. 1992. Non-α_2-adrenoceptor idazoxan binding sites, a new target for drug development. Biochem. Soc. Trans. **20:** 113-118.
5. ERNSBERGER, P., K. WESTBROOKS & L. GRAFF. 1991. Imidazoline binding sites labeled by clonidine analogs (I_1 sites): Putative G-protein-linked receptors. FASEB J. **5:** A1066.
6. MICHEL, M. C. & P. ERNSBERGER. 1992. Keeping an eye on the I site: Imidazoline-preferring receptors. Trends Pharmacol. Sci. **13:** 369-370.
7. MOLDERINGS, G. J., D. MOURA, K. FINK, H. BÖNISCH & M. GÖTHERT. 1993. Binding of [^{3}H]clonidine to I_1-imidazoline sites in bovine adrenal medullary membranes. Naunyn-Schmiedeberg's Arch. Pharmacol. **348:** 70-76.
8. LACHAUD-PETTITI, V., R. A. PODEVIN, Y. CHRÉTIEN & A. PARINI. 1991. Imidazoline-guanidium and α_2-adrenergic binding sites in basolateral membranes from human kidney. Eur. J. Pharmacol. **206:** 23-31.
9. BRICCA, G., H. GRENEY, M. DONTENWILL KIEFFER, J. ZHANG, A. BELCOURT & P. BOUSQUET. 1993. Heterogeneity of the specific imidazoline binding of idazoxan in the human cerebral cortex. Neurochem. Int. **22:** 153-163.
10. FELSEN, D., P. ERNSBERGER, P. M. SUTARIA, R. J. NEJAT, P. NGUYEN, M. MAY, D. S. BRESLIN, D. N. MARION & E. D. J. VAUGHAN. 1994. Identification, localization and functional analysis of imidazoline and alpha adrenergic receptors in canine prostate. J. Pharmacol. Exp. Ther. **268:** 1063-1071.
11. WANG, H., S. REGUNATHAN, M. P. MEELEY & D. J. REIS. 1992. Isolation and characterization of imidazoline receptor protein from bovine adrenal chromaffin cells. Mol. Pharmacol. **42:** 792-801.
12. TIBIRICA, E., J. FELDMAN, C. MERMET, L. MONASSIER, F. GONON & P. BOUSQUET. 1991. Selectivity of rilmenidine for the nucleus reticularis lateralis, a ventrolateral medullary structure containing imidazoline-preferring receptors. Eur. J. Pharmacol. **209:** 213-221.
13. REIS, D. J., S. REGUNATHAN & M. P. MEELEY. 1992. Imidazole receptors and clonidine-displacing substance in relationship to control of blood pressure, neuroprotection, and adrenomedullary secretion. Am. J. Hypertens. **5:** 51S-57S.
14. ERNSBERGER, P., T. H. DAMON, L. M. GRAFF, S. G. SCHÄFER & M. O. CHRISTEN. 1993. Moxonidine, a centrally acting antihypertensive agent, is a selective ligand for I_1-imidazoline sites. J. Pharamcol. Exp. Ther. **264:** 172-182.
15. MUSGRAVE, I. F., R. SEIFERT & G. SCHULTZ. 1994. Maitotoxin activates cation channels distinct from the receptor-activated non-selective cation channels of HL-60 cells. Biochem. J. **301:** 437-441.

16. GRYNKIEWICZ, G., M. POENIE & R. Y. TSIEN. 1985. A new generation of Ca^{2+} indicators with greatly improved fluorescence properties. J. Biol. Chem. **260:** 3440-3450.
17. HAMILL, O. P., A. MARTY, E. NEHER, B. SAKMANN & F. J. SIGWORTH. 1981. Improved patch-clamp techniques for high-resolution current recording from cells and cell-free membrane patches. Pflügers Arch. **391:** 85-100.
18. SCHIEBLER, W. & F. HUCHO. 1978. Membranes rich in acetylcholine receptor: Characterization and reconstitution to excitable membranes from exogenous lipids. Eur. J. Biochem. **85:** 55-63.
19. HARTIG, P. R. & M. A. RAFTERY. 1979. Preparation of right-side-out, acetylcholine receptor enriched intact vesicles from *Torpedo californica* electroplaque membranes. Biochemistry **18:** 1146-1150.
20. MIN, C. K. & G. A. WEILAND. 1993. Effects of substance P on the binding of agonists to the nicotinic acetylcholine receptor of *Torpedo* electroplaque. J. Neurochem. **60:** 2238-2246.
21. GREENE, L. A. & A. S. TISCHLER. 1976. Establishment of a noradrenergic clonal line of rat adrenal pheochromocytoma cells which respond to nerve growth factor. Proc. Natl. Acad. Sci. USA **73:** 2424-2428.
22. MOLDERINGS, G. J., F. HENTRICH & M. GÖTHERT. 1991. Pharmacological characterization of the imidazoline receptor which mediates inhibition of noradrenaline release in the rabbit pulmonary artery. Naunyn-Schmiedeberg's Arch. Pharmacol. **344:** 630-638.
23. GÖTHERT, M. & G. J. MOLDERINGS. 1991. Involvement of presynaptic imidazoline receptors in the α_2-adrenoceptor-independent inhibition of noradrenaline release by imidazoline derivatives. Naunyn-Schmiedeberg's Arch. Pharmacol. **343:** 271-282.
24. OHARA IMAIZUMI, M. & K. KUMAKURA. 1992. Effects of imidazole compounds on catecholamine release in adrenal chromaffin cells. Cell. Mol. Neurobiol. **12:** 273-283.
25. POWIS, D. A. & P. F. BAKER. 1986. α_2-Adrenoceptors do not regulate catecholamine secretion by bovine adrenal medullary cells: A study with clonidine. Mol. Pharmacol. **29:** 134-141.
26. MILLIGAN, G. 1988. Techniques used in the identification and analysis of function of pertussis toxin-sensitive guanine nucleotide binding proteins. Biochem. J. **255:** 1-13.
27. GILMAN, A. G. 1987. G proteins: Transducers of receptor-generated signals. Ann. Rev. Biochem. **56:** 615-649.
28. ROLE, L. W. 1992. Diversity in primary structure and function of neuronal nicotinic acetylcholine receptor channels. Curr. Opinion Neurobiol. **2:** 254-262.
29. KARLIN, A. 1993. Structure of nicotinic acetylcholine receptors. Curr. Opinion Neurobiol. **3:** 299-309.

Calcium Channel-Dependent and I_1-Imidazoline Receptor Binding Properties of 2-(4′-Isothiocyanatobenzyl) Imidazoline Analogs in Vascular and Brain Tissues[a]

L. LEI,[b] M. SLAVICA,[c] P. ERNSBERGER,[d] M. E. GRAVES,[d] P. PATIL,[b] D. MILLER,[c] AND D. FELLER[b]

[b]*College of Pharmacy*
Ohio State University
Columbus, Ohio 43210

[c]*College of Pharmacy*
University of Tennessee
Memphis, Tennessee 38163

[d]*Case Western Reserve University*
School of Medicine
Cleveland, Ohio 44106

The addition of an electrophilic isothiocyanato (-NCS) group to the 4′ position of tolazoline yielded a selective vascular imidazoline agonist, 2-(4′-isothiocyanatobenzyl) imidazoline (IBI). The slow vascular contractile activities of IBI were independent of α-adrenoceptors (α-AR)[1] and were not affected by a variety of receptor antagonists[2]; therefore, this compound represents a unique affinity label for receptor studies. IBI competed for [^{3}H]idazoxan binding in guinea pig cerebral cortex and porcine renal cortex membranes.[3] In the present studies, several IBI analogs (TABLE 1) were prepared to examine the structural requirements for interaction with α-adrenergic and imidazoline receptors in rat aortic and bovine brain tissues. The latter tissue was used because it is a well characterized source of I_1-imidazoline receptors.[4] The role of extracellular calcium in the vascular activities of IBI analogs was also examined.

METHODS

Functional experiments with rat thoracic aorta were conducted as previously described.[1,2] Binding of IBI analogs to α_2-AR and I_1-imidazoline receptor sites was

[a]This work was supported by USPHS grant GM-29358.

TABLE 1. Comparison of EC_{50} Values from Rat Aortic Strips and Ligand Displacing IC_{50} Values from Bovine Ventrolateral Medulla Membranes for IBI-Like Imidazolines[a]

	Compounds	$EC_{50} \pm$ SEM (μM) Control (Max%)	$EC_{50} \pm$ SEM (μM) With NIF (Max%)	$IC_{50} \pm$ SEM (μM) Control (n_H)	$IC_{50} \pm$ SEM (μM) With PBZ (n_H)
IBI	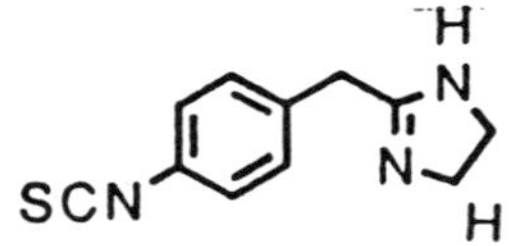	5.8 ± 1.9	23.0 ± 0.95	7.41 ± 0.34	2.34 ± 0.11
		(100 ± 9)	(42 ± 12)	(1.39 ± 0.85)	(0.96 ± 0.87)
II	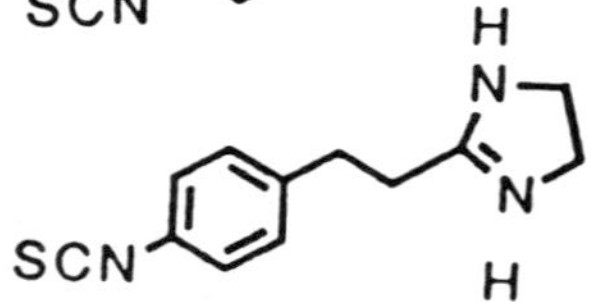	18.3 ± 0.7	67.0 ± 16.5	3.16 ± 0.29	3.38 ± 0.39
		(100 ± 11)	(21 ± 6)	(1.05 ± 0.81)	(0.79 ± 0.85)
III	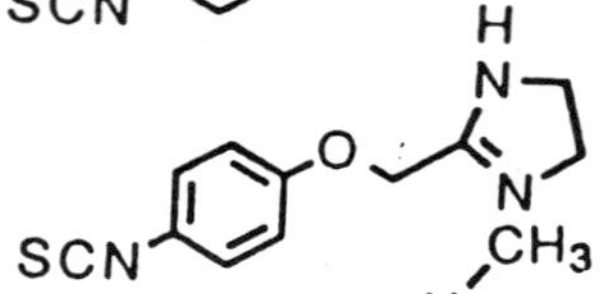	17.2 ± 0.9	52.1 ± 0.0	1.55 ± 0.40	0.42 ± 0.05
		(100 ± 12)	(11 ± 6)	(0.98 ± 0.78)	(0.74 ± 0.87)
IV	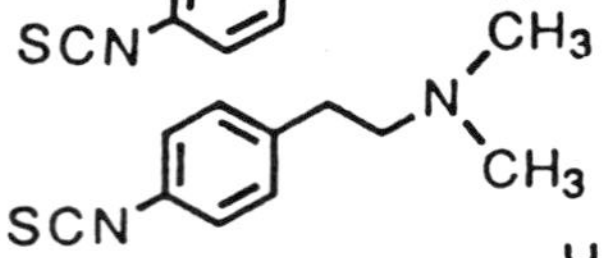	16.8 ± 0.2	48.5 ± 6.1	7.94 ± 0.91	4.17 ± 0.48
		(100 ± 23)	(29 ± 9)	(0.96 ± 0.89)	(0.92 ± 0.81)
V	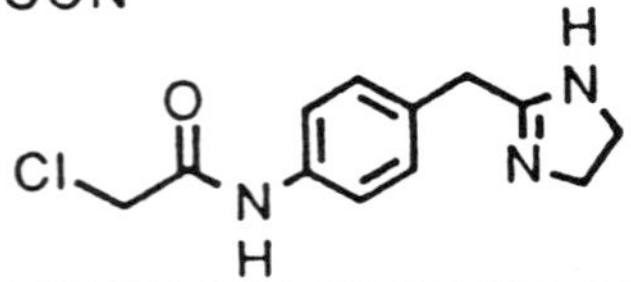	116 ± 8.5	ND	16.2 ± 1.67	8.13 ± 0.58
		(100 ± 9)	ND	(0.84 ± 0.85)	(0.83 ± 89)

NOTE: EC_{50} = concentration at 50% maximal responses; IC_{50} = concentration for 50% maximal binding displacement; NIF = nifedipine (1 μM) for 1 hour; PBZ = phenoxybenzamine (1 μM) with *N*-ethylmaleimide (0.25 mM) to irreversibly inactivate α_2-ARs; Max% = percent maximal contraction where each compound alone is 100%; n_H = Hill number; significantly different from control (p <0.05); ND = not determined.

[a] Geometric mean ± SEM of n = 3-5.

studied in isolated bovine ventrolateral medulla (BVLM) membranes with [^{125}I]*p*-iodoclonidine. Displacement of the radioligand from α_2-adrenergic and I_1-receptors (control) and from I_1-receptors after alkylation of α_2-ARs in BVLM with phenoyxbenzamine (TABLE 1) was studied.[5]

RESULTS AND CONCLUSIONS

IBI analogs contracted rat thoracic aorta with the following rank order of agonist potency: IBI $\geqslant$ IV $\geqslant$ III $\geqslant$ II > V (TABLE 1). The contractile responses of these analogs were not blocked by irreversibly acting α-AR antagonists such as phenoxybenzamine (30 nM), and analog responses in phenoxybenzamine-pretreated tissues were unaffected by imidazoline receptor ligands such as idazoxan (1 μM) and cirazoline (1 μM) (data not presented). In contrast, nifedipine (1 μM) shifted the concentration response curves of these compounds to the right by about two to fourfold with 58-89% suppression of maximal contraction (TABLE 1). Similar effects on the actions of IBI analogs were observed with verapamil (1 μM) or a calcium-deficient buffer (data not presented). Pretreatment of rat aortic strips with R(+)-Bay K 8644 (0.3 μM), another L-type calcium channel blocker, also shifted the concentration response curves of IBI to the right in a competitive fashion. These results indicate that IBI analogs may modulate L-type calcium channels in vascular tissues.

In BVLM membranes, these compounds displaced [^{125}I] *p*-iodoclonidine from I_1-and α_2-AR binding sites, giving the following rank order of potency: III > II > IBI > IV > V (TABLE 1). Alkylation of α_2-ARs did not change the rank order potency of the IBI analogs in BVLM. However, the displacement curves of IBI and analog III were significantly ($p < 0.05$) shifted to the left three to fourfold in phenoxybenzamine-pretreated membranes, implying that these compounds exhibit a modest binding selectivity for I_1-receptors over α_2-AR sites.

The functional effects of IBI and analogs on rat aorta, therefore, are independent of α-AR and are likely coupled to voltage-gated L-type calcium channels. However, these compounds interact with both α-AR and I_1-imidazoline receptor sites in bovine brain stem tissue. Analog III, which bears structural similarity to cirazoline and idazoxan, was the most potent agent for displacement of [^{125}I]*p*-iodoclonidine from I_1-receptor sites. The EC_{50} values in functional assays were in close agreement with the IC_{50} values at I_1-imidazoline receptors. Given the approximate agreement of the concentration range for binding activity and for functional responses, these compounds may be selective irreversible agonists at the I_1-imidazoline receptor; however, further experiments are required to establish this possibility.

REFERENCES

1. VENKATARAMAN, B. V., A. HAMADA, G. SHAMS, D. D. MILLER, D. R. FELLER & P. N. PATIL. 1989. Blood Vessel **26:** 335-346.
2. LEI, L., M. SLAVICA, K. ROMSTEDT, P. N. PATIL, D. D. MILLER & D. R. FELLER. 1992. Pharmacologist (abstract) **34:** 148.
3. HUSSAIN, J. F., P. N. PATIL, D. D. MILLER, D. A. KENDALL & V. G. WILSON. 1992. Fundam. Clin. Pharmacol. (Abstr.) **6:** 56.

4. Ernsberger, P., M. P. Meeley, J. J. Mann & D. J. Reis. 1987. Eur. J. Pharmacol. **134:** 1-13.
5. Felsen, D., P. Ernsberger, P. M. Sutaria, R. J. Nehat, P. Nguyen, M. May, D. S. Breslin, D. N. Marion & D. Vaughan, Jr. 1994. J. Pharmacol. Exp. Ther. **268:** 1063-1071.

Inhibition of Red Cell Ca^{2+}-Activated K^+ Transport by Clotrimazole Does Not Take Place via Imidazoline Binding Sites

I. COUPRY,[a] C. C. ARMSBY, S. ALPER, A. PARINI,[a,b]
AND C. BRUGNARA

Department of Pathology and Clinical Laboratory
Children's Hospital and Molecular Medicine and Renal Units
Beth Israel Hospital
Harvard Medical School
Boston, Massachusetts 02115

[a]*INSERM U388*
Institut Louis Bugnard
CHU Rangueil
31054 Toulouse, France

Recent reports by our group indicate that clotrimazole is a potent and specific inhibitor of the Gardos channel of sickle erythrocytes[1] and prevents Ca^{2+}-dependent dehydration of sickle cells, suggesting a possible therapeutic strategy for sickle cell disease. However, the mechanism by which the effect of clotrimazole is mediated is still unknown; in particular, it is not clear if this effect in red cells takes place by specific interaction with imidazoline binding sites (IBS).

Imidazoline binding sites are a new class of membrane-bound proteins recently described in a variety of tissues and species.[2–4] They are recognized by various imidazoline and guanidinium derivatives, are insensitive to catecholamines, and have been classified as I_1- and I_2BS based on their high or low affinity for clonidine, respectively.[5]

We investigated the effect of imidazoline derivatives on the normal human red cell Gardos channel and present evidence that the effect of clotrimazole is not mediated by imidazoline I_1- or I_2-BS.

Measurement of Ca^{2+}-activated ^{86}Rb influx in human red cells indicates that 10 μM of clotrimazole inhibits 99% of this flux. By contrast, studies with IBS antagonists showed no significant inhibition by either efaroxan, an I_1BS ligand (10 ± 12%), or idazoxan, an I_2BS ligand (5 ± 2%), at concentrations up to 10 μM. These functional studies suggest that the effect of clotrimazole is not mediated by the imidazoline binding site system. However, to further confirm this hypothesis, we investigated whether human red cells possess such binding sites.

[b]Address for correspondence and reprint requests: Dr. Angelo Parini, INSERM U388, Institut Louis Bugnard, CHU Rangueil—Bât. L3, 31054 Toulouse Cedex, France.

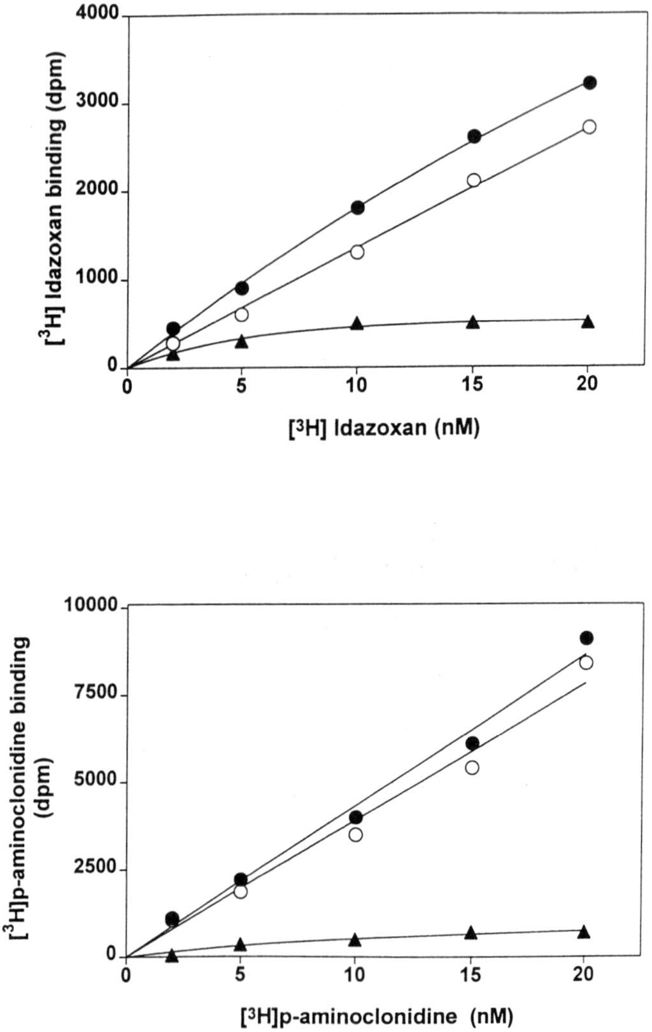

FIGURE 1. Saturation binding experiments on human red cell membranes. Membranes (150–500 μg/ml of protein) were incubated with increasing concentrations of [^{3}H]idazoxan (*left panel*) or [^{3}H]*p*-aminoclonidine (*right panel*) for 45 minutes at 25 °C. Nonspecific binding was defined in the presence of 10 μM cirazoline or idazoxan (for [^{3}H]idazoxan), or 10 μM clonidine or moxonidine (for [^{3}H]*p*-aminoclonidine). Total binding (●), nonspecific binding (○), and Specific binding (▲) are shown.

Saturation binding experiments in human red cell membranes using either [^{3}H]idazoxan (a radioligand for I_2BS) or [^{3}H]*p*-aminoclonidine (a radioligand for I_1BS) showed that in both cases binding was not saturable and that "nonspecific" binding represents 80-90% of total binding (FIG. 1). Similar results were obtained even using different protein concentrations, incubation volumes, and/or times. These results suggest that human red cells do not express imidazoline I_1 or I_2 binding sites.

Taken together, these results show that (1) inhibition of red cell Ca^{2+}-activated K^+ transport induced by clotrimazole is not retrieved with efaroxan and idazoxan, two I_1 and I_2 selective ligands, respectively; and (2) this effect is not mediated by imidazoline I_1 or I_2 binding sites because radioligand binding studies failed to detect any specific binding. However, we cannot rule out possible involvement of another IBS in this effect. Indeed, it was recently shown that stimulation of insulin secretion by imidazoline derivatives is due to a decrease in potassium efflux through ATP-sensitive K^+ channels in the islet β-cells[6] and that this effect occurs via a novel IBS distinct from the already described I_1- and I_2BS. Further experiments are needed to confirm the possible role of such binding sites in the effect of clotrimazole on red cell Ca^{2+}-activated K^+ transport.

REFERENCES

1. BRUGNARA, C., L. DE FRANCESCHI & S. L. ALPER. 1993. J. Clin. Invest. **93:** 520-526.
2. COUPRY, I., D. ATLAS, R. A. PODEVIN, I. UZIELLI & A. PARINI. 1990. J. Pharmacol. Exp. Ther. **252:** 293-299.
3. MALLARD, N. J., H. HUDSON & D. J. NUTT. 1992. Br. J. Pharmacol. **106:** 1019-1027.
4. LANGIN, D. & M. LAFONTAN. 1989. Eur. J. Pharmacol. **159:** 199-203.
5. MICHEL, M. C. & P. ERNSBERGER. 1992. Trends Pharmacol. Sci. **13:** 368-370.
6. CHAN, S. L. F., C. A. BROWN, K. E. SCARPELLO & N. G. MORGAN. 1994. Br. J. Pharmacol. **112:** 1065-1070.

Effects of Rilmenidine on Signal Transduction Mechanisms Associated with α_2-Adrenergic and Imidazoline Receptors in Brain

SOUNDARARAJAN REGUNATHAN,
SONJA BRAMWELL, AND DONALD J. REIS

Division of Neurobiology
Department of Neurology and Neuroscience
Cornell University Medical College
New York, New York 10021

In brain, rilmenidine, an oxazole, binds to imidazoline (I-) receptors of the I_1 subclass and α_2-adrenergic receptors.[1] [^{3}H]rilmenidine binds to a high-affinity site distinct from α_2-adrenergic and I_2 imidazoline sites in rat cerebral cortex.[2] Binding to I_1 imidazoline sites in brain stem is at higher affinity than to I_2 sites in cerebral cortex[1] and exceeds the affinity of rilmenidine for α_2-adrenergic receptors. The interaction of rilmenidine with I-receptors appears functional and responsible for the hypotensive and neuroprotective actions of the drug.[1,3]

The signal transduction pathways related to rilmenidine's interaction with I-receptors are not known. We therefore investigated the effects of rilmenidine on second messenger systems and ion fluxes in slices of rat brain stem that preponderantly expresses I-receptors of the I_1 subclass and in cerebral cortex expressing receptors of the I_2 subclass. Because rilmenidine also binds to α_2-adrenergic sites, studies were undertaken in the presence or absence of rauwolscine, a selective α_2-adrenergic receptor antagonist.

METHODS

The methods used in this study were described in detail in other publications from this laboratory and will only be summarized.

Measurement of Production of cAMP and cGMP and Phosphoinositol Turnover

Rats were stunned and decapitated and brains rapidly removed. Miniprisms (350 μm) of cerebral cortex and brain stem were prepared by described methods[4] with a McIlwain tissue chopper and suspended in Krebs-Ringer bicarbonate buffer. Production of cAMP and cGMP was measured by radioimmunoassay by methods described in detail elsewhere.[5] The production of cAMP and cGMP was determined using radioimmunoassay kits. Protein was assayed by the Coomassie blue method.

Phosphoinositol turnover was measured by estimating the accumulation of inositol-1-phosphate in the presence of lithium in brain slices as described earlier[4] by measuring the production of [^{3}H]inositol-1-phosphate formed from slices prelabeled with [^{3}H]inositol. Results represent the disintegrations per minute (dpm) of inositol-1-phosphate expressed as a percentage of total dpm (inositol-1-phosphate + lipid dpm).

Measurement of ^{45}Ca Influx

The influx of ^{45}Ca into brain minislices was estimated by the method of Geer *et al.*[6] Briefly, rats were killed by decapitation and minislices (200 × 200 μm) were prepared from cerebral cortex and brain stem. These were washed with Krebs-Ringer solution containing 25 mM HEPES (pH 7.4) and preincubated at 25°C for 5 minutes in the same buffer. Tubes containing 1 μCi ^{45}Ca and the testing drug in a volume of 400 μl Krebs-HEPES buffer were prepared and 100 μl of gravity-packed minislices were added to these tubes to start the incubation. After incubating at 25°C for a specified time, the reaction was terminated by rapid filtration over glass fiber filters in a cell harvester (Brandel) and washed once with 5 ml of calcium-free choline chloride-HEPES buffer (pH. 7.3). The radioactivity of the filters was counted in a liquid scintillation system having 90% efficiency for ^{45}Ca.

Measurement of ^{86}Rb Efflux

The activity of K^+ channels was studied by measuring the efflux of the ^{86}Rb that has been used to track the movement of K^+ across the cell membrane.[7] Briefly, minislices (200 × 200 μm), prepared as just described, were preloaded with ^{86}Rb for 2 hours in Krebs-HEPES buffer (pH 7.4) and washed extensively with ice-cold buffer until the washings had negligible radioactivity. They were then incubated at 37°C in Krebs-HEPES buffer in the presence of rilmenidine, KCl (55 mM), or both. After incubating for a specific time, an aliquot of the buffer was removed for counting the radioactivity. The remaining buffer was aspirated and the slices were solubilized in 0.1 N NaOH by sonication. Radioactivity in the solubilized tissue was also determined and the results are expressed as the percentage released to total radioactivity incorporated into the slices.

RESULTS AND DISCUSSION

Effect of Rilmenidine on cAMP Production in Cerebral Cortex and Brain Stem

The effects of rilmenidine on basal and stimulated cAMP production in brain minislices were investigated. In cerebral cortex, rilmenidine did not change basal cAMP concentrations but significantly inhibited the production of cAMP stimulated

TABLE 1. Effect of Rilmenidine on cAMP Production[a]

Drugs	cAMP (pmol/mg protein)	
	Brain Stem	Cerebral Cortex
Control	46.6 ± 4.8	37.5 ± 2.4
Norepinephrine (100 μM)	239 ± 29	233 ± 36
Forskolin (100 μM)	280 ± 41	291 ± 28
Rilmenidine (100 μM)	51.3 ± 5.9	42.3 ± 1.8
Norepinephrine + rilmenidine	246 ± 31	125 ± 18[b]
Norepinephrine + rilmenidine + rauwolscine (100 μM)	220 ± 18	194 ± 31[c]
Forskolin + rilmenidine	297 ± 34	121 ± 21[d]
Forskolin + rilmenidine + rauwolscine	296 ± 44	205 ± 25[e]

[a] Brain slices were incubated for 10 minutes at 37°C in Krebs-Ringer buffer in the presence of 50 μM IBMX.
[b] $p < 0.001$ compared to norepinephrine group.
[c] $p < 0.01$ compared to norepinephrine + rilmenidine group.
[d] $p < 0.001$ compared to forskolin group.
[e] $p < 0.01$ compared to forskolin + rilmenidine group.

by forskolin or norepinephrine (TABLE 1). The response was significantly reversed by rauwolscine, a selective α_2-adrenergic receptor antagonist. By contrast, in brain stem, rilmenidine (100 μM) neither changed basal production nor inhibited the increase in cAMP stimulated by forskolin (10 μM) or norepinephrine (TABLE 1).

The findings suggest that rilmenidine acts as an α_2-adrenergic agonist in cerebral cortex where it inhibits the production of cAMP. By contrast, it is not effective in brain stem, suggesting that stimulation of I_1-receptors in that area may not modify cAMP levels. Interestingly, McAuley *et al.*[8] reported that rilmenidine blocks the inhibitory action of neuropeptide Y on cAMP production in brain stem slices of spontaneous hypertensive rats. The findings raise the possibility that stimulation of I_1 imidazoline sites, although not affecting basal cAMP levels, may prevent the effects of agents that themselves inhibit cAMP production. Further studies with model systems that express only I_1 imidazoline sites are required to verify this hypothesis.

Effect of Rilmenidine on cGMP Production and Phosphoinositol Turnover

Rilmenidine (100 μM) was without any effect on basal or norepinephrine-mediated (100 μM) increases in cGMP in brain stem or cerebral cortex (data not shown). Thus, the binding of rilmenidine to either α_2-adrenergic or I-receptors failed to modify the production of cGMP in brain.

Rilmenidine also failed to modify the basal or epinephrine-stimulated turnover of phosphoinositol (data not shown), suggesting that the interaction of rilmenidine with adrenergic or I-receptors does not modulate hydrolysis of phosphoinositol.

TABLE 2. Effect of Rilmenidine on ^{45}Ca Influx into Rat Brain Slices[a]

Drugs	^{45}Ca Influx (10^3 DPM) Brain Stem	Cerebral Cortex
Control	4.2 ± 1.1	3.4 ± 0.91
KCl (55 mM)	10.4 ± 1.8[b]	8.1 ± 1.2[b]
KCl + $CoCl_2$ (1 mM)	5.3 ± 0.91	4.8 ± 0.85
Rilmenidine (10 μM)	4.7 ± 0.85	3.9 ± 0.75
Rilmenidine (100 μM)	5.1 ± 0.93	5.8 ± 0.84[c]
KCl + Rilmenidine (100 μM)	10.8 ± 1.7[b]	10.4 ± 1.3[b]
Rilmenidine (100 μM) + $CoCl_2$	4.9 ± 0.75	3.3 ± 0.64
Rilmenidine (100 μM) + Nifedipine (10 μM)	5.2 ± 1.2	5.9 ± 0.94[c]
Rilmenidine (100 μM) (no sodium)	5.1 ± 1.1	6.1 ± 1.4[c]

[a] Brain slices (200 × 200 μm) were incubated in Krebs-Ringer buffer for 5 minutes with 1 μCi of ^{45}Ca in the presence of drugs at 37°C. All incubations with rilmenidine also contained 100 μM of SKF-86466. Values are from three experiments, each done in triplicate.

[b] $p < 0.001$ compared to control.

[c] $p < 0.01$ compared to control.

Effect of Rilmenidine on Influx of ^{45}Ca

Depolarization with high K^+ (55 mM KCl) resulted in a significant influx of ^{45}Ca into minislices of cerebral cortex and brain stem, an effect blocked by $CoCl_2$ (TABLE 2). Rilmenidine also elicited a significant and concentration-dependent increase in the uptake of ^{45}Ca but only in cerebral cortical but not brain stem minislices (TABLE 2). Combining rilmenidine with 55 mM KCl resulted in a slightly higher ^{45}Ca influx than that of either agent alone. The increased influx of ^{45}Ca into cerebral cortex slices elicited by rilmenidine was blocked by the nonselective blocker $CoCl_2$ and was not affected by rauwolscine (TABLE 2).

As cerebral cortex expresses I_2 but not I_1 imidazoline sites, the results suggest that the I_2 site may mediate rilmenidine's actions, even though the affinity for the I_2 imidazoline site is low. Moreover, because in the cerebral cortex I_2 receptors are preponderantly expressed in astrocytes,[9] the effect can be largely assigned to these cells. However, in rat cerebral cortex [^{3}H]rilmenidine binds with high affinity to a site that appears to be distinct from α_2-adrenergic and I_2-receptors.[2] Whether the effects of rilmenidine in cerebral cortex are mediated by this site is unknown.

The mechanism accounting for the stimulated influx of ^{45}Ca into cerebral cortex or chromaffin cells is uncertain. The fact that the efflux of ^{86}Rb was not modified by rilmenidine suggests that the influx of ^{45}Ca is not due to inhibition of K^+ channels.

Effect of Rilmenidine on the Efflux of ^{86}Rb

It is conceivable that the increased influx of ^{45}Ca by rilmenidine is secondary to blockade of K^+ channels because I-receptors are sensitive to K^+ and K^+ channel blockers.[10] Therefore, we measured the effects of rilmenidine on ^{86}Rb efflux, a commonly used indicator of K^+ channel activity. The efflux of ^{86}Rb from cerebral cortex and brain stem minislices prelabeled with ^{86}Rb was measured in the presence of rilmenidine. To stimulate efflux, minislices were depolarized by exposure to 55 mM KCl which resulted in significant efflux of ^{86}Rb indicating increased K^+ channel activity.

Rilmenidine, at 10 and 100 μM, failed to modify basal efflux or to inhibit the increase in efflux produced by 55 mM KCl in minislices of brain stem or cerebral cortex (data not shown). Thus, rilmenidine does not interact with K^+ channels, and the observed increase in the influx of ^{45}Ca elicited by the drug in cerebral cortex slices is not due to K^+ channel blockade.

In conclusion, these studies suggest that an interaction of rilmenidine with I_1 imidazoline sites in brain stem may act to maintain cAMP levels under conditions that inhibit its production. In cerebral cortex, rilmenidine increases the influx of calcium by interaction with either I_2 imidazoline sites and/or a novel [^{3}H]rilmenidine binding site. As the majority of I_2 receptors in cortex are neuroglial, the facilitated transport of Ca^{2+} into these cells may, as we suggested,[11] underlie the neuroprotective actions of rilmenidine in focal ischemic infarction.

REFERENCES

1. Ernsberger, P., R. Giuliano, R. N. Willette & D. J. Reis. 1990. J. Pharmacol. Exp. Ther. **253** (1): 408-418.
2. King, P. R., A. L. Gundlach, B. Jarrot & W. J. Louis. 1992. Am. J. Hypertens. **5:** 645-685.
3. Gomez, R. E., P. Ernsberger, G. Feinland & D. J. Reis. 1991. Eur. J. Pharmacol. **195:** 181-191.
4. Regunathan, S. & T. L. Sourkes. 1990. Neurochem. Int. **17:** 481-486.
5. Regunathan, S., M. J. Evinger, M. P. Meeley & D. J. Reis. 1991. Biochem. Pharmacol. **42:** 2011-2018.
6. Geer, J. J., D. J. Dooley & M. E. Adams. 1993. Neurosci. Lett. **158:** 9-100.
7. Takata, Y., F. Shimada & H. Kato. 1992. J. Pharmacol. Exp. Ther. **263:** 1293-1299.
8. McCauley, M. A., J. L. Reid & I. M. Macrae. 1992. J. Cardiovasc. Pharmacol. **19:** 945-952.
9. Regunathan, S., D. L. Feinstein & D. J. Reis. 1993. J. Neurosci. Res. **34:** 681-688.
10. Regunathan, S., M. P. Meeley & D. J. Reis. 1993. Biochem. Pharmacol. **45:** 1667-1675.
11. Reis, D. J., S. Regunathan, E. V. Golanov & D. L. Feinstein. 1994. Am. J. Cardiol. (Suppl.) **74:** 25A-30A.

Endogenous Ligands of Imidazoline Receptors: Classic and Immunoreactive Clonidine-Displacing Substance and Agmatine

DONALD J. REIS, GEN LI, AND
SOUNDARARAJAN REGUNATHAN

Division of Neurobiology
Department of Neurology and Neuroscience
Cornell University Medical College
New York, New York 10021

THE CONCEPT OF THE I-RECEPTOR

Background

The concept of imidazoline (I-) receptors evolved from efforts to understand the receptors through which clonidine, a centrally active antihypertensive agent,[1,2] lowered arterial pressure. While at first assumed that, as in the periphery, clonidine acted centrally as an agonist at α_2-adrenergic receptors,[3] two studies challenged the premise. Karppanen *et al.*[4] showed that the hypotension evoked by intracerebroventricular injection of clonidine resulted from actions at a nonadrenergic site. These, they proposed, were histamine receptors of the H_2 subclass. Bousquet *et al*,[5] analyzing the structure/function activities of a range of agents microinjected into the rostral ventrolateral medulla (RVL, a major site of clonidine's antihypertensive actions[6–8]), demonstrated that the hypotension related to clonidine's *structure* as an imidazoline and not its actions at α_2-adrenergic receptors.

The presence of a nonadrenergic binding site for clonidine in the ventral medulla was directly demonstrated by ourselves in 1987.[9] We observed that while [^{3}H]*para*-aminoclonidine ([^{3}H]PAC) exhibited saturable, reversible, and high affinity binding to membranes of the bovine ventral medulla, only 70% was to α_2-adrenergic receptors (FIG. 1A). The remainder was to a nonadrenergic binding site which recognized a number of imidazoles and imidazolines. This site corresponds to what is now defined as the imidazoline (I-) receptor. The study also indicated that I-receptors were expressed regionally, because they were not present in cerebral cortex.

That the I-receptor related to the hypotensive actions of clonidine and associated drugs (e.g., rilmenidine or moxonidine) is strongly suggested by several observations. First, the hypotensive action of appropriate drugs microinjected into the ventral medulla directly and significantly correlates with their affinity for I- but not α_2-adrenergic receptors.[10] Second, the hypotension elicited by systemic injections of clonidine is not affected by injection into the RVL of selective antagonists of α_2-adrenergic receptors. However, they can be blocked by agents interacting with I-

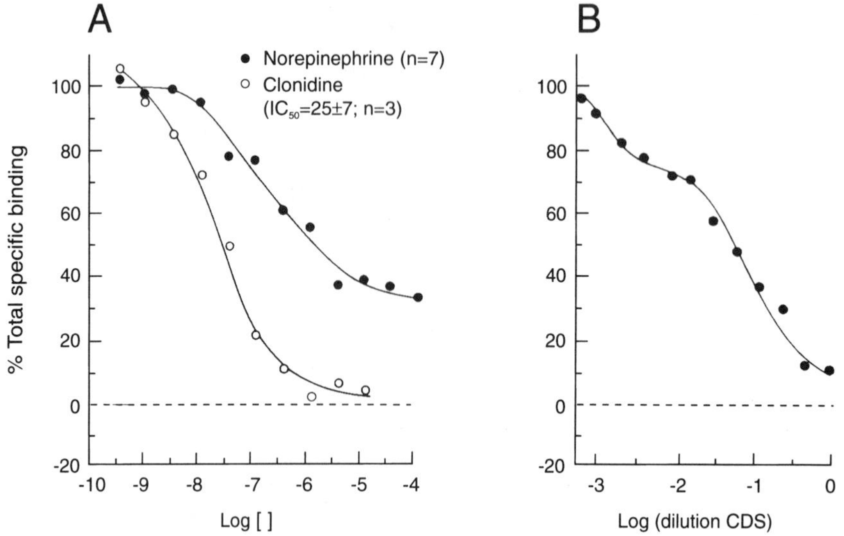

FIGURE 1. Inhibition of [^{3}H]*para*-aminoclonidine ([^{3}H]PAC) binding to membranes isolated from bovine ventrolateral medulla. Crude membrane fractions (P_2) were incubated for 40 minutes at 25°C with 1.0 nM [^{3}H]PAC and 9–12 concentrations of competing drug. Data are expressed as percentage of total specific binding, defined in parallel incubations in the presence of 10 μM phentolamine. Log molar concentrations or dilutions of test compounds are depicted on the x axis. (**A**) Inhibition by clonidine (○) and norepinephrine (●). Note that only 70% of binding is fully displaced by the catecholamine. The remainder of binding is fully displaceable by clonidine. The remaining 30% of binding receptors represent I_1 receptors. (**B**) Displacement by classic clonidine-displacing substance (CDS) isolated from bovine brain. Note that like clonidine, CDS fully displaces the binding of [^{3}H]PAC, indicating that it interacts with both α_2-adrenergic and I-receptors (from ref. 89).

receptors.[11,12] Finally, the clinical efficacy of a range of antihypertensive agents correlates preponderantly with binding to I- but not α_2-adrenergic receptors.[13] However, the fact that all drugs that bind to I-receptors also bind to α_2-adrenergic receptors and that some agents selective for α_2-adrenergic receptors may lower arterial pressure when injected into the RVL clouds the issue.[10]

I-Receptor Family

In 1987 Boyajian *et al.*[14] proposed that [^{3}H]idazoxan ([^{3}H]IDA), an imidazoline and α_2-adrenergic receptor antagonist,[15] labels sites in brain that are distinct from those labeled by [^{3}H]rauwolscine. Subsequently, Wikberg[16] and Wikberg and Uhlen[17] demonstrated that [^{3}H]idazoxan binds with high affinity to nonadrenergic sites in brain, suggesting a class of nonadrenergic receptors in brain differing from those labeled by [^{3}H]PAC.

More extensive analysis of binding profiles, receptor distribution, and biochemical properties has led to a widely accepted view that I-receptors exist in at least two principal classes, now designated as I_1 and I_2.[13,18] The I_1 receptors, relatively uncommon, bind clonidine and idazoxan with approximately equal affinity and are localized to brain stem, kidney, and platelets. Their subcellular distribution and signal transduction processes have yet to be clearly defined, but it has been suggested, on rather weak evidence, that they may be cell-surface receptors possibly linked to G-proteins.

By contrast, the I_2-receptor is much more prevalent and usually identified by its high affinity for [^{3}H]IDA. I_2 receptors are expressed in a number of cells/tissues including basolateral membranes of kidney,[19–21] liver,[21–24] brain[21,25–31] (particularly in cerebral cortex), cultured astrocytes,[31] adipocytes,[32] urethra,[33] platelets,[34] pancreatic β-cells,[35–37] placenta,[38] and adrenal chromaffin cells.[39] The subcellular distribution of I_2 receptors in most tissues is on mitochondrial membranes, and the receptor does not appear to be coupled to G-proteins.[39] Rather, it appears linked with ion channels, possibly for K^+ and/or Na^+.[40–42]

Because I-receptors, particularly those of the I_2 subclass, also bind some non-imidazoles such as the guanidinium guanabenz[21,24] and the oxazole rilmenidine,[11] the term ''imidazoline receptor'' is technically a misnomer. Indeed, it has been proposed that at least one of the receptors be termed an ''imidazoline/guanidinium receptive site'' (IGRS).[19–21] Additionally, evidence that the imidazoline ''receptor'' is more than a ''binding site'' is weak. Nevertheless, the designation ''imidazoline receptor'' has become a convenient and widespread appellation and will be used here with full recognition of the attached imprecision.

ENDOGENOUS LIGANDS FOR I-RECEPTORS

The recognition of the I-receptor as an entity raises the fundamental question: What is its endogenous ligand? The problem is comparable to that posed by the discovery of receptors for drugs such as morphine or marijuana which antedated discovery of their native ligands, the endorphins and anandamide,[43] respectively. Interestingly, the discovery of the leading candidate for an endogenous ligand(s) for I-receptors, clonidine-displacing substance (CDS) (see below), preceded the identification of I-receptors as a unique nonadrenergic binding site for clonidine.

In 1984 Atlas and Burstein[44,45] partially purified a substance from brain that, with high affinity, competitively displaced [^{3}H]clonidine. The material, of unknown structure, was appropriately named ''clonidine-displacing substance'' and was considered a selective ligand for α_2-adrenergic receptors. Its presence was confirmed by Meeley *et al.*[46–49] In 1986[48,49] it was discovered that CDS would also competitively bind with high affinity to I-receptors (FIG. 1B), suggesting that CDS was ''the brain's own clonidine.''[48] It was not until 1994 that a molecule of known structure, agmatine, was characterized as an endogenous ligand at I-receptors.[50] However, the possibility that agmatine was only one of several molecules forming a ''CDS family'' appears likely. The candidates for endogenous ligands for I-receptors in order of their discovery are: (A) ''classic'' CDS (classic CDS); (B) immunoreactive CDS (ir-CDS); and (C) agmatine.

Classic CDS

Isolation

Classic CDS was first isolated by Atlas and Burstein[44,45] from cow and rat brain. Clonidine-displacing substance was measured by a bioassay, specifically displacement of [^{3}H]clonidine from α_2-adrenergic receptors of rat brain membranes. The material was extracted from homogenates of whole brain by methanol precipitation, ion-exchange electrophoresis, high power liquid chromatography, and later[51] by plasma desorption mass spectrometry (PDMS). The findings suggested that the molecule had a molecular mass of ~588 Da and was neither a catecholamine nor a peptide.

The existence of classic CDS was confirmed by Meeley *et al.*[46–48] who modified the extraction procedure by first isolating synaptosomes prior to methanol extraction and dialysis. Activity was tracked by measuring displacement of [^{3}H]PAC from rat cortex. Meeley *et al.*[46] confirmed that the material was not a catecholamine or a peptide and had a molecular weight of approximately 550 Da. Moreover, the material not only displaced ligand from α_2-adrenergic receptors but also competitively bound to I-receptors of the I_1 subclass in ventral medulla[47–50] (FIG. 1B). The subsequent observation that classic CDS would bind to basolateral membranes of kidney[52] and adrenal chromaffin cells[53] indicated that the substance would also bind to I-receptors of the I_2 subclass. Of interest was that the interaction of classic CDS was relatively selective for adrenergic receptors of the α_2-subtype, because it failed to displace radioligands from α_1- or β-adrenergic receptors.[44] The study also suggested that CDS in brain was neuronal.

Distribution

As indicated, classic CDS has been identified in partially purified biological material by radioreceptor assays, usually by assessing displacement of [^{3}H]clonidine or [^{3}H]PAC from α_2-adrenergic receptors of rat cerebrocortical membranes. By this procedure, efforts were made to identify the substance in tissues other than brain. However, interpretation is complicated by the fact that most studies have not measured CDS after chromatographic separation, that binding assays can be variably influenced by the ionic composition of the extract, and, while most extraction procedures eliminate catecholamines (which would compete at the α_2-adrenergic receptor), they do not eliminate agmatine which will also compete for the α_2- and I-receptor,[50] possibly distorting computations. Classic CDS has been detected in human serum,[54,55] cerebrospinal fluid,[55] and placenta[38] and in the neural-glial hybrid line NG-108.[56]

Bioactivity

That classic CDS is biologically active was reported by the Israeli and New York groups using partially purified material.[46,47,53,57] These include actions consistent with CDS's interaction at α_2-adrenergic receptors and with actions possibly at I-receptors. However, the results of these studies must be interpreted with caution; because the structure of classic CDS is unknown, biological responses might result from a molecule(s) other than that responsible for ligand displacement. Moreover, such contaminating moieties, if present, may vary between batches of CDS within and between laboratories.

Several effects of classic CDS have been attributed to actions at α_2-adrenergic receptors which in some, but not all, respects simulate those of clonidine. In human platelets,[57] classic CDS has no effect on basal activity, but it inhibits epinephrine-induced aggregation and potentiates ADP- and collagen-induced aggregation. Unlike clonidine, CDS does not affect the prostacyclin (PGI_2)-stimulated accumulation of cAMP in intact platelets. Classic CDS also inhibits the electrically elicited contraction of rat vas deferens,[58] an effect reversed by yohimbine at an effective dose equivalent to that required to displace specifically bound [^{3}H]clonidine from rat brain membranes. The finding suggests that classic CDS is an α_2-adrenergic receptor agonist.

Like clonidine, classic CDS, partially purified from expired human blood, dose-dependently contracts isolated rat aortic rings,[59] an effect antagonized by the selective α_2-adrenergic receptor antagonist rauwolscine. However, by contrast to clonidine, the α_1-adrenergic antagonist prazosin greatly reduces contractions caused by clonidine but not classic CDS. Moreover, although clonidine dose-dependently relaxed intact but not denuded vessels precontracted with methoxamine, classic CDS did not.

Actions of classic CDS *not* attributable to actions at α_2-adrenergic receptors include dose-dependent contraction of rat gastric fundus[60] (FIG. 2A), an effect not blocked by a range of antagonists to transmitters or peptides contracting the organ. Blockade of the response by verapamil suggests that Ca^{2+} influx is essential for its expression. Comparably, in isolated chromaffin cells of the bovine adrenal medulla, a cell type that expresses I- but not α_2-adrenergic receptors,[39,53] classic CDS elicits a potent dose- and largely Ca^{2+}-dependent release of catecholamines[53] (FIG. 2B and C).

Experiments have been performed by two groups[46,47,61] to assess whether classic CDS microinjected into the ventral medulla replicates the actions of clonidine on arterial pressure. Meeley *et al.*[46,47] reported that classic CDS microinjected into the RVL of the anesthetized rat, like clonidine, dose-dependently and site-specifically reduced arterial pressure and heart rate. By contrast, Bousquet *et al.*[61] found that classic CDS, provided by Atlas, elevated arterial pressure when injected into a comparable site in the medulla of the anesthetized cat. Differences between the groups are inexplicable and might result from differences in the partially purified CDS, in the experimental conditions, and/or in the sites of injection.

Relationship to Pathology

Using a ligand-displacement bioassay, Kreisberg *et al.*[54] reported that methanolic extract from sera of patients with pregnancy-induced hypertension (PIH) contained significantly elevated concentrations of classic CDS compared with the sera of women with normal pregnancies or from pregnant women with chronic hypertension. Goldberg-Stern *et al.*[55] observed elevated levels of classic CDS measured by displacement of [^{3}H]clonidine in the cerebrospinal fluid (CSF) of patients with neoplastic meningitis and stroke, without correlation with age, gender, CSF protein, or number of cells. Clonidine-displacing substance levels in CSF were higher than those in the serum

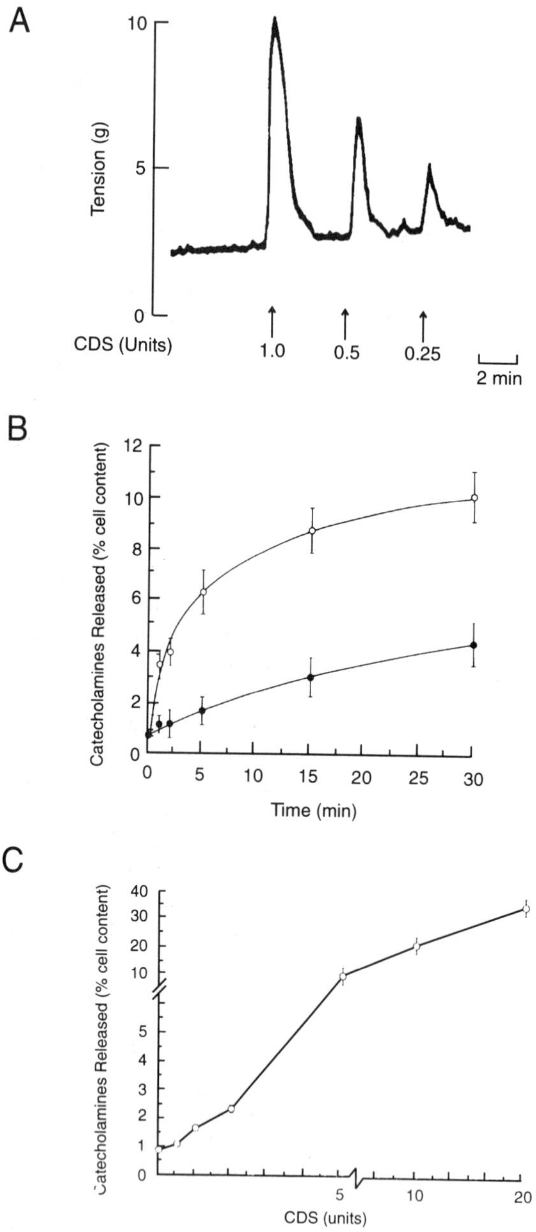

FIGURE 2. Bioactivity of classic clonidine-displacing substance (CDS). (**A**) Dose-dependent contraction of rat gastric fundus by CDS (from ref. 60). (**B**) Time course and (**C**) dose-dependent CDS-stimulated catecholamine (total) release from cultured bovine adrenal chromaffin cells. Values expressed as mean ± SEM. (Modified from refs. 53 and 60.)

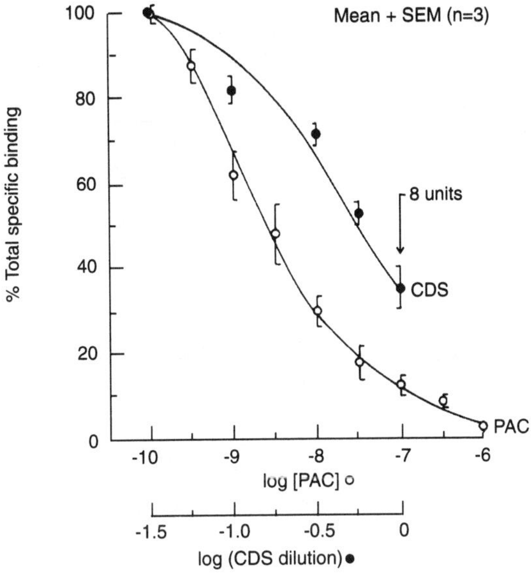

FIGURE 3. Inhibition of [^{3}H]*para*-aminoclonidine ([^{3}H]-PAC) binding to rat anti-PAC immunoglobulin by ir-clonidine-displacing substance (CDS) and PAC. Increasing amounts of partially purified classic-CDS (up to 8 units) or PAC were preincubated in triplicate with anti-PAC, and a displacement assay was performed. Values expressed as mean ± SEM (from ref. 64).

($p < 0.01$), indicating that CDS was present in human CSF and could be elevated in disease.

Summary

In summary, classic CDS appears bioactive at some α_2-adrenergic and I-receptor sites where it appears to act as an agonist. It is expressed in serum and CSF wherein in pathological states it may be elevated. However, absence of a structure for classic CDS and of receptor-selective antagonists for I-receptors limits the certainty of whether the biological responses represent the action of a single molecule and to what extent the responses relate to actions at I-receptors. Also, the fact that agmatine may also interact with the same receptor makes it uncertain how much co-purified agmatine may distort computations.

Immunoreactive Clonidine-Displacing Substance

Discovery and Relationship to Classic Clonidine-Displacing Substance

In 1986 Meeley *et al.*[62] first reported that polyclonal antibodies raised in rats against PAC conjugated with limulus hemocyanin would bind [^{3}H]PAC with high affinity (IC_{50} of 3.9 ± 0.4 nM) and that [^{3}H]PAC binding was inhibited by clonidine and several other imidazolines but not by catecholamines or allied agents. Notable was the observation that the specific binding of [^{3}H]PAC was inhibited in a concentration-dependent manner by CDS partially purified from bovine brain[62–64] (FIG. 3). Compar-

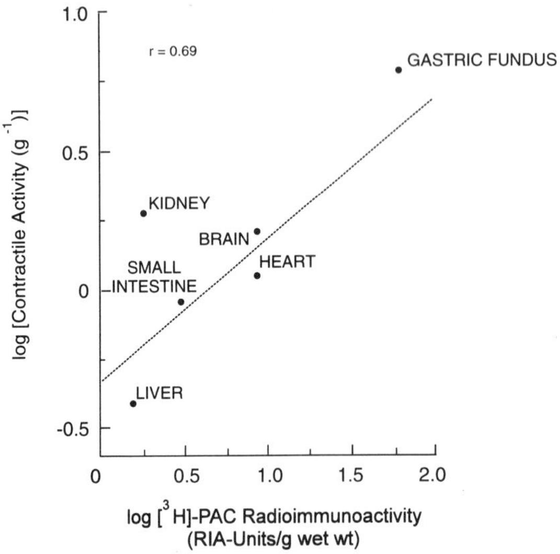

FIGURE 4. Relationship of potency on contraction of rat gastric fundus and immunoreactivity of clonidine-displacing substance (ir-CDS) isolated from different organs of rat. Data represent mean values from 6–10 radioimmunoassays for different tissues performed in triplicate at three dilutions and 6–8 gastric fundus strips tested in at least two trials. The correlation was significant ($p < 0.05$) for the group ($r = 0.69$), and when kidney was removed, it was $r = 0.99$ (from ref. 71).

able binding to classic CDS was described for a polyclonal antibody against idazoxan[65] and clonidine[66,67] and monoclonal antibodies to PAC.[69]

Although the finding might be interpreted to indicate that classic- and ir-CDS are identical and, therefore, need not be distinguished, the possibility exists that partially purified CDS contains different moieties, one responsible for binding displacement, the other for immunoreactivity. Hence, we believe that until structure is established, the two entities should be considered separately.

Distribution

The use of drug-based antibodies has proved useful in the development of radioimmunoassays (RIA) against ir-CDS and has allowed detection in various organs of rat[70,71] and in human serum[65,71] and CSF[65] (FIG. 4). Evaluation by RIA using antibodies to PAC[70,71] showed that the adrenal gland and gastric fundus contained significantly greater amounts of ir-CDS than did the brain; intermediate-to-low activity was present in heart, small intestine, serum, kidney, and liver, whereas lung and skeletal muscle values were near background. Radioimmunoassay-positive extracts elicited contraction of rat gastric fundus with contractions persisting in the presence of antagonists of various transmitters and modulators, but they were abolished by low concentrations of verapamil. Serum levels of CDS were profoundly reduced following removal of the adrenal glands, suggesting that the adrenal was a source of circulating material.[71] In brain, ir-CDS was detected in synaptosomes[62] with regional variations (FIG. 5). The hypothalamus had the highest activity, about 14-fold higher than that of the cerebellar vermis.

The distribution of ir-CDS and agmatine, however, is substantially different. As reported elsewhere in this volume,[72] the rank order of tissues containing the amine

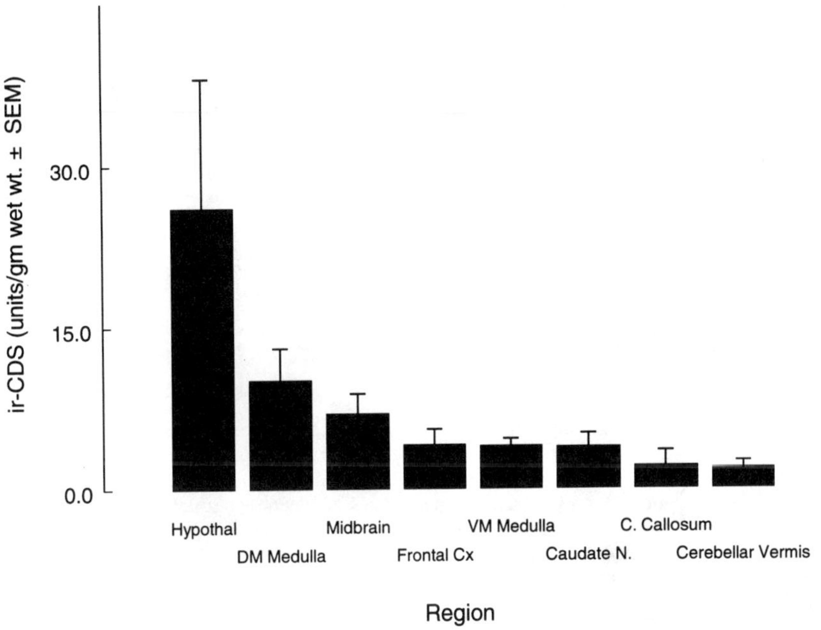

FIGURE 5. Regional concentrations of immunoreactive clonidine-displacing substance (ir-CDS) measured by radioimmunoassay with anti-*para*-aminoclonidine antibodies in rat brain. Data obtained from four male Sprague-Dawley rats (expanded unpublished data from ref. 62). Abbreviations: Hypothal = hypothalamus; DM medulla = dorsomedial medulla; frontal cx = frontal cortex; VM medulla = ventromedial medulla; caudate n. = caudate nucleus; c. callosum = corpus callosum.

are: stomach > small intestine >> adrenal > heart > brain >> plasma. In contrast, ir-CDS is ordered[72,73] adrenal >> stomach >> heart > brain > >> plasma. These findings suggest that ir-CDS and agmatine may be different molecules, a premise supported by observations that agmatine does not cross-react with drug-based antibodies.

Relationship to Disease

In one study, ir-CDS has been measured in the serum of patients with essential hypertension using an anti-clonidine antibody.[69] The study reported that of 32 patients studied, ir-CDS was higher in approximately 30% of cases and positively correlated with values of mean arterial pressure,[69] suggesting some association with the disease. Comparably, elevations of ir-CDS were found in the CSF of a subset of patients with severe intracerebral hemorrhage using an anti-idazoxan antibody with values not correlating with the activity of plasma and in excess of that attributable to leakage of blood into the CSF.[65] As agmatine does not cross-react with either anti-PAC or anti-idazoxan antibody (unpublished), the results suggest that ir-CDS is not agmatine.

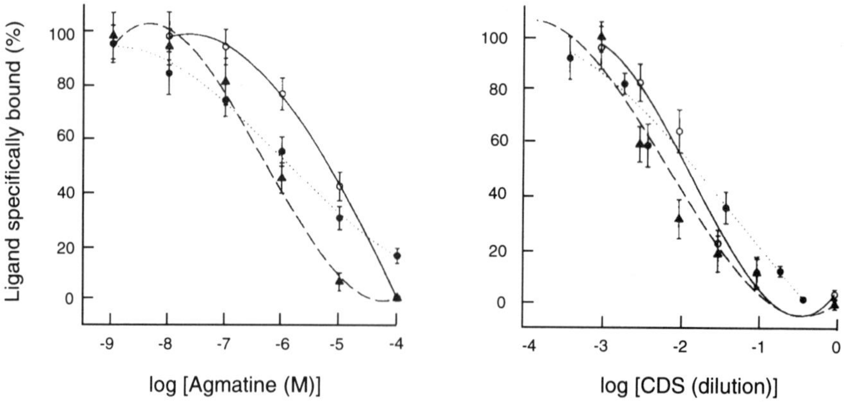

FIGURE 6. Effects of agmatine on α_2-adrenergic and I-receptor binding. Binding to α_2-adrenergic receptors was measured by displacement of [^{3}H]*para*-aminoclonidine (PAC) from membranes prepared from rat cerebral cortex (○); the binding to I_1 imidazoline receptors was measured by competition with [^{3}H]PAC to membranes of bovine ventrolateral medulla in the presence of 10 mM epinephrine to specifically block α_2-adrenergic sites (△) and the interaction with I_2 receptors measured by analysis of competition of [^{3}H]idazoxan binding to membranes of bovine adrenal chromaffin cells (●). Values are mean ± SEM of three experiments done in triplicate (from ref. 50).

Agmatine

Isolation and Structure

In 1994 Li *et al.*[50] demonstrated that a substance that displaced [^{3}H]PAC from α_2-adrenergic receptors of membranes of rat cerebral cortex and hence had CDS-like activity was present in bovine brain. Isolation consisted of homogenization of fresh brain, precipitation of proteins with ethanol and then subjecting the material to a series of chromatographic procedures including passage over a Dowex 50 (H^+) column, elution with 3N hydrochloric acid, C_{18} reversed-phase high performance liquid chromatography (HPLC), elution with acetic acid (25 mM), and size exclusion HPLC with a BioRad SEC 125 column from which CDS was eluted in a region indicating that its molecular weight was less than 300 Da.

When it was noted that the partially purified material did not have a chromophore on ultraviolet spectroscopy, it was suspected to be a guanidinium. By testing several native guanidino substances including guanidino acetic acid, guanidino propionic acid, guanidino succinic acid, and guanidino butylamine (agmatine) in a ligand binding assay, we discovered that agmatine, like classic-CDS, would bind to the α_2-adrenergic receptors and the I_1 and I_2 subclass of I-receptors (FIG. 6); the other substances would not.

To assess whether the CDS was agmatine, it was isolated in derivatized form. Derivatization of the enriched active fraction with 9-fluorenylmethyl chloroformate (Fmoc-Cl) generated a mixture of compounds stable under most purification condi-

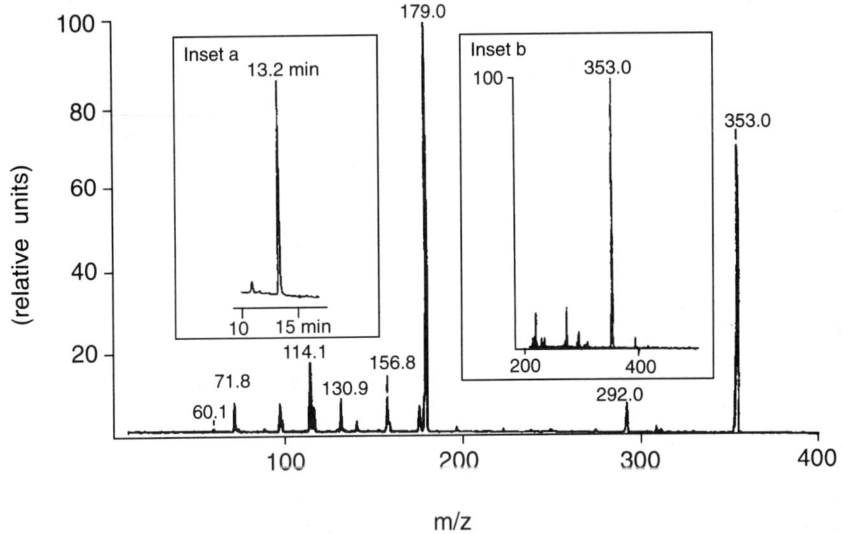

FIGURE 7. Some classic clonidine-displacing substance (CDS) is agmatine. **Inset** shows the HPLC profile of purified CDS derivatized with 9-fluorenylmethoxycarbonyl (Fmoc). FMOC-CDS was isolated using a C18 YMC 20 × 10 mm column with a 65/35 to 20/80 TFA (0.1%) acetonitrile gradient over 20 minutes. Retention time was identical to that obtained for authentic FMOC-agmatine. **Inset b** shows the MS profile of FMOC-CDS with MH^+ ion at 353 Da. The parent ion at 353 Da was passed through a second mass spectrometer, giving the fragmentation pattern shown (main figure). The mass spectral results were identical to those obtained for authentic FMOC-agmatine. MSMS was performed using electrospray ionization with a Finnegan MAT TSQ 700 mass spectrometer (from ref. 50).

tions. Although binding was lost on derivatization, it was regained after base hydrolysis to release the free amine. Sequential C_{18} column chromatography of the derivatized mixture, monitoring biological activity enrichment by release of free amine from a subsample of each fraction, yielded a single peak on C_{18} HPLC that coeluted with authentic Fmoc-agmatine. Subjecting the natural purified Fmoc-CDS derivative to electrospray mass spectroscopy-mass spectroscopy (MSMS) analysis gave identical results to those obtained for authentic Fmoc-agmatine, confirming that agmatine is the isolated CDS compound (FIG. 7).

We quantitated the amount of agmatine in weighed amounts of bovine brain by derivatization to Fmoc-Cl of a fraction (based on 200 U of CDS activity). The quantity of Fmoc agmatine present was determined from a standard curve generated by plotting ultraviolet intensity as a function of Fmoc-agmatine concentration. In this way, the amount of agmatine in bovine brain was found to be 1.5–3.0 nmol per gram of tissue (0.2 to 0.4 μg/g), indicating that agmatine is responsible for most of the CDS activity in the purified sample. Although the molecular weight of classic CDS was estimated to be 580 Da, the molecular weight of agmatine is 130 Da, indicating a major difference from classic CDS.

Distribution

The development of an HPLC method capable of measuring agmatine in mammalian tissue was described elsewhere.[72,73] As noted, we detected agmatine in various tissues of rat in variable concentrations with a rank order of: stomach > small intestine >> adrenal > heart > brain >> plasma. This distribution differs substantially from that of ir-CDS. Agmatine was also localized immunocytochemically with a specific antibody to adrenal chromaffin cells[74] (FIG. 8), to glomus cells of the carotid body,[75] to endothelial and smooth muscles of rat blood vessels,[76] and regionally in specific neurons in brain,[77] suggesting a specific cellular distribution.

It is of importance to note that although agmatine is widely distributed in plants, bacteria, and lower life forms, it has never previously been reported to be a product of mammalian metabolism.[78] However, our discovery that the rat brain expresses arginine decarboxylase[50] indicates that the source of agmatine is not dietary or from enteric bacteria.

Biological Actions

Interestingly, despite its prevalence in nonmammalian forms, agmatine has only infrequently been investigated for biological actions. One interesting feature is that, as a strong cation, it can enter some ionic conductance channels opened by activating nicotinic acetylcholine receptors. This has been shown in sympathetic ganglia,[79–81] in the nudibranch mollusc *Hermissenda crassicornis,*[82] and in the chick retina.[83] It also can be retrogradely transported by nerves.[82] In rat sympathetic ganglia and chick retina, agmatine is a weak antagonist of the nicotinic cholinergic receptor.[83] Of some interest is that agmatine may also have properties related to insulin/glucose metabolism. Thus, it may release insulin from pancreatic islet cells exposed to glucose[84] and exhibit insulin-like properties on isolated adipocytes and other tissues.[85,86] In adrenal chromaffin cells, agmatine is a secretagogue capable of dose-dependently releasing adrenalin and noradrenalin.[50] However, unlike CDS,[59] agmatine does not contract vascular smooth muscle (Estrada, Regunathan, and Reis, unpublished data).

In vivo, far less is known of agmatine's actions. However, what is certain is that it does not act like clonidine. Thus, unlike clonidine, in anesthetized ventilated rats,[87] agmatine administered intracerebroventricularly (400 nmol) increases sympathetic nerve activity and arterial pressure (by 8 ± 2.1 μV and 28.6 ± 2.7 mm Hg, respectively) and blocks, rather than facilitates, arterial baroreflexes. Also, unlike clonidine, microinjection of agmatine into the RVL has no effect on arterial pressure or sympathetic nerve activity, while iontophoresis onto vasomotor neurons of RVL is without effect. Interestingly, systemically agmatine decreases sympathetic nerve activity and arterial pressure largely resulting from blockade of sympathetic ganglionic transmission *in vivo* as *in vitro.*[80,83]

An additional issue of interest is whether agmatine is a precursor of polyamines in mammals as in bacteria.[78,88] Although it is generally assumed that in mammals polyamines are generated from putrescine formed from ornithine by the action of ornithine decarboxylase, the possibility needs to be reconsidered as to whether mam-

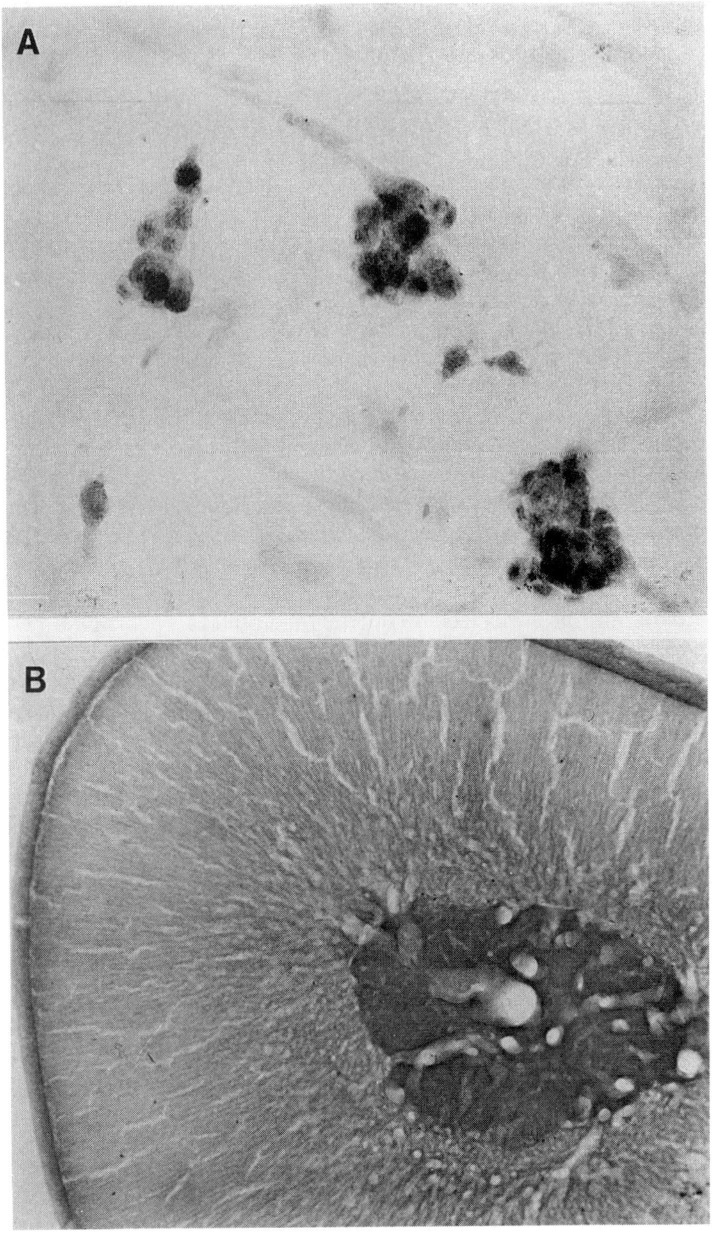

FIGURE 8. Immunoreactive agmatine is contained in adrenal chromaffin cells. (**A**) Staining of clusters of cultured bovine adrenal chromaffin cells (modified from ref. 74). (**B**) Staining of medulla, but not cortex, of rat adrenal gland (courtesy of Dr. C. Youngson).

mals express agmatinase (agmatine uryl-transferase), the enzyme hydrolyzing agmatine to putrescine. Conceivably, agmatine represents part of an alternative pathway for polyamine biosynthesis in mammals.

SUMMARY AND CONCLUSIONS

1. There are several endogenous ligands that bind to I-receptors of both the I_1 and I_2 subclass. These include: (a) classic CDS, a partially purified entity isolated by the criteria that it displaces binding ligands to α_2- and I-receptors; (b) immunoreactive (ir)-CDS, a moiety that binds to antibodies raised against clonidine, *para*-aminoclonidine, or idazoxan; and (c) agmatine.

2. Classic-CDS, not yet defined structurally, binds to I_1, I_2, and α_2-adrenergic receptors, is neither a peptide nor a catecholamine, and has purportedly a molecular weight of 588 Da. By ligand binding assays, it was found in brain, serum, CSF, and placenta and in a neural-glial cell line. Partially purified classic CDS is bioactive. Like clonidine, it contracts aorta and vas deferens and inhibits platelet aggregation, effects largely attributable to agonism at α_2-adrenergic receptors. Unlike clonidine, it contracts rat gastric fundus and releases catecholamines from chromaffin cells, effects attributable to actions at I-receptors. Injected into the RVL, classic CDS alters arterial pressure, but the direction of change of pressure has differed between groups of investigators. However, in the absence of structure, it is possible that ligand binding and bioactivity may be attributable to different molecules.

3. Ir-CDS, also of unknown structure, is a material(s) that binds to antibodies raised against clonidine, PAC, or idazoxan. Ir-CDS, measured by radioimmunoassay, is unevenly distributed in brain with highest concentrations in the hypothalamus, midbrain, and dorsal medulla. It is contained in the gastric fundus, adrenal gland, heart, kidney, and serum in amounts substantially higher than found in brain. Ir-CDS may be elevated in the serum of some patients with hypertension and in the CSF of patients with structural brain disease. The concentration of ir-CDS and bioactivity on gastric fundus directly correlates, suggesting that it may share similarities with classic-CDS. However, until the structure of classic and ir-CDS is determined, the possibility that ligand binding and antibody recognition are properties of different molecules must be considered.

4. Agmatine (decarboxylated arginine) is the only endogenous molecule that, like CDS, binds to α_2- and I-receptors of both classes. It and its biosynthetic enzyme arginine decarboxylase are present in brain, and agmatine is widely distributed throughout the body. However, the distribution of agmatine and ir-CDS differs, whereas the biological actions of agmatine do not mimic those of classic CDS. Its presence raises the possibility of an alternative pathway for polyamine biosynthesis. The fact that its concentration in brain is appropriate, that it is stored in neurons, primarily in vesicles, that it can be locally synthesized and released, and that receptors for agmatine exist suggest that it may be a novel neurotransmitter.

5. It is likely that CDS consists of a "family" of agents that bind to I- and in many, if not all instances, to α_2-adrenergic receptors. The pressing need is to purify these agents, characterize their metabolism, biological activities, and, in concert with investigations aimed at isolating and cloning the I-receptors, determine their action. Whether ligands exist that act *only* at I-receptors is not known.

6. The biological importance of I-receptors and their endogenous ligands remains a promising and challenging issue for contemporary biology and medicine.

REFERENCES

1. Haeusler, G. 1975. Cardiovascular regulation by central adrenergic mechanisms and its alteration by hypotensive drugs. Circ. Res. **36** (Suppl 1): 223-232.
2. Kobinger, W. & L. Pichler. 1978. Differentiation of drugs acting centrally upon the cardiovascular system by means of sympathetic and vagal responses. Clin. Exp. Hypertens. **1:** 229-249.
3. Hieble, J. P. & R. R. Ruffolo, Jr. 1992. Imidazoline receptors: Historical perspective. Fundam. Clin. Pharmacol. **6:** 7S-13S.
4. Karpanen, H., I. Paakkari, P. Paakkari, R. Huotari & A. L. Orma. 1976. Possible involvement of central histamine H2-receptors in the hypotensive effect of clonidine. Nature **259:** 587-588.
5. Bousquet, P., J. Feldman & J. Schwartz. 1984. Central cardiovascular effects of alpha adrenergic drugs: Differences between catecholamines and imidazolines. J. Pharmacol. Exp. Ther. **230:** 232-236.
6. Bousquet, P. & J. Schwartz. 1983. Alpha-adrenergic drugs. Pharmacological tools for the study of the central vasomotor control. Biochem. Pharmacol. **32:** 1459-1465.
7. Reis, D. J., S. Regunathan & M. P. Meeley. 1992. Imidazole receptors and clonidine-displacing substance in relationship to control of blood pressure, neuroprotection and adrenomedullary secretion. Am. J. Hypertens. **5:** 51S-57S.
8. Reis, D. J., E. V. Golanov, D. A. Ruggiero & M.-K. Sun. 1994. Sympathoexcitatory neurons of the rostral medulla are oxygen sensors and essential elements in the tonic and reflex control of the systemic and cerebral circulations. J. Hypertens. (Suppl.) **12:** S159-S180.
9. Ernsberger, P., M. P. Meeley, J. J. Mann & D. J. Reis. 1987. Clonidine binds to imidazole binding sites as well as $alpha_2$-adrenoceptors in the ventrolateral medulla. Eur. J. Pharmacol. **134:** 1-13.
10. Ernsberger, P., R. Giuliano, R. N. Willette & D. J. Reis. 1990. Role of imidazole receptors in the vasodepressor response to clonidine analogs in the rostral ventrolateral medulla. J. Pharmacol. Exp. Ther. **253:** 408-418.
11. Gomez, R. E., P. Ernsberger, G. Feinland & D. J. Reis. 1991. Rilmenidine lowers arterial pressure via imidazole receptors in brainstem C1 area. Eur. J. Pharmacol. **195:** 181-191.
12. Punnen, S., R. Urbanski, A. J. Krieger & H. N. Sapru. 1987. Ventrolateral medullary pressor area: Site of hypotensive action of clonidine. Brain Res. **422:** 336-346.
13. Ernsberger, P., T. H. Damon, L. M. Graff, S. G. Schafer & M. O. Christen. 1993. Moxonidine, a centrally acting antihypertensive agent, is a selective ligand for I_1-imidazoline sites. J. Pharmacol. Exp. Ther. **264:** 172-182.
14. Boyajian, C. L., S. E. Joughlin & F. M. Leslie. 1987. Anatomical evidence for the α_2-adrenoceptor heterogeneity: Differential autoradiographic distributions of ^{3}H-rauwolscine and ^{3}H-idazoxan in rat brain. J. Pharmacol. Exp. Ther. **241:** 1079-1091.
15. Doxey, J. H. C., A. C. Lane, A. G. Roach & N. K. Virdee. 1984. Comparison of the α_2-adrenergic antagonist profiles of idazoxan (RX 781094), yohimbine, rauwolscine, and corynanthine. Naunyn-Schmiedeberg's Arch. Pharmacol. **325:** 136-144.
16. Wikberg, J. E. S. 1989. High affinity binding of idazoxan to a non-catecholaminergic binding site in the central nervous system: Description of a putative idazoxan receptor. Pharmacol. Toxicol. **69:** 152-155.
17. Wikberg, J. E. S. & S. Uhlen. 1990. Further characterization of the guinea pig cerebral cortex idazoxan receptor: Solubilization, distinction from the imidazole site, and demonstration of cirazoline as an idazoxan receptor-selective drug. J. Neurochem. **55:** 192-203.

18. MICHEL, M. C. & P. ERNSBERGER. 1992. Keeping an eye on the I site: Imidazoline-preferring receptors. Trends Pharmacol. Sci. **13:** 369-370.
19. COUPRY, I., D. ATLAS, R. A. PODEVIN, I. UZIELLI & A. PARINI. 1990. Imidazoline-guanidinium receptive site in renal proximal tubule: Asymmetric distribution, regulation by cations and interaction with an endogenous clonidine displacing substance. J. Pharmacol. Exp. Ther. **252:** 293-299.
20. LACHAUD, V., I. LIMON, F. TESSON, I. COUPRY & A. PARINI. 1992. Characterization of imidazoline-guanidinium receptive sites in renal medulla from human kidney. Am. J. Hypertens. **5:** 69S-71S.
21. IVKOVIC, B., V. BAKTHAVACHALAM, W. ZHANG, A. PARINI, D. DIZ, S. BOSCH, J. L. NEUMEYER & S. M. LANIER. 1994. Development of a high-affinity radioiodinated ligand for identification of imidazoline/guanidinium receptive sites (IGRS): Intratissue distribution of IGRS in liver, forebrain, and kidney. Mol. Pharmacol. **46:** 15-23.
22. ATLAS, D. 1991. Clonidine-displacing substance (CDS) and its putative imidazoline receptor. New leads for further divergence of alpha$_2$-adrenergic receptor activity. Biochem. Pharm. **41:** 1541-1549.
23. ZONNENSCHEIN, R., S. DIAMANT & D. ATLAS. 1990. Imidazoline receptors in rat liver cells: A novel receptor or a subtype of alpha$_2$-adrenoceptors? Eur. J. Pharmacol. **190:** 203-215.
24. TESSON, F., C. PRIP-BUUS, A. LEMOINE, J. P. PEGORIER & A. PARINI. 1992. Characterization of mitochondrial imidazoline-guanidinium receptive sites (IGRS) in liver. Am. J. Hypertens. **5:** 80S-82S.
25. TESSON, F. & A. PARINI. 1991. Identification of an imidazoline-guanidinium receptive site in mitochondria from rabbit cerebral cortex. Eur. J. Pharmacol. **208:** 81-83.
26. MALLARD, N. J., A.L. HUDSON & D. J. NUTT. 1992. Characterization and autoradiographical localization of non-adrenoceptor idazoxan binding sites in the rat brain. Br. J. Pharmacol. **106:** 1019-1027.
27. REIS, D. J., S. REGUNATHAN, H. WANG, D. L. FEINSTEIN & M. P. MEELEY. 1992. Imidazoline receptors in the nervous system. Fundam. Clin. Pharmacol. **6** (Suppl 1): 23S-29S.
28. DE VOS, H., G. BRICCA, J. DE KEYSER, J. P. DE BACKER, P. BOUSQUET & G. VAUQUELIN. 1994. Imidazoline receptors, non-adrenergic idazoxan binding sites and alpha$_2$-adrenoceptors in the human central nervous system. Neuroscience **59:** 589-598.
29. REIS, D. J., S. REGUNATHAN, E. V. GOLANOV & D. L. FEINSTEIN. 1994. Protection of focal ischemic infarction by rilmenidine in the animal: Evidence that interactions with central imidazoline receptors may be neuroprotective. Am. J. Cardiol. **74:** 25A-30A.
30. REGUNATHAN, S., S. BRAMWELL & D. J. REIS. 1995. Effects of rilmenidine on signal transduction mechanisms in rat brain slices. Proceedings of the II International Symposium on Imidazoline Receptors. Ann. N.Y. Acad. Sci. This volume.
31. REGUNATHAN, S., D. L. FEINSTEIN & D. J. REIS. 1993. Expression of non-adrenergic imidazoline sites in rat cerebral cortical astrocytes. J. Neurosci. Res. **34:** 681-688.
32. LAFONTAN, M., D. L. LANGIN, M. PORTILLO & H. PARIS. 1992. Imidazoline binding sites in fat cells. Localization and pharmacologic differentiation from α_2-adrenergic receptors. Am. J. Hypertens. **5:** 72S-79S.
33. YABLONSKY, F., J. Y. LACOLLE & J. P. DAUSSE. 1991. Non-adrenergic binding sites for the "α_2-antagonist" ^{3}H-diazoxan in rabbit urethral smooth muscle. J. Recept. Res. **11 (1-4):** 153-162.
34. PILETZ, J. E. & K. SLETTEN. 1993. Nonadrenergic imidazoline binding sites on human platelets. J. Pharmacol. Exp. Ther. **267:** 1493-1502.
35. LACOMBE, C., V. VIALLARD & H. PARIS. 1993. Identification of α_2-adrenoceptors and of non-adrenergic idazoxan binding sites in pancreatic islets from young and adult hamsters. Int. J. Biochem. **25:** 1077-1083.
36. REMAURY, A. & H. PARIS. 1992. The insulin-secreting cell line, RINm5F, expresses an α_{2D}-adrenoceptor and nonadrenergic idazoxan-binding sites. J. Pharmacol. Exp. Ther. **260:** 417-426.

37. BROWN, C. A., A. C. LOWETH, S. A. SMITH & N. G. MORGAN. 1993. Stimulation of insulin secretion by imidazoline compounds is not due to interaction with non-adrenoceptor idazoxan binding sites. Br. J. Pharmacol. **108:** 312-317.
38. DIAMANT, S., T. ELDAR-GEVA & D. ATLAS. 1992. Imidazoline binding sites in human placenta: Evidence for heterogeneity and a search for physiological function. Br. J. Pharmacol. **106:** 101-108.
39. REGUNATHAN, S., M. J. EVINGER, M. P. MEELEY & D. J. REIS. 1991. Effects of clonidine and other imidazoline-receptor binding agents on second messenger systems and calcium influx in bovine adrenal chromaffin cells. Biochem. Pharmacol. **42:** 2011-2018.
40. BIDET, M., P. POUJEOL & A. PARINI. 1990. Effect of imidazolines on Na^+ transport and intracellular pH in renal proximal tubule cells. Biochim. Biophys. Acta **1024:** 173-178.
41. DUNNE, M. J. 1991. Block of ATP-regulated potassium channels by phentolamine and other alpha-adrenoceptor antagonists. Br. J. Pharmacol. **103:** 1847-1850.
42. CHAN, S. L. & N. G. MORGAN. 1990. Stimulation of insulin secretion by efaroxan may involve interaction with potassium channels. Eur. J. Pharmacol. **176:** 97-101.
43. VOGEL, Z., J. BARG, R. LEVY, D. SAYA, E. HELDMAN & R. MECHOULAM. 1993. Anandamide, a brain endogenous compound, interacts specifically with cannabinoid receptors and inhibits adenylate cyclase. J.Neurochem. **61:** 352-355.
44. ATLAS, D. & Y. BURSTEIN. 1984. Isolation and partial purification of a clonidine-displacing endogenous brain substance. Eur. J. Biochem. **144:** 287-293.
45. ATLAS, D. & Y. BURSTEIN. 1984. Isolation of an endogenous clonidine-displacing substance from rat brain. FEBS Lett. **170:** 387-390.
46. MEELEY, M. P., P. ERNSBERGER, A. R. GRANATA & D. J. REIS. 1985. Isolation of clonidine-displacing substance from bovine brain: Functional characterization. Soc. Neurosci. Abstr. **11:** 490.
47. MEELEY, M. P., P. R. ERNSBERGER, A. R. GRANATA & D. J. REIS. 1986. An endogenous clonidine-displacing substance from bovine brain: Receptor binding and hypotensive actions in the ventrolateral medulla. Life Sci. **38:** 1119-1126.
48. MEELEY, M. P., P. R. ERNSBERGER, P. R. ERNSBERGER & D. J. REIS. 1986. An endogenous substance with clonidine-like receptor binding selectivity: The brain's own clonidine? J. Hypertens. **4** (Suppl. 6): S257-S259.
49. ERNSBERGER, P., M. P. MEELEY & D. J. REIS. 1988. An endogenous substance with clonidine-like properties: Selective binding to imidazole sites n the ventrolateral medulla. Brain Res. **441:** 309-318.
50. LI, G., S. REGUNATHAN, C. J. BARROW, J. ESHRAGHI, R. COOPER & D. J. REIS. 1994. Agmatine is an endogenous clonidine-displacing substance in brain. Science **263:** 966-969.
51. ATLAS, D. 1991. Clonidine-displacing substance (CDS) and its putative imidazoline receptor. New leads for further divergence of α_2-adrenergic receptor activity. Biochem. Pharmacol. **41:** 1541-1549.
52. PARINI, A., I. COUPRY, R. M. GRAHAM, I. UZIELLI, D. ATLAS & S. M. LANIER. 1989. Characterization of an imidazoline/guanidinium receptive site distinct from the alpha$_2$-adrenergic receptor. J. Biol. Chem. **264:**11874-11878.
53. REGUNATHAN, S., M. P. MEELEY & D. J. REIS. 1991. Clonidine-displacing substance from bovine brain binds to imidazoline receptors and releases catecholamines in adrenal chromaffin cells. Mol. Pharmacol. **40:** 884-888.
54. KREISBERG, G. A., S. DIAMANT, Y. Z. DIAMANT & D. ATLAS. 1987. Raised levels of an endogenous nonadrenergic substance in the serum of pregnancy-induced hypertension patients. Isr. J. Med. Sci. **23:** 1194-1197.
55. GOLDBERG-STERN, H., D. ATLAS, L. SCHWARTZ, A. ACHIRON, I. ZIV, R. DJALDETTI, Y. ZOLDAN & E. MELAMED. 1993. Detection and measurement of an endogenous clonidine-displacing substance in human cerebrospinal fluid. Brain Res. **601:** 325-328.

56. ERNSBERGER, P., M. P. MEELEY & D. J. REIS. 1989. Neuroblastoma-glioma hybrid cells contain clonidine-displacing substance. Eur. J. Pharmacol. **174:** 135-138.
57. DIAMANT, S., A. ELDOR & D. ATLAS. 1987. A low molecular weight brain substance interacts, similarly to clonidine, with alpha$_2$-adrenoceptors of human platelets. Eur. J. Pharmacol. **144:** 247-255.
58. DIAMANT, S. & D. ATLAS. 1986. An endogenous brain substance, CDS (clonidine-displacing-substance), inhibits the twitch response of rat vas deferens. Biochem. Biophys. Res. Commun. **134:** 184-190.
59. SYNETOS, D., V. G. MANOLOPOULOS, D. ATLAS & E. PIPILLI-SYNETOS. 1991. Human plasma-derived material with clonidine displacing substance (CDS)-like properties contracts the isolated rat aorta. J. Auton. Pharmacol. **11:** 343-351.
60. FELSEN, D., P. ERNSBERGER, M. P. MEELEY & D. J. REIS. 1987. Clonidine displacing substance is biologically active on smooth muscle. Eur. J. Pharmacol. **142:** 453-455.
61. BOUSQUET, P., J. FELDMAN & D. ATLAS. 1986. An endogenous, non-catecholamine clonidine antagonist increases mean arterial blood pressure. Eur. J. Pharmacol. **124:** 167-170.
62. MEELEY, M. P., A. C. TOWLE, L. K. CHAR, P. ERNSBERGER & D. J. REIS. 1986. Polyclonal antiserum to a para-aminoclonidine conjugate recognizes an endogenous clonidine-like substance(s) in brain. Soc. Neurosci. Abstr. **12:** 963.
63. MEELEY, M. P., A. C. TOWLE, P. ERNSBERGER & D. J. REIS. 1988. A specific antiserum recognizes clonidine-displacing substance: Implications for the structure of the brain's own clonidine. Neurosci. Lett. **84:** 84-90.
64. MEELEY, M. P., A. C. TOWLE, P. ERNSBERGER, L. K. CHAR, P. M. MCCAULEY & D. J. REIS. 1989. Clonidine-specific antisera recognize an endogenous clonidine-displacing substance in brain. Hypertension **13:** 341-351.
65. WANG, H., D. MCGOWAN, S. REGUNATHAN, S. BRAMWELL, G. LI & D. J. REIS. 1993. Idazoxan-specific antiserum recognizes an endogenous clonidine-displacing substance (CDS) in human serum and CSF. Soc. Neurosci. Abstr. **19:** 1691.
66. DONTENWILL, M., G. BRICCA, A. MOLINES, A. BELCOURT & P. BOUSQUET. 1987. A polyclonal antibody raised against clonidine: A model for the specific imidazoline receptor. Eur. J. Pharmacol. **137:** 143-144.
67. DONTENWILL, M., G. BRICCA, A. MOLINES, P. BOUSQUET & A. BELCOURT. 1988. Production and characterization of anti-clonidine antibodies not cross-reacting with catecholamines. Eur. J. Pharmacol. **149:** 249-255.
68. DONTENWILL, M., A. MOLINES, G. BRICCA, J. STUTZMAN, J. KEMPF, A. BELCOURT & P. BOUSQUET. 1992. Production and characterization of an iminoimidazolidine specific monoclonal antibody using para-aminoclonidine as antigen. Life Sci. **50:** 1859-1868.
69. DONTENWILL, M., A. MOLINES, A. VERDUN, G. BRICCA, S. LAURENT & P. BOUSQUET. 1993. A circulating substance cross-reacting with antiimidazoline antibodies. Detection in serum in relation to essential hypertension. J. Clin. Invest. **92:** 1068-1072.
70. HENSLEY, M. L., M. P. MEELEY, P. M. MCCAULEY, P. ERNSBERGER & D. J. REIS. 1989. Clonidine-displacing substance is present in peripheral tissues of the rat. Am. J. Hypertens. **2:** 917-919.
71. MEELEY, M. P., M. L. HENSLEY, P. ERNSBERGER, D. FELSEN & D. J. REIS. 1992. Evidence for a bioactive clonidine-displacing substance in peripheral tissues and serum. Biochem. Pharmacol. **44:** 733-740.
72. RAASCH, W., S. REGUNATHAN, G. LI & D. J. REIS. 1995. Agmatine is widely and unequally distributed in rat organs. Proceedings of the II International Symposium on Imidazoline Receptors. Ann. N.Y. Acad. Sci. This volume.
73. RAASCH, W., S. REGUNATHAN, G. LI & D. J. REIS. 1995. Agmatine, the bacterial amine, is widely distributed in mammalian tissues. Life Sci. In press.
74. WANG, H., S. REGUNATHAN, C. YOUNGSON & D. J. REIS. 1995. An antibody to agmatine localizes the amine in bovine adrenal chromaffin cells. Neurosci. Lett. **183:** 1-5.

75. YOUNGSON, C., S. REGUNATHAN, H. WANG, G. LI & D. J. REIS. 1995. Co-expression of imidazoline receptors and agmatine in rat carotid body. Proceedings of the II International Symposium on Imidazoline Receptors. Ann. N.Y. Acad. Sci. This volume.
76. REGUNATHAN, S., C. YOUNGSON, H. WANG & D. J. REIS. 1995. Imidazoline receptors in vascular smooth muscle and endothelial cells. Proceedings of the II International Symposium on Imidazoline Receptors. Ann. N.Y. Acad. Sci. This volume.
77. WANG, H., K. OTAKE, D. A. RUGGIERO, T. A. MILNER, G. LI, W. RAASCH, C. YOUNGSON, S. REGUNATHAN & D. J. REIS. 1994. Regional distribution of agmatine and arginine decarboxylase in rat brain. Soc. Neurosci. Abstr. **20:** 1060.
78. TABOR, C. W. & H. TABOR. 1984. Polyamines. Ann. Rev. Biochem. **53:** 749-790.
79. YOSHIKAMI, D. 1981. Transmitter sensitivity of neurons assayed by autoradiography. Science **212:** 929-930.
80. QUIK, M. 1985. Inhibition of nicotinic receptor mediated ion fluxes in rat sympathetic ganglia by BGT II-S1, a potent phospholipase. Brain Res. **325:** 79-88.
81. QUIK, M. 1987. Evidence for an involvement of membrane lipids in the control of neuronal nicotinic receptor function using bungarotoxin II-S1. J. Neurochem. **48:** 1761-1767.
82. KUZIRIAN, A., E. MEYHOFER, L. HILL, J. T. NEARY & D. L. ALKON. 1986. Autoradiographic measurement of tritiated agmatine as an indicator of physiologic activity in Hermissenda visual and vestibular neurons. J. Neurocytol. **15:** 629-643.
83. LORING, R. H. 1990. Agmatine acts as an antagonist of neuronal nicotinic receptors. Br. J. Pharmacol. **99:** 207-211.
84. SENER, A., P. LEBRUN, F. BLACHIER & W. J. MALAISSE. 1989. Stimulus-secretion coupling of arginine-induced insulin release. Insulinotropic action of agmatine. Biochem. Pharmacol. **38:** 327-330.
85. WEITZEL, G., B. PFEIFFER & W. STOCK. 1980. Insulin-like partial effects of agmatine derivatives in adipocytes. Hoppe-Seyler's Zeitschrift fur Physiologische Chemie. **361:** 51-60.
86. PFEIFFER, B., W. SARRAZIN & G. WEITZEL. 1981. Insulin-like effects of agmatine derivatives *in vitro* and *in vivo*. Hoppe-Seyler's Zeitschrift fur Physiologische Chemie **362:** 1331-1337.
87. SUN, M.-K., S. REGUNATHAN & D. J. REIS. 1995. Cardiovascular responses to agmatine, a clonidine-displacing substance, in anesthetized rats. Clin. Exp. Hypertens. **17:** 115-128.
88. BENDER, D. A. 1985. Amino Acid Metabolism. 2nd Edit. John Wiley & Sons. New York.
89. REIS, D. J., P. ERNSBERGER & M. P. MEELEY. 1987. Clonidine acts on imidazole receptors in brain which bind an endogenous clonidine-displacing substance. *In* Pharmacology. M. J. Rand & C. Raper, Eds. Excerpta Medica International Congress Series.: 41-45. Elsevier. Amsterdam.

Molecular and Physiological Properties of Clonidine-Displacing Substance[a]

DAPHNE ATLAS

Department of Biological Chemistry
The Hebrew University of Jerusalem
Jerusalem 91904, Israel

The rationale for searching for endogenous ligands with biological activity similar to that of alkaloids or synthetic compounds was realized by the discovery of enkephalins and endorphins as the "endogenous morphines" and recently of anandamide as the "endogenous cannabinoids" (marijuana; TABLE 1[1]). The search for an endogenous substance that mimics the many "faces" of the hypotensive drug clonidine led to the identification of a clonidine-displacing substance (CDS).[2,3] Clonidine-displacing substance is a small molecular mass compound that was isolated by screening bovine brain extracts for their ability to displace tritiated clonidine specifically bound to rat brain membranes.[2,3] Its molecular mass (~500 daltons)[3] was determined by size exclusion chromatography, and its precise mass (587.2 daltons) by mass spectrometry.[4,5]

Intracisternal application of CDS to mice, the left cat artery, or directly to the cat ventrolateral medulla (VLM) induces a transient increase in mean arterial pressure without affecting heart rate.[6,7] The presence of another endogenous substance that displaces *p*-aminoclonidine from bovine brain, but is not retained on a reverse phase HPLC column has been reported.[8] By contrast to CDS, this substance induces hypotension when injected into the VLM.[8] Clonidine-displacing substance inhibits the twitch response in rat vas deferens,[9] potentiates ADP- and collagen-induced platelet aggregation,[10] and contracts rat aortic rings.[11] Functionally, CDS action is not identical to that of clonidine; it produces an increase in mean arterial pressure when injected at the VLM and reverses the hypotensive effect of clonidine,[6] it does not reverse a prostaglandin I_2 (PGI_2)-induced rise in cAMP in human platelets,[10] and it does not change its affinity in the presence of GTP or GppNHp in radiolabeled binding studies.[10]

The most intriguing property of CDS is its ability to cause an increase in mean arterial pressure at the VLM.[6] Competition of CDS with clonidine on a common target in the VLM, shown by counteracting the hypotensive effect of clonidine, strongly suggests that CDS mediates hypertension by acting as an agonist at putative and unique clonidine sites in the brain.[4,12] The effectiveness of clonidine, cirazoline, and other imidazoline-like structures as hypotensive agents at the VLM, a region

[a] This work is supported by the H. L. Lauterback Fund for D. A.

TABLE 1. Endogenous Ligands That Compete with Alkaloids and Other Synthetic Substances on Common Receptors

Exogenous/Synthetic Agonist/Antagonist	Endogenous Substance	Receptor (Target)
Morphine	Enkephalins β-endorphins	Opiate (δ, κ, μ)
Cannabinoid (marijuana)	Anandamide	Cannabinoid
Ouabain	Ouabain; 19-norbufalin	Na^+/K^+ ATPase
Clonidine	CDS?	Clonidine sites?
Idazoxan	CDS?	Imidazoline sites?

that is not affected by α_2-adrenergic agonists of the phenylethylamines family (e.g., α-methyl noradrenaline), suggests that CDS is an agonist at imidazoline receptors (IRs), distinct from α_2-adrenergic receptors.[6,13] Indeed, nonadrenergic sites in the VLM were characterized by receptor binding assay in rat brain,[14,15] renal proximal tubule,[16] guinea pig cerebral cortex,[17] rat liver,[18] human platelets,[19] human placenta,[20] and bovine adrenal medulla,[21] as well as by autoradiographical localization.[22,23] These studies indicate that the multiple sites can be classified according to their affinity for clonidine (I_1-IR) and idazoxan (I_2-IR) and further subclassified by their affinity for amiloride (I_{2A} and I_{2B}).[20] Recently, three different putative IRs were visualized by the use of photoaffinity labeling and their apparent molecular weights determined.[24]

Receptor binding assays show that CDS, like many guanido-imidazoline compounds, recognizes the α_2-adrenergic receptors IR-I_1 and IR-I_2. All three receptor types are present together in various proportions, and their individual contribution to the assays confuses the interpretation of their cellular function. Furthermore, no definite function is associated with these sites, partly because the interaction of guanido-imidazoline compounds with their corresponding target sites is determined mostly in ligand-binding assays using synthetic antagonists. Even in the case of α_2-adrenergic receptors, the lack of selective antagonists and functional evidence complicates the assignment of their cellular functions despite the cloning and sequencing of four different receptors subtypes. (For a review see ref. 25.) Human placenta, human liver, and rat liver are unique as all three tissues lack functional α_2-adrenergic receptors and the high affinity clonidine sites (IR-I_1), but display a high density of IR-I_2. Clearly, identification of an endogenous ligand as a selective agonist at IR-I_2 will assign a physiological role for these receptors.

Clonidine sites (IR-I_1) are specifically located in the VLM,[14,15] a nucleus that responds to CDS by increasing mean arterial pressure.[6,7] At the VLM, CDS competes with clonidine, which acts at this site to reduce mean arterial pressure.[6,7] Although CDS might be an agonist-candidate IR-I_1, more studies are required to establish CDS as a specific neurotransmitter/neuroregulator at a specific nucleus in the brain.

A recent study proposed that agmatine, a known compound isolated from bovine brain, is CDS.[26] Although CDS structure is not yet elucidated, we show in this report that the chemical characteristics, molecular mass, and physiological properties of CDS are entirely different from those of agmatine.

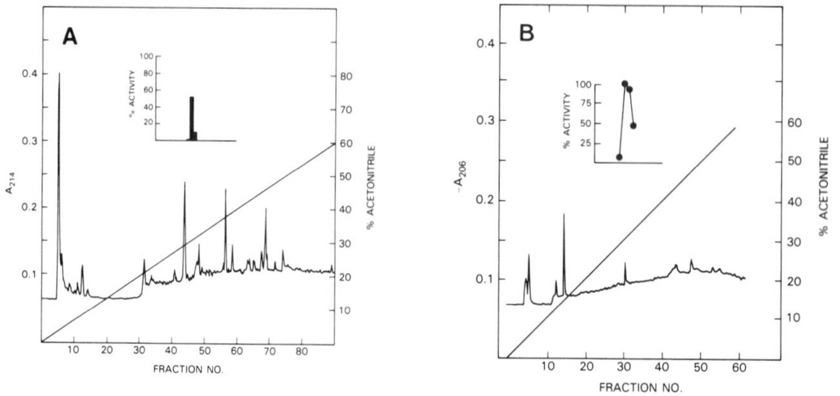

FIGURE 1. Clonidine-displacing substance was extracted from bovine brain and human serum as previously reported, and the two extracts were compared for their elution profile at the third purification step on reverse phase C_{18} HPLC column. All fractions were collected, dried under vacuum, and assayed as previously described by displacement of [^{3}H]clonidine from rat brain membranes.[3,4,10] The active fractions are depicted above the chromatogram and present activity as a percentage of control.

MATERIALS AND METHODS

[^{3}H]Idazoxan (60-80 Ci/mmol) and [^{3}H]rauwolscine (60 Ci/mmol) were purchased from Amersham (UK); clonidine (20-30 Ci/mmol) was from New England Nuclear (US). Rat brain membranes were prepared as previously described.[3] Receptor binding assays and displacement assays of tritiated antagonists from rat brain membranes were carried out according to previously described procedures.[2,10]

RESULTS AND DISCUSSION

Purification of Clonidine-Displacing Substance

Clonidine-displacing substance is a small molecular weight substance isolated from bovine brain and purified in a series of consecutive HPLC columns as previously described.[2-5] Because of its limited availability and lengthy purification procedure, CDS was never purified in enough amounts to allow structure determination by either nuclear magnetic resonance or elemental analysis. Mass spectrometric methods, especially in the FAB-MS mode, which is very helpful in peptide and carbohydrate structure determinations, gave a nonpeptide fragmentation pattern (data not shown). The nonpeptide nature of CDS was confirmed by its resistance to various proteases (e.g., trypsin, chymotrypsin, pronase, and V8 protease) and by negative amino acid analysis (data not shown). The possibility that CDS activity in human serum[27] is associated with another molecule was tested by comparing retention times on HPLC columns with CDS originating from bovine brain (FIG. 1). Following crude extraction procedures of CDS from bovine brain[3] and human serum,[27] the active fractions were chromatographed on a C_{18} reverse-phase column and eluted with CH_3CN/0.1% TFA gradient. As shown in FIGURE 1, CDS originating from either bovine brain or human serum was eluted at a similar retention time, under identical HPLC chromatography, strongly suggesting that CDS, initially defined for bovine brain, is the active species present in human serum. Clonidine-displacing substance recovered from this column is further purified by two consecutive HPLC columns, eventually yielding a pure compound. The final purification step of CDS is achieved on a C_8 reverse-phase HPLC column.[5]

Molecular Mass of Clonidine-Displacing Substance

A tentative molecular mass of 500 daltons was determined for CDS by size exclusion chromatography on a sephadex G-15.[3] For precise molecular mass measurement, a mass spectrometric method, plasma desorption mass spectrometry, was applied.[28] Using this technique the sample is dissolved in methanol and thinly sprayed on aluminized mylar foil and bombarded with high energy ions (fission fragments) from a californium source (^{252}Cf).[28] Pure CDS from its final reverse-phase column shows a single distinct molecular ion in the field desorption of m/z 587.8 ± 2.[4,5,12] The second and smaller peak at m/z 1174 is its corresponding dimer (FIG. 2). The absorbance spectrum of CDS shows two peaks at the ultraviolet range, assigned at 224 and 276 nm (FIG. 3).[5]

Clonidine-Displacing Substance Abundance

Clonidine-displacing substance was identified for the first time in rat brain[2] and later in bovine brain[3] (reviewed in ref. 12). The purification method was developed for bovine brain which has more starting material.[2,3] A unit of activity was defined as the amount needed to displace specifically bound [^{3}H]clonidine (2 nM) to rat brain membranes (250 μg protein) when the assay conditions are 40 minutes at 37°C and 10 μM norepinephrine is used for determining nonspecific binding.[3] Clonidine-displacing substance levels in bovine brain were 100–250 units/100 g wet weight at the methanolic extract. An average of 400 units of CDS per brain is obtained which is calculated to be 4 ng/g wet tissue. The limited availability of CDS in the brain and the fact that CDS is not a peptide might account for some of the obstacles in its purification and structure determination. Clonidine-displacing substance levels in human serum were determined by a method developed by Kreisberg *et al.*[27] and stands at 1.4 ± 0.3 units/ml.[27] Higher CDS levels are found in human cerebrospinal fluid, 4.15 ± 1.2 units/ml,[29,30] which suggests that CDS originates in the central nervous system (TABLE 2).

Clonidine-Displacing Substance and Agmatine Are Two Different Compounds

Clonidine-displacing substance was defined by its ability to displace bound [^{3}H]clonidine from rat brain membranes, specificity, and apparent molecular mass.[2–4]

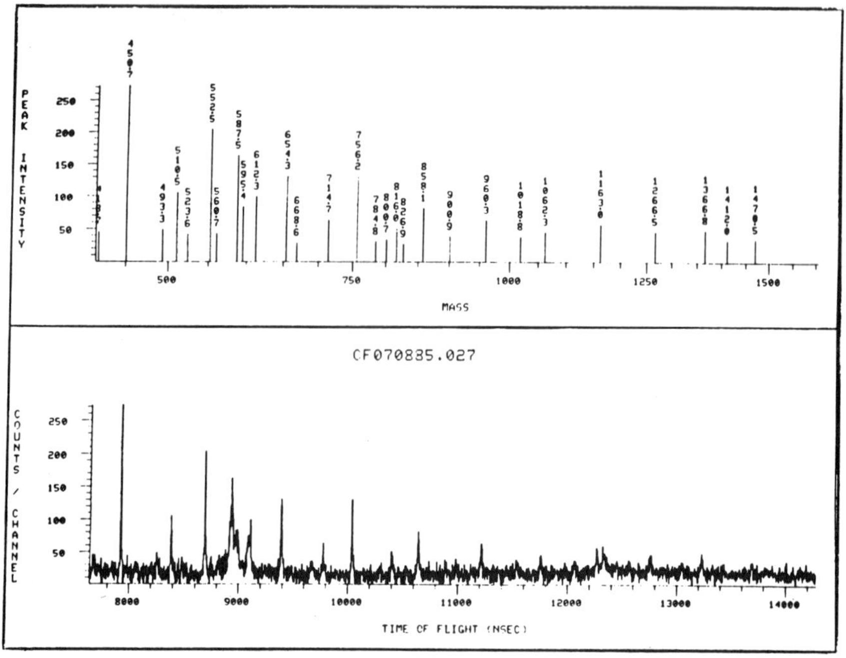

FIGURE 2. Plasma desorption mass spectrometry of CDS. Clonidine-displacing substance eluted from a C_8 reverse phase column[5] was dissolved in methanol, thinly sprayed on aluminized mylar foil, and bombarded with high energy ions (fission fragments) from a californium source (^{252}Cf).[28] Molecular weight measured in field desorption was 587.8 ± 2. The second and smaller peak at m/z1174 is its corresponding dimer. The sharp peaks are the typical background of the mylar foil.

Characterization of its physical and molecular properties showed that CDS is not a peptide; it is heat and acid (pH 2.0) resistant, ninhydrin and fluorescamine negative, hydrophobic by way of its retention time on a C_8 reverse-phase column (FIG. 1),[3,5] and partitions into the aqueous phase in mixtures of either chloroform and water or dioxane and water. These properties were contrasted in a recent paper by Li *et al.*[26] which claims to identify CDS as agmatine. Agmatine is ninhydrin positive by virtue of its primary amino group and is not retained on a reverse-phase column.

Clonidine-displacing substance specificity was determined in receptor binding studies in rat brain membranes, measuring its affinity for α_1-adrenergic receptors using binding of [^{3}H]prazosin to rat brain membranes and β-adrenergic receptors using binding of [^{125}I]hydroxybenzyl pindolol to turkey erythrocyte membranes. Clonidine-displacing substance showed no affinity for either α_1-adrenergic or β-adrenergic receptors.[3,4] The molecular mass of CDS is 587.8 ± 2 (FIG. 2; TABLE 3), whereas agmatine has a low molecular mass of 130 daltons.[26] It displaces tritiated *p*-aminoclonidine from rat brain membranes with ~100-fold lower affinity (μM) and is 100-fold more abundant in bovine brain than is CDS (TABLE 3). The differences between CDS and agmatine are also evident from their biological properties (TABLE

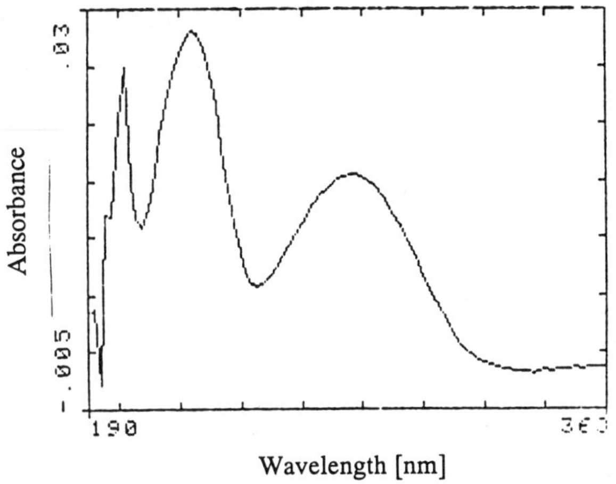

FIGURE 3. Ultraviolet spectrum of the active fraction of pure CDS obtained at the final HPLC step. Peak assignment was given at 224 and 276 nm.

3). During the purification procedure, the amount of CDS recovered is so low that it is spectrally undetectable, yet the small amounts proved potent in receptor binding assays as well as in pharmacological assays (e.g., inhibition of the twitch response in rat vas deferens,[9] potentiation of human platelet aggregation,[10] or increasing mean arterial pressure upon cerebral stereotactic injections[6]). None of these properties was shown for agmatine.

In receptor binding assays the affinity of CDS for α_2-adrenergic receptors in rat brain[3] or human platelets[10] (10 nM) is 100-fold higher than that of agmatine. Similarly, CDS affinity for imidazoline receptors in rabbit kidney (20 nM),[16,31] rat liver (40 nM),[18] and human placenta (23 nM)[20] is ~100-fold higher than that for agmatine.[26]

IMIDAZOLINE SITES IN HUMAN PLACENTA AND RAT LIVER MEMBRANES DISPLAY A SIMILAR AFFINITY PROFILE WITH VARIOUS DRUGS AND CDS

Imidazoline receptor sites (IRs) are present in many tissues, and their function is still not fully understood. As in the α_2-adrenergic receptor family, of which four

TABLE 2. Clonidine-Displacing Substance Abundance in Various Tissues[a]

Tissue	Units
Bovine brain	1–2.5 unit/gr
Human serum	1.4 ± 0.3 unit/ml
Human cerebrospinal fluid	4.14 ± 1.2 unit/ml

[a] One unit of activity is defined as the amount needed to displace 50% of specifically bound [^{3}H]clonidine (2 nM) from rat brain membranes (250 mg protein) in a volume of 250 ml of Tris-HCl buffer, pH 7.5. One unit is approximately 5 pmol (see TABLE 3).

TABLE 3. Physical Properties and Specificity of Clonidine-Displacing Substance (CDS) and Agmatine[a]

Property	CDS	Agmatine[a]
Abundance/brain	3–4 ng/g	200–400 ng/g
Absorbance	224, 276 nm (aromatic)	200 nm (aliphatic)
Ninhydrin	Negative	Positive
Molecular mass	587.8 ± 2 daltons	130 daltons
Affinity for α_2AR	10–12 nM	4 μM
Affinity for IR	20–40 nM	1 μM
Retention time (RP18)	26 minutes	Not retained
Unit of activity[b]	5 pmol	4 μM

Abbreviations: α_2AR = α_2-adrenergic receptors; IR = imidazoline receptors.

[a] Taken from ref. 26.

[b] An estimate for 1 unit of CDS is obtained from the final HPLC separation[5] where the single active peak contains 2,000 units and the amount is less than 5 μg. As the molecular weight is 587.8, it can be calculated that 1 unit of activity corresponds to ~5 pmol.

types have already been cloned and sequenced, lack of a specific functional role makes the four different receptor subtypes difficult to resolve. Even though autoradiographic studies clearly showed that IRs in the CNS are distinct from α_2-adrenergic receptors,[22,23,32] it appears that IR and α_2-adrenergic receptors bear structural similarities because the synthetic ligands belonging to the guanido-imidazoline hypotensive drugs, which act at α_2-adrenergic receptor sites, bind with high affinity to IRs. Imidazoline receptor sites are classified into two distinct types, IR-I_1 with its high affinity for clonidine and IR-I_2 with its low affinity for clonidine. The presence of both IR-I_1 and IR-I_2 in the same cell and their similar pharmacological specificities interfere with assigning functional and specificity for the two imidazoline receptors. Rat liver, human liver cells,[18,33] human placenta, and human placental cells,[20] are unique because they display high IR-I_2 density and are devoid of either IR-I_1 or α_2-adrenergic receptors. Hence, specificity of IR-I_2 is achieved by direct binding of [^{3}H]idazoxan to the single receptor population present in these tissues.[18,20] The functional consequences of interaction between idazoxan and IR-I_2 in placenta and in liver cells are likely to be complex. In this regard, there is recent evidence of a partial sequence of an imidazoline binding site (IR-I_2) in rabbit kidney which is highly homologous to that of monoamine oxidase present on mitochondrial membranes (Parini *et al.*, this volume). These results open a new prospect for sites on plasma membrane which are pharmacologically identical to sites on mitochondrial membrane. It is interesting to speculate that both liver and placenta display the highest density of IR-I_2 binding site and are rich in monoamine oxidase activity.

The affinities of imidazoline and guanidine compounds for human placenta and rat liver membranes, as determined in [^{3}H]idazoxan displacement assays, were correlated, as shown in FIGURE 4. The various drugs display a similar order of potency in both tissues, confirming their interaction with the same IR-I_2 receptors. On the other hand, the affinities of these various drugs for the α_2-adrenergic receptors in

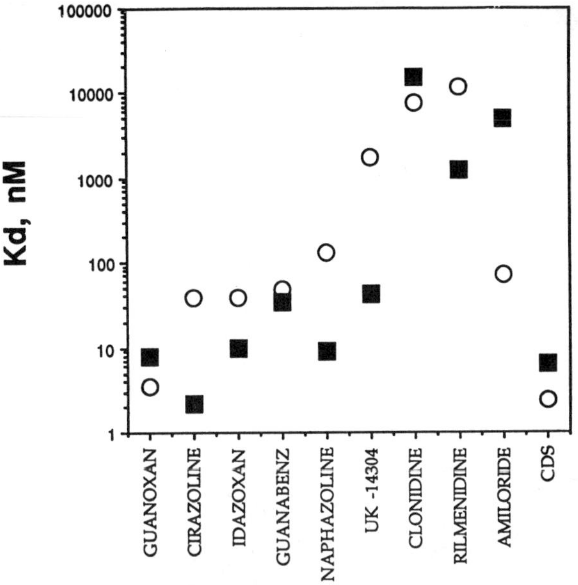

FIGURE 4. The affinity of various ligands for imidazoline sites (IR-I_2) in human placenta and in rat liver membranes. The affinity of various guanidino- and imidazoline ligands which interact with IR-I_2 was compared by plotting the K_d values in decreasing order of potency. These values were obtained in displacement assays using 5 nM of [^{3}H]idazoxan and 10 μM of cirazoline for determining nonspecific binding.

human platelets, as determined by [^{3}H]rauwolscine binding, and human placenta using [^{3}H]idazoxan, clearly show a different order of potency for α_2-adrenergic receptors and IR-I_2 (FIG. 5). Clonidine-displacing substance, like the many imidazoline/guanidine drugs, recognizes both IR-I_1 and IR-I_2 as well as the α_2-adrenergic receptor.[16,18,20,30]

Until a functional assay for these sites is established and CDS structure finally determined, its role at either one of these sites remains elusive. A more detailed understanding of the role of CDS in hypertension awaits the characterization of the functional receptors interacting with CDS at a target nucleus at the CNS.

SUMMARY

An endogenous small molecular mass compound, termed clonidine-displacing substance (CDS), has been isolated and purified from bovine brain. The estimated level of CDS in bovine brain is 400-1,000 units/wet brain, with 1 unit of activity calculated to be approximately 1-2 ng. It is present in human serum, urine, and cerebrospinal fluid. The isolation procedure consists of initial aqueous and methanolic extractions followed by a series of HPLC chromatography steps (reverse phase and TSK sizing columns). The reverse-phase chromatography of CDS extracted under

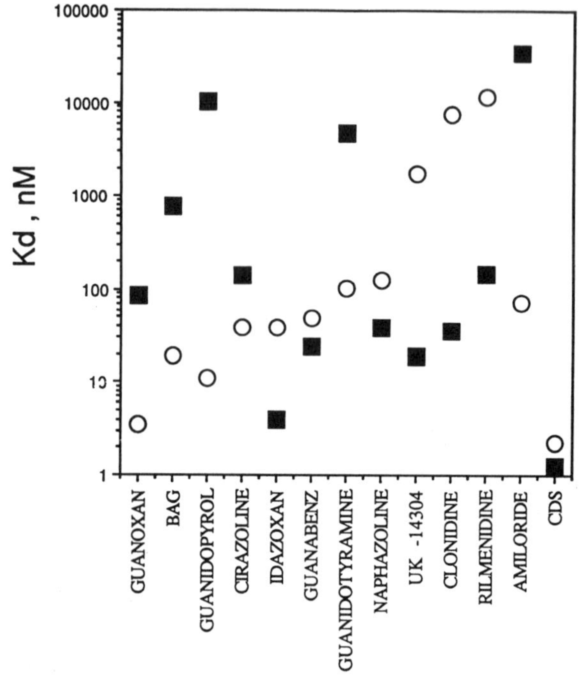

FIGURE 5. The affinities of various ligands that interact with imidazoline sites in human placenta (IR-I_2) and α_2-adrenergic receptors in human platelet membranes were compared. K_d values were determined in displacement assays using 5 nM [^{3}H]idazoxan binding and 10 μM cirazoline for nonspecific binding and plotted in decreasing order for IR-I_2. Binding to α_2-adrenergic receptors was carried out using 5 nM [^{3}H]rauwolscine and 10 μM epinephrine for nonspecific binding.

identical conditions from bovine brain and human serum show similar retention times. The final chromatography step gives a single active peak with a distinct ultraviolet spectrum, a single molecular peak m/z 587.8 ± 2 in plasma desorption mass spectrometry (PDMS), and a unique pharmacological and physiological profile. Clonidine-displacing substance does not partition into organic solvents and it is ninhydrin and fluorescamine negative.

All of these molecular properties clearly distinguish CDS from agmatine, an endogenous 130-dalton compound of far greater abundance which displays lower affinity for *p*-aminoclonidine-labeled sites in rat brain membranes. The ultraviolet spectrum of CDS consists of two aromatic peaks at 224 and 276 nm, whereas agmatine is an aliphatic substance with no ultraviolet absorbance.

Like many antihypertensive drugs of the guanidine and imidazoline family of compounds, CDS recognizes α_2-adrenergic receptors, clonidine sites (IR-I_1), and imidazoline sites (IR-I_2). A good correlation exists between the affinities of various imidazoline/guanidine type ligands for IR-I_2 in both human placenta and rat liver membranes which can be accurately determined because both tissues lack IR-I_1 and

α_2-adrenergic receptors. There is no correlation in the affinities of these ligands for IR-I_2 of human-placental versus α_2-adrenergic receptors of human platelets. By uncovering the role of CDS in the central nervous system we will be able to understand the coupling of IRs to neurotransmission and, in turn, to changes in arterial pressure.

REFERENCES

1. DVANE, W. A., L. HANUS, A. BREUER, R. G. PERTWEE, L. A. STEVENSON, G. GRIFFIN, D. GIBSON, A. MANDELBAUM, A. ETINGER & R. MECHOULAM. 1992. Isolation and structure of a brain constituent that binds to the cannabinoid receptor. Science **256:** 1946-1949.
2. ATLAS, D. & Y. BURSTEIN. 1984. Isolation of an endogenous clonidine displacing substance from rat brain. FEBS Lett. **170:** 887-391.
3. ATLAS, D. & Y. BURSTEIN. 1984. Isolation and partial purification of a clonidine displacing substance. Eur. J. Biochem. **144:** 287-293.
4. ATLAS, D., S. DIAMANT, H. M. FALES & L. PANNELL. 1987. The brain's own clonidine; Purification and characterization of an endogenous clonidine displacing substance. J. Cardiovasc. Pharmacol. **10:** 5122-5127.
5. ATLAS, D. 1994. Identifying clonidine-displacing substance. Science **266:** 462-463.
6. BOUSQUET, P., J. FELDMAN & D. ATLAS. 1986. An endogenous noncatecholamine Clonidine antagonist increases mean arterial pressure. Eur. J. Pharmacol. **124:** 167-170.
7. BOUSQUET, P., J. FELDMAN & D. ATLAS. 1987. Central cardiovascular effects of a non-catecholamine endogenous ligand for clonidine receptors. J. Cardivasc. Pharmacol. **10:** 5167-5171.
8. MEELEY, M. P., P. R. ERNSBERGER, A. R. GRANATA & D. J. REIS. 1986. An endogenous displacing substance from bovine brain: Receptor binding and hypotensive actions at the ventrolateral medulla Life Sci. **38:** 1119-1126.
9. DIAMANT, S. & D. ATLAS. 1986. An endogenous brain substance CDS (clonidine displacing substance) inhibits the twitch response in rat vas deferens. Biochem. Biophys. Res. Commun. **34:** 184-187.
10. DIAMANT, S., A. ELDOR & D. ATLAS. 1987. A low molecular weight brain substance interacts, similarly to clonidine with α_2-adrenergic receptors of human platelets. Eur. J. Pharmacol. **144:** 247-255.
11. SYNETOS, D., V. G. MANOLOPOULOS, D. ATLAS & E. PIPILI-SYNETOS. 1991. Human plasma-derived material with clonidine displacing substance (CDS)-like properties contracts the isolated rat aorta. J. Auton. Pharmacol. **11:** 343-351.
12. ATLAS, D. 1990. Clonidine displacing substance (CDS) and its putative imidazoline receptor-new leads for further divergence of α_2-adrenergic agonists. Biochem. Pharmacol. **41:** 1541-1549.
13. BOUSQUET, P., J. FELDMAN & S. SCHWARTZ. 1984. Central cardiovascular effects of alpha adrenergic drugs: Differences between catecholamines and imidazolines. J. Pharmacol. Exp. Therap. **230:** 232-236.
14. ERNSBERGER, P., M. P. MEELEY, J. J. MANN & D. J. REIS. 1987. Clonidine binds to imidazol binding sites as well as α_2 adrenoceptors in the ventrolateral medulla. Eur. J. Pharmacol. **134:** 1-13.
15. ERNSBERGER, P., M. P. MEELEY & D. J. REIS. 1988. An endogenous substance with clonidine like properties: Selective binding to imidazoline sites in the ventrolateral medulla. Brain Res. **441:** 309-318.
16. PARINI, A., I. COUPRY, R. M. GRAHAM, I. UZIELLI, D. ATLAS & S. M. LANIER. 1989. Characterization of an imidazoline/guanidinium receptive site distinct from the α_2 adrenergic receptor. J. Biol. Chem. **264:** 11874-11878.
17. WIKBERG, J. E. S. & S. UHLEN. 1990. Further characterization of the guinea pig cerebral cortex idazoxan receptor: Solubilization, distinction from the imidazole site, and demon-

stration of cirazoline as an idazoxan receptor-selective drug. J. Neurochem **55:** 192-203.

18. ZONNENSCHEIN, R., S. DIAMANT & D. ATLAS. 1990. Imidazoline receptors in rat liver cells: A novel receptor or a subtype of α_2-adrenergic receptor? Eur. J. Pharmacol **190:** 203-215.
19. MICHEL, M. C., J. W. REAGAN, M. A. GERHARDT, R. R. NEUBIG, P. A. INSEL & H. J. MUTULSKY. 1990. Nonadrenergic [^{3}H]Idazoxan binding sites are physically distinct from α_2 adrenergic receptors. Mol. Pharmacol. **37:** 65-68.
20. DIAMANT, S., T. ELDAR-GEVA & D. ATLAS. 1992. Imidazoline binding sites in human placenta: Evidence for heterogeneity and a search for physiological function. Br. J. Pharmacol. **106:** 101-108.
21. MOLDERINGS, G. J., D. MOURA, K. FINK, H. BONISH & M. GOTHERT. 1993. Binding of [^{3}H]clonidine to I_1-imidazoline sites in bovine adrenal medullary membranes. Naunyn-Schmiedeberg's Arch. Pharmacol. **348:** 70-76.
22. KAMISAKI, Y., T. ISHIKAWA, Y. TAKAO, H. OMODANI, N. KUNO & T. ITOH. 1990. Binding of [^{3}H]*p*-aminoclonodine to two sites, α_2 adrenoceptors and imidazoline binding sites in rat brain. Brain Res. **514:** 15-21.
23. MALLARD, N. J., A. L. HUDSON & D. J. NUTT. 1992. Characterization and autoradiographical localization of non-adrenoceptor idazoxan binding sites in rat brain. Br. J. Pharmacol. **106:** 1019-1027.
24. LANIER, S. M., B. IVKOVITC, I. SINGH, J. L. NEUMEYER & V. BAKTAVACHALAM. 1993. Visualization of multiple imidazoline/guanidinium-receptive sites. J. Biol. Chem. **268:** 16047-16051.
25. MACKINNON, A. C., M. SPEDDING & C. M. BROWN. 1994. α_2 adrenoceptors: More subtypes but fewer functional differences. Trends. Pharmacol. Sci. **15:** 119-123.
26. LI, G., S. REGUNATHAN, C. J. BARROW, J. ESHRAGHI, R. COOPER & D. J. REIS. 1994. Agmatine: An endogenous clonidine-displacing substance in the brain. Science **263:** 966-969.
27. KREISBERG, G. A., S. DIAMANT, Y. DIAMANT & D. ATLAS. 1987. Raised levels of an endogenous non-adrenergic substance in serum of pregnancy-induced hypertension patients. Isr. J. Med. Sci. **23:** 1194-1197.
28. SUNDQUIST, B. & R. D. MACFARLENE. 1985. Plasma desorption mass spectrometry. Mass Spectrom. Rev. **4:** 421-460.
29. GOLDBERG-STERN, H., D. ATLAS, L. SCHWARTZ, A. ACHIRON, I. ZIV, R. DJALDETTI, Y. ZOLDAN & E. MELAMED. 1992. Detection and measurement of an endogenous clonidine-displacing-substance (CDS). Neurology **42** (Suppl. 3): 143.
30. GOLDBERG-STERN, H., D. ATLAS, L. SCHWARTZ, A. ACHIRON, I. ZIV, R. DJALDETTI, Y. ZOLDAN & E. MELAMED. 1993. Detection and measurement of an endogenous clonidine-displacing-substance (CDS) in human cerebrospinal fluid (CSF). Brain Res. **601:** 325-328.
31. COUPRY, I., D. ATLAS, R. A. PODEVIN, I. UZIELLI & A. PARINI. 1990. Imidazoline guanidine receptive site in renal proximal tubule: Asymmetric distribution, regulation by cations and interaction with an endogenous clonidine displacing substance. J. Pharmacol. Exp. Ther. **252:** 293-299.
32. BOYAJIAN, C. L. & F. M. LESLIE. 1987. Pharmacological evidence for α_2-adrenoceptor heterogeneity: Differential binding properties of [^{3}H] rauwolscine and [^{3}H] idazoxan in rat brain. J. Pharmacol. Exp. Ther. **241:** 1092-1099.
33. ATLAS, D., S. DIAMANT & R. ZONNENSCHEIN. 1992. Is imidazoline site a unique receptor? A correlation with clonidine-displacing-substance activity. Am. J. Hypertens. **5:** 83-90S.

Agmatine Is Synthesized by a Mitochondrial Arginine Decarboxylase in Rat Brain

GEN LI, SOUNDARARAJAN REGUNATHAN, AND
DONALD J. REIS

Division of Neurobiology
Department of Neurology and Neuroscience
Cornell University Medical College
New York, New York 10021

We recently discovered that agmatine (decarboxylated arginine) is contained in bovine brain.[1] Moreover, the amine, never before detected in mammals,[2] has properties of a clonidine-displacing substance (CDS); it binds with reasonable affinities (K_i: 1-4 μM) to α_2-adrenergic receptors and to all classes of imidazoline (I-) receptors.[1] In bacteria, fungi, some parasites, and marine animals,[2–5] agmatine is synthesized from arginine by a soluble enzyme, arginine decarboxylase (ADC). Our finding that ADC was expressed in rat brain[1] was the first demonstration of the enzyme in mammals. Moreover, it indicated that mammalian agmatine is a product of local biosynthesis and not attributable to dietary or enteric bacterial sources.

In *Escherichia coli* there are two soluble cytosolic isoforms: one is constitutive or biosynthetic,[6] the other inducible or biodegradative.[2] A third form of ADC which is membrane associated has been found in nematode *C. elegans.*[7] We sought to determine if mammalian brain ADC was similar to any of these other forms.

MATERIALS AND METHODS

Rat brains were homogenized in HEPES-sucrose buffer (5 mM HEPES, pH 7.4; 2 mM DTT; 0.5 mM PMSF; 0.2 mM EDTA) with a Teflon-glass homogenizer and centrifuged at 1,000 × *g* for 10 minutes. The pellet (P1) containing the nuclear fraction was saved for enzyme assays. The supernatant was centrifuged at 12,000 × *g* to produce a crude mitochondrial/synaptosomal pellet (P2). The resulting supernatant was centrifuged at 30,000 × *g* for 20 minutes generating a pellet (P3) containing the plasma/microsomal membrane fraction. This was used for enzyme assays. Separate mitochondrial or synaptosomal fractions were isolated from the P2 pellet by Percoll density gradients.[8] Membranes from each fraction were used for enzyme assays.

Except for studies on enzyme distribution in each subcellular compartment, ADC activity was measured in the P2 pellet which contained mitochondrial/synaptosomal membranes. Arginine decarboxylase was assayed by the method of Wu and Morris[9] measuring conversion of ^{14}C-arginine to $^{14}CO_2$. In brief, the tissue pellet was suspended in pre-chilled subcellular fractionation buffer without sucrose, homogenized by a Polytron (Brinkman, setting 6 for 2 × 15 s), and centrifuged at 100,000 × *g* for

15 minutes. The enzyme reaction was performed in glass tubes with a center well inserted into a tightly closed rubber stopper. The center wells contain strips of filter paper moistened with benzethionium or methylbenzethionium hydroxide to trap the $^{14}CO_2$ produced. The reaction mixture consisted of 500 μl of assay buffer: 10 mM Tris-HCl, pH 8.25, at 30°C; 1 mM DTT; 0.5 mM PMSF; 0.2 mM EDTA; 1 mM $MgSO_4$; 0.2 mM L-arginine and 0.4 μCi of L-[1-^{14}C]-arginine, specific activity 55 mCi/mmol. It was incubated at 30°C for 1 hour in a shaking water bath and the reaction terminated by the addition of 100 μl of 40% trichloroacetic acid injected through the rubber stopper. After further incubation for 20 minutes at 37°C, the filter paper strips were transferred to scintillation vials and radioactivity was determined by liquid scintillation counting. Verification of the product was obtained independently by demonstrating generation of [^{3}H]agmatine from [^{3}H]arginine by HPLC (data not shown).

Ornithine decarboxylase activity was measured by modification of a similar CO_2-trapping method[7] using 0.2 mM L-ornithine and L-[1-^{14}C]-ornithine (specific activity 55 mCi/mmol) as substrates. Protein was measured with a commercial protein assay kit (Bio-Rad) using bovine serum albumin (BSA) as a standard.

RESULTS

Stability. Arginine decarboxylase is labile. Hence, all enzyme activity was lost within a few hours in whole brain when stored on ice, whereas about 50% of activity was lost overnight when stored at 4°C as a membrane suspension. The enzyme was sensitive to detergents. At concentrations of 1%, Triton X-100, NP 40, or CHAPS totally abolished ADC activity, and even at 0.1% concentration, enzyme activity was substantially decreased.

Subcellular Distribution. The subcellular distribution of enzyme was measured after separation into nuclear, mitochondrial/synaptosomal, and plasma membrane fractions. The highest activity, 9.69 nmol CO_2/h/mg protein, or ~89% of total activity was restricted to the mitochondrial/synaptosomal membrane fraction. The activity of the plasma membrane (0.77 nmol. CO_2/h/mg protein) and the nuclear pellet (<0.48 nmol CO_2/h/mg protein) was ~7% and ~4% of total activity, respectively. After further fractionation of the mitochondrial/synaptosomal fraction 80% of total activity (88.67 nmol CO_2/h/mg protein) appeared in mitochondrial membranes. Much of the remaining 20% of ADC activity within the synaptosomal membrane (20.22 nmol CO_2/h/mg protein) was probably attributable to mitochondria, because these organelles are abundant in synaptosomes. Thus, ADC appears to be highly, if not exclusively, associated with mitochondrial membrane.

Kinetics. The activity of ADC was linear with respect to reaction time to 90 minutes and with protein concentrations from 0.5–40 mg/ml. Under optimal reaction conditions, the K_m for arginine, as determined by double-reciprocal (Lineweaver-Burk) plots, was 0.75 mM. The V_{max} was 2.22 nmol/h/mg protein.

pH and Temperature Optima. The optimal pH for brain ADC activity was 8.25, close to that of the biosynthetic ADC of *E. coli.*[10] By contrast to other forms of ADC, the temperature optimum was 30°C. At 37°C, the enzyme was only about one third as active as at 30°C.

TABLE 1. Effect of Substrate Analogs and Polyamines on Arginine Decarboxylase Activity[a]

Compound	Relative Activity (%)
None	100
Amino acids:	
L-Ornithine	18
L-Lysine	59.8
L-Glutamine	79.4
L-Histidine	93.7
D-Arginine	94.7
D-Ornithine	92.3
DFMO	98.5
Polyamines:	
Spermine	18.8
Spermidine	29.4
Agmatine	66.5
Putrescine	66.6
Nitric oxide synthase inhibitors:	
N^G-nitroarginine	79.6
N^G-methylarginine	78.8

[a] The activity of ADC was measured in rat brain mitochondrial membrane for 1 hour at 37°C. Control activity was 88.6 nmol CO_2/h/mg protein. All compounds were tested at 1 mM against the substrate concentration of 0.2 mM arginine.

Inhibition by Substrate Analogs. To determine the effect of substrate analogs on ADC activity, several amino acids were tested for their ability to inhibit the enzyme. As shown in TABLE 1, although ornithine inhibited ADC activity by 82% at 1 mM concentration, lysine inhibited the enzyme by 40%. Therefore, rat brain ADC has properties similar to those of ADC of *C. elegans*[7] but not of *E. coli* and can be considered arginine/ornithine decarboxylase. However, difluoromethylornithine (DFMO), a universal and irreversible inhibitor for all forms of ornithine decarboxylase,[11] could not inhibit rat brain ADC, indicating the distinction between the two forms of the enzyme.

Because in bacteria agmatine is a precursor of polyamines,[12] the effect of several polyamines on ADC activity was examined. Arginine decarboxylase activity was inhibited, in rank order by: spermine > spermidine > agmatine > putrescine. At a concentration of 1 mM, which was fivefold higher than the substrate concentration (TABLE 1), spermidine and spermine blocked 81% and 70% activity, whereas agmatine and putrescine blocked only one third of the enzyme activity. Arginine decarboxylase was not inhibited by the nitric oxide synthase inhibitors N^G-nitroarginine and N^G-methylarginine.

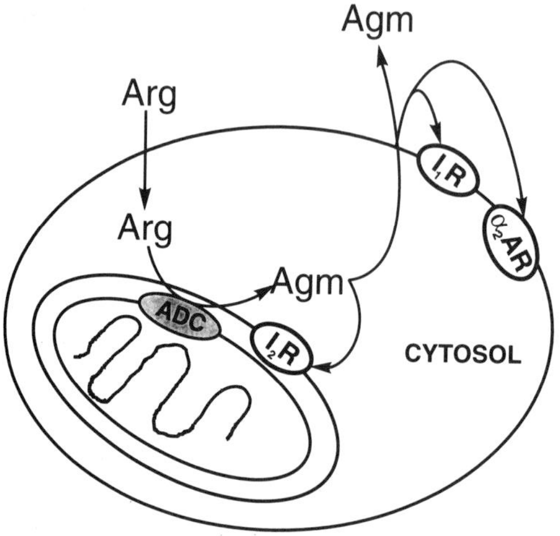

FIGURE 1. Schematic outline of the intracellular biosynthetic pathway for agmatine and its relationship to imidazoline (I-) and α_2-adrenergic receptors. See text for details.

DISCUSSION

This study confirms our previous observation that the rat brain expresses an enzyme that can synthesize agmatine and CO_2 from L-arginine and hence by definition represents an ADC.[1] The enzyme differs from the soluble bacterial forms in that it is associated with mitochondrial membranes. Like the constitutive form of *E. coli*,[10] it is optimally active at a high pH (8.25), but unlike that enzyme it does not require Mg^{2+} (unpublished observations).

Because rat brain ADC also uses ornithine as a substrate (K_m: 0.25 mM), it can be considered an arginine/ornithine decarboxylase. It differs, however, from ODC, a soluble cytoplasmic enzyme, which is irreversibly inhibited by DFMO. By contrast, DFMO is inactive against rat brain ADC. Rat brain ADC is most similar to a membrane-associated ODC isolated from the nematode C. elegans[7] which also uses arginine and ornithine as substrates; however, it is not known if the nematode enzyme is mitochondrial. However, the two enzymes also differ with respect to K_m, pH, and temperature optima and by the fact that the nematode enzyme is inhibited by DFMO.[7] Thus, rat brain ADC appears unique.

The localization of ADC to mitochondrial membranes is of special interest in view of the fact that imidazoline receptors of the I_2 subclass are also localized there.[13,14] FIGURE 1 schematically indicates the relation between the substrate, L-arginine, ADC, agmatine, and I- and α_2-adrenergic receptors in a typical cell. It emphasizes that the substrate L-arginine, which enters the cell by facilitated transport, is converted to agmatine on the mitochondrial membrane. Agmatine, so synthesized, can either bind to I_2 receptors on the mitochondrion, to α_2-adrenergic and possibly I_1 receptors on plasma membranes, or it can be released extracellularly, as evidenced by the presence of agmatine in serum.[14] Within the cell, agmatine may also regulate its own synthesis by feedback inhibition of ADC.

The data therefore suggests that ADC is a component of a complex and highly regulated system within cells generating agmatine, an amine that may act within and beyond the cell as an autocrine, paracrine, and hormonal messenger. We have argued[16] that agmatine may also be a novel neurotransmitter. If so, ADC undoubtedly participates in a potentially new neurotransmitter/neuromodulator network in the CNS.

CONCLUSIONS

1. Rat brain expresses ADC which differs from plant and bacterial forms by its localization to mitochondrial membranes and utilization of ornithine as well as arginine as substrate; yet, it is not a typical ODC as it is not cytosolic or inhibited by DFMO.

2. The colocalization of I_2 receptors and ADC to mitochondria suggests that agmatine may be an important intracellular message possibly regulating, through receptor mediation, its own biosynthesis. The presence of ADC and agmatine in mammalian brain raises questions of whether it might be an alternative pathway for polyamine biosynthesis in mammalian tissue.

REFERENCES

1. Li, G., S. Regunathan, C. J. Barrow, J. Eshraghi, R. Cooper & D. J. Reis. 1994. Science **263:** 966-969.
2. Tabor, C. W. & H. Tabor. 1984. Annu. Rev. Biochem. **53:** 749-790.
3. Yamamoto, S., H. Nakao, K. Yamasaki, K. Takashina, Y. Suemoto & S. Shinoda. 1988. Microbiol. Immunol. **32:** 675-687.
4. Majumder, S., J. J. Wirth, A. J. Bitoni, P. P. McCann & F. Kierszenbaum. 1992. J. Parasitol. **78:** 371-374.
5. Ramakrishna, S. & P. R. Adiga. 1975. Eur. J. Biochem. **59:** 377-386.
6. Buch, J. K. & S. M. Boyle. 1985. J. Bacteriol. **163:** 522-527.
7. Schaeffer, J. M. & M. R. Donatelli. 1990. Biochem. J. **270:** 599-604.
8. Nagy, A. & A. V. Delgado-Escueta. 1984. J. Neurochem. **43:** 1114-1123.
9. Wu, W. H. & D. R. Morris. 1973. J. Biol. Chem. **248:** 1687-1695.
10. Khan, A. J. & S. C. Minocha. 1989. Life Sci. **44:** 1215-1222.
11. Hunter, K. J., C. A. Strobos & A. H. Fairlamb. 1991. Mil. Biochem. Parasitol. **46:** 35-43.
12. Panagiotidis, C. A., S. Blackburn, K. B. Low & E. S. Canellakis. 1987. Proc. Natl. Acad. Sci. USA **84:** 4423-4427.
13. Tesson, F. & A. Parini. 1991. Eur. J. Pharmacol. **208:** 81-83.
14. Regunathan, S., M. P. Meeley & D. J. Reis. 1993. Biochem. Pharmacol. **45:** 1667-1675.
15. Raasch, W., S. Regunathan, G. Li & D. J. Reis. 1995. Proceedings of the II International Symposium on Imidazoline Receptors. Ann. N.Y. Acad. Sci. This volume.
16. Reis, D. J., G. Li & S. Regunathan. 1995. Proceedings of the II International Symposium on Imidazoline Receptors. Ann. N.Y. Acad. Sci. This volume.

Agmatine Is Widely and Unequally Distributed in Rat Organs

WALTER RAASCH,
SOUNDARARAJAN REGUNATHAN, GEN LI, AND
DONALD J. REIS

Division of Neurobiology
Department of Neurology and Neuroscience
Cornell University Medical College
New York, New York 10021

We recently discovered[1] that agmatine, a primary amine generated by decarboxylation of arginine by the enzyme arginine decarboxylase (ADC),[2,3] is stored in bovine brain. As such it is the first clonidine-displacing substance (CDS) isolated whose structure has been determined. (See Reis *et al.*, this volume,[4] for a review.) Like clonidine, agmatine binds to imidazoline (I-) receptors of both I_1 and I_2 subclass and to α_2-adrenergic receptors.[1] While prevalent in bacteria, plants, and invertebrates, neither agmatine nor ADC has been considered present in mammalian tissues.[2,3] Our observation[1] that brain also expresses ADC indicates that the amine is locally synthesized and is not a product of diet or enteric bacteria.

To determine if agmatine is present in rat brain and other organs, we developed a method to allow separation of agmatine from other amines in tissue extracts, permitting its measurement by reversed-phase high pressure liquid chromatography (RP-HPLC). We demonstrate that agmatine is present in this species and widely and unequally distributed throughout the body.

MATERIALS AND METHODS

Male Sprague-Dawley rats weighing about 300 g were stunned and decapitated and tissues were rapidly removed, coarsely minced, and homogenized in 4 volumes (v/w) of 50 mm phosphate buffer, pH 7.5, containing 0.1% Triton X-100, using a tissue homogenizer (Polytron, Brinkman, setting 4, 20 s) and an ultrasonic cell disrupter (Branson sonifier, 25 Watt, 20 s). Protein was precipitated in an aliquot of 300 μl by the addition of 350 μl 0.6 N perchloric acid and 150 μl 0.1 N hydrochloric acid in methanol. After incubation of 30 minutes in an ice bath, samples were centrifuged (5 min, 6,000 *g*) and 650 μl of the supernatant was neutralized by adding 40 μl 5 N sodium hydroxide (NaOH). After derivatization by the addition of 360 μl OPA solution (25 mg *o*-phthalaldehyde [OPA], Sigma) dissolved in 1 ml methanol and mixed with 26.5 μl mercaptoethanol (Sigma) and 9 μl 0.2 M borate buffer, pH 9.2, the solution was loaded onto a Sep Pak Vac cartridge (C18, 1 ml, Millipore) and washed 4 times with 0.8 ml of a mixture of water, methanol, and acetonitrile (1/1/1, v/v/v, pH 13, adjusted with NaOH).

Agmatine was then eluted with 300 μl of a mixture of water, methanol, and acetonitrile (1/3/5, v/v/v, pH 3.3, adjusted with acetic acid). An aliquot of 20 μl was injected onto the HPLC column immediately after sample preparation. Agmatine dilutions were prepared in the same manner for standards or for validation of the whole procedure. The HPLC system consisted of an L 200 Intelligent pump (flow rate 1 ml/min), a manual rheodyne valve (7125), an F-1050 fluorescence spectrophotometer (excitation 338 nm; emission 425 nm), a D-2500 Chromato-Integrator (all from Hitachi Instruments), and a C18 stainless column (5 mm, 250 mm, IBM Instruments, Inc.).

The eluent was mixed with 460 μl 10 mM potassium dihydrogen phosphate in HPLC water (pH 4), 200 μl methanol, and 340 μl acetonitrile, containing 15 mM octylsulfate sodium salt (SOS, Sigma), then filtered through a 0.22 ml nylon membrane filter (Nalgene Company), and degassed by water-pump-vacuum. Agmatine solutions in water with a final concentration of 100 mM were frozen at –20°C until analysis, and a fresh aliquot of this solution was used daily.

RESULTS AND DISCUSSION

The detection of agmatine in tissue samples after derivatization with OPA requires a clean-up step because all primary amines in the homogenate will conjugate to OPA. The problem is illustrated in FIGURE 1, which depicts a chromatogram (Chromatogram A) showing the results of injection of a tissue sample after deproteinization, neutralization, and derivatization with OPA but without prior extraction. The chromatogram is overcrowded with unidentifiable peaks. An agmatine peak with a retention time of about 9.7 minutes as revealed in the standard (Chromatogram C) is lost in the background.

In initial trials we attempted to separate agmatine after deproteinization and derivatization by extraction with different organic liquids. Extraction with butanol or butylacetate resulted in fronting or tailing peaks. Although recovery with chloroform was very small, with ethylacetate or mixtures of ethylacetate with benzene, hexene, or isoamylalcohol the interfering peaks were still present. Moreover, one further disadvantage of such a liquid-liquid extraction is the consequent dilution and decrease in the detection limit. Because of the instability of OPA-agmatine, a reduction in sample volume is not possible. To overcome this disadvantage, the organic phase was re-extracted with some diluted acid, but unfortunately recovery was only about 30% by a clear chromatographic background.

We therefore extracted agmatine with cartridges. Chromatogram B in FIGURE 1 shows that by injecting the same amount of tissue sample onto the HPLC system by identical adjustments of detector and integrator procedure, the separation by solid-liquid extraction results in the appearance on tissue extract of a distinct agmatine peak with a retention time of 9.7 minutes. In a range of 150–400 ml elution buffer (see above), the recovery of agmatine was about 80%. Thus, a small amount of buffer is sufficient for quantitative elution of agmatine from the surface of the cartridge, meaning that a dilution of 300 μl tissue homogenate can be avoided and an even greater concentration of the sample is possible. Because a basic washing buffer was used, most of the other substances were washed from the surface of the C18 material

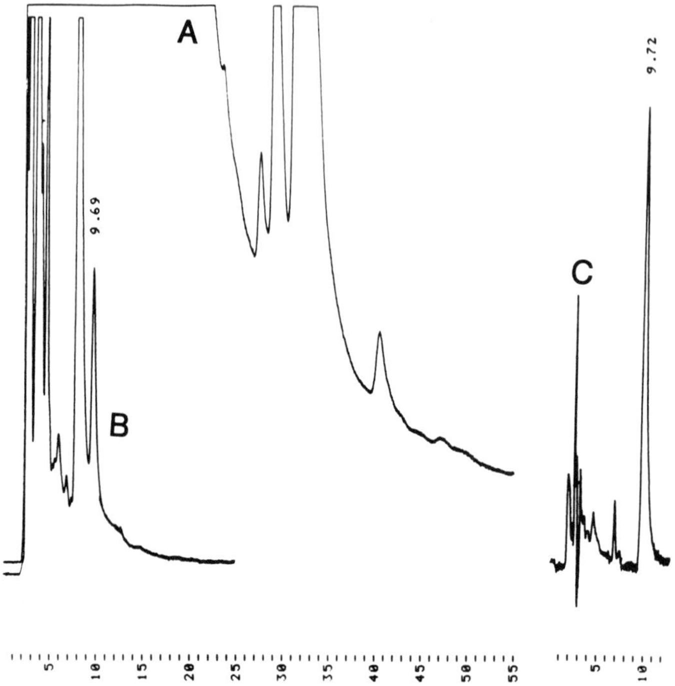

FIGURE 1. Chromatograms of an injection of a stomach sample (prepared as described in Methods) with (**B**) and without (**A**) the clean-up procedure over the C18 cartridge. Chromatogram (**C**) is a standard injection of agmatine (chromatographic system: Octadecyl (C18), 5 mm, 100 Å stainless column, IBM Instruments, Inc.). Eluent: 460 μl 10 mM potassium dihydrogen phosphate in HPLC-water (pH 4), 200 μl methanol, and 340 μl acetonitrile, containing 15 mM octylsulfate sodium salt (SOS, Sigma); the flow: 1 ml/min, flourimetric detection at 338 nm excitation, 425 nm emission.

of the cartridges where agmatine remains bound. By changing the pH to an acidic environment, agmatine, which is charged and hydrophilic, can be eluted. The overall linearity was determined in a range of 37-370 pM on column, the correlation coefficient was 0.99, and overall recovery was 85.2% ± 7.4%. The day-to-day variation was 8.6%.

Agmatine is present in the rat (Table 1), confirming its presence in mammals and indicating its expression in species other than the cow. It is present in various tissues and serum, indicating its wide and uneven distribution in the body. The highest concentration of amine is in the stomach followed by the aorta, small intestine, and large intestine. The concentration in the adrenal gland is approximately 10% and 3% of that in brain and stomach, respectively. Agmatine is also present in serum where it represents a concentration of 6% in brain.

As discussed by Reis *et al.*,[4] while agmatine is an endogenous ligand for I-receptors, it may not be the only one. Two other entities, whose structures have yet to be established, are known. One is classic CDS, the probably bioactive nonpeptide

TABLE 1. Distribution of Agmatine in Various Tissues of Rat

Tissue	Agmatine (ng/g net wt)		Clonidine-Displacing Substances[a] (RIA-Units/g wet wt)	
Stomach	71.00 ± 10.33	(5)	64.0 ± 17.0	(9)
Aorta	57.41 ± 12.74	(5)		
Small intestine	55.35 ± 9.39	(5)	3.0 ± 1.0	(6)
Large intestine	27.86 ± 6.73	(4)		
Spleen	17.38 ± 3.17	(4)		
Lung	10.23 ± 2.82	(5)	0.4 ± 0.2	(4)
Vas deferens	9.45 ± 2.08	(4)		
Adrenal	6.97 ± 3.29	(5)	114.0 ± 16.0	(8)
Kidney	6.45 ± 1.40	(5)	1.8 ± 0.4	(10)
Heart	6.03 ± 0.79	(5)	9.0 ± 3.0	(8)
Liver	5.63 ± 0.87	(5)	1.5 ± 0.3	(10)
Skeletal muscle	5.30 ± 0.72	(5)	0.5 ± 0.5	(5)
Brain	2.40 ± 0.60	(5)	9.0 ± 2.0	(10)
Testes	2.04 ± 0.22	(4)		
Plasma[b]	0.45 ± 0.05	(5)	7.0 ± 2.0	(11)

[a] Data from Meeley *et al.*[13]
[b] ng/ml or RIA-units/ml.
Values are means ± SEM of *n* experiments.

noncatecholamine material of molecular weight 586 Da discovered by Atlas and Burstein[5,6] and whose presence was confirmed by Meeley *et al.*[7] The other is an immunoreactive material, ir-CDS, which is recognized by antibodies to clonidine, *para*-aminoclonidine, or idazoxan[8–13] and whose distribution has been mapped by radioimmunoassay[13] (TABLE 1). Notable is the fact that the tissue distribution of ir-CDS differs from that of agmatine (TABLE 1). Thus, while agmatine is distributed in a rank order of stomach > small intestine >> adrenal > heart > brain >> plasma, ir-CDS is ordered: adrenal >> stomach >> heart > brain > >> plasma. The findings suggest that the two substances may be different. Interestingly, the concentrations of agmatine and ir-CDS in brain are only a small fraction of that in stomach.

The function of agmatine in various organs is unknown. In bacteria, agmatine is a precursor of putrescine and, hence, other polyamines forming a pathway for polyamine biosynthesis parallel to the ornithine-to-putrescine route.[3] Conceivably, an alternate pathway may exist in mammals, but the principal degradative enzyme for agmatine, agmatine uryl-hydrolase, has not been detected in mammals. The presence of agmatine in plasma suggests that it may be a hormone, whereas its distribution in brain has raised the possibility that it may also be a novel neurotransmitter.[1] Investigations aimed at establishing agmatine's function are presently underway.

REFERENCES

1. LI, G., S. REGUNATHAN, C. J. BARROW, J. ESHRAGHI, R. COOPER & D. J. REIS. 1994. Agmatine is an endogenous clonidine-displacing substance in brain. Science **263:** 966–969.

2. TABOR, C. W. & H. TABOR. 1984. Polyamines. Ann. Rev. Biochem. **53:** 749-790.
3. BENDER, D. A. 1985. Amino Acid Metabolism. 2nd Ed. John Wiley and Sons. New York.
4. REIS, D. J., G. LI & S. REGUNATHAN. 1995. Endogenous ligands for imidazoline receptors: Classic and immunoreactive clonidine displacing substance and agmatine. Proceedings of the II International Symposium on Imidazoline Receptors. Ann. N.Y. Acad. Sci. This volume.
5. ATLAS, D. & Y. BURSTEIN. 1984. Isolation and partial purification of a clonidine-displacing endogenous brain substance. Eur. J. Biochem. **144:** 287-293.
6. ATLAS, D. 1991. Clonidine-displacing substance (CDS) and its putative imidazoline receptor; new leads for further divergence of α_2-adrenergic receptor activity. Biochem. Pharmacol. **41:** 1541-1549.
7. MEELEY, M. P., P. R. ERNSBERGER, A. R. GRANATA & D. J. REIS. 1986. An endogenous clonidine-displacing substance from bovine brain: Receptor binding and hypotensive actions in the ventrolateral medulla. Life Sci. **38:** 1119-1126.
8. MEELEY, M. P., A. C. TOWLE, P. ERNSBERGER & D. J. REIS. 1988. A specific antiserum recognizes clonidine-displacing substance: Implications for the structure of the brain's own clonidine. Neurosci. Lett. **84:** 84-90.
9. MEELEY, M. P., A. C. TOWLE, P. ERNSBERGER, L. K. CHAR, P. M. MCCAULEY & D. J. REIS. 1989. Clonidine-specific antisera recognize an endogenous clonidine-displacing substance in brain. Hypertension **13:** 341-351.
10. DONTENWILL, M., G. BRICCA, A. MOLINES, A. BELCOURT & P. BOUSQUET. 1987. A polyclonal antibody raised against clonidine: A model for the specific imidazoline receptor. Eur. J. Pharmacol. **137:** 143-144.
11. DONTENWILL, M., G. BRICCA, A. MOLINES, P. BOUSQUET & A. BELCOURT. 1988. Production and characterization of anti-clonidine antibodies not cross-reacting with catecholamines. Eur. J. Pharmacol. **149:** 249-255.
12. WANG, H., D. MCGOWAN, S. REGUNATHAN, S. BRAMWELL, G. LI & D. J. REIS. 1993. Idazoxan-specific antiserum recognizes an endogenous clonidine-displacing substance (CDS) in human serum and CSF. Soc. Neurosci. Abstr. **19:** 1691.
13. MEELEY, M. P., M. L. HENSLEY, P. ERNSBERGER, D. FELSEN & D. J. REIS. 1992. Evidence for a bioactive clonidine-displacing substance in peripheral tissues and serum. Biochem. Pharmacol. **44:** 733-740.

Immunopurification of a Vasopressive Substance from Human Serum with an Anti-Clonidine Antibody

M. DONTENWILL, A. MOLINES, F. BENNAI, J. FELDMAN, G. BRICCA, A. BELCOURT, AND P. BOUSQUET

Laboratoire de Pharmacologie Cardiovasculaire et Rénale
CNRS URA 589
Faculté de médecine
Université Louis Pasteur
11 rue Humann
67000 Strasbourg, France

The hypotensive effect of clonidine, originating in the central nervous system, has been attributed to stimulation of imidazoline receptors different from adrenoceptors.[1] Binding experiments with [^{3}H]clonidine confirmed the existence of specific imidazoline binding sites insensitive to catecholamines both in the brain and in peripheral tissues from various species.[2,3] It can be assumed that an endogenous ligand exists for these receptors and that this ligand differs from catecholamines. A few years ago a substance extracted from bovine brain was described as capable of displacing [^{3}H]clonidine binding from both α_2-adrenoceptors and imidazoline binding sites.[4] That substance was named clonidine-displacing substance (CDS) and proved hypertensive in animal models.[5] Evidence for the presence of this substance in human serum was presented.[6,7] More recently, agmatine was proposed as an endogenous ligand for imidazoline receptors.[8]

To detect and purify the endogenous ligand for imidazoline receptors, we selected an immunological approach. Assuming that this ligand shares structural analogies with imidazolines, we developed anti-clonidine antibodies to test the eventuality of a cross-reaction between them and the ligand. These anti-clonidine antibodies exhibited similarities to imidazoline receptors in their ability to recognize imidazolines and not catecholamines. With anti-clonidine antibodies, immunoreactivity was detected in sera from normotensive subjects and hypertensive patients. In 30% of the hypertensive patients, immunoreactivity was greater than normal.[9] Various endogenous substances, including neurotransmitters (noradrenaline, adrenaline, histamine, dopamine, and serotonin), amino acids (glycine, arginine, histidine, and tyrosine), urea, or nucleotides, were unable to inhibit [^{3}H]*p*-aminoclonidine (PAC) binding to antibodies. In attempting to purify this so far unknown, endogenous circulating immunoreactive substance, we used the monoclonal anti-clonidine antibody in an immunoaffinity purification strategy.

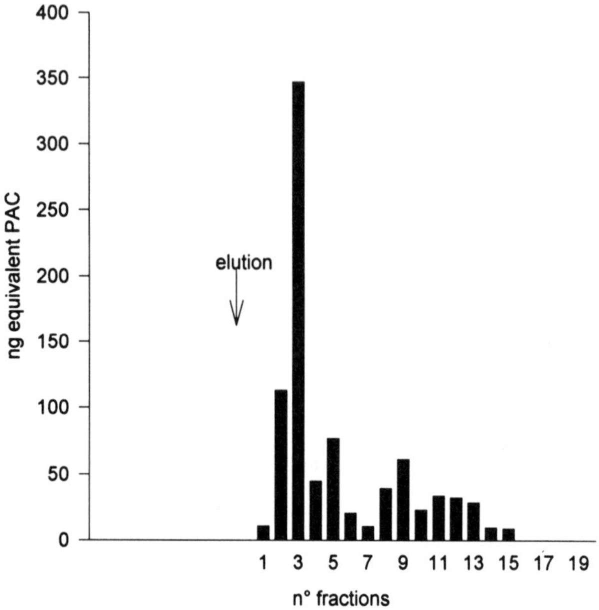

FIGURE 1. Extract of serum samples (pool of 200 subjects) were passed over an mFE7-aminolink column. After extensive washes with phosphate buffer (10 mM, pH 7.2), elution was performed with either an acidic buffer (glycine 0.1M, pH 2.5) or a basic buffer (triethylamine, pH 11.5). A representative experiment with the acidic buffer is shown. Immunoreactivity was assessed by an ELISA test in each fraction.

A monoclonal anti-clonidine antibody (mFE7) was obtained in large amount in ascitic fluids. After purification on an aminoclonidine Sepharose column, it was coupled to an aminolink support. Coupling efficiency was tested with [^{3}H]PAC and eluted with either 0.1 M glycine buffer, pH 2.5, or 0.1 M triethylamine buffer, pH 11.5.

Serum samples were prepared as described elsewhere.[9] Briefly, after ultrafiltration through Amicon membranes (MW cutoff of 1,000 Da) and lyophilization, the ultrafiltrate was methanol extracted three times. The immunoreactivity of the sample was assessed with an enzyme-linked immunosorbent assay (ELISA) developed with PAC-glutaraldehyde-bovine serum albumin (PAC-BSA) as coating antigen and mFE7 as antibody. Standard curves were established with PAC and expressed on a log-logit scale. Immunoreactivity of serum extracts was expressed in nanograms of PAC equivalent/sample. Extract was passed through the mFE7-aminolink column which was then extensively washed with 10 mM phosphate buffer, pH 7.2. Pre-elution buffer, consisting of either 10 mM phosphate buffer, pH 6.8, or 10 mM phosphate buffer, pH 8, was then applied followed, respectively, by either acidic elution with the glycine buffer or basic elution with the triethylamine buffer. Fractions of the acidic elution were neutralized with 1 M of phosphate buffer, pH 8, and their immunoreactivity immediately was tested, whereas fractions of the basic elution were evaporated to dryness and resuspended in distilled water before being tested.

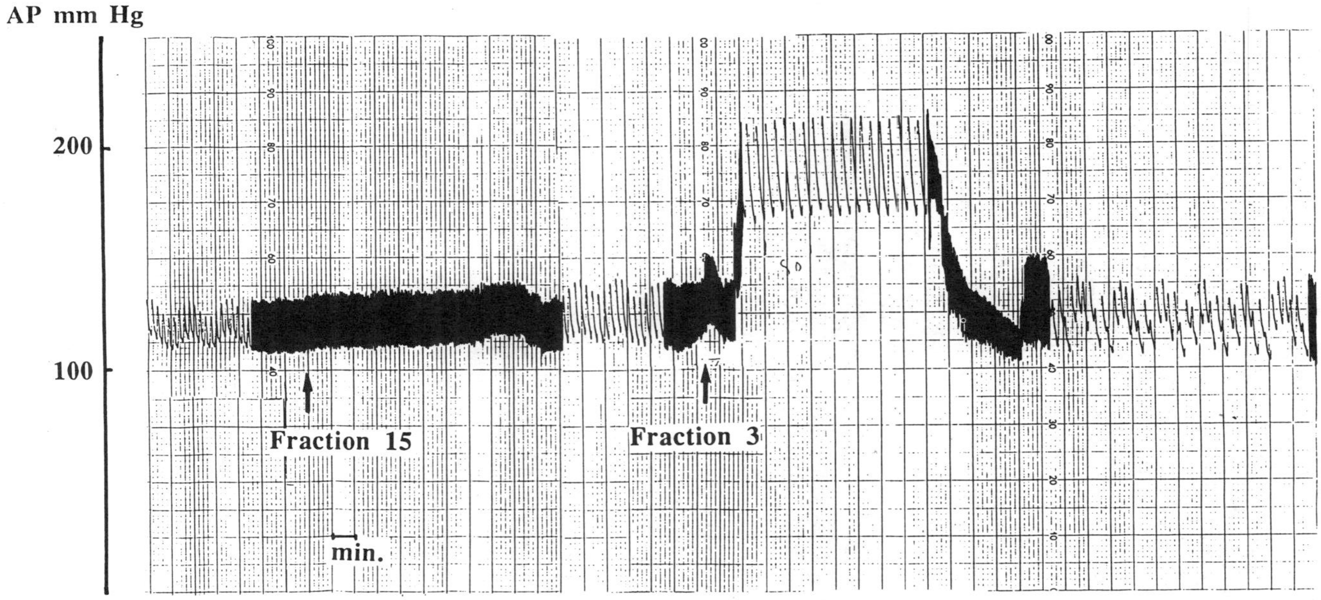

FIGURE 2. The elution fraction exhibiting the highest activity on the ELISA test (fraction 3) was injected into the NRL/RVLM area of anesthetized rabbits. A significant increase in arterial pressure was observed to be rapid and short lasting. Fraction 15, which does not contain a significant level of immunoreactivity, proved inactive when microinjected in the same area.

The ELISA test presented the same characteristics as the radioimmunoassay. The PAC proved capable of displacing mFE7 binding to PAC-BSA complexes with an IC_{50} of 286 ± 54.6 nM (mean ± SD, n = 15). When the serum extract was affinity chromatographed, the inhibitory effect on mFE7 binding to PAC-BSA complexes was undetectable in both effluent and subsequent washings including the pre-elution step, but peak activity was detectable in the elution fractions (FIG. 1). Similar results were obtained when either the acidic or the basic elution buffer was applied. No inhibitory activity was found in blank runs. Activity on arterial blood pressure of the eluted fractions was tested by microinjection of aliquots in the nucleus reticularis lateralis/rostro ventrolateral medulla (NRL/RFLM) area of anesthetized rabbits (FIG. 2). A rapid, short-lasting increase in arterial blood pressure was observed when the fraction containing the largest amount of immunoreactivity was injected and when a fraction devoid of immunoreactivity exhibited no cardiovascular effects. Similar results were obtained with both acidic and basic eluted fractions, thus ruling out a possible effect of the buffer. The presence of free amino-acids in active fractions was analyzed by ion exchange chromatography. Except for glycine in the acidic eluted fractions, no others amino acids were detected. Pools of the most active fractions from 10 runs were reextracted with methanol to remove glycine.

The immunopurified extract dilution curve, in the ELISA test, was not significantly different from the PAC standard curve based on parallel line analysis on log-logit curves. This supports the premise that the same epitope is recognized on both compounds.

In conclusion, with a monoclonal anti-clonidine antibody, we successfully immunopurified a circulating substance exhibiting hypertensive activity when injected in the NRL region of the rabbit. In this area, clonidine proved to be hypotensive by stimulation of imidazoline receptors. Whether this circulating substance is the endogenous ligand of these receptors remains to be established, and only its interaction with cerebral imidazoline-specific binding sites will give a definite answer. Structural analysis of this substance is under process in the laboratory.

REFERENCES

1. BOUSQUET, P., J. FELDMAN & J. SCHWARTZ. 1984. Central cardiovascular effects of alpha-adrenergic drugs. Differences between catecholamines and imidazolines. J. Pharmacol. Exp. Ther. **230:** 232-236.
2. BRICCA, G., H. GRENEY, J. ZHANG, M. DONTENWILL, J. STUTZMANN, A. BELCOURT & P. BOUSQUET. 1994. Human brain imidazoline receptors: Further characterization with [^{3}H]clonidine. Eur. J. Pharmacol. Mol. Pharm. Sect. **266:** 25-33.
3. HAMILTON, C. A., M. A. YAKUBU, E. JARDINE & J. L. REID. 1991. Imidazole binding sites in rabbit kidney and forebrain membranes. J. Auton. Pharmacol. **11:** 277-283.
4. ATLAS, D., S. DIAMANT, H. M. FALES & L. PANNELL. 1987. The brain's own clonidine: Purification and characterization of endogenous clonidine displacing substance from brain. J. Cardiovasc. Pharmacol. **10:** s122-s127.
5. BOUSQUET, P., J. FELDMAN & D. ATLAS. 1986. An endogenous, noncatecholamine clonidine antagonist increases mean arterial blood pressure. Eur. J. Pharmacol. **124:** 167-170.
6. KREISBERG, G. A., S. DIAMANT, Y. Z. DIAMANT & D. ATLAS. 1987. Raised levels of an endogenous nonadrenergic substance in the serum of pregnancy-induced hypertension patients. Isr. J. Med. Sci. **23:** 1194-1197.

7. Synetos, D., V. G. Manopoulos, D. Atlas & E. Pipili-Synetos. 1991. Human plasma derived material with clonidine displacing substance (CDS)-like properties contracts the isolated rat aorta. J. Auton. Pharmacol. **11:** 343-351.
8. Li, G., S. Regunathan, C. J. Barrow, J. Eshraghi, R. Cooper & D. J. Reis. 1994. Agmatine: An endogenous clonidine-displacing substance in the brain. Science **263:** 966-969.
9. Dontenwill, M., A. Molines, A. Verdun, G. Bricca, S. Laurent & P. Bousquet. 1993. A circulating substance cross-reacting with antiimidazoline antibodies. Detection in serum in relation to essential hypertension. J. Clin. Invest. **92:** 1068-1072.

Imidazoline Receptors and Sodium Excretion in the Rat Kidney[a]

DONALD D. SMYTH,[b,c] FRANCIS K. DARKWA,[b] AND S. BRIAN PENNER[b,c]

Departments of [b]Pharmacology & Therapeutics and [c]Internal Medicine
University of Manitoba
Winnipeg, Manitoba, Canada R3E 0W3

In the last few years, it has become apparent that many effects previously attributed to the α_2-adrenoceptor may in fact involve a novel class of receptors, the imidazoline receptor. This has led to a better understanding of the antihypertensive actions of centrally acting compounds such as clonidine which have been reported to lower blood pressure primarily by imidazoline receptors and not solely by α_2-adrenoceptors.[1,2] This has resulted in studies delineating differences between these two receptor classes in regard to binding distribution[3,4] and possible function.[5]

Both receptor classes (α_2-adrenoceptors and imidazoline receptors) have been identified in the rat kidney[6] with a clear difference in distribution.[7] In the present study, we report similarities and differences between the renal actions of these two receptor systems. Clonidine was chosen as an α_2-adrenoceptor agonist due to radioligand binding studies which have shown that clonidine has a fivefold greater selectivity for these receptors over imidazoline receptors in the kidney.[8] Also, previous studies documented the direct action of clonidine in the kidney, at least in part, by interaction of this compound with the renal actions of vasopressin.[9,10] Moxonidine, 2,6-dimethyl clonidine, and agmatine were chosen as imidazoline receptor agonists. Moxonidine possess 600-fold selectivity for imidazoline receptors over α_2-adrenoceptors in the kidney.[6] 2,6-Dimethyl clonidine has different effects on the kidney as compared to clonidine[11,12] and possesses fivefold selectivity for imidazoline receptors over α_2-adrenoceptors in the kidney.[8] Finally, agmatine was proposed to be the endogenous agonist for the imidazoline receptor.[13] The present results indicate that although both classes of compounds increase urine flow rate, this most conceivably occurs by two unique mechanisms/sites in the kidney.

METHODS

Experimental Animals

Standard procedures were previously reported by our laboratory.[9,10] Male Sprague-Dawley rats (200–225 g) were obtained from the University of Manitoba (Charles

[a]This work was supported by the Medical Research Council of Canada. D.D.S. is a Scientist of the Medical Research Council of Canada.

River Breeding Stock) and cared for as per regional animal care standards protocol. Animals were housed at 22°C with environmental humidity maintained at 50% with a light/dark cycle from 7:00 AM to 7:00 PM. Rats were fed Purina rat chow and received tap water for drinking. Seven to 10 days prior to the experiment, all animals underwent unilateral nephrectomy (right kidney) during ether anesthesia.

Renal Effects of Clonidine, 2,6-Dimethyl Clonidine, Agmatine, or Moxonidine

On the day of the experiment, rats were anesthetized with pentobarbitone (BDH Chemicals, Poole, England; 50 mg/kg, ip). Additional anesthetic was administered in a bolus dose of 3 mg/kg iv as needed. Rats were placed on a thermostatically controlled heating blanket. A rectal thermometer was connected to a Harvard Animal Blanket Control Unit which was used to maintain body temperature at 37.5 °C. The trachea was cannulated with polyethylene tubing (PE-240), and the animal was allowed to breathe spontaneously. The left carotid artery was cannulated with polyethylene tubing (PE-50), and blood pressure was measured with a Statham pressure transducer (model P23Dc) connected to a Grass Polygraph Model V. The left jugular vein was cannulated (PE-160) and connected to an infusion pump (Syringe pump model 355) for the infusion of saline solution. The left kidney was exposed by a flank incision, and the left ureter was cannulated (PE-50) for the collection of urine. A 30-gauge stainless steel needle was inserted into the aorta and advanced into the origin of the left renal artery. A single drop of glue (Superglue, LePage's Limited) secured the base of the needle to the surrounding tissue. This was connected by Tygon tubing (size 0.25 mm) to an infusion pump (Harvard infusion/withdrawal pump model 600-900) for infusion of either saline solution or study drug at 3.4 μl/min.

Following the surgical procedure, saline solution was administered by intravenous infusion (97 μl/min). Animals were allowed to stabilize for a 45-minute period. After the stabilization period, a 15-minute control urine collection was obtained. This was followed by infusion of saline solution (vehicle), clonidine (1.12 or 3.75 nmol/kg/min), agmatine (3, 10, or 30 nmol/kg/min), 2,6-dimethyl clonidine (1.59 or 5.29 nmol/kg/min), or moxonidine (1, 3, or 10 nmol/kg/min) directly into the renal artery for the duration of the experiment at 3.4 μl/min. During this infusion, four additional urine collections of 15 minutes each were obtained in preweighed tubes.

Sodium concentration in plasma and urine was determined with a Beckman KLiNa flame photometer. Creatinine concentration was determined with a Beckman creatinine analyzer model 2. Urine and plasma osmolality was analyzed with a MicroOsmette (Precision Systems). Urine volume was determined gravimetrically.

Effects of Pertussis Toxin

The effects of pertussis toxin on the natriuretic response to clonidine and moxonidine were determined in another group of rats. Animals were treated with either vehicle or pertussis toxin injected directly into the tail vein (3 μg/kg). Five days later, the natriuretic response to an intrarenal infusion of clonidine (1.12 nmol/kg/min) or moxonidine (1 nmol/kg/min) was determined.

Effects of V_2 Vasopressin Receptor Antagonism

The effects of V_2 vasopressin receptor antagonism on the diuretic and natriuretic response to clonidine and moxonidine were determined. Immediately following the first urine collection, 30 nmol/kg of the V_2 vasopressin receptor antagonist ([$d(CH_2)_5$,D-Ile^2, Ile^4]AVP) was administered. After 15 minutes, the response to an intrarenal infusion of clonidine (3.75 nmol/kg/min) or moxonidine (3 nmol/kg/min) was determined.

Effects of Indomethacin

In another group of rats, animals were pretreated with indomethacin (5 mg/kg) or vehicle in a total volume of 0.3 ml given by intraperitoneal injection immediately following administration of the anesthetic. The natriuretic and diuretic response to clonidine (1.12 nmol/kg/min) or moxonidine (1 nmol/kg/min) was determined.

1 Kidney-1 Clip Hypertension

Animals were divided into two groups. In one group (1K-1C), animals were anesthetized (ether), a silver clip (0.254 mm) was placed around the left renal artery, and the right kidney was removed via an abdominal incision. In the other group (1K-sham), animals were treated identically except that no silver clip was placed around the artery. Animals were kept for 12-16 days after surgery and subsequently underwent the protocol just described. The natriuretic response to 2,6-dimethyl clonidine (1.59 and 5.29 nmol/kg/min) was determined in the normotensive sham group and the hypertensive group that had received the silver clip.

Data Analysis

Each experimental group consisted of 6-8 rats. Data are presented as the mean ± standard error of the mean. Data were analyzed with Repeated Measures ANOVA using SAS version 6.07. Significant interactions were further analyzed using least square means difference. Data from the final collection period were presented because this point was representative of differences observed between groups. In the figures, -*- denotes $p < 0.05$ between groups and * denotes $p < 0.05$ as compared with the control within the same group receiving the vehicle infusion.

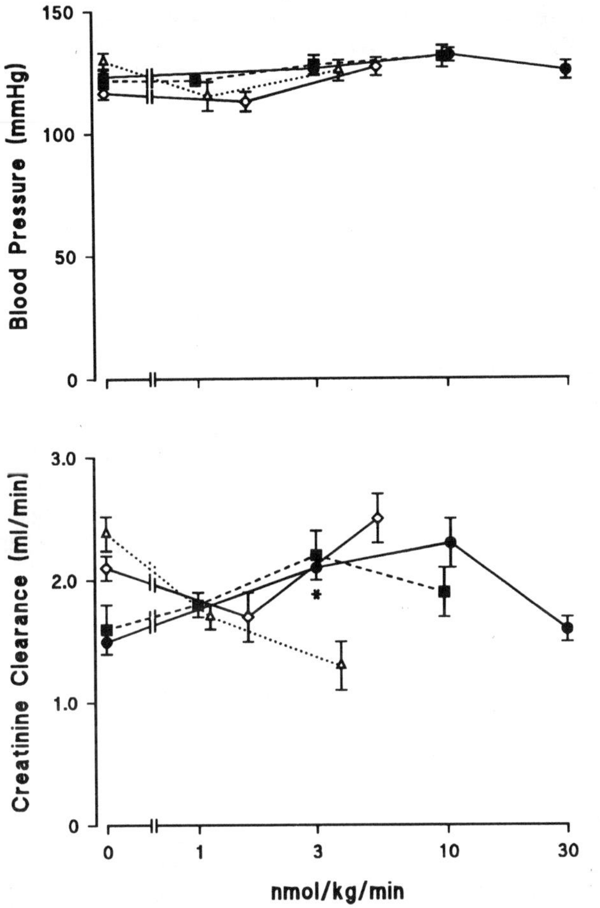

FIGURE 1. Effects of clonidine (*open triangles*), 2,6-dimethyl clonidine (*open diamonds*), moxonidine (*closed squares*), and agmatine (*closed circles*) on blood pressure and creatinine clearance in the anesthetized rat. Data are presented as the mean ± standard error of the mean. * denotes $p < 0.05$ as compared with the respective control group (0 nmol/kg/min).

RESULTS

Hemodynamic Effects of Intrarenal Infusions

The effects of intrarenal infusions of clonidine, 2,6-dimethyl clonidine, moxonidine, and agmatine on blood pressure and creatinine clearance were recorded (FIG. 1). Infusion rates of the different drugs used in this study did not change blood pressure as compared to that in the respective control groups (0 nmol/kg/min) which only received saline vehicle. Similarly, creatinine clearance was also not altered by the infusion rates used in this study. The only exception was agmatine 3 nmol/kg/min which had a significantly greater creatinine clearance than did the control group. Higher infusion rates of these different compounds did change blood pressure and are not included in the present study.

Renal Effects of Clonidine, 2,6-Dimethyl Clonidine, Agmatine, or Moxonidine

Clonidine, 2,6-dimethyl clonidine, moxonidine, and agmatine all increased urine flow rate (FIG. 2). Clonidine, 2,6-dimethyl clonidine, and moxonidine appeared to all have similar potencies. Agmatine had to be administered at a much higher infusion rate to increase urine flow rate. The mechanism for this increase in urine flow rate, however, differed between the compounds. With moxonidine, 2,6-dimethyl clonidine, and agmatine, the increase in urine flow rate was secondary to an increase in osmolar clearance, but higher infusion rates did increase free water clearance (data not shown). Clonidine, on the other hand, increased urine flow rate secondary to an increase in free water clearance. We also found that infusion rates of clonidine which were higher than those reported in the present study would also increase osmolar clearance.

Effects of Pertussis Toxin

We previously documented that pretreatment of animals with pertussis toxin attenuated the natriuretic action of neuropeptide Y,[14] indicating that the effects of this compound may be mediated through a G_i protein. In the present study animals were pretreated with 3 μg/kg pertussis toxin 5 days prior to the experiment. In the control animal, clonidine significantly increased urine flow rate (FIG. 3). However, in the animals that received the pertussis toxin, clonidine failed to increase urine flow. Similarly, moxonidine increased urine flow in the control animals. In the animals pretreated with pertussis toxin, moxonidine failed to increase urine flow rate and sodium excretion. The natriuretic response to furosemide (0.1 mg/kg) was not altered by the pertussis toxin (data not shown).

Effects of V_2 Vasopressin Receptor Antagonism

Previous studies have indicated the importance of vasopressin in the diuretic action of α_2-adrenoceptor agonists.[10] We therefore removed the renal actions of endogenous vasopressin by pretreatment with a specific V_2 vasopressin antagonist. As previously reported, clonidine, moxonidine, or the V_2 antagonist alone increased urine flow rate (FIG. 4). In the presence of the V_2 receptor antagonist, clonidine failed to significantly increase urine flow rate as compared to the group that received the V_2 receptor antagonist alone. Moxonidine in the presence of V_2 receptor antagonism, on the other hand, increased urine flow rate significantly as compared to the group that received the V_2 receptor antagonist alone. Thus, the renal actions of clonidine but not moxonidine were dependent on the renal actions of vasopressin being intact.

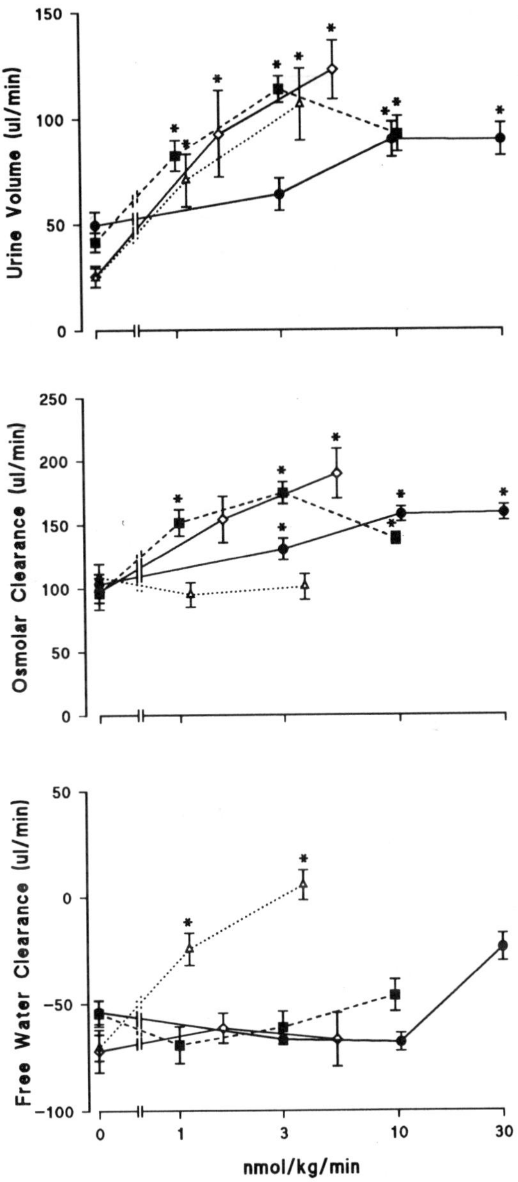

FIGURE 2. Effects of clonidine (*open triangles*), 2,6-dimethyl clonidine (*open diamonds*), moxonidine (*closed squares*), and agmatine (*closed circles*) on urine volume, osmolar clearance, and free water clearance in the anesthetized rat. Data are presented as the mean ± standard error of the mean. * denotes $p < 0.05$ as compared with the respective control group (0 nmol/kg/min).

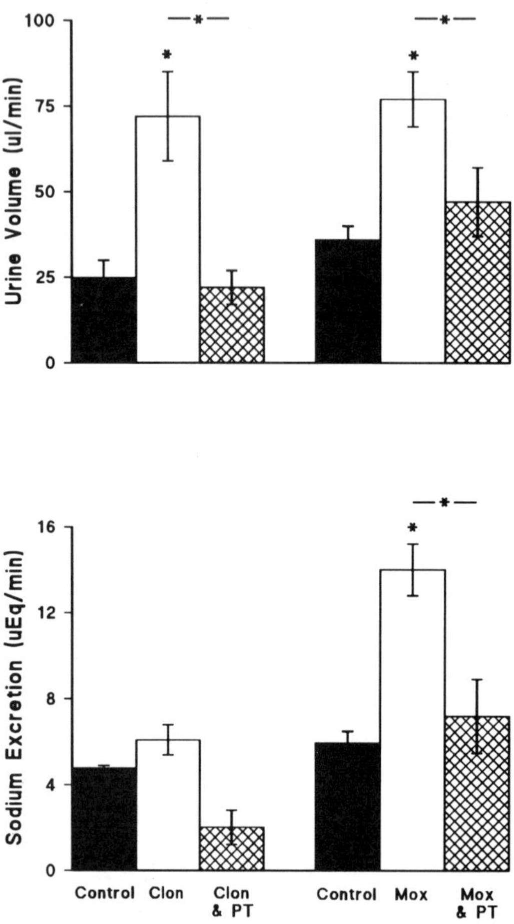

FIGURE 3. Effects of pertussis toxin on the natriuretic and diuretic response to clonidine or moxonidine in the anesthetized rat. Control represents the group receiving vehicle throughout the experiment. Clon = clonidine 1.12 nmol/kg/min; Mox = moxonidine 1 nmol/kg/min; PT = pertussis toxin (3 μg/kg) 5 days prior to the experiment. * denotes $p < 0.05$ as compared with the respective control group (0 nmol/kg/min). -*- denotes $p < 0.05$ between groups.

Effects of Indomethacin

In another series of experiments, indomethacin was used to determine the role of endogenous prostaglandin production in the diuretic response to clonidine and moxonidine (FIG. 5). As previously documented, clonidine increased urine flow rate. In animals pretreated with indomethacin (5 mg/kg), the diuretic response to clonidine was significantly increased. Moreover, this dose of clonidine also increased sodium excretion in indomethacin-pretreated animals. Conversely, in animals pretreated with indomethacin, moxonidine failed to increase urine flow rate or sodium excretion.

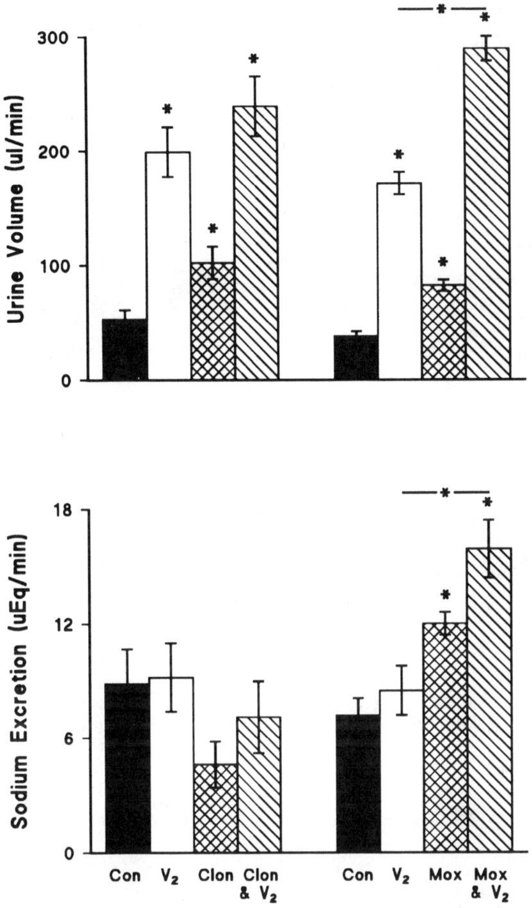

FIGURE 4. Effects of a specific V_2 vasopressin receptor antagonist on the natriuretic and diuretic response to clonidine or moxonidine in the anesthetized rat. Con = control group receiving vehicle throughout the experiment; Clon = clonidine 3.75 nmol/kg/min; Mox = moxonidine 3 nmol/kg/min; V_2 = the V_2 vasopressin receptor antagonist (30 nmol/kg). * denotes $p < 0.05$ as compared with the respective control group (0 nmol/kg/min). -*- denotes $p < 0.05$ between groups.

1 Kidney-1 Clip Hypertension

We previously reported that imidazoline-based compounds such as rilmenidine and moxonidine have a decreased natriuretic response in 1 kidney-1 clip hypertension. 2,6-Dimethyl clonidine also displayed a decreased response in animals with this model of acquired hypertension (FIG. 6). Interestingly, the response to clonidine is not altered in this model of hypertension.[12]

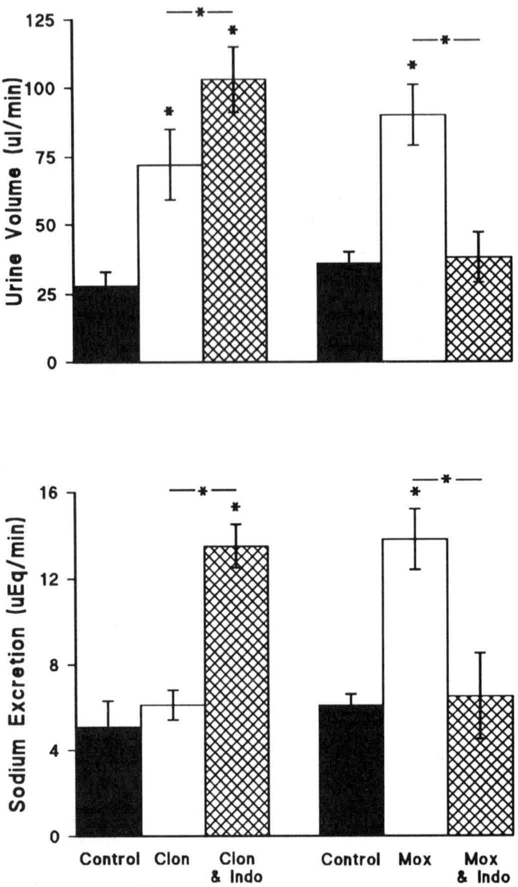

FIGURE 5. Effects of indomethacin on the natriuretic and diuretic response to clonidine or moxonidine in the anesthetized rat. Control represents the group receiving vehicle throughout the experiment. Clon = clonidine 1.12 nmol/kg/min; Mox = moxonidine 1 nmol/kg/min; Indo = indomethacin (5 mg/kg). * denotes $p < 0.05$ as compared with the respective control group (0 nmol/kg/min). -*- denotes $p < 0.05$ between groups.

DISCUSSION

In the present study, clonidine was chosen as a representative α_2-adrenoceptor agonist and moxonidine, 2,6-dimethyl clonidine, and agmatine as imidazoline receptor agonists. All compounds increased urine flow rate, and this increase was attenuated by pretreatment with pertussis toxin. However, although the increase in urine flow rate following clonidine was secondary to an increase in free water clearance, the increase following moxonidine, 2,6-dimethyl clonidine, or agmatine was secondary to an increase in osmolar clearance. The change in free water clearance after clonidine

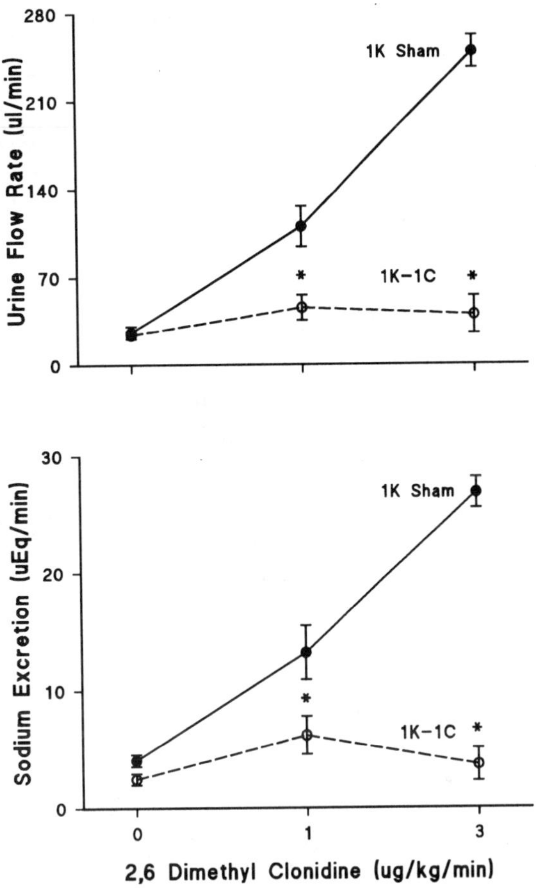

FIGURE 6. Natriuretic and diuretic response to 2,6-dimethyl clonidine (1 and 3 μg/kg/min) in sham-operated (1K Sham) and rats with 1 kidney-1 clip (1K-1C)-induced hypertension.

was conceivably due to an interaction with the renal actions of vasopressin. Consistent with this, pretreatment with a V_2 vasopressin receptor antagonist attenuated the diuretic response to clonidine but not moxonidine. The studies with indomethacin indicated that prostaglandins may antagonize the renal actions of clonidine and play a significant role in the actions of moxonidine. Finally, the response to 2,6-dimethyl clonidine but not clonidine was attenuated in an acquired model of hypertension. These studies collectively indicated a clear difference between the renal actions of compounds reported to be α_2-adrenoceptor selective (clonidine) or imidazoline receptor selective (moxonidine, 2,6-dimethyl clonidine, and agmatine).

The effects of renal α_2-adrenoceptor stimulation have long been known to be mediated through the G_i protein of the adenylate cyclase system.[15] In microdissected collecting ducts, α_2-adrenoceptor stimulation decreased vasopressin-induced cAMP production.[16] This ability of α_2-adrenoceptor stimulation to inhibit the renal actions of vasopressin has been confirmed in the isolated perfused rat kidney[17] and in whole animal preparations.[9,18] The decrease in the diuresis mediated by clonidine following pertussis toxin would be consistent with this effect being mediated in part through the G_i protein. Suppression of the renal actions of moxonidine, however, was somewhat surprising, because previous studies failed to link the imidazoline receptors with the adenylate cyclase system. This was not secondary to general toxicity of the treatment, as the response to furosemide remained intact. An alternate explanation may be related to the lack of specificity of pertussis toxin for the G_i protein. Although the dose of pertussis toxin used in the present study was less than one tenth that used by others to inhibit the G_i protein,[19] other G proteins may possibly have been affected by this treatment. Thus, attenuation of the effects of clonidine and moxonidine by pertussis toxin implies a similarity but does not confirm this possibility.

The ability of a V_2 vasopressin receptor antagonist to attenuate the actions of clonidine but not moxonidine clearly delineates two sites of action. Previous studies reported on the importance of vasopressin in mediating the diuretic effect of α_2-adrenoceptor agonists.[9,18] This interaction with vasopressin would also be consistent with the increase in free water clearance observed after clonidine. The increase in osmolar clearance, and not free water clearance, observed following imidazoline agonists such as moxonidine and agmatine also suggest independence from the renal actions of vasopressin. In fact, the ability of clonidine to increase free water clearance and 2,6-dimethyl clonidine to increase osmolar clearance was interpreted to indicate that some purported α_2-adrenoceptor agonists such as 2,6-dimethyl clonidine may in fact be acting at a unique receptor, the imidazoline receptor.[11]

The ability of indomethacin to potentiate the renal actions of clonidine and attenuate those of moxonidine also indicates a clear difference in these receptor systems. However, the mechanism(s) by which this dichotomy occurs is not clear. Depending on the basal situation (presence of clonidine or moxonidine), prostaglandins apparently may mediate either natriuresis or antinatriuresis. This difference is consistent with previous studies which reported both natriuretic and antinatriuretic actions of prostaglandins in different renal preparations.[20,21] In preliminary experiments, we also observed that indomethacin attenuates the renal actions of agmatine; however, the inhibition was not as complete as that observed with moxonidine (unpublished data).

Attenuation of the renal actions of 2,6-dimethyl clonidine but not clonidine[12] in 1 kidney-1 clip hypertension also points to two unique sites of action and indicates that these two receptor systems may be independently regulated. We also found that the renal response to moxonidine and rilmenidine was attenuated in this model of acquired hypertension.[22] Although not conclusive, this supports 2,6-dimethyl clonidine as an imidazoline agonist. The mechanism for this decrease in response has not been determined, but it may be related to the decrease in [^{3}H]idazoxan-labeled binding sites in kidneys obtained from 1 kidney-1 clip hypertensive animals.[12]

In summary (TABLE 1), the increase in urine flow rate observed following α_2-adrenoceptor stimulation appears secondary to an increase in free water clearance, mediated through a G protein (G_i?), dependent on vasopressin and attenuated by endogenous prostaglandins. The increase in urine flow rate following imidazoline receptor stimulation is secondary to an increase in osmolar clearance, mediated

TABLE 1. Summary of the Effects of Pertussis Toxin, V_2 Vasopressin Receptor Inhibition, Indomethacin, and an Acquired Model of Hypertension on the Renal Actions of Clonidine (α_2-Adrenoceptor Selective Agonist) and Moxonidine, 2,6-Dimethyl Clonidine, and Agmatine (Imidazoline-Selective Agonists) in the Anesthetized Rat Preparation

	α_2-Adrenoceptors	Imidazoline Receptors
Urine flow	⇑ Free water clearance	⇑ Osmolar clearance
G Protein couples	Yes	Yes
Vasopressin	Dependent	Independent
Prostaglandins	Attenuate	Mediate?
1 Kidney-1 clip	⇔ Response	⇓ Response

through a G protein, independent of vasopressin, and stimulated (mediated?) by endogenous prostaglandin production. The ability of these interventions to have differing effects depending on agonists used clearly indicates that the α_2-adrenoceptor and the imidazoline receptor in the rat kidney represent two separate and unique receptor systems.

ACKNOWLEDGMENTS

The authors wish to express their gratitude to Donald R. Jones and Marilyne I. Vandel for their expert technical assistance and Mary Cheang for expert assistance with the statistical analysis.

REFERENCES

1. Bousquet, P., J. Feldman & J. Schartz. 1984. Central cardiovascular effects of the alpha adrenergic drugs: Differences between catecholamines and imidazolines. J. Pharmacol. Exp. Ther. **230:** 232-236.
2. Ernsberger, P., H. D. Damon, L. M. Graff, S. G. Schafer & M. O. Christen. 1993. Moxonidine, a centrally acting antihypertensive agent, is a selective ligand for I_1-imidazoline sites. J. Pharmacol. Exp. Ther. **264:** 172-182.
3. Michel, M. C., O.-E. Brodde, B. Schnepel, J. Behrendt, R. Tschada, H. J. Motulsky & P. A. Insel. 1989. [^{3}H]Idazoxan and some other alpha-2 adrenergic drugs also bind with high affinity to a non-adrenergic site. Mol. Pharmacol. **35:** 324-330.
4. Michel, M. C., J. W. Regan, M. A. Gerhardt, R. R. Neubig, P. A. Insel & H. J. Motulsky. 1990. Nonadrenergic [^{3}H]idazoxan binding sites are physically distinct from alpha-2 adrenergic receptors. Mol. Pharmacol. **37:**65-68.
5. Tibirica, E., J. Feldman, C. Mermet, F. Gonon & P. Bousquet. 1991. An imidazoline-specific mechanism for the hypotensive effect of clonidine: A study with yohimbine and idazoxan. J. Pharmacol. Exp. Ther. **256:** 606-613.
6. Ernsberger, P., K. L. Westbrooks, M. O. Christen & S. G. Schäfer. 1992. A second generation of centrally acting antihypertensive agents act on putative I_1-imidazoline receptors. J. Cardiovasc. Pharmacol. **20** (Suppl 4): S1-S10.

7. ZAKEIL, N., P. ERNSBERGER & J. G. DOUGLAS. 1993. Distribution of I_1-imidazoline and α_2-adrenergic binding sites in renal tubule epithelial cells. FASEB J., Part I: A145 (Abstr. 837).
8. SMYTH, D. D., P. LI & P. ERNSBERGER. 1994. 2,6 Dimethyl clonidine: α_2 Adrenoceptor agonist or I_1 imidazoline preferring agonist? Submitted.
9. BLANDFORD, D. E. & D. D. SMYTH. 1988. Dose selective dissociation of water and solute excretion after renal alpha2-adrenoceptor stimulation. J. Pharmacol. Exp. Ther. **247:** 1181-1186.
10. BLANDFORD, D. E. & D. D. SMYTH. 1990. Role of vasopressin in the response to intrarenal infusions of α_2-adrenoceptor agonists. J. Pharmacol. Exp. Ther. **255:** 264-270.
11. SMYTH, D. D., P. LI, D. E. BLANDFORD & S. B. PENNER. 1992. Opposite rank order of potency for clonidine, 2,6 dimethyl clonidine and UK 14,304 on water and solute excretion in the rat. Two sites and/or receptors? J. Pharmacol. Exp. Ther. **261:** 21-25.
12. LI, P. & D. D. SMYTH. 1993. Suppressed renal response to 2,6-dimethyl clonidine but not clonidine in one kidney-one clip hypertensive rats. J. Pharmacol. Exp. Ther. **267:** 1395-1400.
13. LI, G., S. REGUNTHAN, C. J. BARROW, J. ESHRAGHI, R. COOPER & D. REIS. 1994. Agmatine, an endogenous clonidine-displacing substance in the brain. Science **263:** 966-969.
14. SMYTH, D. D., D. E. BLANDFORD & S. THOM. 1988. Disparate effects of neuropeptide Y and clonidine on the excretion of sodium and water in the rat. Eur. J. Pharmacol. **152:** 157-162.
15. PETTINGER, W. A., S. UMEMURA, D. D. SMYTH & W. B. JEFFRIES. 1987. Renal α_2-adrenoceptors and the adenylate cyclase-cAMP system: Physiological significance. Am. J. Physiol. **252:** F199-F208.
16. UMEMURA, S., D. MARVER, D. D. SMYTH & W. A. PETTINGER. 1985. α_2-Adrenoceptors and cellular cAMP levels in isolated single nephron segments in the rat. Am. J. Physiol. **249:** F28-F33.
17. SMYTH, D. D., S. UMEMURA & W. A. PETTINGER. 1985. Alpha$_2$-adrenoceptor antagonism of vasopressin-induced changes in sodium excretion. Am. J. Physiol. **248:** F767-F772.
18. GELLAI, M. 1990. Modulation of vasopressin antidiuretic action by renal α_2-adrenoceptors. Am. J. Physiol. **28:** F1-F8.
19. PEDRAZA-CHAVERRI, J., M. C. ALTORRE-GONZALEZ, M. E. IBARRA-RUBIO, J. C. PENA & J. A. GARCIA-SAINZ. 1984. Effect of pertussis toxin on the adrenergic regulation of plasma renin activity. Life Sci. **35:** 1638-1642.
20. FULGRAFT, G. & A. MEIFORTH. 1971. Effects of prostaglandin E_2 on excretion and reabsorption of sodium and fluid in rat kidneys (micropuncture studies). Pfluegers Arch. **330:** 243-256.
21. FASSINA, G., F. CARPENEDO & R. SANTI. 1976. Effect of prostaglandin E1 on isolated short-circuited frog skin. Life Sci. **8:** 181-187.
22. LI, P., S. B. PENNER & D. D. SMYTH. 1994. Attenuated response to moxonidine and rilmenidine in one kidney-one clip hypertensive rats. Br. J. Pharmacol. **112:** 200-206.

Rilmenidine Alters Renal Function When Administered Intracerebroventricularly or Intrarenally[a]

S. BRIAN PENNER AND DONALD D. SMYTH

Department of Pharmacology & Therapeutics
Faculty of Medicine
University of Manitoba
Winnipeg, Canada, R3E 0W3

Radioligand binding studies have demonstrated the presence of imidazoline receptors in different tissues including the brain and the kidney.[1] Previously, utilizing doses of moxonidine, an I_1-imidazoline selective compound, that did not alter blood pressure or heart rate, our laboratory showed an increase in urine volume associated with an increase in sodium excretion when administered directly into the renal artery (ir).[2] The increase in urine flow was secondary to an increase in osmolar clearance, not a change in free water clearance.[2] These changes could be inhibited by the imidazoline antagonist idazoxan, but not rauwolscine, a selective α_2-adrenoceptor antagonist,[2] thus implying direct activation of imidazoline receptors in the kidney. Recently, moxonidine administered into the lateral cerebral ventricle (icv) at doses that did not change systemic hemodynamics also increased urine volume through changes in osmolar clearance.[3] However, the highest dose of icv moxonidine (1 nmol) resulted in an increase in free water clearance.[3] Previously, increases in free water clearance have been postulated to be secondary to activation of the α_2-adrenoceptor.[4] Therefore, in the present study we compared the effect of a highly selective I_1-imidazoline compound, rilmenidine, or renal function when administered ir or icv to determine if the site of injection altered the effect observed.

METHODS

All studies used Sprague-Dawley rats subjected to unilateral nephrectomy 7-10 days prior to the experimental day. When necessary, central cannulas were implanted 3-5 days before the study. Rats weighed approximately 300 g on the day of the experiment.

Experimental Procedure. On the day of the study rats were anesthetized with pentobarbital, the trachea cannulated (PE-240), the carotid and jugular veins cannulated (PE-50 and PE-160, respectively), and the ureter cannulated following a flank

[a] This work was supported by the Heart and Stroke Foundation of Canada.

incision (PE-50). Continuous infusion of 97 μl min^{-1} of saline solution was given intravenously throughout the experiment. After completion of surgery, a 30-minute stabilization period was allowed. Infusion or bolus of vehicle or study drug was given after a 15-minute control collection, and for the ir animals it was continued throughout the next two 30-minute urine collections. Blood pressure and heart rate were recorded throughout the experiment. Urine volume was measured gravimetrically. Sodium and creatinine concentration and osmolality of urine and plasma was measured.

Intrarenal Experiments. Four groups of animals were studied, a control group and three groups each receiving rilmenidine 3, 10, and 30 nmol kg^{-1} min^{-1}. A 30-gauge needle was inserted into the aorta and advanced into the left renal artery, through which an infusion of saline solution or rilmenidine at a rate of 3.4 μl min^{-1} was given.[5]

Intracerebroventricular Experiments. Four groups of animals were studied, a control group and three groups each receiving rilmenidine 30, 100, and 300 nmol. Following anesthesia with pentobarbital, the tip of a modified 23-gauge needle, with a solid obturator, was placed into the lateral cerebral ventricle using a stereotaxic apparatus. Coordinates for implantation in relation to the skull were as follows: 0.3 mm posterior to the bregma, 1.4 mm lateral to the midline, and 3.5 mm below the surface of the skull. A 10-μl Hamilton syringe was used for the bolus injection which was given over 2 minutes in a volume of 5 μl.[6]

Statistical Analysis. Data are presented as the mean ± standard error of the mean (SEM). TABLE 1 shows the difference between the second 30-minute collection and the respective baseline collection. Comparisons were made between the control and the different doses within either the intrarenal or intracerebroventricular group. Each group contained five to seven animals. Repeated measures of analysis of variance were done. When differences were found, the least squares means difference test was performed. Analysis was carried out on both absolute values as well as delta results so that the same comparisons could be made. A p value of ≤ 0.05 was considered significant.

RESULTS

In both the icv and ir baseline values (data not shown), no differences were found between groups for blood pressure, heart rate, creatinine clearance, or urine flow. In animals that received rilmenidine 30 nmol icv the baseline sodium excretion, osmolar clearance, and free water clearance were significantly higher than those in the other groups. Changes from the respective control for the second postintervention collection are given in TABLE 1. Animals receiving ir rilmenidine did not demonstrate any changes in blood pressure or creatinine clearance. Only the highest dose lowered heart rate. Urine flow was increased by 3 and 30 nmol kg^{-1} min^{-1}, whereas sodium excretion was greater than that in control animals with all doses. The 3- and 30-nmol kg^{-1} min^{-1} doses of rilmenidine caused an increase in osmolar clearance. Rilmenidine ir did not alter free water clearance. However, with icv rilmenidine the two highest doses caused a decrease in both the blood pressure and heart rate. Rilmenidine administered icv resulted in an increase in urine flow only with the

TABLE 1. Differences between the Second 30-Minute Collection after Intervention and the Baseline in the Different Groups[a]

	Intracerebroventricular Route				Intrarenal Route			
Dose	0 (n = 7)	30 (n = 5)	100 (n = 7)	300 (n = 6)	0 (n = 5)	3 (n = 6)	10 (n = 5)	30 (n = 5)
BP	1 ± 3	2 ± 3	−13 ± 3*	−20 ± 3*	−2 ± 3	5 ± 3	7 ± 3	3 ± 3
HR	−4 ± 7	−2 ± 8	−47 ± 7*	−58 ± 8*	−10 ± 11	−12 ± 10	−28 ± 11	−50 ± 11*
UV	32 ± 6	22 ± 7	33 ± 6	70 ± 7*	33 ± 8	64 ± 7*	47 ± 8	65 ± 8*
UNa	7 ± 1	5 ± 1	5 ± 1	9 ± 1	5 ± 1	10 ± 1*	9 ± 1*	9 ± 1*
CrCl	−0.4 ± 0.2	−0.4 ± 0.2	−0.2 ± 0.2	−0.1 ± 0.2	−0.2 ± 0.1	−0.1 ± 0.1	0 ± 0.1	−0.1 ± 0.1
C_{OSM}	53 ± 7	25 ± 8*	35 ± 7	63 ± 7	47 ± 6	76 ± 6*	65 ± 6	76 ± 6*
C_{H2O}	−20 ± 4	−3 ± 5*	−2 ± 4*	7 ± 5*	−14 ± 5	−11 ± 4	−18 ± 5	−11 ± 5

[a] Each group contains at least five animals. Dose administered intracerebroventricularly was in nanomoles and dose administered intrarenally was in nmol kg^{-1} min^{-1}; BP = blood pressure (mm Hg); CrCL = creatinine clearance (ml min^{-1}); HR = heart rate (beats min^{-1}); UV = urine flow (μl min^{-1}); UNa = urinary sodium excretion (μmol min^{-1}); C_{OSM} = osmolar clearance (μl min^{-1}); C_{H_2O} = free water clearance (μl min^{-1}). The mean ± SE are given. *p value ≤0.05 between the groups for each parameter.

highest dose, whereas sodium excretion remained unaltered. The 30-nmol dose of icv rilmenidine, which did not alter blood pressure or heart rate, decreased osmolar clearance, but resulted in an increase in free water clearance. The two highest doses did not change osmolar clearance; however, similar to the lowest dose, free water clearance increased.

DISCUSSION

Rilmenidine administered either directly into the renal artery or intracerebroventricularly results in an increase in urine flow. Unlike ir rilmenidine, a dose of icv rilmenidine that increased urine output without a hemodynamic change could not be found. Previously, we showed that ir infusions of 2,6-dimethyl clonidine[7] or moxonidine,[2] which are both imidazoline compounds, caused an increase in sodium excretion which was reflected in an increase in osmolar clearance. Similarly in the current study, ir rilmenidine increased urine output through an increase in sodium excretion. When moxonidine was given intracerebroventricularly at doses that did not alter systemic hemodynamics, an increase in urine flow, baseline sodium excretion, and osmolar clearance was observed,[3] while icv rilmenidine increased only free water clearance, with and without any changes in blood pressure or heart rate. Thus, rilmenidine resulted in different changes in kidney function, depending on the site of administration. This suggests either a different site or mechanism of action between both the route as well as the different imidazoline compounds we have studied. To further elucidate the effect of central administration of imidazoline compounds on renal function, site-specific injections will have to be utilized.

REFERENCES

1. Ernsberger, P. R., K. L. Westbrooks, M. O. Christen & S. G. Schafer. 1992. J. Cardiovasc. Pharmacol. **20**(suppl. 4): S1-S10.
2. Allan, D. R., S. B. Penner & D. D. Smyth. 1993. Br. J. Pharmacol. **108:** 870-875.
3. Penner, S. B. & D. D. Smyth. 1994. Br. J. Pharmacol. In press.
4. Blandford, D. E. & D. D. Smyth. 1988. J. Pharmacol. Exp. Ther. **261:** 1080-1086.
5. Penner, S. B. & D. D. Smyth. 1988. Br. J. Pharmacol. **93:** 446-452.
6. Penner, S. B., D. D. Smyth & G. B. Glavin. 1990. Neurobiology of stress ulcers. Ann. N.Y. Acad. Sci. **597:** 43-50.
7. Smyth, D. D., P. Li, D. E. Blandford & S. B. Penner. 1992. J. Pharmacol. Exp. Ther. **261:** 1080-1086.

α_2 Adrenoceptor- and Imidazoline-Preferring Binding Sites in the Dog Kidney

ROGER G. EVANS[a] AND JOHN M. HAYNES[b,c]

Emily E. E. Stewart Renal Laboratory
Baker Medical Research Institute
Prahran, Victoria, Australia

[b]*University of Melbourne Department of Medicine*
Repatriation Hospital
Heidelberg West, Victoria, Australia

Certain imidazoline and guanidinium α_2 adrenoceptor (α_2AR) ligands have high affinity for nonadrenoceptor binding sites. Heterogeneous populations of these imidazoline binding sites have been found in a range of tissues and species.[1] Those imidazoline binding sites that display a high affinity for clonidine and its derivatives have been termed I_1, while those with high affinity for idazoxan have been termed I_2.[2] Imidazoline binding sites have been identified in the kidneys of all species in which it has been tested. In pigs and humans only the I_2-subtype has been identified,[3,4] whereas both subtypes coexist in the kidneys of rats and rabbits.[5,6]

Our long-term aim is to investigate the functional significance of these binding sites in the kidney. We have chosen the dog as our experimental model, because these animals can be surgically prepared for repeated and simultaneous measurement of systemic and renal hemodynamic, renal excretory, and hormonal function. Furthermore, these observations can be made without the confounding effects of anesthesia. The aim of the present, preliminary study was to characterize the nature of α_2AR and imidazoline binding sites in the dog kidney. A more complete report of these experiments was published previously.[7]

METHODS

The right kidneys (89 ± 11 g) of three male Rottweiler dogs (28–34 kg) were removed under halothane anesthesia. These animals were being surgically prepared for other studies. The kidneys were perfused with 100 ml of ice cold heparinized saline solution and a membrane homogenate was prepared by conventional methods.[8]

[a] Address for correspondence: Dr. Roger Evans, Baker Medical Research Institute, P.O. Box 348, Prahran, Victoria 3181, Australia.

[c] Present address: Department of Physiology and Pharmacology, University of Nottingham, Queen's Medical Centre, Nottingham, UK.

Frozen (–70 °C) aliquots were later resuspended at a protein concentration of ~200 μg/mg protein for receptor binding studies.

Binding reactions were performed as described previously[7,8] at 37°C in a final volume of 200 μl of Krebs phosphate buffer. Saturation and competition experiments were performed with [^{3}H]rauwolscine (α_2AR binding sites), [^{3}H]*p*-aminoclonidine (I_1 binding sites), and [^{3}H]idazoxan (I_2 binding sites). Nonspecific binding was defined by 10 μmol/L idazoxan, 10 μmol/L phentolamine, and 1 mmol/L phentolamine, respectively. Incubation times were 10, 10, and 30 minutes, respectively. Data were analyzed as previously described.[7,8]

RESULTS

[^{3}H]Rauwolscine. Scatchard analysis demonstrated binding to a single site with affinity (pK_D; negative Log mol/L) 8.7 ± 0.1 and maximum binding density (B_{max}) 59 ± 5 fmol/mg protein. [^{3}H]Rauwolscine binding was displaced from an apparent single site ($p > 0.05$ for a two-site fit) by 2-methoxyidazoxan, idazoxan, phentolamine, methysergide, and amiloride, but in a complex manner indicative of competition at two binding sites ($p < 0.05$ for a two-site fit) by prazosin (FIG. 1; TABLE 1).

[^{3}H]p-Aminoclonidine. Scatchard analysis demonstrated binding to a single site (pK_D 8.3 ± 0.2; B_{max} 72 ± 26 fmol/mg protein). No specific binding could be detected when this experiment was performed in the presence of 1 μmol/L of the α_2AR-selective ligand 2-methoxyidazoxan.

[^{3}H]Idazoxan. Scatchard analysis demonstrated higher affinity (pK_D 9.0 ± 0.1; B_{max} 44 ± 13 fmol/mg protein) and lower affinity (pK_D 8.0 ± 0.2; B_{max} 94 ± 9 fmol/mg protein) components of [^{3}H]idazoxan binding. In the presence of 10 μmol/L of guanosine 5′-0-(3-thiotriphosphate) (GTPγS), maximum binding at the higher affinity site was reduced by 79%, but the lower affinity site was unaffected. [^{3}H]Idazoxan binding was displaced in a monophasic manner (slope factors not less than unity) but with low potency by amiloride, 2-methoxyidazoxan, and phentolamine (FIG. 1; TABLE 1). By contrast, idazoxan itself displaced [^{3}H]idazoxan binding with relatively high potency, but the displacement curve was shallow, with a slope factor less than unity (0.61 ± 0.03) (FIG. 1; TABLE 1).

DISCUSSION

We compared the binding of [^{3}H]rauwolscine, [^{3}H]*p*-aminoclonidine, and [^{3}H]idazoxan in the dog kidney, to characterize the nature of α_2AR and imidazoline binding sites in this organ. Our major finding is that under our experimental conditions [^{3}H]rauwolscine and [^{3}H]idazoxan appear to bind to different sites.

[^{3}H]Rauwolscine was shown previously to bind to 5-hydroxytryptamine (5-HT) receptors as well as α_2ARs.[9] We observed that [^{3}H]rauwolscine binding was displaced with high affinity by the selective α_2AR ligand 2-methoxyidazoxan, but with low affinity by the 5-HT receptor ligand methysergide (FIG. 1; TABLE 1), indicating that [^{3}H]rauwolscine binds with high affinity to α_2ARs but not to 5-HT receptors in dog kidney. The complex displacement profile of prazosin provides evidence that these α_2ARs are heterogeneous, perhaps representing α_{2A}-like and α_{2B}-like ARs.[10]

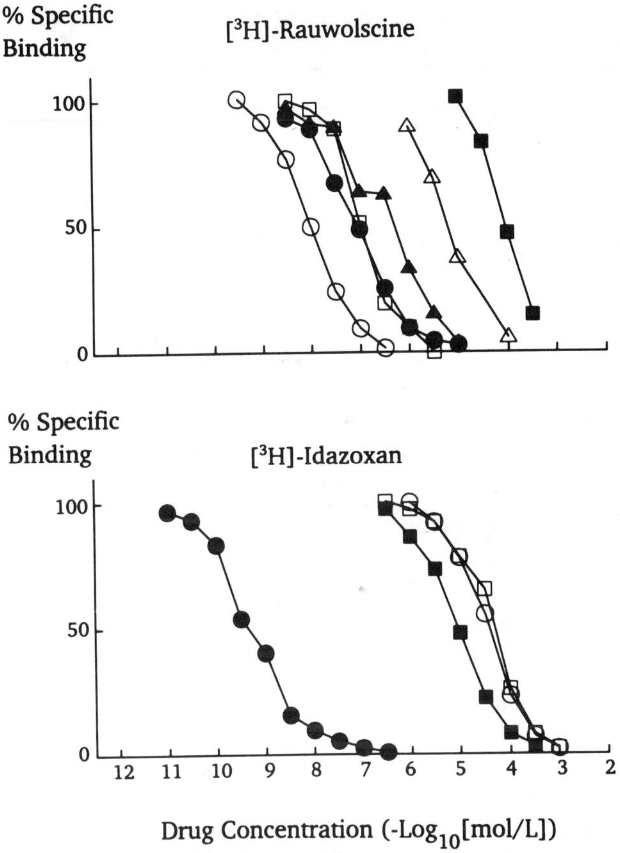

FIGURE 1. Inhibition of specific binding of [^{3}H]rauwolscine (2 nmol/L) and [^{3}H]idazoxan (3 nmol/L) to dog kidney membranes in a representative experiment. (○) 2-methoxyidazoxan, (●) idazoxan, (□) phentolamine, (■) amiloride, (△) methysergide, and (▲) prazosin. Slope factors were not significantly less than unity except for prazosin displacement of [^{3}H]rauwolscine (0.77 ± 0.06) and idazoxan displacement of [^{3}H]idazoxan (0.61 ± 0.03).

[^{3}H]*p*-Aminoclonidine selectively labels I_1 binding sites in a range of tissues and species,[2,3] but appears to exclusively label α_2ARs in the dog kidney, because all specific binding was eliminated by 1 μmol/L 2-methoxyidazoxan. Thus, we can find no evidence for the existence of I_1 binding sites in dog kidney.

[^{3}H]Idazoxan appears to label a heterogeneous population of binding sites in the dog kidney that are not α_2ARs, inasmuch as 2-methoxyidazoxan displaced [^{3}H]idazoxan binding with very low potency (FIG. 1; TABLE 1). The low affinity of amiloride for these binding sites suggests that they are pharmacologically similar to the putative I_{2B} binding sites found in the kidneys of pigs, rats, and humans.[1,2]

We also observed that the soluble guanosine triphosphate analog GTPγS reduced maximum binding at the higher affinity but not the lower affinity component of

TABLE 1. Inhibition of $[^3H]$Rauwolscine and $[^3H]$Idazoxan Binding to Dog Kidney Membranes[a]

Drug	$[^3H]$Rauwolscine pK_i		(% sites)	$[^3H]$Idazoxan pIC_{50}
Phentolamine		7.6 ± 0.1		4.5 ± 0.1
Idazoxan		7.3 ± 0.1		8.8 ± 0.1
2-Methoxyidazoxan		8.4 ± 0.1		4.9 ± 0.1
Methysergide		5.6 ± 0.2		Not determined
Amiloride		4.1 ± 0.2		5.3 ± 0.1
Prazosin	8.1 ± 0.4		(31%)	Not determined
	6.4 ± 0.1		(69%)	

[a]All values are the mean ± SEM of three experiments. The concentrations of $[^3H]$rauwolscine and $[^3H]$idazoxan were 2 and 3 nmol/L, respectively. pK_i values are given for $[^3H]$rauwolscine because Scatchard analysis demonstrated that $[^3H]$rauwolscine bound to a single population of sites, giving a unique K_D value. pK_i values were calculated by applying the correction of Cheng and Prusoff[11] to the pIC_{50}. Only pIC_{50} values are given for $[^3H]$idazoxan, because Scatchard analysis demonstrated that it bound to two distinct populations of binding sites.

$[^3H]$idazoxan binding. GTP and its analogs were shown previously to inhibit the binding of radioligands, but only in situations where the radioligand is an agonist and the receptor belongs to the guanine nucleotide regulatory protein (G-protein)-coupled family.[9] Thus, our data are consistent with the hypothesis that the higher affinity component of $[^3H]$idazoxan binding represents a G-protein-coupled receptor at which idazoxan behaves as an agonist.

In summary, we conclude that the dog kidney contains two α_2AR subtypes that can be selectively labeled with $[^3H]$rauwolscine and a heterogeneous population of imidazoline binding sites similar to the I_{2B} binding sites described in pig, rat, and human kidneys, that can be selectively labeled with $[^3H]$idazoxan.

REFERENCES

1. MICHEL, M. C. & P. A. INSEL. 1989. Trends Pharmacol. Sci. **10:** 342-344.
2. REIS, D. J., S. REGUNATHAN, H. WANG, D. L. FEINSTEIN & M. P. MEELEY. 1992. Fundam. Clin. Pharmacol. **6**(Suppl. 1): 23s-30s.
3. MICHEL, M. C., O. E. BRODDIE, B. SCHNEPEL, J. BEHRENDT, R. TSCHADA, H. J. MOTULSKY & P. A. INSEL. 1989. Mol. Pharmacol. **35:** 324-330.
4. VIGNE, P., M. LAZDUNSKI & C. FRELIN. 1989. Eur. J. Pharmacol. **160:** 295-298.
5. COUPRY, I., D. ATLAS, R. A. PODEVIN, I. UZIELLI & A. PARINI. 1989. J. Pharmacol. Exp. Ther. **252:** 293-299.
6. ERNSBERGER, P., G. FEINLAND, M. P. MEELEY & D. J. REIS. 1990. Am. J. Hypertens. **3:** 90-97.
7. EVANS, R. G. & J. M. HAYNES. 1994. Clin. Exp. Pharmacol. Physiol. **21:** 649-658.
8. EVANS, R. G., J. M. HAYNES & J. LUDBROOK. 1993. Br. J. Pharmacol. **109:** 37-47.
9. DE VOS, H., E. CZERWIEC, J. P. DE BACKER, W. DE POTTER & G. VAUQUELIN. 1991. Eur. J. Pharmacol.-Mol. Pharmacol. Sect. **207:** 1-8.
10. BYLUND, D. B., C. RAY-PRENGER & T. J. MURPHY. 1988. J. Pharmacol. Exp. Ther. **245:** 600-607.
11. CHENG, Y. C. & W. H. PRUSOFF. 1973. Biochem. Pharmacol. **22:** 3099-3108.

Characterization of the Imidazoline Binding Site Involved in Regulation of Insulin Secretion[a]

NOEL G. MORGAN, SUSAN L. F. CHAN, COLIN A. BROWN, AND EFTHYMIA TSOLI

Cellular Pharmacology Group
Department of Biological Sciences
Keele University
Keele, Staffs ST5 5BG, UK

Insulin is the principal anabolic hormone in mammals, and its level in the circulation is regulated by a diverse range of molecules which reflect the metabolic status of the animal. These include nutrients (such as sugars, amino acids, and fatty acids), hormones, and neurotransmitters, all of which can impinge upon the pancreatic B-cell to elicit changes in the insulin secretion rate.[1] To respond to this array of signals, the B-cell is equipped with numerous receptor and signal transduction mechanisms that allow the various inputs to be sensed and integrated.

The primary determinant of the rate of insulin secretion is the prevailing glucose concentration, and it is well established that oxidative metabolism of the sugar is used to generate the signals that initiate secretion.[2,3] In contrast, hormones and neurotransmitters utilize more classical receptor-mediated signaling mechanisms to elicit their effects in the pancreatic B-cell. Thus, there must be certain critical components of the stimulus-secretion coupling pathway that can be controlled by both metabolic and receptor-mediated events.

Very few of these common components have been identified; however, potassium permeability of the B-cell plasma membrane is one of the most important events controlled by metabolic stimuli.[4–6] Under basal conditions potassium permeability is high and efflux of potassium ions maintains the cell in a hyperpolarized state. An increase in glucose metabolism results in a reduction in potassium efflux and leads to depolarization of the plasma membrane, gating of voltage-sensitive calcium channels, and a subsequent increase in insulin secretion. The central regulator of these events is a class of potassium channels whose open state varies according to the ratio of ATP to ADP in the cell and which close in response to a net increase in this ratio (i.e., in response to increased metabolic flux). Therefore, these channels (K_{ATP} channels) perform a critical role in coupling metabolic changes to electrical activity in the B-cell.[4–6]

K_{ATP} channels also serve as targets for pharmacological agents in that they mediate the insulin secretory effects of hypoglycemic sulfonylureas.[7,8] In addition, there is

[a] This work was supported by grants from the Wellcome Trust, Medical Research Council, British Diabetic Association, and the Royal Society.

increasing evidence that the activity of K_{ATP} channels can also be regulated by a newly defined class of antihyperglycemic agents that possess an "imidazoline" ring.[9–12] At present, the best characterized compounds in this class are synthetic reagents, but evidence has already begun to emerge that endogenous compounds possessing similar activity may exist. Thus, on this basis, K_{ATP} channels may represent a common site at which both metabolic and certain receptor-generated signals are integrated.

In the present paper, evidence is reviewed which supports the concept that imidazoline insulin secretagogues may interact with a physiologically relevant receptor site that is coupled to K_{ATP} channel activity.

IMIDAZOLINE INSULIN SECRETAGOGUES

The concept that certain imidazoline compounds might possess insulin secretagogue activity has developed over a number of years and has been the subject of several recent reviews.[13–15] Much of the evidence was initially derived from studies with antagonists of α_2-adrenoceptors (such as phentolamine and midaglizole) which were observed to elicit an increase in circulating insulin when administered to experimental animals or to man.[16–18] At first, this was attributed solely to blockade of a tonic inhibitory control of insulin secretion mediated by endogenous catecholamines acting on islet α_2-adrenoceptors. However, it is difficult to reconcile this mechanism with the lack of efficacy of the potent α_2-antagonist idazoxan[19] and with the finding that certain imidazoline α-antagonists can stimulate insulin secretion directly, in isolated islet preparations.[9,10,20,21] One possible explanation comes from evidence that adrenoceptors can express a basal level of signaling activity even in the absence of agonist. Binding of some (but not all) antagonists to the receptor leads to inhibition of this activity.[22] When this occurs, it gives rise to a phenomenon known as "intrinsic antagonist activity," in which the antagonists appear to elicit direct responses that are independent of agonists.[22] If this were the case for α_2-adrenoceptors in the endocrine pancreas, then the ability of selected antagonists to stimulate insulin secretion could be mediated by such a mechanism.

There are, however, several additional pieces of evidence which indicate that α_2-adrenoceptors are not involved in the action of imidazoline insulin secretagogues. First, the dose-response characteristics for stimulation of secretion by imidazoline α_2-antagonists *in vitro* are not consistent with such a mechanism.[9] In the second place, irreversible α_2-blockade does not prevent stimulation of insulin secretion by imidazoline α_2-antagonists.[23] Third, some imidazoline compounds that are not α_2-antagonists can elicit an insulin secretory response in islets.[24]

Thus, the weight of evidence favors the view that the ability of imidazoline compounds to act as insulin secretagogues is not due primarily to interaction with islet α_2-adrenoceptors. Therefore, an important new question arises: Is there a second site in the pancreatic B-cell at which these drugs act?

As just outlined, radiochemical and electrophysiological experiments have shown that stimulation of insulin secretion by imidazoline compounds correlates with closure of K_{ATP} channels.[9–15] Because this effect can be demonstrated in isolated plasma membrane patches,[10,25–27] this suggests that the binding site and the channel must both be present in the plasma membrane. The nature of the interaction between the two is still unresolved, and it is not yet clear if they exist as separate molecular entities or if the K_{ATP} channel carries the imidazoline binding site. To understand how imidazoline binding leads to an alteration in channel activity, it is clearly crucial to identify and characterize the binding site and to establish its relationship with the K_{ATP} channel.

EVIDENCE FOR AN ISLET IMIDAZOLINE RECEPTOR

During the time that the work just described, in the endocrine pancreas, was proceeding, a number of important parallels emerged with observations made in other systems. For example, it has become clear that a dissociation exists between α_2-adrenoceptor binding and the actions of certain imidazoline α_2-ligands in tissues as diverse as brain, kidney, and adrenal gland.[28,29] In some cases (e.g., adrenal gland and ventrolateral medulla) functional distinctions have been made, whereas in others differences in pharmacology have been noted.[28,29] These have now been examined in some detail, and the functional and pharmacological effects have been attributed to the existence of a class of binding sites that are distinct from α_2-adrenoceptors and that have become known as imidazoline binding sites (imidazoline receptors). The nomenclature associated with these sites has still to be standardized, but they are now often subclassified as either I_1 or I_2 sties on the basis of pharmacological criteria.[28]

In the presence of increasing evidence for the existence of functional imidazoline receptors in other tissues, it is entirely reasonable to consider the possibility that the effects of imidazoline compounds on insulin secretion might be due to interaction with a member of this class of binding site. This hypothesis is still under scrutiny, but increasing evidence suggests that the islet imidazoline binding site may be different from those found elsewhere.

The principal pharmacological distinction between I_1 and I_2 sites relates to their affinity for specific radioligands. I_1 sites bind clonidine with high affinity and can be labeled with [^{3}H]clonidine (or a derivative such as *p*-amino- or iodoclonidine).[30,31] I_2 sites are labeled most readily with [^{3}H]idazoxan.[32,33] Initial experiments with the pancreatic B-cell line RINm5F revealed the presence of I_2 sites (labeled with [^{3}H]idazoxan),[34] and this has been confirmed for normal islets of rat[23] and hamster.[35] However, strong evidence indicates that I_2 sites are not involved in mediating the effects of imidazoline compounds on insulin secretion.[23] Firstly, idazoxan is, at best, weakly effective as an insulin secretagogue and K_{ATP} channel blocker, and it does not antagonize the response to other imidazoline compounds.[9,10] Thus, although idazoxan is a potent I_2 ligand, this does not correlate with its activity in islets. Furthermore, one of the most effective imidazoline insulin secretagogues is efaroxan,[9,10,13–15,37] a compound that displays very low affinity for I_2 binding sites.[23,36] In this context, the emerging consensus suggests that I_2 sites are predominantly localized on mitochondria,[37] whereas imidazoline α_2-antagonists are able to regulate K_{ATP} channel activity in isolated patches of plasma membrane.[13–15] Therefore, this evidence also argues against involvement of I_2 sites in mediating their activity.

Data obtained with the isolated perfused rat pancreas have substantiated these conclusions and have provided evidence that the pancreatic vasculature may contain

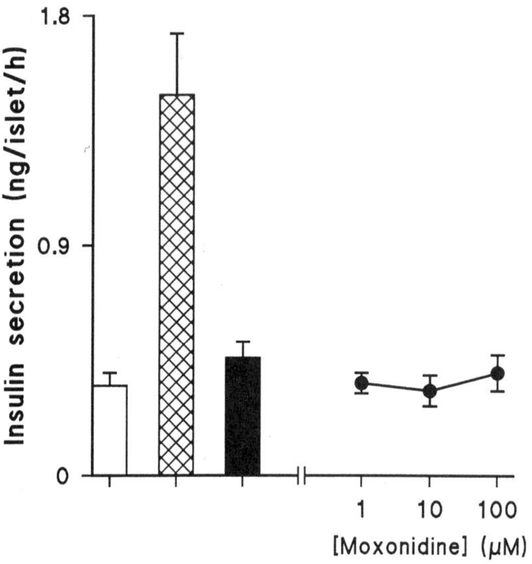

FIGURE 1. Effects of moxonidine on inhibition of insulin secretion mediated by diazoxide. Isolated rat islets were incubated with 2 mM glucose (*open bar*), 18 mM glucose (*hatched bar*), 18 mM glucose plus 250 μM diazoxide (*solid bar*) or 18 mM glucose, 250 μM diazoxide, and increasing concentrations of moxonidine (*closed circles*). Yohimbine (100 μM) was present in all tubes to block α_2-adrenoceptors.

functional imidazoline receptors.[38] These have a pharmacology similar to that of I_2 sites but different from that of the imidazoline binding site involved in direct control of insulin secretion. Because, however, changes in islet blood flow may contribute to alterations in insulin secretion, I_2 sites on blood vessels could play an indirect role in control of islet secretory activity *in vivo.*

The situation with respect to islet I_1 sites has proved more difficult to clarify. Efaroxan is a potent and selective I_1 ligand[36] which might suggest that an I_1 site could be a target for this molecule in islets. However, in the ventrolateral medulla of the brain, efaroxan exhibits the characteristics of an I_1 antagonist, whereas the ligand moxonidine has even higher affinity and shows agonist activity.[36] We compared the effects of these two compounds in rat islets and found that efaroxan acts as an agonist in that it directly stimulates insulin secretion and antagonizes the inhibition of insulin secretion in response to the K_{ATP} channel agonist diazoxide.[9,10] By contrast, moxonidine fails to alter the effects of diazoxide on insulin secretion from rat islets (Fig. 1), and it does not modify the response to efaroxan, suggesting that it is neither an agonist nor an antagonist at the islet site. Hence, apparently the islet imidazoline binding site does not have the same functional pharmacology as does the I_1 site in the ventrolateral medulla.

To clarify this situation further we performed ligand binding experiments with the I_1-selective ligand *p*-iodoclonidine (PIC). Because of limited tissue availability, experiments were carried out using clonal pancreatic B-cells rather than normal islets. Membranes from the cells displayed nonadrenergic binding sites which could be labeled with ^{125}I-PIC, but the label was displaced more effectively by idazoxan than by efaroxan (Fig. 2). This finding suggested that the majority of sites labeled by ^{125}I-PIC were more closely related to I_2 than to I_1 sites and that the number of I_1 sites present on pancreatic B-cells is, at best, very low.

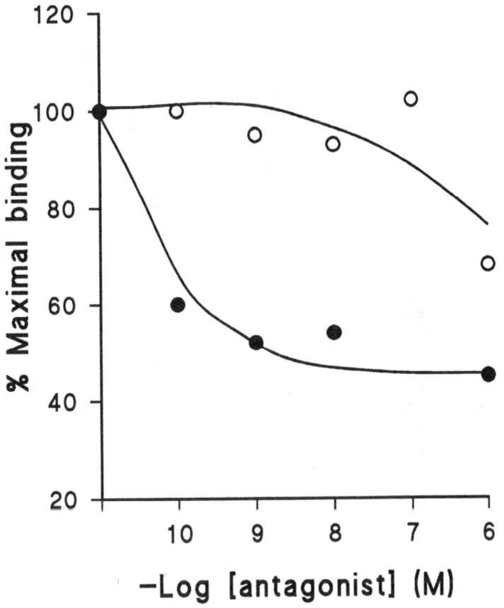

FIGURE 2. Displacement of ^{125}I-iodoclonidine binding to HIT-T15 cell membranes. Membranes from cultured HIT-T15 pancreatic B-cells were incubated with 1 nM ^{125}I-iodoclonidine under conditions of α_2-blockade. Binding was displaced with increasing concentrations of either efaroxan (*open circles*) or idazoxan (*closed circles*).

Thus, the current weight of evidence suggests that the imidazoline binding site involved in regulation of insulin secretion has pharmacological characteristics that distinguish it from the I_1 and I_2 sites described in other tissues.

IS THERE A FUNCTIONAL IMIDAZOLINE BINDING SITE IN ISLETS?

Because of mounting evidence that the effects of imidazoline insulin secretagogues cannot be attributed to interaction with either I_1 or I_2 imidazoline binding sites, the question arises as to whether their effects in islets are relatively nonspecific or whether a novel imidazoline binding site is involved. We employed a variety of approaches to address this issue and obtained several independent lines of evidence that support the existence of an imidazoline receptor that controls K_{ATP} channel activity in islets.

ISLET RESPONSES TO EFAROXAN SHOW STEREOSELECTIVITY

For many receptors that bind small ligand molecules, binding of the ligand occurs in a stereospecific manner with one enantiomeric form having a higher binding affinity. This has been shown for I_2 receptors,[39] but the degree of enantiomeric selectivity with the high affinity ligand idazoxan is small.

In rat islets, responses to the imidazoline efaroxan exhibited marked stereoselectivity.[40] The (–) enantiomer of the drug was much more potent as a direct insulin secretagogue than the (+) enantiomer, and a similar separation of activities was observed in relation to their abilities to antagonize the inhibitory effects of diazoxide

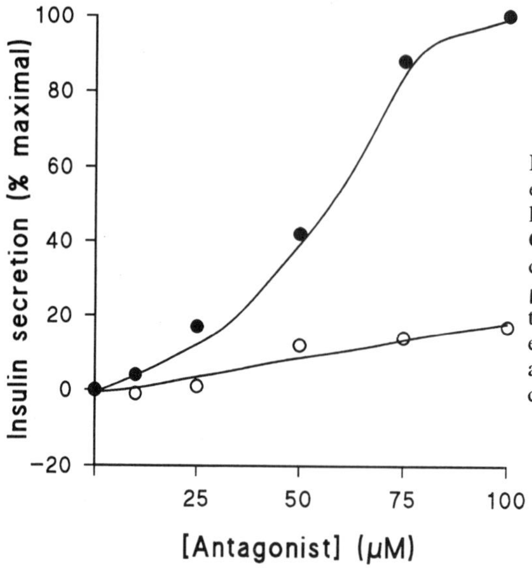

FIGURE 3. Effects of the enantiomers of efaroxan on inhibition of insulin secretion mediated by diazoxide. Groups of isolated rat islets were incubated with 20 mM glucose and 250 μM diazoxide. Increasing concentrations of either the (+) enantiomer of efaroxan (*open circles*) or the (–) enantiomer (*closed circles*) were included, as shown.

in islets. This can be seen clearly in FIGURE 3 in which data have been normalized to allow direct comparison of the results from different experimental strategies. The (–)enantiomer was considerably more potent than the (+) form. This contrasts with the α_2-adrenoceptor affinities of these two ligands in that the (+)enantiomer is the more potent α_2-ligand.

Thus, at least with efaroxan, stimulation of insulin secretion apparently displays a stereospecific preference which is consistent with involvement of a putative receptor molecule. We have currently been unable to obtain any other imidazoline insulin secretagogues in optically pure form, and the extent to which the separation of activity of the enantiomers of efaroxan is more generally applicable remains to be established.

Although this evidence for the stereospecificity of efaroxan is striking, it is more difficult to define other structural criteria that are required for insulin secretagogue activity. With I_1 and I_2 imidazoline binding sites the possession of a closed ring forming the imidazoline structure is not an absolute requirement. Indeed, certain guanidinium compounds are potent ligands at these receptors.[28,29,36] To date, little work has been performed in islets to test the effects of guanidinium compounds as ligands at imidazoline binding sites, but such compounds have been known as effective antihyperglycemic agents for many years.[41,42] One such reagent, amiloride, has the capacity to antagonize the inhibitory effect of diazoxide on secretion (FIG. 4), suggesting that it may be active at the "imidazoline" binding site in islets. However, inasmuch as amiloride can also interact with other binding sites in cells,[43] its effects on insulin secretion may be complex, and further studies are required before definitive conclusions can be drawn.

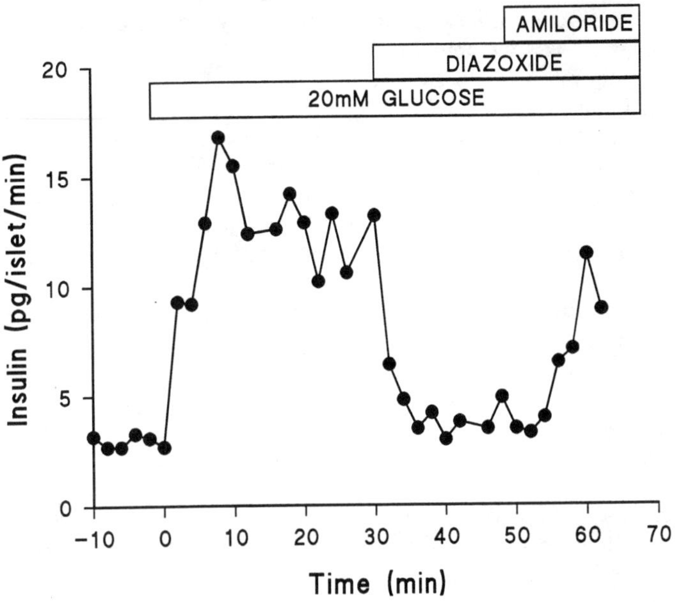

FIGURE 4. Effects of amiloride on inhibition of insulin secretion mediated by diazoxide. Groups of isolated rat islets were perfused in medium containing 4 mM glucose for 30 minutes, after which (t = 0) 20 mM glucose was introduced. Diazoxide (250 μM) was added 30 minutes later, followed by amiloride (100 μM).

IMIDAZOLINE RESPONSES IN ISLETS CAN BE DOWNREGULATED

Many receptor-mediated responses are subject to desensitization phenomena which result in loss of responsiveness to agonists.[44] This can occur on a very short time scale (seconds to minutes), when desensitization results from functional uncoupling of the receptor from its effector system or it may require longer term agonist exposure. In the latter case the loss of response is usually associated with removal of receptors from the cell surface.

In isolated rat islets, the stimulatory effects of efaroxan and other imidazolines are rapid in onset and sustained for periods in excess of 30 minutes without any evidence of short-term loss of responsiveness.[10,20,21,38] Thus, in functional terms there is no evidence that imidazoline responses are subject to rapid desensitization in islets. This is also true in isolated patches of B-cell plasma membrane where the effects of efaroxan are rapid, sustained, and reversible.[10] With certain other imidazoline compounds (e.g., SL 84.014) the effects on K_{ATP} channel activity are prolonged and not readily reversed, even following agonist withdrawal from the medium.[45] The reasons for this are not yet clear, but observations confirm that inhibition of K_{ATP} channel activity by imidazolines does not exhibit rapid desensitization.

By contrast with the results just described, exposure of rat islets to imidazolines for up to 18 hours in culture leads to significant loss of responsiveness.[40,46] This was first shown for efaroxan, but similar results were also obtained with phentolamine. In these experiments, the ability of imidazoline drugs to antagonize the effects of diazoxide on insulin secretion was used as the assay system. Culture of islets in the absence of imidazoline did not alter their responsiveness per se, but treatment with an imidazoline "agonist" resulted in apparent desensitization. The effect was not agonist specific in that culture of islets in the presence of efaroxan resulted in subsequent loss of response to either efaroxan or phentolamine. In contrast, culture of islets in the presence of imidazolines that do not elicit significant increases in insulin secretion (e.g., idazoxan and UK14,304) failed to affect the subsequent response of islets to efaroxan.[46]

These results suggest several conclusions. First, they indicate that the islet imidazoline receptor can be controlled independently of K_{ATP} channels. Thus, culture of islets in the presence of an imidazoline compound resulted in selective loss of response, as evidenced by the observation that the secretory changes mediated by either glucose or diazoxide (both of which require functional K_{ATP} channel activity) were not modified. This selectivity implies that the imidazoline binding site may be physically separate from the K_{ATP} channel. Second, the results indicate that islet cells have the capacity to modulate their response to imidazoline compounds according to the prevailing conditions. This implies that imidazoline drugs interact with a receptor molecule which is under dynamic regulation in islet cells.

IDENTIFICATION OF AN ISLET IMIDAZOLINE ANTAGONIST

So far, a number of pieces of data have been discussed which support the possibility that islet cells may express an "imidazoline receptor." In an effort to substantiate this concept further, we sought to identify possible imidazoline antagonists that can block the effects of agonists such as efaroxan and phentolamine. A number of structural analogs of efaroxan were screened, and one in particular (an unsubstituted derivative, RX801080[47]) exhibited interesting properties. This compound is inactive as a direct insulin secretagogue; it does not modify the inhibition of insulin secretion mediated by diazoxide; and it fails to alter the rate of insulin secretion in response to 20 mM of glucose.[47] However, inclusion of increasing concentrations of RX801080 together with efaroxan resulted in progressive loss of the response to efaroxan. This antagonism was observed when direct stimulation of insulin secretion was measured and also when the ability of efaroxan to overcome the inhibitory effect of diazoxide on secretion was assayed. Thus, RX801080 appears to act as an efaroxan antagonist in rat islets.[47]

The mechanism by which RX801080 antagonizes the actions of efaroxan has not been elucidated, but as it is a structural analog of efaroxan, the most likely mechanism would involve interaction at a common binding site. This cannot yet be proven directly, but the finding that RX801080 also blocks the stimulatory effects of other imidazoline compounds in islets (including phentolamine and midaglizole) is consistent with the concept that these agents all share a common binding site.[47]

While studying the effects of RX801080 in islets, some very surprising results emerged, in that this agent was also found to antagonize the stimulatory effects of a sulfonylurea (glibenclamide) on insulin secretion. This was unexpected, because no clear structural homology exists between imidazolines and sulfonylureas. The result, however, is significant in that both types of drug exert their influence by regulating the activity of K_{ATP} channels. Use of [^{3}H]glibenclamide to label the sulfonylurea receptor confirmed that RX801080 does not displace sulfonylurea binding, indicating that the mode of antagonism in this case does not involve a common binding site.[47]

In light of these data it would seem appropriate to propose a model in which RX801080 exerts two distinct, but related, actions in the pancreatic B-cell. These derive from an initial binding event in which RX801080 interacts with the islet "imidazoline binding site" which is associated with K_{ATP} channels. This leads to competitive antagonism of the binding of other imidazoline reagents and, as a consequence, inhibits their effects on insulin secretion. At the same time, occupation of the imidazoline binding site must also influence the coupling of the sulfonylurea receptor to K_{ATP} channels. This second effect is not mediated by direct competition for a common binding site, and the mechanisms involved remain uncertain. However, there is additional evidence for the existence of functional interactions between these two components. For example, coadministration of efaroxan and glibenclamide to rats leads to synergistic potentiation of circulating insulin levels beyond that achieved with either agent alone[48] (Brown, Smith, and Morgan, unpublished observations).

LABELING OF THE ISLET IMIDAZOLINE RECEPTOR

It is clear that one of the most definitive means to demonstrate the presence of an islet imidazoline receptor would be to label it with a suitable radioligand. This objective has proved very difficult to achieve, because the compounds that have been used to characterize the site in functional experiments are active at relatively high concentrations (in the μM range), suggesting that their binding affinities are low. As such, they are not well suited for use in radioligand binding studies. Despite this, some success was recently achieved with the ligand ^{3}H-RX821002.[46] This molecule is an imidazoline α_2-antagonist which was originally considered to be highly selective for α_2-adrenoceptors; however, there is now evidence that it may also bind weakly to 5-HT_{1A} receptors.[49]

RX821002 can stimulate insulin secretion from isolated rat islets[46] and perfused rat pancreas preparations.[37] Therefore, we investigated the possibility that high concentrations of a radiolabeled form of this ligand ([^{3}H]RX821002) might be useful for labeling the islet imidazoline binding site. At concentrations below 20 nM the ligand binds exclusively to α_2-adrenoceptors in pancreatic B-cells,[34,35,46] but as the ligand concentration is increased beyond this level, a second, much lower affinity site also becomes labeled (FIG. 5). Binding to the second site can be displaced completely by those imidazoline compounds that are effective insulin secretagogues (e.g., efaroxan, phentolamine) but not by imidazolines that fail to affect secretion (e.g., idazoxan, UK14,304).[46] This evidence indicates that the low affinity [^{3}H]RX821002 binding site may represent the functional islet imidazoline receptor.

[^{3}H]RX821002 is certainly not an ideal radioligand with which to label the islet imidazoline binding site, because it has a very low affinity for the site. It is difficult to estimate the affinity constant with precision, but curve fitting analysis suggests

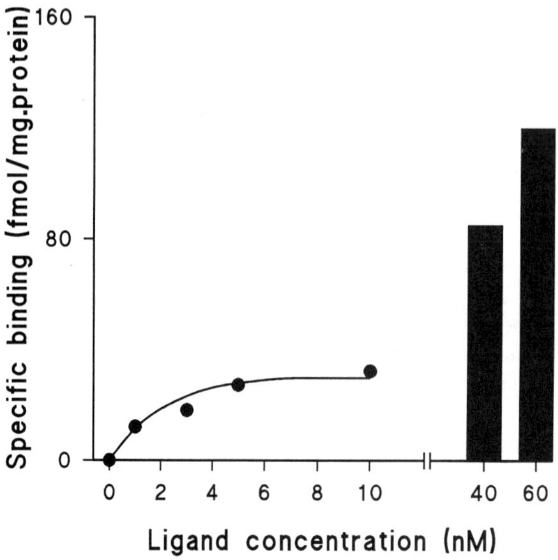

FIGURE 5. Binding of [^{3}H]RX821002 to RINm5F pancreatic B-cell membranes. Membranes from cultured RINm5F cells were incubated with increasing concentrations of [^{3}H]RX821002, as shown. Specific binding was assessed by displacement of the radioligand with 100 μM phentolamine.

that it is likely to be in excess of 140 nM. Thus, the search for a ligand that binds with much higher affinity is still an important objective, but these results provide the first encouraging indication that radiolabeling of the site can be achieved.

IS THERE AN ENDOGENOUS LIGAND FOR THE ISLET IMIDAZOLINE RECEPTOR?

The evidence just described is entirely consistent with the possibility that islet cells express some type of imidazoline receptor involved in control of insulin secretion, but this molecule still awaits detailed characterization. If such a receptor does exist, then it seems likely that an endogenous ligand may also be produced in species expressing the receptor. In this context, Li *et al.*[50] recently identified a molecule in extracts of mammalian brain which appears to possess many of the characteristics expected of an endogenous imidazoline receptor ligand. This molecule is a decarboxylated derivative of arginine, known as agmatine.[50]

Agmatine is an intermediate in the synthesis of polyamines in certain plants and bacteria, but it had not previously been found in mammalian tissues. Interestingly, it is known to possess hypoglycemic properties and has even been used in clinical practice as an oral antidiabetic agent.[41]

Despite this encouraging background information, experiments with isolated rat islets have not provided strong support for the hypothesis that agmatine may be an agonist at the islet imidazoline receptor. The compound certainly stimulates insulin secretion (Fig. 6), but the time course of this stimulation is slow, indicating that the reagent has to accumulate inside cells in order to exert its effects.[51] In addition, stimulation of secretion is elicited only by high (>mM) concentrations of agmatine

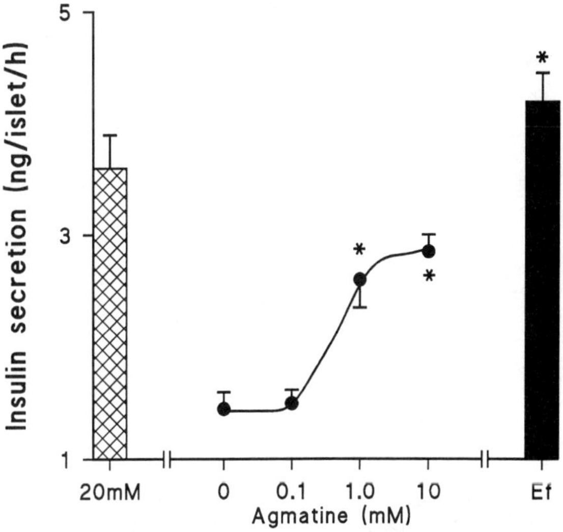

FIGURE 6. Effects of agmatine on inhibition of insulin secretion mediated by diazoxide. Isolated rat islets were incubated with 20 mM glucose alone (*hatched bar*) or 20 mM glucose plus 250 μM diazoxide (all other conditions). Either efaroxan (100 μM; *solid bar*) or agmatine (*closed circles*) was also included, as shown. $*p < 0.01$ relative to zero agmatine.

and the compound is weakly potent as an antagonist of the effects of diazoxide in islets (Chan and Morgan, unpublished observations). None of these characteristics supports the possible involvement of an imidazoline receptor in mediation of the response to agmatine.

Thus, it remains an open question as to whether a different endogenous imidazoline receptor ligand awaits characterization, that is more active at the islet receptor.

CONCLUSIONS

Overall, evidence reviewed in the present paper supports the view that the pancreatic B-cell expresses a binding site for imidazoline compounds that is functionally associated with K_{ATP} channels. This site has characteristics which indicate that it is closely related to the I_1 and I_2 imidazoline receptor subtypes now being characterized in other tissues, but it exhibits certain features which appear to distinguish it from these receptors. Evidence that the site can be radiolabeled is now emerging, but synthesis of a high affinity radioligand remains an important priority in order to facilitate further characterization of the site. Despite the uncertainties that still exist, there is encouraging evidence that the islet imidazoline binding site may represent an important target for design of new antihyperglycemic agents, perhaps formulated in combination with sulfonylureas.

REFERENCES

1. Flatt, P. R. (Ed.) 1992. Nutrient Regulation of Insulin Secretion. Portland Press. Colchester, UK.
2. Malaisse, W. J. 1984. Diabete & Metab. **9:** 313–320.

3. MALAISSE, W. J. 1992. D-glucose metabolism in islet cells. *In* Nutrient Regulation of Insulin Secretion. P. R. Flatt, Ed.: 83-100. Portland Press. Colchester, UK.
4. ASHCROFT, F. M. & S. J. H. ASHCROFT. 1990. Cell Signalling **2:** 197-214.
5. COOK, D. L., L. S. SATIN, M. L. J. ASHFORD & C. N. HALES. 1988. Diabetes **37:** 495-498.
6. BOYD, A. E. 1992. J. Cell. Biochem. **48:** 234-241.
7. BOYD, A. E. 1988. Diabetes **37:** 847-850.
8. NELSON, D. A., L. AGUILER-BRYAN, H. RAEF & A. E. BOYD. 1992. Molecular mechanisms of sulphonylurea action in the pancreatic B-cell. *In* Nutrient Regulation of Insulin Secretion. P. R. Flatt, Ed.: 319-339. Portland Press. Colchester, UK.
9. CHAN, S. L. F. & N. G. MORGAN. 1990. Eur. J. Pharmacol. **176:** 97-101.
10. CHAN, S. L. F., M. J. DUNNE, M. R. STILLINGS & N. G. MORGAN. 1991. Eur. J. Pharmacol. **204:** 41-48.
11. PLANT, T. D. & J.-C. HENQUIN. 1990. Br. J. Pharmacol. **101:** 115-120.
12. JONAS, J.-C., T. D. PLANT & J.-C. HENQUIN. 1992. Br. J. Pharmacol. **107:** 8-14.
13. CHAN, S. L. F. 1993. Clin. Sci. **85:** 671-677.
14. MORGAN, N. G. 1994. Exp. Opin. Invest. Drugs **3:** 561-569.
15. MORGAN, N. G., S. L. F. CHAN, R. J. LACEY & C. A. BROWN. 1995. Pharmacology and molecular biology of islet cell adrenoceptors. *In* Frontiers of Insulin Secretion and Pancreatic B-Cell Research. P. R. Flatt & S. Lenzen, Eds. Smith-Gordon Ltd. London UK. In press.
16. ROBERTSON, R. P., J. B. HALTER & D. PORTE. 1976. J. Clin. Invest. **57:** 791-795.
17. BROADSTONE, V. L., M. A. PFEIFER, J. I. BAJAJ, J. I. STAGNER & E. SAMOLS. 1987. Diabetes **36:** 932-937.
18. KAWAZU, S., M. SUZUKI, K. NEGISHII, T. WATANABE & J. ISHII. 1987. Diabetes **36:** 216-220.
19. BROWN, M. J., A. D. STRUTHERS, L. DI SILVIO, T. YEO, M. GHATEI & J. M. BURRIN. 1985. Clin. Sci. **68:** (Suppl. 10): 137s-139s.
20. SCHULZ, A. & A. HASSELBLATT. 1988. Naunyn-Schmiedeberg's Arch. Pharmacol. **337:** 637-643.
21. SMITH, M. & FURMAN, B. L. 1988. Acta Endocrinol. **188:** 89-96.
22. SAMAMA, P., G. PEI, T. COSTA, S. COTECCHIA & R. J. LEFKOWITZ. 1994. Mol. Pharmacol. **45:** 390-394.
23. BROWN, C. A., A. C. LOWETH, S. A. SMITH & N. G. MORGAN. 1993. Br. J. Pharmacol. **108:** 312-317.
24. SCHULZ, A. & A. HASSELBLATT. 1989. Naunyn-Schmiedeberg's Arch. Pharmacol. **340:** 305-309.
25. DUNNE, M. J. 1991. Br. J. Pharmacol. **103:** 1847-1850.
26. PLANT, T. D., J.-C. JONAS & J.-C. HENQUIN. 1991. Br. J. Pharmacol. **104:** 385-390.
27. EDWARDS, G. & A. H. WESTON. 1993. Ann. Rev. Pharmacol. Toxicol. **32:** 597-637.
28. MICHEL, M. C. & P. ERNSBERGER. 1992. Trends Pharmacol. Sci. **13:** 369-370.
29. KILPATRICK, A. T., C. C. BROWN & A. C. MACKINNON. 1992. Biochem. Soc. Trans. **20:** 113-118.
30. LANGIN, D. & M. LAFONTAN. 1989. Eur. J. Pharmacol. **159:** 199-203.
31. RENOUARD, A., P. S. WIDDOWSON & A. CORDI. 1993. Br. J. Pharmacol. **109:** 625-631.
32. MACKINNON, A. C., M. STEWART, H. J. OLVERMAN, M. SPEDDING & C. M. BROWN. 1993. Br. J. Pharmacol. **232:** 79-87.
33. BRICCA, G., H. GRENEY, J. ZHANG, M. DONTENWILL, J. STUTZMANN, A. BELCOURT & P. BOUSQUET. 1994. Eur. J. Pharmacol. **266:** 25-33.
34. REMAURY, A. & H. PARIS. 1992. J. Pharmacol. Exp. Ther. **260:** 417-426.
35. LACOMBE, C., V. VIALLARD & H. PARIS. 1993. Int. J. Biochem. **25:** 1077-1083.
36. ERNSBERGER, P. R., K. L. WESTBROOKS, M. O. CHRISTEN & S. G. SCHAFER. 1992. J. Cardiol. Pharmacol. **20** (Suppl. 4): S1-S10.
37. TESSON, F., I. LIMON & A. PARINI. 1992. Eur. J. Pharmacol. **219:** 335-338.

38. BERDEU, D., R. GROSS, G. RIBES, M.-M. LOUBATIERES-MARIANI & G. BERTRAND. 1994. Eur. J. Pharmacol. **254:** 119-125.
39. TESSON, F., C. PRIP-BUUS, A. LEMOINE, J.-P. PEGORIER & A. PARINI. J. BIOL. CHEM. **266:** 155-160.
40. CHAN, S. L. F., C. A. BROWN & N. G. MORGAN. 1993. Eur. J. Pharmacol. **230:** 375-378.
41. AYNSLEY-GREEN, A. & K. G. M. M. ALBERTI. 1973. Horm. Metab. Res. **5:** 55.
42. ALSEVER, R. N., R. GEORG & K. E. SUSSMAN. 1970. Endocrinology **86:** 332-336.
43. NOVOTNY, W. F., O. CHASSANDE, M. BAKER, M. LAZDUNSKI & P. BARBRY. 1994. J. Biol. Chem. **269:** 9921-9925.
44. SIBLEY, D. R. & R. J. LEFKOWITZ. 1985. Nature **317:** 124-129.
45. JONAS, J.-C., T. D. PLANT, I. ANGEL, S. Z. LANGER & J.-C. HENQUIN. 1994. Eur. J. Pharmacol. **254:** 27-33.
46. CHAN, S. L. F., C. A. BROWN, K. E. SCARPELLO & N. G. MORGAN. 1994. Br. J. Pharmacol. **112:** 1065-1070.
47. BROWN, C. A., S. L. F. CHAN, M. R. STILLINGS, S. A. SMITH & N. G. MORGAN. 1993. Br. J. Pharmacol. **110:** 1017-1022.
48. BERRIDGE, T. L., J. C. DOXEY, A. G. ROACH & C. F. C. SMITH. 1992. Eur. J. Pharmacol. **213:** 205-212.
49. MIRALLES, A., G. OLMOS, M. SASTRE, F. BARTUREN, I. MARTIN & J. A. GARCIA-SEVILLA. 1993. J. Pharmacol. Exp. Ther. **264:** 1187-1197.
50. LI, G., S. REGUNATHAN, C. J. BARROW, J. ESHRAGHI, R. COOPER & D. J. REIS. 1994. Science **263:** 966-969.
51. SENER, A., P. LEBRUN, F. BLACHIER & W. J. MALAISSE. 1989. Biochem. Pharmacol. **38:** 327-330.

Imidazolines Stimulate Release of Insulin from RIN-5AH Cells Independently from I_1- and I_2-Imidazoline Receptors

GABRIEL OLMOS,[a,b] ROHIT N. KULKARNI,[c]
MUNIRUL HAQUE,[c] AND JOHN MacDERMOT[a]

[a]*Department of Clinical Pharmacology*
Royal Postgraduate Medical School
Du Cane Road
London W12 ONN, UK

[c]*Department of Medicine*
Francis Fraser Laboratories
Hammersmith Hospital
Du Cane Road
London W12 ONN, UK

The activity of an ATP-dependent K^+ channel (K^+_{ATP} channel) is well established as a major determinant of the resting potential of β-cells. Plasma glucose inhibits this channel via metabolic production of ATP, which leads to depolarization of the cell, opening of voltage-sensitive calcium channels, and an increase in the rate of insulin release.[1,2]

K^+_{ATP} channels in β-cells are also subject to regulation by a variety of pharmacological agents. Certain compounds with an imidazoline structure stimulate insulin release from β-cells by inducing closure of K^+_{ATP} channels.[3,4] However, the nature of the interaction between those compounds and the channels has not been established. It has been proposed that the effects of imidazolines on K^+_{ATP} channels in β-cells are mediated through imidazoline receptors, because their effects are stereoselective and subject to downregulation.[5] However, there is currently no direct evidence to support the proposal. The present study was designed (1) to assess if imidazoline drugs with selectivity for the I_1- or I_2-imidazoline receptors could stimulate insulin release in the insulinoma cell line RIN-5AH and (2) to examine the possible contribution of the I_1- and I_2-imidazoline receptors to this effect.

The rat insulinoma RIN-5AH cell line was routinely cultured in RPMI 1640 medium supplemented with 10% (v/v) fetal calf serum, 100 μg/ml streptomycin, 100 U/ml penicillin, 5.5 mM glucose, and 2 mM L-glutamine. The cells were grown in

[b]Address for correspondence: Dr. Gabriel Olmos, Lab. Neurofarmacologia, Dept. Biologia Fonamental i Ciències de la Salut, Universitat de les Illes Balears, Cra. Valldemossa km 7.5, E-07071 Palma de Mallorca, Spain.

an atmosphere of 5% CO_2 in air at 100% humidity. For insulin release experiments, cells were plated 48 hours before assay at a density of 1×10^6 cells/well in 24-well plates. Thereafter, the medium was discarded and the cells were preincubated for 2 hours in RPMI 1640 medium alone. The medium was then discarded and replaced by RPMI 1640 containing 0.6 mM glucose. Test substances were dissolved in phosphate-buffered saline solution and added in a volume of 50 μl per well. Experiments were carried out for 2 hours in a final volume of 1,050 μl/well. At the end of each experiment, culture medium was collected and assayed for insulin content by radioimmunoassay.[6]

Radioligand binding assays with $[^{125}I]p$-iodoclonidine (0.5×10^{-9} M), a high affinity ligand for I_1-imidazoline receptors, were performed in membranes (P_2 fractions) of RIN-5AH cells as previously described by others.[7] Analyses of competition experiments as well as the fitting of data to the appropriate binding models were performed by computer-assisted nonlinear regression using the EBDA-LIGAND programs.[8]

One- or two-way analysis of variance (ANOVA) followed by Scheffé's test and two-tailcd Student's *t* test were used for statistical evaluations. The level of significance was chosen as $p = 0.05$.

Exposure of RIN-5AH cells to glucose (0–5 mM) stimulated insulin release in a dose-dependent manner. The imidazoline and α_2-adrenoceptor antagonist efaroxan (100 μM) caused a significant additional release of insulin that was more pronounced at low glucose concentrations (0.3–1.2 mM). The concentration of 0.6 mM of glucose was chosen for all further experiments. Irreversible blockade of the α_2-adrenoceptor population by preincubation with 10 μM benextramine did not significantly modify insulin release from RIN-5AH cells (104 ± 6% of insulin release in control cells: $p > 0.05$) and did not block the stimulatory effect of 100 μM efaroxan (155 ± 10% of insulin release in control cells; $p < 0.05$). Moreover, antazoline, an imidazoline drug that is not an α_2-adrenoceptor antagonist, also stimulated insulin release from these cells in a dose-dependent manner (200 ± 11% of insulin release in control cells at 100 μM; $p < 0.05$). Thus, the primary determinant of the secretagogue activity of efaroxan in RIN-5AH cells is not α_2-antagonism per se, but rather the possession of an imidazoline ring within the molecule. Epinephrine completely displaced the binding of $[^{125}I]p$-iodoclonidine to membranes of RIN-5AH cells from a single site and with high potency (IC_{50} = 5.2 nM). Efaroxan, clonidine, rilmenidine, and moxonidine also displaced the total binding of the radioligand to the same extent as did epinephrine and from only one binding site (IC_{50s} = 3.2, 6.2, 28.7, and 62.4 nM, respectively). Imidazole-4 acetic acid, a compound that binds to I_1-imidazoline receptors but not to α_2-adrenoceptors,[7] did not displace the binding of $[^{125}I]p$-iodoclonidine.

The order of potency for seven other imidazolines stimulating insulin release was cirazoline > idazoxan ~ naphazoline > clonidine > RX821002 ~ tolazoline > moxonidine. The guanidines guanoxan (200 μM) and amiloride (200 μM) but not aganodine also stimulated insulin release from RIN-5AH cells (154 and 151% of insulin release in control cells, respectively; $p < 0.05$). Cirazoline, idazoxan, and guanoxan are high affinity ligands for I_2-imidazoline receptors.[9] Preincubation of RIN-5AH cells for 40 minutes with 10 μM clorgyline, a compound that irreversibly binds to I_2-imidazoline receptors,[10] did not block the stimulatory effects of cirazoline or idazoxan on insulin release. However, cirazoline (100 μM) and idazoxan (500 μM) completely reversed the inhibition by diazoxide (250 μM), an opener of K^+_{ATP} channels, of glucose-induced insulin release. Several conclusions can be deduced from the present data: (1) several agents that possess either an imidazoline or a guanidine moiety within their structure stimulate insulin release from RIN-5AH insulinoma cells; (2) epinephrine completely displaced the binding of [^{125}I]*p*-iodoclonidine to membranes of RIN-5AH cells from a single high affinity binding site, indicating that these cells express α_2-adrenoceptors but not I_1-imidazoline receptors; (3) irreversible blockade of the I_2-imidazoline receptors with clorgyline did not prevent the stimulatory effects on insulin release of cirazoline or idazoxan; however, these compounds completely reversed the inhibition by diazoxide of glucose-induced insulin release; and (4) these results suggest that certain imidazoline/guanidine compounds might interact with a low affinity site at or near the K^+_{ATP} channel that does not represent any of the known subtypes of imidazoline receptors.

REFERENCES

1. Ashcroft, F. M., D. E. Harrison & S. J. H. Ashcroft. 1984. Nature **312:** 446-448.
2. Cook, D. L. & C. N. Hales. 1984. Nature **311:** 271-273.
3. Plant, T. D. & J. C. Henquin. 1990. Br. J. Pharmacol. **101:** 115-120.
4. Chan, S. L. F., M. J. Dunne, M. R. Stillings & N. G. Morgan. 1991. Eur. J. Pharmacol. **204:** 41-48.
5. Chan, S. L. F., C. A. Brown & N. G. Morgan. 1993. Eur. J. Pharmacol. **230:** 375-378.
6. Albano, D. J. M., R. P. Ekins, G. Maritz & R. C. Turner. 1972. Acta Endocrinol. **70:** 487-494.
7. Ernsberger, P. R., T. H. Damon, L. M. Graff, S. G. Schäfer & M. O. Christen. 1993. J. Pharmacol. Exp. Ther. **264:** 172-182.
8. Munson, P. J. & D. Rodbard. 1980. Anal. Biochem. **107:** 220-239.
9. Miralles, A., G. Olmos, M. Sastre, F. Barturen, I. Martin & J. A. García-Sevilla. 1993. J. Pharmacol. Exp. Ther. **264:** 1187-1197.
10. Olmos, G., A. M. Gabilondo, A. Miralles, P. V. Escriba & J. A. García-Sevilla. 1993. Br. J. Pharmacol. **108:** 597-603.

Influence of Clonidine on Alloxan-Induced Diabetes in Rats

MIHAI NECHIFOR,[a] CRISTIANA FILIP,
ELENA TESLARIU, BOGDAN I. NEUGHEBAUER,
MARCEL COSTULEANU, AND ADRIANA NEGRU

Department of Pharmacology
University of Medicine and Pharmacy
16 University Street
RO-6600, Iasi, Romania

Accumulated data indicate that one or both populations of imidazoline receptors are present in various organs and tissues and that some substances exert actions on these sites.[1–3] This work investigates the possible involvement of these receptors in the regulation of glycemia and in the pathogenesis of alloxanic diabetes.

MATERIAL AND METHODS

Experiment 1. We studied the influence of clonidine on experimental alloxan-induced diabetes in rats. The series consisted of three groups of 10 adult male Wistar rats each, bred under identical laboratory conditions. Twenty hours before the start of the experiment, animals were only allowed water ad libitum.

Group I served as the controls and received no substances. Groups II, III, and IV received alloxan 175 mg/kg subcutaneously to induce diabetes. Group III also received clonidine 0.5 mg/kg orally (single dose) 3 hours before the administration of alloxan. This dose was repeated at 24 and at 44 hours. Forty-eight hours after the induction of diabetes and 4 hours after the last dose of clonidine, we determined the level of glycemia.

Group IV was treated the same as group III, but, in addition, rats received yohimbine 1 mg/kg intraperitoneally 1 hour after alloxan administration. These doses were repeated after 24 hours and 4 hours before the determination of glycemia.

After the induction of alloxanic diabetes, all animals received water ad libitum and 3 g/kg of glucose (3% aqueous solution through an endogastric catheter) every 6 hours. The last dose of glucose was given 6 hours before clonidine administration. The obtained data were statistically interpreted by Student's *t* test.

Experiment 2. Two groups of 10 adult male Wistar rats each, bred under identical laboratory conditions, were studied. Twenty hours before the experiment was started, animals were not fed, but were allowed water ad libitum. Alloxanic diabetes was induced by the subcutaneous administration of 175 mg/kg of alloxan in both series.

[a] Address for correspondence: Prof. Dr. Mihai Nechifor, Department of Pharmacology, University of Medicine and Pharmacy, 16 University Street, RO-6600, Iasi, Romania.

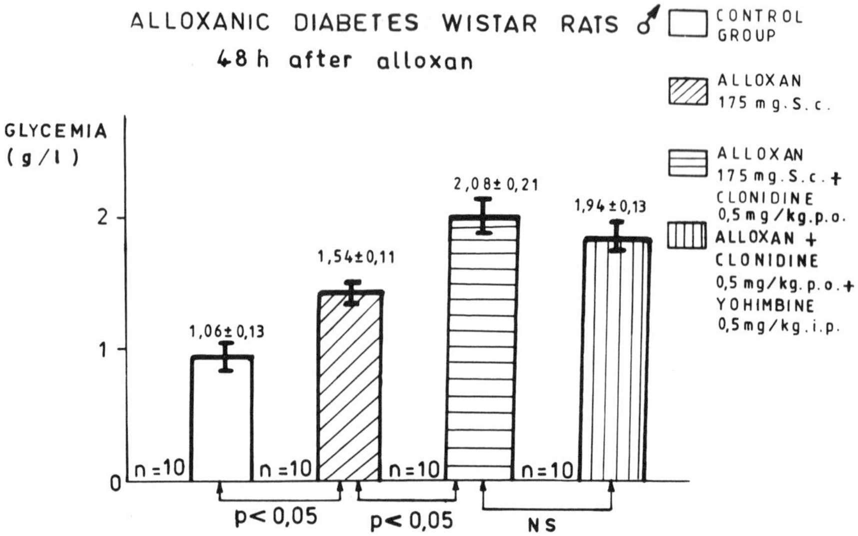

FIGURE 1. Effect of clonidine alone or both clonidine and yohimbine on glycemia in alloxanic diabetes.

Seventy-two hours after the induction of alloxanic diabetes, Group I received insulin Actrapid (Nova) 3 IU/kg intraperitoneally; Group II received Actrapid (Nova) 3 IU/kg intraperitoneally and clonidine 0.5 mg/kg orally 3 hours before insulin administration.

Blood for determination of glycemia was taken from the retroorbital plexus 2 and 5 hours after insulin administration. The obtained data were statistically interpreted by means of Student's *t* test.

RESULTS

The influence of clonidine alone or both clonidine and yohimbine on glycemia in alloxanic diabetes is shown in FIGURE 1. The effect of clonidine on the action of insulin in alloxanic diabetes is shown in FIGURE 2.

DISCUSSION AND CONCLUSIONS

These data show that clonidine increases hyperglycemia from alloxanic diabetes and reduces the hypoglycemic effect of insulin in experimental diabetes in rats.

The effect of clonidine was obtained at high doses, which cannot be achieved in therapeutic practice. It is not significantly influenced by yohimbine. That is why it cannot be mediated by the action of clonidine to stimulate α_2 presynaptic receptors.

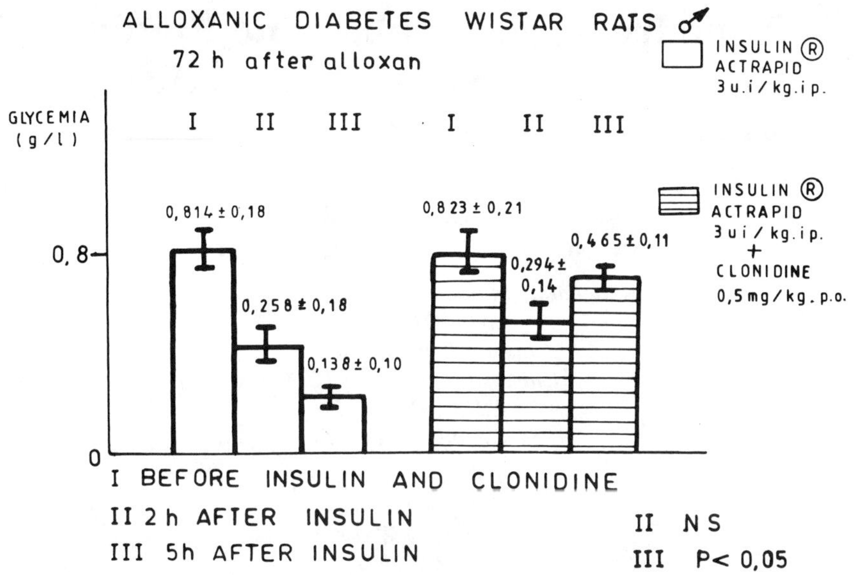

FIGURE 2. Effect of clonidine on the action of insulin in alloxanic diabetes.

We believe that this effect is mediated by the action of clonidine on imidazoline receptors. Which kind of receptor (I_1 or I_2) we cannot say precisely. Clonidine is known to affect, but to have different affinities for, both I_1 and I_2 sites.

REFERENCES

1. Gomez, R., P. Ernsberger, G. Feinland & D. J. Reis. 1990. Eur. J. Pharmacol. **183:** 168.
2. Hamilton, C. A., M. A. Yakubu, E. Jardine & J. L. Reid. 1990. Eur. J. Pharmacol. **183:** 1150.
3. Piletz, J. E. & K. Sletten. 1993. J. Pharmacol. Exp. Ther. **267:** 1493-1502.

Nonadrenergic Imidazoline Binding Sites and Amine Oxidase Activities in Fat Cells

CHRISTIAN CARPÉNÉ,[a] LUC MARTI,
ALAN HUDSON,[b] AND MAX LAFONTAN

Institut National de la Santé et de la Recherche Médicale
(INSERM U 317)
Institut Louis Bugnard
Université Paul SABATIER
C.H.U. Rangueil
31054 Toulouse Cedex, France

[b]*Psychopharmacology Unit*
School of Medical Sciences
University WALK
Bristol, BS8 LTD, UK

Numerous organs, tissues, and cell types possess binding sites for imidazoline drugs (for a review see refs. 1-3). These sites have generally been characterized by their binding parameters rather than their biochemical properties or biological activity. At present, sites labeled with high affinity by [^{3}H]*para*-aminoclonidine[4] or other clonidine analogs[5] are clearly distinct from those labeled by [^{3}H]idazoxan.[6] The former type, now called the I_1-site, is mainly involved in the hypotensive effects of imidazolines administered in animals (reviewed in ref. 7 and in this volume). The latter type, called the I_2-site, was previously termed the "nonadrenergic [^{3}H]idazoxan binding site" (NAIBS),[6] the "imidazoline-guanidinium receptive site" (IGRS),[2] or other acronyms. In early studies, I_2-sites were described as pharmacologically[8-10] and physically[6,11] distinct from adrenergic receptors. Recently, imidazoline bindings sites were demonstrated to be located essentially in the mitochondrial outer membrane, in the liver,[12] but also in the kidney, brain,[2] and adrenal medulla.[13] The molecular weights (55-70 kD) of the I_2-sites in the kidney,[14] liver,[15] and chromaffin cells[16] were also reported. However, the previous implications that I_2-sites are involved in the regulation of biological events such as K^+ transport,[17] noradrenaline release,[3] or insulin secretion[18] were in apparent disagreement with the mitochondrial location of such sites. Therefore, the search for a biological effect of I_2-site ligands remains open. Recent reports indicate a parallel between I_2-sites and another mitochondrial protein, monoamine oxidase (MAO); some inhibitors of MAO competed *in vitro* for [^{3}H]idazoxan binding in rat[19] and rabbit[20] brain membranes. However, the authors

[a]Address for correspondence: Dr. Christian Carpéné, INSERM U 317, Institut Louis Bugnard, C.H.U. Rangueil, Bât. L3, 31054 Toulouse Cedex, France.

did not argue for a close relation between I_2-sites and MAO because: (1) the affinity of MAO inhibitors (MAOIs) was 10-fold lower for imidazoline binding sites than for MAO[20]; (2) idazoxan was only a weak competitor of the binding of labeled MAOIs[19,21]; and (3) in the brain the density of I_2-sites was much lower than that of the two MAO isoenzymes. In accordance with these pharmacological observations, a similar conclusion was drawn from the differences observed in the isoelectric points of MAO-A, MAO-B, and I_2-sites purified from kidney.[14] Nevertheless, MAO-B and I_2-sites were found to be co-regulated in human brain; both exhibited increasing densities with age.[21] The increased density of I_2-sites was also observed in the brain of depressed suicide victims.[22] Finally, chronic treatment with the MAOI clorgyline downregulates [^{3}H]idazoxan binding sites in rat brain.[19] When it is noted that MAOIs are largely used for their antidepressive properties, it is also suspected that interactions between I_2-sites and MAO have functional implications. Such interactions between I_2-sites and MAO-A or MAO-B could follow from their common location in the outer mitochondrial membrane, originally described in liver by Tesson and coworkers.[12]

All these observations prompted us to determine if the subcellular co-location of MAO and I_2-sites could be extended to a functional relationship and even to a closer homology between the two entities. To obtain such evidence we first tested the capacity of some inhibitors of MAO to compete for [^{3}H]idazoxan binding on the imidazoline receptors of white adipose tissue which were previously described in various animal species[10,23] including man.[24] Secondly, the effects of imidazoline derivatives were compared to those of various MAO blockers (including MAO-A or MAO-B selective inhibitors) on the MAO activity present in adipose tissue. The same approach was used in rat liver, a model well known for its I_2-sites[17,25] and MAO activity.[26] Finally, we tested if the inhibition of amine oxidases could influence adipocyte responses to lipolytic amines.

The results indicate that among the tested agents, some MAO inhibitors and imidazoline molecules share displacing capacities of [^{3}H]idazoxan binding and inhibition of MAO activity. Some I_2-selective imidazoline compounds were evident (RX 801077, S 15674, and BU 224); they exhibited 100- to 2,000-fold less affinity for α_2-adrenoceptors than did idazoxan itself. These I_2-site ligands inhibited amine oxidation in both liver cells and adipocytes. In addition, [^{3}H]idazoxan binding was also inhibited in both models by several inhibitors of MAO which are not imidazoline or guanidinium derivatives: clorgyline, tranylcypromine, and harmaline. Indeed, inhibition of MAO activity represents a novel biological effect of ligands selective for I_2-sites and poses the question of the functional interplay or even the homology between I_2-sites and MAO.

CHARACTERIZATION OF NONADRENERGIC IMIDAZOLINE BINDING SITES IN FAT CELLS

[^{3}H]Idazoxan total binding on adipocyte membranes can be separated into at least three components, as shown by the competition studies depicted in FIGURE 1. The first component was adrenergic in nature and corresponded to the portion of binding that was displaced by 100 μM of epinephrine. The second portion was recognized by imidazoline derivatives such as UK 14304 and cirazoline but not by epinephrine

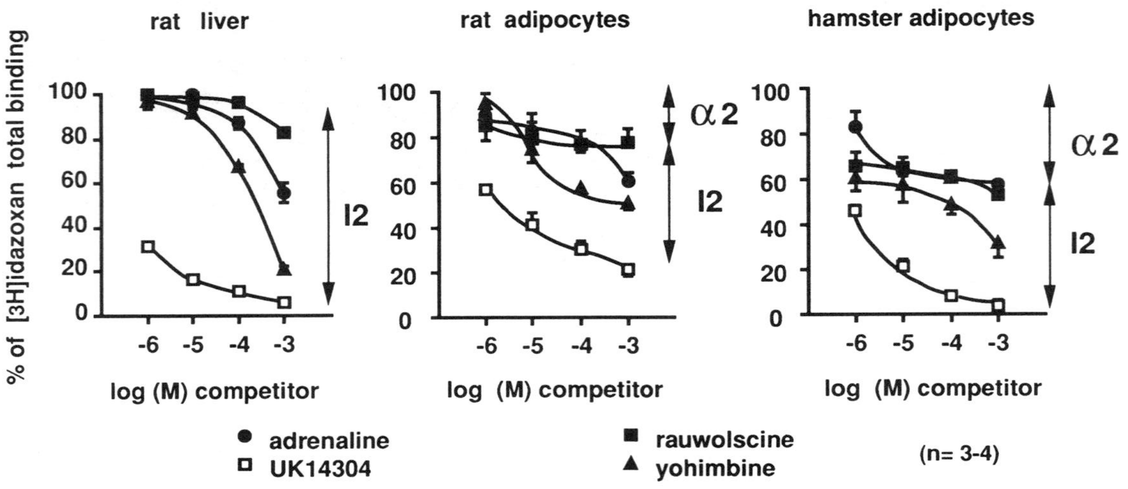

FIGURE 1. Inhibition of total binding of [^{3}H]idazoxan to crude membranes of rat liver, rat adipocytes, and hamster adipocytes. Membranes were incubated for 45 minutes in the presence of a fixed concentration of [^{3}H]idazoxan (6 nM for liver and 10 nM for adipocytes) and with increasing concentrations of the following competitors: adrenaline (●), UK 14304 (□), which is an α_2-agonist with an imidazoline moiety, and the non-imidazoline α_2-antagonists rauwolscine (■) and yohimbine (▲). Binding of [^{3}H]idazoxan to adrenergic receptors displaced by micromolar concentrations of epinephrine is delimited by the α_2-portion, whereas binding to nonadrenergic sites is delimited by the I_2 portion.

which, at 1 mM, displaced only a minor part of this binding. The third portion was not competed for by any of the compounds tested. This nonspecific binding, when defined in the presence of 100 μM of cirazoline, averaged 5 and 12% of the total binding at [^{3}H]idazoxan 10 nM in hamster and rat adipocytes, respectively. The same percentages of nonspecific binding were obtained with millimolar concentrations of other imidazoline drugs, such as UK 14304, as illustrated in FIGURE 1. In liver membranes, 5% of [^{3}H]idazoxan total binding was nonspecific. Moreover, [^{3}H]idazoxan mainly labeled nonadrenergic sites in rat liver because 100 μM of epinephrine displaced less than 10% of total binding (FIG. 1). A greater part of the binding was competed for by 1 mM of epinephrine but not by the α_2-antagonist rauwolscine. Among the α_2-antagonists tested in competition for [^{3}H]idazoxan binding, rauwolscine was more selective for α_2-adrenergic receptors than for nonadrenergic sites. Another α_2-antagonist, yohimbine, recognized part of the nonadrenergic binding sites especially at millimolar concentrations, that is, at doses higher than those needed to block α_2-receptors, because the K_I value of yohimbine was 3 nM for the α_2-receptors of human adipocytes.[24] Analysis of the inhibition curves of rauwolscine evidenced a plateau between 1 μM and 1 mM of this agent; the percentage of total [^{3}H]idazoxan binding resistent to displacement by rauwolscine ranged between 100 and 82% in the liver, 87 and 80% in rat fat cells, and 65 and 55% in hamster fat cells (FIG. 1). This result indicates that the relative proportion of α_2-adrenoceptors to imidazoline binding sites was negligible in the rat liver, about one fourth in rat adipocytes, and two thirds in hamster adipocytes. These proportions were confirmed by saturation analyses. For these direct quantifications of nonadrenergic binding sites, increasing concentrations of [^{3}H]idazoxan were used in the presence of 10 μM of rauwolscine to preclude binding to α_2-adrenoceptors. The α_2-adrenoceptors were quantified along with the α_2-antagonist [^{3}H]RX 821002. This agent, which is 2-methoxy-idazoxan, did not interfere with the nonadrenergic sites, at least in the nanomolar range, as previously reported in hamster,[27] rat,[23] and human fat cells.[24] In rat fat cells, nonadrenergic idazoxan binding sites were characterized by an apparent dissociation constant (K_D of 15.7 ± 2.8 nM and a maximal binding capacity (B_{max}) of 290 ± 93 fmol/mg protein (n = 5). In hamster adipocyte membranes, the nonadrenergic sites were more numerous: B_{max} = 564 ± 129 fmol/mg protein (K_D = 9.3 ± 2.2 nM, n = 4). However, fat cell membranes from the hamster also possessed more α_2-adrenoceptors than did those from the rat (B_{max} of [^{3}H]RX 821002 = 776 ± 60 and 39 ± 2 fmol/mg protein, respectively). In rat liver membranes, [^{3}H]idazoxan labeled 450 ± 46 fmol/mg protein of nonadrenergic sites (K_D = 13.0 ± 3.4, n = 14), whereas [^{3}H]RX 821002 labeled only a small amount of binding sites (B_{max} = 43 fmol/mg protein) of relatively low affinity (K_D around 13 nM).

THE FAT CELL: A MODEL FOR DETERMINING THE SELECTIVITY OF LIGANDS FOR NONADRENERGIC IMIDAZOLINE BINDING SITES

Because white fat cells express both α_2-adrenoceptors and nonadrenergic binding sites, it was convenient to verify the distinct pharmacologies of both types of sites on the same cell type. Competition studies of [^{3}H]idazoxan binding confirmed that in rat and hamster fat cells, most of the imidazoline derivatives possessing α_2-adrenergic properties were able to bind to nonadrenergic binding sites (e.g., cirazoline, naphazoline, UK 14304, and idazoxan itself), whereas other imidazolines, such as RX 821002, phentolamine, rilmenidine, and clonidine (not shown), were weak competitors. This was in agreement with previous findings in human adipose tissue.[24] The low affinity of clonidine allowed nonadrenergic binding sites to be classified as I_2-sites according to the recently proposed classification.[28,29] The presence of I_1-sites on white adipocytes was not tested directly by binding analyses of labeled clonidine derivatives (*p*-aminoclonidine or *p*-iodoclonidine) but was excluded from the pattern of competition curves of [^{3}H]clonidine binding. At 100 μM, idazoxan, clonidine, UK 14304, RX 821002, phentolamine, yohimbine, and epinephrine maximally inhibited to the same extent the [^{3}H]clonidine binding to hamster fat cell membranes; 76 to 88% of the total binding of 8 nM of ligand was inhibited in these conditions. In addition, the maximal binding capacity of [^{3}H]clonidine was 468 ± 76 fmol/mg protein in hamster adipocytes, i.e., slightly less than that reported for [^{3}H]RX 821002. This was in agreement with the respective agonist and antagonist properties of these two radioligands towards α_2-adrenoceptors.[27] Accordingly, [^{3}H]clonidine binding to rat adipocytes was at the limits of detection because of the small population of α_2-adrenoceptors in this model (see above and ref. 23).

To discriminate new compounds more selective for I_2 sites than for α_2-adrenoceptors, we compared the binding properties of commonly used imidazoline ligands (such as idazoxan or cirazoline) to those of various molecules that can be classified according to their origin: (1) derivatives of idazoxan that were first developed by Reckitt and Colman Laboratories (Hull, England) and now acquired by Pierre Fabre Laboratories (Labège, France), namely, RX 821029 or 2-(1,3-benzodioxanyl)-2-imidazoline, and RX 801077 or 2-benzofuran-2-yl-imidazoline; (2) heptyl imidazole derivatives such as S 15 674 (1-methyl 5-*n*-heptyl imidazole) and S 15430 (2-*n*-heptylimidazoline), synthesized in the Institut de Recherches Servier (Suresnes, France), in addition to S 15898 (2-(3,3-dicyclopropylpropyl)imidazoline); and (3) BU 224 (2-(4,5-dihydroimidaz-2-yl)-quinoline) and BU 239 (2-(4,5-dihydroimidaz-2-yl)-quinoxaline) synthesized by Dr. S. Husbands (Bristol, England).

RX 821002 up to 10 μM did not bind to I_2-sites of rat or hamster adipocytes and was the least I_2-selective of the tested compounds, justifying its use as an α_2-adrenergic ligand (FIG. 2). When also tested for their selectivity, neither RX 821029 nor RX 801077 exhibited a significantly higher affinity than did idazoxan itself (RX 781094) for adipocyte I_2-sites, whatever the animal model used (FIG. 2). However, they exhibited much lower affinity in competition for [^{3}H]RX 821002 binding, particularly RX 801077, which demonstrated the lowest affinity for α_2-adrenoceptors (TABLE 1). The most I_2-site-selective agent of the tested compounds was BU 224; it displayed the highest affinity for I_2-sites but had an affinity comparable to that of S 15430 or S 15898 with respect to α_2-adrenoceptors and was thus endowed with the greatest ratio of K_I α_2/K_I I_2. The data in TABLE 1 indicate that RX 801077, S 15674, and BU 224 display the highest selectivity for I_2-sites. By contrast, cirazoline, naphazoline, and bromoxidine (UK 14304), which exhibited high affinity for both I_2-sites and α_2-adrenoceptors, cannot be considered further as highly I_2-site-selective.

The use of RX 801077, S 15674, and BU 224 as imidazoline ligands, therefore, is recommended because they offer considerably improved selectivity for I_2-sites, at

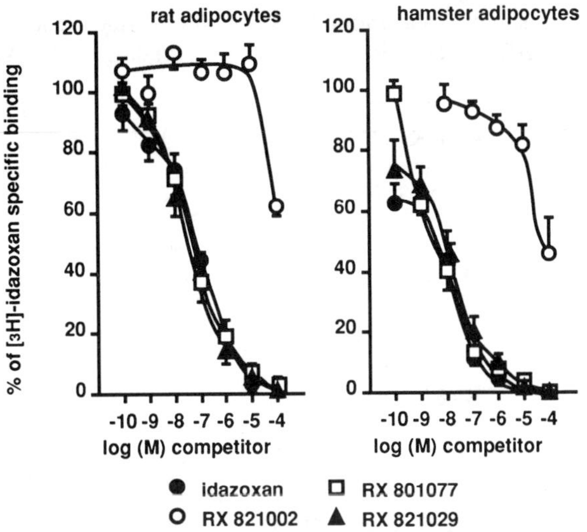

FIGURE 2. Inhibition of [^{3}H]idazoxan binding by idazoxan derivatives. To prevent the binding to α_2-adrenoceptors, binding studies were performed with 8–11 nM [^{3}H]idazoxan in the presence of 10 μM of rauwolscine (rat adipocytes, *left panel*) or 100 μM of epinephrine (hamster adipocytes, *right panel*). Data are expressed as a percentage of specific [^{3}H]idazoxan bound after subtraction of the binding not displaced by 100 μM cirazoline. Mean ± SEM of four (hamster) or six (rat) experiments.

least when compared with α_2-adrenoceptors. However, the need for ligands possessing high affinity and selectivity for nonadrenergic imidazoline binding sites remains crucial in order to label only I-sites, because many investigators still need to mask the α-adrenergic receptors with high concentrations of adrenergic agents mixed with radiolabeled idazoxan or clonidine derivatives. The adipocyte could be a convenient and predictive model to test the selectivity of these ligands. In this regard, the relatively high affinities of RX 801077 and S 15674 described here on adipocytes were also found for rat liver I_2-sites.[30]

DISTINCT PHARMACOLOGIES OF ADIPOCYTE α_2-ADRENERGIC RECEPTORS AND IMIDAZOLINE BINDING SITES

The use of newly synthesized imidazoline derivatives with higher affinity for I_2-sites than for α_2-adrenoceptors could also be used to confirm that these proteins are distinct, as already evidenced by biochemical approaches.[11] As expected, the pharmacological profile of adipocyte I_2-sites was not superimposed on that of α_2-adrenoceptors (FIG. 3). When 16 different drugs including α-agonists, α-antagonists, and imidazoline compounds were analyzed in competition experiments, no correlation was found between their respective affinities for I_2-sites and α_2-adrenoceptors. Only two compounds, idazoxan and UK 14304, were on the identity line of the correlational

TABLE 1. Comparison of Competition Parameters of Various Imidazolinic Derivatives on [^{3}H]Idazoxan and [^{3}H]RX 821002 Binding to Rat Adipocyte Membranes[a]

Competitors	Mean K_i (nM) on [^{3}H]Idazoxan Binding K_i I_2-Sites		Mean K_i (nM) on [^{3}H]RX 821002 Binding $K_i\alpha_2$-ARs		Ratio of $K_i\alpha_2/K_i$ I_2
Naphazoline	157 ± 91	(3)	0.2 ± 0.1	(3)	0.001
Bromoxidine	48.1 ± 11.3	(6)	74.9 ± 11.8	(3)	1.6
Cirazoline	3.8 ± 0.9	(3)	165.5 ± 0.5	(3)	43.6
RX 821002	97,721 ± 17,143	(6)	0.3 ± 0.1	(3)	0.000003
RX 821029	39.2 ± 33.4	(6)	1,028 ± 202	(3)	26.2
RX 801077	55.0 ± 38.3	(6)	84,257 ± 22,977	(3)	1532
S 15430	122 ± 44	(3)	8,940 ± 5,839	(3)	73.3
S 15898	56.1 ± 17.9	(3)	6,263 ± 1,522	(3)	111
S 15674	314.6 ± 63.4	(3)	70,050 ± 8,156	(3)	223
BU 239	68 ± 49.7	(3)	14,773 ± 9,516	(4)	217
BU 224	3.0 ± 0.4	(4)	7,333 ± 1,736	(4)	2444

[a] Competitions of [^{3}H]idazoxan binding were performed in the presence of 10 μM of rauwolscine to preclude binding on α_2-adrenoceptors. Pseudo-inhibition constants (mean K_i, in nM) were calculated from the IC_{50} values following the equation: $K_i = IC_{50}/(1 + (L/KD))$ where L, concentration of radioligand, was 7-12 nM for [^{3}H]idazoxan and 6-8 nM for [^{3}H]RX 821002. KD were 15 and 6 nM, respectively. Concentrations of competitor reducing the specific binding (IC_{50}) by one half were calculated from Hill plot of the competition dose response even when Hill coefficients (nHill) were different from unity, that is, even when each individual competition curve was shallow or biphasic. The ratio of mean K_i values was an estimate of the selectivity ratio between α_2-adrenoceptors and imidazoline binding sites. Values are means ± SEM of (*n*) separate determinations.

plot. With hindsight, this can explain the previously observed similarities of the adrenergic and nonadrenergic binding components of these two ligands in hamster[27] and rat[23] adipocytes.

In parallel studies, the affinity constants of 20 agents for adipocyte α_2-adrenoceptors were compared in the rat and hamster. This approach showed that although less numerous, rat adipocyte α_2-adrenoceptors shared the same pharmacological profile as did those of the hamster (FIG. 4a). Only prazosin and chlorpromazine shared the low affinity of S 15430, S 15674, S 15430, RX 821029, and RX 801077 for α_2-adrenoceptors. When the affinities for I_2-sites were compared between the two species, the regression line drawn from the correlational plot did not demonstrate any interspecific difference (FIG. 4b). However, this overall homology does not imply that any further imidazoline drugs to be tested will display the same affinity for I_2-sites in both species. In fact, numerous competitors of [^{3}H]idazoxan binding demonstrate shallow or biphasic competition curves, arguing for heterogeneity of the I_2-sites as already detailed in guinea pig kidney.[31] It therefore cannot be excluded that in future studies, a novel molecule may exhibit a higher affinity for one subclass of I_2-sites

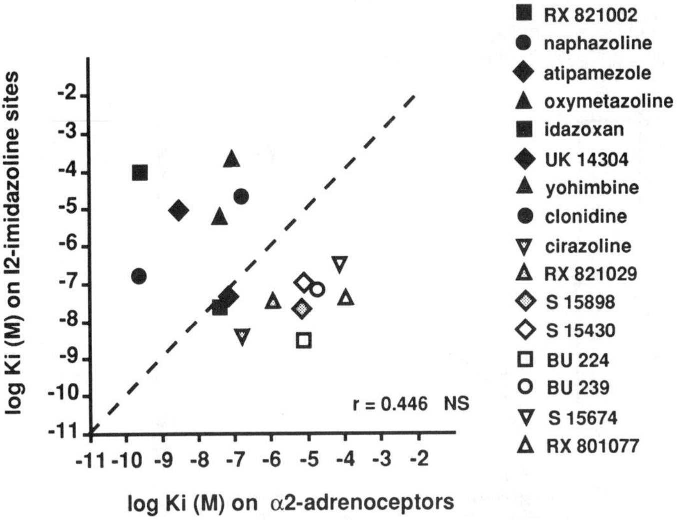

FIGURE 3. Correlational plot of affinity constants (K_i) for I_2-sites versus K_i for α_2-adrenoceptors in rat adipocyte membranes. Abscissa values designate the logarithmic values of K_i (M) for the compounds tested in competition with [^{3}H]RX 821002 binding. Ordinate values designate log(K_i) values of the same compounds tested in competition with [^{3}H]idazoxan binding (in the presence of rauwolscine). In the legend, adrenergic agents or imidazoline derivatives are presented in decreasing rank order of affinity for α_2-adrenoceptors. The correlation line was not drawn, because there was not a significant relation between the two variables; correlation coefficient *r* was 0.446, nonsignificant (NS) at 14 degrees of freedom. Each point is the mean of at least three determinations. *Dotted line:* line of identity.

and present a different overall affinity between the models used, especially if the proportion of each subclass of I_2-sites, as yet unknown, differs from one species to another. Despite the heterogeneity of I_2-sites, the present correlation analyses, which used pseudo-K_I values derived from IC_{50} values even when Hill coefficients were different from unity, remain a valid tool to assess the overall selectivity of I_2-ligands. It has been successfully used to discriminate the I_1- and I_2-sites from the α_2-adrenoceptors present on human platelets.[5] The same analysis did not evidence major differences between rat liver and rat adipocyte I_2-sites (not shown).

In these correlation analyses, RX 821002 was one of the idazoxan derivatives that exhibited a high selectivity for α_2-adrenoceptors in rat and hamster adipocytes (TABLE 1, FIGS. 2-4), as previously reported in human and rabbit adipoctyes.[24,32] However, it was the imidazoline with the lowest affinity in competition of binding to I_2-sites in both liver and adipocytes. This observation is sufficient to illustrate the difficulty in establishing a structure-activity relationship with the compounds tested. However, among the imidazolines tested, several compounds possess a cyclopropyl ring (S 15898, cirazoline). One of the two non-imidazoline drugs reported in FIGURE 4b also possessed a cyclopropyl ring. This compound, namely, tranylcypromine (*trans*-2-phenylcyclopropanamine), displayed an intermediate affinity for both rat and hamster I_2-sites. The other non-imidazoline agent reported in FIGURE 4b was

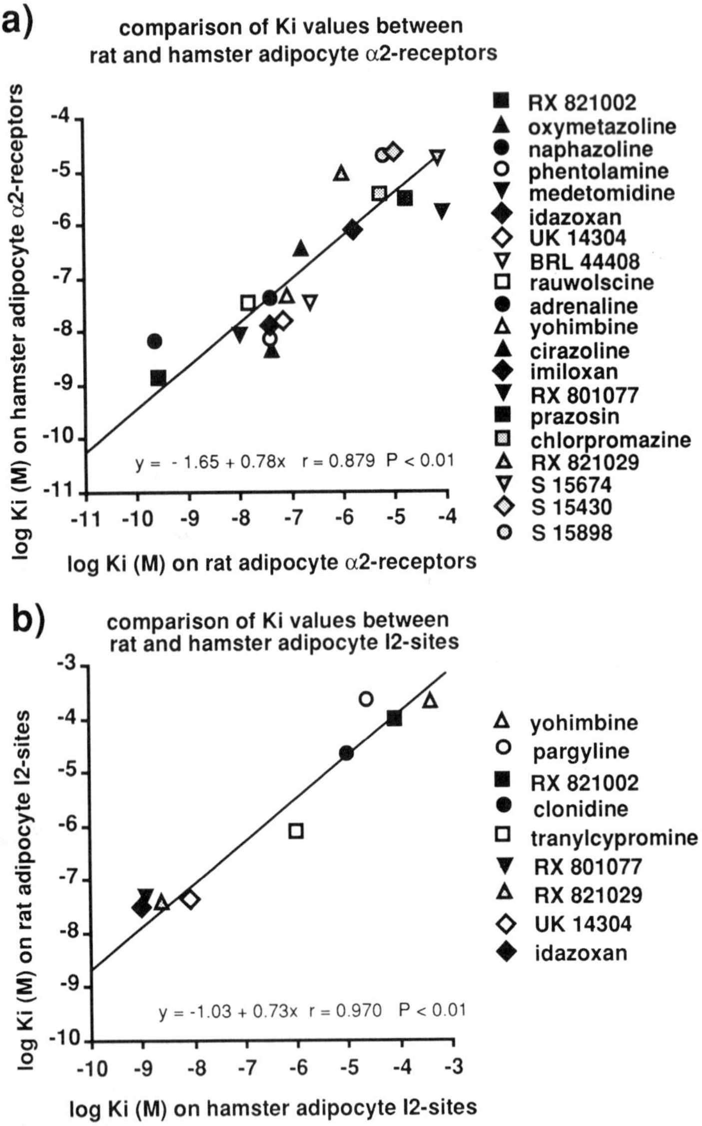

FIGURE 4. Correlational plots of affinity constants (K_i) for α_2-adrenoceptors (**a**) or for I_2-sites (**b**) in rat versus hamster adipocyte membranes. (**a**) Logarithms of K_i values of competitors were derived from competition studies of [^{3}H]RX 821002 binding to adipocyte membranes from hamsters (ordinate) or rats (abscissa). In the legend, adrenergic agents or imidazoline derivatives are in decreasing rank order of affinity for hamster α_2-adrenoceptors. (**b**) Log (K_i) values are derived from competition studies of [^{3}H]idazoxan binding (with a rauwolscine mask) to adipocyte membranes of rat (ordinate) or hamster (abscissa). Correlation lines were drawn by linear regression analysis (equation and significance of correlation coefficient are reported). Each point is the mean of at least three determinations.

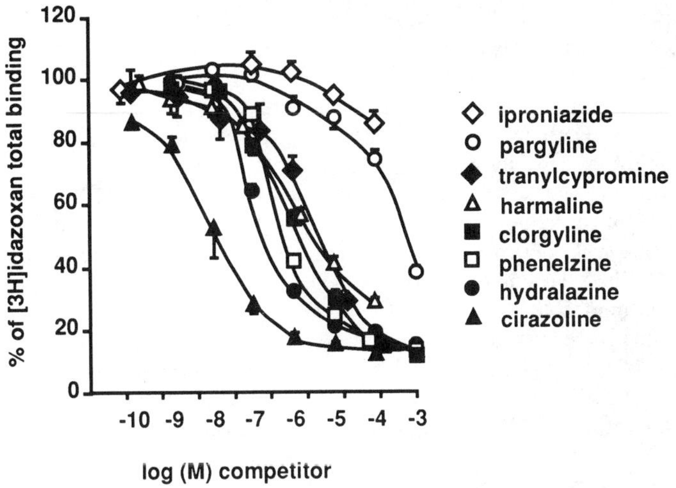

FIGURE 5. Inhibition of [^{3}H]idazoxan binding to rat fat cell membranes (n = 3–4) by monoamine oxidase inhibitors. Inhibition curve for cirazoline is given as a reference. Data are expressed as a percentage of total [^{3}H]idazoxan bound, in the presence of 10 μM rauwolscine, which ranged from 180–250 fmol/mg protein.

pargyline, a classical inhibitor of MAO activity. Further comments on the inhibition of [^{3}H]idazoxan binding by non-imidazoline drugs follow.

INHIBITION OF [^{3}H]IDAZOXAN BINDING TO I_2-SITES BY MONOAMINE OXIDASE INHIBITORS

Several MAO inhibitors induced dose-dependent inhibition of [^{3}H]idazoxan binding to I_2-sites in rat liver, hamster, and rat adipocytes (FIG. 5). In this last model, clorgyline exhibited a shallow inhibition curve (K_I = 460 ± 138 nM, n_H = 0.60 ± 0.08, n = 3) that was intermediate between the clearly biphasic curve previously observed in rat cerebral cortex by Garcia-Sevilla and coworkers[19] (K_{IH} was 0.04 nM and K_{IL} was 10,600 nM) and the monophasic inhibition reported for rabbit cerebral cortex[20] (K_I around 5,930 nM and n_H about 0.88). A biphasic competition curve was also obtained with clorgyline on liver I_2-sites (n_H = 0.47 ± 0.04, n = 3). Despite these discrepancies, clorgyline was a more efficient inhibitor of [^{3}H]idazoxan binding than was pargyline in the five models just mentioned. Only iproniazide possessed a lower inhibiting capacity than did pargyline, whereas the other unselective and irreversible MAOIs, phenelzine and tranylcypromine (*trans*-2-phenylcyclopropylamine), displayed only a 150- to 200-fold lower affinity than did the better imidazoline competitor tested, cirazoline. Inhibition of [^{3}H]idazoxan binding was not limited to the irreversible blocking properties of the inhibitors tested, because harmaline (4,9-dihydro-7-methoxy-1-methyl-^{3}H-pyrido[3,4-b]indole), which is a reversible inhibitor selective for MAO-A, shared the relatively high affinity of the irreversible MAOIs

(FIG. 5). It must be noted that the most efficient MAOI was hydralazine (K_I = 116 ± 46 nM, n = 3) which also possesses antihypertensive properties.

Although these MAOIs do not possess an imidazoline ring in their chemical structure and also are devoid of any imidazole moiety or guanidium group, some of them exhibited a relatively high affinity for [^{3}H]idazoxan binding sites and complete maximal inhibitory capacity comparable to that of some imidazoline derivatives.

Moreover, the MAO inhibitors competed for [^{3}H]RX 821002 binding on α_2-adrenoceptors of rat adipocytes with a noticeably lower affinity than that observed for I_2-sites (mean K_I > 1 mM for hydralazine). The same observation was made in hamster fat cells.[30] To our knowledge, MAOIs are the first molecules that do not belong to the imidazole or guanidium family of derivatives but exhibit affinity for the imidazoline-preferring sites. Because these competitors of [^{3}H]idazoxan are known to block the MAO[33] and because MAO activity and I_2-sites are colocalized in the mitochondrial fraction of various tissues,[2] we compared the effect of some imidazoline derivatives with that of MAOIs on amine oxidation in liver and fat cells.

INHIBITION OF MONOAMINE OXIDASE ACTIVITY BY MAOIs AND I_2-SITE LIGANDS

In white adipose tissue, as in liver, amine oxidase activity was measured with crude membrane preparations according to the method of Yu[26] using [^{14}C]tyramine as substrate (which is oxidized by both MAO-A and MAO-B). Under the conditions previously described,[30] the mean K_m values for tyramine were 0.33, 0.35, and 0.39 mM for rat adipocytes, hamster adipocytes, and liver cells, respectively. In the three preparations studied, basal oxidation of tyramine was totally abolished in the presence of tranylcypromine (irreversible inhibitor of MAO-AB) (FIG. 6). Pargyline (irreversible inhibitor of MAO-B) and harmaline (reversible inhibitor of MAO-A) were less potent inhibitors of enzyme activity. Their respective dose-dependent inhibition curves were biphasic or shallow, as expected for the presence of both MAO-A and MAO-B acting on the nonspecific substrate [^{14}C]tyramine.[33] In hamster and rat adipocytes, the isoform of MAO involved in tryamine oxidation remains to be characterized, but the biphasic inhibition curve for pargyline suggests that two MAO subtypes can coexist in adipose tissue. Furthermore, at least one third of tyramine oxidation by fat cell membranes was resistant to blockade by pargyline or harmaline and was due to an amine oxidase distinct from MAO-A or MAO-B. This enzyme belongs to the semicarbazide-sensitive amine oxidase family (SSAO), as previously described in brown[34] and white[35] adipocytes. It was not evident in liver cells. Strikingly, in all tissue preparations, various imidazolinic drugs were able to inhibit MAO activity. Among the drugs tested, the classical imidazoline ligand cirazoline was the most potent (mean IC_{50} values ranged between 20 and 180 μM), whereas RX 821029 and naphazoline exhibited lower affinity and intrinsic activity (total tyramine oxidation was maximally inhibited by 30–40%, 50–60%, and 60–70% in the presence of 1 mM of these compounds in rat fat cells, hamster fat cells, and liver cells, respectively) (FIG. 6). In addition, S15 674 and RX 801077 totally abolished tyramine oxidation by liver cells in a noncompetitive manner,[30] probably by acting on a site distinct from the catalytic one. In addition, the inhibitory action of BU 224 and BU 239 on

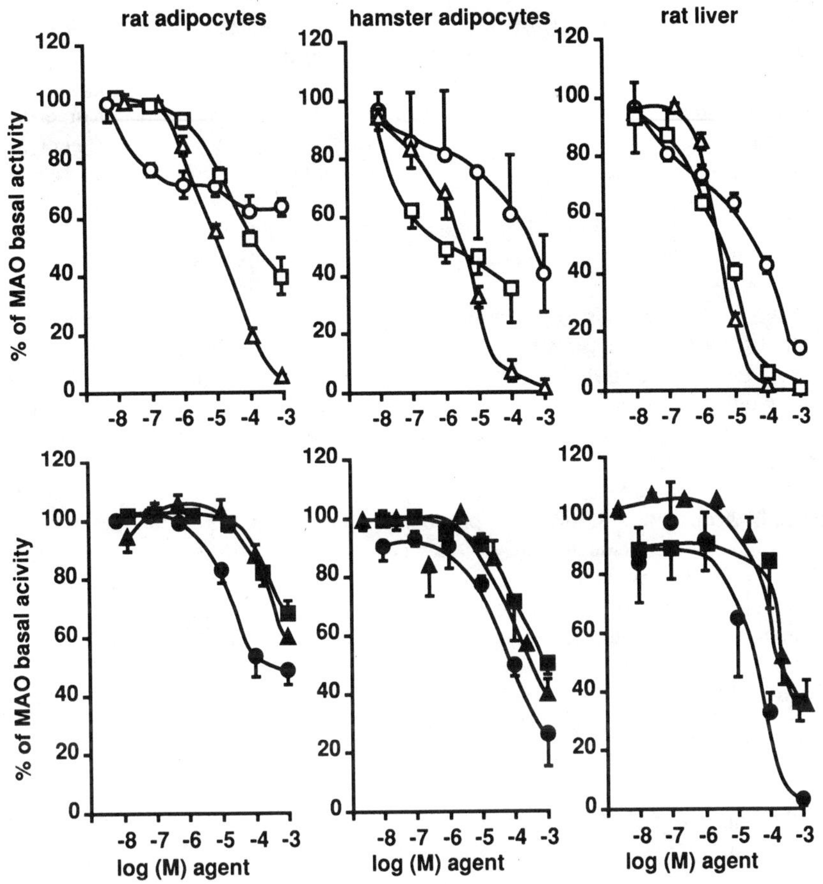

FIGURE 6. Inhibition of monoamine oxidase activity by various MAO inhibitors and imidazoline derivatives in rat (*left*) or hamster (*middle*) fat cell membranes and in rat liver membranes (*right*). The different preparations were incubated for 30 minutes in the absence (basal activity) or presence of increasing concentrations of the following MAO inhibitors: tranylcypromine (△), pargyline (□), and harmaline (○). Imidazolines RX 821029 (▲), cirazoline (●), and naphazoline (■) were added under identical conditions. Results are expressed as a percentage of basal MAO activity which represented 4.9 ± 0.7, 4.4 ± 0.3, and 5.4 ± 0.5 nmol of [^{14}C]tyramine oxidized/min/mg protein for hamster adipocyte (n = 6), rat adipocyte (n = 6), and rat liver (n = 9), respectively. The 0% corresponds to residual tyramine oxidation observed without or with inactivated biological material; values greater than 100% designate slight stimulation of basal oxidation.

liver MAO was reversible; after extensive washing, inhibition of basal MAO activity was lost, whereas it was maintained in the case of the irreversible MAOI clorgyline (data not shown). In this regard, the I_2-site ligands behaved differently from substrate suicide inhibitors which interact with the FAD group of the MAO catalytic site.[36]

Idazoxan itself was a poor inhibitor of liver MAO; at 1 mM it reduced basal oxidation of tyramine only by half. Because of a recent report on pancreatic islets in which agonist and antagonist properties were attributed to several imidazolines according to their effect on insulin secretion,[37] we verified if idazoxan could behave as an antagonist of the inhibitory effect of imidazolines on amine oxidase activity. However, complementary studies failed to demonstrate that idazoxan impaired the complete inhibitory action of other imidazolines.[30] Therefore, idazoxan cannot be considered a blocker of the imidazoline-induced inhibition of MAO, but rather it is a partial agonist. Clonidine (100 μM) was unable to inhibit more than 20% of MAO activity, a finding that, in addition to the low affinity of RX 821029 for α_2-adrenoceptors, implies that the effects observed were independent of α_2-adrenergic activation.

When the compounds just mentioned were tested on amine oxidase preparations purified from bovine plasma, tranylcypromine and RX 821029 shared similar IC_{50} values (23 and 17 μM) and maximal inhibitory capacity (less than 8% of the basal activity remained in the presence of 1 mM of inhibitor). Naphazoline and cirazoline blocked a large part of the enzyme activity at submillimolar doses. Surprisingly, pargyline and harmaline were ineffective on these preparations.[30]

Further studies are required for complete definition of this novel biological effect induced by imidazolines, inhibition of amine oxidase. From the present preliminary studies, it can be concluded that I_2-site ligands can inhibit MAO-A and/or MAO-B, because some of them totally counteracted the oxidation of a nonselective substrate, tyramine, in rat liver which is commonly known to possess equal proportions of both isoenzymes.[26] Furthermore, kinetic studies demonstrated that the site of action of imidazolines involved in inhibition of oxidase activity seems to be distinct from that of classical "substrate suicide" inhibitors of MAO. Whether the MAO enzyme itself, whose two isoforms, A and B, have already been sequenced,[38] harbors the binding site(s) of imidazolines remains to be elucidated. Such a hypothesis is relevant, because in liver both proteins are located in the mitochondrial outer membrane,[12,39] and share a similar molecular weight of around 60 kD.[15,38]

INFLUENCE OF THE INHIBITION OF AMINE OXIDATION ON FAT CELL LIPOLYTIC ACTIVITY

Beyond the functional relationship between MAO and I_2-sites, which is being intensively investigated in the brain,[19–21] it was of interest to define the physiological role of amine oxidation in peripheral organs. Involvement of peripheral MAO in the degradation of hyperpressive amines was essentially evident during the therapeutic use of some MAOIs as antidepressants. These nonselective inhibitors blocked peripheral MAO-A mainly in the liver and small intestine and prevented the degradation of amines assimilated from foodstuffs. During MAO blockade, exogenous amines, including tyramine, were not catabolized, released norepinephrine, and led to excessive, even fatal, hypertension crises, also called the "cheese effect."[36] Whereas liver and intestinal MAO activities are associated with a protective function against hypertensive attacks induced by endogenous or exogenous amines, very little is known about the role of MAO and SSAO in adipose tissue. Because amines of neuronal or systemic origin are involved in the regulation of lipolysis, we first tested if inhibition of their oxidation could influence their lipolytic capacities.

Briefly, blockade of MAO by 1 mM of pargyline did not influence lipolytic responses to maximal norepinephrine, octopamine (which acts via β_3-adrenoceptors[40]), or histamine (which essentially stimulates H_2-receptors[41]) (FIG. 7). As mentioned here, another adipocyte enzyme distinct from MAO is also responsible for the oxidation of amines, SSAO. The SSAO activity was thus blocked by 1 mM of semicarbazide. This failed to modify the lipolytic responses to the three amines studied. This last observation contrasted with the findings of Raimondi *et al.*[41] who reported an increase in the moderate lipolytic effect of histamine when rat adipocytes were incubated in the presence of 1 mM of semicarbazide.[41] Simultaneous blockade of both MAO and SSAO did not further modify the lipolytic dose-responses reported in FIGURE 7. Catecholamines can also induce antilipolytic responses through activation of α_2-adrenoceptors, but no influence of MAO blockade was evident on α_2-adrenergic antilipolysis. Finally, idazoxan itself did not modify basal or stimulated lipolysis. It counteracted the antilipolytic effects of various α_2-agonists but had no effect on the antilipolytic responses to adenosine analogs, nicotinic acid, and prostaglandin E1 (not shown). In this regard, it apparently behaved as a pure α_2-antagonist, at least when lipolysis was measured, as previously reported on the same cell model.[23]

Despite the lack of influence of inhibition of amine oxidase activity on the lipolytic response, it is necessary to verify if other aspects of adipocyte metabolism could be influenced by the regulation of amine oxidation. For instance, catecholamines influence not only lipolytic responses but also glucose metabolism and lipogenesis. Another question must therefore be posed: why do fat cells, whose mitochondrial apparatus is less developed than is that of liver or kidney cells, express a similar number of I_2-sites and amine oxidase activity? It can be assumed that adipose cells need to exert a scavenging function by eliminating excess circulating amines in order to preserve their fat stores. This may agree with the finding that orthosympathetic activity is reduced in various models of obesity.

CONCLUSION

Our findings indicate that some imidazoline derivatives and MAO inhibitors can inhibit both [^{3}H]idazoxan binding and MAO activity. Competition of [^{3}H]idazoxan binding by various MAO inhibitors belonging to the propargylamine, cyclopropylamine, or hydrazine families shows that I_2-sites not only bind imidazoline or guanidinium derivatives but also exhibit affinity for other families of molecules. Inhibition of MAO activity represents a novel biological effect of I_2-site ligands which is common to a variety of anatomical locations (adipose tissue and liver). However, the site of action of the imidazolines involved in MAO inhibition seems distinct from that of "substrate suicide" MAO inhibitors. Localization of the I_2-sites on the MAO itself remains to be demonstrated.

ACKNOWLEDGMENTS

We are grateful to M. Leynaert and P. Collon for their assistance and to A. Cordi (Servier Labs.), D. Nutt (Bristol University), and P. Ladure (P. Fabre Labs.) for providing the imidazoline compounds.

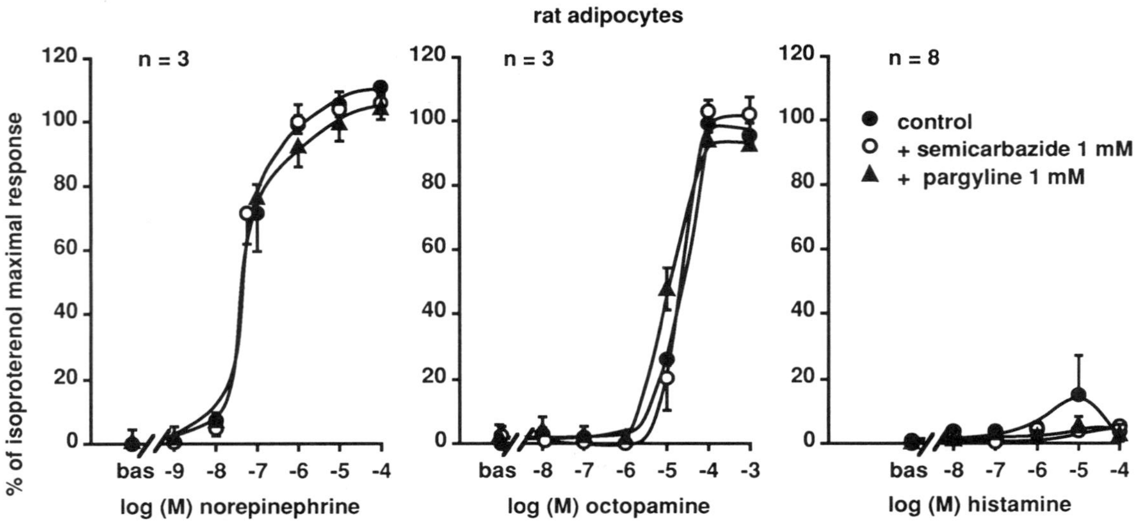

FIGURE 7. Lack of influence of inhibitors of amine oxidase on the lipolytic activity of rat adipocytes. Isolated adipocytes were incubated for 90 minutes without (bas) or with the indicated concentrations of norepinephrine (*left*), octopamine (*middle*), or histamine (*right*). Incubations were carried out in the absence (control, ●) or presence of the SSAO inhibitor semicarbazide (1 mM, ○) or the MAO inhibitor pargyline (1 mM, ▲). The amount of glycerol released in the medium was taken as an index of lipolysis. Data are expressed as a percentage of the isoproterenol maximal response which was observed with 1–10 μM of the β-agonist. Basal and isoproterenol-stimulated lipolysis averaged 0.25 ± 0.04 and 2.33 ± 0.16 μmol glycerol/100 mg cell lipid/90 minutes, respectively (n = 8).

REFERENCES

1. Hieble, J. P. & R. R. Ruffolo. 1992. Imidazoline receptors: Historical perspective. Fundam. Clin. Pharmacol. **6** (suppl 1): 7-13.
2. Parini, A., I. Coupry, F. Tesson, I. Limon & C. Gargalidis-Moudanos. 1992. Imidazoline-guanidinium receptive site (IGRS): Mitochondrial distribution and regulation properties. Fundam. Clin. Pharmacol. **6** (suppl 1): 35-40.
3. Reis, D. J., S. Regunathan, H. Wang, D. L. Feinstein & M. P. Meeley. 1992. Imidazoline receptors in the nervous system. Fundam. Clin. Pharmacol. 6 (suppl. 1): 23.
4. Ernsberger, P., M. P. Meeley, J. J. Mann & D. J. Reis. 1987. Clonidine binds to imidazole binding sites as well as alpha2-adrenoceptors in the ventrolateral medulla. Eur. J. Pharmacol. **134:** 1-13.
5. Piletz, J. E. & K. Sletten. 1993. Nonadrenergic imidazoline binding sites on human platelets. J. Pharmacol. Exp. Ther. **267:** 1493-1502.
6. Michel, M. C., J. W. Regan, M. A. Gerhardt, R. R. Neubig, P. A. Insel & H. J. Motulsky. 1990. Nonadrenergic [^{3}H]-Idazoxan binding sites are physically distinct from α_2-adrenergic receptors. Mol. Pharmacol. **37:** 65-68.
7. Bousquet, P., G. Bricca, M. Dontenwill, J. Feldman, H. Greney, A. Belcourt, J. Stutzmann & J. Tibriça. 1992. From the α_2-adrenoceptors to the imidazoline preferring receptors. Fundam. Clin. Pharmacol. **6:** (suppl. 1): 15S-22S.
8. Coupry, I., R. A. Podevin, J. P. Dausse & A. Parini. 1987. Evidence for imidazoline binding sites in basolateral membranes from rabbit kidney. Biochem. Biophys. Res. Comm. **147:** 1055-1060.
9. Hamilton, C. A., J. L. Reid & M. A. Yakabu. 1988. [^{3}H]yohimbine and [^{3}H]idazoxan bind to different sites on rabbit forebrain and kidney membranes. Eur. J. Pharmacol. **146:** 345-348.
10. Langin, D. & M. Lafontan. 1989. [^{3}H]idazoxan binding at non-alpha2-adrenoceptors in rabbit adipocyte membranes. Eur. J. Pharmacol. **159:** 199-203.
11. Parini, A., I. Coupry, R. M. Graham, I. Uzielli, D. Atlas & S. Lanier. 1989. Characterization of an imidazoline/guanidinium receptive site distinct from the α_2-adrenergic receptor. J. Biol. Chem. **264:** 11874-11878.
12. Tesson, F., C. Prip-Buus, A. Lemoine, J. P. Pegorier & A. Parini. 1991. Subcellular distribution of imidazoline-guanidinium-receptive sites in human and rabbit liver. Major localization to the mitochondrial outer membrane. J. Biol. Chem. **266:** 155-160.
13. Regunathan, S., M. P. Meeley & D. J. Reis. 1993. Expression of non-adrenergic imidazoline sites in chromaffin cells and mitochondrial membranes of bovine adrenal medulla. Biochem. Pharmacol. **45:** 1667-1675.
14. Limon, I., I. Coupry, S. M. Lanier & A. Parini. 1992. Purification and characterization of mitochondrial imidazoline-guanidinium receptive site from rabbit kidney. J. Biol. Chem. **267:** 21645-21649.
15. Lanier, S. M., B. Ivkovic, I. Singh, J. L. Neumeyer & V. Bakthavachalam. 1993. Visualization of multiple imidazoline/guanidinium receptive sites. J. Biol. Chem. **268:** 16047-16051.
16. Wang, H., S. Regunathan, M. P. Meeley & D. J. Reis. 1992. Isolation and characterization of imidazoline receptor protein from bovine adrenal chromaffin cells. Mol. Pharmacol. **42:** 792-801.
17. Zonnenschein, R., S. Diamant & D. Atlas. 1990. Imidazoline receptors in rat liver cells: A novel receptor or a subtype of α_2-adrenoceptors? Eur. J. Pharmacol. **190:** 203-215.
18. Chan, S. L. F. 1993. Role of α_2-adrenoceptors and imidazoline-binding sites in the control of insulin secretion. Clin. Sci. **85:** 671-677.
19. Olmos, G., A. M. Gabilondo, A. Miralles, P. V. Escriba & J. A. Garcia-Sevilla. 1993. Chronic treatment with the monoamine oxidase inhibitors clorgyline and pargyline down regulates non-adrenoceptor [^{3}H]idazoxan binding sites in the rat brain. Br. J. Pharmacol. **108:** 597-603.

20. RENOUARD, A., P. S. WIDDOWSON & A. CORDI. 1993. [^{3}H]idazoxan binding to rabbit cerebral cortex recognises multiple imidazoline (I2-type receptors). Pharmacological characterization and relationship to monoamine oxidase. Br. J. Pharmacol. **109:** 625-631.

21. SASTRE, M. & J. A. GARCIA-SEVILLA. 1993. Opposite age-dependent changes of alpha2A-adrenoceptors and nonadrenoceptor [^{3}H]idazoxan binding sites (I2-imidazoline sites) in the human brain. Strong correlation of I2 with monoamine oxidase-B sites. J. Neurochem. **61:** 881-889.

22. MEANA, J. J., F. BARTUREN, I. MARTIN & J. A. GARCIA-SEVILLA. 1993. Evidence of increased non-adrenoceptor [^{3}H]-idazoxan binding sites in the frontal cortex of depressed suicide victims. Biol. Psychiatry **34:** 498-501.

23. CARPÉNÉ, C., J. GALITZKY, D. LARROUY, D. LANGIN & M. LAFONTAN. 1990. Nonadrenergic sites for imidazolines are not directly involved in the α_2-adrenergic antilipolytic effect of UK14304 in rat adipocytes. Biochem. Pharmacol. **40:** 437-445.

24. LANGIN, D., H. PARIS & M. LAFONTAN. 1990. Binding of [^{3}H]idazoxan and of its methoxy derivative [^{3}H]RX 821002 in human fat cells: [^{3}H]idazoxan but not [^{3}H]RX 821002 labels additional non-α_2-adrenergic binding sites. Mol. Pharmacol. **37:**876-885.

25. ATLAS, D. 1991. Clonidine-displacing substance (CDS) and its putative imidazoline receptor. New leads for further divergence of alpha2-adrenergic receptor activity. Biochem. Pharmacol. **41:** 1541-1549.

26. YU, P. H. 1986. Monoamine oxidase. *In* Neuromethods, Neurotransmitter Enzymes. A. A. Boulton & G. B. Baker, Eds. Vol. **5:** 235-272. Humana Press. Clifton, N.J.

27. SAULNIER-BLACHE, E., C. CARPÉNÉ, D. LANGIN & M. LAFONTAN. 1989. Imidazolinic radioligands for the identification of hamster adipocytes alpha2-adrenoceptors. Eur. J. Pharmacol. **171:** 145-157.

28. ERNSBERGER, P. 1992. Heterogeneity of imidazoline binding sites: Proposed I1 and I2 subtypes. Fund. Clin. Pharmacol. **6:** (suppl. 1): 55S.

29. MICHEL, M. C. & P. ERNSBERGER. 1992. Keeping an eye on the I site: Imidazoline preferring receptors. Trends Pharmacol. Sci. **13:** 369-370.

30. CARPÉNÉ, C., P. COLLON, A. REMAURY, A. CORDI, A. HUDSON, D. NUTT & M. LAFONTAN. 1995. Inhibition of amine oxidase activity by derivatives that recognize imidazoline I_2 sites. J. Pharmacol. Exp. Ther. **272:** 681-688.

31. WIKBERG, J. E. S., S. UHLÉN & V. CHAJLANI. 1992. Evidence that drug binding to non-adrenergic [^{3}H]-idazoxan binding sites (I-receptors) occurs to interacting or interconvertible affinity forms of the receptor. Pharmacol. Toxicol. **70:** 208-219.

32. LANGIN, D., H. PARIS, M. DAUZATS & M. LAFONTAN. 1990. Discrimination between α_2-adrenoceptors and [^{3}H]idazoxan non-adrenergic sites in rabbit fat cells. Eur. J. Pharmacol. **188:** 261-272.

33. YOUDIM, M. B. H. & J. P. M. FINBERG. 1991. New directions in monamine oxidase A and B selective inhibitors and substrates. Biochem. Pharmacol. **41:** 155-162.

34. BARRAND, M. A. & B. A. CALLINGHAM. 1984. Solubilization and some properties of a semi carbazide-sensitive amine oxidase in brown adipose tissue of the rat. Biochem. J. **222:** 467-475.

35. RAIMONDI, L., R. PIRISINO, G. IGNESTI, S. CAPECCHI, G. BANCHELLI & F. BUFFONI. 1991. Semicarbazide-sensitive amine oxidase activity (SSAO) of rat epididymal white adipose tissue. Biochem. Pharmacol. **41:** 467-469.

36. KYBURZ, E. 1990. New developments in the field of MAO inhibitors. Drugs News Perspectives **3:** 592-599.

37. BROWN, C. A., S. L. CHAN, M. R. STILLINGS, S. A. SMITH & N. G. MORGAN. 1993. Antagonism of the stimulatory effects of efaroxan and glibenclamide in rat pancreatic islets by the imidazoline, RX 801080. Br. J. Pharmacol. **110:** 1017-1022.

38. BACH, A. W. J., N. C. LAN, D. L. JOHNSON, C. W. ABELL, M. E. BEMBENEK, S. W. KWAN, P. H. SEEBURG & J. C. SHIH. 1988. cDNA cloning of human liver monoamine oxidase

A and B. Molecular basis of differences in enzymatic properties. Proc. Natl. Acad. Sci. USA **85:** 4934-4938.

39. SCHNAITMAN, C., G. ERWIN & J. W. GREENAWALT. 1967. The submitochondrial localization of monoamine oxidase. J. Cell Biol. **32:** 719-735.

40. GALITZKY, J., C. CARPÉNÉ, M. LAFONTAN & M. BERLAN. 1993. Stimulation spécifique des récepteurs β3-adrénergiques du tissu adipeux par l'octopamine. C. R. Acad. Sci. (Paris) **316:** 519-523.

41. RAIMONDI, L., G. CONFORTI, G. BANCHELLI, G. IGNESTI, R. PIRISINO & F. BUFFONI. 1993. Histamine lipolytic activity and semicarbazide-sensitive amine oxidase (SSAO) of rat white adipose tissue (WAT). Biochem. Pharmacol. **46:** 1369-1376.

Evidence for a Role for Imidazoline I_1 Binding Site in Rat Brown Adipocytes

G. MORY, F. TESSON, M. COMBES-GEORGE,
M. NECHAD, AND A. PARINI[a]

CNRS URA 1488
UPMC (boîte 14) F-75252
Paris Cédex 05, France

[a]INSERM U 388
F-31054, France

Brown adipose tissue (BAT) is a tissue that produces heat in response to cold exposure or to overfeeding in most neonatal mammals and some adults such as rat. Its thermogenic capacity is due to the expression of a unique protein, called the uncoupling protein (UCP) and located in mitochondrial inner membranes, which uncouples respiration from ATP synthesis. As the rat is exposed to cold, the sympathetic nervous system triggers lipolysis and heat production within minutes through activation of β-adrenoceptors. When rats are chronically exposed to cold, a trophic response can also be observed: cell number, mitochondrial mass per cell, and UCP concentration in mitochondrial protein increase. Trophic response is also triggered by sympathetic nervous system and β-adrenoceptors.[2,3] Rat BAT also contains α_2-adrenoceptors, determined by rauwolscine binding.[4] In isolated brown fat cells, 10^{-6} M α_2-agonist clonidine (which induces decreased cAMP production) counteracts the stimulation of respiration and lipolysis by the β-adrenergic agonist isoproterenol (which induces an increase in cAMP production).[5] However, the present data show that under different experimental conditions, clonidine can also trigger non-α_2 effects in rat BAT.

A 9-day chronic administration of clonidine (10^{-9} mol·h^{-1} delivered by ALZETT osmotic minipumps implanted subcutaneously upon interscapular BAT of 40-day-old Wistar rats) had no effect on UCP mitochondrial concentration and BAT thermogenic capacity (determined by the capacity of isolated mitochondria to bind *in vitro* GDP, a ligand of UCP).[6] However, clonidine reinforced the effect of isoproterenol administration (TABLE 1).

This paradoxical effect of clonidine was further studied *in vitro*. Clonidine 10^{-6} M did not modify basal or isoproterenol-stimulated respiration of isolated rat brown

TABLE 1. Effects of a Nine-Day Chronic Administration of Clonidine on Interscapular Brown Adipose Tissue of Rats Bred at 25°C and Receiving Isoproterenol Simultaneously or Not and Effects of a Nine-Day Exposure to 5°C

Treatment	Tissue Weight (mg)	GDP Binding $(pmol{\cdot}mg^{-1})^a$	UCP $(\%)^b$
Vehicle	212 ± 17 (8)	21.9 ± 0.4	4.7 ± 0.4
Clonidine $10^{-8}M{\cdot}h^{-1}$	193 ± 19 (8)	20.7 ± 1.6	4.7 ± 0.4
Isoproterenol $10^{-8}M{\cdot}h^{-1}$	241 ± 13 (8)	58.7 ± 0.4	6.6 ± 0.2
Isoproterenol + clonidine	237 ± 22 (8)	93.0 ± 8.0	8.6 ± 0.5
5°C, 9 days	343 ± 45 (6)	127.0 ± 12	9.2 ± 0.7

[a] pmol GDP·mg mitochondrial protein^{-1}.
[b] Percentage of mitochondrial proteins. Number of determinations is in brackets (first column). Bars link values that are statistically similar (Games-Howell ANOVA and test for multiple comparisons) with $p < 0.05$.

adipocytes (TABLE 2A). However, a lower concentration of clonidine (10^{-8} M) did not modify basal respiration (TABLE 2B), but significantly reinforced respiration stimulated by the β-adrenergic agonist isoproterenol used at a concentration of 10^{-9} M (TABLE 2C). α_2-Blocking by yohimbine (10^{-7} M) was responsible for an even larger effect of the isoproterenol + clonidine combination (TABLE 2C). These results suggest that the net effect of clonidine is the result of inhibition of respiration through the α_2-adrenoceptor and of stimulation of respiration via another clonidine binding site. As clonidine is known to bind a novel family of binding sites, the imidazoline binding sites (IBS), this hypothesis was tested. I_1-ligand rilmenidine (SERVIER Laboratories) did not modify basal respiration of isolated brown fat cell as clonidine (TABLE 2D), and reinforced isoproterenol-stimulated respiration using as low a concentration as 10^{-10} M. Rilmenidine respiratory effect was insensitive to α_2-blocking by yohimbine (TABLE 2E).

These results suggest that the activating site sensitive to clonidine is an I_1BS. A preliminary study of displacement of [^{125}I]*para*-iodoclonidine binding on purified BAT plasma membrane by clonidine and rilmenidine supplied evidence for the presence of I_1BS in rat brown adipocyte.

Two observations suggest that the effects of clonidine and rilmenidine on BAT require second-messenger generation: (1) clonidine or rilmenidine did not exhibit any respiratory effect on isolated BAT mitochondria in our experimental conditions and (2) clonidine 10^{-8} M also reinforces the respiration of isolated brown fat cells prestimulated by forskolin, dibutyryl-cyclic AMP, or butyrate (TABLE 2F, G, and H).

In conclusion, these results suggest that rat BAT contain I_1BS which modulate respiration and UCP synthesis through a still unknown transduction pathway.

TABLE 2. Effect of Clonidine and Rilmenidine on Respiration of Rat Isolated Brown Fat Cells[a]

	Added Drugs	Respiration in % of Basal Value, taken as 100%
A	Basal (8)	100
	Isoproterenol 10^{-9} M	159.4 ± 15.9
	Isoproterenol + clonidine 10^{-6} M	152.4 ± 17.7
B	Basal (8)	100
	Clonidine 10^{-8} M	101.2 ± 11.1
C	Basal (36)	100
	Isoproterenol 10^{-9} M	237.8 ± 20.7
	Isoproterenol + clonidine 10^{-8} M	297.1 ± 29.0
	Isoproterenol + clonidine + yohimbine 10^{-7} M	341.9 ± 32.9
D	Basal	100
	Rilmenidine 10^{-10} M	99.6 ± 6.6
E	Basal (15)	100
	Isoproterenol 10^{-9} M	182.1 ± 6.8
	Isoproterenol + rilmenidine 10^{-10} M	211.5 ± 21.8
	Isoproterenol + rilmenidine 10^{-8} M	225.0 ± 21.9
	Isoproterenol + rilmenidine + yohimbine 10^{-8} M	205.2 ± 17.2
F	Basal (8)	100
	Forskoline 10^{-7} M	161.1 ± 17.2
	Forskoline + clonidine 10^{-8} M	211.3 ± 21.2
	Forskoline + clonidine + yohimbine 10^{-7} M	213.8 ± 29.3
G	Basal (8)	100
	Dibutyryl cAMP 10^{-6} M	130.3 ± 14.6
	dbcAMP + clonidine 10^{-8} M	176.7 ± 22.9
H	Basal (8)	100
	Butyrate 10^{-4} M	178 ± 18.1
	Butyrate + clonidine 10^{-8} M	240 ± 22.9

[a] Results are expressed as means ± SEM; number of determinations is in brackets. Each series shows the additive effects of drugs on same cell preparations. Statistics were analyzed with original values, not with percentages using Neuman-Keuls test for paired values. Bars link values that are statistically similar, with p <0.05.

REFERENCES

1. TRAYHURN, P. & D. G. NICHOLLS. Brown Adipose Tissue. 1986. Arnold E.
2. MORY, G., F. BOUILLAUD, M. COMBES-GEORGE & D. RICQUIER. 1984. FEBS Lett. **166:** 393-396.
3. RICQUIER, D., F. BOUILLAUD, P. TOUMELIN, G. MORY, R. BAZIN, J. ARCH & L. PÉNICAUD. 1986. J. Biol. Chem. **261:** 13905-13910.
4. MORY, G., B. WOESSMER & D. RICQUIER. 1990. Path. Biol. **38:** 908.
5. SUNDIN, U. & J. FAIN. 1983. Biochem. Pharmacol. **32:** 3117-3221.
6. RIAL, E. & D. G. NICHOLLS. 1984. Biochem. J. **222:** 685-693.

Regulation of Electrolyte Transport in Rabbit Tracheal Epithelial Cells by the I_1-Imidazoline Agonist Moxonidine[a]

CAROLE M. LIEDTKE AND PAUL ERNSBERGER

Departments of Pediatrics and Physiology
and
Biophysics and Medicine
Case Western Reserve University
Cleveland, Ohio 44106

In previous studies, we demonstrated that rabbit tracheal epithelial cells bind clonidine, an α_2-adrenergic (AR) agent, to α_2AR sites and to I_1 sites with a similar affinity.[1] Clonidine and the α_{2A}AR agent guanabenz stimulate NaCl(K) cotransport through activation of pertussis toxin-sensitive phosphatidylinositol-4,5-bisphosphate (Ptdins(4,5)P_2)-dependent phospholipase C with the transient generation of messengers, inositol trisphosphate and diaclyglycerol.[2]

The I_1 agent moxonidine (MOX)[1,2] increases Na, but not Cl, transport in rabbit tracheal epithelial cells through a mechanism that is independent of cyclic nucleotides and activation of cAMP-dependent protein kinase A. The intracellular mechanism of signal transduction for MOX-stimulated Na transport is not known. However, elevated Na transport is insensitive to loop diuretics and hence is mediated by a transport carrier other than NaCl cotransport.

I_1 agents have been linked to modulation of diuresis and natriuresis in kidney epithelium,[4] inhibition of Na/H exchange in renal proximal tubule cells[5] and K channels in pancreatic islet cells.[8] Our goal in these studies is to determine if MOX regulates vectorial Na transport in rabbit tracheal epithelium and to determine the intracellular second messenger system that mediates the effects of MOX. We examined the effects of MOX in reconstituted epithelium using an electrophysiological approach and investigated MOX-induced generation of a lipid mediator, diacylglycerol.

METHODS

Rabbit tracheal epithelial cells were isolated and grown in *in vitro* cell culture on tissue culture plastic dishes or Millipore culture inserts, as previously described.[7]

[a] This work was done during the tenure of an Established Investigatorship of the American Heart Association for C.M.L. C.M.L. received support for this work from the NHLBI (HL-43907) and P.E. from NHLBI (HL-44514).

TABLE 1. Effect of Moxonidine on Short Circuit Current of Rabbit Tracheal Transmonolayer Cultures

Additive	Concentration	Change in Short Circuit Current ($\mu A/cm^2$)
Amiloride	20 μM	−3.0 ± 0.4 (14)
Bumetanide	10 μM	−5.4 ± 0.7 (12)
Moxonidine	10 μM	−3.4 ± 0.9 (6)
Amiloride + moxonidine		−5.0 ± 0.9 (5)
Bumetanide + moxonidine		−3.0 ± 0.5 (3)
Ba^{2+}	1 mM	−9.3 ± 1.7 (3)
Ba^{2+} + moxonidine		−1.2 ± 0.4 (5)
A23187		11.8 ± 1.4 (17)

Electrophysiological measurements were performed on transmonolayer cultures with resistances of 500 ohms · cm^2 or greater. Cultures were inserted into a Ussing chamber for continuous measurement of short circuit current. Apical and basolateral surfaces were continuously perfused with DMEM : F12. Cultures were pulsed at intervals to determine the transmonolayer voltage (V_{tm}). Transmonolayer conductance (G_{tm}) was calculated using Ohm's Law. Agents of interest were added to apical or basolateral perfusate.

Diacylglycerol levels were quantitated in cultures prelabeled with [^{3}H]arachidonic acid, which radiolabels the glycerol backbone of Ptdins(4,5)P_2.[2] [^{3}H]Diacylglycerol was separated by thin layer chromatography against authentic standards. [^{3}H]Diacylglycerol levels were calculated as [^{3}H]dpm/μg DNA and normalized to basal levels determined in cells treated with vehicle.

RESULTS AND DISCUSSION

Rabbit tracheal epithelial cells were grown as transmonolayer culture to reconstitute the epithelium and thus study the role of MOX in a polarized epithelium. Short circuit current of unstimulated transmonolayers was 19.8 ± 3.0 (10) $\mu A/cm^2$. V_{tm} and G_{tm} were 9.98 ± 0.8 (9) mV and 2.0 ± 0.4 (10) mS/cm^2, respectively. These baseline bioelectric properties are consistent with those of other mammalian tracheal epithelial cells.

We used electrolyte transport blockers to characterize active transport mechanisms that contributed to short circuit current in transmonolayer cultures. The results are shown in TABLE 1. Apical application of amiloride, an epithelial Na channel blocker, reduced short circuit current, indicating decreased uptake of Na ions. Basolateral application of bumetanide also reduced short circuit current. These results indicate that apical Na channels and basolateral NaCl cotransport are active in unstimulated transmonolayer cultures. To assess the ability of transmonolayers to respond to stimulatory agents, we exposed cultures to the Ca^{2+}-ionophore A23187 after each

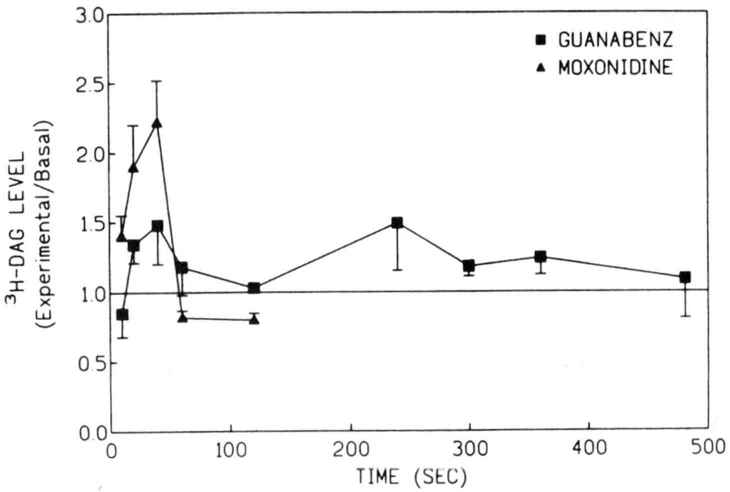

FIGURE 1. Diacylglycerol generation in the presence of 10 μM moxonidine or guanabenz. Values represent the mean of 3-5 experiments. Error bars are 1 SE unit.

experiment. In intact tracheal epithelium, A23187 stimulates Cl secretion and thus increases short circuit current. In transmonolayers, A23187 increased short circuit current, indicating induction of Cl secretion and confirming the ability of cells to respond to stimulatory agents.

We next tested whether MOX modulated Na transporters. Basolateral application of MOX decreased short circuit current, indicating loss of charge transfer from apical to basolateral perfusates (TABLE 1). The addition of MOX to the apical perfusate did not alter short circuit current (data not shown). Sequential stimulation of transmonolayers with amiloride and then MOX lowered short circuit current to a value not significantly different from that of MOX alone or an additive response of the two agents. This suggests that MOX and amiloride act on different transporters. Basolateral application of MOX to transmonolayers pretreated with bumetanide decreased short circuit current to a level observed with MOX alone. These results indicate that MOX does not modulate NaCl cotransport and are in agreement with our earlier studies showing that MOX did not activate bumetanide-sensitive NaCl cotransport.[1]

In the absence of MOX-mediated modulation of apical Na channels and basolateral NaCl cotransport, we explored a role for K channels in the I_1-mediated response. Basolateral K channels include leak pathways for the exit of K from epithelial cells. These channels are sensitive to Ba^{2+}. Basolateral perfusion with Ba^{2+} deceased short circuit current significantly (TABLE 1). This effect is likely due to accumulation of positive charges in the cell which prevent uptake of Na. Sequential treatment of monolayers with Ba^{2+} and MOX did not lower short circuit current to the extent found with MOX alone. These results suggest that Ba^{2+} prevents, at least partially, the effects of MOX on short circuit current. One explanation for this result is that Ba^{2+} blocked activated basolateral K channels.

To ascertain the intracellular mechanism of signal transduction, we investigated a role for lipid mediators as secondary messengers in I_1-mediated modulation of vectorial electrolyte transport. FIGURE 1 illustrates the kinetics of [^{3}H]diacylglycerol generated by MOX and guanabenz. Both agonists transiently increased diacylglycerol levels. Guanabenz stimulated biphasic generation of diacylglycerol with peak levels attained at 40 seconds and 4 minutes. Stimulation with MOX, however, induced a single diacylglycerol peak at 20 seconds that returned to baseline within 1 minute. The contrast between guanabenz and MOX may be due to the action of MOX but not guanabenz at I_1-receptors. In addition, transient generation of [^{3}H]diacylglycerol in response to MOX has been observed in pheochromocytoma cells, which lack α_2AR receptors (Separovic and Ernsberger, unpublished observations). Pretreatment of rabbit tracheal epithelial cells with 100 ng pertussis toxin/ml blunted the effects of MOX and guanabenz (data not shown). These findings indicate that the I_1-receptor is linked to activation of pertussis toxin-sensitive phospholipase(s) that generates diacylglycerol.

ACKNOWLEDGMENT

We thank Linda Thomas for excellent technical assistance.

REFERENCES

1. LIEDTKE, C. M., J. FURIN & P. ERNSBERGER. 1993. Am. J. Physiol. **264** (Cell Physiol. 33): C568-C576.
2. LIEDTKE, C. M. & M. KESTER. 1994. J. Cellular Physiol. **15**(Suppl. 18D): 35.
3. ERNSBERGER, P., T. H. DAMON, L. M. GRAFF, S. G. SCHÄFER & M. O. CHRISTEN. 1993. J. Pharmacol. Exp. Ther. **264:** 172-182.
4. ALLAN, M., S. B. PENNER & D. D. SMYTH. 1993. Br. J. Pharmacol. **208:** 870-875.
5. BIDET, M., P. POUJEOL & A. PARINI. 1991. J. Hypertens. **9**(Suppl 6): S216-S217.
6. CHAN, S. L. F. & N. G. MORGAN. 1990. Eur. J. Pharmacol. **176:** 97-101.
7. DAVIS, P. B., C. L. SILSKI & C. M. LIEDTKE. 1992. Am. J. Respir. Cell Mol. Biol. **6:** 140-145.

α_2-Adrenoceptor-Independent Inhibition by Imidazolines and Guanidines of Noradrenaline Release from Peripheral, but Not Central Noradrenergic Neurons

M. GÖTHERT,[a] G. J. MOLDERINGS, K. FINK, AND E. SCHLICKER

Institute of Pharmacology and Toxicology
University of Bonn
Reuterstr. 2b
53113 Bonn, Germany

PUTATIVE PRESYNAPTIC IMIDAZOLINE RECEPTORS IN THE RABBIT PULMONARY ARTERY

On the basis of previous experiments mainly carried out in the rabbit pulmonary artery and heart it was claimed that the sympathetic axon terminals in these tissues are endowed with inhibitory presynaptic imidazoline receptors[1–4] (FIG. 1) in addition to the well-known α_2-autoreceptors.[5] Several isoindoline derivatives substituted with aminoimidazoline (BDF 6100, BDF 6143, and BDF 7572) or guanidine (BDF 7579, aganodine; for chemical structures, see FIG. 2) and some important ligands at I_1- and I_2-imidazoline binding sites (IBS)[6–13] such as clonidine, cirazoline, and idazoxan were applied as pharmacological tools in those studies. The α_2-autoreceptor-blocking imidazolines BDF 6143, idazoxan (FIG. 3), BDF 6100, and BDF 7572 increased the electrically (2 Hz) evoked norepinephrine (NE) release in the rabbit pulmonary artery due to the interruption of the inhibitory feedback loop maintained by the endogenous NE released from the sympathetic nerves. However, in contrast to the concentration-response curve of rauwolscine, an α_2-adrenoceptor antagonist devoid of an imidazoline structure, the curves of the imidazolines were bell-shaped (FIG. 3). The biphasic character of the concentration-response curve was explained by an α_2-adrenoceptor-independent inhibitory effect of the imidazolines which was unmasked (i.e., reversed to a monophasic inhibition) by previous reversible or irreversible blockade of α_2-adrenoceptors by rauwolscine (FIG. 4) and phenoxybenzamine, respectively.[1,2] The inhibitory component of BDF 6143 was also disclosed even in the absence of rauwolscine when the frequency of electrical stimulation was decreased from 2 to 0.66 Hz,

[a] Author for correspondence.

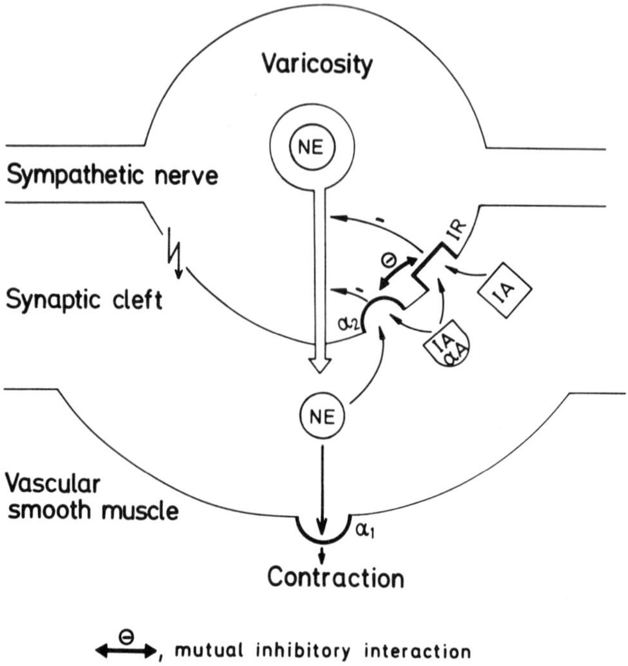

FIGURE 1. Schematic representation of a synapse between a noradrenergic varicosity of a sympathetic axon terminal and a smooth muscle cell of the rabbit or human pulmonary artery or rabbit aorta with a presynaptic imidazoline receptor (IR), a presynaptic α_2-adrenoceptor (α_2), and a postsynaptic α_1-adrenoceptor (α_1). NE = norepinephrine; IA = imidazoline receptor agonist; IAαA = imidazoline receptor agonist with affinity for α_2-adrenoceptors; – = inhibition of release; ϟ = action potential. Scheme based on results published in refs. 1 and 2 and in the present report.

leading to a considerably decreased degree of α_2-autoreceptor activation by NE and, accordingly, an only negligible disinhibition of NE release due to the α_2-blocking component of BDF 6143.[1,3]

The suggestion that so far unknown imidazoline receptors are involved in this inhibitory effect was supported by experiments with certain imidazolines and guanidines such as clonidine, BDF 7579, and aganodine (FIG. 2) which are known not to block, but to activate α_2-adrenoceptors. After complete or nearly complete α_2-adrenoceptor blockade with phenoxybenzamine or rauwolscine, these imidazolines and guanidines still inhibited noradrenaline release.[2] Furthermore, cirazoline, an imidazoline derivative devoid of intrinsic activity at α_2-adrenoceptors, also induced an inhibitory effect on NE release in the rabbit pulmonary artery[1,3] (FIG. 3). In the search for a compound that acts as an antagonist at these presynaptic imidazoline receptors, rauwolscine was found to produce a rightward shift of the concentration response curve for the inhibitory effect of BDF 6143 on the norepinephrine release evoked by electrical impulses at 0.66 Hz, suggesting that rauwolscine exerts such

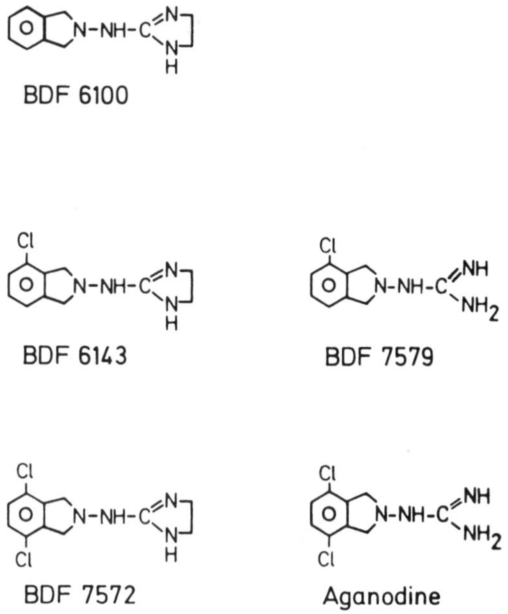

FIGURE 2. Chemical structure of aminoimidazoline- and guanidine-substituted isoindoline derivatives.

an antagonism in addition to its α_2-adrenoceptor blocking property.[3] Determination of apparent pA_2 values against further appropriate imidazoline receptor agonists and α_2-adrenoceptor agonists supported this conclusion and revealed that rauwolscine was clearly less potent in antagonizing imidazoline receptors than α_2-adrenoceptors (ref. 2 and unpublished data).

Whether such presynaptic imidazoline receptors also occur in human vascular tissues and whether they are also present on the noradrenergic neurons of the central nervous system are open questions. In addition, it is of interest to characterize the presynaptic imidazoline receptors on the sympathetic axon terminals of rabbit blood vessels in more detail by application of further relevant drugs including the recently identified endogenous imidazoline receptor ligand agmatine.[14] It is also desirable to compare the pharmacological properties of these receptors with I_1 and I_2 imidazoline binding sites (IBS) and to examine whether the presynaptic imidazoline receptors interact with the presynaptic α_2-autoreceptors. The following sections of this report will focus on these problems.

FAILURE TO IDENTIFY PRESYNAPTIC IMIDAZOLINE RECEPTORS IN THE RABBIT AND RAT CEREBRAL CORTEX

In superfused slices of the rabbit cerebral cortex, BDF 6143 (10 nM–10 μM) produced a monophasic, concentration-dependent increase in the electrically (3 Hz)

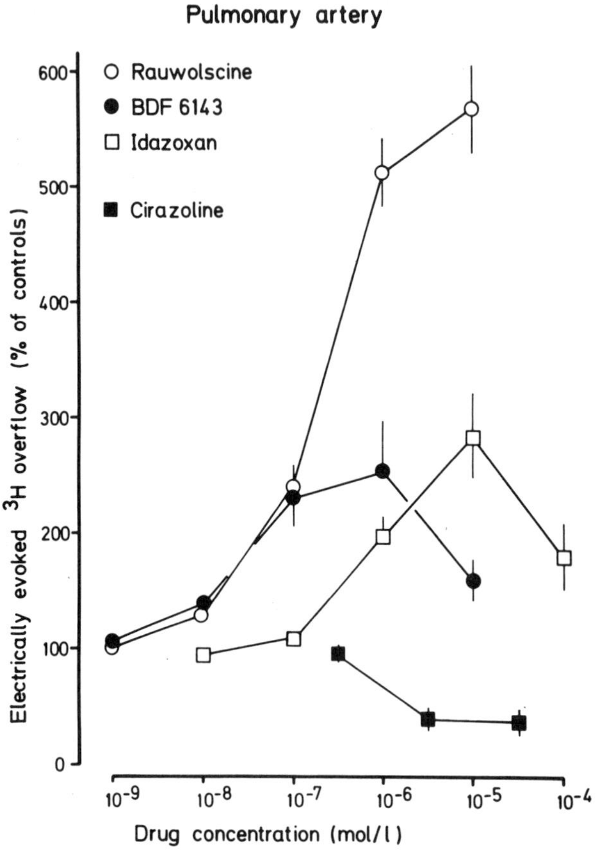

FIGURE 3. Effects of cirazoline, rauwolscine, and two imidazolines with α-adrenoceptor-blocking property, BDF 6143 and idazoxan, on the electrically (2 Hz) evoked ^3H overflow from superfused strips of the rabbit pulmonary artery preincubated with [^3H]norepinephrine. Tritium overflow (mainly consisting of [^3H]norepinephrine) was expressed as a percentage of ratios in control experiments. Values are given as mean ± SEM of 4-7 experiments; in some cases, SEM was smaller than symbols. All changes in evoked ^3H overflow that exceeded 30% of the corresponding controls were statistically significant (at least $p < 0.05$). Modified from reference 1 with permission.

evoked [^3H]NE release (by about 300% in the concentration range of 0.1-10 μM; details of the method applied are given elsewhere[15]), whereas cirazoline (1-30 μM) was ineffective. In slices preincubated with 10 μM phenoxybenzamine, that is, after irreversible blockade of the α_2-autoreceptors, the facilitatory effect of BDF 6143 on [^3H]NE release was abolished, but not reversed to an inhibition. In slices not pretreated with phenoxybenzamine, the evoked [^3H]NE release was concentration-dependently inhibited by NE, aganodine, and clonidine (0.01-1 μM). Preincubation with phenoxybenzamine led to abolition of the inhibitory effect of aganodine at all concentrations

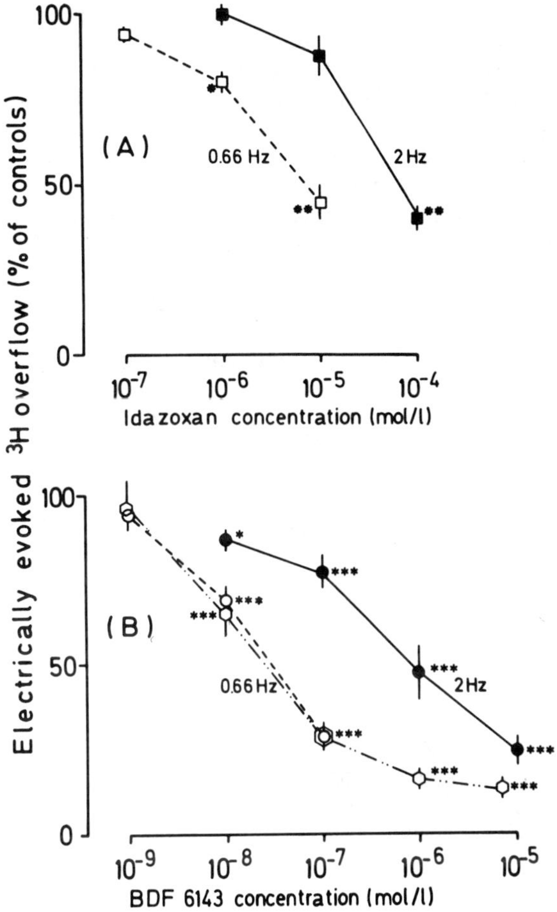

FIGURE 4. Frequency dependence of the effect of (**A**) idazoxan and (**B**) BDF 6143 on the electrically evoked ^{3}H overflow from the rabbit pulmonary artery (preincubated with [^{3}H]norepinephrine) after α_2-adrenoceptor blockade with 3 μM rauwolscine. Tritium overflow (mainly consisting of [^{3}H]norepinephrine) from the strips was stimulated by transmural electrical impulses at either 0.66 Hz for 3 (○) or 9 minutes (□, ⬡) or at 2 Hz for 3 minutes (■, ●). Tritium overflow was expressed as a percentage of the ratios in the control experiments with rauwolscine alone at the relevant frequency of stimulation. In agreement with a typical feature of presynaptic receptors,[5,32] the drugs were more potent in inhibiting tritium overflow when the frequency of electrical stimulation was decreased. Means ± SEM of 4–7 experiments. *$p < 0.05$; **$p < 0.025$; ***$p < 0.001$ (compared with the corresponding controls). Reproduced from reference 2 with permission.

applied and of that of clonidine and NE up to 0.1 μM; the inhibition induced by 1 μM clonidine or 1 μM NE was considerably attenuated, but a slight inhibitory effect was left, suggesting that the irreversible blockade of the presynaptic α_2-adrenoceptors was not complete. The endogenous imidazoline receptor ligand agmatine[14] (at concentrations of 0.1-10 μM) did not affect the evoked [^{3}H]NE release, whether or not 3.2 μM rauwolscine was present in the superfusion fluid.

In rat cerebrocortical slices with intact α_2-adrenoceptors, noradrenaline, moxonidine, and clonidine inhibited the electrically evoked [^{3}H]NE release. This effect was abolished in slices preexposed to 10 μM phenoxybenzamine. Basically the same results were obtained in superfused synaptosomes from the rat cerebral cortex, in which stimulation of [^{3}H]NE release was carried out with 15 mM K^+ (for methodological details, see ref. 16).

These results indicate that the noradrenergic axon terminals of the rabbit and rat cerebral cortex are not endowed with inhibitory imidazoline receptors. Accordingly, imidazoline derivatives devoid of intrinsic activity at α_2-adrenoceptors, such as cirazoline and BDF 6143, which have been shown to stimulate presynaptic imidazoline receptors in the rabbit pulmonary artery, did not inhibit NE release in the cerebral cortex.

CHARACTERISTICS OF PRESYNAPTIC IMIDAZOLINE RECEPTORS IN RABBIT BLOOD VESSELS AND COMPARISON WITH I_1- AND I_2-IMIDAZOLINE BINDING SITES

In view of the failure to identify presynaptic imidazoline receptors on central noradrenergic neurons of the rabbit, it was of interest to provide further evidence that such receptors exist not only in the rabbit pulmonary artery but also in another blood vessel of this species, namely, the aorta. Moreover, it was desirable to characterize the vascular presynaptic imidazoline receptors in more detail.

The rabbit aorta, in which the α_2-adrenoceptors were strongly blocked by 3 μM rauwolscine, resembled the pulmonary artery in that the imidazolines BDF 6143, BDF 6100, clonidine, cirazoline, ST 587, idazoxan, and tolazoline and the guanidine BDF 7579 concentration-dependently inhibited the electrically (2 Hz) evoked [^{3}H]NE release; norepinephrine (up to 1 μM) and amiloride were ineffective. As a potency measure, the negative logarithms of the concentrations that produced 30% inhibition of [^{3}H]NE release (pIC_{30}) were determined from the concentration-response curves of the effective compounds. It is evident from the list of the pIC_{30} values (TABLE 1) that the rank order of potency of drugs in the aorta is very similar to that in the pulmonary artery. Accordingly, there is a highly significant correlation between the pIC_{30} values in both blood vessels (FIG. 5).

Interestingly, in the rabbit aorta the inhibitory effect of the imidazolines and of the guanidine BDF 7579 was shared by agmatine ($pIC_{30} = 6.19$), which was recently proposed to be an endogenous ligand at imidazoline receptors.[14] This finding certainly supports the suggestion that imidazoline receptors are involved in the inhibitory effect of the compound on NE release, because an important criterion for the mediation of a given effect via a newly discovered receptor is the identification of an endogenous agonist.

TABLE 1. Inhibitory Potencies (expressed as pIC_{30} values) of Imidazolines and the Guanidine BDF 7579 on the Electrically (2 Hz) Evoked ^3H Overflow from the Rabbit Pulmonary Artery and Aorta[a]

Compound	Pulmonary Artery[b]	Aorta
BDF 6143[c]	7.05	6.77
BDF 7579[c]	6.98	6.92
Clonidine	6.22	5.89
BDF 6100[c]	6.10	6.32
Cirazoline	5.89	5.30
ST 587[d]	5.26	5.26
Idazoxan	4.62	4.50
Tolazoline	4.56	4.82

[a] Strips of blood vessels were incubated with [^{3}H]norepinephrine and subsequently superfused with physiological salt solution containing 3 μM rauwolscine. For further details see reference 2.

[b] From reference 2.

[c] See FIGURE 2 for chemical structures.

[d] 2-(2-Chloro-5-trifluoro-methylphenylimino)imidazoline.

In the rabbit pulmonary artery evidence was presented that rauwolscine also possesses antagonistic properties at presynaptic imidazoline receptors (see above), but its potency was about 30 times lower in this respect than its potency in acting as an antagonist at α_2-adrenoceptors.[2] To examine whether rauwolscine also exhibits antagonistic properties at the presynaptic imidazoline receptors on the rabbit aorta, the same key agonists were applied for the interaction experiments with rauwolscine as previously used in the rabbit pulmonary artery.[2] Apparent pA_2 values of at least 8.0 should be expected for rauwolscine when it actually interacts with a certain agonist at presynaptic α_2-adrenoceptors (which in the rabbit pulmonary artery belong to the α_{2A}-subtype[17]). Such an interaction at presynaptic α_2-adrenoceptors has been found with oxymetazoline and moxonidine, whose concentration-response curves were strongly shifted to the right by rauwolscine, yielding apparent pA_2 values of 8.33 and 8.12, respectively. These values are similar to those previously determined in the pulmonary artery and, as in that tissue,[2] it may be concluded that oxymetazoline and moxonidine do not appear to activate presynaptic imidazoline receptors, but exclusively presynaptic α_2-adrenoceptors. On the other hand, the fact by itself that BDF 6143 and cirazoline (in the absence of further drugs) inhibited the [^{3}H]NE release evoked by electrical impulses at 0.66 and 2 Hz provides evidence that these drugs, which are devoid of agonistic property at α_2-adrenoceptors, act via an α_2-adrenoceptor-independent mechanism, probably by activating presynaptic imidazoline receptors (BDF 6143 at remarkably high potency: $pIC_{30} = 8.96$). Rauwolscine induced rightward shifts of the concentration-response curves for BDF 6143, cirazoline, and clonidine (apparent pA_2 values 6.43-7.32) which were clearly less pronounced than the shifts of the concentration response curves for oxymetazoline and than those that would have been expected if α_2-adrenoceptors were involved. (For a complete list

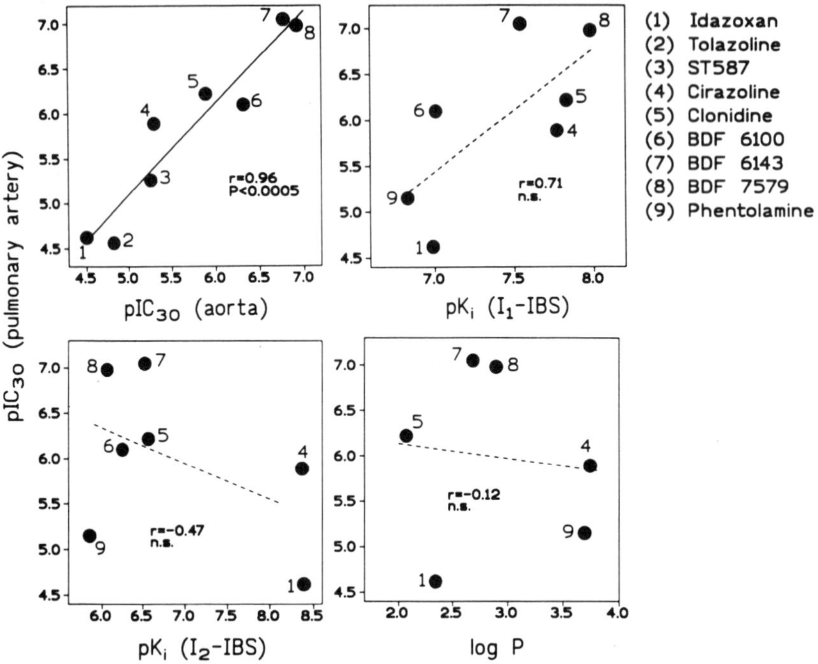

FIGURE 5. Comparison of the inhibitory potency of imidazolines and guanidines in the rabbit pulmonary artery with that in the rabbit aorta, with the affinity for I_1- and I_2-imidazoline binding sites and the lipophilicity (log P values) of the drugs. Significant correlation of the pIC_{30} values in the rabbit pulmonary artery[2] with those in the rabbit aorta (left upper panel; values from TABLE 1), but not with pK_i values for the high affinity [^{3}H]clonidine binding site[12] (I_1-imidazoline binding site [I_1-IBS]; right upper panel), the high affinity [^{3}H]idazoxan binding site[13] (I_2-imidazoline binding site [I_2-IBS]; left lower panel), and the log P values (unpublished; right lower panel). r = correlation coefficient, n.s. = not significant.

of apparent pA_2 values of rauwolscine, see TABLE 2.) This finding suggests that in addition to its well-known α_2-adrenoceptor-blocking property, rauwolscine acts as a relatively weak antagonist at presynaptic imidazoline receptors.

The possibility had to be examined that the presynaptic imidazoline receptors in the rabbit blood vessels might have similar pharmacological properties as the I_1- and I_2-imidazoline binding sites identified in various tissues including the adrenal medulla (see above). For this purpose, the pIC_{30} values of the imidazolines and of BDF 7579 in the rabbit pulmonary artery were compared with the K_i values determined in competition experiments for the I_1- and I_2-imidazoline binding sites in the bovine adrenal medulla.[12,13] No correlation was found between the potencies of the drugs in inhibiting [^{3}H]NE release and their K_i values at I_1- or I_2-imidazoline binding sites (FIG. 5), suggesting that the presynaptic imidazoline receptors differ from I_1- and I_2-imidazoline binding sites with respect to their pharmacological properties. A nonspecific inhibitory effect due to a hydrophobic interaction with lipids and proteins of

TABLE 2. Apparent pA_2 Values of Rauwolscine at Presynaptic Receptors Mediating Inhibition of Norepinephrine Release in the Rabbit Pulmonary Artery and Aorta

Agonist	pA_2 (Pulmonary Artery)	pA_2 (Aorta)
Oxymetazoline	8.08	8.33
Moxonidine	8.22	8.12
Clonidine	6.77	6.55
BDF 6143	—	6.43
BDF 7579	6.37	—
Aganodine	6.67	—
Cirazoline	—	7.32

the cell membrane could also be excluded, inasmuch as no correlation was found between the inhibitory potency of the drugs and their lipophilicity (log P values determined by an HPLC method[18,19]).

The possibility that the imidazolines and guanidines might inhibit [^{3}H]NE release by activating the presynaptic inhibitory histamine H_3 receptors recently identified in blood vessels was ruled out by the results of a radioligand binding study of [^{3}H]N^α-methylhistamine to the H_3 receptors of rat cerebrocortical membranes. The imidazolines and aganodine had either no affinity or only negligible affinity for H_3 binding sites. Above all, the failure of histamine to inhibit the release of [^{3}H]NE in the rabbit pulmonary artery indicates that the sympathetic axon terminals of the rabbit blood vessels are obviously not endowed with inhibitory presynaptic histamine receptors.

A typical feature of imidazoline recognition sites is their particularly high sensitivity to K^+ ions. These ions markedly reduce radioligand binding to I_1- and I_2-imidazoline binding sites,[4,12,13,20–22] whereas radioligand binding to α_2-adrenoceptors is at best moderately decreased by K^+ ions.[23–25] Therefore, the possibility of similar differences between presynaptic imidazoline receptors and α_2-adrenoceptors was examined by investigating the release of [^{3}H]NE in response to membrane depolarization by high K^+. In the rabbit aorta, a K^+ concentration of 150 mM induced a [^{3}H]NE release of similar magnitude as electrical stimulation at 2 Hz. BDF 6143 in the presence of 3 μM rauwolscine was applied as a presumably selective imidazoline receptor agonist and moxonidine as an α_2-adrenoceptor agonist devoid of intrinsic activity at the presynaptic imidazoline receptor (see above). BDF 6143 was 17 times less potent in inhibiting the K^+- than the electrical impulse-evoked [^{3}H]NE release, whereas the degree of moxonidine-induced inhibition of the K^+-evoked release was almost identical to that of the electrically induced release. According to these findings, the presynaptic imidazoline receptor resembles the I_1- and I_2-imidazoline binding site in that it is more sensitive to the inhibitory effect of high K^+ than is the α_2-adrenoceptor.

INTERACTION OF PRESYNAPTIC IMIDAZOLINE RECEPTORS WITH α_2-AUTORECEPTORS IN THE RABBIT AORTA

Evidence has been presented in various tissues that different presynaptic receptor systems located on the same noradrenergic axon terminals can interact with each other.[26–30] Therefore, it was of interest to investigate whether such interaction also occurs between the presynaptic imidazoline receptors and the α_2-autoreceptors in the rabbit aorta. Cirazoline and moxonidine, respectively, were used as presumably selective agonists at these presynaptic receptors.

In experiments shown in the left panel of FIGURE 6, the inhibitory effect of moxonidine was investigated first in the absence and then in the presence of cirazoline. In the absence of this drug, moxonidine 0.3 μM decreased the [^{3}H]NE overflow evoked by S_3 at 0.66 Hz by 99% (empty column). To investigate the interaction of moxonidine with 3 μM cirazoline, the latter drug (which by itself decreased the evoked [^{3}H]NE overflow) was present during S_4 and S_5, and moxonidine, in addition, was present during S_5. The frequency of electrical stimulation was doubled at S_4 and S_5 (1.32 Hz) to compensate for the inhibitory effect of cirazoline. Under this condition, moxonidine-induced inhibition of the evoked [^{3}H]NE overflow (at S_5) was considerably decreased to 37% (hatched column).

Conversely, 3 μM cirazoline, given alone at S_3, decreased the electrically (0.66 Hz) evoked [^{3}H]NE overflow by 63% (right panel of FIG. 6, empty column). In the presence of 0.1 μM moxonidine at S_4 and S_5, the stimulation frequency was increased to 2 Hz to compensate for the inhibitory effect of the interacting drug moxonidine, which was present at a concentration of 0.1 μM during these stimulation periods. In addition, 3 μM cirazoline was given at S_5. In the presence of moxonidine, cirazoline-induced inhibition of evoked [^{3}H]NE overflow was significantly reduced to 17% (hatched column).

In conclusion, the interaction experiments revealed that whenever one of the two receptor systems is in an activated state, the effect mediated by the other is attenuated (FIG. 1). The mechanism underlying this mutual interaction between the presynaptic imidazoline receptor and the α_2-autoreceptor, which was postulated previously[1,2] and which may occur, for example, at the level of the receptor proteins or one of the steps of signal transduction, is unknown.

PRESYNAPTIC IMIDAZOLINE RECEPTORS IN THE HUMAN PULMONARY ARTERY

In view of the potential clinical implications, it is worthwhile to investigate whether presynaptic imidazoline receptors are also present in human blood vessels. Preliminary studies designed to examine this possibility were carried out in two blood vessels, namely, the human saphenous vein and the pulmonary artery. BDF 6143 and cirazoline were applied as pharmacological tools.

In strips of the saphenous vein, 0.01–1 μM BDF 6143 produced a monophasic concentration-dependent facilitation of the electrically evoked [^{3}H]NE release[31]; in preparations superfused with solution containing rauwolscine, this facilitatory effect was abolished but not reversed to an inhibition. These results argue against the

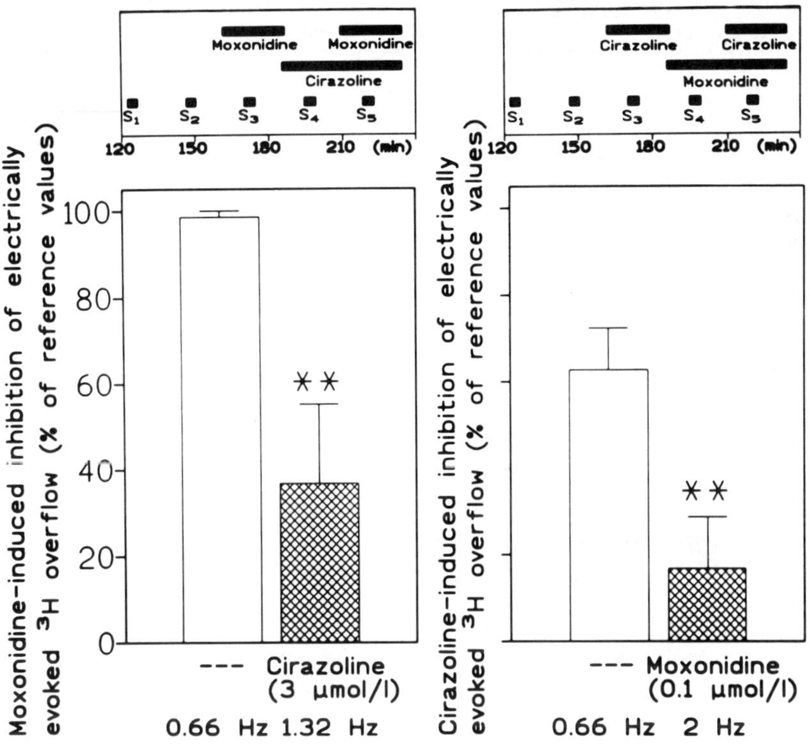

FIGURE 6. Moxonidine-induced inhibition (*left lower panel*) and cirazoline-induced inhibition (*right lower panel*) of the electrically evoked ^{3}H overflow from superfused strips of the rabbit aorta (preincubated with [^{3}H]norepinephrine) in the absence (*open columns*) and presence (*hatched columns*) of an interacting drug. Five periods of stimulation (S_1–S_5) were applied at the times after onset of superfusion indicated in the *upper panels*. Stimulation parameters are indicated below the columns. In the experiments in which the moxonidine (0.3 µM)-induced inhibition of ^{3}H overflow was investigated in the absence and presence of cirazoline (*left panel*), moxonidine was added to the superfused fluid 9 minutes before, during, and after S_3 and S_5; cirazoline was administered from 9 minutes before S_4 until the end of the experiment (see *left upper panel* for time schedule; 360 impulses at S_1–S_3, 475 impulses at S_4 and S_5). Conversely, an analogous time schedule was used in the experiments in which the cirazoline (3 µM)-induced inhibition of ^{3}H overflow was determined in the absence and presence of moxonidine (*right panel;* cirazoline applied at S_3 and S_5; moxonidine administered from 9 minutes before S_4 until the end of the experiment; see *right upper panel* for time schedule; 360 impulses at S_1–S_5). The values for moxonidine- or cirazoline-induced inhibition of evoked ^{3}H overflow in the absence and presence of the respective interacting compound are given as percentages of the overflows evoked by the corresponding reference stimulation periods S_2 and S_4, respectively. Means ± SEM of 5–7 experiments. In control experiments (n = 7–9) carried out in the presence of cirazoline and moxonidine as interacting drugs from 9 minutes before S_4 until the end of the experiments, the ratios S_3/S_2 and S_5/S_4 were close to unity. **$p < 0.025$ (compared with the effect of moxonidine and cirazoline in the absence of the respective interacting compound).

TABLE 3. Drugs Acting at Presynaptic Imidazoline Receptors and/or α_2-Adrenoceptors in the Rabbit Pulmonary Artery and Aorta[a]

Drug	Presynaptic Imidazoline Receptor	Presynaptic α_2-Adrenoceptor
Cirazoline	Agonist	0
Oxymetazoline	0	Agonist
Moxonidine	0	Agonist
Clonidine	Agonist	Agonist
BDF 6143	Agonist	Antagonist
Rauwolscine	Antagonist[b] <	Antagonist[c]

[a] 0 = no effect; < = lower potency.
[b] pA_2: 6.37-6.77 (–7.32).
[c] pA_2: 8.08-8.33.

possibility that the sympathetic nerves of the human saphenous vein are endowed with inhibitory presynaptic imidazoline receptors.

In contrast, the human pulmonary artery resembled the rabbit pulmonary artery in that BDF 6143 induced a biphasic increase in electrically evoked [^{3}H]NE release; up to 0.1 μM BDF 6143, facilitation increased with the concentration applied, whereas facilitation caused by 1 μM BDF 6143 was less pronounced than was that induced by the 10 times lower concentration (bell-shaped concentration-response curve). In the presence of 3 μM rauwolscine, BDF 6143 concentration-dependently (0.1-10 μM) inhibited the electrically evoked [^{3}H]NE release; under this condition (as in the absence of rauwolscine) the evoked [^{3}H]NE release was also inhibited by cirazoline (1 and 10 μM). These findings are similar to those in the rabbit pulmonary artery and aorta and are compatible with the suggestion that inhibitory presynaptic imidazoline receptors exist in the human pulmonary artery.

GENERAL CONCLUSIONS

Experiments with slices and synaptosomes of the rabbit and rat cerebral cortex revealed that presynaptic imidazoline receptors do not appear to exist in this brain region, but that the imidazolines clonidine and moxonidine as well as the guanidine aganodine inhibit NE release by stimulating presynaptic α_2-autoreceptors only. However, the postganglionic sympathetic axon terminals of the rabbit pulmonary artery and aorta and the human pulmonary artery are endowed with presynaptic imidazoline receptors in addition to the α_2-autoreceptors. Both types of inhibitory presynaptic receptors mutually interact with each other. However, the results reported here cannot only be interpreted in terms of the possibility just discussed that presynaptic imidazoline receptors and α_2-autoreceptors are different proteins in the cell membrane, but it is also conceivable that the imidazoline receptor represents a modulatory site of the α_2-adrenoceptor. In any case the pharmacological properties of the presynaptic imidazoline receptors differ from those of the I_1- and I_2-imidazoline binding site. Important pharmacological features of presynaptic imidazoline receptors are summarized in TABLE 3; cirazoline is a "pure" imidazoline receptor agonist, whereas BDF 6143 is a mixed imidazoline receptor agonist/α_2-adrenoceptor antagonist, and clonidine seems to be an agonist at both types of receptors; rauwolscine blocks not only the presynaptic α_2-adrenoceptors but at lower potency also the presynaptic imidazoline receptors.

REFERENCES

1. GÖTHERT, M. & G. J. MOLDERINGS. 1991. Involvement of presynaptic imidazoline receptors in the α_2-adrenoceptor-independent inhibition of noradrenaline release by imidazoline derivatives. Naunyn-Schmiedebergs Arch. Pharmacol. **343:** 271-282.
2. MOLDERINGS, G. J., F. HENTRICH & M. GÖTHERT. 1991. Pharmacological characterization of the imidazoline receptor which mediates inhibition of noradrenaline release in the rabbit pulmonary artery. Naunyn-Schmiedebergs Arch. Pharmacol. **344:** 630-638.
3. GÖTHERT, M. & G. J. MOLDERINGS. 1992. Modulation of norepinephrine release in blood vessels: Mediating by presynaptic imidazoline receptors and α_2-adrenoceptors. J. Cardiovasc. Pharmacol. **20:** S16-S20.
4. FUDER, H. & P. SCHWARZ. 1993. Desensitization of inhibitory prejunctional α_2-adrenoceptors and putative imidazoline receptors on rabbit heart sympathetic nerves. Naunyn-Schmiedebergs Arch. Pharmacol. **348:** 127-133.
5. STARKE, K., M. GÖTHERT & H. KILBINGER. 1989. Modulation of neurotransmitter release by presynaptic autoreceptors. Physiol. Rev. **69:** 864-989.
6. ERNSBERGER, P., M. P. MEELEY, J. J. MANN & D. J. REIS. 1987. Clonidine binds to imidazole binding sites as well as alpha$_2$-adrenoceptors in the ventrolateral medulla. Eur. J. Pharmacol. **134:** 1-13.
7. MICHEL, M. C., O. E. BRODDE, B. SCHNEPEL, J. BEHRENDT, R. TSCHADA, H. J. MOTULSKY & P. A. INSEL. 1989. [^{3}H]Idazoxan and some other α_2-adrenergic drugs also bind with high affinity to a noradrenergic site. Mol. Pharmacol. **35:** 324-330.
8. BRICCA, G., M. DONTENWILL, A. MOLINES, J. FELDMAN, E. TIBIRICA, A. BELCOURT & P. BOUSQUET. 1989. Rilmenidine selectivity for imidazoline receptors in human brain. Eur. J. Pharmacol. **163:** 373-377.
9. LACHAUD-PETTITI, V., R. A. PODEVIN, Y. CHRETIEN & A. PARINI. 1991. Imidazoline-guanidinium and alpha$_2$-adrenergic binding sites in basolateral membranes from human kidney. Eur. J. Pharmacol.-Mol. Pharmacol. Sect. **206:** 23-31.
10. ERNSBERGER, P. R., K. L. WESTBROOKS, M. O. CHRISTEN & S. G. SCHÄFER. 1992. A second generation of centrally acting antihypertensive agents act on putative I_1-imidazoline receptors. J. Cardiovasc. Pharmacol. **20:** S1-S10.
11. MIRALLES, A., G. OLMOS, M. SASTRE, F. BARTUREN, I. MARTIN & J. A. GARCIA-SEVILLA. 1993. Discrimination and pharmacological characterization of I_2-imidazoline sites with [^{3}H]idazoxan and alpha$_2$-adrenoceptors with [^{3}H]RX821002 (2-methoxy-idazoxan) in the human and rat brains. J. Pharmacol. Exp. Ther. **264:** 1187-1197.
12. MOLDERINGS, G. J., D. MOURA, K. FINK, H. BÖNISCH & M. GÖTHERT. 1993. Binding of [^{3}H]clonidine to I_1-imidazoline sites in bovine adrenal medullary membranes. Naunyn-Schmiedeberg's Arch. Pharmacol. **348:** 70-76.
13. MOLDERINGS, G. J., L. KUNDT & M. GÖTHERT. 1994. [^{3}H]Idazoxan binding to bovine adrenal medullary membranes: Identification and pharmacological characterization of I_2-imidazoline sites. Naunyn-Schmiedeberg's Arch. Pharmacol. **350:** 252-257.
14. LI, G., S. REGUNATHAN, C. J. BARROW, J. ESTRAGHI, R. COOPER & D. J. REIS. 1994. Agmatine: An endogenous clonidine-displacing substance in the brain. Science **263:** 966-969.

15. SCHLICKER, E., M. GÖTHERT, F. KÖSTERMANN & R. CLAUSING. 1983. Effects of α-adrenoceptor antagonists on the release of serotonin and noradrenaline from rat brain cortex slices. Influence of noradrenaline uptake inhibition and determination of pA_2 values. Naunyn-Schmiedeberg's Arch. Pharmacol. **232:** 106-113.
16. FINK, K. & M. GÖTHERT. 1992. Presynaptic site of action underlying the ethanol-induced inhibition of norepinephrine release evoked by stimulation of *N*-methyl-D-aspartate (NMDA) receptors in rat cerebral cortex. Brain Res. **572:** 27-32.
17. MOLDERINGS, G. J. & M. GÖTHERT. 1993. Pharmacological characterization of the inhibitory α_2-autoreceptors on the sympathetic nerves of human, rat and rabbit blood vessels. Fundam. Clin. Pharmacol. **7:** 372.
18. HAKY, J. E. & A. M. YOUNG. 1984. Evaluation of a simple HPLC correlation method for the estimation of the octanol-water partition coefficient of organis compounds. J. Liq. Chromatogr. **7:** 675-689.
19. KOJDA, G., W. KLAUS, G. WERNER & U. FRICKE. 1991. The influence of 3-ester side chain variation on the cardiovascular profile of nitrendipine in porcine isolated trabeculae and coronary arteries. Naunyn-Schmiedeberg's Arch. Pharmacol. **344:** 488-494.
20. LANGIN, D. & M. LAFONTAN. 1989. [^{3}H]Idazoxan binding at non-α_2-adrenoceptors in rabbit adipocyte membranes. Eur. J. Pharmacol. **159:** 199-203.
21. ZINNENSCHEIN, R., S. DIAMANT & D. ATLAS. 1990. Imidazoline receptors in rat liver cells: A novel receptor or a subtype of alpha$_2$-adrenoceptors? Eur. J. Pharmacol. **190:** 203-215.
22. YABLONSKY, F. & J. P. DAUSSE. 1991. Non-adrenergic binding sites for the "alpha$_2$-antagonist" [^{3}H]idazoxan in the rabbit urethral smooth muscle. Pharmacological and biochemical characterization. Biochem. Pharmacol. **41:** 701-707.
23. GLOSSMANN, H. & P. PRESEK. 1979. Alpha noradrenergic receptors in brain membranes: Sodium, magnesium and guanyl nucleotides modulate agonist binding. Naunyn-Schmiedeberg's Arch. Pharmacol. **306:** 67-73.
24. COUPRY, I., D. ATLAS, R. A. PODEVIN, I. UZIELLI & A. PARINI. 1990. Imidazoline-guanidinium receptive site in renal proximal tubule: Asymmetric distribution, regulation by cations and interaction with an endogenous clonidine displacing substance. J. Pharmacol. Exp. Ther. **252:** 293-299.
25. YAKUBU, M. A., N. M. DEIGHTON, C. A. HAMILTON & J. L. REID. 1990. Differences in the regulation of [^{3}H]idazoxan and [^{3}H]yohimbine binding sites in the rabbit. Eur. J. Pharmacol. **176:** 305-311.
26. MASSINGHAM, R., M. L. DUBOCOVICH & S. Z. LANGER. 1980. The role of presynaptic receptors in the cardiovascular actions of *N,N,*-di-*n*-propyldopamine in the cat and dog. Naunyn-Schmiedeberg's Arch. Pharmacol. **314:** 17-28.
27. LOIACONO, R. E., M. J. RAND & D. F. STORY. 1985. Interaction between the inhibitory action of acetylcholine and the α_2-adrenoceptor autoinhibitory feedback system on release of [^{3}H]noradrenaline from rat atria and rabbit ear artery. Br. J. Pharmacol. **84:** 697-705.
28. RAMME, D., P. ILLES, L. SPÄTH & K. STARKE. 1986. Blockade of α_2-adrenoceptors permits the operation of otherwise silent opioid k-receptors at the sympathetic axons of rabbit jejunal arteries. Naunyn-Schmiedeberg's Arch. Pharmacol. **334:** 48-55.
29. LIMBERGER, N., L. SPÄTH & K. STARKE. 1988. Presynaptic α_2-adrenoceptor, opioid k-receptor and adenosine A_1-receptor interactions on noradrenaline release in rabbit brain cortex. Naunyn-Schmiedeberg's Arch. Pharmacol. **338:** 53-61.
30. MOLDERINGS, G. J. & M. GÖTHERT. 1990. Mutual interaction between presynaptic α_2-adrenoceptors and 5-HT_{1B} receptors on the sympathetic nerve terminals of the rat inferior vena cava. Naunyn-Schmiedeberg's Arch. Pharmacol. **341:** 391-397.

31. GÖTHERT, M., E. SCHLICKER, F. HENTRICH, N. ROHM & H.-R. ZERKOWSKI. 1984. Modulation of noradrenaline release in human saphenous vein via presynaptic α_2-adrenoceptors. Eur. J. Pharmacol. **102:** 261-267.

32. STARKE, K. 1977. Regulation of noradrenaline release by presynaptic receptor systems. Rev. Physiol. Biochem. Pharmacol. **77:** 1-124.

No Relationship of I_1- and I_2-Imidazoline Binding Sites to Inhibitory Effects of Imidazolines on Ligand-Gated Ion Channels

An Investigation in the Adrenal Medulla and in Neuroblastoma Cells

G. J. MOLDERINGS,[a] K. RUPPERT, H. BÖNISCH, AND M. GÖTHERT

Institute of Pharmacology and Toxicology
University of Bonn
Reuterstr. 2b
D-53113 Bonn, Germany

It has been established that imidazoline derivatives and structurally related compounds bind not only to α_2-adrenoceptors but also to imidazoline binding sites (IBS). The latter differ from α_2-adrenoceptors in structure, function, and distribution among and within organs. (For a review, see ref. 1.) The different rank orders of affinity of ligands for these sites indicate the existence of at least two classes of IBS denoted as I_1- and I_2-IBS. [^{3}H]Clonidine is preferentially bound to I_1-IBS,[2,3] whereas [^{3}H]idazoxan preferentially labels I_2-IBS.[4–6] In addition, we recently identified a presynaptic imidazoline receptor mediating inhibition of norepinephrine release from postganglionic sympathetic nerve fibers in rabbit pulmonary artery and aorta with pharmacological properties different from those of both I_1- and I_2-IBS.[7–9]

Recent studies of binding of [^{3}H]clonidine and [^{3}H]idazoxan to membranes prepared from bovine adrenal chromaffin cells revealed that these cells lack α_2-adrenoceptors but appear to possess I_1-IBS[3,10] and I_2-IBS.[11,12] Doubts were expressed over the coexistence of both I_1- and I_2-IBS on chromaffin cells,[13] but in the first part of the present account we report on recent investigations[14,15] that provide evidence for the coexistence of I_1- and I_2-IBS on bovine adrenomedullary membranes.

It has been speculated that IBS on chromaffin cells might be involved in the modification of catecholamine (CA) release from the adrenal medulla. Certain imidazolines cause an increase in *basal* CA release,[16] whereas CA release evoked by acetylcholine or nicotine was inhibited by a variety of imidazolines including clonidine and related compounds at very high concentrations[17–19] ($\geq$10 μM). However, CA release evoked by either histamine or a high K^+ concentration was not modified by the imidazolines.[17–19] On the basis of these findings, we analyze in the second part

[a] To whom correspondence should be addressed.

of this report whether the inhibitory potency of imidazolines on the acetylcholine-evoked CA release from bovine chromaffin cells is correlated with their affinity for I_1 or I_2-IBS as a potential hint in the involvement of I_1 or I_2 recognition sites in this inhibition.

In the third part of the present account, we analyze whether this inhibitory effect of imidazolines is restricted to the nicotinic receptor channel or whether it also comes into play at other ligand-gated ion channels such as the 5-HT_3-receptor. For this purpose, N1E-115 neuroblastoma cells were used to study the influence of imidazolines on 5-HT-induced [^{14}C]guanidinium influx through the 5-HT_3 receptor channel.[20–22] Because, in this context, it was of interest to investigate if ion influx through voltage-gated Na^+ channels is also modulated by imidazoline derivatives, additional investigations were carried out in which the influence of BDF 6143 (4-chloro-2-[2-imidazolin-2-ylamino]-isoindoline), an imidazoline with high affinity for IBS, on the veratridine-induced [^{14}C]guanidinium influx into N1E-115 cells was examined.

RADIOLIGAND BINDING TO BOVINE CHROMAFFIN CELL MEMBRANES

Details of the preparation of the adrenomedullary membranes and of the [^{3}H]clonidine and [^{3}H]idazoxan binding assays were described elsewhere.[14,15] Our binding studies[14,15] confirmed the results of previous investigations in membranes from bovine chromaffin cells, namely, that [^{3}H]idazoxan[12,13] and [^{3}H]clonidine[3,10] bind to different populations of nonadrenoceptor IBS. However, the existence of both types of IBS in bovine adrenal medullary membranes has previously never been demonstrated in the same laboratory. As a further extension of previous investigations, our competition experiments of imidazolines and guanidines with the radioligands also included a series of isoindolines substituted with aminoimidazoline or guanidine. These compounds were used as key drugs for the characterization of presynaptic imidazoline receptors in rabbit blood vessels (see above). The possibilities that [^{3}H]clonidine and [^{3}H]idazoxan label a histamine receptor or an α_2-adrenoceptor were excluded, because in competition experiments histamine and the α_2-adrenoceptor ligands rauwolscine and (–)-epinephrine exhibited no or very low affinity for [^{3}H]clonidine and [^{3}H]idazoxan sites. Moreover, the rank order of potency of the α_2-adrenoceptor ligands in displacing [^{3}H]clonidine and [^{3}H]idazoxan (TABLE 1) differs from the rank order expected for binding to one of the α_2-adrenoceptor subtypes.[24]

We found shallow inhibition curves with Hill coefficients significantly different from unity for most of the compounds investigated in competition experiments with [^{3}H]clonidine (FIG. 1) and for two compounds (idazoxan and cirazoline) in competition experiments with [^{3}H]idazoxan[14,15] (FIG. 2). Similar results were obtained in radioligand binding experiments with [^{3}H]idazoxan in the guinea pig kidney.[25] The authors suggested that IBS may exist in two interconvertible forms which are induced in different proportions by the imidazolines or related compounds.[25]

The pharmacological properties of [^{3}H]clonidine and [^{3}H]idazoxan binding sites in the bovine adrenal medulla are summarized in TABLE 1. The affinities of competing drugs for the [^{3}H]clonidine binding site were similar to those found in another recently

TABLE 1. Affinity (K_i values) of Imidazolines (I) and Guanidines (G) for High Affinity [^{3}H]Clonidine and [^{3}H]Idazoxan Binding Sites in Bovine Adrenal Medullary Membranes[a]

Compound	[^{3}H]Clonidine K_i (nM)[b]	[^{3}H]Idazoxan K_i (nM)[c]
Naphazoline (I)	8.4	1,451
Clonidine (I)	14.8	272
Cirazoline (I)	17.0	4.3
BDF 6143 (I)	28.9	299
Moxonidine (I)	77.8	–
Idazoxan (I)	103	4.1
Phentolamine (I)	148	1,424
Amiloride (G)	2,350	>100,000
Histamine	8,935	>100,000
(–)Epinephrine	>100,000	>100,000

[a] Results are from computer analysis of the competition of various concentrations of the compounds with [^{3}H]clonidine and [^{3}H]idazoxan 10 nM.
[b] Taken from ref. 14.
[c] Taken from ref. 15.

published study of I_1-IBS in the adrenal medulla.[3,10] Because the K_i value of clonidine for the high affinity binding site was in the nanomolar range, this site is unlikely to represent an I_2-IBS, for which clonidine has an affinity in the micromolar range.[16] The K_i value of clonidine of 16 nM found in our experiments for the high affinity site is in the same range as its K_1 values at the I_1-IBS (1-17 nM) previously described in bovine and rat ventrolateral medulla,[2,3,26–28] rat kidney, and human platelet membranes.[29,30] Moreover, the rank order of affinity observed in our experiments for the high affinity [^{3}H]clonidine site (TABLE 1) was very similar to that of I_1-IBS in other tissues, which was clonidine = naphazoline = moxonidine ≥ BDF 6143 > cirazoline = idazoxan ≫ ≫ rauwolscine and norepinephrine.[27,31,32]

The rank order of affinity of the competing drugs for the [^{3}H]idazoxan binding sites, in particular the affinity of clonidine, is clearly different from that observed in the experiments with [^{3}H]clonidine (TABLE 1) and at other I_1-IBS (see above). The rank order of affinity of imidazolines and guanidines for I_2-IBS in various tissues was cirazoline ≥ idazoxan > guanabenz > tolazoline > rilmenidine > clonidine > phentolamine[5,6,33,34] and, hence, conforms to the rank order found in our experiments with [^{3}H]idazoxan. Accordingly, [^{3}H]idazoxan binding sites in the bovine adrenal medullary membranes may be denoted as I_2-IBS. The latter can be further subclassified, depending on its sensitivity to displacement by the guanidine amiloride. An amiloride-sensitive I_2-IBS has been found in rabbit kidney, liver, and fat cells[35–37] as well as in human placenta,[38] whereas amiloride-insensitive I_2-IBS have been detected in kidney (man,[39] pig,[40] rat[4]) liver (man,[37] rat[33]) guinea pig ileum,[38] and brain (man,[6] guinea pig[41]). Obviously, the I_2-IBS on bovine adrenal chromaffin cells identified in our study belong to the amiloride-insensitive subclass.

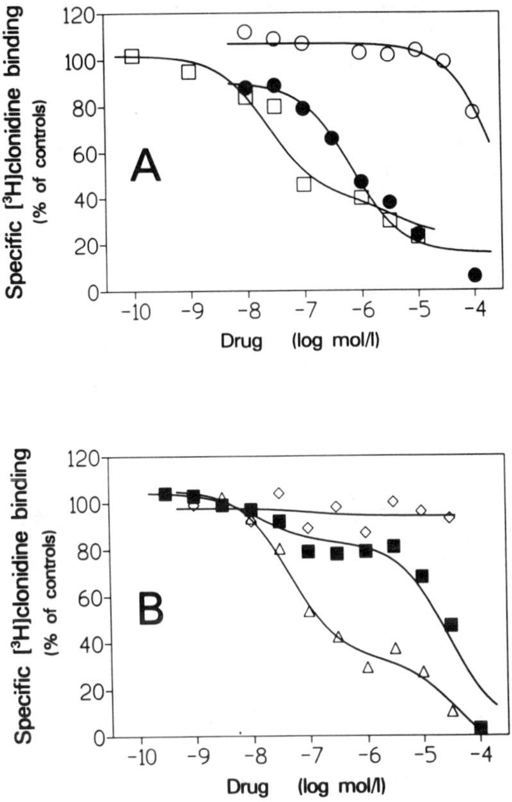

FIGURE 1. Competition of six drugs with [^{3}H]clonidine for its specific binding sites in bovine adrenal medullary membranes. Membranes were incubated for 40 minutes with [^{3}H]clonidine 10 nM and increasing concentrations of the competitor. **(A)** ○, (–)-epinephrine; □, clonidine; ●, guanabenz; **(B)** ◇, rauwolscine; ■, naphazoline; △, BDF 6143. Each point is the mean of 5-7 experiments performed in triplicate. SEM amounted to up to 20% of the respective mean (reproduced with permission from ref. 14).

Taken together, our data provide evidence that [^{3}H]clonidine and [^{3}H]idazoxan label different specific IBS in the bovine adrenal medulla which resemble the I_1- and I_2-subtype, respectively. Accordingly, I_1- and I_2-IBS coexist in this tissue.

EFFECTS OF IMIDAZOLINES ON BOVINE CHROMAFFIN CELLS

On the basis of these results we examined whether I_1- and I_2-IBS play a functional role in the adrenal medulla. Activation of I_2-imidazoline receptors in bovine adrenal chromaffin cells has been suggested to cause an increase in basal release of catecholamines.[12,16] In contrast, it is well known that certain (but not all) imidazolines such as clonidine, phentolamine, and naphazoline at high concentrations (≥10 μM) inhibit acetylcholine- or nicotine-evoked catecholamine release; by contrast, phenylethylamines and rauwolscine were ineffective.[17–19,42,43] However, the discrepancy between the nanomolar affinities of the compounds in the radioligand binding studies and their micromolar potencies in the functional model sheds doubts on the hypothesis that I_1- or I_2-IBS might be involved in the imidazoline-induced inhibition of evoked

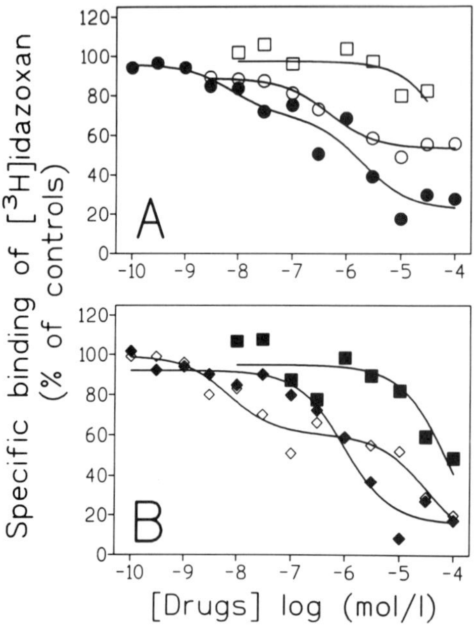

FIGURE 2. Competition of six drugs with [^{3}H]idazoxan for its specific binding sites in bovine adrenal medullary membranes. Membranes were incubated for 45 minutes with [^{3}H]idazoxan 10 nM and increasing concentrations of the competitor. (**A**) □, (–)-epinephrine; ○, clonidine; ●, cirazoline; (**B**) ■, rauwolscine; ◇, idazoxan; ◆, BDF 6100. Each point is the mean of 5–8 experiments performed in triplicate. SEM amounted to up to 20% of the respective mean (reproduced with permission from ref. 15).

CA release. Similar inconsistencies between the binding affinities and the concentrations necessary to inhibit ion conductance in chromaffin cells and PC 12 cells were observed in electrophysiological experiments. Whole-cell patch-clamp experiments in bovine chromaffin cells revealed that clonidine concentrations of 2–30 μM had to be applied to reduce the acetylcholine-induced current, whereas epinephrine up to 60 μM was without influence.[17,44] In PC 12 cells, a rat pheochromocytoma cell line that shares important pharmacological properties with adrenal chromaffin cells, it was recently shown by means of the whole-cell patch-clamp technique that several imidazolines (clonidine, cirazoline, naphazoline, oxymetazoline, idazoxan, and moxonidine) at concentrations of 10–100 μM inhibit the Na^+ current induced by nicotine.[45] However, nicotine-evoked Na^+ current was not influenced by epinephrine and yohimbine up to 100 μM.[45] In view of the specific action of imidazolines (in contrast to non-imidazoline α_2-adrenoceptor ligands) on ion currents, it is tempting to speculate that the nicotinic receptor channel is a site of action of imidazolines, which is independent of α_2-adrenoceptors and I_1- or I_2-IBS.

Nevertheless, the possibility that I_1- and I_2-IBS might play a role in the imidazoline-induced inhibition of catecholamine release certainly is a particularly attractive hypothesis. Therefore, we reanalyzed this possibility more carefully. We compared a series of compounds with respect to their inhibitory potencies, that is, data reported in the literature (for references, see TABLE 2), with their affinities determined in our radioligand binding studies. In detail, the rank order of potencies of relevant drugs (imidazolines and non-imidazolines; TABLE 2) in inhibiting CA release from bovine adrenal medulla evoked by nicotine was compared to the rank order of affinity of these substances at I_1- and I_2-IBS on bovine adrenomedullary membranes (TABLE 1).

TABLE 2. Potency (pIC_{50}) of Drugs for Inhibiting Nicotine-Evoked Catecholamine Release from Bovine Chromaffin Cells (values reported in the literature) and Lipophilicity (log P) of Drugs[a]

Compound	Potency (pIC_{50})	Lipophilicity (log P)
1. Clonidine[b]	4.5	1.65
2. Phentolamine[b]	4.5	3.66
3. Oxymetazoline[b]	5	4.08
4. Naphazoline[c]	4.6	4.52
5. Guanabenz[b]	5	2.95
6. Tolazoline[b]	Facilitation	2.95
7. Norepinephrine[d]	<4	1.26
8. Epinephrine[d]	<4	1.84
9. Yohimbine[e]	~4	2.63
10. Chlorpheniramine[b]	5	4.25

[a] Determined in our laboratory; unpublished.
[b] Taken from ref. 19.
[c] Taken from ref. 42.
[d] Taken from ref. 17.
[e] Taken from ref. 18.

Potency in inhibiting CA release decreased in the order oxymetazoline = guanabenz > naphazoline ≥ clonidine = phentolamine > yohimbine, whereas the rank orders of affinity for binding to I_1- and I_2-IBS of bovine chromaffin cells were naphazoline > clonidine ≫ phentolamine > guanabenz ≫ ≫ yohimbine (I_1-IBS) and clonidine > phentolamine = naphazoline ≫ ≫ yohimbine (I_2-IBS), respectively. Thus, the rank order of potency of the imidazolines for their ability to induce an inhibitory effect does not conform to the pharmacological properties of I_1- or I_2-IBS, and it also does not fit in with the pharmacological properties of the presynaptic imidazoline receptors on the sympathetic nerves of the rabbit pulmonary artery[8] (clonidine > phentolamine > tolazoline > oxymetazoline). Tolazoline, which inhibited evoked norepinephrine release from sympathetic nerves, even facilitated evoked CA release from bovine chromaffin cells.[19]

In addition, the possibility was considered that the inhibitory effect of the imidazolines might be due to an interaction with membrane lipids or hydrophobic regions of membrane proteins such as the nicotinic receptor. Therefore, we determined the lipophilicity (log P values) of the relevant drugs with a simple but reliable HPLC method[46,47] (TABLE 2). The potency of the imidazolines in inhibiting acetylcholine-evoked CA release was not correlated with their log P values (FIG. 3), suggesting that their inhibitory effect is not due to a hydrophobic interaction with lipid components of the cell membrane or hydrophobic regions of the receptor protein.

Taken together, the inhibitory effect of imidazolines on CA release from bovine chromaffin cells evoked by stimulation of nicotinic receptors is due neither to activation of I_1- or I_2-IBS or of imidazoline receptors with the pharmacological characteris-

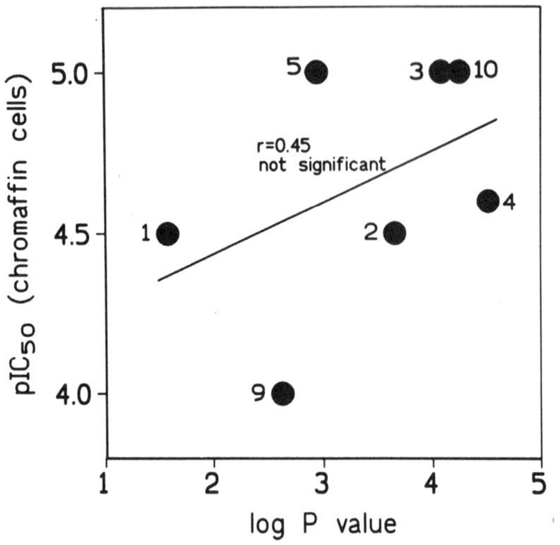

FIGURE 3. Comparison of the pIC_{50} values for inhibition of nicotine-induced catecholamine release from bovine chromaffin cells by imidazolines and related substances with their lipophilicity (log P values; TABLE 2). The two parameters were not correlated (r, correlation coefficient). Numbers refer to the numbers of the compounds in TABLE 2.

tics of the presynaptic imidazoline receptors nor to interaction between the drugs and lipids or hydrophobic regions of proteins of the cell membrane. Patch-clamp experiments revealed that several imidazolines differ from the catecholamine epinephrine in that they inhibit ion conductance through the nicotinic receptor channel, suggesting that this ligand-gated ion channel may be a site of action of these drugs.

EFFECTS OF IMIDAZOLINES ON N1E-115 CELLS

Since ligand-gated ion channels represent a superfamily of several membrane receptors, it was of interest to study whether the inhibitory effect of the imidazolines occurs not only at the nicotinic receptor but also at other ligand-gated ion channels such as the 5-HT_3 receptor. Cation influx via this receptor's channel can be investigated in mouse neuroblastoma cell line N1E-115 which is well known to be endowed with 5-HT_3 receptors at particularly high density (for a review, see ref. 48). In these cells, the organic cation [^{14}C]guanidinium can be used to measure ion influx through the 5-HT_3 receptor channel.[20–22,49] In studies of the influence of imidazolines on [^{14}C]guanidinium flux through the 5-HT_3 receptor channel, an incubation buffer containing 139 mM Na^+ (and other cations such as K^+ and Mg^{2+}) was applied. It has been established that under these ionic conditions 5-HT_3 receptor stimulation with 5-HT has to be carried out in the presence of 10 μM of substance P in order to obtain reliably high [^{14}C]guanidinium influx.[20,22,49] The [^{14}C]guanidinium influx evoked by 5-HT under this condition occurs via 5-HT_3 receptor channels, because it is inhibited by 5-HT_3 receptor antagonists such as ondansetron but not by tetrodotoxin.[20,22,49]

Basal [^{14}C]guanidinium influx into N1E-115 cells was not or only weakly inhibited by the compounds investigated. BDF 6143 (FIG. 4), idazoxan, cirazoline, naphazoline, and the endogenous ligand of imidazoline receptors, agmatine,[50] concentration-depen-

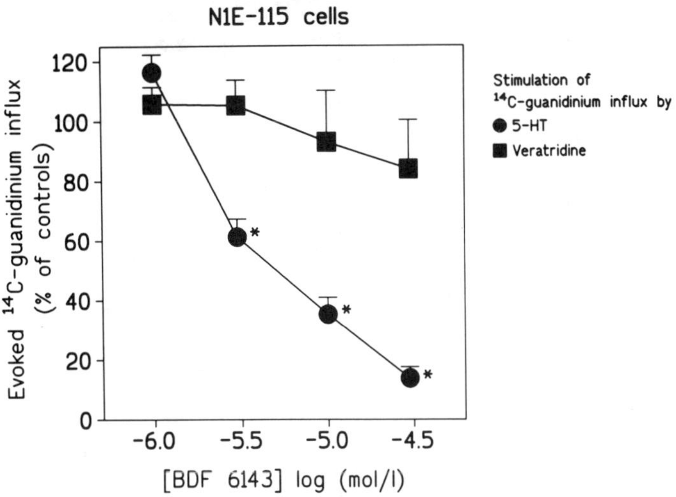

FIGURE 4. Inhibition of 5-HT (1 μM for 2 minutes in the presence of 10 μM of substance P)-induced (*closed circles*) and veratridine (1 mM for 2 minutes)-induced (*closed squares*) influx of [^{14}C]guanidinium into N1E-115 cells by BDF 6143. Shown are mean values ± SEM ($n = 4$) for inhibition of [^{14}C]guanidinium influx (expressed as percent of the respective control). $^*p < 0.05$.

dently inhibited the [^{14}C]guanidinium influx induced by 1 μM of 5-HT (in the presence of 10 μM of substance P), whereas clonidine and, above all, epinephrine were ineffective up to the highest concentration investigated (100 μM). Accordingly, it is obvious that the 5-HT_3 receptor resembles the nicotinic receptor in that ion influx through the respective channel is inhibited by certain imidazolines (and agmatine) but not by the catecholamine epinephrine. The potencies of the imidazolines in inhibiting [^{14}C]guanidinium influx are listed in TABLE 3 together with their log P values. Comparison of these potencies with their lipophilicity, and their pK_i values at I_1-

TABLE 3. Potency (pIC_{50}) of Drugs in Inhibiting 5-HT (1 μM; in the presence of 10 μM substance P)-Evoked [$_{14}$C]Guanidinium Influx in Mouse N1E-115 Neuroblastoma Cells and Lipophilicity (log P) of the Drugs

Compound	Potency (pIC_{50})	Lipophilicity (log P)
1. Agmatine	3.85	—
2. BDF 6143	5.30	3.24
3. Cirazoline	5.15	4.13
4. Naphazoline	4.77	4.52
5. Idazoxan	5.18	2.62
6. Clonidine	<4	1.65

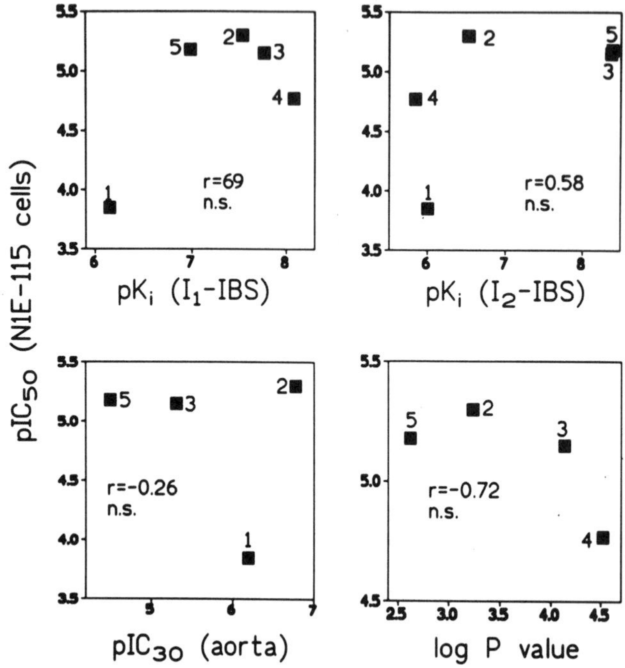

FIGURE 5. Comparison of pIC_{50} values for inhibition of 5-HT (1 μM, in the presence of 10 μM of substance P)-induced [^{14}C]guanidinium influx into N1E-115 cells by imidazolines and agmatine (TABLE 3) with their affinity (pK_i values) at I_1- and I_2-IBS (TABLE 1), their potency (pIC_{30} values) in inhibiting norepinephrine release from the sympathetic nerves of the rabbit aorta (from ref. 56), and their lipophilicity (log P values; TABLE 3). The pIC_{50} values in N1E-115 cells and the other parameters were not significantly correlated (r, correlation coefficient; n.s., not significant). Numbers refer to the numbers of the compounds in TABLE 3.

and I_2-IBS and their potencies in inhibiting norepinephrine release from the sympathetic nerves of the rabbit aorta revealed no correlation between the inhibitory effect on the 5-HT-induced [^{14}C]guanidinium influx and the other properties of the drugs (FIG. 5). These findings suggest that the imidazolines might act at certain sites of the 5-HT_3 receptor protein different from I_1- and I_2-IBS and from presynaptic imidazoline receptors, and not by interaction with membrane lipids or hydrophobic regions of the receptor protein.

N1E-115 cells are endowed not only with 5-HT_3 receptor channels but also with fast, voltage-dependent sodium channels.[51] In N1E-115 cells incubated in a solution in which Na^+ was omitted and replaced by cholinium$^+$, activation of the sodium channel by veratridine induced [^{14}C]guanidinium influx; the latter was not affected by the 5-HT_3 receptor antagonist ondansetron but was abolished by tetrodotoxin, indicating that the veratridine-induced guanidinium influx occurred exclusively via voltage-gated sodium channels.[21] Interestingly, the imidazoline BDF 6143 failed to decrease the veratridine-induced [^{14}C]-guanidinium influx under these ionic conditions (Na^+ replaced by cholinium$^+$), although, as already mentioned, it inhibited 5-HT-induced [^{14}C]guanidinium influx (FIG. 4). This finding suggests that the imidazoline selectively modulates the ligand-gated sodium channel, leaving the voltage-gated sodium channel unaffected.

CONCLUSIONS

In the bovine adrenal medulla, evidence was presented that I_1- and I_2-IBS labeled by [^{3}H]clonidine and [^{3}H]idazoxan, respectively, coexist in the same tissue. Neither the I_1- nor the I_2-IBS appear to be involved in the inhibitory effect of imidazolines on the CA release evoked by nicotinic receptor activation. This inhibitory effect is probably also not due to a hydrophobic interaction with lipid constituents or proteins of the cell membrane. Imidazolines have been shown to decrease ion conductance through the nicotinic receptor channel. In N1E-115 neuroblastoma cells, imidazolines and the endogenous imidazoline receptor ligand agmatine also inhibited the ion influx through the 5-HT_3 receptor, that is, another ligand-gated sodium channel. These findings are compatible with the suggestion that ligand-gated ion channels are sites of action of these drugs. In this context it is of interest that the ATP-dependent K^+ channel is modulated by imidazolines in the same concentration range as is the nicotinic and 5-HT_3 receptor channel.[52–55]

REFERENCES

1. BOUSQUET, P., G. BRICCA, M. DONTENWILL, J. FELDMAN, H. GRENEY, A. BELCOURT, J. STUTZMANN & E. TIBIRICA. 1992. From the α_2-adrenoceptors to the imidazoline preferring receptors. Fundam. Clin. Pharmacol. **6** (Suppl 1): 15S-21S.
2. ERNSBERGER, P., M. P. MEELEY, J. J. MANN & D. J. REIS. 1987. Clonidine binds to imidazoline binding sites as well as alpha$_2$-adrenoceptors in the ventrolateral medulla. Eur. J. Pharmacol. **134:** 1-13.
3. ERNSBERGER, P. R., K. L. WESTBROOKS, M. O. CHRISTEN & S. G. SCHÄFER. 1992. A second generation of centrally acting antihypertensive agents act on putative I_1-imidazoline receptors. J. Cardiovasc. Pharmacol. **20** (Suppl 4): S1-S10.
4. MICHEL, M. C., O. E. BRODDE, B. SCHNEPEL, J. BEHRENDT, R. TSCHADA, H. J. MOTULSKY & P. A. INSEL. 1989. [^{3}H]idazoxan and some other α_2-adrenergic drugs also bind with high affinity to a nonadrenergic site. Mol. Pharmacol. **35:** 324-330.
5. LACHAUD-PETTITI, V., R. A. PODEVIN, Y. CHRETIEN & A. PARINI. 1991. Imidazoline-guanidinium and alpha$_2$-adrenergic binding sites in basolateral membranes from human kidney. Eur. J. Pharmacol.-Mol. Pharmacol. Sect. **206:** 23-31.
6. MIRALLES, A., G. OLMOS, M. SASTRE, F. BARTUREN, I. MARTIN & J. A. GARCIA-SEVILLA. 1993. Discrimination and pharmacological characterization of I_2-imidazoline sites with [^{3}H]idazoxan and alpha-2 adrenoceptors with [^{3}H]RX821002 (2-methoxy idazoxan) in the human and rat brains. J. Pharmacol. Exp. Ther. **264:** 1187-1197.
7. GÖTHERT, M. & G. J. MOLDERINGS. 1991. Involvement of presynaptic imidazoline receptors in the α_2-adrenoceptor-independent inhibition of norepinephrine release by imidazoline derivatives. Naunyn-Schmiedeberg's Arch. Pharmacol. **343:**271-282.
8. MOLDERINGS, G. J., F. HENTRICH & M. GÖTHERT. 1991. Pharmacological characterization of imidazoline receptor which mediates inhibition of noradrenaline release in the rabbit pulmonary artery. Naunyn-Schmiedeberg's Arch. Pharmacol. **344:**630-638.

9. Göthert, M. & G. J. Molderings. 1992. Modulation of norepinephrine release in blood vessels: Mediation by presynaptic imidazoline receptors and α_2-adrenoceptors. J. Cardiovasc. Pharmacol. **20** (Suppl. 4): S16-S20.

10. Ernsberger, P., T. H. Damon, M. Graff, S. G. Schäfer & O. Christen. 1993. Moxonidine, a centrally acting antihypertensive agent, is a selective ligand for I_1-imidazoline sites. J. Pharmacol. Exp. Ther. **264:** 172-182.

11. Regunathan, S., M. J. Evinger, M. P. Meeley & D. J. Reis. 1991. Effects of clonidine and other imidazole-receptors binding agents on second messenger systems and calcium influx in bovine adrenal chromaffin cells. Biochem. Pharmacol. **42:** 2011-2018.

12. Regunathan, S., M. P. Meeley & D. J. Reis. 1991. Clonidine-displacing substance from bovine brain binds to imidazoline receptors and releases catecholamines in adrenal chromaffin cells. Mol. Pharmacol. **40:** 884-888.

13. Reis, D. J., S. Regunathan, H. Wang, D. L. Feinstein & M. P. Meeley. 1992. Imidazoline receptors in the nervous system. Fundam. Clin. Pharmacol. **6** (Suppl. 1): 23S-20S.

14. Molderings, G. J., D. Moura, K. Fink, H. Bönisch & M. Göthert. 1993. Binding of [^{3}H]clonidine to I_1-imidazoline sites in bovine adrenal medullary membranes. Naunyn-Schmiedeberg's Arch. Pharmacol. **348:** 70-76.

15. Molderings, G. J., L. Kundt & M. Göthert. 1994. [^{3}H]Idazoxan binding to bovine adrenal medullary membranes: Identification and pharmacological characterization of I_2-imidazoline sites. Naunyn-Schmiedeberg's Arch. Pharmacol. **350:** 252-257.

16. Reis, D. J., S. Regunathan & M. P. Meeley. 1992. Imidazoline receptors and clonidine-displacing substance in relationship to control of blood pressure, neuroprotection, and adrenomedullary secretion. Am. J. Hypertens. **5:** 51S-57S.

17. Powis, D. A. & P. F. Baker. 1986. α_2-Adrenoceptors do not regulate catecholamine secretion by bovine medullary cells: A study with clonidine. Mol. Pharmacol. **29:** 134-141.

18. Wan, D. C. C., D. A. Powis, P. D. Marley & B. G. Livett. 1988. Effects of α- and β-adrenoceptor agonists and antagonists on ATP and catecholamine release and desensitization of the nicotinic response in bovine adrenal chromaffin cells. Biochem. Pharmacol. **37:** 725-736.

19. Ohara-Imaizumi, M. & K. Kumakura. 1992. Effects of imidazole compounds on catecholamine release in adrenal chromaffin cells. Cell. Mol. Neurobiol. **12:** 273-283.

20. Reiser, G. & B. Hamprecht. 1989. Substance P and serotonin act synergistically to activate a cation permeability in a neuronal cell line. Brain Res. **479:** 40-48.

21. Bönisch, H., M. Barann, J. Graupner & M. Göthert. 1993. Characterization of 5-HT_3 receptors of N1E-115 neuroblastoma cells by use of the influx of the organic cation [^{14}C]-guanidinium. Br. J. Pharmacol. **108:** 436-442.

22. Emerit, M. B., M. Riad, C. M. Fattaccini & M. Hamon. 1993. Characteristics of [^{14}C]guanidinium accumulation in NG 108-15 cells exposed to 5-HT_3 receptor ligands and substance. P. J. Neurochem. **40:** 2059-2067.

23. Wang, H., S. Regunathan, M. P. Meeley & D. J. Reis. 1992. Isolation and characterization of imidazoline receptor protein from bovine adrenal chromaffin cells. Mol. Pharmacol. **42:** 792-801.

24. Simonneaux, V., M. Ebadi & D. B. Bylund. 1991. Identification and characterization of alpha$_{2D}$-adrenergic receptors in bovine pineal gland. Mol. Pharmacol. **40:** 235-241.

25. Wikberg, J. E. S., S. Uhlén & V. Chhajlani. 1992. Evidence that drug binding to non-adrenergic [^{3}H]-Idazoxan binding sites (I-receptors) occurs to interacting or interconvertible affinity forms of receptor. Pharmacol. Toxicol. **70:** 208-219.

26. Ernsberger, P., R. Giuliano, R. N. Willette & D. J. Reis. 1990. Role of imidazole receptors in the vasodepressor response to clonidine analogs in the rostral ventrolateral medulla. J. Pharmacol. Exp. Ther. **253:** 408-418.

27. Kamisaki, Y., T. Ishikawa, Y. Takao, H. Omodani, N. Kuno & T. Itoh. 1990. Binding of [^{3}H]*p*-aminoclonidine to two sites, α_2-adrenoceptors and imidazoline binding sites: Distribution of imidazoline binding sites in rat brain. Brain Res. **514:** 15-21.

28. Gomez, R. E., P. Ernsberger, G. Feinland & D. J. Reis. 1991. Rilmenidine lowers arterial pressure via imidazole receptors in brainstem C1 area. Eur. J. Pharmacol. **195:** 181-191.
29. Piletz, J. E., A. C. Andorn, J. R. Unnerstall & A. Halaris. 1991. Binding of [^{3}H]-*p*-aminoclonidine to alpha$_2$-adrenoceptor states plus a non-adrenergic site on human platelet plasma membranes. Biochem. Pharmacol. **42:** 569-584.
30. Muramatsu, I. & S. Kigoshi. 1992. Tizanidine may discriminate between imidazoline-receptors and α-adrenoceptors. Japan. J. Pharmacol. **59:** 457-459.
31. Bricca, G., M. Dontenwill, A. Lonies, J. Feldman, A. Belcourt & P. Bousquet. 1989. The imidazoline preferring receptor: Binding studies in bovine, rat and human brainstem. Eur. J. Pharmacol. **162:** 1-9.
32. Felsen, D., P. Ernsberger, P. M. Sutaria, R. J. Nejat, P. Nguyen, M. May, D. S. Breslin, D. N. Marion & E. D. Vaughan. 1993. Identification, localization and functional analysis of imidazoline and alpha adrenergic receptor in canine prostrate. J. Pharmacol. Exp. Ther. **268:** 1063-1071.
33. Zonnenschein, R., S. Diamant & D. Atlas. 1990. Imidazoline receptors in rat liver cells: A novel receptor or a subtype of alpha$_2$-adrenoceptors? Eur. J. Pharmacol. **190:** 203-215.
34. Yablonsky, F. & J. P. Dausse. 1991. Non-adrenergic binding sites for the "alpha$_2$-antagonist" [^{3}H]idazoxan in the rabbit urethral smooth muscle. Pharmacological and biochemical characterization. Biochem. Pharmacol. **41:** 701-707.
35. Langin, D. & M. Lafontan. 1989. [^{3}H]Idazoxan binding at non-α_2-adrenoceptors in rabbit adipocyte membranes. Eur. J. Pharmacol. **159:** 199-203.
36. Coupry, I., D. Atlas, R. A. Podevin, I. Uzielli & A. Parini. 1990. Imidazoline-guanidinium receptive site in renal proximal tubule: Asymmetric distribution, regulation by cations and interaction with an endogenous clonidine displacing substance. J. Pharmacol. Exp. Ther. **252:** 293-299.
37. Tesson, F., A. Lemoine, M. Johann & A. Parini. 1990. Identification of a nonadrenergic [^{3}H]-idazoxan binding site in rabbit and human liver. Eur. J. Pharmacol. **183:** 736.
38. Diamant, S., T. Eldar-Geva & D. Atlas. 1992. Imidazoline binding sites in human placenta: Evidence for heterogeneity and a search for physiological function. Br. J. Pharmacol. **106:** 101-108.
39. Michel, M. C. & P. A. Insel. 1989. Are there multiple imidazoline binding sites? Trends Pharmacol. Sci. **10:** 342-344.
40. Vigne, P., M. Lazdunski & C. Frelin. 1989. Guanabenz, guanochlor, guanoxon and idazoxan bind with high affinity to non-adrenergic sites in pig kidney membranes. Eur. J. Pharmacol. **160:** 295-298.
41. Wikberg, J. E. S. & S. Uhlén. 1990. Further characterization of the guinea pig cerebral cortex idazoxan receptor: Solubilization, distinction from the imidazole site, and demonstration of cirazoline as an idazoxan receptor-selective drug. J. Neurochem. **55:** 192-203.
42. Sakurai, S., A. Wada, F. Izumi, H. Kobayashi & N. Yanagihara. 1983. Inhibition by α_2-adrenoceptor agonists of the secretion of catecholamines from isolated adrenal medullary cells. Naunyn-Schmiedeberg's Arch. Pharmacol. **324:** 15-19.
43. Ohara-Imaizumi, M., Y. Miyakawa & K. Kumakura. 1988. Pertussis toxin attenuates clonidine inhibition of catecholamine release in adrenal chromaffin cells. Neurosci. Lett. **93:** 294-299.
44. Cull-Candy, S. G., A. Mathie & D. A. Powis. 1988. Acetylcholine receptor channels and their block by clonidine in cultured bovine chromaffin cells. J. Physiol. **402:** 255-278.
45. Musgrave, I. F., D. Krautwurst, J. Hescheler & G. Schultz. 1994. Clonidine and cirazoline inhibit activation of nicotinic channels on PC-12 cells. Ann. N.Y. Acad. Sci., this issue.

46. HAKY, J. E. & A. M. YOUNG. 1984. Evaluation of a simple HPLC correlation method for the estimation of the octanol-water partition coefficient of organic compounds. J. Liq. Chromatogr. **7:** 675-689.
47. KOJDA, G., W. KLAUS, G. WERNER & U. FRICKE. 1991. The influence of 3-ester side chain variation on the cardiovascular profile of nitrendipine in porcine isolated trabeculae and coronary arteries. Naunyn-Schmiedeberg's Arch. Pharmacol. **344:** 488-494.
48. KILPATRICK, G. J., K. T. BUNCE & M. B. TYERS. 1990. 5-HT_3 receptors. Med. Res. Rev. **10:** 441-475.
49. BARANN, M., K. RUPPERT, M. GÖTHERT & H. BÖNISCH. 1993. Influence of buffer conditions on 5-HT_3-receptor mediated cation influx in cultured neural cell lines. Fundam. Clin. Pharmacol. **7:** 346.
50. LI, G., S. REGUNATHAN, C. J. BARROW, J. ESHRAGHI, R. COOPER & D. J. REIS. 1994. Agmatine: An endogenous clonidine-displacing substance in the brain. Science **263:** 966-969.
51. HONERJÄGER, P., C. FRELIN & M. LAZDUNSKI. 1982. Actions, interactions, and apparent affinities of various ceveratum alkaloids at sodium channels of cultured neuroblastoma and cardiac cells. Naunyn-Schmiedeberg's Arch. Pharmacol. **321:** 123-129.
52. SCHULZ, A. & A. HASSELBLATT. 1989. An insulin-releasing property of imidazoline derivatives is not limited to compounds that block α-adrenoceptors. Naunyn-Schmiedeberg's Arch. Pharmacol. **340:** 321-327.
53. CHAN, S. L. & N. G. MORGAN. 1990. Stimulation of insulin secretion by efaroxan may involve interaction with potassium channels. Eur. J. Pharmacol. **176:** 97-101.
54. JONAS, J. C., T. D. PINT & J. C. HENQUIN. 1992. Imidazoline antagonists of α_2-adrenoceptors increase insulin release *in vitro* by inhibiting ATP-sensitive K^+ channels in pancreatic β-cells. Br. J. Pharmacol. **107:** 8-14.
55. BERDEU, D., R. GROSS, G. RIBES, M. M. LOUBATIERES-MARIANI & G. BERTRAND. 1994. Effects of imidazolines and derivates on insulin secretion and vascular resistance in perfused rat pancreas. Eur. J. Pharmacol. **254:** 119-125.
56. GÖTHERT, M., G. J. MOLDERINGS, K. FINK & E. SCHLICKER. 1994. α_2-Adrenoceptor-independent inhibition by imidazolines and guanidines of noradrenaline release from peripheral, but not central noradrenergic neurones. Ann. N.Y. Acad. Sci., this issue.

Moxonidine Decreases Gastric Secretion and Gastric Mucosal Injury in Rats

G. B. GLAVIN AND D. D. SMYTH

Departments of Pharmacology & Therapeutics
Surgery and Internal Medicine
Faculty of Medicine
University of Manitoba
Winnipeg, Manitoba, Canada, R3E 0W3

Previous reports of the effects of α_2-adrenoceptor stimulation on gastric secretion have been inconclusive.[1,2] On the basis of observations in other systems, the α_2-adrenoceptor agonist (clonidine) that was utilized in these studies may have been activating the newly described imidazoline receptor. For example, it is now widely appreciated that the antihypertensive action of clonidine was mediated primarily by the stimulation of central I_1-imidazoline receptors (not α_2-adrenoceptors) located in the nucleus reticularis lateralis.[3] As studies have reported a close association between hypertension and gastroduodenal ulcer disease,[4] an antihypertensive agent with beneficial gastric effects would be of interest clinically. In the present experiments, because of inconsistency in the literature regarding the gastrointestinal actions of traditional centrally-acting antihypertensive agents (α_2-adrenoceptor or imidazoline receptor agonists), we tested the effects of moxonidine (I_1-imidazoline receptor agonist) in two models of gastric secretion and a model of experimental gastric mucosal injury.

METHODS

Briefly, experimental preparations were as follows. For the study of gastric acid secretion, male Sprague-Dawley rats were implanted with chronic indwelling gastric cannulas, as previously described.[5] Secretory testing was performed 14 days later. Each session involved a 1-hour baseline collection followed by the injection (ip) of vehicle or moxonidine and two subsequent 1-hour posttreatment collections. The sequence of the sessions was: vehicle (saline solution 1.0 ml/kg), moxonidine 0.01 mg/kg, 0.1 mg/kg, 1.0 mg/kg ip, and vehicle again, and sessions were separated by a 96-hour period. Acid output was determined by titration to pH 7.0 with 0.01 M NaOH and expressed as a percentage decrease from baseline. For ethanol-induced gastric mucosal injury, food-deprived rats (24 h) were given 1.0 ml of 75% (v/v) ethyl alcohol orally by gavage.[6] Vehicle or moxonidine (0.01, 0.1, or 1.0 mg/kg ip) was given 5 minutes prior to ethanol. Two hours later, stomachs were removed, everted, washed, and fixed in 10% (v/v) buffered formalin and the number and cumulative length of gastric glandular mucosal lesions determined. Preparation for ligation of the pylorus, under ether anesthesia, involved a midline laparotomy to

TABLE 1. Effects of Moxonidine on Gastric Secretion and Gastric Mucosal Injury (Mean + SEM)

Dose (mg/kg ip)	% Decrease in Basal Acid Output in Conscious Rats	Intragastric pH	Ethanol-Induced Gastric Lesions (mm)	Acid Output in Pylorus-Ligated Rats (μmol/h)	Pepsin Output (mg)
0	0	2.2	62.5 (5.8)	33.0 (2.1)	15.6 (1.9)
0.01	41[a]	5.3[a]	37.2 (9.3)[a]	13.2 (0.8)[a]	19.3 (2.3)
0.10	63[a]	3.9[a]	56.2 (13.4)	11.3 (1.1)[a]	13.3 (2.2)
1.0	65[a]	5.5[a]	16.4 (4.2)[a]	10.5 (1.2)[a]	11.5 (2.5)[a]

[a] $p < 0.05$ vs. vehicle

expose the pylorus, which was then ligated with 2-0 silk. The incision was closed and vehicle (saline solution) or moxonidine (0.01, 0.1, or 1.0 mg/kg ip) administered. Four hours later, the stomachs were removed and the total volume of secretion was recorded. Acid output was determined as just described and expressed as micromoles per hour. Pepsin output was determined as described previously[7] and expressed as milligrams of pepsin secretion.

RESULTS

Moxonidine (0.01, 0.1, and 1 mg/kg ip) potently inhibited basal acid output (BAO) in conscious rats, with an $ED_{50\%}$ of 0.04 mg/kg (TABLE 1). At the largest dose of moxonidine tested (1 mg/kg ip), anacidity was observed 2 hours after administration. Moxonidine at 0.1 and 1 mg/kg ip significantly increased intragastric pH. Moxonidine at the smallest dose (0.01 mg/kg ip) decreased the length and at the largest dose (1 mg/kg ip) also reduced the number (data not shown) and length of ethanol-induced gastric mucosal lesions. In pylorus-ligated rats, moxonidine significantly decreased both acid output (all doses) and pepsin secretion (1 mg/kg ip).

DISCUSSION

In the present study, moxonidine beneficially altered gastric function at doses that were less than those previously shown to be required to decrease blood pressure in spontaneously hypertensive rats. Gastric acid secretion was inhibited with an ED_{50} of about 0.04 mg/kg ip. Moxonidine also reduced both acid and pepsin secretion in pylorus-ligated rats and reduced ethanol-induced gastric mucosal injury. Interestingly, moxonidine not only reduced aggressive, pro-ulcerogenic factors such as gastric acid and pepsin, but also appeared to enhance defensive, anti-ulcerogenic mechanisms. In particular, intragastric pH was significantly elevated by moxonidine, suggesting a possible action on gastric bicarbonate secretion that remains to be investigated.

The results of this study with moxonidine are in contrast to comparable experiments that used clonidine. For example, Al-Bekairi *et al.*[8] observed that clonidine exacerbated ethanol-induced gastric damage. In these studies, clonidine also decreased gastric acid secretion to a maximum of 63% at a dose of 0.3 mg/kg. Moxonidine, at doses of 0.1 and 1.0 mg/kg, initially resulted in 63% and 65% inhibition of basal acid output, respectively, but with the 1.0 mg/kg dose, this inhibition persisted into the subsequent hour of testing during which 100% suppression of gastric secretion was observed. The contrasting gastrointestinal actions of clonidine and moxonidine thus lend further support to the hypothesis that these compounds act at different sites, α_2-adrenoceptors and imidazoline receptors, respectively. The relative selectivity of these agonists for their respective receptors, as reported by others, supports this contention. Clonidine displayed fivefold selectivity for the α_{2B}-adrenoceptor over the I_1-imidazoline receptor. Conversely, moxonidine displayed 600-fold selectivity for the I_1-imidazoline receptor.[9]

In summary, the I_1-imidazoline receptor agonist moxonidine suppressed basal gastric acid output in both conscious and anesthetized rats, decreased pepsin output, and decreased ethanol-induced gastric lesions. These studies indicate a potential role for I_1-specific imidazoline receptor agonists in the management of gastroduodenal diseases in addition to their blood pressure lowering activity. The relative contribution of the central and peripheral effects of moxonidine to these gastrointestinal actions remains to be determined.

REFERENCES

1. Del Soldato, P. 1986. Japan. J. Pharmacol. **41:** 257-259.
2. Kunchandy, J., S. Khanna & K. Kulkarni. 1985. Arch. Int. Pharmacodyn. Ther. **275:** 123-138.
3. Ernsberger, P., T. H. Damon, L. M. Graff, S. G. Schafer & M. O. Christen. 1993. J. Pharmacol. Exp. Ther. **264:** 172-182.
4. Mitana, S., J. Yoshida, M. Saitoh, M. Asaka, T. Matsushima & T. Miyazaki. 1989. J. Gastroenterol. & Hepatol. **4**(suppl. 1): 146-150.
5. Paré, W., K. Isom, G. Vincent & G. Glavin. 1977. Lab. Animal Sci. **27:** 244-247.
6. Robert, A., J. Nezamis, C. Lancaster & A. Hanchar. 1979. Gastroenterology **77:** 433-443.
7. Dupuy, D. & S. Szabo. 1986. Gasteroenterology **91:** 966-974.
8. Al-Bekairi, A., A. Al-Rajhi & M. Tariq. 1993. Arch. Int. Pharmacodyn. Ther. **323:** 97-113.
9. Ernsberger, P., K. L. Westbrooks, M. O. Christen & S. G. Schafer. 1992. J. Cardiovasc. Pharmacol. **20**(suppl. 4): S1-S10.

Role of the Epithelium in the Release of Contractile Agents from the Rat Vas Deferens by Clonidine

AFONSO CARICATI-NETO,[a] ANDRE S. PUPO,[a]
ALMIR G. WANDERLEY,[a]
MANUEL NUÑEZ-VERGARA,[a] IVAN HO JUN KOH,[b]
NEIDE H. JURKIEWICZ,[a] AND ARON JURKIEWICZ [a,c]

[a]Department of Pharmacology
and
[b]Department of Surgery
Escola Paulista de Medicina
São Paulo 04034-970, Brazil

In the rat vas deferens, clonidine was shown to induce the release of endogenous substances that can contract the rat uterus and guinea pig ileum.[1] This clonidine-induced release is partially blocked by denervation, by the serotonin antagonists ketanserin and cyproheptadine, and by naloxone,[1] suggesting that it depends on neuronal integrity and involves different substances. However, the localization of these released endogenous substances and the corresponding releasing mechanisms are unknown. Therefore, we investigated the influence of the epithelium of the rat vas deferens mucosa in this phenomenon.

METHODS

Vasa deferentia with the attached epididymus and testis were removed from adult (4-5-month old) male Wistar rats from our own colony (BAW-2), carefully cleared of the surrounding tissues, and placed under a dissecting microscope. The epithelium of the vas deferens was removed microsurgically,[2] as shown in FIGURE 1. To check if the epithelium was completely removed, light and electron microscopic examinations were performed on some preparations (not shown).

Uteri were obtained from female rats (1-2 months old, 100-150 g body weight) pretreated with a single subcutaneous dose of stilbestrol (15 mg/100 g, 24 hours before). Isolated uterus preparations for recording of isotonic contractions were mounted in 10-ml organ chambers containing a nutrient solution with the same composition as that used for the vas.[3]

[c] Address for correspondence: Prof. Aron Jurkiewicz, Department of Pharmacology, Escola Paulista de Medicina, Caixa Postal 20372, São Paulo 04034-970, Brazil.

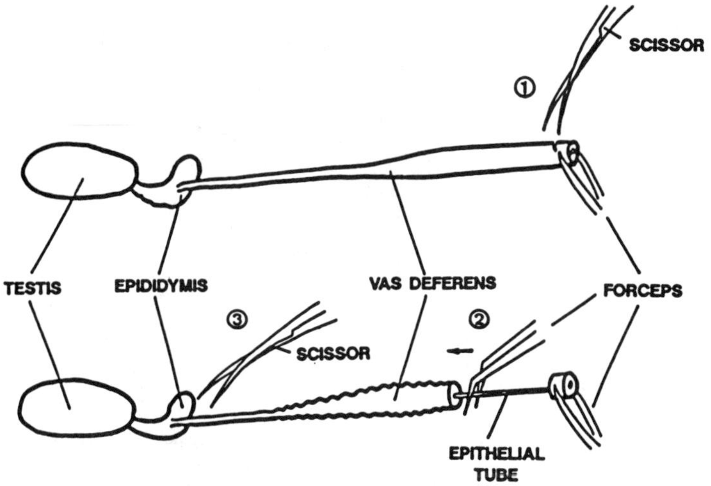

FIGURE 1. Technique for removal of epithelium of the rat vas deferens. (1) The muscle layer is carefully transected about 5 mm from the prostatic end, leaving the mucosa intact. (2) The prostatic stump is firmly gripped with a forceps and the muscles are pushed with a second forceps, resulting in detachment of the epithelial tube. (3) Muscles and mucosa are completely separated by transecting the vas deferens close to the epididymus and subsequently sliding the entire muscle tube around the mucosal tube.

Clonidine (3×10^{-3} M) was incubated with isolated epithelium or vasa deferentia with or without epithelium, in 10 ml of nutrient solution. After a 10-minute incubation, tissues were withdrawn and the entire liquid contents transferred at once to another 10-ml organ bath containing the rat uterus to be tested for its contractile activity.

RESULTS

Light and electron microscopy revealed that the epithelium from rat vas deferens was completely removed without significant damage to muscle and muscular nerve terminals (not shown).

FIGURE 2 shows a typical contractile response of rat isolated uterus induced by clonidine alone (3×10^{-3} M) or by incubates of clonidine plus isolated epithelium or plus a vas with or without epithelium. As previously described,[1] clonidine incubated with intact vas deferens induced a fast sustained contraction of the uterus (22.8 ± 7.4 mm) in contrast to the negligible effects observed when clonidine was incubated alone (7.4 ± 3.6 mm). When the epithelium-free vas deferens was incubated instead of the normal organ, a striking reduction in contractile activity was observed in the corresponding incubate (2.6 ± 1.4 mm). In addition, the incubate of clonidine plus isolated epithelium induced a contraction (38.2 ± 2.3 mm) similar to that produced by the incubate of clonidine plus intact vas deferens.

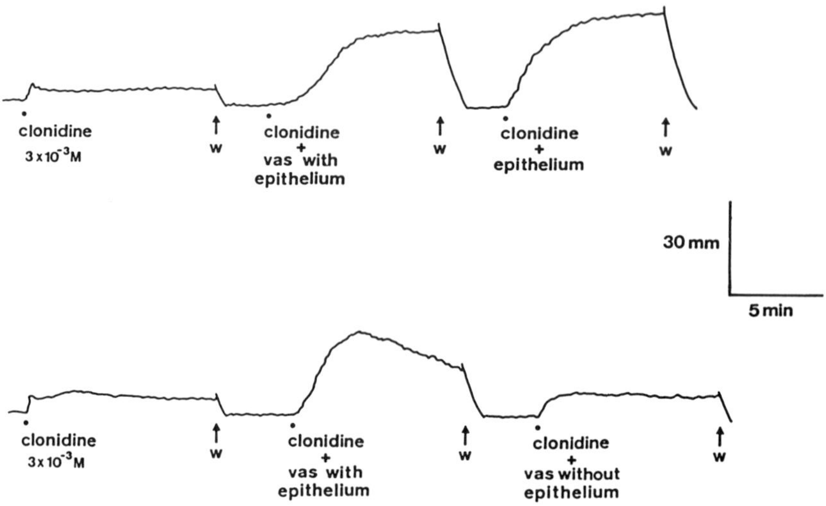

FIGURE 2. Typical contractile response of rat isolated uterus induced by clonidine (3×10^{-3} M) alone or incubated in the presence of rat vas deferens with or without epithelium or with epithelium alone.

DISCUSSION

The present study shows that clonidine-induced release of contractile agents from rat vas deferens[1] depends on the presence of the epithelium of the mucosa, because this imidazoline induced a negligible release when used in vasa deferentia without epithelium. In addition, the role of epithelium was corroborated by the effect observed after incubation of clonidine with the isolated epithelial tube. The later result also shows that the lack of an effect after incubation with epithelium-free vas deferens is not due to organ damage caused by the surgical procedure, because the epithelium should be equally damaged. It is known that clonidine is able to release various endogenous substances, such as opioids and serotonin in nervous pathways,[4] atrial natriuretic factor,[5] beta-endorphin in rat brain,[6] neuropeptide Y in guinea pig cardiac nerves,[7] dynorphin in rat spinal cord,[8] and growth hormone in various species including man.[9] Additionally, it was shown that the endogenous clonidine displacing substance (CDS) competes for the binding of clonidine to its receptors and has a contractile activity in smooth muscle.[10] In several tissues, the epithelium represents a source of released endogenous substances with pharmacological activity,[11] and therefore it is plausible to assume that the epithelium in rat vas deferens, as in other tissues, is a source of release of contractile agents and that clonidine can trigger this release. It is still unknown how this release is interrelated with our previous finding that denervation, naloxone, and serotonin antagonists can reduce clonidine-induced release of contractile agents in this preparation.[1]

REFERENCES

1. JURKIEWICZ, N. H. & A. JURKIEWICZ. 1991. Life Sci. **49:** 1643-1649.
2. JURKIEWICZ, A., M. NUÑEZ-VERGARA, I. KOH, A. CARICATI-NETO, A. S. PUPO, A. G. WANDERLEY & N. H. JURKIEWICZ. 1994. An. Acad. Bras. Cienc. **66:** 124.
3. JURKIEWICZ, A. & N. H. JURKIEWICZ. 1976. Br. J. Pharmacol. **56:** 169-178.
4. LIN, M. T., M. L. CHI, A. CHANDRA & B. L. TSAY. 1980. Pharmacology **20:** 323-328.
5. PAN, L. & J. GUTKOWSKA. 1988. Endocrinology **123:** 1259-1263.
6. VAN GIERSBERGEN, P. L. M., S. A. V. TIERNEY, V. M. WIEGANT & W. JONG. 1989. Hypertension **13:** 83-90.
7. NAGATA, M., A. FRANCO-CERECEDA, T. H. SVENSSON & J. M. LUNDBERG. 1986. Acta Physiol. Scand. **128:** 321-322.
8. XIE, C. W., J. TANG & J. S. HAN. 1986. Neurosci. Lett. **65:** 224-228.
9. PINTOR, C., R. CORDA, R. PUGGIONI, S. G. CELLA, V. LOCATELLI, S. LOCHE & E. E. MULLER. 1985. Lancet **1:** 1482-1485.
10. FELSEN, D., P. ERNSBERGER, M. P. MEELEY & D. J. REIS. 1987. Eur. J. Pharmacol. **142:** 453-455.
11. VANHOUTTE, P. M. 1988. Am. Rev. Resp. Dis. **138:** S24-S30.

Coexpression of Imidazoline Receptors and Agmatine in Rat Carotid Body

CHARLOTTE YOUNGSON, SOUNDARARAJAN REGUNATHAN, HONG WANG, GEN LI, AND DONALD J. REIS

Division of Neurobiology
Department of Neurology and Neuroscience
Cornell University Medical College
New York, New York 10021

The recent discovery that agmatine (decarboxylated arginine) is an endogenous ligand for imidazoline (I-) as well as α_2-adrenergic receptors[1] raises questions as to whether agmatine and I-receptors are coexpressed, a feature that might suggest an action of agmatine as an auto- or paracrine message. The fact that adrenal chromaffin cells express I-receptors[2] and immunoreactive agmatine[3] are consistent with the possibility. In this study we investigated whether glomus (type I) cells of the carotid body, a homogeneous population of cells developmentally related to chromaffin cells as neural crest derivatives, also express I-receptors and contain agmatine. We report that glomus but not sustentacular cells of rat carotid bodies express I-receptors of the I_2 subclass but that both cells types express agmatine. The data are consistent with agmatine acting as a local signal for I-receptors on glomus cells in this organ.

METHODS

Glomus cells of the rat carotid body were prepared after enzymatic and mechanical dissociation by modification of the method of Nurse.[4] Rat pups (Wistar; Hilltop) of postnatal age 6–10 days were deeply anesthetized with pentobarbital and sacrificed by cardiac puncture. The carotid bodies were removed and enzymatically dissociated for 1 hour at 37°C in HBSS containing 0.1% collagenase, 0.01% trypsin, 0.01% DNAase, and 1% penicillin-streptomycin. The enzyme solution was removed and replaced with F12 growth medium to inactivate any remaining enzyme, and the cells were centrifuged at 500 rpm for 5 minutes. The growth medium contained 10% fetal calf serum, 1% fungizone, and 80 U/L insulin. The carotid bodies were slowly triturated before plating on poly-L-lysine-coated culture dishes. The cells were maintained in an incubator at 37°C in a humidified atmosphere of 95% air–5% CO_2 for 3–7 days before staining.

For ligand binding analysis, membranes were prepared from 80–100 carotid bodies for ligand binding assay by methods described elsewhere.[5] Briefly, tissue was homogenized with a Polytron (Brinkman Instruments) in Tris-HCl buffer (pH 7.4)

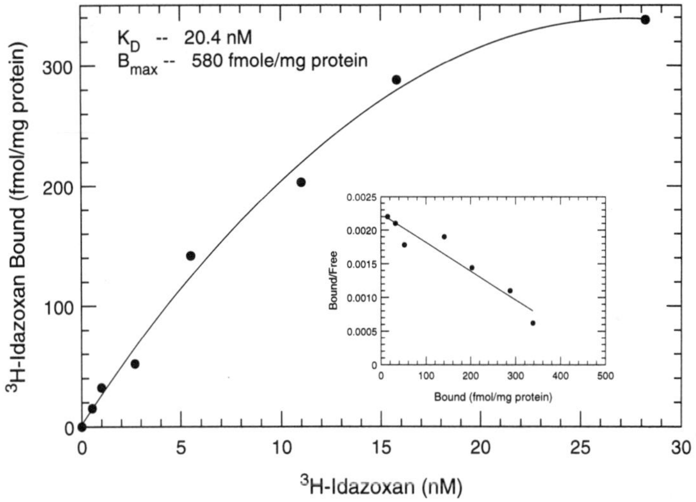

FIGURE 1 Saturation binding of [^{3}H]idazoxan to membranes of carotid body in the presence of epinephrine (10 μM). Scatchard analysis (**inset**) showed that binding is high affinity (K_D: 20 nM) and saturable (B_{max}: 810 fmol/mg protein).

and centrifuged at 1,000 × *g* for 10 minutes. The pellets were suspended in ice-cold Tris-HCl buffer (pH 7.7) with 5 mM EDTA and homogenized (Polytron) for 10 seconds. The homogenate was centrifuged at 40,000 × *g* for 30 minutes; the resulting membrane pellet was suspended in buffer and recentrifuged. The final washed pellet was suspended in fresh Tris-HCL buffer (pH. 7.4) to provide approximately 50 μg protein per assay tube. Binding assays were performed in Tris-HCl buffer (pH. 7.7) using a total volume of 1 ml. Membranes were incubated with 10 nM [^{3}H]idazoxan in the presence of various concentrations of idazoxan or epinephrine. Following incubation at 25°C for 30 minutes, the reaction was terminated by rapid vacuum filtration over Whatman GF/B filters and washed with 10 ml of ice-cold buffer. Filters were suspended in scintillation fluid and radioactivity counted in a liquid scintillation counter. The binding data were analyzed using LIGAND (Elsevier Biosoft), a parametric nonlinear regression analysis. Membrane protein was assayed by the Coomassie blue method (Pierce) using bovine serum albumin as standard.

The production and characterization of polyclonal antibodies raised in rabbits to the imidazoline receptor binding protein of bovine adrenal chromaffin cells and to agmatine covalently linked to keyhole limpet hemocyanin have been published elsewhere.[2,6] Primary cultures were fixed with 4% paraformaldehyde and blocked with 0.5% bovine serum albumin in 0.1M Tris-saline solution. Cultures were then incubated with antiserum (IRP, 1:1000; agmatine, 1:500 for carotid body cultures in 0.1% BSA containing Tris-saline solution) at 4°C overnight. Immunoreactivity was visualized by the sequential addition of: (a) biotinylated secondary IgG, 1:50 for 30 minutes; (b) avidin-biotin complex, 1:100 for 30 minutes; and (c) 3,3′-diaminobenzidine, 0.44 mg/100 ml Tris-saline solution with 10 μl of 30% hydrogen peroxide. To

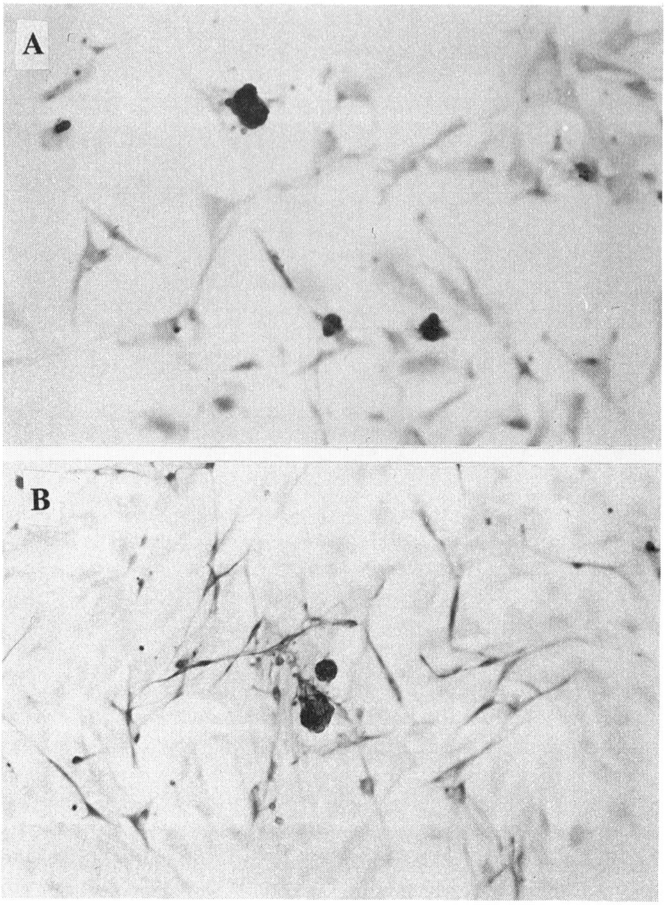

FIGURE 2 **(A)** Cultured glomus cells of the carotid body are immunoreactive for IRP, whereas all other cells in the culture remain unstained. Preimmune serum showed no specific staining (not shown). **(B)** Cultured glomus cells as well as sustentacular cells of the carotid body are immunoreactive for agmatine. Preimmune serum showed no specific staining (not shown).

permeabilize tissue, 0.1–0.3% Triton-X was included in most of these steps to improve penetration of the antibody. Cultures were then dehydrated and mounted.

RESULTS AND CONCLUSIONS

The binding of [^{3}H]idazoxan to membranes of rat carotid body was specific, of high affinity (FIG. 1) (K_D: 20 nM), saturable (B_{max}: 810 fmol/mg protein), and not inhibited by catecholamines (data not shown). Cirazoline and naphazoline potently displaced binding, whereas clonidine and phentolamine were ineffective, indicating that the binding site conforms to the I_2-imidazoline site.[7]

The cellular distribution of I-receptors was examined in cultured bodies immunostained with antibodies to an imidazoline receptor binding protein. Almost all glomus cells contained immunoreactivity (FIG. 2A). On the other hand, sustentacular cells were unstained (FIG. 2A). Cultured carotid bodies were also immunostained with a polyclonal antiserum against agmatine.[6] Virtually all glomus and sustentacular cells stained with antiserum (FIG. 2B) but not with preimmune serum (not shown).

The present study demonstrates that the rat carotid body expresses I-receptors. The affinity of the site for various ligands suggests that the receptors are of the I_2 subclass which, as in other tissues, are probably localized to mitochondrial membranes.[8] Immunocytochemistry studies performed with a polyclonal antibody to an I-receptor binding protein revealed that I-receptors are expressed within glomus but not sustentacular cells. Thus, the distribution of agmatine within the carotid body is cell-specific. Whether agmatine is synthesized in either or both of these cell types is not certain, as the presence of the biosynthetic enzyme for agmatine, arginine decarboxylase, has yet to be determined.

The fact that I-receptors and the endogenous ligand agmatine are contained in the same cells in the carotid body is of interest and suggests that agmatine may act in glomus cells of the carotid body as a paracrine and/or autocrine message. Moreover, since agmatine is present in serum,[3] it also possible that given the high basal blood flow of the carotid bodies, it may also act as a hormone. The function of agmatine and I-receptors in the glomus cells is not known. The fact that I-receptors are mitochondrial and closely positioned to elements of the respiratory chain suggests that they may serve a function in oxygen sensing.[9]

REFERENCES

1. LI, G., S. REGUNATHAN, C. J. BARROW, J. ESHRAGHI, R. COOPER & D. J. REIS. 1994. Agmatine: An endogenous clonidine-displacing substance in the brain. Science **263:** 966–969.
2. WANG, H., S. REGUNATHAN, D. A. RUGGIERO & D. J. REIS. 1993. Production and characterization of antibodies specific for the imidazoline receptor protein. Mol. Pharmacol. **43:** 509–515.
3. RAASCH, W., S. REGUNATHAN, G. LI & D. J. REIS. 1995. Agmatine is widely and unequally distributed in rat organs. Proceedings of the II International Symposium on Imidazoline Receptors. Ann. N.Y. Acad. Sci. This volume.
4. NURSE, C. A. 1987. Localization of acetylcholinersterase in dissociated cell cultures of the carotid body of the rat. Cell Tiss. Res. **250:** 21–27.
5. REGUNATHAN, S., M. P. MEELEY & D. J. REIS. 1993. Expression of nonadrenergic imidazoline sites in chromaffin cells and mitochondrial membranes of bovine adrenal medulla. Biochem. Pharmacol. **45:** 1667–1675.
6. WANG, H., S. REGUNATHAN, C. YOUNGSON & D. J. REIS. 1995. An antibody to agmatine localizes the amine in bovine adrenal chromaffin cells. Neurosci. Lett. **183:** 1–5.
7. MICHEL, M. C. & P. R. ERNSBERGER. 1992. Keeping an eye on the I site: Imidazoline-preferring receptors. Trends Pharmacol. Sci. **13:** 369–370.

8. Tesson, F., C. Prip-Buus, A. Lemoine, J.-P. Pergorier & A. Parini. 1991. Subcellular distribution of imidazoline-guanidinium-receptive sites in human and rabbit liver: Major localization to the mitochondrial outer membrane. J. Biol. Chem. **260:** 155–160.
9. Wilson, D. F., A. Mokashi, D. Chugh, S. Vinogradov, S. Osanai & S. Lahiri. 1994. The primary oxygen sensor of the cat carotid body is cytochrome a3 of the mitochondrial respiratory chain. FEBS Lett. **351:** 370–374.

Effect of I_1-Imidazoline Receptor Activation on Responses of Hypoglossal and Phrenic Nerve to Chemical Stimulation[a]

M. A. HAXHIU,[b] I. A. DRESHAJ, B. EROKWU,
L. A. COLLINS, AND P. ERNSBERGER

Departments of Medicine and Neuroscience
Case Western Reserve University
Cleveland, Ohio 44106

Hypertension is associated with well described cardiovascular, morphologic, and hemodynamic changes. Recently, a strong association was observed between hypertension and sleep obstructive apneas. Epidemiological studies have demonstrated a surprisingly high prevalence (~40 %) of sleep-related respiratory disturbances (SRRD) in hypertensive subjects.[1] Conversely, patients with severe sleep apnea are more likely to be hypertensive than is the general population and are at an increased risk of adverse cardiovascular events.[2–6] Furthermore, hypertensive subjects with SRRD commonly experience severely fragmented sleep and manifest profound sleep deprivation which leads to impaired vigilance, memory, and learning and deficiencies in executive functions.[7,8]

We hypothesize that several pathophysiological processes may be a consequence of hypertension and lead to SRRD. Blood pressure fluctuations acting through carotid and aortic baroreceptors influence breathing. Elevated blood pressure preferentially inhibits pharyngeal dilator muscle activity.[9] Furthermore, during sleep baroreceptor inputs triggered by blood pressure increases may preferentially reduce the responses of upper airway respiratory muscles to hypoxia, hypercapnia, and other stimuli, causing failure of the dilating forces to counterbalance the constricting luminal forces generated by the inspiratory thoracic muscles. In addition, changes in brain blood flow associated with cardiovascular disease may directly affect the acid-base environment of medullary chemoreceptors and neurotransmitter metabolism. Repetitive closure of the pharyngeal airway segment is followed by hypoxic episodes which lead to further elevation of blood pressure and hormonal and metabolic changes, including insulin resistance and lipid abnormalities.

[a] This work was supported by grants from Solvay Pharma Deutchland (Hanover) and Biersdorf-Lilly (Hamburg) and by the National Institutes of Health grant R29-HL-44514.

[b] Address for correspondence: Musa A. Haxhiu, MD, PhD, Pulmonary Division and Critical Care Medicine, Department of Medicine, University Hospitals of Cleveland, 11000 Euclid Avenue, Cleveland, OH 44106.

Because elevated sympathetic activity may underlie hypertension in persons with sleep apneas, the treatment of choice is a sympathoinhibitory agent. The use of centrally acting antihypertensive drugs has been limited by their sedative action,[10–12] which through depression of respiratory activity[13,14] may contribute to the appearance of occlusive apneas.

Obstructive sleep apnea is a common disorder affecting 1-4% of the male population.[15] Most studies implicate the pharynx as the predominant site of airway closure in patients with the obstructive sleep apnea syndrome. The pharynx is a compliant structure subject to extrinsic forces that alter its size and shape. Contraction of the thoracic muscles during inspiration produces negative pressures in the upper airway lumen; when these negative intraairway pressures are excessive relative to the forces that maintain airway stability, the airway will close. Contraction of muscles that stiffen or dilate the upper airways is required to oppose the collapsing forces generated by thoracic muscle contraction. The upper airway muscles, such as the genioglossus, are more sensitive to changes in chemoreceptor and mechanoreceptor input than is the diaphragm; therefore, reduction in the intensity of these afferent inputs during sleep may cause a greater reduction in upper airway activity relative to diaphragm muscle activity. These specific effects of sleep on the respiratory system become more pronounced during rapid eye movement (REM) sleep.

It was recently observed that clonidine treatment reduces the number of apneic episodes, suggesting that it may attenuate nocturnal breathing abnormalities. However, in a subgroup of patients, possibly because of clonidine's significant sedative action, an increase in complete occlusive attacks occurred.[16] An agent with a reduced sedative action relative to clonidine and with REM suppressant activities might be of significant benefit in hypertension associated with sleep apneas and in the obstructive sleep apnea syndrome.

Moxonidine is a centrally acting antihypertensive agent with a high affinity for I_1-imidazoline sites and a weak tendency to interact with α_2-adrenergic receptors.[17–22] Therefore, moxonidine in small doses acting centrally may be expected to produce hypotension comparable to that from clonidine,[23] but respiratory inhibitory effects may be significantly less than those of clonidine.

Moxonidine and rilmenidine represent a second generation of central antihypertensive drugs with the reduced side effects of sedation and dry mouth.[17–21] Extensive studies comparing the cardiovascular effects of first- and second-generation agents found them to be very similar. Few studies have systematically compared side effect profiles, which should clearly distinguish first- and second-generation agents. Sedation is a difficult endpoint to quantify, and studies in humans are, of necessity, subjective. We chose to study respiratory activity, because it is readily quantifiable by direct nerve recordings in anesthetized animals and is predictably influenced by substances that possess hypnotic and sedative properties.

The present study tested the hypothesis that respiratory depressive effects of the centrally acting hypotensive drugs are due to preferential activation of α_2-receptors rather than imidazoline sites in the rostral ventrolateral medulla, the final common pathway for excitatory vasomotor activity descending to the spinal cord.[24] Moxonidine in therapeutic doses is expected to decrease sympathetic outflow without influencing the activity of upper airway and chest wall muscles. By contrast, drugs like clonidine, in comparable concentrations to those of moxonidine, will inhibit respiratory neuronal output, particularly genioglossus muscle activity or the nerve that innervates it. Evidence for or against this hypothesis is provided by examining the effect of activation of imidazoline receptors by moxonidine on neuronal output of the hypoglossal and phrenic nerves and the responses of these nerves to changes in chemical drive. Furthermore, the effects of I_1-receptor activation by moxonidine on cervical sympathetic responses to hypercapnia and hypoxic loading were examined. Results were compared with the effects of clonidine given systemically. In addition, we attempted to determine if moxonidine exerts its hypotensive effects on I_1-imidazoline receptors of superficially located neurons or dendrites in the ventrolateral medulla.

METHODS

Autoradiographic Studies. Autoradiographic studies were performed using [^{125}I]*p*-iodoclonidine in the presence of 10 μm epinephrine to block α_2-adrenergic sites or 100 nM moxonidine to mask I_1-imidazoline sites, as described earlier.[19] Briefly, brains were quickly removed from anesthetized cats, rinsed in ice-cold saline solution, and flash-frozen in isopentane chilled on dry ice. Blocks of cat medulla were mounted in a cryostat and sectioned at a thickness of 15 μm at -18°C and thaw-mounted onto glass slides coated twice with 1% gelatin and 0.1% chrome alum. Slides were kept at 4°C during sectioning, dried in a vacuum desiccator for 1 hour, and stored at -70°C. Every twelfth section was fixed in buffered formalin without drying and stained with thionin for anatomical reference. Sections were incubated for 45 minutes at 20°C in 60 ml of assay buffer consisting of (in mM): sucrose (200), HEPES (25), Tris base (16.5), disodium EDTA (1), (1,10)-phenanthroline (0.1), and phenylmethylsulfonylfluoride (0.05). Sections were washed for 15 minutes in 60 ml of the same buffer without EDTA and containing one of the following drug additions: to define total binding, vehicle (0.1 mM ascorbic acid in 1.0% dimethylsulfoxide); to define nonspecific binding, 10 μM BDF-6143; to define α_2-adrenergic binding, 10 μM epinephrine supplemented with 10 μM pargyline to block its degradation and 10 μM desipramine to block its uptake; and to define I_1-imidazoline sites, 100 nM moxonidine. After this second preincubation, sections were incubated for 60 minutes in 13.3 ml of fresh buffer containing the same drug additions just described and 0.6 nM [^{125}I]*p*-iodoclonidine. Labeled sections were washed for 7 minutes in two changes of ice-cold assay buffer, rinsed for 3 seconds in ice-cold water to remove buffer salts, rapidly dried under a stream of dry cold air passed through calcium sulfate desiccant and a cold trap, and then dried under vacuum overnight and placed in X-ray cassettes in contact with Hyperfilm (Amersham) at 4°C for 4 days. Films were developed with Kodak D19 and fixed.

Physiological Studies. Physiological investigations of the effects of moxonidine were performed on healthy cats of either gender. Animals (2.0-4.0 kg body weight) were anesthetized with α-chloralose using dosages (40 mg/kg intraperitoneally with an additional 5-10 mg/kg doses given intravenously) that we previously used in both nonparalyzed and paralyzed cats.[25,26] Cats were placed supine on a heating pad while rectal temperature was maintained at 37°C. A femoral artery and vein were cannulated for measurement of arterial pressure and for systemic administration of saline solution, anesthetics, and paralyzing agents. The neck area was widely exposed through a ventral midline incision, and a tracheal cannula was inserted. After intravenous administration of gallamine triethiodide (Flaxedil, 4 mg/kg), animals were ventilated with a Harvard respiratory pump. End-tidal concentration of CO_2 was monitored with an infrared gas analyzer (Beckman LB2). Guidelines of the American Physiological Society regarding anesthetized, curarized animals were respected. The effects of the paralyzing agent were allowed to dissipate periodically, and the anesthetic level was monitored by assessing corneal reflexes and responses to tactile stimuli. In addition, blood pressure, heart rate, and rate of phrenic nerve firing were continuously monitored as indices of the depth of anesthesia. Vagus nerves were isolated in all but one animal and cut to eliminate lung mechanoreceptor inputs and prevent eventual entrainment of respiratory output to the ventilator. In one group of nine cats the carotid sinus nerves were isolated free from surrounding tissue and severed before beginning the experiment.

The fifth root of the phrenic nerve was dissected and exposed. In addition, the hypoglossal and cervical sympathetic nerves were also separated, cut, and prepared for recording of mass electrical activity. The central ends of the nerves were each placed on pairs of stainless steel hook electrodes (250 μm). Electrical activity was recorded using Grass amplifiers (band pass 10 to 3 kHz), which was rectified and processed through filters (Charles Ward Enterprises, Philadelphia) to obtain a moving average activity. The processed nerve activities, end-tidal CO_2, and blood pressure were recorded on a stripchart recorder, before, during, and after activation of I_1-imidazoline receptors, by intravenous administration of moxonidine.

In one group of cats, a wide craniotomy was performed and the ventral surface of the medulla was exposed for topical applications of drugs. After surgery was completed, animals were allowed to recover for at least 30 minutes before blood gas status, analyzing arterial blood gases, was monitored. Changes in arterial pH during preparation was corrected by bicarbonate infusion.

In the first series of experiments, we examined the effects of moxonidine administered intravenously on responses to hyperoxic hypercapnia of (1) phrenic nerve (in four completely chemodenervated and in two animals with carotid sinus nerves intact), (2) hypoglossal nerve (in three animals with carotid sinus nerves cut and in two cats with carotid sinus nerves intact), and (3) cervical sympathetic nerve discharge (in three chemodenervated animals and in two with carotid sinus nerve intact). Complete peripheral chemodenervation was obtained to eliminate baroreceptor influences on CO_2 and hypoxic response of recorded nerves. Cats were hyperventilated with O_2 until phrenic and hypoglossal activities disappeared. At the apneic point, arterial PCO_2 was 33.0 ± 1 mm Hg, PaO_2 463 ± 23 mm Hg, and pH 7.415. After complete cessation of phrenic activity, animals were switched from O_2 to a rebreathing mixture of 3% CO_2 in O_2 until peak expired CO_2 reached 10.5% (arterial $PCO_2 \approx 70$ mm Hg). In each of these animals CO_2 rebreathing was instituted before and after intravenous administration of moxonidine (30-100 μ/kg). Responses of phrenic, hypoglossal, and cervical activities over three breath epochs with every 0.5% increment in end-tidal CO_2 were analyzed.

In the second series of experiments, we determined the effects of clonidine on the phrenic (six cats), hypoglossal (four cats), and cervical sympathetic nerve discharge (four cats) responses to progressive hypercapnia. In five of six animals, the CO_2 response was repeated after administration of SK&F-86466, a selective α_2-adrenergic receptor blocker. These animals were vagotomized, five of them with carotid sinus nerves intact.

In the third series of experiments, we examined the effects of moxonidine on responses of phrenic nerve (nine cats), hypoglossal nerve (eight cats), and cervical sympathetic nerve (four cats) to stimulation of peripheral chemoreceptors by hypoxemic hypoxia. To avoid superposition of secondary effects induced by vagal afferent inputs, cats were vagotomized while sinus nerves were intact. Animals were ventilated, employing a constant tidal volume of 10 ml/kg. The rate of ventilation was adjusted to provide end-tidal CO_2 above the apneic point and magnitude of phrenic nerve activity around 30% of that observed while the animal was ventilated with 7% CO_2 in O_2. Just before hypoxic exposure, the following blood gases were recorded: $PaCO_2$ 37 ± 3 mm Hg; PaO_2 498 ± 27 mm Hg, and pH 7.402.

In the fourth series of experiments, we examined whether moxonidine influences the centrally mediated effects of hypoxia. Experiments were performed in vagotomized cats after transection of carotid sinus nerves. In a steady-state period, inspired gas was switched from an oxygen to a hypoxic gas mixture (12–10% O_2 in N_2), reducing arterial PO_2 from 498 ± 27 to 30 ± 2 mm Hg. During hypoxic exposure, on average $PaCO_2$ was 38 ± 4 mm Hg and pH 7.395.

The effect of topical application of efaroxan (a selective blocker for I_1-imidazoline over α_2-adrenergic receptors[27] to the ventral medullary surface upon changes induced by administration of moxonidine was studied in six cats. Vehicle and chemical substances were administered locally by a circle-shaped pledget (2 mm in diameter) made of Gel-Foam soaked with 10 μl of solution. Filter paper wicks were placed at the lateral and rostral margins of the exposed ventral medullary surface to prevent cerebrospinal fluid accumulation. Drugs were applied bilaterally on the area rostral to the hypoglossal roots and 3 mm lateral to the midline. The field corresponds approximately to the intermediate or Schlaefke's area.[28] Control applications of the vehicle in which test substances were dissolved had no effect. In these studies we did not attempt to compare the effects obtained by application of these drugs to a described field with those of adjacent areas. Efaroxan was applied immediately after the full effects of moxonidine were expressed.

Moxonidine and clonidine were given intravenously by gradual injection (up to 2 minutes). Drugs were administered at a point at which phrenic nerve activity reached steady state and the amplitude was approximately 50% of that during hyperoxic hypercapnia ($PaCO_2 \approx$ 56–60 mm Hg).

Data Collection and Analyses. Records were analyzed to determine activities of phrenic, hypoglossal, and cervical sympathetic nerves, arterial blood pressure, and heart rate before and after moxonidine administration. In addition, all recorded variables were analyzed before and after topical application of efaroxan to the ventral medullary surface after intravenous injection of moxonidine. Data were analyzed by a paired or unpaired t test, as appropriate. The criterion for statistical significance was $p < 0.05$.

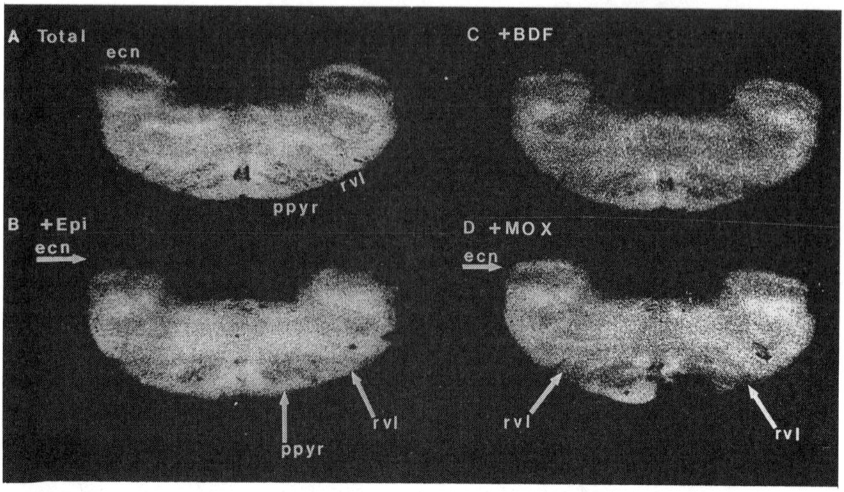

FIGURE 1. Autoradiography of [^{125}I]*p*-iodoclonidine binding sites in the cat medulla. Shown are photographic prints made under identical conditions using the autoradiograms as negatives. Light areas correspond to the highest intensity of labeling. Adjacent 15-μm sections of cat brain stem were incubated with 0.5 nM [^{125}I]*p*-iodoclonidine under one of four different conditions. (**A**) This section incubated with antioxidant vehicle alone shows total binding of [^{125}I]*p*-iodoclonidine. (**B**) Section incubated with 10 μM epinephrine to mask α_2-receptors shows the distribution of I_1 binding. (**C**) Section incubated with 10 μM BDF-6143 to block both I_1 and α_2 binding shows the distribution of nonspecific sites. (**D**) Section incubated with 100 nM moxonidine to selectively mask I_1-imidazoline sites mainly shows the distribution of α_2-adrenergic sites. ecn = dorsal (external) cochlear nucleus; ppyr = parapyramidal region; rvl = rostral ventrolateral medulla; Epi = epinephrine; BDF = BDF-6143; MOX = moxonidine.

RESULTS

Autoradiographic Studies. Autoradiographic studies with [^{125}I]*p*-iodoclonidine in the presence of 10 μM epinephrine to block α_2-adrenergic sites or 100 nM moxonidine to mask I_1-imidazoline sites show that both α_2 and I_1 are localized in putative chemosensory areas of the ventral medulla oblongata in the cat (FIG. 1), resembling that observed in the rat.[19] In the ventral medulla, [^{125}I]*p*-iodoclonidine binding sites were present in both the parapyramidal region (ppyr) and the rostral ventrolateral medulla (rvl) (FIG. 1A), whereas in the dorsal medulla there was a moderate density of sites in the nucleus tractus solitarii and a high density in the dorsal (or external) cochlear nucleus (ecn). The addition of 10 μM epinephrine eliminated labeling of the ecn and nucleus tractus solitarii (FIG. 1B), but labeling persisted in the ppyr and rvl (see arrows), indicating that portions of the binding sites for [^{125}I]*p*-iodoclonidine in these regions were nonadrenergic and thus probably were I_1-imidazoline sites. These I_1-imidazoline sites were also distributed throughout the reticular and raphé regions (FIG. 1B). Nonspecific binding was much sparser than the total or epinephrine mask conditions (FIG. 1C). The addition of 100 nM moxonidine to selectively mask a portion of the I_1-imidazoline sites resulted in loss of labeling

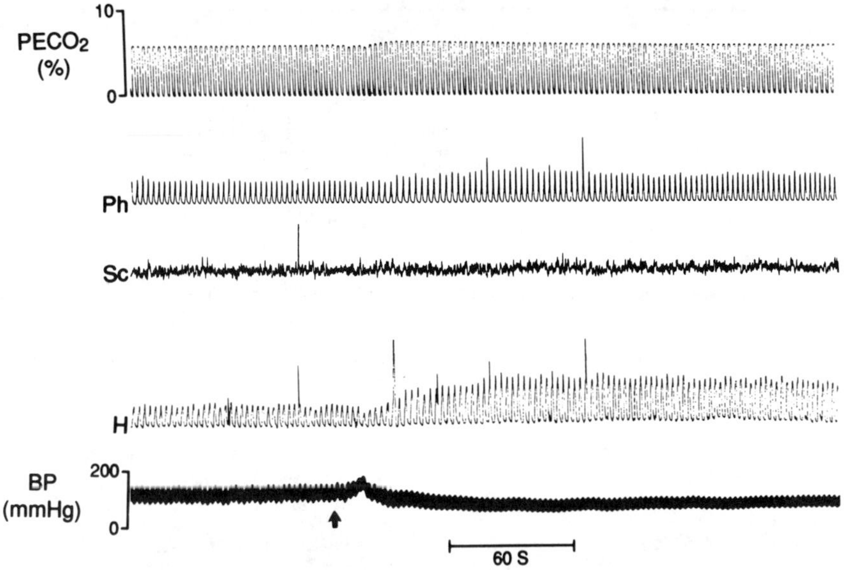

FIGURE 2. Effect of intravenous administration of moxonidine (30 μg/kg) on phrenic, hypoglossal, and cervical sympathetic nerve activities in a vagotomized, paralyzed, and mechanically ventilated cat with 100% O_2 at end-tidal CO_2 above the apneic point. *Arrow* indicates moxonidine administration.

in the ppyr and rvl (FIG. 1D, arrows), but labeling in regions containing mainly α_2-adrenergic receptors, such as ecn and the nucleus tractus solitarii, persisted in the presence of 100 nM moxonidine. The anatomically selective occupancy of I_1-imidazoline sites by moxonidine is revealed by a comparison of FIGURE 1B and 1D.

Effects of Moxonidine and Clonidine on Respiratory Activity and Blood Pressure. FIGURE 2 shows the effect of intravenous administration of 30 μg/kg of moxonidine on phrenic, hypoglossal, and cervical sympathetic activity in a vagotomized cat, ventilated with O_2 at end-tidal CO_2 above the apneic point ($PaCO_2$ 39.1 mm Hg; PO_2 510 mm Hg; pH 7.407). Moxonidine given systemically consistently caused an initial transient decrease in peak phrenic and hypoglossal nerve activities and breathing rate, which paralleled a transient initial increase in arterial blood pressure. The transient pressor response was followed by hypotension and a return to basal respiratory activity. On average in all animals studied, mean arterial pressure decreased from 139 ± 7 to 91 ± 6 mm Hg ($p = 0.001$), and heart rate decreased from 228 ± 7 to 190 ± 8 beats/min ($p = 0.02$). Under steady-state conditions, 10 to 15 minutes after the administration of moxonidine (30 μg/kg) respiratory rate tended nonsignificantly to increase due to a slight shortening of inspiratory and expiratory duration. The duration of the respiratory cycle before moxonidine administration was 5.1 ± 0.85 seconds; and 10-15 minutes after moxonidine the duration was 4.5 ± 0.4 seconds ($p = 0.41$). Peak phrenic activity, expressed in arbitrary units, was not different from

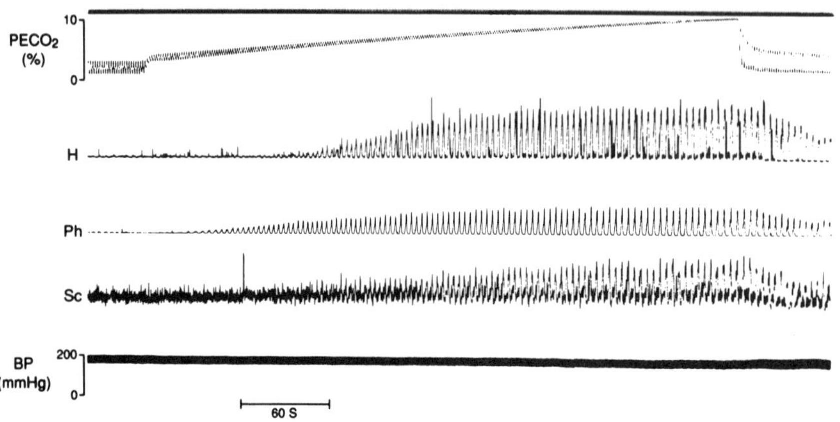

FIGURE 3. Time course of phrenic, hypoglossal, and cervical sympathetic nerve responses to progressive hyperoxic hypercapnia in a vagotomized, carotid sinus nerves cut, paralyzed, and mechanically ventilated cat before moxonidine administration.

control activity. However, hypoglossal discharge tended to increase slightly with administration of lower concentrations of moxonidine (3-10 μg/kg). Moxonidine-induced hypotension was not invariably associated with a comparable decrease in respiratory related cervical sympathetic activity. However, moxonidine in a concentration of 100 μg/kg or in higher doses, in some experiments, resulted in complete cessation of phrenic and hypoglossal nerve activities which developed rapidly and was often associated with a decrease in CO_2 output, resembling the effect of 10-30 μg clonidine. The time course and respiratory pattern of responses to administration of moxonidine were similar in cats with intact and carotid sinus nerves cut; therefore, data were combined.

Intravenous administration of clonidine (30-100 μg/kg) in five of six studied cats caused inhibition of phrenic nerve activity and a biphasic response of blood pressure, that is, initial elevation followed by depression and slowing of the heart rate. On average, systolic arterial blood pressure decreased from 146 ± 5 to 112 ± 8 mm Hg (p = 0.007) and heart rate from 210 ± 8 to 164 ± 10 beats/min (p = 0.034).

Effects of Moxonidine and Clonidine on Respiratory Response to Progressive Hyperoxic Hypercapnia. Following hyperventilation-induced respiratory neural apnea, activities of phrenic, hypoglossal, and cervical sympathetic nerves increased in parallel with elevation of CO_2. Administration of moxonidine in a concentration of 30-100 μg/kg had insignificant effects on phrenic and hypoglossal nerve responses to progressive hyperoxic hypercapnia in animals with carotid sinus nerves intact and in peripherally chemodenervated cats. FIGURES 3 and 4 illustrate the response of phrenic, hypoglossal, and cervical sympathetic nerves in a completely chemodenervated cat to a gradual increase in inhaled CO_2 before and after the administration of moxonidine, respectively. Although I_1-imidazoline receptor activation by moxonidine decreased blood pressure and attenuated the response of cervical sympathetic nerve

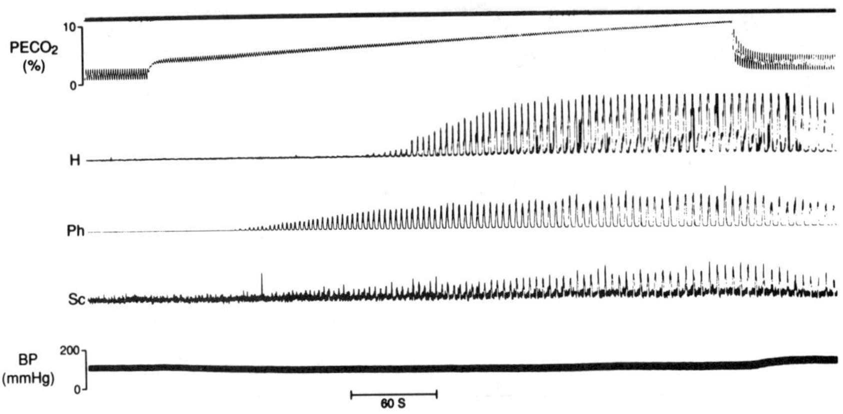

FIGURE 4. Time course of phrenic, hypoglossal, and cervical sympathetic nerve responses to progressive hyperoxic hypercapnia in the same cat after moxonidine administration.

discharge to CO_2, CO_2 responses of the phrenic and hypoglossal nerves were not affected. Average results are presented in FIGURE 5. After moxonidine administration phrenic nerve activity reappeared at 5.8 ± 0.2% versus 5.6 ± 0.3% CO_2 in the control period. Following moxonidine, the slope of the steep portion of the CO_2 response tended to increase (10.3 ± 1.8 vs 7.3 ± 0.9). Peak phrenic nerve activity was comparable to control values at any given end-tidal CO_2. For example, at 7.5% CO_2 average peak phrenic activity was 20 ± 2 in a control period versus 20 ± 3 postmoxonidine administration, and at 9.5% it was 27 ± 2 versus 30 ± 3 units (FIG. 5, upper panel). Similarly, hypoglossal nerve activity was not altered by moxonidine administration (FIG. 5, middle panel). However, cervical sympathetic response decreased, and at any given end-tidal CO_2 average activity of cervical sympathetic nerve was lower after moxonidine administration than in a control period (FIG. 5, lower panel). Differences were significant at PCO_2 6.5-7.5% ($p = 0.012$). The relative suppression of sympathetic activity by moxonidine tended to decrease as CO_2 increased.

By contrast to moxonidine, clonidine (30-100 μg/kg) affected CO_2 responses of phrenic, hypoglossal, and cervical sympathetic activities. Average data, analyzed at 0.5 increments in end-tidal CO_2, are presented in FIGURE 6. As can be seen, following clonidine administration, CO_2 sensitivity was reduced, and the slope of the curve shifted to the right. The threshold for the appearance of phrenic nerve discharge was elevated from 5.3 ± 0.5 to 6.7 ± 0.4% ($p = 0.01$), and at any level of end-tidal CO_2, peak phrenic nerve activity was lower after clonidine administration than in a control period (FIG. 6, upper panel). For example, at 7% CO_2 peak phrenic nerve activity in a control period was 23 ± 2 versus 8 ± 4 units after clonidine administration ($p =$ 0.0036), and at 9% CO_2 it was 32 ± 1 versus 18 ± 4 units ($p = 0.0036$). Similarly, the hypoglossal nerve response to CO_2 was inhibited by clonidine administration. After clonidine the peak hypoglossal nerve response at the highest level of CO_2 (10%) was significantly lower than the control condition, 8 ± 2 versus 23 ± 3 units

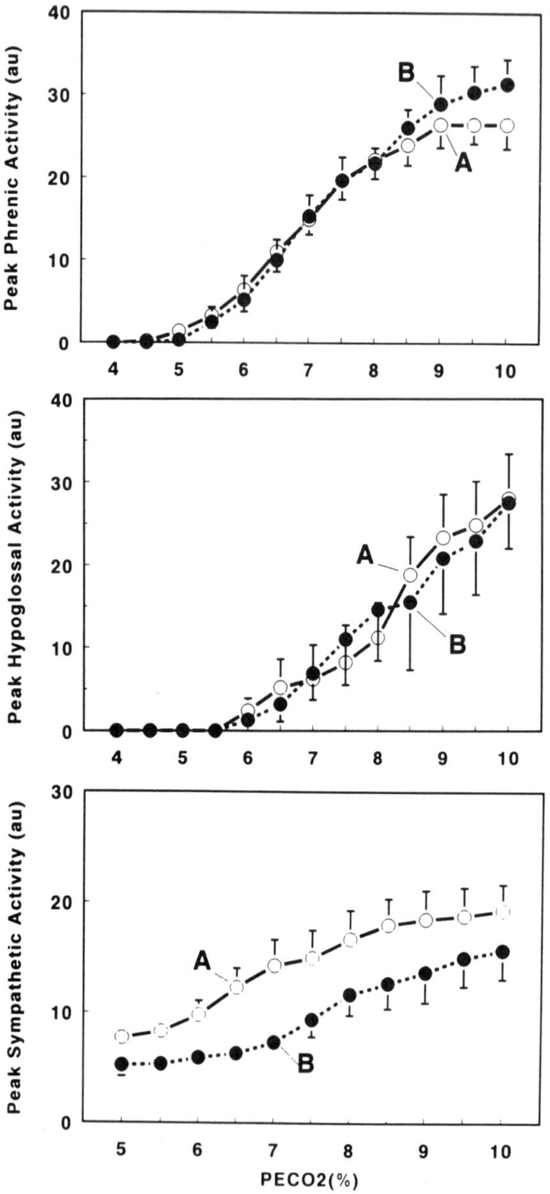

FIGURE 5. Average data (X ± SEM) of phrenic (*upper panel*), hypoglossal (*middle panel*), and cervical sympathetic (*lower panel*) nerve responses to progressive hyperoxic hypercapnia in a control period (**A**) and after moxonidine administration (**B**).

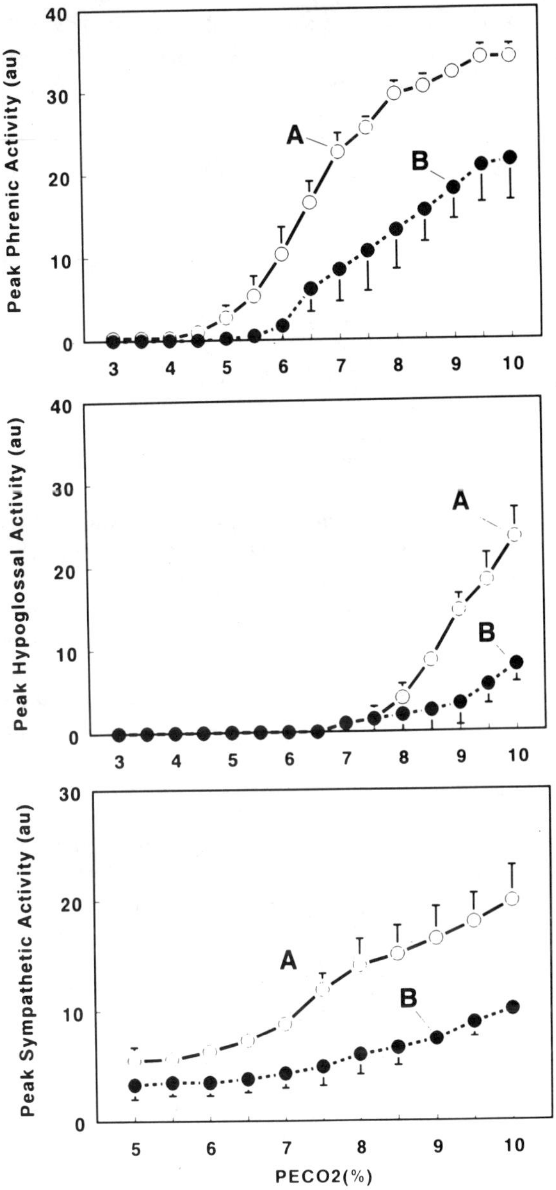

FIGURE 6. Average data of phrenic (*upper panel*), hypoglossal (*middle panel*), and cervical sympathetic (*lower panel*) nerve responses to progressive hyperoxic hypercapnia before (**A**) and after (**B**) clonidine administration.

($p = 0.01$). Furthermore, clonidine significantly diminished the cervical sympathetic response to CO_2 (FIG. 6, lower panel). At end-tidal CO_2 of 10%, cervical activity in a control period was 20 ± 3 versus 10 ± 0.3 units after clonidine administration ($p = 0.02$).

In five of six cats, effects of SK&F-86466 on clonidine-induced changes were examined. SK&F-86466 given intravenously (0.5-1 mg/kg) reversed the effects of clonidine on CO_2 responses of phrenic and hypoglossal nerves. Respiratory related cervical sympathetic activity also increased. In response to SK&F-86466, arterial blood pressure transiently increased and then decreased close to levels before SK&F-86466 administration.

Effect of Moxonidine on the Response to Hypoxia in Cats with Carotid Sinus Nerves Intact. The effect of moxonidine on respiratory response to stimulation of peripheral chemoreceptors by decreasing partial pressure of oxygen in an inhaled gas mixture was also examined. Phrenic nerve response was studied in nine cats, hypoglossal nerve response in eight animals, and the response of cervical sympathetic discharge in four cats. FIGURE 7 presents average time courses of phrenic (upper panel), hypoglossal (middle panel), and cervical sympathetic nerve responses (lower panel) to isocapnic hypoxia. Transition from a hyperoxic to a hypoxic condition was associated with an increase in electrical activities of all three studied nerves, reaching a peak response at the third minute of exposure.

Moxonidine given in concentrations of 30-100 μg/kg had no effect on phrenic and hypoglossal nerve responses to hypoxia, but it reduced the response of cervical sympathetic nerve activity. The response of cervical sympathetic nerve discharge to stimulation of peripheral chemoreceptors by low arterial PO_2 was suppressed by lower doses of moxonidine than was the activity of the phrenic and hypoglossal nerves. At any given time of on-response curve to isocapnic hypoxia, values of cervical sympathetic activity were significantly lower after moxonidine administration. For example, before treatment at the third minute of exposure to hypoxia cervical sympathetic activity was higher than activity after moxonidine administration (20 ± 3 versus 12.5 ± 2.5 units; $p = 0.02$).

Switching the animal from a hypoxic to a hyperoxic gas mixture caused activity of all nerves to decrease rapidly, below the pre-hypoxic exposure levels, and then activities gradually increased on average above the control values. Posthypoxic cervical sympathetic nerve discharge in the pretreatment period was higher than that in a period before hypoxic exposure (14 ± 2 versus 8.5 ± 0.9 units; $p = 0.04$). Moxonidine administration did not alter the response of phrenic and hypoglossal nerves, but it significantly reduced the posthypoxic increase in cervical sympathetic nerve discharge. Peak cervical sympathetic activity by switching the animal from a hypoxic to a hyperoxic condition insignificantly increased, only from 4.2 ± 0.9 to 5.7 ± 1.8 units.

Effect of Moxonidine on Response to Hypoxic Loading in Cats with Carotid Sinus Nerves Cut. In five vagotomized, carotid body denervated cats ventilated with 100% O_2 at end-tidal CO_2 above apneic point, the effects of moxonidine administration on the responses to hypoxic loading were studied. As expected, exposure to hypoxia in peripherally chemodenervated cats caused a gradual decrease in phrenic nerve activity (on-response), which was reversed by switching the animal from 12-10% O_2 to 100% O_2. In the posthypoxic period (off- response), phrenic nerve output tended to increase,

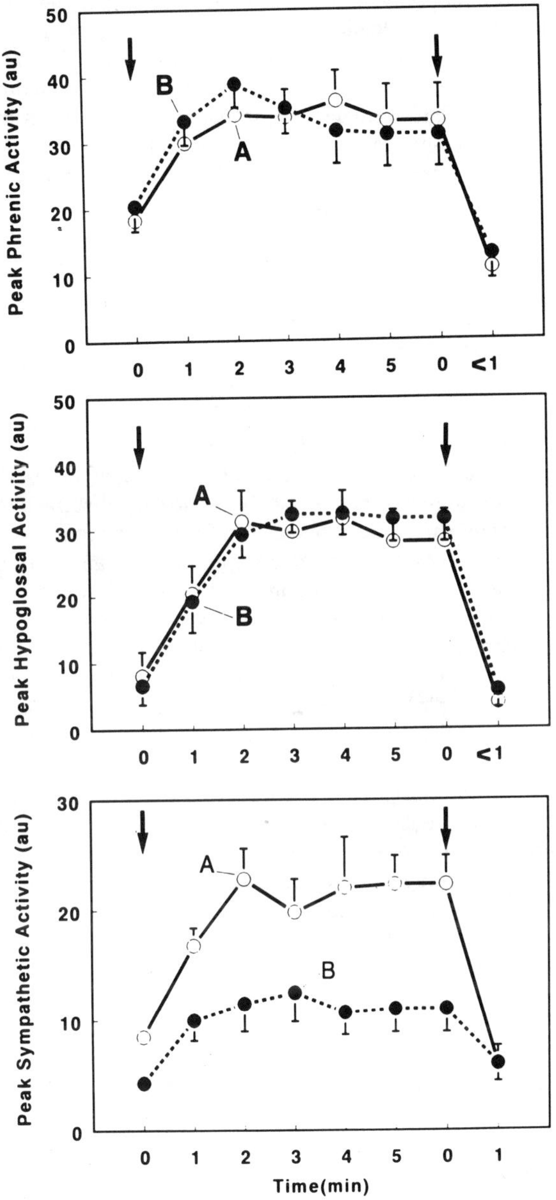

FIGURE 7. Time course of average phrenic (*upper panel*), hypoglossal (*middle panel*), and cervical sympathetic (*lower panel*) responses to hypoxic exposure in a control period (**A**) and after moxonidine administration (**B**). At the *first arrow,* the animal was switched from O_2 to a hypoxic gas mixture and at the *second arrow* back to O_2.

reaching values above the prehypoxic exposure level. Moxonidine had no effect on off- or on-response to hypoxic loading.

Effect of Mixed I_1-Imidazoline and α_2-Adrenergic Receptor Blockade on Hypotensive Action of Moxonidine. The reversal effect of the I_1/α_2-antagonist efaroxan[27] applied bilaterally to the intermediate area on hypotension induced by intravenous administration of moxonidine was tested in six cats. Efaroxan (10 μl of 50 mM in a pledget/site) caused blood pressure to increase from 106 ± 9 to 146 ± 9 ($p = 0.01$) and heart rate from 183 ± 16 to 230 ± 22 beats/min ($p = 0.03$). Furthermore, efaroxan locally administered to chemosensitive areas induced an increase in peak phrenic nerve activity (20 ± 2 versus 31 ± 4 units; $p = 0.052$) and in peak hypoglossal nerve discharge (12 ± 2 to 33 ± 5 units; $p = 0.02$).

DISCUSSION

The results of this study confirm earlier findings that moxonidine given intravenously effectively decreases arterial blood pressure and sympathetic outflow. Furthermore, the present study suggests that in cats the ventrolateral surface of the medulla oblongata is an important site of the hypotensive effects of moxonidine given intravenously. This statement is based on the finding that topical application of efaroxan to intermediate area reverses moxonidine-induced hypotension and bradycardia.

The ventral aspect of the medulla oblongata plays a major role in central chemosensitivity and is an important site for coordination of respiratory related activities of different groups of respiratory muscles, including chest wall pumping and upper airway dilating muscles,[25,26,28,29] by maintaining balanced activity and upper airway patency. Superficial areas of the ventrolateral medulla are the sites of action of many substances such as the benzodiazepines[30] and centrally acting antihypertensive agents.[31–35] Many of these compounds elicit parallel changes in blood pressure and respiratory drive. We have localized I_1-imidazoline as well as α_2-adrenergic receptor sites to the chemosensory regions of the ventral medulla of the cat and have further shown that moxonidine demonstrates anatomical selectivity in the occupancy of [^{125}I]*p*-iodoclonidine binding sites, because moxonidine selectively occupies binding sites in regions such as the RVL and parapyramidal regions which are enriched in I_1-imidazoline sites relative to α_2-adrenergic sites. The regions near the ventral surface of the medulla expressing the highest densities of I_1-imidazoline sites correspond to regions sensitive to applications of I_1-imidazoline agents to the ventral surface by pledgets.

Agents that have a generalized depressive effect on breathing usually have a greater inhibitory influence on the activity of the genioglossus muscle (tongue protrussor) or the nerve that innervates this muscle (hypoglossal nerve). Low concentrations of inhibitory substances acting on the ventrolateral medullary surface can cause an appreciable decrease in hypoglossal nerve activity with little change in phrenic nerve output.[36] Disproportional changes in upper and lower airway musculature appear to increase the likelihood of obstructive apneas.[37]

In the studies reported here, moxonidine in concentrations that effectively lowered blood pressure and suppressed sympathetic tone had practically no depressive effect on phrenic and hypoglossal nerve responses to progressive stimulation of central and peripheral chemoreceptor inputs which are needed to produce functionally appropriate discharges. Furthermore, the pattern of hypoglossal nerve discharge during normocapnia, progressive hypercapnia, or isocapnic hypoxia was not influenced by the administration of moxonidine in therapeutic doses, even in concentrations of 30 μg/kg. In higher doses of ≥100 μg/kg, systemically given moxonidine induced rapid transient decreases in phrenic and hypoglossal nerve activities, progressing to central apnea as long as 1 minute or more. In animals with intact carotid sinus nerves this inhibitory response can be a consequence of transient increases in arterial blood pressure, changes in blood flow and CO_2 output, and activation of α_2-adrenergic receptors. Moxonidine partly activates α_2-adrenergic receptors[19,20] that might elicit respiratory inhibition. However, the α_2-contribution could be weak, because it had no discernible effect on the response of phrenic and hypoglossal nerves to increases in chemical drive.

Recently, the effects of blood pressure on the respiratory activity of cranial nerves as well as the phrenic nerve were evaluated. It was found that at all levels of chemical drive, increases in blood pressure elicit inhibition of respiration activity, whereas decreases in blood pressure increase respiratory output. Fluctuations in arterial blood pressure resulted in more pronounced changes in hypoglossal than in phrenic nerve discharge.[9] Preserved respiratory output response to changes in chemical drive after moxonidine cannot be fully explained by the facilitatory influence of moxonidine-induced hypotension, which might overcome central inhibitory influences. In peripherally chemodenervated cats, the response to activation of the central chemosensory system was not different from the responses in animals with intact carotid sinus nerves.

By contrast to moxonidine, clonidine inhibited responses to CO_2. Specifically, the response of the hypoglossal nerve was strongly inhibited. These effects were reversed by intravenous administration of SK&F-86466, a selective blocker of the α_2-adrenergic receptor, suggesting that activation of α_2-adrenergic receptors is responsible for clonidine-induced respiratory depression. Preferential inhibition of hypoglossal nerve activity may explain clonidine-produced episodes of an obstructive pattern associated with a decrease in oxygen saturation in patients receiving this drug as preoperative medication.[38] It is believed that moxonidine in therapeutic doses has less sedative effects than does clonidine.[19,20,39]

Recently, clonidine given orally was shown to influence sleep architecture, causing suppression of REM sleep and improvement in oxygen saturation. In some patients, clonidine induced conversion of obstructive hypopneas during REM sleep in complete airway occlusion, suggesting that clonidine in particular patients worsens nocturnal breathing activities,[16] probably because of its preferential inhibitory effects on hypoglossal nerve activity. Differences between the effects of clonidine and moxonidine on the CO_2 threshold for activation of phrenic and hypoglossal nerves are probably caused by differences in the affinity for α_2-adrenergic receptors. Because the general pharmacology and central effects of this novel, centrally acting antihypertensive agent are well known,[40–42] our studies indicate that moxonidine may be of special value in hypertensive patients with sleep-related respiratory disorders and in combination with other approaches in patients with nocturnal oxygen desaturation attacks.[43]

In summary, activation of I_1-imidazoline receptors in the ventrolateral medulla decreases blood pressure and reduces sympathetic outflow without affecting the response of upper airway dilating and chest wall pumping muscles to peripheral and central chemosensory inputs.

SUMMARY

Sedation elicited by some centrally acting antihypertensive agents may interfere with respiratory control, and by selectively inhibiting upper airway dilating muscle activity it may facilitate obstructive sleep apnea. Autoradiographic studies with [^{125}I]*p*-iodoclonidine in the presence of 10 μM epinephrine to block α_2-adrenergic sites or 100 nM moxonidine to mask I_1-imidazoline sites show that both I_1- as well as α_2-sites are localized in putative chemosensory areas of the rostral ventrolateral medulla in the cat. We sought to determine the effect of activating I_1 and α_2-receptors on central chemosensitivity by using moxonidine as a selective I_1 agonist, clonidine as a mixed I_1/α_2 agonist, SK&F-86466 as a specific α_2-antagonist, and efaroxan as a mixed I_1/α_2 antagonist.

We recorded responses of phrenic, hypoglossal, and cervical sympathetic nerve activities to progressive hypercapnia after hyperventilation to apnea. Moxonidine (3-100 μg/kg iv) caused dose-dependent decreases in tonic cervical sympathetic nerve activity and blood pressure, but had no effect on the CO_2 threshold (after 30 or 100 μg/kg moxonidine, phrenic nerve activity reappeared at 5.8 ± 0.2% CO_2 versus 5.6 ± 0.3% CO_2 in control). Following moxonidine, the slope of the steep portion of the CO_2 response tended to increase (10.3 ± 1.8 versus 7.3 ± 0.9). Peak phrenic nerve activity was comparable to control at 7.5% CO_2 (20 ± 2 U in control) and at 9.5% CO_2 (30 ± 3 versus 27. ± 2 U). Similarly, the response of hypoglossal and inspiratory phasic cervical sympathetic nerve activity to a progressive CO_2 rise was not affected by moxonidine. By contrast, clonidine in the same doses decreased CO_2 sensitivity, because the CO_2 threshold was elevated from 5.3 ± 0.5% to 6.7 ± 0.4% ($p < 0.001$). The slope of the CO_2 response was decreased from 9.7 ± 1.9 to 7.4 ± 1.3 ($p = 0.05$). Peak phrenic nerve activity was reduced at 7.5% CO_2 (11 ± 5 versus 25 ± 2 U; $p < 0.05$) and at 9.5% CO_2 (21 ± 4 versus 33 ± 2 U; $p = 0.06$). Clonidine selectively inhibited the response of hypoglossal nerve activity to CO_2. The depressive effects of clonidine were reversed by α_2-blockade with SK&F-86466 (0.5 or 1 mg/kg). Inspiratory phasic cervical sympathetic nerve activity increased after SK&F-86466 in parallel with phrenic and hypoglossal nerve activity, but the tonic component of cervical sympathetic nerve activity and blood pressure increased only transiently. Efaroxan applied topically to the ventrolateral medulla reversed the effects of moxonidine on arterial blood pressure and heart rate and tended to increase respiratory output.

We conclude that α_2 activation impairs respiratory output responses to central chemosensory stimulation, whereas I_1- imidazoline activation lacks this adverse effect. Therefore, I_1 agonists such as moxonidine may be indicated in hypertension with sleep-related respiratory disorders.

ACKNOWLEDGMENT

We thank Cecily Lewis for secretarial assistance.

REFERENCES

1. KALES, A., R. J. CADIEUX, L. C. SHAW, A. VELA-BUENO, E. O. BIXLER, D. SCHNECK, T. LOCKE & C. R. SOLDATOS. 1984. Lancet **2:** 1005-1008.
2. TILIKIAN, A. G., C. GUILLEMINAULT, J. S. SCHRODER, K. L. LEHRMAN & W. C. DEMENT. 1976. Ann. Intern. Med. **85:** 714-719.
3. MOTTA, J., C. GUILLEMINAULT, J. S. SCHRODER & W. C. DEMENT. 1978. Ann. Intern. Med. **89:** 454-458.
4. BUDA, A. J., J. S. SCHRODER & C. GUILLEMINAULT. 1981. Internat. J. Cardiol. **1:** 67-74.
5. WILLIAMS, A. J., D. HOUSTON, S. FINBERG, C. MA, J. L. KINNEY & S. SANTIAGO. 1985. Am. J. Cardiol. **55:** 1019-1022.
6. MILLMAN, R. P., S. REDLINE, C. C. CARLISLE, A. L. ASSAF & P. LEVISON. 1991. Chest **99:** 861-866.
7. FINDELY, L. J., M. E. UNVERZAGT & P. SURRAT. 1988. Am. Rev. Resp. Dis. **138:** 337-340.
8. HORNE, J. A., N. R. ANDERSON & R. T. WILKINSON. 1983. Sleep **6:** 347-358.
9. SALAMONE, J. A., K. P. STROHL, D. M. WEINER, J. MITRA & N. S. CHERNIACK. 1983. J. Appl. Physiol. **55:** 61-68.
10. TIMMERMANS, P. B. M. W. M., A. M. C. SSCHOOP, H. Y. KWA & P. A. ZWIETEN. 1981. Eur. J. Pharmacol. **70:** 7-15.
11. TIMMERMANS, P. B. M. W. M. & P. A. VAN ZWIETEN. 1982. J. Med. Chem. **25:** 1389-1401.
12. ZORNOW, M. H. 1991. Neuropharmacology **30:** 1065-1071.
13. MCCRIMMON, D. R. & P. M. LALLEY. 1982. J. Pharmacol. Exp. Ther. **222:** 771-777.
14. BAMFORD, O. S., R. L. HAWKINS & C. E. BLANCO. 1990. Am. J. Obstet. Gynecol. **163:** 661-668.
15. GISLASON, T., M. ALQUIST, G. ERIKSSON, A. TAUBE & G. BOMCM. 1988. J. Clin. Epidemiol. **61:** 571-576.
16. ISA, F. G. 1992. Am. Rev. Resp. Dis. **145:** 435-439.
17. ERNSBERGER, P. R., K. L. WESTBROOKS, M. O. CHRISTEN & S. G. SCHÄFER. 1992. J. Cardiovasc. Pharmacol. **20:** (Suppl. 4): S1-S10.
18. ERNSBERGER, P., T. H. DAMON, L. M. GRAFF, M. O. CHRISTEN & S. G. SCHÄFER. 1993. J. Pharmacol. Exp. Ther. **264:** 172-182.
19. ERNSBERGER, P., M. A. HAXHIU, L. M. GRAFF, L. A. COLLINS, I. DRESHAJ, D. L. GROVE, M. E. GRAVES, S. G. SCHÄFER & M. O. CHRISTEN. 1994. Cardiovasc. Drugs Ther. **8:** 27-41.
20. ERNSBERGER, P., H. L. ELIOTT, H.-J. WEINMANN, A. RAAP, M. A. HAXHIU, E. HOFFERBER, A. LOW-KROGER, J. L. REID & H.-J. MEST. 1993. Cardiovasc. Drug Rev. **11:** 411-431.
21. HAXHIU, M. A., I. DRESHAJ, S. G. SCHÄFER & P. ERNSBERGER. 1994. J. Cardiovasc. Pharmacol. **241** (Suppl. 1): S1-S8.
22. ERNSBERGER, P., M. A. HAXHIU, T. H. DAMON & L. A. WENDORFF. 1992. FASEB J. **6:** A1874.
23. HAXHIU, M. A., I. DRESHAJ, B. EROKWU, S. G. SCHÄFER, M. O. CHRISTEN & P. R. ERNSBERGER. 1992. J. Cardiovasc. Pharmacol. **20** (Suppl. 4): S11-S15.
24. REIS, D. J., S. MORRISON & D. A. RUGGIERO. 1988. Hypertension **11** (Suppl. 1): I8-I13.
25. HAXHIU, M. A., J. MITRA, E. VAN LUNTEREN, E. W. BRUCE & N. S. CHERNIACK. 1984. Am. J. Physiol. **247:** R939.
26. HAXHIU, M. A., J. MITRA, E. VAN LUNTEREN, N. R. PRABHAKAR & N. S. CHERNIACK. 1985. Am. J. Physiol. **249:** R266-R273.

27. Ernsberger, P., D. L. Grove, M. E. Graves, I. Dreshaj & M. A. Haxhiu. 1994. Am. J. Hypertens. **7:** 139A.
28. Schläefke, M. E. 1981. Rev. Physiol. Biochem. Pharmacol. **90:** 171-244.
29. Cherniack, N. S., E. von Euler, I. Homma & F. F. Kao. 1979. J. Physiol. **287:** 191-211.
30. Haxhiu, M. A., E. Van Lunteren, N. S. Cherniack & E. C. Deal. 1989. Am. J. Physiol. **257:** R810-R815.
31. Ernsberger, P., R. Giuliano, R. N. Willette, A. R. Granata & D. J. Reis. 1988. J. Hypertens. **6** (Suppl. 4): S554-S557.
32. Ernsberger, P., R. N. Giuliano, R. N. Willette & D. J. Reis. 1990. J. Pharmacol. Exp. Ther. **253:** 408-418.
33. Punnen, S., R. Urbanski, A. J. Krieger & H. N. Sapru. 1987. Brain Res. **422:** 336-346.
34. Bousquet, P. & P. G. Guertzenstein. 1973. Br. J. Pharmacol. **49:** 573-579.
35. Weber, M. A., W. F. Graettinger & D. G. Cheung. 1990. Centrally acting sympathetic inhibitors. *In* Hypertension: Pathophysiology, Diagnosis, and Management. J. H. Laragh & B. M. Brenner, Eds.: 2251-2261. Raven Press. New York.
36. Haxhiu, M. A., J. Mitra, E. Van Lunteren, N. Prabhakar, E. N. Bruce & N. S. Cherniack. 1986. Respiration **50:** 130-138.
37. Remmers, J. E., W. J. Degroot, E. K. Sauerland & A. M. Anch. 1978. J. Appl. Physiol. **44:** 931-938.
38. Benhamou, O., Y. Veuieillette, P. Narchi & C. Ecoffey. 1991. Anesth. Analg. **73:** 799-803.
39. Kirch, W., H.-J. Hutt & V. Planitz. 1990. J. Clin. Pharmacol. **30:** 1088-1095.
40. Armah, B. I., E. Hofferber & W. Stenzel. 1988. Arzneimittelforschung **38:** 1426-1434.
41. Haxhiu, M. A., I. Dreshaj, S. G. Schäfer & P. Ernsberger. 1994. J. Cardiovasc. Pharmacol. **241** (Suppl. 1): S1-S8.
42. Udvary, E., S. G. Schäfer, A. Vegh & L. Szekeres. 1984. Eur. J. Pharmacol. **27:** 147-152.
43. Strohl, K. P., N. S. Cherniack & B. Gothe. 1986. Am. Rev. Respir Dis. **134:** 791-802.

Centrally Mediated Ocular Hypotension: Potential Role of Imidazoline Receptors[a]

W. RICHARD CAMPBELL[b] AND DAVID E. POTTER[c]

[b]*Department of Biology*
Morris Brown College
Atlanta, Georgia 30314-4140

[c]*Department of Pharmacology and Toxicology*
Morehouse School of Medicine
Atlanta, Georgia

Glaucoma is a major public health problem,[1] causing blindness in a significant number of Americans, particularly African Americans.[2] One of the characteristic, measurable risk factors most consistently associated with the disease is a chronic increase in intraocular pressure (IOP).[3] It is therefore logical that reducing IOP could have therapeutic benefit. Clinical studies suggest a significant reduction in the rate of initial injury when eyedrops are used to lower IOP.[4–6]

In recent years, therapeutic and research interest in α_2-agonists as potential antiglaucoma drugs has increased.[7] Drugs such as clonidine,[8] *para*-aminoclonidine (iopidine),[9] UK-14,304 (brimonidine),[10] moxonidine,[11–13] and oxymetazoline,[14,15] which have been classified as α_2-agonists, lower IOP when given systemically or applied topically to eyes of animals[16,17] or humans.[18,19] Agents that have either a phenylethylamine structure (norepinephrine, epinephrine, and dopamine) or an imidazoline structure (clonidine, moxonidine, and oxymetazoline) have been demonstrated to act as α_2-agonists. However, recently some actions of imidazoline compounds are believed to be mediated by a site termed an imidazoline-preferring receptor (IPR). This receptor was found in a number of tissues, cell types, and species.[20–23] Recently, the centrally mediated systemic antihypertensive actions of clonidine have been attributed to interactions at imidazoline-preferring receptors[24] rather than to actions on α_2-adrenergic receptors.[20,25] The centrally mediated antihypertensive actions of other imidazolines, such as moxonidine (MOX), when injected into the rostral ventrolateral medulla (RVLM) area of the brain stem, have also been attributed, in part, to actions on an imidazoline-preferring receptor rather than to actions on α_2-adrenergic receptors.[22,26] The central nervous system is a key organ in which imidazoline receptors, particularly I_1, are located.[22,23]

An intact sympathetic nervous system is required for some aspects of IOP regulation, and drugs that suppress sympathoadrenal function are effective antiglaucoma

[a]This research was supported in part by EY06338 from the National Eye Institute.

agents. Superior cervical ganglionectomy eliminates the elevation of IOP at night in rabbits,[27] indicating that intact ocular sympathetic innervation is required for maintenance of the normal circadian rhythm of IOP in this species.[28,29] In patients with open-angle glaucoma, blocking the stellate ganglion with local anesthesia was reported to lower IOP.[30,31] Other clinical reports have postulated a relation between the central nervous system and the control of IOP.[32,33] Past workers, using electrical stimulation in the brain of the cat, sought a center that would affect IOP selectively[34–39] (for a review see ref. 39). Stimulation in some areas, such as the posterior hypothalamus, induced an IOP change simultaneous with a change in blood pressure, whereas stimulation in other areas, such as the dorsal hypothalamus and ventral thalamus, induced changes in IOP without parallel blood pressure changes.[40]

Substances have been introduced into the cerebrospinal fluid by intraventricular perfusion or by single intraventricular injections to study their effects on IOP. Effects on IOP were demonstrated to be independent of actions on related physiologic parameters such as blood pressure.[39] These substances include hypo- and hyperosmotic agents, prostaglandins, arachidonic acid, calcium, clonidine, substance P, thyrotropin-releasing hormone, and vasopressin.[33,41–46] When perfused intracisternally in conscious rabbits,[46] clonidine lowered IOP. Administration of agents into cerebrospinal fluid, however, delivers these agents to many areas of the brain, and multiple sites of action are therefore possible. Like the studies with diencephalic stimulation, intraventricular perfusion and injection studies have not identified a specific brain locus for the regulation of aqueous humor dynamics or an integrating neural mechanism for IOP regulation.[39]

Recently, the imidazolines MOX and oxymetazoline (OXY)[47] were shown by this laboratory to lower IOP bilaterally when applied topically to the eyes of rabbits[13,15,48,49] or when injected intracerebroventricularly (icv) in rabbits.[12,15,48] Additionally, the α_2-agonist UK-14,304-18 (UK; brimondine),[50] defined chemically as an imidazolidine, was shown by this laboratory to induce ocular hypotension when applied topically to the eyes of the rabbit[10,48] or injected into the lateral ventricle of the rabbit.[48] This study seeks to: (1) further assess the contribution of central nervous system mechanisms to the ocular actions of MOX, OXY, and UK; (2) determine the magnitude of involvement of central imidazoline (I_1) receptors and α_2-adrenoceptors in the ocular hypotensive responses to MOX, OXY, and UK; (3) investigate an integrating neural mechanism for IOP regulation; and (4) gain evidence for the brain as a target site for potential antiglaucoma drugs. Efaroxan, which has been classified as a relatively selective I_1 imidazoline receptor antagonist,[20] and rauwolscine, a relatively selective α_2-adrenoceptor antagonist,[51] were selected to help characterize the centrally mediated ocular hypotensive actions of MOX, OXY, and UK.

MATERIALS AND METHODS

Experimental Animals

New Zealand white male rabbits (2-4 kg), used in IOP experiments, were maintained on a reversed 12-hour dark/light cycle. All experiments were conducted between 9:00 AM and 5:00 PM. All animals were housed individually at normal room temperature and maintained on Purina rabbit chow and tap water ad libitum. Animal care and treatment in this investigation were in accordance with the National Institutes of Health and Association for Research in Vision and Ophthalmology (ARVO) guidelines on the use of animals for research and were approved by the Institutional Animal Care and Use Committee of Morehouse School of Medicine.

Implantation of Chronic Indwelling Cannulas

Male New Zealand white rabbits (2-4 kg) were anesthetized with ketamine/xylazine, 25/3 mg/kg (im), respectively, and placed in a stereotaxic apparatus (Kopf). A small incision was made in the dorsal cranial surface to reveal skull sutures lambda and bregma, which were used to align the skull and served as stereotaxic reference points. Trephination of the skull was performed to produce entry ports for the stainless steel guide cannula and three anchor screws. After the anchor screws were placed and the guide cannula stereotaxically implanted, cranioplastic dental cement was built up around the guide cannula and the anchor screws to solidly hold the cannula in place. The surface of the hardened dental cement cap was filed to remove any rough edges that could irritate the skin of the animal. The guide cannula was sealed with a removable dummy cannula to prevent foreign material from entering the guide cannula and subsequently the CNS. The incision was closed using stainless steel wound clips, which are removed after 1 week. Animals were allowed a minimum of 5 days to recuperate prior to experimentation. Coordinates for cannula placement were obtained using the "Atlas of the Rabbit Brain and Spinal Cord."[52] Injection sites were verified by macroscopic and microscopic methods. In each case trypan blue dye was injected into the injection site to aid in visualizing the area.

Intraocular Pressure and Pupil Diameter Measurements

Intraocular pressure was measured using an applanation pneumatonograph (Digilab Inc., Division of Bio-Rad, Cambridge, Massachusetts), which was calibrated manometrically for rabbit eyes and values were expressed in millimeters of mercury (mm Hg). Pupil diameter (PD) was measured horizontally in millimeters under constant illumination using an Optistick (Allergan, Irvine, California). Before tonometry, 25 μl of 0.05% tetracaine (Sigma, St. Louis, Missouri) was applied to the corneas to minimize discomfort to the animal. Two baseline readings were taken at –0.5 and 0 hour after which the test drug and vehicle were administered, as will be described. Subsequent IOP and PD measurements were taken at 0.5, 1, 2, 3, and 4 hours after drug administration. At the end of each day's measurement, stability of the tonometer was confirmed using the verifier supplied by Digilab.

Drug Preparation and Administration

The drugs used and their suppliers were: moxonidine (4-chloro-*N*-[4,5 dihydro-1H-imidazol-2-yl]-6-methoxy-2 methyl-pyridinamine hydrochloride (Beiersdorf, Hamburg, Germany); oxymetazoline (2-[3-hydroxy-2,6-dimethyl-4-*t*-butylbenzyl]-2-

imidazoline) hydrochloride (Sigma Chemical Co., St. Louis, Missouri); UK-14,304-18 (brimonidine) 5-bromo-*N*-(4,5-dihydro-1H-imidazol-2-yl)-6-quinoxalinamine (Research Biochemicals, Inc.); efaroxan (2-ethyl-2-[imidazolin-2-yl]-2,3-dihydrobenzofuran)hydrochloride; and rauwolscine (17 alpha-hydroxy-20 alpha yohimban-16 beta-carboxylic acid methy ester hydrochloride) (Sigma Chemical Co., St. Louis, Missouri); and 6-hydroxydopamine (Sigma Chemical Co., St. Louis, Missouri). The salt of each agent was dissolved in distilled water on each day of the experiment; concentrations are expressed as the salt.

Drug and vehicle solutions used in IOP experiments were administered: (1) topically to the cornea of the right eye in 50-μl volumes using a Gilman pipetman microliter pipeter; (2) intracerebroventricularly into the anterior lateral ventricle and third ventricle in 20-μl volumes; and (3) into the RVLM in 1.0-μl volumes using cannulas (26 g) and injectors (33 g) supplied by Plastic Products, Inc and Hamilton microliter syringes. MOX, OXY, and UK were administered at time $t = 0$; efaroxan and rauwolscine were administered intracerebroventricularly or into the RVLM 30 minutes prior to the addition of MOX, OXY, or UK. Because the drugs were salts, saline solution was used as placebo treatment for control.

Chemical Sympathectomy

Animals were pretreated with 400 μg/40 μl of 6-hydroxydopamine (6-OH-DA) injected into the lateral ventricle 48 hours prior to administration of MOX and OXY to determine if the site of action of these α_2/I_1 agonists was a presynaptic site.

Presentation and Statistical Analysis of Data

On the basis of the experimental design, Student's *t* test for paired data was used to evaluate drug-induced changes. A probability level of 0.05 was selected as the minimum level of significance; data presented in tables represent the mean ± standard error (SE).

RESULTS

Ocular Responses to Topical Administration of Agonists and Central Administration of Antagonists

Antagonists (efaroxan and rauwolscine) administered into the CNS were used to assess the activity of agonists (MOX, OXY, and UK) administered topically using IOP and PD as the surrogate endpoints of action. Moxonidine (25 μg, 50 μl, n =

TABLE 1. Topical Application of Moxonidine and Intracerebroventricular Injection of Antagonist[a]

	Ipsilateral Eye			Contralateral Eye		
Time (h)	MOX (25 μg)	EFAR (3 μg) + MOX	RAUW (3 μg) + MOX	Opposite Eye to MOX	EFAR (3 μg) + MOX	RAUW (3 μg) + MOX
–0.5	0	0	0	0	0	0
0	0	0	0	0	0	0
0.5	–0.22 ± 0.6	–0.99 ± 0.4	–1.8 ± 0.6	1.5 ± 0.9	4.8 ± 1.2*	–0.16 ± 0.9
1.0	–3.6 ± 0.5	–0.93 ± 0.4*	–1.1 ± 0.7*	1.6 ± 1.6	3.6 ± 1.3	–4.96 ± 1.4*
2.0	–3.5 ± 0.5	0.93 ± 0.4*	–6.3 ± 0.7	–3.8 ± 0.9	–.46 ± 1.3*	–7.5 ± 0.9
3.0	–2.7 ± 0.6	2.2 ± 1.2*	–2.6 ± 1.1	–5.4 ± 0.5	–1.7 ± 1.1*	–5.7 ± 0.9
4.0	0.99 ± 0.6	2.4 ± 1.3	–3.03 ± 0.7	–4.2 ± 0.6	–1.3 ± 1.1*	–4.7 ± 1.4
5.0	1.6 ± 0.7	3.3 ± 0.9	–1.7 ± 1.1	–2.8 ± 0.8	–0.9 ± 0.9	–2.5 ± 1.1

[a] Effect of unilateral topical moxonidine (MOX) on intraocular pressure (IOP) with and without icv efaroxan and rauwolscine (n = 5 for all groups: values represent mean ± SE). Starting IOPs for treatment groups were: MOX alone, 28.8 ± 0.7 mm Hg; efaroxan (EFAR) plus MOX, 27.8 ± 1.3 mm Hg; and rauwolscine (RAUW) plus MOX, 27.7 ± 1.1 mm Hg. Individual significance levels were determined by Student's t test: MOX was compared to saline controls: efaroxan (icv) plus MOX (topical) data were compared to data of MOX alone; rauwolscine plus MOX data were compared to those of MOX alone; * p <0.001.

5), applied unilaterally and topically to the cornea of the right eye, lowered IOP over a 4-hour period, as measured by applanation pneumatonometry (TABLE 1). From a starting IOP of 28.8 ± 0.7 mm Hg, MOX (25 μg/50 μl) lowered IOP bilaterally, 3.6 ± 0.54 (p <0.001) and 3.8 ± 0.53 (p <0.001) mm Hg, at 2 hours in the ipsilateral and contralateral eye, respectively, as compared to saline controls. Intraocular pressure returned to control levels after 5 hours. Moxonidine increased pupil diameter at this dose (data not shown) in the ipsilateral eye only and produced no observable change in rabbit behavior. Efaroxan, an I_1-imidazoline receptor antagonist, injected into the lateral ventricle (LV, 3.0 μg, 20 μl icv, n = 5) 30 minutes prior to unilateral topical MOX, inhibited bilateral IOP reductions by 73.4 and 87.9% (p <0.001) in the ipsilateral and contralateral eye (TABLE 1), respectively. By contrast, rauwolscine, an α_2-receptor antagonist, injected (3.0 μg, 20 μl icv, n = 5) 30 minutes prior to unilateral topical MOX did not inhibit bilateral reductions in IOP in either the ipsilateral or the contralateral eye (TABLE 1).

Unilateral, topical application of OXY (50 μg, 50 μl, n = 5) from a starting IOP of 28.2 ± 1.2 mm Hg produced maximal, bilateral hypotensive responses of 6.4 ± 0.88 (p <0.001) and 8.7 ± 1.6 (p <0.001) mm Hg, at 2 hours, in the contralateral and ipsilateral eyes, respectively (TABLE 2). Efaroxan (3.0 μg, 20 μl icv, n = 5), injected 30 minutes prior to unilateral, topical OXY, inhibited maximal IOP decreases by 83.4 and 39.6% in the contralateral and ipsilateral eyes, respectively (TABLE 2). Topical OXY increased PD at this dose (data not shown) in the ipsilateral eye

TABLE 2. Topical Application of Oxymetazoline and Intracerebroventricular Injection of Antagonist

	Ipsilateral Eye			Contralateral Eye		
Time (h)	OXY (50 μg)	EFAR (3 μg) + OXY	RAUW (3 μg) + OXY	Opposite Eye to OXY	EFAR (3 μg) + OXY	RAUW (3 μg) + OXY
–0.5	0	0	0	0	0	0
0	0	0	0	0	0	0
0.5	0.8 ± 1.4	2.5 ± 1.1	–1.1 ± 1.1	–0.9 ± 0.9	–0.7 ± 1.0	–1.4 ± 0.8
1.0	–4.4 ± 1.7	–0.7 ± 0.6*	–1.9 ± 0.8*	–4.1 ± 0.6	–0.9 ± 0.6*	–1.3 ± 1.7*
2.0	–8.7 ± 1.6	–5.3 ± 1.4*	–7.2 ± 1.7	–6.4 ± 0.8	–1.1 ± 0.6*	–4.4 ± 1.4
3.0	–9.3 ± 1.5	–6.4 ± 1.3*	–7.9 ± 1.7	–6.2 ± 1.1	–4.2 ± 0.9	–5.8 ± 0.9
4.0	–7.9 ± 1.9	–4.7 ± 0.5*	–8.8 ± 1.4	–4.9 ± 0.6	–2.4 ± 0.3*	–2.5 ± 1.3*
5.0	–7.4 ± 1.8	–3.6 ± 0.8*	–7.3 ± 1.9	–3.6 ± 0.8	–3.6 ± 0.8	–1.8 ± 1.2

[a] Effect of unilateral topical oxymetazoline (OXY) on intraocular pressure (IOP) with and without icv efaroxan and rauwolscine (n = 5 for all groups: values represent mean ± SE). Starting IOPs for treatment groups were: OXY alone 28.2 ± 1.2 mm Hg; efaroxan (EFAR) plus OXY, 26.8 ± 1.4 mm Hg; and rauwolscine (RAUW) plus OXY, 29.1 ± 1.7 mm Hg. Individual significance levels were determined by Student's t test: OXY was compared to saline controls; efaroxan (icv) plus OXY (topical) data were compared to those of OXY alone; rauwolscine (icv) plus OXY (topical) data were compared to those of OXY alone; * p <0.001.

only and produced no observable change in rabbit behavior. Efaroxan administered centrally did not inhibit ipsilateral eye mydriasis induced by topically administered OXY.

Unilateral, topical application of UK (50 μg/50 μl), from a starting IOP of 26.8 ± 1.3 mm Hg, lowered IOP bilaterally, 5.6 ± 0.6 (p <0.001) and 6.2 ± 1.9 (p <0.001) mm Hg in the ipsilateral and contralateral eye, respectively, as compared to saline controls (TABLE 3). Intraocular pressure returned to control levels after 5 hours. UK increased PD at this dose (data not shown) in the ipsilateral eye only and produced no observable change in rabbit behavior. Efaroxan, an I_1-imidazoline receptor antagonist, injected (3.0 μg, 20 μl icv, n = 5) 30 minutes prior to unilateral topical UK, did not inhibit bilateral IOP reductions in the ipsilateral or contralateral eye (TABLE 3). By contrast to the MOX-induced IOP effect, rauwolscine injected (3.0 μg, 20 μl, icv, n = 5) 30 minutes prior to unilateral topical UK inhibited bilateral reductions in IOP by 47.4% in the contralateral eye (TABLE 3). Although rauwolscine did not inhibit the decrease in IOP at 2 hours in the ipsilateral eye, it did inhibit the increase in IOP that occurred at 0.5 and 1 hour in this eye. Neither saline (20 μl, icv, n = 5), efaroxan (3 and 10 μg, 20 μl icv, n = 5), nor rauwolscine (3.0 μg, 20 μl icv, n = 5) alone, injected into the anterior lateral ventricle, produced any significant change in IOP (FIG. 1).

TABLE 3. Topical Application of UK-14,304-18 and Intracerebroventricular Injection of Antagonist

	Ipsilateral Eye			Contralateral Eye		
Time (h)	UK (25 μg)	EFAR (3 μg) + UK	RAUW (3 μg) + UK	Opposite Eye to UK	EFAR (3 μg) + UK	RAUW (3 μg) + UK
−0.5	0	0	0	0	0	0
0	0	0	0	0	0	0
0.5	1.7 ± 1	−1.5 ± 0.8	−0.9 ± 0.3	−4.2 ± 0.6	−8.1 ± 0.8	−1.9 ± 1.1
1.0	1.1 ± 2.1	−4.3 ± 0.8	−3.7 ± 0.5	−7.9 ± 0.9	−11.3 ± 0.6	−5.2 ± 1.7*
2.0	−5.6 ± 0.6	−10.6 ± 0.5	−6.8 ± 1.3	−6.3 ± 1.9	−6.1 ± 1.4	−3.3 ± 1.4*
3.0	−7.3 ± 2.2	−11.3 ± 0.3	−6.4 ± 1.8	−4.2 ± 1.9	−2.1 ± 0.7	−3.11 ± 0.9
4.0	−5.5 ± 1.5	−9.2 ± 0.7	−5.7 ± 0.8	−0.1 ± 2.3	−2.6 ± 0.7	−0.5 ± 1.5
5.0	−3.1 ± 2.4	−3.7 ± 1.7	−3.2 ± 1	2.7 ± 1.5	−2.3 ± 1.5	−0.6 ± 0.2
6.0	−3.9 ± 1.2	−1.9 ± 1.8	−2.1 ± 0.8	−0.4 ± 1.1	0.8 ± 0.9	−0.16 ± 0.5

[a] Effect of unilateral topical UK on intraocular pressure (IOP) with and without icv efaroxan and rauwolscine ($n = 5$ for all groups; values represent mean ± SE). Starting IOPs for treatment group were: UK alone 26.8 ± 1.3 mm Hg; efaroxan (EFAR) plus UK, 26.1 ± 1.5 mm Hg; and rauwolscine (RAUW) plus UK, 25.4 ± 0.9 mm Hg. Individual significance levels were determined by Student's t test: OXY was compared to saline controls; efaroxan (icv) plus OXY (topical) data were compared to those of OXY alone; rauwolscine (icv) plus OXY (topical) data were compared to those of OXY alone; * $p < 0.001$.

Ocular Responses to Central Administration of Agonists and Antagonists

Lateral Ventricle

MOX, OXY, UK, efaroxan, and rauwolscine were administered icv into the CNS to further assess their actions on the modulation of aqueous humor dynamics, using IOP and PD as the surrogate endpoints of activity. MOX (0.3, 0.9, 3 μg, 20 μl, n = 5) injected icv into the anterior lateral ventricle lowered IOP over a 4–6-hour period, as measured by applanation pneumatonometry (FIG. 2). From a starting IOP of 28.8 ± 0.7 mm Hg, MOX (0.3 μg) lowered IOP bilaterally, 7.3 ± 0.8 ($p < 0.001$) and 7.4 ± 0.6 ($p < 0.001$) mm Hg at 30 and 120 minutes, respectively, as compared to saline controls. Intraocular pressure returned to control levels after 4 hours (FIG. 2). MOX at all doses tested had no effect on PD (data not shown) and produced no observable change in rabbit behavior. From a starting IOP of 27.8 ± 0.6 mm Hg, MOX (0.9 μg) lowered IOP bilaterally, 7.1 ± 0.9 ($p < 0.001$) and 10.1 ± 0.8 ($p < 0.001$) mm Hg at 30 and 120 minutes, respectively, as compared to saline controls. Intraocular pressure returned to control levels after 4 hours (FIG. 2). From a starting IOP of 26.8 ± 0.7 mm Hg, MOX (3 μg) lowered IOP bilaterally, 3.4 ± 0.9 ($p < 0.05$) and 10.9 ± 0.6 ($p < 0.001$) mm Hg at 30 and 120 minutes, respectively, as compared to saline controls. Intraocular pressure returned to control levels after 6 hours (FIG. 2). Efaroxan, injected (3.0 μg, 20 μl icv, n = 5) 30 minutes prior to MOX icv, inhibited bilateral IOP reductions at 30 and 120 minutes by 71.2 and

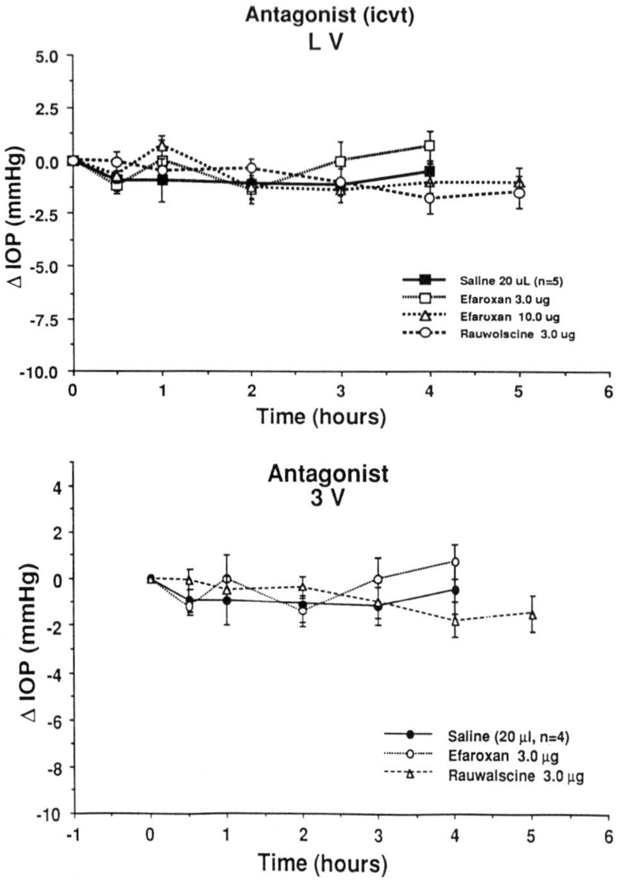

FIGURE 1. **(Top)** Effects of saline, efaroxan (3.0 and 10.0 μg icv), and rauwolscine (3.0 μg icv), injected into the anterior lateral ventricle (LV; 20 μl) alone on intraocular pressure (IOP); values represent mean ± SE. Starting IOPs for the treatment groups were: 24.5 ± 0.9 for saline; 26.5 ± 0.8 for 3.0 μg and 27.2 ± 0.7 for 10.0 μg efaroxan, respectively; and 28.7 ± 0.3 for rauwolscine. **(Bottom)** Effects of saline, efaroxan (3.0 μg), and rauwolscine (3.0 μg), alone (20 μl), injected into the third ventricle, on IOP. Starting IOPs for treatment groups were: saline, 27.5 ± 0.9 mm Hg; efaroxan, 26.8 ± 0.9); and rauwolscine (3.0 μg), 28.7 ± 0.3 mm Hg.

34.3% (p <0.001), respectively (FIG. 2). Rauwolscine, injected (3.0 μg, 20 μl icv, n = 5) 30 minutes prior to MOX, inhibited bilateral reductions at 30 and 120 minutes by 47 and 42% (p <0.05), respectively (FIG. 2).

Oxymetazoline (0.3, 0.9, 3 μg, 20 μl, n = 5), injected icv into the anterior lateral ventricle, lowered IOP over a 4-6 hour period, as measured by applanation pneumatonometry (FIG. 3). Oxymetazoline (0.3 μg, 20 μl icv, n = 5), from a starting pressure of 30.3 ± 0.7 mm Hg, lowered IOP bilaterally, 2.2 ± 0.9 (p <0.05) and

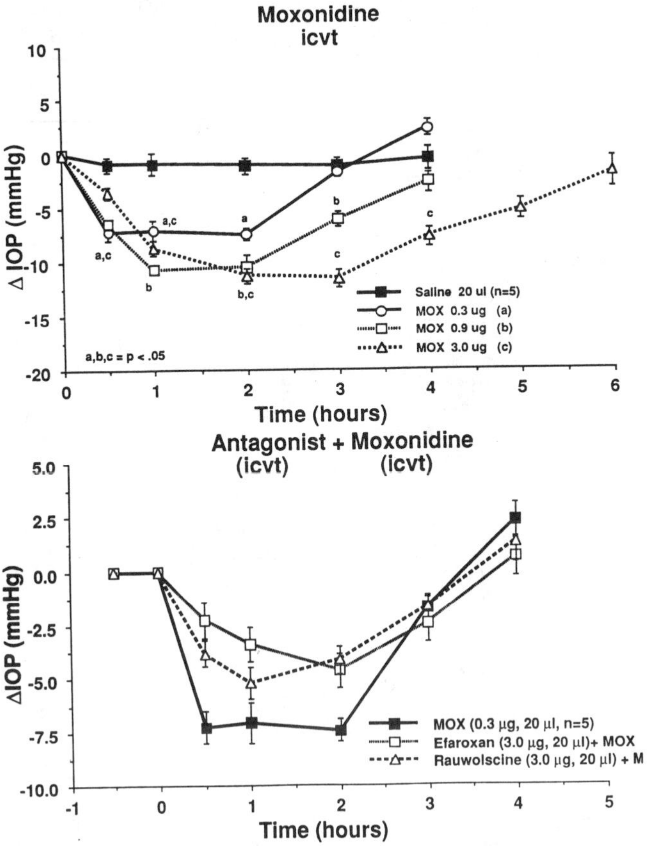

FIGURE 2. Effect of icv moxonidine (MOX) injected into the lateral ventricle on IOP, with and without efaroxan (icv) and rauwolscine (icv), as compared to saline (icv) controls ($n = 5$ for all groups; values represent mean ± SE). Starting IOPs for treatment groups were: MOX alone, 28.8 ± 0.7 mm Hg; efaroxan plus MOX, 29.8 ± 0.9 mm Hg; rauwolscine plus MOX, 29.3 ± 0.5 mm Hg. Individual significance levels were determined by Student's *t* test: MOX was compared to saline controls; efaroxan (icv) plus MOX data were compared to those of MOX alone; rauwolscine (icv) plus MOX data were compared to those of MOX alone; $^{*}p < 0.001$.

6.7 ± 0.94 (p <0.001) mm Hg, at 30 and 120 minutes, respectively, as compared to saline controls (FIG. 3). Oxymetazoline (0.3 μg icv) had no observable effect on PD at this dose (data not shown) and produced no observable change in rabbit behavior. Oxymetazoline (0.9 μg, 20 μl icv, n = 5), from a starting pressure of 24.4 ± 0.2 mm Hg, lowered IOP bilaterally, 4.02 ± 0.64 (p <0.001) and 7.6 ± 1.2 (p <0.001) mm Hg, at 30 and 120 minutes, respectively, as compared to saline controls (FIG. 3). At a dose of 0.9 μg icv there was no observable effect on PD (data not shown) and no observable change in rabbit behavior. OXY (3.0 μg, 20 μl icv, n = 5), from

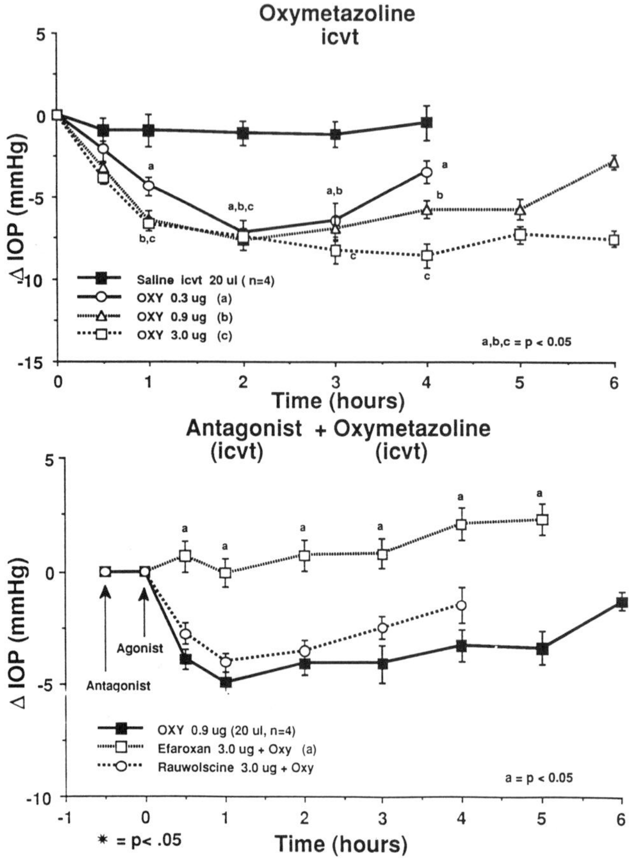

FIGURE 3. Effect of icv oxymetazoline (OXY) injected into the lateral ventricle on IOP, with and without efaroxan (icv) and rauwolscine (icv), as compared to saline (icv) controls ($n = 5$ for all groups; values represent mean ± SE). Starting IOPs for treatment groups were: OXY alone, 24.4 ± 0.2 mm Hg; efaroxan plus OXY, 25.1 ± 0.7 mm Hg; rauwolscine plus MOX, 27.6 ± 0.9 mm Hg. Individual significance levels were determined by Student's t test: MOX was compared to saline controls; efaroxan (icv) plus MOX data were compared to those of MOX alone; rauwolscine (icv) plus MOX data were compared to those of MOX alone; *p <0.001.

a starting pressure of 26.3 ± 1.2 mm Hg, lowered IOP bilaterally, 4.3 ± 1.1 (p <0.05) and 8.6 ± 0.77 (p <0.001) mm Hg, at 30 and 120 minutes, respectively, as compared to saline controls (FIG. 3). The antagonist efaroxan (3.0 μg, 20 μl icv, n = 5), given 30 minutes prior to OXY (0.9 μg), inhibited OXY-induced ocular hypotension by 100% at 30 minutes (p <0.001) and by 100% at 120 minutes (p <0.001) (FIG. 3). Rauwolscine, an α_2-receptor antagonist, injected (3.0 μg, 20 μl icv, n = 5) 30 minutes prior to OXY, did not statistically inhibit the bilateral reductions in IOP (FIG. 3).

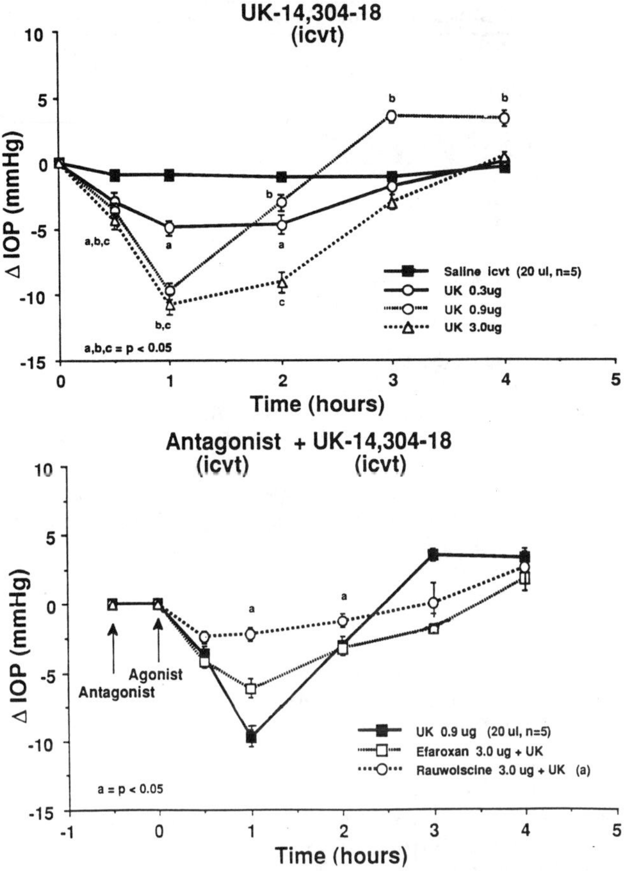

FIGURE 4. Effect of icv UK-14,304 injected into the lateral ventricle on IOP with and without icv rauwolscine and efaroxan (*n* = 5 for all groups; values represent mean ± SE). Starting IOPs for treatment groups were: UK-14,304 alone, 31.7 ± 0.5 mm Hg; efaroxan plus UK, 31.6 ± 0.8 mm Hg; rauwolscine plus UK, 32.0 ± 0.7 mm Hg. Individual significance levels were determined by Student's *t* test: UK was compared to saline controls; rauwolscine (icv) plus UK data were compared to those of UK alone; efaroxan (icv) plus UK data were compared to those of UK alone; **p* <0.001.

UK (0.3, 0.9, 3 μg, 20 μl, *n* = 5), injected icv into the anterior lateral ventricle, lowered IOP over a 4-6-hour period, as measured by applanation pneumatonometry (FIG. 4). UK (0.3 μg, 20 μl icv, *n* = 5), from a starting pressure of 29.7 ± 0.8 mm Hg, lowered IOP bilaterally, 2.9 ± 0.7 (*p* <0.001) and 4.7 ± 0.7 (*p* <0.05) mm Hg, at 30 and 120 minutes, respectively, as compared to saline controls (FIG. 4). UK (0.9 μg, 20 μl icv, *n* = 5), from a starting pressure of 27.7 ± 0.7 mm Hg, lowered IOP bilaterally, 3.6 ± 0.5 (*p* <0.001) and 9.7 ± 0.7 (*p* <0.001) mm Hg, at 30 and 60 minutes, respectively, as compared to saline controls (FIG. 4). UK (3 μg, 20 μl icv,

TABLE 4. Injections of Moxonidine and Antagonists into the Third Ventricle

Time (h)	Saline (20 μl)	MOX (0.3 μg)	EFAR (3 μg) + MOX	RAUW (3 μg) + MOX
−0.5	0	0	0	0
0	0	0	0	0
0.5	−0.99 ± 0.4	−5.2 ± 0.7	−2.4 ± 0.8*	−3.8 ± 0.8
1.0	−0.93 ± 0.4	−7.4 ± 0.8	−0.4 ± 0.7*	−4.6 ± 1.5
2.0	0.93 ± 0.4	−6.7 ± 0.7	−0.1 ± 0.7*	−2.9 ± 1.9**
3.0	2.2 ± 1.2	−4.4 ± 0.6	−3.01 ± 0.8	−0.9 ± 2.2
4.0	2.4 ± 1.3	−1.3 ± 0.4	−2.4 ± 0.9	0.56 ± 0.6

[a] Effect of icv moxonidine (MOX) injected into the third ventricle on intraocular pressure (IOP) with and without efaroxan (icv) and rauwolscine (icv), as compared to saline (icv) controls ($n = 4$ for all groups; values represent mean ± SE). Starting IOPs for treatment groups were: MOX alone, 31.7 ± 0.5 mm Hg; efaroxan (EFAR) plus MOX, 28.8 ± 0.9 mm Hg; and rauwolscine (RAUW) plus MOX, 29.3 ± 0.7 mm Hg. Individual significance levels were determined by Student's t test: MOX was compared to saline controls; efaroxan (icv) plus MOX data were compared to those of MOX alone. * $p < 0.01$; rauwolscine (icv) plus MOX data were compared to those of MOX alone; ** $p < 0.05$.

$n = 5$), from a starting pressure of 27.2 ± 0.8 mm Hg, lowered IOP bilaterally, 4.4 ± 0.5 ($p < 0.001$) and 9.1 ± 0.7 ($p < 0.001$) mm Hg, at 30 and 120 minutes, respectively, as compared to saline controls (FIG. 4). At all doses tested, UK had no observable effect on PD (data not shown) and produced no observable change in rabbit behavior. Rauwolscine (3.0 μg icv, $n = 5$), given 30 minutes prior to UK, inhibited UK-induced ocular hypotension by 54.6% at 30 minutes ($p < 0.001$) and by 95% at 120 minutes ($p < 0.001$) (FIG. 4). Unlike rauwolscine, efaroxan (3.0 μg, 20 μl icv, $n = 5$), given 30 minutes prior to UK (0.9 μg), had no effect on UK-induced ocular hypotension (FIG. 4). Neither saline (20 μl icv, $n = 4$) efaroxan (3 and 10 μg, 20 μl icv, $n = 4$), nor rauwolscine (3.0 μg, 20 μl icv, $n = 4$) alone produced any significant change in IOP (FIG. 1).

Third Ventricle

Single doses of MOX, OXY, UK, efaroxan, and rauwolscine were used in experiments to assess responses of drugs administered into the third ventricle, using IOP and PD as the surrogate endpoints of activity. Moxonidine (0.3 μg, 20 μl, $n = 5$), injected icv into the third ventricle, lowered IOP over a 4-hour period, as measured by applanation pneumatonometry (TABLE 4). From a starting IOP of 31.7 ± 0.5 mm Hg, MOX (0.3 μg) lowered IOP bilaterally, 5.2 ± 0.7 ($p < 0.001$) and 6.7 ± 0.7 ($p < 0.001$) mm Hg at 30 and 120 minutes, respectively, as compared to saline controls (TABLE 4). Intraocular pressure returned to control levels after 4 hours. Moxonidine (0.3 μg) had no effect on PD at this dose (data not shown) and produced no observable change in rabbit behavior. Efaroxan, an I_1-imidazoline receptor antagonist, injected (3.0 μg, 20 μl icv, $n = 4$) 30 minutes prior to icv MOX, inhibited bilateral IOP

TABLE 5. Injections of Oxymetazoline and Antagonists into the Third Ventricle

Time (h)	Saline (20 μl)	OXY (0.9 μg)	EFAR (3 μg) + OXY	RAUW (3 μg) + OXY
–0.5	0	0	0	0
0	0	0	0	0
0.5	–0.91 ± 0.6	–3.2 ± 0.4	–0.7 ± 0.7*	–3.9 ± 0.5
1.0	–0.93 ± 0.9	–6.5 ± 0.7	–0.1 ± 0.6*	–4.9 ± 0.5**
2.0	–1.09 ± 0.7	–7.6 ± 0.6	0.7 ± 0.7*	–4.1 ± 0.5**
3.0	–1.13 ± 0.7	–6.9 ± 0.7	0.8 ± 0.6*	–3.3 ± 0.7**
4.0	–0.45 ± 1.1	–5.8 ± 0.6	2.1 ± 0.7*	–3.4 ± 0.8**
5.0		–5.7 ± 0.6	2.3 ± 0.7*	–1.3 ± 0.4**
6.0		–2.8 ± 0.5		

[a] Effect of icv OXY injected into the third ventricle on intraocular pressure (IOP) with and without efaroxan (icv) and rauwolscine (icv), as compared to saline (icv) controls (n = 4 for all groups; values represent mean ± SE). Starting IOPs for treatment groups were: OXY alone, 30.2 ± 1.3 mm Hg; efaroxan (EFAR) plus OXY, 27.9 ± 0.9 mm Hg; and rauwolscine (RAUW) plus OXY, 29.1 ± 0.5 mm Hg. Individual significance levels were determined by Student's t test: OXY was compared to saline controls; efaroxan (icv) plus OXY data were compared to those of OXY alone; *p <0.001; rauwolscine (icv) plus OXY data were compared to those of OXY alone; **p <0.05.

reductions at 30 and 120 minutes by 54 and 100% (p <0.001), respectively (TABLE 4). Injection of rauwolscine (3.0 μg, 20 μl icv, n = 4) 30 minutes prior to MOX, inhibited bilateral reductions at 30 and 120 minutes by 27.3 and 56.7% (p <0.05), respectively (TABLE 4).

Oxymetazoline (0.9 μg, 20 μl icv, n = 5), from a starting pressure of 30.2 ± 1.3 mm Hg, lowered IOP bilaterally, 3.2 ± 0.4 (p <0.001) and 7.6 ± 0.6 (p <0.001) mm Hg, at 30 and 120 minutes, respectively, as compared to saline controls (TABLE 5). At a dose of 0.9 μg icv it had no observable effect on PD (data not shown) and produced no observable change in rabbit behavior. Efaroxan (3.0 μg icv, n = 5), given 30 minutes prior to OXY, inhibited OXY-induced ocular hypotension by 100% at 30 minutes (p <0.001) and by 100% at 120 minutes (p <0.001) (TABLE 9). Injection of rauwolscine (3.0 μg, 20 μl icv, n = 4) 30 minutes prior to OXY inhibited bilateral reductions at 30 and 120 minutes by 0 and 46.5% (p <0.05), respectively (TABLE 5).

UK (0.9 μg, 20 μl icv, n = 5), from a starting pressure of 31.7 ± 0.5 mm Hg, lowered IOP bilaterally, 4.8 ± 0.9 (p <0.001) and 6.2 ± 1.7 (p <0.001) mm Hg, at 30 and 120 minutes, respectively, as compared to saline controls (TABLE 6). UK (0.9 μg icv) had no observable effect on PD at this dose (data not shown) and produced no observable change in rabbit behavior. Rauwolscine (3.0 μg icv, n = 5), given 30 minutes prior to UK, inhibited UK-induced ocular hypotension by 55% at 30 minutes (p <0.001) and by 100% at 120 minutes (p <0.001) (TABLE 6). Efaroxan (3.0 μg, 20 μl icv, n = 5) given 30 minutes prior to UK (0.9 μg), had no effect on UK-induced ocular hypotension (TABLE 6). Neither saline (20 μl icv, n = 5), efaroxan

TABLE 6. Injections of UK-14,304 and Antagonists into the Third Ventricle

Time (h)	Saline (20 μl)	UK (0.9 μg)	EFAR (3 μg) + UK	RAUW (3 μg) + UK
−0.5	0	0	0	0
0	0	0	0	0
0.5	−0.9 ± 0.6	−4.8 ± 0.9*	−3.6 ± 1.9	−2.2 ± 1.3*
1.0	−0.9 ± 1	−6.6 ± 0.7*	−5.3 ± 1.7	1.6 ± 1*
2.0	−1.1 ± 0.7	−6.1 ± 1.7*	−6.4 ± 1.3	0.3 ± 1*
3.0	−1.2 ± 0.8	−3.1 ± 1.2	−5.3 ± 0.9	0.87 ± 1.6*
4.0	−0.5 ± 1.1	−3.3 ± 1*	−4.3 ± 2.2	−1.3 ± 1.2

[a] Effect of icv UK-14,304 injected into the third ventricle on intraocular pressure (IOP) with and without icv rauwolscine and efaroxan ($n = 4$ for all groups; values represent mean ± SE). Starting IOPs for treatment groups were: UK-14,304 alone, 31.7 ± 0.5 mm Hg; efaroxan (EFAR) plus UK, 31.6 ± 0.8 mm Hg; and rauwolscine (RAUW) plus UK, 32.0 ± 0.7 mm Hg. Individual significance levels were determined by Student's t test: UK was compared to saline controls; rauwolscine (icv) plus UK data were compared to those of UK alone; efaroxan (icv) plus UK data were compared to those of UK alone; $*p < 0.001$.

(3.0 μg, 20 μl icv, $n = 5$), nor rauwolscine (3.0 μg, 20 μl icv, $n = 5$) alone, produced any significant change in IOP (FIG. 1).

Rostral Ventrolateral Medulla

Single doses of MOX, OXY, and UK (0.9 μg/1.0 μl) were unilaterally injected in the RVLM region and possible changes in IOP were evaluated (FIG. 5). From a starting pressure of 29.0 ± 0.9 mm Hg ($n = 4$), MOX lowered IOP, −3.8 ± 0.4 ($p < 0.05$) mm Hg at 30 minutes and −6.9 ± 1.3 ($p < 0.001$) mm Hg at 120 minutes. The I_1 antagonist efaroxan, injected unilaterally into the RVLM region, inhibited the MOX-induced hypotensive response (100% at 30 minutes and 60% at 120 minutes) more than did the α_2-antagonist rauwolscine (1% at 30 minutes and 31.4% at 120 minutes) (FIG. 5). There was no observable change in PD as a result of MOX (0.9 μg) injection into the RVLM region.

From a starting pressure of 32.4 ± 1.6 mm Hg ($n = 4$), OXY lowered IOP, 4.2 ± 0.0 ($p < 0.001$) mm Hg at 30 minutes and 6.7 ± 0.8 ($p < 0.001$) mm Hg at 120 minutes (TABLE 7). The I_1-antagonist efaroxan, injected unilaterally into the RVLM region, inhibited the OXY-induced hypotensive response (86% at 30 minutes and 65% at 120 minutes) (TABLE 7). There was no observable change in PD as a result of OXY (0.9 μg) injection into the RVLM region.

When UK was injected into the RVLM region, IOP was not lowered significantly (FIG. 5). From an initial IOP reading of 33.8 ± 0.4 mm Hg, the change following UK was only −0.9 mm Hg at 30 minutes and only −1.8 mm Hg at 120 minutes. Neither saline (1.0 μl, $n = 4$), efaroxan (3.0 μg, 20 μl icv, $n = 5$), nor rauwolscine

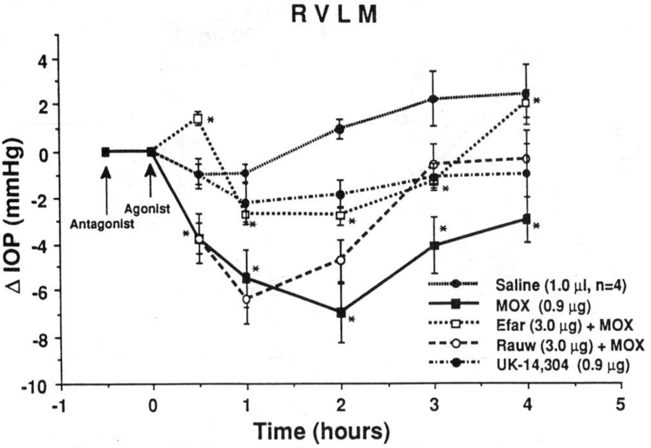

FIGURE 5. Effect of unilateral injection of (1) MOX into the RVLM with and without efaroxan and rauwolscine, and (2) UK into the RVLM alone as compared to saline controls (n = 4 for all groups; values represent mean ± SE). Starting IOPs for treatment groups were: MOX alone, 29.0 ± 0.9 mm Hg; efaroxan plus MOX, 27.8 ± 1.3 mm Hg; rauwolscine plus MOX, 29.9 ± 0.5 mm Hg; UK alone, 33.8 ± 0.4 mm Hg. Individual significance levels were determined by Student's t test: MOX was compared to saline controls; efaroxan plus MOX data were compared to those of MOX alone; rauwolscine plus MOX data were compared to those of MOX alone; *p <0.001.

TABLE 7. Injections of Oxymetazoline and Efaroxan into the Rostral Ventrolateral Medulla

Time (h)	Saline (1 μl)	OXY (0.9 μg)	EFAR (3 μg) + OXY
−0.5	0	0	0
0	0	0	0
0.5	−1.0 ± 0.5	−4.2 ± 0.9*	−0.55 ± 0.9*
1.0	−0.9 ± 0.4	−6.7 ± 1*	−3.3 ± 0.8*
2.0	−0.93 ± 0.4	−6.7 ± 0.8*	−2.3 ± 0.8*
3.0	2.2 ± 1.2	−7.5 ± 0.8*	−2.5 ± 0.9*
4.0	2.4 ± 1.3	−6.1 ± 1.7	−0.8 ± 0.8*
5.0		−4.9 ± 1.6*	

[a] Effect of unilateral injection of (1) oxymetazoline (OXY) into the rostral ventrolateral medulla with and without efaroxan (n = 4 for all groups; values represent mean ± SE). Starting intraocular pressures for treatment groups were: OXY alone, 28.9 ± 0.9 mm Hg; efaroxan (EFAR) plus OXY, 27.8 ± 1.3 mm Hg. Individual significance levels were determined by Student's t test: OXY was compared to saline control data; efaroxan plus OXY data were compared to those of OXY alone; *p < 0.001.

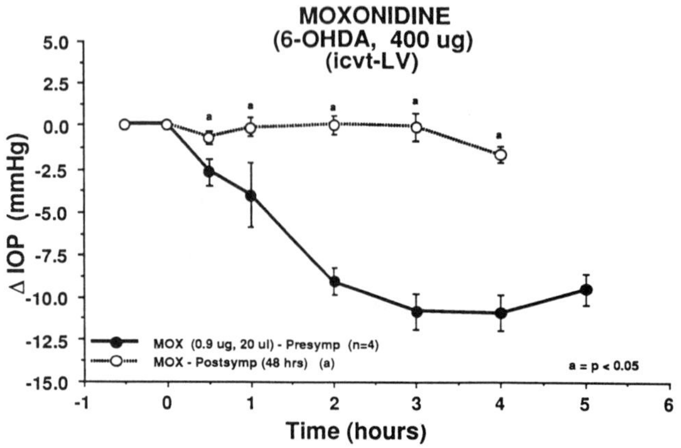

FIGURE 6. Effect of icv MOX injected into the lateral ventricle on IOP with and without 6-OH-DA (400 μg, 40 μl icv) ($n = 4$ for all groups; values represent mean ± SE). Starting IOPs for treatment groups were: MOX alone, 28.8 ± 0.7 mm Hg; 6-OH-DA plus MOX, 23.8 ± 0.8 mm Hg; Individual significance levels were determined by Student's t test: MOX was compared to MOX plus 6-OH-DA; *p <0.001.

(3.0 μg, 20 μl icv, $n = 5$) alone produced any significant change in IOP when injected unilaterally into the RVLM region.

Chemical Sympathectomy

Some animals were pretreated with 400 μg/40 μl of 6-hydroxydopamine (6-OH-DA) injected into the lateral ventricle 48 hours prior to administration of MOX and OXY to determine the site of action of these α_2/I_1 agonists (FIGS. 6 and 7, respectively). Pretreatment with 6-OH-DA lowered the initial starting IOP of the rabbits 5 mm Hg, but completely eliminated MOX-induced ocular hypotension (FIG. 6). In another group of rabbits, pretreatment with 6-OH-DA lowered the initial starting IOP of the rabbits 3 mm Hg, but eliminated only about 50% of OXY-induced ocular hypotension (FIG. 7).

DISCUSSION

The objectives of this study were to assess the role of centrally located α_2-adrenoceptors and I_1-imidazoline-preferring receptors in the mediation of changes in intraocular pressure. To explore a possible central site of action of α_2- and I_1-imidazoline receptors in mediating ocular hypotension, drugs (MOX, OXY, and UK) were administered topically to the cornea of one eye, given icv into the lateral and third ventricles, and microinjected into the RVLM region of the brain stem. MOX, OXY, and UK, administered unilaterally and topically to the right eye, or injected icv into the lateral or third ventricle, significantly lowered IOP in conscious New

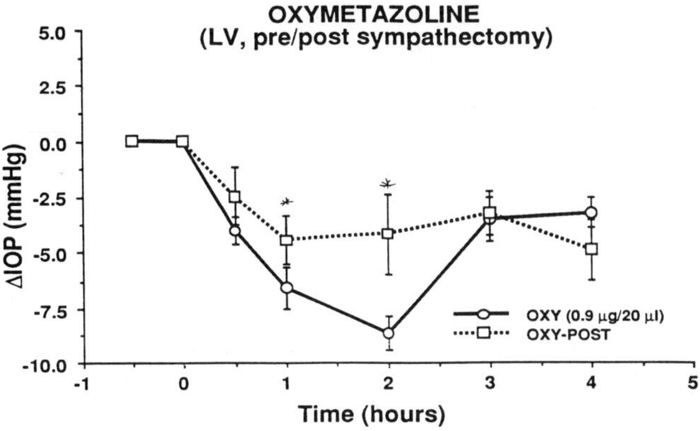

FIGURE 7. Effect of icv OXY injected into the lateral ventricle on IOP with and without 6-OH-DA (400 μg, 40 μl icv) ($n = 4$ for all groups; values represent mean ± SE). Starting IOPs for treatment groups were: OXY alone, 25.4 ± 0.7 mm Hg; 6-OH-DA plus OXY, 22.3 ± 0.4 mm Hg. Individual significance levels were determined by Student's t test: OXY was compared to OXY plus 6-OH-DA; $^*p < 0.001$.

Zealand white male rabbits. Topical MOX-, OXY-, and UK-induced ocular hypotension was significantly inhibited by an antagonist injected into the anterior lateral cerebral ventricle. The most inhibited response of both eyes was the contralateral eye decrease in IOP. Efaroxan, an I_1-imidazoline antagonist, was significantly more potent in inhibiting both topical and icv MOX- and OXY-induced ocular hypotension, whereas rauwolscine, an α_2-antagonist, was significantly more potent in inhibiting both topical and icv UK-induced ocular hypotension. MOX and OXY, but not UK, significantly lowered IOP when microinjected unilaterally into the RVLM region of the brain stem. Efaroxan was significantly more effective than rauwolscine in inhibiting both MOX- and OXY-induced ocular hypotension when pretreated into the same site. These results are consistent with a role for I_1 receptors in MOX- and OXY-induced ocular hypotension. Chemical sympathectomy completely eliminated MOX-induced ocular hypotension, but eliminated only 50% of OXY-induced ocular hypotension.

Moxonidine is a clinically utilized second generation antihypertensive agent that was originally classified as an α_2-adrenergic agonist.[53] The recent findings of several laboratories have questioned the limitations of this classification, however, and have proposed the idea that MOX may activate an IPR[12,21,22,24,26,54,55] rather than an α_2 receptor. This IPR has been subdivided into subtypes, I_1 and I_2,[23] as suggested by binding experiments[21,22,26] and by functional alterations in cardiovascular parameters.[55] On the basis of recent binding studies by Ernsberger and coworkers, MOX was shown to be one of the most selective IPR (I_1) agonists available at this time. To determine the actions of MOX on cardiovascular parameters, workers injected MOX into the brain stem area called the RVLM[55] in different animal species. Moxonidine (2-10 nmol), an imidazoline, significantly lowered blood pressure in rabbits,[55] while α-methylnorepinephrine (0.1-10 μg/kg), a catecholamine, injected into rats was unable to affect blood pressure in this brain stem region and clonidine produced dose-dependent hypotension.[20] Therefore, not only MOX but also a number of imidazoline structured compounds, including clonidine and rilmenidine, another second generation antihypertensive agent, have been shown to preferentially induce systemic hypotension when injected into the RVLM. Moreover, nonimidazoline compounds traditionally classified as α_2-agonists, such as α-methylnorepinephrine[20] and guanabenz,[26] are much less effective in reducing systemic blood pressure when administered into the RVLM. Oxymetazoline, an oxazoline classified as an α_2-receptor agonist, is the active ingredient in a number of over-the-counter nasal and ocular decongestant preparations. Oxymetazoline also lowered IOP when applied topically to the eyes of rabbits.[14] UK-14,304-18 (UK; brimonidine), classified as a relatively selective α_2-adrenergic agonist,[50] lowered IOP bilaterally when applied unilaterally to the eyes of rabbits[10,48] or injected icv into the anterior lateral ventricle of rabbits.[48]

To ascertain the receptor type(s) and site(s) of action that mediate(s) the ocular hypotension induced by topically and centrally administered MOX, OXY, and UK, relatively selective antagonists were used to inhibit the effects of these agonists. Efaroxan, described as a relatively selective I_1-imidazoline-preferring antagonist,[26] and rauwolscine, described as a relatively selective α_2-adrenoceptor antagonist,[51] were used to evaluate their effects on MOX, OXY, and UK. Efaroxan injected icv significantly inhibited ocular hypotension induced by unilateral topical MOX and OXY and icv MOX and OXY, whereas rauwolscine was a much less effective antagonist when injected icv into the anterior lateral ventricle prior to MOX or OXY. The decrease in IOP induced by both unilateral topical MOX and OXY and icv MOX and OXY was inhibited markedly by pretreatment with icv efaroxan, whereas rauwolscine only inhibited topical and icv MOX to a moderate degree. Rauwolscine (icv) did not inhibit either topical or icv OXY. The ocular hypotension induced by UK administered unilaterally and topically was significantly inhibited by icv pretreatment with rauwolscine, but not efaroxan. Intraocular pressure responses to UK injected icv into the anterior lateral or third ventricle was significantly inhibited by icv rauwolscine. By contrast, efaroxan was relatively ineffective against icv UK. The marked inhibition by icv efaroxan of the actions of centrally applied MOX and OXY on IOP indicates that central I_1-imidazoline receptors function to mediate ocular hypotension. The ability of icv rauwolscine to markedly inhibit the IOP effects of UK demonstrates that central α_2-receptors also play a role in ocular hypotension. It was previously suggested that the sympathetic nervous system be intact for MOX to induce ocular hypotension when applied topically to the eyes of rabbits[13]; however, this former study was not designed to identify central site(s) of action for MOX and therefore did not use centrally mediated responses to assess the activity of MOX. The results of chemical sympathectomy indicate that MOX may act through a presynaptic receptor-mediated mechanism, but that the effects of OXY may be due to actions at both pre- and postsynaptic receptor-mediated mechanisms.

To explore a possible central site of action of α_2- and I_1-imidazoline receptors in mediating ocular hypotension, MOX, OXY, and UK in a smaller volume (1 μl) were injected unilaterally into the RVLM region of the brain stem. The RVLM is a critical component of an area in the brain stem called the medullary intermediate reticular zone (MIRZ). The MIRZ contains premotor autonomic and respiratory neurons as well as interneurons.[56] The RVLM is a critical region of the MIRZ which contains vasomotor,[57–59] cardiovagal,[60] and respiratory neurons[61] that form a cardiorespiratory network which controls input to spinal, respiratory, and preganglionic motoneurons.[62] The MIRZ and its associated RVLM are components of the central autonomic network (CAN).[63] The CAN is a network that receives and integrates viscerosensory, humoral, and exteroceptive information and, in turn, controls and coordinates preganglionic, neurosecretory, and respiratory output. The main output of the CAN is mediated through preganglionic sympathetic and parasympathetic neurons. (For a review see ref. 63.) Although a number of midbrain regions in the hypothalamus and thalamus have been shown to induce alterations in IOP upon electrical stimulation,[40] there is a paucity of reports describing modulation of IOP by specific sites in the brain stem. MOX and OXY injected unilaterally into the RVLM region significantly lowered IOP. This MOX- and OXY-induced ocular hypotension was significantly and preferentially inhibited by unilateral pretreatment with efaroxan into the RVLM. Efaroxan inhibited MOX-induced ocular hypotension significantly (approximately 60%), whereas rauwolscine inhibited the MOX-induced decrease in IOP by a modest amount (approximately 30%). Efaroxan inhibited OXY-induced ocular hypotension significantly (approximately 70%). UK injected unilaterally into the RVLM did not induce a change in IOP.

The inhibition of efaroxan, injected into the LV, 3V, and RVLM, of MOX- and OXY-induced ocular hypotension implicates the involvement of a centrally mediated I_1-imidazoline receptor mechanism in the ocular action of MOX and OXY. The inhibition of UK-induced ocular hypotension by rauwolscine, when injected into the third ventricle demonstrates that central α_2-adrenoceptors also play a role in ocular hydrodynamics. The inability of UK to induce a decrease in IOP when injected into the RVLM region suggests that this region contains a predominance of IPRs rather than α_2-receptors. Additionally, however, recent data from our laboratory[12,15] suggested that icv MOX can also induce ocular hypotension via an α_2-adrenergic mechanism. For example, the relatively selective α_2-adrenergic receptor antagonists L-659,066[64] and L-657,743[65] significantly inhibited the MOX-induced decrease in IOP when injected into the anterior lateral ventricle, whereas rauwolscine again only partially inhibited the icv MOX-induced decrease in IOP. It is not known, however, if L-659,066 and L-657,743 also have IPR antagonist activity. This latter possibility is currently being explored.

CONCLUSION

In conclusion, the findings of this study indicate that (1) activation of central α_2-adrenoceptors and I_1 imidazoline receptors influences aqueous humor dynamics; (2) MOX and OXY act primarily via I_1 imidazoline receptors to lower IOP when applied topically to the eye and when injected into the LV, 3V, or RVLM; (3) UK acts primarily via α_2-receptors to lower IOP when applied topically to the eye and when injected into the LV or 3V, but it was ineffective in changing IOP when injected into the RVLM region; (4) I_1 imidazoline receptors are present in the RVLM region as demonstrated by the ability of MOX and OXY, but not UK-14,304-18, to induce ocular hypotension and by antagonism of MOX and OXY effects by efaroxan; and

(5) the receptors mediating the effects of MOX may be located presynaptically, whereas OXY may be acting through both pre- and postsynaptic receptors to lower ocular pressure.

SUMMARY

These experiments sought to: (1) determine if α_2/I_1 agonists that are topically active on the eye have similar effects on intraocular pressure when applied to the CNS and (2) ascertain whether these agents lower IOP, in part, via central α_2 receptors and/or imidazoline (I_1) receptors.

New Zealand White rabbits were fitted with chronic indwelling stainless-steel guide cannulas in several brain regions including the lateral ventricle, third ventricle (3V), or medullary intermediate reticular zone. Animals were allowed 5 days' recovery time prior to experiments measuring the effects of drugs on IOP via applanation pneumatonometry. Some animals were also pretreated with 400 μg of 6-hydroxydopamine injected into the lateral ventricle to determine the site of action of these α_2/I_1 agonists.

In initial experiments involving microinjection into the lateral ventricle, UK-14,304-18 evoked ocular hypotension that was inhibited by the α_2-antagonist rauwolscine but not by the I_1-receptor antagonist efaroxan. Conversely, moxonidine and oxymetazoline were preferentially inhibited by efaroxan rather than by rauwolscine. Subsequently, experiments have shown that moxonidine and oxymetazoline, but not UK-14,304-18 will lower intraocular pressure when microinjected into the medullary intermediate reticular zone region and that efaroxan, but not rauwolscine, will inhibit ocular hypotension induced by moxonidine and oxymetazoline. Pretreatment with 6-hydroxydopamine (48 hours) completely eliminated the ocular hypotension induced by moxonidine.

These preliminary data demonstrate that α_2- and I_1-receptors in the brain mediate ocular hypotension induced by UK-14,304-18 and moxonidine/oxymetazoline, respectively. Moreover, the medullary intermediate reticular zone area of the brain stem is the probable presynaptic site mediating ocular hypotension induced by moxonidine and oxymetazoline.

REFERENCES

1. Wilson, M. R. 1992. Primary open-angle glaucoma: Magnitude of the problem in the United States. J. Glaucoma **1:** 64-68.
2. Sommer, A., J. M. Tielsch, J. Katz, H. A. Quigley, J. D. Gottsch, J. Javitt, K. Singh & Baltimore Eye Survey Research Group. 1991. Relationship between intraocular pressure and primary open angle glaucoma among white and black Americans. Arch. Opthalmol. **109:** 1090-1095.
3. Quigley, H. A. 1993. Open angle glaucoma. Medical progress. N. Engl. J. Med. **328:** 1097-1106.
4. Epstein, D. L., J. H. Krug, Jr., E. Hertzmark, L. L. Remis & D. J. Edelstein. 1989. A long-term clinical trial of timolol therapy versus no treatment in the management of glaucoma suspects. Ophthalmology **96:** 1460-1467.
5. Kass, M. A., M. O. Gordon, M. R. Hoff, *et al.* 1989. Topical timolol administration reduced the incidence of glaucomatous damage in ocular hypertensive individuals: A

randomized, double-masked, long-term clinical trial. Arch. Ophthalmol. **107:** 1590-1598.
6. Schulzer, M., S. M. Drance & G. R. Douglas. 1991. A comparison of treated and untreated glaucoma suspects. Ophthalmology **98:** 301-307.
7. Potter, D. E., C. E. Crosson, A. J. Heath & M. J. Ogidigben. 1990. Alpha-2 and DA-2 agonists as antiglaucoma agents. Comparative pharmacology and clinical potential. J. Ocular Pharmacol. **6:** 252-257.
8. Makabe, R. 1966. Ophthalmologische Untersuckunger unit Dichlorophenylaminoimidazolin. Dtsch. Med. Woschenschr. **91:** 1686-1688.
9. Robin, A. L., I. P. Pollack, B. House & C. Eager. 1987. Effects of AL02145 on ocular pressure following argon laser trabeculoplasty Arch. Ophthalmol. **105:**646-650.
10. Burke, J. A. & D. E. Potter. 1986. Ocular effects of a relatively selective α_2 agonist (UK 14,304-18) in cats, rabbits and monkeys. Curr. Eye Res. **5:** 665-676.
11. Campbell, W. R. & D. E. Potter. 1992. The central effects of moxonidine on intraocular pressure and its antagonism by L-659,066, in the rabbit. American College of Neuropsychopharmacology, p. 143.
12. Campbell, W. R. & D. E. Potter. 1994. The central effects of moxonidine on intraocular pressure and its antagonism by L-659,066 and L-657,743 in the rabbit. J. Prog. Neuro-Psychopharmacol. Biol. Psy **18:** 1051-1061.
13. Ogidigben, M. J. & D. E. Potter. 1991. Moxonidine: Effects on intraocular pressure and pupil diameter and contraction of the nictitating membrane. Invest. Ophthalmol. Vis. Sci. (ARVO Suppl.) **32:** 867.
14. Wang, R. F., P. Y. Lee, T. T. Taniguchi, B. Becker, S. M. Podos, J. B. Serle & T. W. Mittag. 1993. Effect of oxymetazoline on aqueous humor dynamics and ocular blood flow in monkeys and rabbits. Arch. Ophthalmol. **111:** 535-538.
15. Campbell, W. R., C. J. Wilson & D. E. Potter. 1993. Ocular hypotensive effects of moxonidine and oxymetazoline: Involvement of central imidazoline (I1) receptors. Invest. Ophthalmol. Vis. Sci. (ARVO Suppl.) **34:** 928.
16. Allen, R. C. & M. E. Langham. 1976. The intraocular pressure response of conscious rabbits to clonidine. Invest. Ophthalmol. **15:** 815-823.
17. Campbell, W. R. & D. E. Potter. 1993. Potential role of imidazoline (I_1) receptors in modulating aqueous humor dynamics. Int. Symp. Ocular Pharm., Novi, Michigan, #53.
18. Hodapp, E., A. E. Kolker, M. A. Kass, I. Goldberg, B. Becker & M. Gordon. 1981. The effect of topical clonidine on intraocular pressure. Arch. Ophthalmol. **99:** 1208-1212.
19. Chacko, D. M. & C. B. Camras. 1994. The potential of α_2-adrenergic agonists in the medical treatment of glaucoma. Curr. Opin. Ophthalmol. **5:** 76-84.
20. Bousquet, P., J. Feldman & J. Schwartz. 1984. Central cardiovascular effects of alpha adrenergic drugs: Differences between catecholamines and imidazolines. J. Pharmacol. Exp. Ther. **230:** 232-236.
21. Ernsberger, P., R. Giuliano, R. N. Willette, A. R. Granata & D. J. Reis. 1988. Hypotensive action of clonidine analogues correlates with binding affinity at imidazoline and not alpha-2-adrenergic receptors in the rostral ventrolateral medulla. J. Hypertens. **6:** S554-557.
22. Ernsberger, P., T. H. Damon, L. M. Graff, S. G. Schafer & M. O. Christen. 1993. Moxonidine, a centrally acting antihypertensive agent, is a selective ligand for I1-imidazoline sites. J. Pharmacol. Exp. Ther. **264:** 172-182.
23. Reis, D. J., S. Regunathan, H. Wang, D. L. Feinstein & M. P. Meeley. 1992. Imidazoline receptors in the nervous system. Fundam. Clin. Pharmacol. **6:** 23S-29S.
24. Lehman, J., E. Koenig-Berard & P. Vitou. 1989. The imidazoline-preferring receptor. Life Sci. **45:** 1609-1615.
25. Karppanen, H., I. Paakkari & P. Paakkari. 1977. Studies on the nature of cerebral receptors mediating the hypotensive effects of clonidine in rats. *In* Histamine Receptors. T. O. Yellin, Ed.: 25. S.P. Medical and Scientific. New York.

26. Ernsberger, P., K. L. Westbrooks, M. O. Christen & S. G. Schafer. 1992. A second generation of centrally acting antihypertensive agents act on putative I_1-imidazoline receptors. J. Cardiovasc. Pharmacol. **20:** S1-S10.
27. Gregory, D. S., D. G. Aviado & M. L. Sears. 1985. Cervical ganglionectomy alters the circadian rhythm of intraocular pressure in New Zealand rabbits. Curr. Eye Res. **4:** 1273-1279.
28. Rowland, J. M., D. E. Potter & R. J. Reiter. 1981. Circadian rhythm in an intraocular pressure model. Curr. Eye Res. **1:** 169-173.
29. Braslow, R. A. & D. S. Gregory. 1987. Adrenergic decentralization modifies the circadian rhythm of intraocular pressure. Invest. Ophthalmol. Vis. Sci. **28:** 1730.
30. Miller, S. J. H. 1953. Stellate ganglion block in glaucoma. Br. J. Ophthalmol. **37:** 70.
31. Endo, Y. 1961. Fluctuation of intraocular pressure by blockage of the stellar ganglion. Acta Soc. Ophthalmol. Jpn **70:** 926.
32. Waitzman, M. B. 1971. Hypothalmus and ocular pressure. Surv. Ophthalmol. **16:** 161.
33. Schmerl, E. & B. Steinberg. 1949. The role of the ciliary and superior cervical ganglion in ocular tension. Am. J. Ophthalmol. **32:** 947.
34. Schmerl, E. & B. Steinberg. 1950. Separation of diencephalic centers concerned with pupillary motility and ocular tension. Am. J. Ophthalmol. **33:** 1379.
35. Nagai, M., T. Ban & T. Kurotsu. 1951. Studies on the changes of intraocular pressure induced by electrical stimulation of the hypothalamus. Med. J. Osaka Univ. **2:** 87.
36. von Sallmann, L. & O. Lowenstein. 1955. Responses of intraocular pressure, blood pressure and cutaneous vessels to electric stimulation in the diencephalon. Am. J. Ophthalmol. **39:** 11.
37. Gloster, J. & D. P. Greaves. 1957. Effect of diencephalic stimulation upon intra-ocular pressure. Br. J. Ophthalmol. **41:** 513.
38. Gloster, J. 1960. Responses of the intra-ocular pressure to diencephalic stimulation. Br. J. Ophthalmol. **44:** 649.
39. Stone, R. A. & Y. Kuwayama. 1989. The nervous system and intraocular pressure. *In* The Glaucomas. R. Ritch, M. B. Shields & T. Krupin, Eds.: 257-279. C. V. Mosby. St. Louis.
40. von Sallmann, L. 1959. The role of the central nervous system in the regulation of the intraocular pressure. Doc. Ophthalmol. **13:** 93.
41. Krupin, T., S. M. Podos & B. Becker. 1973. Alteration of intraocular pressure after third ventricle injections of osmotic agents. Am. J. Ophthalmol. **76:** 948-952.
42. Krupin, T., C. J. Oestrich, S. M. Podos & B. Becker. 1976. Increased intraocular pressure after third ventricle injections of prostaglandins E1 and arachidonic acid. Am. J. Ophthalmol. **81:** 346-350.
43. Krupin, T. *et al.* 1978. Increased intraocular pressure and hypothermia following injection of calcium into the rabbit third ventricle. Exp. Eye Res. **27:** 129-134.
44. Krupin, T. *et al.* 1982. Increased intraocular pressure and hypothermia following injection of substance P into the rabbit third ventricle. Exp. Eye Res. **34:** 319-324.
45. Krupin, T. *et al.* 1984. Central effects of thyrotropin-releasing hormone and arginine vasopressin on intraocular pressure in rabbits. Invest. Ophthalmol. Vis. Sci. **25:** 932-937.
46. Liu, J. H. K. & A. H. Neufeld. 1985. Study of central regulation of intraocular pressure using ventriculocisternal perfusion. Invest. Ophthalmol. Vis. Sci. **26:** 136-143.
47. Bylund, D. B. C. Ray-Prenger & T. J. Murphy. 1988. Alpha-2A and alpha-2B adrenergic receptor subtypes: Antagonists binding in tissues and cell lines containing only one subtype. J. Pharmacol. Exp. Ther. **245:** 600-607.
48. Campbell, W. R., S. Long, S. Strothers & D. E. Potter. 1993. Centrally-induced ocular hypotension in the rabbit: Involvement of alpha-2 and imidazoline receptor subtypes. Soc. Neurosci. Abstr. **19:** 1790.

49. Potter, D. E., M. J. Ogidigben & T. C. Chu. 1992. Oxymetazoline: An atypical alpha-2 agonist with ocular hypotensive activity. Invest. Ophthalmol. Vis. Sci. (ARVO Suppl.) **33:** 1116.
50. Cambridge, D. 1981. UK-14,304-18, a potent and selective α_2-agonist for the characterization of α-adrenoceptor subtype. Eur. J. Pharmacol. **72:** 413-415.
51. Perry, B. D. & D. C. U'Prichard. 1981. [^{3}H]Rauwolscine (α-yohimbine), a specific antagonist radioligand for brain α_2-adrenergic receptors. Eur. J. Pharmacol. **76:** 461-464.
52. Shek, J. W., G. Y. Wen & H. M. Wisniewski. 1986. Atlas of the Rabbit Brain and Spinal Cord. Karger. New York.
53. Armah, I. B. & W. Stenzel. 1981. BE 5895, a new clonidine-type antihypertensive aminopyrimidine derivative. Naunyn-Schmiedeberg's Arch Pharmacol. **316:** R42.
54. Ernsberger, P., R. Giuliano, R. N. Willette & D. J. Reis. 1990. Role of imidazole receptors in the vasodepressor response to clonidine analogs in the rostral ventrolateral medulla. J. Pharmacol. Exp. Ther. **253:**408-418.
55. Haxhiu, M. A., I. Dreshaj, B. Erokwu, S. G. Schafer, M. O. Christen & P. R. Ernsberger. 1992. Vasodepression elicited in hypertensive rats by the selective I_1-imidazoline agonist moxonidine administered into the rostral ventrolateral medulla. J. Cardiovasc. Pharmacol. **20:** S11-S15.
56. Tork, I., D. A. McRitchie, G. C. Rikard-Bell & G. Paxinos. 1990. Autonomic regulatory centers in the medulla oblongata. *In* The Human Nervous System. G. Paxinos, Ed.: 221-259. Academic Press. San Diego.
57. Guyenet, P. G., J. R. Haselton & M.-K. Sun. 1989. Sympathoexcitatory neurons of the rostroventrolateral medulla and the origin of the sympathetic vasomotor tone. Prog. Brain Res. **81:** 105-116.
58. Arango, V., D. A. Ruggiero, J. L. Callaway, M. Anwar, J. J. Mann & D. J. Reis. 1988. Catecholaminergic neurons in the ventrolateral medulla and nucleus of the solitary tract in the human. J. Comp. Neurol. **273:** 224-240.
59. McAllen, R. M. & R. A. L. Dampney. 1990. Vasomotor neurons in the rostral ventrolateral medulla are organized topographically with respect to type of vascular bed but not body region. Neurosci. Lett. **110:** 91-96.
60. Loewy, A. D. & K. M. Spyer. 1990. Vagal preganglionic neurons. *In* Central Regulation of Autonomic Functions. A. D. Loewy & K. M. Spyer, Eds.: 68-87. Oxford University Press. New York.
61. Feldman, J. L. 1986. Neurophysiology of breathing in mammals. *In* Handbook of Physiology. Section 1: The Nervous System. V. B. Mountcastle, F. E. Bloom & S. R. Greiger, Eds.: 463-524. American Physiological Society. Bethesda, MD.
62. Richter, D. W. & K. M. Spyer. 1990. Cardiorespiratory control. *In* Central Regulation of Autonomic Functions. A. D. Loewy & K. M. Spyer, Eds.: 189-207. Oxford University Press. New York.
63. Benarroch, E. E. 1993. The central autonomic network: Functional organization, dysfunction and perspective. Mayo Clin. Proc. **68:** 988-1001.
64. Clineschmidt, B. V., D. J. Pettibone, V. J. Lotti, H. B. Hucker, B. M. Sweeney, D. R. Reis, E. V. Lis, J. R. Huff & J. Vacca. 1988. A peripherally acting alpha-2 adrenoceptor antagonist: L-650,066. J. Pharmacol. Exp. Ther. **245:** 32-40.
65. Pettibone, D. J., B. V. Clineschmidt, V.J. Lotti. J. R. Huff, W. C. Randall, J. Vacca & S. D. Young. 1987. Pharmacological profile of a new potent and specific α2-adrenoceptor antagonist, L-657,743. Naunyn-Schmiedeberg's Arch. Pharmacol. **336:** 169-175.

Chronic Imidazoline Drug Treatment Increases the Immunoreactivity of Glial Fibrillary Acidic Protein in Rat Brain

LSL 60101 as a Novel and Selective Ligand for I_2-Imidazoline Receptors[a]

JESÚS A. GARCÍA-SEVILLA, REGINA ALEMANY, GABRIEL OLMOS, ANGEL MENARGUES,[b] AND ROSEND OBACH[b]

Laboratory of Neuropharmacology
Department of Fundamental Biology and Health Sciences
University of the Balearic Islands
E-07071 Palma de Mallorca, Spain

[b]*Research Department LASA Laboratories*
Barcelona, Spain

The density of I_2-imidazoline receptors (I_2-IR) increases with age in the human brain, and the regional distribution of these sites correlates well with that of the B isoform of the enzyme monoamine oxidase (MAO-B),[1] which is mainly located in serotonergic and glial cells. Cultured cortical astrocytes, but not cortical neurons, are reported to express I_2-IR.[2] These facts suggest that the binding of [^{3}H]idazoxan to nonadrenoceptor sites in the brain is preponderantly, if not exclusively, located on glial cells. Glial fibrillary acidic protein (GFAP) is the main constituent of intermediate filaments of astrocytes[3] which accumulates in response to brain injury. Idazoxan was shown to induce a marked concentration-dependent (1-100 μM) increase in mRNA levels for GFAP in cultured rat cortical astrocytes, suggesting that occupancy of astrocytic I_2-IR may have some functional activity in regulating the expression of this protein.[2]

Recently, the novel and selective I_2-IR ligand LSL 60101 (2-[2-benzofuranyl]imidazole hydrochloride) was shown to induce hyperphagia in satiated rats involving activity only at I_2-IR.[4] The present study was designed to assess whether chronic treatment with LSL 60101 and other imidazoline drugs, selective for I_1- or I_2-IR, could modulate *in vivo* the density of GFAP in rat brain and also to relate possible changes in GFAP levels with parallel increases in the density of brain I_2-IR.

[a]This study was supported by grants DGICYT (PM 91-0069), FIS (93-0641), and CDTI-S.A. LASA Laboratories (92-0198). R.A. was supported by a fellowship from FIS (Fondo de Investigaciones Sanitarias, Spain).

Male Sprague-Dawley rats (250-300 g) were used. In chronic treatment, animals received intraperitoneally every 12 hours for 7 days either 0.9% saline vehicle, LSL 60101 (10 mg/kg), cirazoline (1 mg/kg), idazoxan (10 mg/kg), RX821002 (2-methoxy idazoxan, 10 mg/kg), yohimbine (10 mg/kg), or efaroxan (10 mg/kg). The time course for the effect of LSL 60101 (10 mg/kg) was studied at days 1 (acute), 3, 7, 14, and 21 after treatment. In both acute and chronic treatments, rats were killed 24 hours after the last injection. Solubilized GFAP (0.5% Triton X-100) was submitted to polyacrylamide gel electrophoresis (SDS-PAGE), transferred to nitrocellulose membranes (immunoblotting, western blotting), incubated with the primary antibody, rabbit anti-cow GFAP (1:5000 dilution, DAKO A/S, Denmark), and then with the secondary antibody, horseradish peroxidase-linked donkey anti-rabbit IgG. Immunoreactivity was detected with the Enhanced Chemiluminescence (ECL) Western Blot Detection system (Amersham International plc, UK). Films were scanned in the image analyzer Bio Image (Millipore, Ann Arbor, Michigan). After scanning, standard curves of increasing GFAP content were constructed using samples from saline-treated rats. In these curves, the total protein loaded in, at least, four points of increasing protein content were plotted against integrated optical density (IOD). The percentage of GFAP content of a given sample was referred to the standard curve. Radioligand binding assays for [^{3}H]idazoxan to I_2-IR in membrane fractions from rat parietooccipital cortex were carried out as previously described.[5]

In receptor binding studies, LSL 60101 had a lower affinity than cirazoline and idazoxan for brain I_2-IR (K_i values: 350, 3.4, and 14 nM, respectively), but a much higher α_2R/I_2-IR selectivity (K_i ratio: 280, 90, and 0.8, respectively). Chronic (7-21 days) but not acute (1-3 days) treatment with LSL 60101 (10 mg/kg) induced a significant increase in the levels of GFAP in the cerebral cortex (44-47%). Similar chronic treatment (ip every 12 hours for 7 days) with the mixed I_2-IR/$\alpha_{1/2}$R ligands cirazoline (1 mg/kg) and idazoxan (10 mg/kg) also increased GFAP immunoreactivity in the cerebral cortex (47 ± 9% and 45 ± 12%, respectively) (FIG. 1). Chronic treatment with selective α_2R antagonists RX821002 and yohimbine (10 mg/kg) did not significantly modify the levels of GFAP. Treatment with the mixed I_1-IR/α_{2B} ligand efaroxan (10 mg/kg) failed to increase GFAP immunoreactivity significantly (FIG. 1).

The same chronic treatments with LSL 60101, cirazoline, and idazoxan but not with RX821002, yohimbine, or efaroxan were also associated with upregulation of I_2-IR in the cerebral cortex (B_{max} values increased by 25-39%) (FIG. 1). A strong correlation ($r = 0.95$; $p < 0.01$) was found between changes in the density of I_2-IR and GFAP immunoreactivity after the various treatments.

Several conclusions can be deduced from the present data: (1) chronic, but not acute treatments with the selective I_2-IR ligand LSL 60101 and other imidazoline drugs that possess affinity for brain I_2-IR are associated with increased levels of the astrocyte marker GFAP; (2) increased GFAP immunoreactivity after chronic treatment with imidazole/imidazoline drugs occurs in parallel with upregulation of the density of I_2-IR in the rat cerebral cortex; this appears to be a specific parallel phenomenon, because also during aging and in Alzheimer's disease both increased GFAP expression and density of I_2-IR in the brain have been described,[1,6–8] and (3) increased GFAP expression and astrocytic activation may function as a protective mechanism in the maintenance of neural function following brain injury and, in this sense, idazoxan

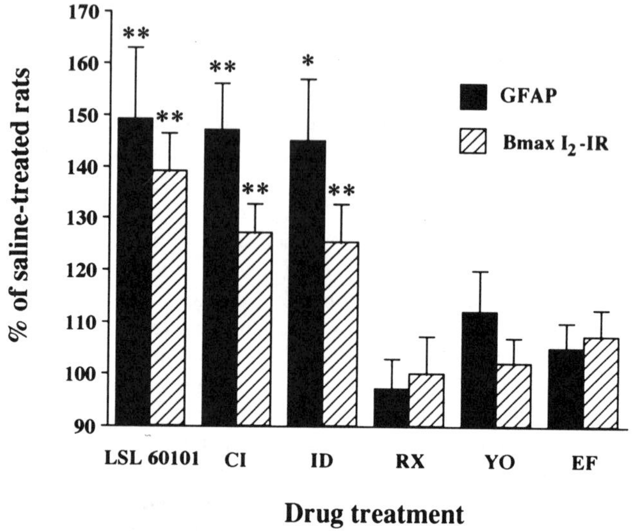

FIGURE 1. Effect of chronic treatment (ip every 12 hours for 7 days) with LSL 60101 (10 mg/kg), cirazoline (CI, 1 mg/kg), idazoxan (ID, 10 mg/kg), RX821002 (RX, 10 mg/kg), yohimbine (YO, 10 mg/kg), or efaroxan (EF, 10 mg/kg) on GFAP immunoreactivity and on the density (B_{max}) of I_2-IR in the rat cerebral cortex. Saline-treated rats were used as controls. Columns represent the mean ± SEM of 3-12 experiments with an animal per experiment. One-way ANOVA followed by a multiple comparison test detected a significant increase in GFAP immunoreactivity and B_{max} for I_2-IR after chronic treatment with LSL 60101, cirazoline, and idazoxan, but not after chronic RX821002, yohimbine, or efaroxan treatments. $*p < 0.01$ and $**p < 0.001$ as compared with saline-treated group.

and rilmenidine have been shown to provide protection from global cerebral ischemia.[9] These findings suggest important roles for the glial I_2-IR.

REFERENCES

1. Sastre, M. & J. A. García-Sevilla. 1993. Opposite age-dependent changes of α_{2A}-adrenoceptors and non-adrenoceptor [^{3}H]idazoxan binding sites (I_2-imidazoline sites) in the human brain: Strong correlation of I_2 with MAO-B sites. J. Neurochem. **61:** 881-889.
2. Regunathan, S., D. L. Feinstein & D. J. Reis. 1993. Expression of non-adrenergic imidazoline sites in rat cerebral cortical astrocytes. J. Neurosci. Res. **34:** 681-688.
3. Eng, L. F. 1985. Glial fibrillary acidic protein (GFAP): The major protein of glial intermediate filaments in differentiated astrocytes. J. Neuroimmunol. **8:** 203-214.
4. Menargues, A., M. Cedo, O. Artiga, R. Obach & J. A. García-Sevilla. 1994. Modulation of food intake by α_2-adrenoceptor antagonists and I_2-imidazoline drugs in rats: LSL 60101 as a novel and selective ligand for I_2-imidazoline sites. Br. J. Pharmacol. **111:** 298P.
5. Olmos, G., A. M. Gabilondo, A. Miralles, P. V. Escriba & J. A. García-Sevilla. 1993. Chronic treatment with the monoamine oxidase inhibitors clorgyline and pargyline

down-regulates non-adrenoceptor [^{3}H]-idazoxan binding sites in the rat brain. Br. J. Pharmacol. **108:** 597-603.
6. O'Callaghan, J. P. & D. B. Miller. 1991. The concentration of glial fibrillary acidic protein increases with age in the mouse and rat brain. Neurobiol. Aging **12:**171-174.
7. Duffy, P. E., M. Rapport & L. Graf. 1980. Glial fibrillary acid protein and Alzheimer-like senile dementia. Neurology **30:** 778-782.
8. Ruiz, J., I. Martín, L. F. Callado, J. J. Meana, F. Barturen & J. A. García-Sevilla. 1993. Increased non-adrenoceptor [^{3}H]idazoxan binding (I_2-imidazoline sites) in post-mortem brain from patients with Alzheimer's disease. Neurosci. Lett. **160:** 109-112.
9. Maiese, K., L. Pek, S. B. Berger & D. J. Reis. 1992. Reduction in focal cerebral ischemia by agents acting at imidazole receptors. J. Cereb. Blood Flow Metab. **12:** 53-63.

In Vivo Modulation of Norepinephrine and Glutamate Release through Imidazoline Receptors in the Rat Central Nervous System[a]

J. J. MEANA,[b,c] M. HERRERA-MARSCHITZ,[c] AND M. GOINY[c]

[b]*Department of Pharmacology*
University of the Basque Country
Leioa, Bizkaia, Spain

[c]*Department of Physiology and Pharmacology*
Karolinska Institute
Stockholm, Sweden

Different α_2-adrenergic compounds with an imidazoli(di)ne chemical structure, such as the antagonist idazoxan and the agonist clonidine, can also recognize with high affinity nonadrenergic binding sites. The nonadrenergic binding site identified by [^{3}H]idazoxan has been termed an I_2-imidazoline site[1,2] and the nonadrenoceptor site labeled by [^{3}H]*p*-aminoclonidine has been termed the I_1-imidazoline site.[1,3] Clonidine inhibits the firing of locus coeruleus neurons by acting on inhibitory α_2 adrenoceptors.[4] However, a paradoxical stimulatory effect of clonidine and other imidazoline/oxazoline drugs on locus coeruleus activity was recently demonstrated after inactivation of α_2 adrenoceptors with the alkylating agent *N*-ethoxycarbonyl-2-ethoxy-1,2-dihydroquinoline (EEDQ).[5] This effect seems to be mediated by I_1-imidazoline receptors and suggests that EEDQ could be a tool to investigate the modulation of norepinephrine (NE) release by the new putative receptor in the CNS.

MATERIALS AND METHODS

Control and EEDQ-treated rats were anesthetized using a mixture of air and halothane and placed in a Kopf stereotaxic frame. Two microdialysis probes were implanted in the cingulate cortex (B + 2.8, L ± 1.0, V − 4.5) and perfused with a modified cerebrospinal fluid solution containing the NE reuptake blocker desipramine (10^{-6} M). Samples of 35 μl were collected and analyzed by high performance liquid chromatography coupled to electrochemical or fluorometric detection.[6] *In vitro*

[a]This study was supported by the Basque Government, the Spanish CICYT (SAF 93-0459) and CDTI (92-0178), the Swedish MRC (8669), and the Foundation Wenner-Grenska Samfundet.

recoveries of NE and glutamate (GLU) were approximately 45%. Clonidine (0.3-2.4 mg/kg ip) and the α_2-adrenoceptor antagonist RX821002 (2-5 mg/kg ip) were administered systemically. EEDQ (6 mg/kg ip) was administered 6 hours before starting the collection of dialysates. For local administration, clonidine (10^{-5} M) was perfused through the probe, and the irreversible α_2-adrenoceptor antagonist benextramine (10^{-4} M) was included from the beginning in the perfusion medium. For each experimental group, four to six rats were used. Statistical analysis was carried out by one-or two-way ANOVA. Experimental data of each animal were pooled and analyzed for the best, simple, nonlinear fit to the three-parameter equations of Parker and Waud.[7]

RESULTS

Basal levels of NE were 4.79 ± 0.60 nM (n = 7) and of GLU were 2.20 ± 1.09 μM (n = 7). RX821002 (2 mg/kg ip) produced an increase in NE efflux, reaching a maximum of 299 ± 36% of basal levels (n = 8; p <0.01). Administration of clonidine (0.3-1.2 mg/kg ip) led to a dose-dependent decrease in cortical NE levels ($F[3,28] = 4.42$; p <0.05) with a maximal effect of 107 ± 39%. The administration of RX821002 (2 mg/kg ip) reversed the effects of clonidine and increased NE release over basal values. The GLU output was enhanced by 71 ± 17% after clonidine administration. However, a significant and clear dose-dependent effect of clonidine on GLU could not be demonstrated ($F[3,96] = 1.85$; $p = 0.14$). Following RX821002 administration, GLU levels did not return to basal values.

Basal extracellular levels of NE evaluated after EEDQ treatment (6 mg/kg injected ip 6 hours before sample collection) tended to be higher (6.38 ± 0.37 nM; n = 76) than those in control groups. In contrast, GLU levels after EEDQ treatment (2.61 ± 0.49 μM; n = 53) were similar to GLU control values. Systemic injection of clonidine (0.6-2.4 mg/kg ip) resulted in a significant dose-dependent increase in NE levels ($F[2,48] = 5.07$; p <0.05). The ED_{50} and E_{max} values for the stimulatory effect of clonidine, estimated from simultaneous analysis in a dose-effect curve of the individual maximal effect, were 1.03 ± 0.58 mg/kg and 88 ± 37%, respectively. Systemic administration of RX821002 (5 mg/kg ip) failed to reverse or prevent the stimulatory effect of clonidine and even produced a significant increase of 70 ± 14% in NE concentration. Glutamate output was enhanced after clonidine injection in EEDQ-pretreated rats with a maximal effect of 190 ± 18%. In this group, RX821002 administration (5 mg/kg ip) did not affect GLU levels.

Local administration of clonidine (10^{-5} M) induced a decrease in GLU levels to 45 ± 18%. This effect was reversed by local RX821002 administration (200 ± 36% of basal values). Following inclusion of benextramine (10^{-4} M) in the medium, basal levels of NE and GLU were increased by 57 ± 13% and 43 ± 15%, respectively (n = 24; p <0.05). In α_2-adrenoceptor alkylating conditions, clonidine did not modify NE levels ($F[4,27] = 1.99$; $p = 0.12$). In similar conditions, GLU levels were similar after clonidine administration (113 ± 20% of basal values; n = 7) to those in the contralateral control side (91 ± 14%; n = 7). Systemic injection of RX821002 (5 mg/kg ip) induced a significant bilateral increase in NE levels to 128 ± 4% of baseline (n = 2), but it failed to produce an observable effect on GLU efflux (92 ± 7% of basal values; n = 2).

DISCUSSION

The present results show functional activity of the new entity termed an imidazoline receptor. Imidazoline derivatives as clonidine have shown stimulatory effects on firing of locus coeruleus noradrenergic neurons.[5] As this effect is normally masked by the more prominent α_2-adrenoceptor-mediated inhibition of noradrenergic neurons, irreversible blockade of α_2-adrenoceptors is necessary. Such blockade has been achieved by administration of EEDQ or benextramine[8–11] which lack effect on imidazoline receptors.[9,12] Under conditions of irreversible α_2-adrenoceptor blockade, clonidine induced a dose-dependent increase in NE levels that was not abolished or reversed by RX821002 administration. Thus, the results suggest a nonadrenergic mechanism of clonidine that could be related to the imidazolinic properties of clonidine. Although systemic intraperitoneal administration of clonidine after α_2-adrenoceptor alkylation with benextramine induced release of NE, the drug failed to produce a similar response following local administration via the dialysis probe. The observation suggests that the stimulatory effect may be mediated by an indirect mechanism unrelated to presynaptic action of the drug on noradrenergic neurons. It was previously shown that an important afferent pathway to the locus coeruleus is an excitatory input from the ventrolateral medulla[13,14] which involves the release of excitatory amino acids. In this context, preliminary results suggest that the nonspecific excitatory amino acid antagonist kynurenic acid blocks the stimulatory effect of clonidine on the firing of locus coeruleus noradrenergic neurons.[15]

In vitro studies have documented the regulation of GLU levels by α_2-adrenoceptors in the rat cortex.[16] The administration of clonidine induced a decrease, whereas RX821002 increased GLU levels. However, decreased GLU levels after local clonidine treatment were not observed after systemic injection of the drug. Moreover, RX821002 was unable to reverse clonidine-induced increases in GLU levels, suggesting that α_2-adrenoceptor-mediated inhibition of GLU release reflects a local mechanism. By contrast, clonidine seems to play a more important functional role, acting indirectly on GLU cortical release through nonadrenergic receptors.

According to binding experiments the imidazoline receptor responsible for the nonadrenergic activity of clonidine has been termed an I_1-imidazoline receptor.[1,3] The I_1-imidazoline receptor has been implicated in the hypotensive effect of clonidine and other imidazoline/oxazoline agents[17–19] and seems to be located in the ventrolateral medulla.[20,21] Therefore, findings strongly suggest that a substantial component of the noradrenergic stimulation elicited by clonidine could be dependent on a neural pathway originating in or projecting through the ventrolateral medulla and involving an I_1-imidazoline receptor. The exact location of this receptor has yet to be determined.

In conclusion, clonidine can not only inhibit NE and GLU release in rat cerebral cortex involving α_2-adrenoceptor mechanisms but also induce paradoxical NE and GLU release through an indirect, nonpresynaptic mechanism. This last effect does not appear to be related to α_2-adrenoceptors but may involve activity of I_1-imidazoline receptors.

REFERENCES

1. MICHEL, M. C. & P. ERNSBERGER. 1992. Trends Pharmacol. Sci. **13:** 369-370.
2. MIRALLES, A., G. OLMOS, M. SASTRE, F. BARTUREN, I. MARTIN & J. A. GARCÍA-SEVILLA. 1993. J. Pharmacol. Exp. Ther. **264:** 1187-1197.
3. ERNSBERGER, P. R., K. L. WESTBROOKS, M. O. CHRISTEN & S. G. SCHÄFER. 1992. J. Cardiovasc. Pharmacol. **20**(Suppl. 4): S1-S10.
4. CEDARBAUM, J. M. & G. K. AGHAJANIAN. 1976. Brain Res. 113: 413-419.
5. PINEDA, J., L. UGEDO & J. A. GARCÍA-SEVILLA. 1993. Naunyn-Schmiedeberg's Arch. Pharmacol. **348:** 134-140.
6. HERRERA-MARSCHITZ, M., J. J. MEANA, W. T. O'CONNOR, M. GOINY, M. S. REID & U. UNGERSTEDT. 1992. Amino Acids **2:**157-179.
7. PARKER, R. B. & D. R. WAUD. 1971. J. Pharmacol. Exp. Ther. **177:** 1-12.
8. TAOUIS, M., M. BERLAN & M. LAFONTAN. 1987. Mol. Pharmacol. **31:** 89-96.
9. SCHULTZ, A. & A. HASSELBLATT. 1989. Naunyn-Schmiedberg's Arch. Pharmacol. **340:** 712-714.
10. ADLER, C. H., E. MELLER & M. GOLDSTEIN. 1985. Eur. J. Pharmacol. **116:**175-178.
11. BARTUREN, F. & J. A. GARCÍA-SEVILLA. 1992. Mol. Pharmacol. **42:** 846-855.
12. MIRALLES, A., C. RIBAS, G. OLMOS & J. A. GARCÍA-SEVILLA. 1993. J. Neurochem. **61:** 1602-1610.
13. ASTON-JONES, G., M. ENNIS, V. A. PIERIBONE, W. T. NICKELL & M. T. SHIPLEY. 1986. Science **234:** 734-737.
14. ENNIS, M. & G. ASTON-JONES. 1988. J. Neurosci. **8:** 3644-3657.
15. PINEDA, J., J. A. RUIZ-ORTEGA, L. UGEDO & J. A. GARCÍA-SEVILLA. 1994. Ann. N.Y. Acad. Sci., this volume.
16. KAMIASAKI, Y., T. HAMAHASHI, C. M. OKADA & T. ITOH. 1991. Brain Res. **568:** 193-198.
17. FELDMAN, J., E. TIBIRIÇA, G. BRICCA, M. DONTENWILL, A. BELCOURT & P. BOUSQUET. 1990. Br. J. Pharmacol. **100:** 600-604.
18. TIBIRIÇA, E., J. FELDMAN, C. MERMET, F. GONON & P. BOUSQUET. 1991. J. Pharmacol. Exp. Ther. **256:** 606-613.
19. ERNSBERGER, P., T. H. DAMON, L. M. GRAFF, S. G. SCHÄFER & O. CHRISTEN. 1993. J. Pharmacol. Exp. Ther. **264:**172-182.
20. ERNSBERGER, P., M. P. MEELEY, J. J. MANN & D. J. REIS. 1987. Eur. J. Pharmacol. **134:** 1-13.
21. BRICCA, G., J. ZHANG, H. GRENEY, M. DONTENWILL, J. STUTZMANN, A. BELCOURT & P. BOUSQUET. 1993. Br. J. Pharmacol. **110:** 1537-1543.

Effects of the I_2-Imidazoline Receptor Ligand LSL 60101 on Various Models of Anorexia in Rats[a]

ANGEL MENARGUES, MAR CEDÓ, OLGA ARTIGA, ROSSEND OBACH, AND JESÚS A. GARCÍA-SEVILLA[b]

Research Department
LASA Laboratories
E-08980 Sant Feliu de Llobregat
Barcelona, Spain

[b]*Laboratory of Neuropharmacology*
Department of Fundamental Biology & Health Sciences
University of the Balearic Islands
E-07071 Palma de Mallorca, Spain

Alpha$_2$-adrenoceptor antagonists and I_2-imidazoline receptor ligands, such as yohimbine, idazoxan, and RX821002 (2-methoxy idazoxan), induce acute hyperphagic effects in freely-feeding rats, which appear to be mediated by activation of α_2-adrenoceptors and/or I_2-imidazoline receptors.[1,2] Recently, the novel and selective I_2-imidazoline receptor ligand LSL 60101 (2-(2-benzofuranyl)imidazole hydrochloride) was also demonstrated to induce hyperphagia in satiated rats.[2] In receptor binding studies, LSL 60101 had a lower affinity than idazoxan for brain I_2-imidazoline receptors (K_i values 350 nM and 14 nM, respectively) but a much higher α_2-adrenoceptor/I_2-imidazoline receptor selectivity (K_i ratio 280 and 0.8, respectively). Although idazoxan-induced feeding may involve activation of both α_2-adrenoceptors and I_2-imidazoline receptors, the hyperphagic effect induced by LSL 60101 apparently involves activity only at I_2-imidazoline receptors.[2] This study extends these observations by evaluating the effects of LSL 60101 on food intake and body weight of rats in various models of anorexia.

In male Sprague-Dawley rats (360–450 g) and during the light period (minimal food intake), LSL 60101 induced in a dose-related manner (3–30 mg/kg ip) significant increases (at least $p < 0.05$) in food intake over 4 hours (4 ± 1, 5 ± 2, and 13 ± 1 g/kg body weight, respectively; $n = 7$) compared to freely-feeding control rats over the same light period (2 ± 1 g/kg body weight; $n = 6$). However, similarly to idazoxan, food intake and body weight over 24 hours were unaffected by LSL 60101 (control

[a]This work was supported by Centro para el Desarrollo Tecnológico e Industrial (CDTI P-920178 grant) from the Ministerio de Industria y Energía, Madrid, Spain.

rats: food intake 100 g/kg body weight; LSL 60101-treated rats: 98 g/kg body weight; $n = 15$).

In food-deprived (20 h/day for 10 days) rats (200-270 g at the start of experiments), LSL 60101 (30 mg/kg ip) did not induce hyperphagia over 4 hours (72 ± 9 g/kg body weight; $n = 5$) compared to control rats (68 ± 3 g/kg body weight; $n = 5$). In this set of experiments body weight of rats was maintained or slightly reduced (200-250 g body weight). In another set of experiments in food-deprived rats (20 h/day for 10 days; accumulated food intake 32-68 g/kg body weight over 1-4 hours), fluoxetine (10 mg/kg ip; $n = 5$) reduced food intake (43-65% over 1-4 hours, $p < 0.05$) as expected,[3] and the anorectic effect of the antidepressant drug was not reversed by the coadministration of LSL 60101 (30 mg/kg ip; $n = 6$) or idazoxan (10 mg/kg ip; $n = 5$) (food intake reduced by 38-50% or 28-52% over 1-4 hours, respectively).

The acute administration of lipopolysaccharide of *Escherichia coli* (1 mg/kg ip) to freely-feeding rats (360-450 g body weight) caused significant decreases (12-76%) in food intake over 12-60 hours (accumulated food intake was markedly reduced from 31-127 g/kg to 8-78 g/kg body weight; $n = 10$, $p < 0.001$) with a concomitant decrease in body weight. In this model of anorexia, LSL 60101 (30 mg/kg ip every 12 hours for 2.5 days) potentiated the effect of lipopolysaccharide of *E. coli* over 24-60 hours (accumulated food intake further reduced by 52-36%, respectively; $n = 10$; $p < 0.001$).

Chronic immobilization (2 h/day for 10 days) of rats (320-420 g body weight) also induced, as expected,[4] severe anorexia at 24 hours postimmobilization (accumulated food intake was reduced from 32 to 23 g/kg body weight; $n = 9$; $p < 0.05$) with a concomitant reduction in body weight during the same period (about 65 ± 4 g body weight; $n = 6$; $p < 0.001$). In this stress-induced model of anorexia, chronic treatment with LSL 60101 (30 mg/kg for 10 days; $n = 6$) or idazoxan (10 mg/kg for 10 days; $n = 6$) also potentiated the anorexia (17% and 5%, $p < 0.05$, respectively) and loss of body weight (17% and 15%, $p < 0.05$, respectively).

The results indicate that LSL 60101, similarly to idazoxan, induces an acute hyperphagic effect in satiated rats. The preferential alkylation of brain α_2-adrenoceptors with *N*-ethoxycarbonyl-2-ethoxy-1,2-dihydroquinoline (EEDQ, 1 mg/kg ip for 6 hours)[5] did not alter the increase in food intake induced by LSL 60101, but it attenuated that of idazoxan, which suggested that the effect of LSL 60101 was most probably mediated only through I_2-imidazoline receptors.[2] However, results also indicate that drug-, infection-, and stress-induced anorexia are not reversed (and even potentiated) by idazoxan or LSL 60101, which argues about the possible therapeutic effectiveness of these drugs as orectics in humans.

REFERENCES

1. Jackson, H. C., I. J. Griffin & D. J. Nutt. 1991. The effects of idazoxan and other α_2-adrenoceptor antagonists on food and water intake in the rat. Br. J. Pharmacol. **104:** 258-262.
2. Menargues, A., M. Cedó, O. Artiga, R. Obach & J. A. García-Sevilla. 1994. Modulation of food intake by α_2-adrenoceptor antagonists and I_2-imidazoline drugs in rats: LSL 60101 as a novel and selective ligand for I_2-imidazoline sites. Br. J. Pharmacol. **111:** 298P.

3. ROWLAND, N. E., S. M. ANTELMAN & T. J. BARTNESS. 1985. Comparison of the effects of fenfluramine and other anorectic agents in different feeding and drinking paradigms in rats. Life Sci. **36:** 2295–2300.
4. MARTI, O., J. MARTI & A. ARMARIO. 1994. Effect of chronic stress on food intake in rats: Influence of stressor intensity and duration of daily exposure. Physiol. Behav. **55:** 747–753.
5. MIRALLES, A., C. RIBAS, G. OLMOS & J. A. GARCÍA-SEVILLA. 1993. Differential effects of the alkylating agent *N*-ethoxycarbonyl-2-ethoxy-1,2-dihydroquinoline on brain α_2-adrenoceptors and I_2-imidazoline sites *in vitro* and *in vivo*. J. Neurochem. **61:** 1602–1610.

Spinal Imidazoline Receptors Do Not Mediate the Antinociceptive Action of Intrathecal Clonidine in the Rat

PHILIP J. MONROE, DEBORAH L. SMITH, AND DAVID J. SMITH

Departments of Anesthesiology and Pharmacology & Toxicology
The Robert C. Byrd Health Sciences Center of
West Virginia University
3618 Health Sciences South
Morgantown, West Virginia 26506-9134

Noradrenergic projections that descend from the midbrain to the spinal cord have been reported to play a prominent role in mediating the analgesic effects of morphine and other narcotics. The intrathecal application of noradrenergic receptor antagonists blocks the antinociception produced by morphine administration into the midbrain.[1,2] Moreover, spinal application of norepinephrine produces a profound antinociceptive effect.[3] These effects are apparently mediated via the α_2-receptor, because antagonists specific for this site attenuate the antinociceptive actions of morphine and norepinephrine,[3] and spinal administration of α_2-selective agonists such as ST-91 and clonidine also produces a profound antinociceptive effect.[4,5] However, clonidine, idazoxan, and other α_2-selective compounds with imidazoline structures are capable of interacting with nonadrenergic imidazoline receptors. These receptors have been identified on the basis of their insensitivity to catecholamines, using radioligand binding techniques,[6,7] and are apparently responsible for the hypotensive effect of imidazolines following their microinjection into the ventrolateral medulla of the rat.[8] These sites have been subdivided into I_1 and I_2 subtypes on the basis of their preferential affinities for clonidine and idazoxan, respectively.[9] Furthermore, recent results suggest that multiple subtypes of I_1 sites may exist.[10]

Because the presence of nonadrenergic imidazoline-preferring receptors in the spinal cord has not previously been assessed, it could be that the antinociceptive effect of spinally administered clonidine, an imidazoline, is mediated (at least in part) through these sites. Consequently, radioligand binding studies examining the abilities of norepinephrine, idazoxan, clonidine, and yohimbine to compete for spinal sites labeled with [^{3}H]clonidine were performed to determine if imidazoline sites are present in rat spinal cord. Then, using the tail flick reflex, the potential involvement of these sites in mediating spinal antinociception was evaluated by comparing the abilities of idazoxan and yohimbine (a non-imidazoline) to antagonize the action of spinally administered clonidine.

METHODS

Radioligand Binding Assays. To determine the existence and pharmacological characteristics of spinal imidazoline receptors, [^{3}H]clonidine binding studies were performed using a whole spinal cord homogenate preparation. Assays were performed in 50 mM TRIS-HCl buffer (pH 7.4) and were incubated at 25°C for 45 minutes. The final radioligand concentration was 4 nM. Nonspecific binding was defined using 10 μM clonidine.

Antinociceptive Studies. Rats were fitted with chronic intrathecal catheters and allowed to recover for 1 week. Antinociception was produced using a 400-nmol dose of clonidine and quantitated using the radiant heat-induced tail flick reflex. Dose antagonist relationships for yohimbine and idazoxan were determined following administration of the antagonists at times such that their peak effects coincided with that of clonidine.

RESULTS

Idazoxan and clonidine completely inhibited the binding of [^{3}H]clonidine to rat spinal cord membranes (FIG. 1). In contrast, norepinephrine and yohimbine did not compete for all spinal sites labeled by 4 nM of [^{3}H]clonidine, inhibiting only 80% of total specific binding at a 10-μM concentration (FIG. 1). These results demonstrate that [^{3}H]clonidine labels multiple spinal sites. The competition of *p*-aminoclonidine for spinal sites labeled by [^{3}H]clonidine was characterized by a two-site model (TABLE 1), also suggesting the interaction of the radioligand with multiple spinal sites.

To further characterize the nature of the nonadrenergic binding component, additional competition experiments were conducted in the presence of a 10 μM concentration of norepinephrine (to prevent the radioligand labeling spinal adrenergic sites). Yohimbine failed to compete for [^{3}H]clonidine binding to spinal nonadrenergic sites (TABLE 1), demonstrating that these sites display little affinity for the antagonist. On the other hand, at micromolar concentrations clonidine, *p*-aminoclonidine, and idazoxan completely inhibited [^{3}H]clonidine binding to spinal nonadrenergic sites (TABLE 1). In addition, the IC_{50} of *p*-aminoclonidine for nonadrenergic sites was comparable to that of the low affinity component determined in the absence of norepinephrine (TABLE 1). Concentrations of histamine, cimetidine, and imidazole-4-acetic acid up to 30 μM did not significantly inhibit [^{3}H]clonidine binding in either the presence or the absence of a 10 μM concentration of norepinephrine.

Intrathecal administration of clonidine (400 nmol) resulted in antinociception that peaked 15 minutes after injection and persisted at least 60 minutes. Idazoxan dose-dependently inhibited the antinociceptive effect of clonidine, producing maximal inhibition at a 12.5-nmol dose ($ID_{50} = 1.27 \pm 0.86$ nmol). Similarly, the alkaloid antagonist yohimbine inhibited the effect of clonidine, producing maximal inhibition at a 30-nmol dose ($ID_{50} = 3.39 \pm 0.94$ nmol).

DISCUSSION

The results of radioligand binding studies demonstrate the existence of nonadrenergic, imidazoline sites in rat spinal cord. Approximately 20% of specific spinal sites

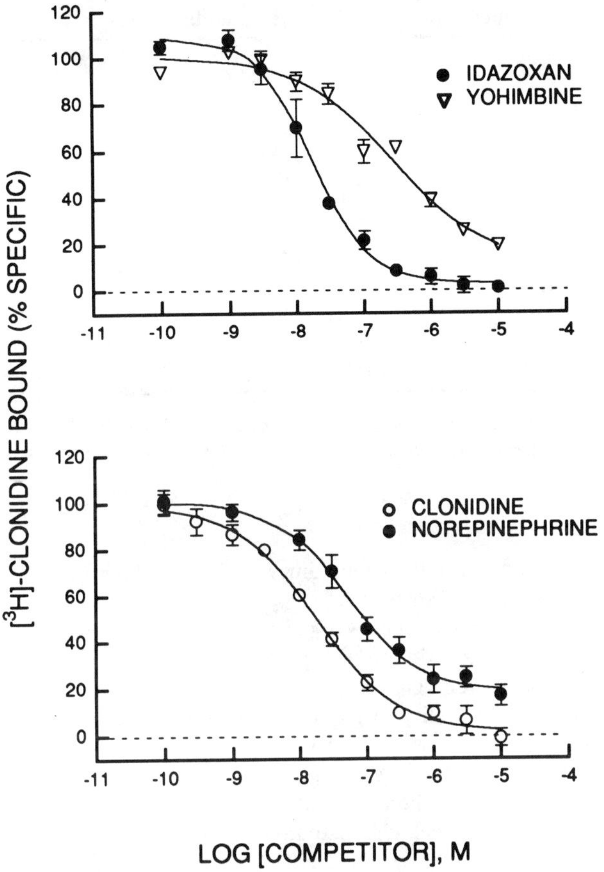

FIGURE 1. Competition of idazoxan, yohimbine, clonidine, and norepinephrine for spinal sites labeled with 4 nM [^{3}H]clonidine. All curves were best fit to a one-site model. While clonidine and idazoxan completely inhibited binding, norepinephrine and yohimbine plateaued at 22 ± 5 and 23 ± 4%, respectively.

labeled by 4 nM [^{3}H]clonidine display no affinity for norepinephrine. An equivalent proportion of spinal sites also displayed no affinity for yohimbine. Also, yohimbine did not compete for any of the binding that remained in the presence of a 10 μM concentration of norepinephrine, suggesting the two compounds compete for the same sites.

Based on the affinities determined for the competitors used in the current study, spinal nonadrenergic sites are analogous to the imidazoline sites described in the human nucleus reticularis lateralis.[10] These sites have micromolar affinities for clonidine, *p*-aminoclondine, and idazoxan and display little or no affinity for histamine. This finding is in contrast to that for imidazoline sites identified in bovine brain stem

TABLE 1. Competitor Interaction at Spinal [^{3}H]Clonidine Sites[a]

Competitor	Control Site 1	Control Site 2	+10 μM Norepinephrine Site 1
Norepinephrine	47 ± 10		nd
Yohimbine	121 ± 35		>10,000
Clonidine	19.4 ± 2.8		1,726 ± 657
p-NH_2-clonidine	2.7 ± 0.1	821 ± 5	3,750 ± 1,653
Idazoxan	17.1 ± 4.5		2,006 ± 665
Histamine	>10,000		>10,000
Cimetidine	>10,000		>10,000
Imidazole-4-acetic acid	>10,000		>10,000

[a] Concentration (nM) of competitors that inhibited 50% of total specific [^{3}H]clonidine binding (IC_{50} ± SEM). Only the *p*-aminoclonidine (*p*-NH_2-clonidine) curve could be best fit to a two-site model. nd = not determined.

which have high affinity for *p*-aminoclonidine and micromolar affinity for histamine, imidazole-4-acetic acid, and cimetidine.[6]

Despite the existence of two classes of receptors with which clonidine can interact in the spinal cord, antinociception resulting from spinal administration is apparently mediated solely by α_2-adrenergic receptors. Idazoxan and yohimbine, compounds that displayed little or no ability to interact with spinal imidazoline sites, blocked all of the antinociception resulting from intrathecally administered clonidine. This observation is consistent with the results of Takano and Yaksh[11] who demonstrated cirazoline (an imidazoline α_1-adrenergic receptor agonist) to have no antinociceptive activity following its intrathecal administration.

REFERENCES

1. YAKSH, T. L. 1979. Brain Res. **160:** 180-185.
2. CAMARATA, P. J. & T. L. YAKSH. 1985. Brain Res. **336:** 133-142.
3. CRISP, T & D. J. SMITH. 1989. Neuropharmacology **28:** 1047-1053.
4. MONASKY, M. S., A. R. ZINSMEISTER, C. W. STEVENS & T. L. YAKSH. 1990. J. Pharmacol. Exp. Ther. **254:** 383-392.
5. REDDY, S. V. R. A., J. L. MALDERDRUT & T. L. YAKSH. 1980. J. Pharmacol. Exp. Ther. **213:** 525-533.
6. ERNSBERGER, P., M. P. MEELEY, J. J. MANN & D. J. REIS. 1987. Eur. J. Pharmacol. **134:** 1-13.
7. MACKINNON, A. C., M. STEWART, H. J. OLVERMAN, M. SPEDDING & C. M. BROWN. 1993. Eur. J. Pharmacol. **232:** 79-87.
8. ERNSBERGER, P., R. GIULIANO, R. N. WILLETTE & D. J. REIS. 1990. J. Pharmacol. Exp. Ther. **253:** 408-418.
9. MICHEL, M. C. & P. ERNSBERGER. 1992. Trends Pharmacol. Sci. **213:** 369-370.
10. BRICCA, G., H. GRENEY, J. ZHANG, M. DONTENWILL, J. STUTZMANN, A. BELCOURT & P. BOUSQUET. 1994. Eur. J. Pharmacol. **266:** 25-33.
11. TAKANO, Y. & T. L. YAKSH. 1993. J. Pharmacol. Exp. Ther. **264:** 327-335.

The Stimulatory Effect of Clonidine on Locus Coeruleus Noradrenergic Neurons through Imidazoline Receptors Is Modulated by Excitatory Amino Acids[a]

JOSEBA PINEDA, J. ANGEL RUIZ-ORTEGA, LUISA UGEDO, AND JESÚS A. GARCÍA-SEVILLA [b]

Department of Pharmacology
University of the Basque Country
Leioa, Bizkaia, E-48940, Spain

[b]*Laboratory of Neuropharmacology*
Department of Fundamental Biology and Health Sciences
University of the Balearic Islands
Palma de Mallorca, Spain

We have demonstrated that large doses of clonidine produce a stimulatory effect on locus coeruleus noradrenergic (LC-NA) neurons after irreversible inactivation of α_2-adrenoceptors with *N*-ethoxycarbonyl-2-ethoxy-1,2-dihydroquinoline (EEDQ). This effect suggests that I_1-imidazoline receptors can mediate the effect of clonidine on LC-NA neurons. It was also observed that pretreatment with EEDQ plus reserpine enhances this stimulatory effect.[1] This study investigates the mechanism underlying these effects.

Male albino Sprague-Dawley rats (200-300 g) were treated 6 hours before cell recording with 6 mg/kg of EEDQ. In addition, the nonselective excitatory amino acid antagonist kynurenic acid (1-3 μmol in 10-30 μl icv 2-5 minutes before doses of clonidine) was administered to one group of rats. The other groups of rats were pretreated with reserpine (5 mg/kg sc 18 hours before the experiment) or *p*-chlorophenylalanine (PCPA; 400 mg/kg ip 24 hours before the experiment) or α-methyl-*p*-tyrosine (αMPT; 250 mg/kg ip 24 hours before the experiment). Rats were anesthetized with chloral hydrate (400 mg/kg ip) and with additional injections given intravenously when needed. A tracheal cannula and jugular vein catheter were inserted before the rats were mounted in a stereotaxic frame. A 23-gauge steel cannula was placed in the left lateral ventricle 1.0 mm posterior and 1.3 mm lateral to bregma and 4-5 mm deep to the skull surface. Omegadot glass electrode filled with a 2% solution of Pontamine sky blue in 0.5% sodium acetate was situated 1.1 mm lateral

[a]This study was supported by the Spanish CYCYT (SAF93-0459) and CDTI (92-0178). J. Pineda was supported by a postdoctoral fellowship from GV. J. A. Ruiz-Ortega was supported by a fellowship from MEC.

TABLE 1. Parameters of Dose-Effect Curves for Stimulatory Effects of Clonidine on the Firing Rate of Locus Coeruleus Noradrenergic (LCNA) Neurons in Anesthetized Rats[a]

Pretreatment	Dose Range (μg/kg)	ED_{50} (μg/kg)	E_{max} (%)	*n*
EEDQ	10-5,120	671	86	13
Reserpine + EEDQ	320-5,210	767	147[b]	5
αMPT + EEDQ	320-5,120	479	84	4
PCPA + EEDQ	320-2,560	707	215[b]	6
EEDQ + kynurenic acid	320-5,120	—	—	7

[a] For dose-effect curves, LC-NA neuron firing rates were recorded before (for 3-8 min) and after the cumulative administration at 1-minute intervals of increasing doses of clonidine until maximal response was reached. Increases in firing rate were expressed as a percentage of the average basal firing rate. Experimental data from *n* rats in each group were pooled and analyzed for the best simple nonlinear fit to the three-parameter equation of Parker and Waud,[11] using the fitting program GraFit.

[b] $p < 0.05$ when compared with the value obtained EEDQ-pretreated rats using the goodness of fit (determined by an F-test) of a model with and without a constrained E_{max}.

and 3.7 mm caudal to the lambda and 5.5-6.5 mm below the cortical surface. Signal from the recording electrode were amplified, discriminated, and monitored on an oscilloscope and fed into the pen chart recorder and computer, creating interspike time interval histograms. Only one cell was used in each animal. LC-NA cells were identified by criteria described by Cedarbaum and Aghajanian.[2] Changes in the firing rate of LC-NA neurons after cumulative doses of clonidine (320-5,120 μg/kg × 2 iv) were evaluated at 1-minute intervals, and dose-effect curves for clonidine-induced stimulation were analyzed using the nonlinear least squares fitting program GraFit.[3] The stimulatory effect of clonidine was transformed as a percentage of the spontaneous firing rate, and ED_{50} and E_{max} values were estimated.

To study the effect of clonidine not mediated by α_2-adrenoceptors, rats were pretreated with EEDQ (6 mg/kg ip). As shown previously,[1] the inhibitory effect of clonidine which is mediated through α_2-adrenoceptors was completely abolished, and clonidine dose dependently stimulated LC-NA neurons (TABLE 1, FIG. 1).

The stimulatory effect of clonidine (320-5,120 μg/kg iv) on LC-NA cell firing was blocked by the administration of kynurenic acid (1-3 μmol in 10-30 μl icv, 7/7) (FIG. 1). Kynurenic acid also abolished the biphasic excitation-inhibition response to noxious stimuli as reported by Ennis and Aston-Jones.[4]

After reserpine (5 mg/kg sc 18 hours) or PCPA (400 mg/kg ip 24 hours), the stimulatory effect of high doses of clonidine was enhanced. While the E_{max} value in rats pretreated only with EEDQ was 86% ($n = 13$), the E_{max} values in rats pretreated with EEDQ plus reserpine and in rats pretreated with EEDQ plus PCPA were 147% ($n = 5$) and 215% ($n = 6$), respectively (TABLE 1, FIG. 1). However, the ED_{50} values

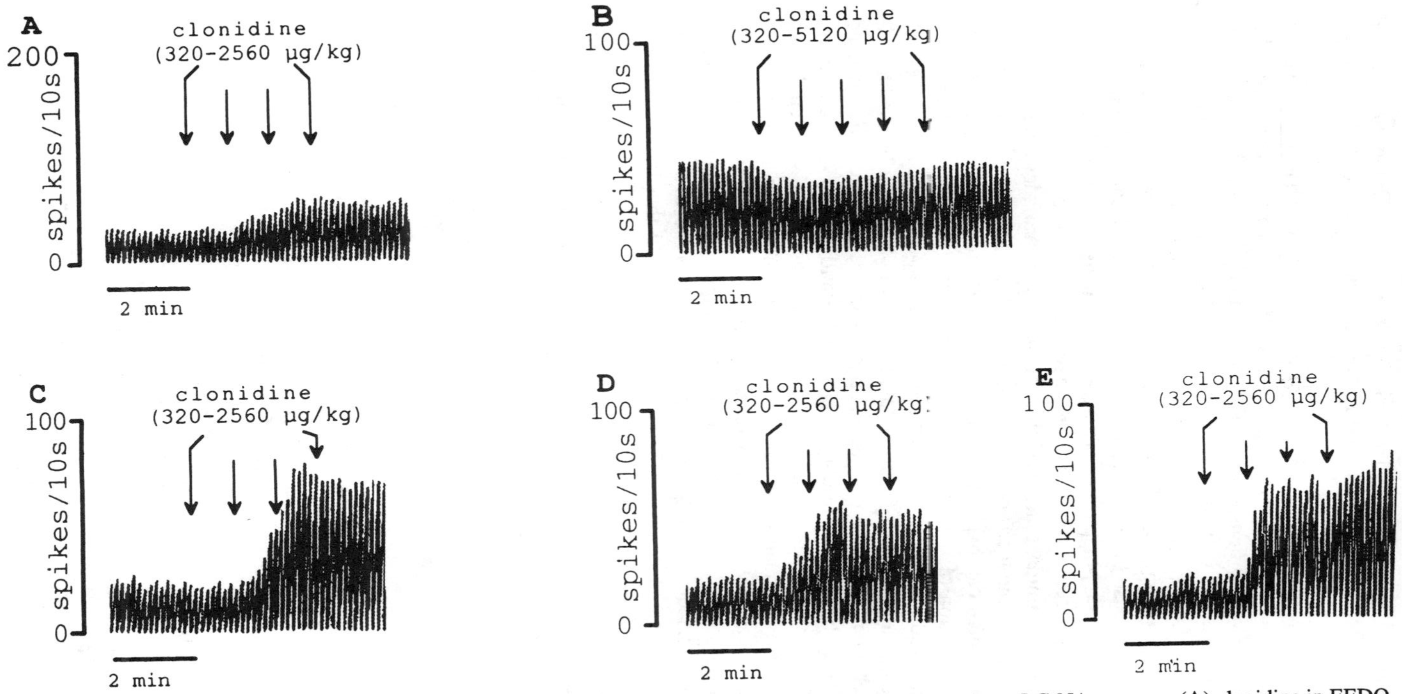

FIGURE 1. Ratemeter records showing the effect of clonidine (320–5,120 μg/kg x2, iv) on firing rate from LC-NA neurons: (**A**) clonidine in EEDQ-pretreated rats; (**B**) clonidine in EEDQ-pretreated rats after the administration of kynurenic acid; (**C**) clonidine in rats pretreated with EEDQ and reserpine; (**D**) clonidine in rats pretreated with EEDQ and αMPT; and (**E**) clonidine in rats pretreated with EEDQ and PCPA.

were not modified when compared with the values of the EEDQ group. Also, αMPT (250 mg/kg, ip, 24 hours) did not modify the stimulatory effect of clonidine (TABLE 1, FIG. 1).

Pretreatment with reserpine, PCPA, or αMPT did not change the basal firing rates of LC-NA neurons when compared with the basal firing rates from EEDQ-pretreated rats. The low density of imidazoline binding sites in the locus coeruleus[5] and the large doses of clonidine needed to stimulate LC-NA neurons suggest that the stimulatory effect is not produced by a direct action on LC-NA. In this context, preliminary results suggest that local administration of clonidine via the dialysis probe failed to induce release of norepinephrine, whereas intraperitoneal administration did induce release of norepinephrine.[6] The locus coeruleus receives afferents from the nucleus paragigantocellularis[7] located in the rostral ventrolateral medulla, where there is a high density of imidazoline receptors.[8] This nucleus provides a potent excitatory amino acid input to the locus coeruleus and controls its activity by excitatory amino acid release,[4] which can explain the blockade of the stimulatory effect of clonidine by icv administration of kynurenic acid, an excitatory amino acid antagonist. Thus, this antagonist did not block the effect mediated by α_2-adrenoceptors. These results suggest that the stimulatory effect of clonidine on LC-NA may be mediated by an indirect mechanism in which the nucleus paragigantocellularis can be involved.

On the other hand, the stimulatory effect of clonidine on LC-NA neurons was enhanced after 5-hydroxytryptamine (5-HT) synthesis inhibition, which indicated that endogenous 5-HT attenuates the response to imidazoline receptor activation. In this context, Aston-Jones *et al.*[9] demonstrated that 5-HT reduces the excitatory amino acid response of locus coeruleus neurons. This input of endogenous 5-HT is not provided from the dorsal raphe; instead, it was observed that 5-HT cells in the pericoerulear region innervate to the locus coeruleus and that this region provides the endogenous 5-HT to the locus coeruleus.[10]

In summary, these results suggest that the imidazoline effect of clonidine on LC-NA neurons is mediated through excitatory amino acids and can be modulated by endogenous 5-HT.

REFERENCES

1. PINEDA, J., L. UGEDO & J. A. GARCÍA-SEVILLA. 1993. Naunyn-Schmiedebergs Arch. Pharmacol. **348:** 134–140.
2. CEDARBAUM, J. M. & G. K. AGHAJANIAN. 1976. Brain Res. **112:** 413–419.
3. LEATHERBARROW, R. J. 1990. Trends Biochem. Sci. **15:** 455–458.
4. ENNIS, M. & G. ASTON-JONES. 1988. J. Neurosci. **8:** 3644–3657.
5. KAMISAKI, Y., T. ISHIKAWA, Y. TAKAO, H. OMODANI, N. KUNO & T. ITOH. 1990. Brain. Res. **514:** 15–21.
6. MEANA, J. J., M. HERRERA-MARSCHITZ & M. GOINY. 1994. Ann. N.Y. Acad. Sci., this volume.
7. ASTON-JONES, G., M. ENNIS, V. A. PIERIBONE, W. T. NICKELL & M. T. SHIPLEY. 1986. Science **234:** 734–737.
8. ERNSBERGER, P., M. P. MEELEY, J. J. MANN & D. J. REIS. 1987. Eur. J. Pharmacol. **134:** 1–13.

9. Aston-Jones, G., H. Akaoka, P. Charléty & G. Chouvet. 1991. J. Neurosci. **11:** 760–769.
10. Pieribone, V. A., E. Van Bockstaele, M. T. Shipley & G. Aston-Jones. 1989. Soc. Neurosci. Abstr. **15:** 420.
11. Parker, R. B. & D. R. Waud. 1971. J. Pharmacol. Exp. Ther. **177:** 1–12.

Chronic Treatment with Phenelzine and Other Irreversible Monoamine Oxidase Inhibitors Downregulates I_2-Imidazoline Receptors in the Brain and Liver[a]

REGINA ALEMANY, GABRIEL OLMOS, AND JESÚS A. GARCÍA-SEVILLA

Laboratory of Neuropharmacology
Department of Fundamental Biology and Health Sciences
University of the Balearic Islands
E-07071 Palma de Mallorca, Spain

The I_2-imidazoline receptors (I_2-IR) were increased in the frontal cortex of depressed suicide victims,[1] and recently, downregulation of these receptors after chronic treatment with the irreversible monoamine oxidase (MAO) inhibitor clorgyline was reported in the rat brain.[2] On the other hand, interactions of irreversible MAO inhibitors with cytochrome P-450 enzymes[3] have been described. The present study was designed (1) to further assess and extend the effects of chronic treatment with irreversible MAO inhibitors on rat brain and liver I_2-IR, (2) to investigate *in vitro* the interaction of these drugs with the receptors, and (3) to seek a possible link between I_2-IR and cytochrome P-450 enzymes.

Male Sprague-Dawley rats (250-300 g) were used. For chronic treatment the animals received ip every 12 hours either 0.9% saline vehicle or the irreversible MAO inhibitors phenelzine (0.3-20 mg/kg for 7 days), isocarboxazid (10 mg/kg for 7 days), clorgyline (3 mg/kg for 14 days), and tranylcypromine (10 mg/kg for 7 days). The time course for the effect of phenelzine (3 mg/kg) was studied at days 1 (acute), 7, 14, and 21 after treatment. In another series of experiments rats were treated chronically with cytochrome P-450-inducing agents phenobarbital (40 or 80 mg/kg ip every 24 hours for 7 days), 3-methylcholanthrene (20 mg/kg ip every 12 hours for 7 days), or 2-methylimidazole (40 mg/kg ip, every 24 hours for 7 days). In both acute and chronic treatment, rats were killed 24 hours after the last injection. The parietooccipital cortex and portions of liver were used to prepare cortical and liver membranes (P_2 fractions). Total [^{3}H]idazoxan binding to brain and liver I_2-IR was measured in 1.1-ml aliquots of these membranes as previously described.[4] Binding of [^{3}H]idazoxan to brain I_2-IR was always done in the presence of 10^{-6}M (–)-

[a]This study was supported by grants DGICYT (PM 91-0069), FIS (93-0641), and CDTI-S.A. LASA Laboratories (92-0198). R.A. was supported by a fellowship from FIS (Fondo de Investigaciones Sanitarias, Spain).

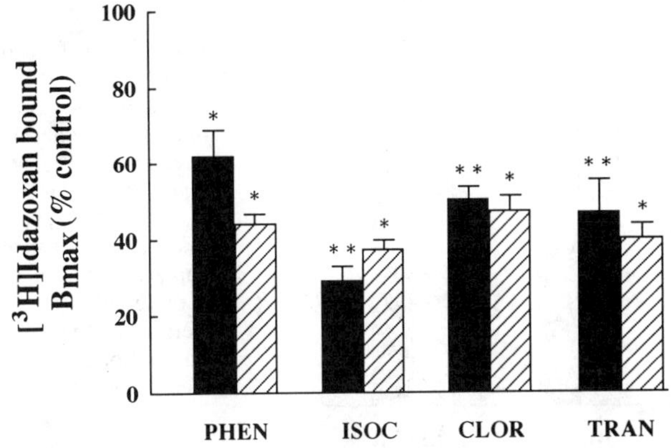

FIGURE 1. Effect of chronic treatment (ip every 12 h) with the irreversible MAO inhibitors phenelzine (PHEN, 3 mg/kg for 7 days), isocarboxazid (ISOC, 10 mg/kg for 7 days), clorgyline (CLOR, 3 mg/kg for 14 days), and tranylcypromine (TRAN, 10 mg/kg for 7 days) on the density (B_{max}) of I_2-imidazoline receptors (IR) in the rat cerebral cortex (*solid columns*) and liver (*hatched columns*). I_2-IR density (B_{max}) after treatment is expressed as a percentage of that from saline-treated rats (mean control values: K_d = 14.4 ± 0.7 nM; B_{max} = 68 ± 3 fmol/mg protein in the cerebral cortex and K_d = 9.0 ± 1.0 nM; B_{max} = 335 ± 21 fmol/mg protein in the liver; *n* = 24). *Columns* represent the mean ± SEM of 3-6 experiments with one animal per experiment. One-way ANOVA followed by a multiple comparison test detected a significant decrease in B_{max} for I_2-IR after treatments. No significant changes in K_d values were found after treatments. $*p < 0.05$ and $**p < 0.001$ as compared with saline-treated group (ANOVA followed by Scheffé's test).

epinephrine. Analyses of saturation isotherms and competition experiments were performed by computer-assisted nonlinear regression using the EBDA-LIGAND programs.[5]

Chronic treatment (7 days) with phenelzine (1-20 mg/kg) dose-dependently decreased the density of brain and liver I_2-IR (B_{max} values reduced by 21-60%). The effect of phenelzine (3 mg/kg) was rapid (1 day) in the liver but not in the brain (7 days), and in both tissues the downregulation of I_2-IR was maintained after 14 and 21 days of repeated treatment (B_{max} values reduced by 35% in brain and 60% in liver). Chronic treatment with isocarboxazid (10 mg/kg), a hydrazine structurally related to phenelzine, also resulted in downregulation of brain and liver I_2-IR (71 and 63% reduction in B_{max} values, respectively) (FIG. 1). Similar results were obtained in both tissues after chronic treatment with the propargylamine clorgyline (3 mg/kg) or the cyclopropylamine tranylcypromine (10 mg/kg) (50-60% reduction in B_{max} values) (FIG. 1). *In vitro,* phenelzine and tranylcypromine displayed low affinity (Ki_H = 0.3 – 6 μM) for rat brain and liver I_2-IR, whereas clorgyline displayed very high affinity for rat brain (Ki_H = 40 pM) but not rat liver I_2-IR (Ki_H = 169 nM). The hydrazine isocarboxazid did not compete against [^{3}H]idazoxan binding to I_2-IR in the two tissues studied.

Preincubation of cortical or liver membranes (30 minutes at 25 °C) with phenelzine (10^{-4} M) did not alter the total density of I_2-IR, indicating that this irreversible MAO

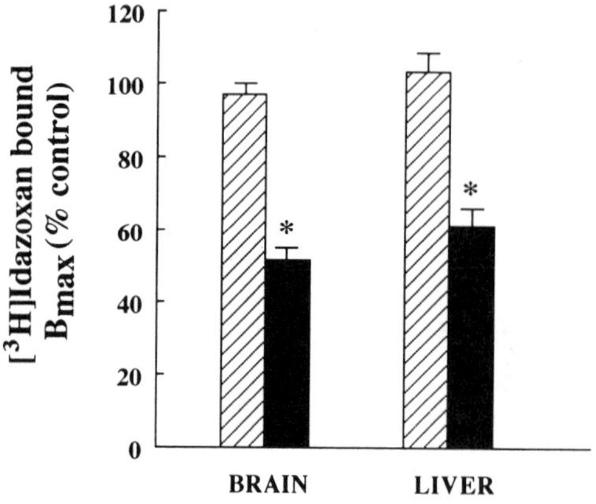

FIGURE 2. Effects of preincubation (30 min at 25°C) of cortical or liver membranes in buffer containing 10^{-4} M phenelzine (*hatched columns*) or 10^{-6} M clorgyline (*solid columns*) on the total density (B_{max}) of I_2-IR. B_{max} values are expressed as a percentage of that preincubated in buffer only (*control*) and represent mean ± SEM of 3-6 experiments. $*p < 0.05$ as compared with control group (ANOVA followed by Scheffé's test).

inhibitor does not irreversibly bind to I_2-IR. In contrast, preincubation with 10^{-6} M clorgyline reduced by 40% the B_{max} of [^{3}H]idazoxan to brain and liver I_2-IR (FIG. 2).

The compound SKF 525A, a potent inhibitor of cytochrome P-450 enzymes, completely displaced with moderate affinity the specific binding of [^{3}H]idazoxan to brain and liver I_2-IR. However, chronic treatment (7 days) with various inducers of these enzymes including phenobarbital, 3-methylcholanthrene, and 2-methylimidazole (20-80 mg/kg) did not alter the density of I_2-IR in the liver or rat brain.

Several conclusions can be deduced from the present data: (1) Chronic treatment with the MAO inhibitor antidepressant phenelzine and other clinically effective irreversible MAO inhibitors including clorgyline and tranylcypromine is associated with downregulation of brain and liver I_2-IR; (2) there is no relation between the *in vitro* potency of irreversible MAO inhibitors competing against [^{3}H]idazoxan binding to brain and liver I_2-IR and their ability *in vivo* to downregulate these receptors; (3) the downregulation of brain and liver I_2-IR after chronic treatment with MAO inhibitors is not related to irreversible binding of these compounds to I_2-IR; (4) chronic treatment with inducers of cytochrome P-450 enzymes including phenobarbital, 3-methylcholanthrene, and 2-methylimidazole did not modulate the density of I_2-IR in brain and liver. These results discounted a possible relation between I_2-IR and cytochrome P-450 enzymes; and (5) together these findings suggest an indirect mechanism for the *in vivo* downregulation of I_2-IR found after chronic treatment with irreversible MAO inhibitors that could be of relevance in the treatment of depression.

REFERENCES

1. MEANA, J. J., F. BARTUREN, I. MARTIN & J. A. GARCÍA-SEVILLA. 1993. Evidence of increased non-adrenoceptor [^{3}H]idazoxan binding sites in the frontal cortex of depressed suicide victims. Biol. Psychiatry **34:** 498-501.
2. OLMOS, G., A. M. GABILONDO, A. MIRALLES, P. V. ESCRIBA & J. A. GARCÍA-SEVILLA. 1993. Chronic treatment with the monoamine oxidase inhibitors clorgyline and pargyline down-regulates non-adrenoceptor [^{3}H]-idazoxan binding sites in the rat brain. Br. J. Pharmacol. **108:** 597-603.

3. Dupont, H., D. S. Davies & M. Strolin-Benedetti. 1987. Inhibition of cytochrome P-450-dependent oxidation reactions by MAO inhibitors in rat liver microsomes. Biochem. Pharmacol. **36:** 1651-1657.
4. Olmos, G., R. Alemany, P. V. Escriba & J. A. García-Sevilla. 1994. The effects of chronic imidazoline drug treatment on glial fibrillary acidic protein concentrations in rat brain. Br. J. Pharmacol. **111:** 997-1002.
5. Munson, P. J. & D. Rodbard. 1980. LIGAND: A versatile computerized approach for the characterization of ligand binding systems. Anal. Biochem. **107:** 220-239.

Involvement of I_1-Imidazoline Receptors in Mood Disorders

JOHN E. PILETZ AND ANGELOS HALARIS

Departments of Psychiatry and Human Behavior and of Pharmacology and Toxicology
University of Mississippi Medical Center
Jackson, Mississippi 39216–4505

Mood disorders, such as depression, are widely assumed to involve dysregulation of monoaminergic neurotransmission in the central nervous system (CNS).[1] Despite this widely held assumption, an exact defect has been elusive.[2] Owing to their physiological importance in regulating the release of monoamines, presynaptic α_2-adrenoceptors (α_2ARs) have attracted considerable attention as a possible site of the primary defect in depression. Stimulation of the presynaptic α_2AR inhibits neuronal firing and release.[3] Accordingly, antidepressant drugs have been proposed[4] to act, at least in part, by desensitizing a presumably supersensitive α_2AR in depression, rendering the normalization of neurotransmission and consequential clinical improvement. Indeed, idazoxan, an imidazoline α_2-antagonist, was shown to possess antidepressant efficacy.[5] However, other α_2-antagonists, such as yohimbine, which are not imidazolines, have been proved to lack antidepressant efficacies. Indeed, the role of α_2ARs in depression remains controversial.

In attempting to unravel some of the discrepancies that exist in the α_2AR literature on depression, it is apparent that most of the evidence implicating an aberrant α_2AR in depression derives from studies with imidazoline compounds. Imidazolines, such as idazoxan and clonidine, have widely been assumed to be pure α_2-selective ligands and have accordingly been used to probe the human depressive condition either as *in vivo* pharmacological challenge agents or, in the radioligand form, to radiolabel tissue receptors. In this article we suggest that some of the arguments used to hypothesize supersensitivity of presynaptic α_2ARs in depression might better be applied to imidazoline receptors (either subtype I_1 or I_2 receptors).

As already mentioned, the rationale linking α_2ARs to depression is based on the presynaptic inhibitory function of these receptors. Interestingly, the inhibitory function of α_2ARs encompasses two major monoamine systems of importance to depression: norepinephrine[6] and serotonin.[7] Thus, a supersensitive state of α_2ARs could explain a depletion of both norepinephrine and serotonin in depression. Accordingly, one rationale for the efficacy of idazoxan as an antidepressant agent is that it stimulates (i.e., restores) norepinephrine and serotonin release by toning down a supersensitive presynaptic α_2AR in the CNS of depressed patients. But it is also interesting that recent studies indicate that I-receptors might also act in an inhibitory fashion. For instance, clonidine has been shown to act through *nonadrenergic* mechanisms to mediate similar inhibitory properties as the α_2ARs. In the brain, clonidine inhibits dopaminergic activity in the nucleus reticularis lateralis in a nonadrenergic manner.[8]

Clonidine also has been reported to inhibit acetylcholine-evoked catecholamine (CA) release in adrenal chromaffin cells[9–12] and to inhibit electrically evoked CA release in pulmonary arteries[13,14] and heart sympathetic nerves[15] via imidazoline receptors. Furthermore, [^{3}H]clonidine binding sites, presumably I_1-imidazoline receptors, were identified in brain regions of relevance to depressive illness (i.e., hypothalamus and hippocampus).[16] In light of these findings, the previous use of imidazoline agents as selective α_2AR compounds for studies of depression may need to be reevaluated carefully. Indeed, the rationale for implicating inhibitory α_2ARs in depression might equally well apply to inhibitory I-receptors.

Besides these theoretical considerations, experimental evidence has begun to link imidazoline receptors with depression. Foremost among these observations is the report by Meana *et al.*[17] showing a 133% increase in cortical I_2 sites from the brain of depressed suicide victims relative to controls. By comparison, the same authors reported a more modest 30% increase in the density of cortical α_2ARs in depressed suicide victims.[18] Furthermore, monoamine oxidase inhibitors (clorgyline, pargyline, and deprenyl), known to be effective antidepressants, display reasonable affinities for I_2 sites.[19,20] In fact, clorgyline displays a component of very high affinity (K_i = 40 pM) in competition with ^{3}H-idazoxan at rat brain I_2 sites. Furthermore, chronic treatment (14 days) of rats with either clorgyline or pargyline has been shown to downregulate I_2 sites in brain.[20] Lastly, in depressed patients, there are numerous reports of blunted growth hormone responses to clonidine challenge.[21] In fact, the consistency of this finding throughout a number of laboratories makes it probably the most reproducible finding in the study of depression.[22] Although there has been little doubt heretofore that this hyposensitive growth hormone response to clonidine in depression is α_2-mediated, to our knowledge this has never been proved. This issue needs to be revisited in view of a possible aberration in imidazoline receptors in depression.

Human platelets have also provided evidence for I-receptor aberrations in depression. Because of the ethical infeasibility of obtaining brain biopsies from living patients, platelets have been studied as a peripheral marker of depression. Platelets possess an array of serotonin-like neuronal properties, including neuron-specific enolase.[23] Additionally, α_2AR binding sites on platelets are upregulatable following reduced sympathetic nervous system activation,[24] indicating that this platelet receptor is regulated in conjunction with overall sympathetic activity. Furthermore, the density of platelet α_2AR binding sites (defined with the non-imidazoline antagonist yohimbine) has been shown to be under genetic control.[25] Because platelets are enucleated, megakaryocytic precursors may be involved in this level of gene expression. In any case, substantial evidence exists for a common level of α_2AR gene control in both platelets and neurons. Whether the I_1 site is under similar regulatory control in platelets and neurons is unknown.

In this article we review earlier studies with depressed patients and healthy subjects that have compared platelet α_2AR and I_1 binding sites. New data are also presented, drawing attention to a unique aspect of our data, which is that α_2AR and I_1 binding site parameters are correlated on platelets. This finding may have ramifications for linking supersensitive α_2AR and I_1 receptors to depression.

METHODS

A detailed description of the criteria for subject selection and the biochemical procedures used to analyze platelets was previously presented.[26–28] Therefore, only methods relevant to the premises of this manuscript (i.e., identification of correlations between I_1 and α_2AR binding parameters) will be elaborated. A subsequent report will deal with the actual clinical findings of the study.

Platelets were collected from 25 depressed patients and 25 healthy subjects. In depressed patients, blood was drawn prior to treatment ($n = 25$) and again after a 6-week treatment course of desipramine or fluoxetine (21 completed; 4 dropped out). In healthy subjects, blood was collected one to three times over a 3-6-week period. Six healthy male subjects each gave blood three times, and one randomly gave blood twice throughout the period. In healthy females, blood drawings were spaced over different stages of the menstrual cycle. Healthy female subjects were required to (1) have regular menstrual cycles, and (2) record their menstrual stage during the study. Five females gave blood during their proliferative, ovulatory, and midluteal phases; five gave blood during their proliferative and midluteal phases; and eight gave blood only during their proliferative phase. Phases of the menstrual cycle were defined as follows: proliferative, days 2-13 after the onset of menstruation; ovulatory, days (–12) to (–13) before the next menses; luteal, days (–0.5) to (–11) before the next menses.

Diagnoses were made by two experienced psychiatrists. Patients met Research Diagnostic Criteria for primary Major Depressive Disorder. The health of the control subjects was confirmed by both physical and psychiatric examination.[29] None of the subjects reported a history of alcohol or drug abuse, and none was taking known platelet-altering medications during the week preceding the drawing of blood. Blood was collected between 9 and 10 AM. A total population of platelets was obtained from 100 ml of fresh blood according to the method of Corash.[30]

The procedure for radioligand ([^{125}I]*p*-iodoclonidine, 2,200 Ci/mmol, New England Nuclear) binding allowed a distinction between α_2AR and I_1 sites on human platelets. Purified plasma membranes were prepared in a low concentration of magnesium (0.5 mM) plus 5 mM HEPES and 0.5 mM EGTA, which produced high affinity radioligand binding to I_1 sites, while still being sufficient to form stable α_2AR-agonist complexes.[26] α_2AR sites were assessed between 0.01 and 2.5 nM radioligand concentrations and the I_1 sites were assessed between 0.5 and 10.0 nM. To selectively assess I_1 sites, an epinephrine mask (10 μM, EPI mask) was used. Norepinephrine (10 μM) was used to measure nonspecific binding at the α_2AR, and 10 μM moxonidine (Solvay Pharmaceuticals, Germany) was used to measure nonspecific binding at the I_1 site. B_{max} and K_D parameters were analyzed by single-site modeling of the saturation data using the LIGAND computer program.[31] Additional measurements were made of platelet-poor plasma concentrations of norepinephrine, epinephrine, and dopamine using HPLC with a Coulochem detector (ESA Corp., Bedford, Massachusetts).

REVIEW OF PREVIOUS PLATELET STUDIES

There have been at least 50 reports concerning platelet α_2ARs in depression. A review of this field[32] indicates that, by and large, studies using non-imidazoline

TABLE 1. Platelet Imidazoline Binding Studies in Depression

Measurement Used	Number of Labs Reporting: No Changes	Changes	Findings in Depression up to 1993[a]
[^{3}H] Clonidine binding	2	6	Increased B_{max}
[^{3}H] *p*-Aminoclonidine binding	0	1	Increased binding
	0	1	Decreased B_{max}

[a] See refs. 28 and 32 for reviews of these data.

radioligands demonstrated no differences from those in healthy subjects. By contrast, as shown in TABLE 1, most studies using radiolabeled imidazolines [^{3}H]clonidine or [^{3}H]*p*-aminoclonidine demonstrated increased B_{max} values in depressed patients than in healthy control subjects.[28] These data were interpreted solely as reflecting α_2AR effects, because I_1 sites were not known to exist on platelets at the time. Nevertheless, the inconsistency across studies and the failure to demonstrate an increased B_{max} value using purer α_2AR ligands, such as [^{3}H]UK,14304,[33] have made this field difficult to interpret. One suggestion, based on the results to be described, is that the earlier studies with [^{3}H]clonidine and [^{3}H]*p*-aminoclonidine may have inadvertently radiolabeled both α_2ARs and I_1 sites, leading to the inconsistencies.

REVIEW OF THREE PLATELET STUDIES OF α_2AR/I_1 SITES IN DEPRESSION

Thus far, ours is the only laboratory to distinguish α_2AR and I_1 sites on platelets of depressed patients. This was achieved by using [^{3}H]*p*-aminoclonidine binding in one study[29] and [^{125}I]*p*-iodoclonidine binding in two additional studies[27,28] of depression. A good close correlation ($r = 0.888$) was observed between the findings with either radioligand, and therefore the results were merged in TABLE 2. Therefore, it is of interest that not withstanding occasional interstudy exceptions,[28] our results demonstrate no significant differences between α_2AR B_{max} values in depressed patients and those in controls. By contrast, platelet I_1 site B_{max} values in depressed patients in all three studies have tended to be increased compared to those in healthy subjects. Moreover, following treatment of the patients with desipramine for 6 weeks, normal I_1 B_{max} values were highly consistent. Detailed studies can be found in our earlier publications.[27–29] Thus, our findings indicate increased platelet I_1 sites in depression with no or only minor changes in the α_2AR.

PRIMARY DATA CORRELATING α_2AR AND I_1 SITES IN DEPRESSION

Our recent data provide evidence for a possible interrelationship between α_2AR and I_1 binding sites on human platelets. This had been speculated to occur, because

TABLE 2. Summary of Three Platelet α_2AR and I_1 Studies in Depression

	α_2AR Site Study Means		I_1 Site Study Means	
Subjects	K_D (nM)	B_{max} (fmol/mg)	K_D (nM)	B_{max} (fmol/mg)
Depressed, pretreatment	0.2	121	3.2	246[a]
	(total $n = 70$)		(total $n = 66$)	
Depressed, post desipramine	0.2	95[b]	1.6	136[c]
	(total $n = 43$)		(total $n = 40$)	
Controls, healthy volunteers	0.3	98	2.8	137
	(total $n = 56$)		(total $n = 46$)	

[a] I_1 B_{max} in depressed (untreated) patients versus I_1 B_{max} in healthy subjects: study 1, $p = 0.005$; study 2, $p = 0.001$; study 3, $p = 0.078$, two-tailed.

[b] α_2AR B_{max} in depressed (treated) patients versus α_2AR B_{max} in depressed patients (untreated): study 1, $p = 0.04$; study 2, $p = 0.006$; study 3, $p > 0.5$. Similarly, the α_2AR B_{max} in patients did not significantly differ from that in healthy controls in any of the studies.

[c] I_1 B_{max} in depressed (treated) patients versus I_1 B_{max} in depressed patients (untreated): study 1, $p = 0.04$; study 2, $p = 0.0004$; study 3, $p = 0.01$. (The primary data for these studies are in refs. 27 to 29.)

both of these sites were reported to be inhibitory receptors,[13] and it was felt that certain relationships might exist. Therefore, correlations in the [^{125}I]iodoclonidine binding parameters are described.

Binding site density (B_{max}) was the first parameter under consideration. B_{max} values at α_2AR and I_1 binding sites in the overall group of subjects (patients and controls) were not correlated ($r = -0.017$, $p = 0.892$). Accordingly, the α_2AR and I_1 B_{max} values were not correlated in pretreated patients ($r = -0.218$, $p = 0.295$), posttreatment patients ($r = -0.129$, $p = 0.581$), or all healthy subjects ($r = 0.352$, $p = 0.152$). However, subgrouping of the sexes among the healthy subjects revealed a robust correlation between α_2AR and I_1 B_{max} values in the males (FIG. 1, top panel) ($r = 0.653$, $p = 0.002$). A nonsignificant trend in the same direction was also seen in the healthy female subjects ($r = 0.263$, $p = 0.133$), but further subgrouping by phases of the menstrual cycle revealed little (FIG. 1, bottom panel). It should be noted that this finding among males was confined to healthy subjects and did not appear among the depressed male patients. Although this is not an overwhelming finding, one interpretation is that a relationship between α_2AR and I_1 B_{max} values exists in only healthy males because this is the less confounded group. For instance, it may be possible that female hormonal variation and/or the presence of depression in the other groups has disrupted an underlying relationship between the receptors.

Binding site affinity (K_D) was the second parameter under consideration. The α_2AR and I_1 site K_D values were highly correlated in all subject groups (FIG. 2). K_D values at α_2AR and I_1 binding sites in the overall group of subjects (both patients and controls) were correlated ($r = 0.546$, $p = 0.0002$), as were the α_2AR and I_1 site

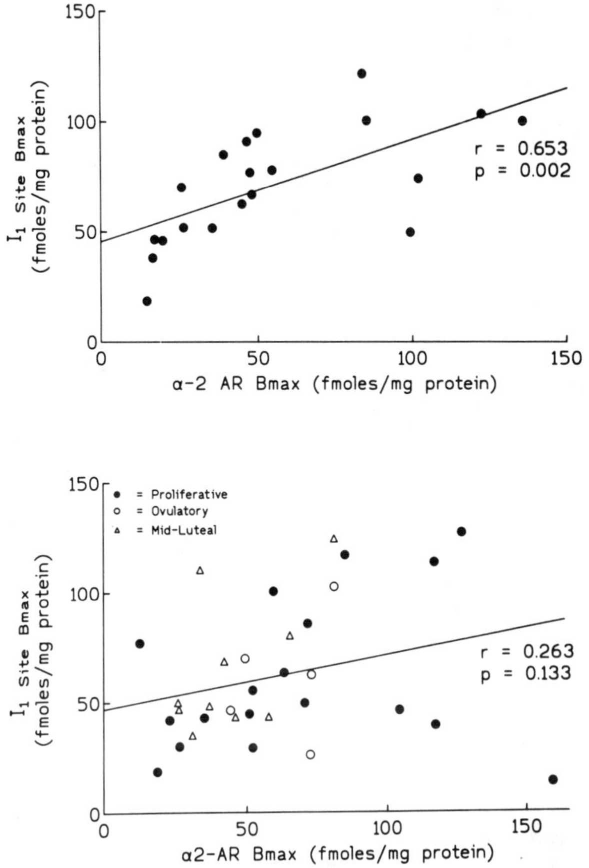

FIGURE 1. Correlation of B_{max} values using [^{125}I]*p*-iodoclonidine binding to platelets from healthy males (*top panel*) and healthy females (*lower panel*).

K_D values in the subgroups: pretreated patients ($r = 0.588$, $p = 0.002$), posttreatment patients ($r = 0.539$, $p = 0.012$), and healthy subjects ($r = 0.511$, $p = 0.030$) were all significantly correlated. Males and females were both correlated, and their K_D values did not differ significantly in the degree of correlation. This high degree of correlation between the binding affinity of these two sites implies some interrelationship.

Plasma catecholamines were measured next to establish additional interrelationships within the data. As expected, B_{max} values (but not K_D values) at platelet α_2AR sites were negatively correlated with plasma epinephrine concentrations. By comparison, no correlations were found between either norepinephrine or dopamine concentrations and binding parameters, and no correlations between catecholamines were evident at the I_1 site. Epinephrine concentrations were clearly correlated with α_2AR B_{max} values in the overall group of subjects (both patients and controls; $r =$

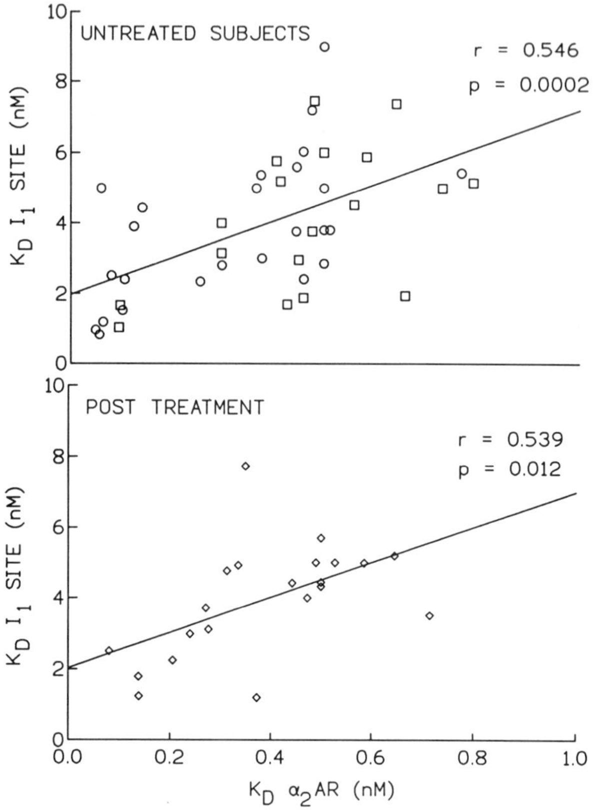

FIGURE 2. Correlation of K_D values using [^{125}I]*p*-iodoclonidine binding to platelets from untreated subjects (*top panel*) and posttreatment subjects (*lower panel*). Untreated subjects included healthy subjects (○) and depressed patients (□). Depressed patients, posttreatment with either desipramine or fluoxetine, are plotted together (◇).

−0.341, $p = 0.005$). Subgroups also demonstrated trends in the same direction; pretreated patients ($r = -0.502, p = 0.009$), posttreatment patients ($r = -0.329, p = 0.135$), and healthy subjects ($r = -0.099, p = 0.695$) showed correlations between α_2AR B_{max} values and epinephrine concentrations. Thus, the platelet α_2AR site displayed the predicted autologous downregulation by its endogenous agonist epinephrine. However, neither of the relevant B_{max} or K_D correlations observed earlier between the α_2AR and I_1 sites (FIGS. 1 and 2) were interrelated with catecholamine concentrations. Thus, although plasma epinephrine concentrations appear to regulate platelet α_2AR density, epinephrine does not explain the interrelationships we have observed between α_2AR and I_1 sites.

DISCUSSION

A review of the literature (TABLE 1) and of our own findings (TABLE 2) indicates that platelet I_1 sites are increased in depression. By comparison, little, if any, aberration in platelet α_2AR is noted in depression. Furthermore, platelet I_1 sites appear to be downregulated (normalized) following antidepressant therapy with desipramine. Taken together with findings from brain autopsy tissue,[17] these results provide considerable evidence for I_1 receptor involvement in depression. The nature of that involvement and its relevance to the pathophysiology of depression remain open issues. However, the strong correlation observed in this study between the affinity of I_1 and α_2AR binding sites may provide some clues.

One possibility that cannot be disproved at this time is that α_2AR and I_1 sites are structurally linked. However, our results tend to rule out that possibility on the grounds that these two sites appeared to be differentially regulated. First, only the I_1 site was affected in depressed patients (TABLE 2; only I_1 sites were increased). Second, only I_1 sites were downregulated by desipramine therapy (TABLE 2). Third, as expected since I_1 sites have low affinity for epinephrine, only α_2AR sites displayed the characteristic downregulatory pattern expected in conjunction with plasma epinephrine. Thus, it is unlikely that α_2AR and I_1 sites are structurally linked.

Another factor that links α_2AR and I_1 receptors is that they both have been ascribed inhibitory functions on neuronal firing.[13,15] Thus, these receptors might be functionally interrelated. It is our current working hypothesis that a functional interrelationship underlies our observed correlations between K_D values for $[^{125}I]$iodoclonidine binding at these two receptors. If this is the case, it is tempting to speculate further that a supersensitive inhibitory function of I_1 receptors (perhaps predicted by the increase in platelet I_1 sites in depression) might underlie the depletion of monoamine release in depression. Obviously, verification of this hypothesis awaits further study.

REFERENCES

1. SCHILDKRAUT, J. J. 1965. The catecholamine hypothesis of affective disorders: A review of the supporting evidence. Am. J. Psychiatry **122:** 509-522.
2. CALDECOTT-HAZARD, S., D. G. MORGAN, F. DELEON-JONES, D. H. OVERSTREET & D. JANOWSKY. 1991. Clinical and biochemical aspects of depressive disorders. II. Transmitter/receptor theories. Synapse **9:** 251-301.
3. STARKE, K. 1987. Presynaptic α-autoreceptors. Rev. Physiol. Biochem. Pharmacol. **107:** 73-146.
4. GARCIA-SEVILLA, J. A., A. P. ZIS, P. J. HOLLINGSWORTH, J. F. GREDEN & C. B. SMITH. 1981. Platelet alpha-2 adrenergic receptors in major depressive disorder. Arch. Gen. Psychiatry **38:** 1327-1333.
5. OSMAN, O. T., M. V. RUDORFER & W. Z. POTTER. 1989. Idazoxan: A selective α2-antagonist and effective sustained antidepressant in two bipolar depressed patients. Arch. Gen. Psychiatry **46:** 958-959.
6. LANGER, S. Z. 1981. Presynaptic regulation of the release of catecholamines. Pharmacol. Rev. **32:** 337-362.
7. RAITERI, M., G. MAURA & P. VERSACE. 1983. Functional evidence for two stereochemically different alpha-2 adrenoceptors regulating central norepinephrine and serotonin release. J. Pharmacol. Exp. Ther. **224:** 679-684.

8. Tibirica, E., J. Feldman, C. Mermet, F. Gonon & P. Bousquet. 1991. An imidazoline-specific mechanism for the hypotensive effect of clonidine: A study with yohimbine and idazoxan. J. Pharmacol. Exp. Ther. **256:** 606.
9. Powis, D. A. & P. F. Baker. 1986. α_2-Adrenoceptors do not regulate catecholamine secretion by bovine adrenal medullary cells: A study with clonidine. Mol. Pharmacol. **29:** 134-141.
10. Nguyen, T. T. & A. DeLean. 1987. Nonadrenergic modulation by clonidine of the cosecretion of catecholamines and enkephalins in adrenal chromaffin cells. Canad. J. Physiol. & Pharmacol. **65:** 823-827.
11. Ohara-Imaizumi, M. & K. Kumakura. 1992. Effects of imidazole compounds on catecholamine release in adrenal chromaffin cells. Cell. Molec. Neurobiol. **12:** 273-283.
12. Orts, A., C. Orellana, T. Canto, V. Cena, C. Gonzalez-Garcia & A. G. Garcia. 1987. Inhibition of adrenomedullary catecholamine release by propranolol isomers and clonidine involving mechanisms unrelated to adrenoceptors. Br. J. Pharmacol. **92:** 795-801.
13. Gothert, M. & G. J. Molderings. 1991. Involvement of presynaptic imidazoline receptors in the alpha2-adrenoceptor independent inhibition of noradrenaline release by imidazoline derivatives. Naunyn-Schmiedeberg's Arch. Pharmacol. **343:** 271-282.
14. Molderings, G. J., F. Hentrich & M. Gothert. 1991. Pharmacological characterization of the imidazoline receptor which mediates inhibition of noradrenaline release in the rabbit pulmonary artery. Naunyn-Schmiedeberg's Arch. Pharmacol. **344:** 630-638.
15. Fuder, H. & P. Schwarz. 1993. Desensitization of inhibitory prejunctional alpha-2 adrenoceptors and putative imidazoline receptors on rabbit heart sympathetic nerves. Naunyn-Schmiedeberg's Arch. Pharmacol. **348:** 127-133.
16. Devos, H., G. Bricca, J. DeKeyseer, J. DeBacker, P. Bousquet & G. Vauquelin. 1994. Imidazoline receptors, non-adrenergic idazoxan binding sites and alpha-2 adrenoceptors in the human central nervous system. Neuroscience **59:** 589-598.
17. Meana, J. J., F. Barturen, I. Martin & J. A. Garcia-Sevilla. 1993. Evidence of increased non-adrenoceptor [^{3}H]Idazoxan binding sites in the frontal cortex of depressed suicide victims. Biol. Psychol. **34:** 498-501.
18. Meana, J. J., F. Barturen & J. A. Garcia-Sevilla. 1992. α2-Adrenoceptors in the brain of suicide victims: Increased receptor density associated with major depression. Biol. Psychiatry **31:** 471-490.
19. Renouard, A., P. S. Widdowson & A. Cordi. 1993. [^{3}H]-idazoxan binding to rabbit cerebral cortex recognizes multiple imidazoline I2-type receptors: Pharmacological characterization and relationship to monoamine oxidase. Br. J. Pharmacol. **109:** 625-631.
20. Olmos, G., A. M. Gabilondo, A. Miralles, P. V. Escriba & J. A. Sevilla. 1993. Chronic treatment with the monoamine oxidase inhibitors clorgyline and pargyline down-regulates non-adrenoceptor [^{3}H]-idazoxan binding sites in the rat brain. Br. J. Pharmacol. **108:** 597-603.
21. Siever, L. J. & T. W. Uhde. 1984. New studies and perspectives on the noradrenergic receptor system in depression: Effects of the alpha-2 adrenergic agonist clonidine. Biol. Psychol. **19:** 131-156.
22. Manji, H. K., M. V. Rudorfer & W. Z. Potter. 1994. Affective disorders and adrenergic function. *In* Adrenergic Dysfunction and Psychobiology. O. G. Cameron, Ed.
23. Da Prada, M., A. M. Cesura, J. M. Launay & J. G. Richards. 1988. Platelets as a model for neurons? Experientia **44:** 115-126.
24. Hollister, A. S., G. A. Fitzgerald, J. H. J. Nadeau & D. Robertson. 1983. Acute reduction in human platelet α2-adrenoreceptor affinity of agonist by endogenous and exogenous catecholamines. J. Clin. Invest. **72:** 1498-1505.
25. Propping, P., W. Friedl, J. Hebebrand & K.-U. Lentes. 1986. Genetic studies at the receptor level: Investigations in human twins and experimental animals. *In* Antidepressants and Receptor Function. Vol. 42-56. Wiley. Chichester.

26. Piletz, J. E. & K. Sletten. 1993. Nonadrenergic imidazoline binding sites on human platelets. J. Pharmacol. Exp. Ther. **267:** 1493-1502.
27. Piletz, J. E., D. Chikkala, L. Khaitan, K. Jackson & Y. S. Qu. 1993. Delayed desensitization of alpha-2 adrenoceptor-mediated platelet aggregation in depressed patients. Neuropsychopharmacology **9:** 55-66.
28. Piletz, J. E., P. R. Ernsberger & A. Halaris. 1995. Imidazoline and alpha-2 adrenergic receptors in depression. CRC Crit. Rev. Neurobiol. In press.
29. Piletz, J. E., A. Halaris, A. Saran & M. R. Marler. 1991. Desipramine lowers 3H-para-aminoclonidine binding in platelets of depressed patients. Arch. Gen. Psychiatry **48:** 813-820.
30. Corash, L. 1980. Platelet heterogeneity: Relevance to the use of platelets to study psychiatric disorders. Schizophrenia Bull. **6:** 254-258.
31. McPherson, C. A. 1985. Analysis of radioligand binding experiments. A collection of computer programs for the IBM PC. J. Pharmacol. Methods **14:** 213-228.
32. Grossman, F., H. K. Manji & W. Z. Potter. 1993. Platelet alpha-2 adrenoceptors in depression: A critical examination. J. Psychopharmacol. **7:** 4-18.
33. Theodorou, A. E., A. S. Hale, S. L. Davies, C. L. E. Katona, Y. Yamaguchi, R. W. Horton, J. S. Kelly & E. S. Paykal. 1986. Platelet high affinity adrenoceptor binding sites labelled with the agonist [^{3}H]UK-14,304 in depressed patients and matched controls. Eur. J. Pharmacol. **126:** 329-332.

Decreased Number and Immunoreactivity of I_2-Imidazoline Receptors in the Frontal Cortex of Suicide Victims[a]

MAGDALENA SASTRE, PABLO V. ESCRIBÁ, DONALD J. REIS,[b] AND JESÚS A. GARCÍA-SEVILLA

Laboratory of Neuropharmacology
Department of Fundamental Biology and Health Sciences
University of the Balearic Islands
E-07071 Palma de Mallorca, Spain

[b]*Division of Neurobiology*
Cornell University Medical College
New York, New York 10021

In recent pilot studies, imidazoline sites labeled by [^{3}H]*p*-aminoclonidine (proposed I_1-IRs) were increased and downregulated after desipramine treatment in platelets from depressed patients.[1] Moreover, competition experiments with norepinephrine/[^{3}H]idazoxan revealed an increase in nonadrenoceptor [^{3}H]idazoxan binding sites (proposed I_2-IRs) in the frontal cortex of depressed suicide victims.[2] In the rat brain, I_2-IRs do not appear to be modulated by desipramine, but these putative receptors are markedly downregulated by chronic treatment with various monoamine oxidase inhibitor antidepressant drugs.[3] The present study was designed to further assess the status of IRs in postmortem brains from suicide victims by saturation experiments with [^{3}H]idazoxan and the immunological detection of IRs with a specific antibody.

Human brains were obtained from 25 control subjects (17 men, 8 women; age: 50 ± 4 years) and from 26 subjects (21 men, 5 women; age: 52 ± 4 years) who had died by suicide. Frontal cortex-Brodmann's area 9 was dissected at necropsy, frozen, and stored at −70°C until assay. After thawing, preparation of neural membranes (P_2 fractions) was done as previously described.[4] Binding of [^{3}H]idazoxan to brain I_2-IR was always done in the presence of 10^{-6} M (−)epinephrine. Specific binding, defined as the difference between total and nonspecific binding (determined with 10^{-4} M naphazoline), ranged from 97–30% at 6×10^{-10} M to 3×10^{-8} M [^{3}H]idazoxan. Analysis of saturation isotherms was performed by computer-assisted nonlinear regression using EBDA-LIGAND programs.[5] Levels of IRs were also assessed by immuno-(western) blot using a specific antiserum against the IRs purified from bovine adrenal chromaffin cells.[6] Tissue (50 mg) was homogenized in 2 ml of 50 mM Tris-HCl buffer, pH 7.5, containing 1 mM EDTA, 2 mM $MgCl_2$, 1 mM

[a]This study was supported by grants DGICYT PM 91-0069 and FIS 93-0641 to J.A.G.S.

phenylmethylsulfonylfluoride (1 mM), and iodoacetamide (5 mM). Aliquots of 1.4 ml of this suspension were centrifuged at 4°C for 30 minutes at 40,000 × *g*. The resulting pellets were resuspended in 250 μl of 40 mM Tris-HCl buffer, pH 6.8, and 4% sodium dodecyl sulfate (SDS) and incubated at 75°C for 5 minutes. After the addition of 120 μl of loading buffer (62.5 mM Tris-HCl, pH 6.8, 3% SDS, 20% glycerol, and 0.005% bromophenol blue), samples were boiled for 3 minutes and submitted to 10% polyacrylamide gel electrophoresis (SDS-PAGE). Proteins were then transferred to nitrocellulose membranes (immunoblot, western blot) and incubated at room temperature for 1 hour in phosphate-buffered saline solution containing 10% nonfat dry milk and 0.05% Tween 20 (blocking solution). Membranes were then incubated in fresh blocking solution containing the primary antibody (1 : 5,000 dilution) at 4°C overnight. The secondary antibody, a horseradish peroxidase-linked donkey anti-rabbit IgG (1 : 5,000 dilution), was incubated for 2 hours at room temperature in blocking solution. Immunoreactivity was detected with the Enhanced Chemiluminescence (ECL) Western Blot Detection System (Amersham) followed by exposure to Amersham ECL-hyperfilm. The integrated optical densities of the bands were measured with the Bio Image (Millipore) image analyzer. Samples were evaluated using standard curves (i.e., total protein loaded versus integrated optical density), consisting of at least four points of different protein content (usually 5-50 μg of protein) of a control subject, all loaded in the same gel. The mean intraassay coefficient of variation was 10% (same subject). The quantitation procedure was performed as described for other proteins.[7]

The specific binding of [^{3}H]idazoxan in the suicide group (B_{max} = 45 ± 2 fmol/mg protein) was decreased compared to that in the control group (B_{max} = 75 ± 8 fmol/mg protein) (40% reduction, p <0.001). K_d values were similar in both groups (suicide, K_d: 9.4 ± 0.8 nM; control, K_d: 10.5 ± 0.9 nM). The binding parameters of [^{3}H]idazoxan in the frontal cortex in healthy men were not different from those obtained in healthy women.

On the other hand, the immunological detection of IRs in the human frontal cortex revealed a double band of 29-30 kD, a less intense band of 45 kD, and in some subjects one more band of 66 kD. In control and suicide subjects together, the density of I_2-IR labeled by [^{3}H]idazoxan correlated with the total immunoreactivity of the various bands (r = 0.64, n = 18, p <0.005) and with the 29-30 kD band immunoreactivity (r = 0.69, n = 18, p <0.005), suggesting that this latter protein detected by immunoblot analysis was related to I_2-IR. However, there was no correlation between the specific binding of [^{3}H]idazoxan and the immunoreactivity of the 45-kD band (r = –0.26, n = 10, p >0.05).

In line with [^{3}H]idazoxan binding data, a reduction in the total immunoreactivity of IRs (14 ± 4%, n = 13, p < 0.01) and in the 29-30 kD band (18 ± 3%, n = 13, p <0.001) was observed in the frontal cortex of suicides. By contrast, the 45-kD band immunoreactivity was increased in the same suicide victims (47 ± 20%, n = 8, p <0.05).

Several conclusions can be deduced from the current data: (1) the density of I_2-imidazoline receptors (I_2-IR) quantitated by means of [^{3}H]idazoxan saturation binding assays is decreased (40%) in the frontal cortex of suicide victims; (2) the correlation between immunoreactivity of the 29-30-kD band and data for [^{3}H]idazoxan binding in the same brains suggests that this protein is related to I_2-IRs; (3) the levels of immunoreactivity of the 29-30-kD protein are also reduced (18%) in the frontal cortex of suicide subjects compared with matched controls; and (4) other forms of IRs (e.g., a 45-kD protein) can be upregulated in suicide.

REFERENCES

1. PILETZ, J. E., A. HALARIS, A. SARAN & M. R. MARLER. 1991. Desipramine lowers tritiated *para*-aminoclonidine binding in platelets of depressed patients. Arch. Gen. Psychiatry **48:** 813-820.
2. MEANA, J. J., F. BARTUREN, I. MARTIN & J. A. GARCÍA-SEVILLA. 1993. Evidence of increased non-adrenoceptor [^{3}H]idazoxan binding sites in the frontal cortex of depressed suicide victims. Biol. Psychiatry **34:** 498-501.
3. OLMOS, G., A. M. GABILONDO, A. MIRALLES, P. V. ESCRIBÁ & J. A. GARCÍA-SEVILLA. 1993. Chronic treatment with the monoamine oxidase inhibitors clorgyline and pargyline down-regulates non-adrenoceptor [^{3}H]-idazoxan binding sites in the rat brain. Br. J. Pharmacol. **108:** 597-603.
4. SASTRE, M. & J. A. GARCÍA-SEVILLA. 1993. Opposite age-dependent changes of α_2-adrenoceptors and nonadrenoceptor [^{3}H]idazoxan binding sites (I_2 imidazoline sites) in the human brain: Strong correlation with monoamino oxidase-B sites. J. Neurochem. **61:** 881-889.
5. MUNSON, P. J. & D. RODBARD. 1980. LIGAND: A versatile computerized approach for the characterization of ligand binding systems. Anal. Biochem. **107:** 220-239.
6. WANG, H., S. REGUNATHAN, D. A. RUGGIERO & D. J. REIS. 1993. Production and characterization of antibodies specific for the imidazoline receptor protein. Mol. Pharmacol. **43:** 509-515.
7. ESCRIBÁ, P. V., M. SASTRE & J. A. GARCÍA-SEVILLA. 1994. Increased density of Guanine nucleotide-binding proteins in the postmortem brains of heroin addicts. Arch. Gen. Psychiatry **51:** 494-501.

Human Cerebral Imidazoline Binding Sites

Further Biochemical Investigations

H. GRENEY, M. DONTENWILL, G. BRICCA,
A. BELCOURT, AND P. BOUSQUET

Laboratoire de Pharmacologie Cardiovasculaire et Rénale
CNRS URA 589
Faculté de Médecine
Université Louis Pasteur
11, rue Humann
67000 Strasbourg, France

The central hypotensive action of imidazolines and related drugs involves specific receptors distinct from the α-adrenoceptors and located in the rostral ventrolateral medulla (RVLM).[1] Binding studies revealed the existence of [^{3}H]clonidine and [^{3}H] idazoxan binding sites insensitive to catecholamines in human brain membranes prepared from this region.[2,3] The aim of this study was to purify these imidazoline binding sites from the human brain stem. Two [^{3}H]idazoxan binding sites have already been purified from the rabbit kidney[4] and from bovine chromaffin cells.[5] Nevertheless, these two binding sites seem different from the human cerebral imidazoline binding site. Indeed, first, both are sensitive to amiloride, whereas the human central one proved insensitive, and second, the rabbit imidazoline binding site is sensitive (IC_{50} = 22.5 nM) to a derivative of cirazoline, (2[3-aminophenoxy]methyl imidazoline), whereas the human central one is insensitive to this molecule ($IC_{50} > 1$ μM). On the basis of different sensitivity to these drugs,[6] heterogeneity of imidazoline binding sites is thus confirmed.

We decided to purify the human brain imidazoline binding sites to determine how many proteins exist and also if [^{3}H]clonidine and [^{3}H]idazoxan binding proteins are different. [^{3}H]Clonidine and [^{3}H]idazoxan binding sites have being solubilized from the RVLM region with 13 mM CHAPS.[7] These solubilized imidazoline binding sites kept their binding properties similar to that observed in membrane preparation: saturability, high affinity, stereoselectivity, and insensitivity to catecholamines and to amiloride. It is important to note that the affinity of clonidine for the solubilized [^{3}H]idazoxan binding site was higher than that for membranes. This was already observed by Wang *et al.*[5] with the [^{3}H]idazoxan binding site purified from bovine chromaffin cells. Analysis of solubilized imidazoline binding sites by gel filtration on Sephacryl S200 columns showed that [^{3}H]clonidine and [^{3}H]idazoxan binding activity eluted in the same fractions, that is, the void volume and a fraction corresponding to an apparent molecular weight of 35 kD.

Purification was started according to two procedures. In one, the solubilized imidazoline binding sites were passed through an affinity column, namely, NH_2-

idazoxan linked on an epoxy-Sepharose 6B matrix. About 75% of the [^{3}H]idazoxan binding activity was found in the pass-through when the 25% remaining was eluted with 1 M KCl. This last eluted fraction was then submitted to gel filtration chromatography, allowing (1) further purification of this [^{3}H]idazoxan binding site and (2) desalting of the fraction eluted from affinity chromatography. The fraction containing the binding activity was then analyzed by SDS-PAGE, and the silver-stained gel revealed bands with apparent molecular weights of 65, 55, 45, and 20 kD.

In a second procedure, another affinity chromatography allowed isolation of both [^{3}H]clonidine and [^{3}H]idazoxan binding sites.[8] Western blot analysis of this affinity chromatography-purified fraction was performed with an antiidiotypic antibody. These antibodies inhibited [^{3}H]clonidine and [^{3}H]idazoxan binding to human membranes.[9] These antibodies labeled one 43-kD major band on immunoblots when the preimmune serum did not reveal any band.

In conclusion, we have shown that: (1) based on binding studies, human central imidazoline binding sites seem to be different from those of rabbit kidney and bovine chromaffin cells described elsewhere[4,5]; (2) after solubilization, [^{3}H]clonidine and [^{3}H]idazoxan binding sites from human brain keep their binding properties as compared to the membrane sites; (3) proteins bearing [^{3}H]clonidine and [^{3}H]idazoxan binding activity coeluted whatever chromatography procedure was used; these results seem to indicate that either [^{3}H]clonidine or [^{3}H]idazoxan binding was located on the same site or that these two sites are closely related; and (4) a protein of about 40 kD molecular weight represents a good candidate for the imidazoline binding protein as assessed by SDS-PAGE or Western blot analysis. Further characterization of this purified fraction is underway in this laboratory.

REFERENCES

1. Bousquet, P., J. Feldman & J. Schwartz. 1984. Central cardiovascular effects of alpha-adrenergic drugs. Differences between catecholamines and imidazolines. J. Pharmacol. Exp. Ther. **230:** 232-236.
2. Bricca, G., H. Greney, M. Dontenwill-Kieffer, J. Zhang, A. Belcourt & P. Bousquet. 1993. Heterogeneity of the specific imidazoline binding of [^{3}H]idazoxan in the human cerebral cortex. Neurochem. Int. **22:** 153-163.
3. Bricca, G., H. Greney, J. Zhang, M. Dontenwill, J. Stutzmann, A. Belcourt & P. Bousquet. 1994. Human brain imidazoline receptors: Further characterization with [^{3}H]clonidine. Eur. J. Pharmacol. **266:** 25-33.
4. Limon, I., I. Coupry, S. M. Lanier & A. Parini. 1992. Purification and characterization of mitochondrial imidazoline-guanidinium receptive site from rabbit kidney. J. Biol Chem. **267:** 21645-21649.
5. Wang, H., S. Regunathan, M. P. Meeley & D. J. Reis. 1992. Isolation and characterization of imidazoline receptor protein from bovine adrenal chromaffin cells. Mol. Pharmacol. **42:** 792-801.
6. Greney, H., A. Molines, P. Bousquet & M. Dontenwill. 1995. Heterogeneity of imidazoline binding sites revealed by AMPI, a cirazoline derivative. Eur. J. Pharmacol. **271:** 533-536.
7. Greney, H., G. Bricca, M. Dontenwill, J. Stutzmann, P. Bousquet & A. Belcourt. 1994. Characterization of imidazoline binding protein(s) solubilized from human brainstem: Studies with [^{3}H]idazoxan and [^{3}H]clonidine. Neurochem. Int. **25:** 183-191.

8. GRENEY, H., F. BENNAI, M. DONTENWILL, A. BELCOURT & P. BOUSQUET. 1995. Purification of imidazoline binding sites for the human brainstem. **265:** R1-R2.
9. BENNAI, F., H. GRENEY, J. STUTZMANN, A. BELCOURT, P. BOUSQUET & M. DONTENWILL. 1994. Polyclonal anti-idiotypic antibodies to idazoxan and their interaction with imidazoline binding sites: Revelation of a 43 Kdalton protein in human brain preparations. Submitted.

Imidazoline Receptors and Cardiovascular Regulations

A Statement

PASCAL BOUSQUET,[a] HUGUES GRENEY,
FATIMA BENNAI, JOSIANE FELDMAN,
JEANNE STUTZMANN, ALAIN BELCOURT, AND
MONIQUE DONTENWILL

Laboratoire de Pharmacologie Cardiovasculaire et Rénale
CNRS URA 589
Faculté de Médecine
Université Louis Pasteur
67000, Strasbourg, France

The relation between the imidazoline receptors and the regulation of cardiovascular function is approached from two aspects, the first one being pathophysiological and functional and the second one truly biochemical.

PATHOPHYSIOLOGICAL AND FUNCTIONAL ASPECTS

Primary Hypertension

With poly- and monoclonal antibodies, the presence of an immunoreactive substance was detected in human sera. These antibodies recognized very well the imidazolines and more particularly the amino-imidazolines when they did not cross-react with any of a whole series of endogenous neuromediators and hormones, among which were the endogenous catecholamines as well as histamine, adenine, or purine.[1–3] In fact, two antibodies, one polyclonal and one monoclonal, exhibited close specificity. Radioimmunoassays developed with both antibodies have had similar results. With these radioimmunoassays, the presence of a circulating substance was first detected in the sera of normotensive subjects; its concentration in the sera of normotensive and hypertensive subjects was then quantified with the aid of a unit defined on the basis of calibration curves of the radioimmunoassays. Under these conditions and from measurements taken in 26 normotensive subjects, an upper limit of normal was set at around 75 units. In sera from a series of 32 patients with primary hypertension, about 30% exhibited values exceeding, sometimes greatly, this upper limit; some even reached 400 units.[3]

[a] To whom correspondence should be addressed.

In a short series of secondary hypertension, such as renovascular hypertension or Conn's syndrome, levels of that circulating immunoreactive substance were always within the limits established with the sera of normotensive subjects. Although that series is still of too short duration to allow any conclusions, our study strongly suggests that in a significant proportion of these subjects, primary hypertension may be associated with high levels of that circulating immunoreactive substance, provisionally called "imidazoline-like." This result led to questions about the identity of this circulating immunoreactive substance and the endogenous ligand of the imidazoline receptors or one of its metabolites. If they are identical, questions will arise about the significance of the cause-and-effect relation between the high levels of that substance present in hypertensive patients and the hypertension itself. Indeed, it must be demonstrated that this excess is involved in the genesis of the hypertension itself or at least in the maintenance of high blood pressure. To make progress in this area, it is necessary to identify the relation between the immunoreactive circulating substance and the neuromediator specific for imidazoline receptors, and this cannot be done until both substances are purified. This work is now in progress.

In fact, human blood extract was already partially purified. It exhibits some exciting characteristics; on the one hand, it cross-reacts with the antibodies in the ELISA tests; on the other hand, it raises arterial blood pressure significantly when injected directly into the ventrolateral area of the rabbit's medulla where imidazolines and related compounds induce their hypotensive effect.[4,5] However, we are still faced with technical difficulties in completely purifying that molecule.

When that substance is purified, antibodies raised directly against it will be developed to confirm our results before identifying a category of hypertension for which it is possible to claim that dysfunction in the imidazoline system, at least of the endogenous ligand, is involved. The possibility of dysfunction of the receptors themselves must also be investigated; for instance, is it "a down or even more likely an up regulation?"

An Additional Antiarrhythmic Effect?

Another recent aspect in the development of functional research in the field of imidazoline receptors in connection with the cardiovascular system concerns an additional property of imidazoline-type substances that is of potential therapeutic interest. In fact, an antiarrhythmic property of imidazoline-type or related substances was recently observed. In more than 30% of cases, hypertension was frequently associated with left ventricular hypertrophy which itself constitutes a risk factor for ventricular arrhythmia, which is frequently responsible for sudden death. In addition, it is extensively documented that reduction in left ventricular mass in hypertensive patients is unfortunately insufficient to decrease that arrhythmogenic risk. The existence of an antiarrhythmic property with a powerful hypotensive effect would be a major asset in the specification of a substance that could be proposed as a modern antihypertensive drug.

This problem is approached in our laboratory mainly from the point of view of ventricular arrhythmias induced by intracerebral manipulations. An extremely efficient model of ventricular arrhythmias of central origin results from the central injection of bicuculline.[6] Immediately after its intracisternal injection, bicuculline produces either isolated extrasystoles or salvos or even ventricular tachycardia which often proceeds to fibrillation and death. These experiments were performed in anesthetized rabbits. The effect of bicuculline is directly associated with sympathetic hyperactivity. When animals were pretreated with 15 μg/kg of clonidine or 1 mg/kg of rilmenidine injected intravenously 10 minutes before the injection of bicuculline, interesting qualitative and quantitative prevention of these arrhythmias was observed. These results strengthen the general idea that the central nervous system represents a target for the development of antiarrhythmic drugs and also that drugs with central antihypertensive effects mediated by the imidazoline receptors present a non-negligible interest from this point of view. This concept still has to be validated in other models of ventricular tachyarrhythmias. In this connection, an interesting effect of imidazolines and related drugs, particularly rilmenidine, was observed in a model of arrhythmias induced by a general anesthetic.[7]

BIOCHEMICAL ASPECTS

The relation between imidazoline receptors and cardiovascular function from a biochemical point of view is based on a theoretical assumption: imidazoline receptors that are more clearly involved in regulating cardiovascular function are logically those involved in the central hypotensive effect of clonidine and related substances. These specific sites will thus be sensitive to clonidine, insensitive to catecholamines. These are also the receptors that are sensitive to idazoxan and that Bricca *et al.*[9] showed insensitive to GppNHp.[8] In fact, these sites are almost insensitive to *p*-aminoclonidine. In the ventrolateral area of the human medulla where we performed all of our experiments of characterization and purification *p*-aminoclonidine competed only weakly with the binding of tritiated clonidine.[9] No imidazoline-specific binding of tritiated *p*-aminoclonidine was observed in membrane preparations sampled from that area of the human brain (unpublished data). In addition, there are sites sensitive to other imidazolines such as idazoxan which recognized clonidine and its derivatives either poorly or not at all.[8]

The existence of these receptors, which perhaps will be called I_1-receptors from now on, was proposed at the conclusion of the structure/activity relationship study performed in 1984, which indicated that only substances with an imidazoline structure were likely to induce a hypotensive effect from the nucleus reticularis lateralis (NRL)/rostroventrolateral medulla (RVLM) region when the phenylethylamines or catecholamines proved ineffective.[4] These receptors are those involved in clonidine- and rilmenidine-induced inhibition of catecholaminergic neurons in the same area as that demonstrated by the voltammetric technique.[10–12]

Specific binding proteins sensitive to clonidine are difficult to isolate from human brain or even from other tissues or cell lines. Isolation and purification of the clonidine-sensitive receptor from the ventrolateral region of the human medulla were carried out.

Briefly, after solubilization of these proteins in 3[-(3 cholamidopropyl dimethyl-ammonio-1-propane sulfonate)] CHAPS and the observation that dilution and the addition of glycerol led to a significant increase in the solubilization rate, [^{3}H]clonidine binding proteins were largely purified. The effect of dilution of solubilized membrane proteins in CHAPS was impressive, as at a dilution of 1 : 20 the binding of [^{3}H] clonidine to soluble protein was increased by 200%. In membrane preparations, similar observations were made. A dilution of 1 : 50 did not change [^{3}H] clonidine binding, because the amount of proteins was divided by 50. In this assay, it means that binding was, in fact, increased by the dilution. Centrifugation performed after dilution of the membrane preparation further increased the specific binding of [^{3}H] clonidine. One way to interpret these data is to speculate on the existence of a binding inhibitor, that is, an endogenous ligand present in the soluble fraction that could mask the specific binding of tritiated clonidine to its sites. This possibility was already mentioned by Parini's group[13] in their work on the purification of imidazoline specific binding sites from kidney.[13] After solubilization, [^{3}H] clonidine binding was saturable and reversible.[14] Various chromatographic techniques were applied to the solubilized protein and fractions exhibiting 3-5 electrophoretic bands on SDS-PAGE. Among these, a unique band at 43 kD was always revealed with a specific imidazoline antiidiotypic antibody recently developed in our laboratory.[15] Up to now, this 43-kD protein represents the best candidate for a clonidine-sensitive binding protein specific to imidazolines in the human brain. It must be noted that this site is also a high affinity site for idazoxan and cirazoline.

Thus, a 43-kD protein candidate to be a binding site specific to imidazolines must exhibit high affinities for clonidine, idazoxan, and cirazoline simultaneously. Its molecular mass distinguishes that protein from those previously purified from peripheral tissues by Parini and Reis' groups.[13,16] It looks similar to a protein recently identified by Garcia-Sevilla and co-workers in the rat brain.

CONCLUSION

Concerning the functional aspects, the presence of a circulating immunoreactive substance was detected in human serum with anti-imidazoline antibodies. In some patients with essential hypertension, levels were very high. The question of identity between that substance and the endogenous ligand of the imidazoline receptors remains unresolved.

An additional antiarrhythmic effect of the action of central hypotensive drugs on the imidazoline receptors was also identified. The mechanism of that action remains to be explained. Clinical trials should demonstrate the particular interest of that kind of drug in hypertensive patients with left ventricular hypertrophy associated with ventricular arrhythmias.

In biochemistry, the insensitivity to catecholamines and the sensitivity to both clonidine and idazoxan are proposed as the most reliable features in defining the cerebral imidazoline receptors involved in cardiovascular regulation at the time. From this point of view, a 43-kD protein was recently identified and largely purified from the human brain with the aid of antiidiotypic antibodies. That protein binds clonidine and idazoxan with high affinity. One remaining question is to find out if this protein represents the whole receptor or a subunit. This problem will remain until complete purification is achieved.

REFERENCES

1. DONTENWILL, M., G. BRICCA, A. MOLINES, P. BOUSQUET & A. BELCOURT. 1988. Production and characterization of anti-clonidine antibodies not cross-reacting with catecholamines. Eur. J. Pharmacol. **149:** 249-255.
2. DONTENWILL, M., A. MOLINES, G. BRICCA, J. STUTZMANN, A. BELCOURT & P. BOUSQUET. 1992. Production and characterization of an iminoimidazolidine specific monoclonal antibody using *p*-aminoclonidine as antigen. Life Sci. **50:** 1859-1868.
3. DONTENWILL, M., A. MOLINES, A. VERDUN, G. BRICCA, S. LAURENT & P. BOUSQUET. 1993. A circulating substance cross-reacting with anti-imidazoline antibodies: Detection in serum in relation to essential hypertension. J. Clin. Invest. **92:** 1068-1072.
4. BOUSQUET, P., J. FELDMAN & P. BOUSQUET. 1984. Central cardiovascular effects of alpha-adrenergic drugs: Difference between catecholamines and imidazolines. J. Pharmacol. Exp. Ther. **230:** 232-236.
5. FELDMAN, J., E. TIBIRICA, G. BRICCA, M. DONTENWILL, A. BELCOURT & P. BOUSQUET. 1990. Evidence for the involvement of imidazoline receptors in the central hypotensive effect of rilmenidine in the rabbit. Br. J. Pharmacol. **100:** 600-604.
6. SEGAL, A. S., T. JACOB & R. A. GILLIS. 1984. Blockade of central nervous system GABAergic tone causes sympathetic-mediated increases in coronary vascular resistance in cats. Circ. Res. **55:** 404-415.
7. HAYASHI, Y., T. KAMIBAYASHI, M. MAZE, A. YAMATODANI, K. SUMIKAWA, M. KURO & I. YOSHIYA. 1993. Role of imidazoline preferring receptors in the genesis of epinephrine-induced arrhythmias in halothane anesthetized dogs. Anesthesiology **78:** 524-530.
8. BRICCA, G., H. GRENEY, J. ZHANG, M. DONTENWILL, J. STUTZMANN, A. BELCOURT & P. BOUSQUET. 1994. Human brain imidazoline receptors: Further characterization with [^{3}H] clonidine. Eur. J. Pharmacol.-Mol. Pharmacol. **266:** 25-33.
9. BRICCA, G., H. GRENEY, M. DONTENWILL-KIEFFER, J. ZHANG, A. BELCOURT & P. BOUSQUET. 1993. Heterogeneity of the specific imidazoline binding of [^{3}H]idazoxan in the human cerebral cortex. Neurochem. Int. **22:** 153-163.
10. TIBIRICA, E., C. MERMET, J. FELDMAN, F. GONON & P. BOUSQUET. 1989. Correlation between the inhibitory effect of catecholaminergic ventrolateral medullary neurons and the hypotension evoked by clonidine: A voltammetric approach. J. Pharmacol. Exp. Ther. **250:** 642-647.
11. TIBIRICA, E., C. MERMET, J. FELDMAN, F. GONON & P. BOUSQUET. 1991. An imidazoline specific mechanism for the hypotensive effect of clonidine: A study with yohimbine and idazoxan. J. Pharmacol. Exp. Ther. **256:** 606-613.
12. TIBIRICA, E., J. FELDMAN, C. MERMET, L. MONASSIER, F. GONON & P. BOUSQUET. 1991. Selectivity of rilmenidine for the nucleus reticularis lateralis, a ventrolateral medullary structure containing imidazoline-preferring receptors. Eur. J. Pharmacol. **209:**213-221.
13. LIMON, I., I. COUPRY, S. M. LANIER & A. PARINI. 1992. Purification and characterization of mitochondrial imidazoline-guanidinium receptive site from rabbit kidney. J. Biol. Chem. **267:** 21645-21649.
14. GRENEY, H., G. BRICCA, M. DONTENWILL, J. STUTZMANN, P. BOUSQUET & A. BELCOURT. 1994. Characterization of imidazoline binding protein(s) solubilized from human brainstem: Studies with [^{3}H]idazoxan and [^{3}H]clonidine. Neurochem. Int. In press.
15. BENNAI, F., H. GRENEY, J. STUTZMANN, A. BELCOURT, P. BOUSQUET & M. DONTENWILL. 1994. Polyclonal anti-idiotypic antibodies to idazoxan and their interaction with imidazoline binding sites: Revelation of a 43 Kdalton protein in human brain preparations. Submitted.
16. WANG, H., S. REGUNATHAN, M. P. MEELEY & D. J. REIS. 1992. Isolation and characterization of imidazoline receptor protein from bovine adrenal chromaffin cells. Mol. Pharmacol. **42:** 792-801.

Importance of Imidazoline Receptors in the Cardiovascular Actions of Centrally Acting Antihypertensive Agents[a]

GEOFFREY A. HEAD[b]

Baker Medical Research Institute
Department of Neuropharmacology
Prahran, Victoria 3181, Australia

It has been suggested that baroreflex control of blood pressure by the autonomic nervous system is limited to the short term, while the kidney plays a major role in determining the long-term level of blood pressure.[1] However, much evidence now supports the view that the sympathetic nervous system is activated in the early phases of hypertension and that this may be a contributory factor to the disease.[2] The increase in sympathetic drive is not generalized but is confined to renal and cardiac beds. Recently it was observed that central monoamine pathways are also disturbed in human essential hypertension. Esler and colleagues[3,4] have shown increased noradrenaline spillover and decreased dopamine spillover from the subcortical regions of the central nervous system (CNS) in hypertensive patients. An association exists between subcortical catecholaminergic neuronal activity and renal sympathetic activation.[3] Our own studies and those from other groups have suggested a disturbance in central noradrenergic pathways in the spontaneously hypertensive rat (SHR).[5–7] The mechanism of the antihypertensive action of clonidine-like drugs involves an action mainly within the CNS to reduce sympathetic tone. They appear to modulate central pathways regulating vasomotor tone to specific organs such as the kidney[8,9] and heart[10] but not to muscle.[11,12] Therefore, these agents provide a unique therapeutic opportunity to target sympathetic tone to the organs that are selectively activated in early hypertension. In support of a selective influence on monoamine pathways in the CNS, we also found a dependence of centrally acting agents on the monoamine pathways themselves. Destruction of the noradrenergic pathways with 6-hydroxydopamine or the serotonergic pathways with 5,6-dihydroxytryptamine leads to attenuation of the hypotension and bradycardia produced by α-methyldopa, clonidine, and also rilmenidine.[13,14] The latter agent is an oxazoline derivative recently introduced into therapeutic use as an antihypertensive agent.

[a] These studies were supported by a block institute grant from the National Health and Medical Research Council of Australia and by a grant from Institut de Recherches Internationales SERVIER & Compagnie Developpement.

[b] Address for correspondence and reprint requests (*please use air mail*): Dr. Geoffrey A. Head, Baker Medical Research Institute, Commercial Road, Prahran, Victoria, 3181, Australia.

The receptors that are involved in the antihypertensive actions of drugs such as clonidine and their location in the CNS have been the subject of much attention. Central and peripheral interaction with α_2-adrenoceptors has long been suggested because these agents are potent agonists.[15,16] However, this view has changed dramatically in recent years with the observations that these agents may also interact with a nonadrenergic imidazoline receptor (IR)[17] and that this interaction at the level of the rostral ventrolateral medulla oblongata may be important for their hypotensive actions.[18–21] However, numerous studies have now either supported[20–22] or rejected[23–26] this view on the basis of evidence in various species, strains, routes of administration, and drugs used. Thus, it still remains to be demonstrated that at "clinically relevant" antihypertensive doses in conscious animals with reflex autonomic control systems intact, IR rather than α_2-adrenoceptors are mediating the antihypertensive action of these drugs. This review outlines our studies in conscious rabbits in which this question was examined.

METHODS

All studies described were performed in male and female cross-bred conscious rabbits bred at the Baker Medical Research Institute and chronically implanted with a vinyl intracisternal catheter under halothane anesthesia as described previously.[13] In some rabbits a renal nerve electrode and balloon cuffs for performing baroreflexes were also implanted.[27] We used conscious rabbits to avoid anesthesia and central administration to eliminate possible peripheral actions of the drugs.

On each experimental day the central ear artery of each rabbit was catheterized and the end of the intracisternal catheter was retrieved from under the skin at the back of the neck under local anesthesia. Cardiovascular variables were recorded on a Grass (model 7D) polygraph. Mean arterial pressure was measured with a Statham P23 transducer and heart rate determined by a heart rate meter triggered from the arterial pulse. The renal nerve activity was rectified, filtered, and integrated as described previously.[27] Baroreflex blood pressure-renal sympathetic activity relationships were assessed with slow balloon-induced ramp rises and falls in blood pressure with subsequent sigmoidal curve fitting.[27]

In microinjection experiments, rabbits with previously implanted renal nerve electrodes were anesthetized with intravenous urethane (1.4 g/kg), mounted in a stereotaxic frame and the head flexed 90 degrees. The dorsal medulla was exposed through the atlanto-occipital membrane. The obex was used as stereotaxic zero and microinjections were made with a fine (20–40 μm) tip glass micropipette using a volume of 100 nl per injection.

ROLE OF α_2-ADRENOCEPTORS AND IMIDAZOLINE RECEPTORS

We sought to determine the relative importance of IR and α_2-adrenoceptors in the antihypertensive effects of rilmenidine and moxonidine and to compare them with clonidine. Rilmenidine and moxonidine have a greater selectivity for IR over α_2-adrenoceptors than does clonidine.[28,29] From cumulative dose-response curves, we determined the relative potency of the three agents given intracisternally in conscious rabbits with implanted intracisternal catheters. Rilmenidine, moxonidine, and clonidine produced a similar maximum reduction in mean arterial pressure. Rilmenidine was 11 times less potent than clonidine ($p < 0.001$) in lowering blood pressure, while moxonidine was 3 times more potent than clonidine ($p < 0.05$). From these data we chose a dose expected to produce 90% of the maximum hypotensive effect of the three antihypertensive agents. All the antagonists have affinity for α_2-adrenoceptors; therefore, it is important to allow for this effect using "pure" α_2-adrenoceptor agonists and antagonists. Thus, we compared the reversal of rilmenidine- and α-methyldopa-induced hypotension by idazoxan (a mixed IR/α_2-adrenoceptor antagonist) and 2-methoxy-idazoxan (α_2-adrenoceptor antagonist). The hypotension after a near-maximal intracisternal dose of rilmenidine was reversed by lower doses of idazoxan than was the hypotension produced by intracisternal α-methyldopa. By contrast, hypotension produced by α-methyldopa was reversed more readily by 2-methoxy-idazoxan.[30,31] The antagonists given by themselves had no effect on blood pressure. The preferential reversal effect of idazoxan therefore suggests that rilmenidine acts principally through IR. We did observe that higher doses of 2-methoxy-idazoxan completely reversed rilmenidine-induced hypotension. This finding was surprising, because it suggested that blockade of α_2-adrenoceptors could reverse the hypotension produced by activation of IR. One explanation for this is that α_2-adrenoceptors and IR may be in series along the same autonomic pathway.[30,31]

In a further study we examined the effects of intracisternal clonidine, but because of the short time course of action of this agent, we were only able to use single doses of idazoxan (30 μg/kg) and 2-methoxy-idazoxan (1 μg/kg).[32] These doses had a similar α_2-adrenoceptor blocking effect as determined from earlier experiments with α-methyldopa. The hypotension produced by clonidine and α-methyldopa was similarly reversed by the two antagonists, suggesting that both of these agents act mainly on α_2-adrenoceptors (FIG. 1). By contrast, the rilmenidine-induced decrease in blood pressure was not affected by 2-methoxy-idazoxan, but was markedly reduced by idazoxan, confirming a preferential effect on IR (FIG. 1).[32]

It was recently suggested that there are at least two subtypes of IR, I_1 (clonidine site) and I_2 (idazoxan site), and that it is the former that is important in mediating hypotensive responses.[33,34] Because idazoxan has greater selectivity for I_2 than for I_1 receptors, we also examined the effects of the highly selective I_1 antagonist efaroxan (100,000-fold selectivity for I_1- than for I_2-imidazoline sites).[28] The doses of the two antagonists were chosen to have equal antagonist effects against the hypotension produced by α-methyldopa and therefore equal α_2-adrenoceptor blockade. Efaroxan was more effective at reversing the hypotension produced by rilmenidine and moxonidine than was 2-methoxy-idazoxan, suggesting involvement mainly of IR (FIG. 2). By contrast, 2-methoxy-idazoxan was more effective against clonidine than efaroxan, suggesting that the hypotension was mainly due to stimulation of α_2-adrenoceptors (FIG. 2). Using therapeutically relevant doses, these studies indicate an important difference between the receptors mediating the antihypertensive actions of the newer agents with higher selectivity for IR and the less selective drugs such as clonidine. Efaroxan which we used has no I_2 affinity, suggesting that principally I_1 subtype rather than the I_2 receptors are involved. Thus, the effects of idazoxan which were similar to those of efaroxan were also probably through I_1 receptors despite its 10-

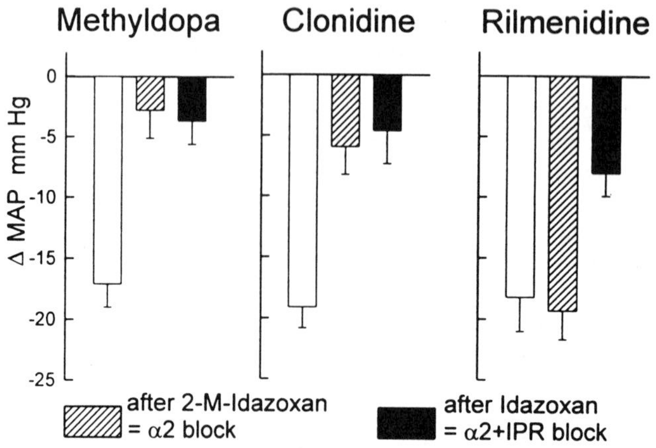

FIGURE 1. Average decreases in blood pressure (ΔMAP) following α-methyldopa, clonidine, and rilmenidine after vehicle treatment (*open bars*), 2-methoxy-idazoxan (*shaded bars*), or idazoxan (*black bars*). Error bars indicate SEM. *p <0.05 compared with control. Reproduced with permission from Head *et al.*[32]

fold higher affinity for the I_2 subtype. The second-generation agents appear to activate different receptors in the CNS from those of the first-generation agents such as clonidine.

EFFECTS OF RILMENIDINE ON RENAL SYMPATHETIC REFLEXES

Relatively little is known about baroreflex control of sympathetic function with these agents. We therefore examined the effects of α_2-adrenoceptors/imidazoline-preferring antagonists on the renal sympathetic effects of central administration of rilmenidine in conscious rabbits. The animals were chronically instrumented with an intracisternal catheter, an aortic and vena caval cuff occluder, and a renal nerve electrode. We recorded renal sympathetic nerve activity, heart rate, and mean arterial pressure and assessed baroreflex relationships with balloon-induced ramp increases and decreases in blood pressure.[35] Rabbits were treated with intracisternal rilmenidine followed by intracisternal idazoxan or intracisternal 2-methoxy-idazoxan at doses similar to those used in the aforementioned studies. Rilmenidine induced significant falls in blood pressure, heart rate, and sympathetic nerve activity. Sympathetic baroreflex curves were shifted to the left, with reductions in both upper and lower sympathetic plateaus. Thus, sympathetic range (difference between plateaus) was reduced. 2-Methoxy-idazoxan caused only partial recovery of blood pressure, renal sympathetic nerve activity, and sympathetic baroreflex range. By comparison, idazoxan induced greater reversal of hypotension and completely restored basal sympathetic nerve activity as well as baroreflex sympathetic range.[35] The renal sympathetic baroreflex responses produced by α-methyldopa were qualitatively similar to those

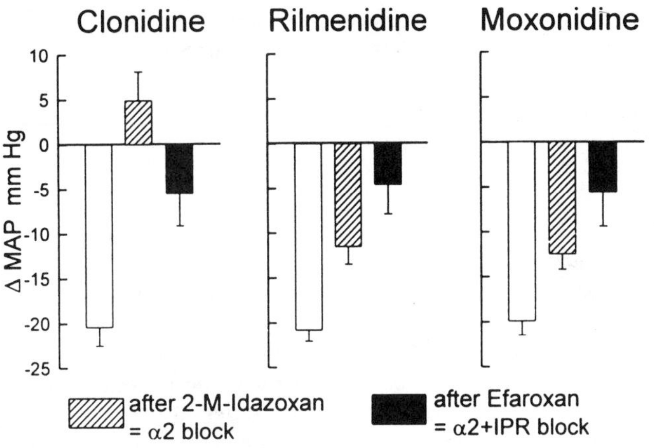

FIGURE 2. Average decreases in blood pressure (ΔMAP) following clonidine, rilmenidine, and moxonidine after vehicle treatment (*open bars*), 2-methoxy-idazoxan (*shaded bars*), or efaroxan (*black bars*). Error bars indicate SEM. *$p < 0.05$ compared with control. Reproduced with permission from Head.[42]

with rilmenidine but equally affected by the two antagonists (doses were chosen to have equal α_2-adrenoceptor blockade). These findings suggest that in conscious rabbits, both IR and central α_2-adrenoceptors can influence the renal baroreflex, but the former are relatively more important for the actions of rilmenidine.

SITE OF BAROREFLEX EFFECTS OF RILMENIDINE

Studies by Gomez and colleagues[21] provided convincing evidence to suggest that the subretrofacial nucleus of the rostral ventrolateral medulla was the site of the hypotensive action of rilmenidine. We determined whether the subretrofacial nucleus was also involved in the renal sympathoinhibitory and renal sympathetic baroreflex effects of rilmenidine or whether there was a contribution from the nucleus of the solitary tract.[36] Cumulative dose-response curves to intracisternal and bilateral microinjection of rilmenidine into the subretrofacial nucleus were compared. Rabbits were anesthetized with urethane or were studied conscious with an indwelling intracisternal catheter. Rilmenidine was equally effective intracisternally in reducing mean arterial pressure in conscious and anesthetized rabbits, but was 1.5 times more potent in the anesthetized than in the conscious state. Bilateral cumulative dose-response curves to microinjection of rilmenidine into the subretrofacial nucleus of anesthetized rabbits produced maximum hypotension which was similar to that produced by intracisternal administration at sixfold lower doses. We then compared the effects of a half-maximal subretrofacial nucleus dose on renal sympathetic baroreflexes in anesthetized rabbits. Rilmenidine into the subretrofacial nucleus reduced mean arterial pressure, renal sympathetic activity, and baroreflex maximum sympathetic range.[36] These effects were qualitatively similar to those observed after intracisternal rilmenidine. By contrast, rilmenidine injected into the nucleus of the solitary tract had no effect on mean arterial pressure, sympathetic activity, or baroreflex parameters. These studies suggest that the subretrofacial nucleus is a major site for rilmenidine-induced inhibition of the renal sympathetic baroreflex.

AGMATINE

An exciting recent development in the field has been the characterization of "clonidine-displacing substance" (CDS) by Li and colleagues[37] as agmatine. Clonidine-displacing substance was known to be a nonpeptide, noncatecholamine, low molecular weight substance that was isolated and purified from animal brains and could displace radiolabeled imidazolines.[38–41] Agmatine is decarboxylated arginine that possesses a guanidine group structurally similar to the imidazoline ring of clonidine and moxonidine. Binding assays in rat cerebral cortex demonstrated that both agmatine and CDS have high affinity for the I_1-subtype of IR.[37] However, little information on the cardiovascular effects of agmatine is available. The current study was undertaken to investigate the effects of agmatine on blood pressure and heart rate when injected intracisternally to conscious rabbits. Initial experiments involved performing a dose-response curve to agmatine. In the second series of experiments we determined if agmatine would affect the hypotensive response to clonidine or moxonidine. Agmatine at doses up to 0.01 - 10 μg/kg had little effect on blood pressure but did cause a dose-dependent moderate bradycardia. At a higher dose of 100 μg/kg agmatine produced a marked increase in blood pressure and reversed the bradycardia. Intracisternal administration of 10 μg/kg of agmatine 10 minutes after a single dose of clonidine (0.9 μg/kg intracisternally) did not further alter hypotension but produced greater bradycardia. Similarly, hypotension induced by moxonidine was not altered by intracisternal administration of agmatine (10 μg/kg) at 15 minutes, but displayed significantly greater bradycardia. These results suggest that agmatine possesses the bradycardic actions of clonidine but does not mimic the hypotensive action of clonidine or moxonidine. Moreover, agmatine does not appear to be an antagonist of α_2-adrenoceptors or IR.

CONCLUSION

Our studies show that the second generation of centrally acting agents such as rilmenidine and moxonidine appear to have a selectivity for IR and that this is important in mediating hypotension and also for their actions on renal sympathetic baroreflexes. By contrast, clonidine mainly acts through α_2-adrenoceptors. This finding suggests that not all agents that purport to have a selectivity for IR act in the same manner. This has not been so apparent, because both IR and α_2-adrenoceptors, when activated, produce similar hemodynamic and cardiovascular reflex effects. Our conscious rabbit preparation with intracisternal administration to eliminate the systemic influences, in combination with careful calibration of IR and non-IR antagonists, has proved most useful in distinguishing the various agents. Thus, it appears to require a greater degree of selectivity for IR than is possessed by clonidine before the IR action can become dominant. Part of the controversy that surrounds the importance of IR is based on the lack of recent evidence to suggest that clonidine lowers blood pressure through IR.[23,25,26] However, this is perhaps not so surprising given our current findings.

SUMMARY

Increasing evidence indicates that the hypotensive effect of centrally acting antihypertensive drugs is not due to stimulation of α_2-adrenoceptors but to action on imidazoline receptors (IR). This has led to the development and recent clinical use of second generation agents such as rilmenidine and moxonidine that possess a much greater selectivity toward these nonadrenergic receptors. However, relatively few studies have examined the role of these receptors in conscious animals or have adequately accounted for the α_2-adrenoceptor antagonist properties of IR antagonists such as idazoxan.

We have taken the approach of initially calibrating the α_2-adrenoceptor antagonist potency of intracisternally (ic) administered idazoxan and the IR-1 receptor antagonist efaroxan against 2-methoxyidazoxan, a highly selective α_2-adrenoceptor antagonist with little or no imidazoline antagonist effect. This was done using α-methyldopa, a hypotensive agent affecting only α_2-adrenoceptors. Thus, we chose doses of the antagonists with equal α_2-adrenoceptor blocking action such that differences in the ability of idazoxan or efaroxan compared to 2-methoxy-idazoxan to reverse the hypotension produced by rilmenidine, moxonidine, or clonidine indicate an interaction with IR. By this method we found that the hypotensive effects of rilmenidine and moxonidine at moderate intracisternal doses were more readily reversed by the imidazoline antagonists than by 2-methoxy-idazoxan, indicating that IR were largely responsible for their hypotensive actions. By contrast, clonidine's effects were equally reversed by all antagonists, suggesting interaction mainly with α_2-adrenoceptors.

In conscious rabbits with chronic renal sympathetic nerve electrodes we examined the effect of rilmenidine and α-methyldopa on the renal sympathetic baroreflex. Both drugs reduced renal sympathetic nerve activity and sympathetic baroreflex responses, but only the effect of rilmenidine was preferentially reversed by idazoxan. Thus, both IR and central α_2-adrenoceptor receptors can influence the renal baroreflex, but the former are relatively more important for the actions of rilmenidine.

We recently examined the possible sites of action of rilmenidine in anesthetized rabbits and showed that sixfold lower doses were required to reduce blood pressure when the drug was injected into the rostral ventrolateral medulla compared to intracisternal administration. At this site rilmenidine also reduced renal sympathetic tone and inhibited renal sympathetic baroreflex responses. By contrast, rilmenidine was relatively ineffective when injected into the nucleus of the solitary tract. These experiments support the view that rilmenidine acts primarily at IR in the rostral ventrolateral medulla to reduce sympathetic tone and modulate sympathetic baroreflexes. Initial studies with agmatine, the putative clonidine-displacing substance, had relatively little cardiovascular effects but did produce bradycardia. Agmatine had little effect on the actions of clonidine or moxonidine. These studies suggest that agmatine is neither an agonist nor an antagonist at IR.

ACKNOWLEDGMENTS

The assistance of Shirley Godwin and Sandra Burke and the contribution by Frederick Sannajust and Candy Chan are greatly appreciated. The gift of moxonidine from Solvay Pharma is acknowledged.

REFERENCES

1. GUYTON, A. C. 1991. Blood pressure control: Special role of the kidneys and body fluids. Science **252:** 1813-1816.
2. ESLER, M., G. JACKMAN, A. BOBIK, P. LEONARD, D. KELLEHER, H. SKEWS *et al.* 1981. Norepinephrine kinetics in essential hypertension. Defective neuronal uptake of norepinephrine in some patients. Hypertension **3:** 149-156.
3. FERRIER, C., G. L. JENNINGS, G. EISENHOFER, G. LAMBERT, H. S. COX, V. KALFF *et al.* 1993. Evidence for increased noradrenaline release from subcortical brain regions in essential hypertension. J. Hypertens. **11:** 1217-1227.
4. LAMBERT, G. W., C. FERRIER, D. M. KAYE, V. KALFF, M. J. KELLY, H. S. COX *et al.* 1994. Monoaminergic neuronal activity in subcortical brain regions in essential hypertension. Blood Pressure **3:** 55-66.
5. HEAD, G. A. & W. DE JONG. 1986. Differential blood pressure responses to intracisternal clonidine, α-methyldopa, and 6-hydroxydopamine in conscious normotensive and spontaneously hypertensive rats. J. Cardiovasc. Pharmacol. **8:** 735-742.
6. TIBIRICA, E., J. FELDMAN & P. BOUSQUET. 1992. Contribution of catecholaminergic neurons of the dorsomedial and ventrolateral medulla-oblongata to the hypotensive effect of clonidine in spontaneously hypertensive rats: *In vivo* voltammetric studies. J. Hypertens. **10:** 1327-1334.
7. QUALY, J. M. & T. C. WESTALL. 1993. Age-dependent overflow of endogenous norepinephrine from paraventricular hypothalamic nucleus of hypertensive rats. Am. J. Physiol. **265:** H39-H46.
8. GARTY, M., A. DEKA-STAROSTA, P. CHANG, I. J. KOPIN & D. S. GOLDSTEIN. 1990. Effects of clonidine on renal sympathetic activity and norepinephrine spillover. J. Pharmacol. Exp. Ther. **254:** 1068-1075.
9. SANNAJUST, F., C. BARRES, E. KOENIG-BERARD & J. SASSARD. 1992. Sympathoinhibitory effects of rilmenidine may be mediated by sites located below the brainstem. Br. J. Pharmacol. **105:** 535-541.
10. BADOER, E., G. A. HEAD & P. I. KORNER. 1983. Effects of intracisternal and intravenous α-methyldopa and clonidine on haemodynamics and baroreceptor-heart rate reflex properties in conscious rabbits. J. Cardiovasc. Pharmacol. **5:** 760-767.
11. SHAW, J., S. N. HUNYOR & P. I. KORNER. 1971. The peripheral circulatory effects of clonidine and their role in the production of arterial hypotension. Eur. J. Pharmacol. **14:** 101-111.
12. WALLIN, B. G. & M. FRISK-HOLMBERG. 1981. The antihypertensive mechanism of clonidine in man: Evidence against a general reduction of sympathetic activity. Hypertension **3:** 340-346.
13. HEAD, G. A., P. I. KORNER, S. L. LEWIS & E. BADOER. 1983. Contribution of noradrenergic and serotonergic neurons to the circulatory effects of centrally acting clonidine and α-methyldopa in' rabbits. J. Cardiovasc. Pharmacol. **5:** 945-953.
14. HEAD, G. A. & S. L. BURKE. 1991. Importance of central noradrenergic and serotonergic pathways in the cardiovascular actions of rilmenidine and clonidine. J. Cardiovasc. Pharmacol. **18:** 819-826.
15. SCHMITT, H. 1977. The pharmacology of clonidine and related products. *In* Handbook of Experimental Pharmacology. Vol. **39:** 299. Springer-Verlag. Berlin.

16. Van Zwieten, P. A. 1988. Pharmacology of the alpha-2 adrenoceptor agonist rilmenidine. Am. J. Cardiol. **61:** 6D-14D.
17. Ruffolo, R. R., B. S. Turowski & P. N. Patil. 1977. Lack of cross desensitization between structurally dissimilar alpha-adrenoceptor agonists. J. Pharm. Pharmacol. **29:** 378-380.
18. Bousquet, P., J. Feldman & J. Schwartz. 1984. Central cardiovascular effects of alpha adrenergic drugs: Differences between catecholamines and imidazolines. J. Pharmacol. Exp. Ther. **230:** 232-236.
19. Ernsberger, P., R. Guiliano, R. N. Willette & D. J. Reis. 1990. Role of imidazole receptors in the vasodepressor response to clonidine analogs in the rostral ventrolateral medulla. J. Pharmacol. Exp. Ther. **253:** 408-418.
20. Feldman, J., E. Tibirica, G. Bricca, M. Dontenwill, A. Belcourt & P. Bousquet. 1990. Evidence for the involvement of imidazoline receptors in the central hypotensive effect of rilmenidine in the rabbit. Br. J. Pharmacol. **100:** 600-604.
21. Gomez, R. E., P. Ernsberger, G. Feinland & D. J. Reis. 1991. Rilmenidine lowers arterial pressure via imidazole receptors in brainstem C1 area. Eur. J. Pharmacol. **195:** 181-191.
22. Mayorov, D., M. Chernobelski & O. Medvedev. 1993. Sympathoinhibitory action of rilmenidine in conscious sinoaortically denervated rats. J. Cardiovasc. Pharmacol. **22:** 314-320.
23. Szabo, B., R. Urban & K. Starke. 1993. Sympathoinhibition by rilmenidine in conscious rabbits: Involvement of alpha(2)-adrenoceptors. Naunyn-Schmiedeberg's Arch. Pharmacol. **348:** 593-600.
24. Guyenet, P. G., K. R. Lynch, A. M. Allen, D. L. Rosin & R. L. Stornetta. 1993. Alpha2-adrenergic receptors rather than "Imidazoline" binding sites mediate the sympatholytic effect of clonidine in the rostral ventrolateral medulla: A new look at the evidence. *In* Ventral Brainstem Mechanisms and Control Functions. Marcel Dekker, Inc. New York.
25. Allen, A. M. & P. G. Guyenet. 1993. Alpha2-adrenoceptor-mediated inhibition of bulbospinal barosensitive cells of rat rostral medulla. Am. J. Physiol. **265:** R1065-R1075.
26. Hieble, J. P. & D. C. Kolpak. 1993. Mediation of the hypotensive action of systemic clonidine in the rat by alpha(2)-adrenoceptors. Br. J. Pharmacol. **110:** 1635-1639.
27. Dorward, P. K., W. Riedel, S. L. Burke, J. Gipps & P. I. Korner. 1985. The renal sympathetic baroreflex in the rabbit. Arterial and cardiac baroreceptor influences, resetting, and effect of anesthesia. Circ. Res. **57:** 618-633.
28. Ernsberger, P. R., K. L. Westbrooks, M. O. Christen & S. G. Schafer. 1992. A second generation of centrally acting antihypertensive agents act on putative I(1)-imidazoline receptors. J. Cardiovasc. Pharmacol. **20:** s1-s10.
29. Bricca, G., M. Dontenwill, A. Molines, J. Feldman, E. Tibirica, A. Belcourt *et al.* 1989. Rilmenidine selectivity for imidazoline receptors in human brain. Eur. J. Pharmacol. **163:** 373-377.
30. Sannajust, F. & G. A. Head. 1994. Rilmenidine induced hypotension involves imidazoline preferring receptors in conscious rabbits. J. Cardiovasc. Pharmacol. **23:** 42-50.
31. Head, G. A. & F. Sannajust. 1992. Importance of imidazoline receptors in the cardiovascular responses to clonidine and rilmenidine in conscious rabbits. Fundam. Clin. Pharmacol. **6** (Suppl. 1): 31s-43s.
32. Head, G. A., S. J. Godwin & F. Sannajust. 1993. Differential receptors involved in the cardiovascular effects of clonidine and rilmenidine in conscious rabbits. J. Hypertens. **11** (suppl. 5): S322-S323.
33. Michel, M. C. & P. A. Insel. 1989. Are there multiple imidazoline binding sites? Trends Pharmacol. Sci. **10:** 334-342.
34. Hieble, J. P. & R. R. Ruffolo. 1992. Imidazoline receptors—historical perspective. Fundam. Clin. Pharmacol. **6:** S7-S13.

35. SANNAJUST, F. & G. A. HEAD. 1994. Evidence for imidazoline receptors mediating the inhibition of the renal sympathetic baroreflex produced by rilmenidine in rabbits. J. Hypertens. **12:** S188.
36. HEAD, G. A., S. L. BURKE & F. SANNAJUST. 1994. Subretrofacial nucleus (SN) mediates rilmenidine-induced inhibition of the renal sympathetic baroreflex in rabbits. Can. J. Physiol. Pharmacol. **72:** 369.
37. LI, G., S. REGUNATHAN, C. J. BARROW, J. ESHRAGHI, R. COOPER & D. J. REIS. 1994. Agmatine—an endogenous clonidine-displacing substance in the brain. Science **263:** 966-969.
38. ERNSBERGER, P., M. P. MEELEY & D. J. REIS. 1988. An endogenous substance with clonidine-like properties: Selective binding to imidazole sites in the ventrolateral medulla. Brain Res. **441:** 309-318.
39. ATLAS, D., S. DIAMANT, H. M. FALES & L. PANNELL. 1987. The brain's own clonidine: Purification and characterization of endogenous clonidine displacing substance from brain. J. Cardiovasc. Pharmacol. **10** (suppl. 12): S122-S127.
40. ATLAS, D. & Y. BURNSTEIN. 1984. Isolation of an endogenous clonidine-displacing substance from rat brain. FEBS Lett. **170:** 387-390.
41. ATLAS, D. 1991. Clonidine-displacing substance (CDS) and its putative imidazoline receptor: New leads for further divergence of alpha$_2$-adrenergic receptor activity. Biochem. Pharmacol. **41:** 1541-1549.
42. HEAD, G. A. 1995. Central monoamine systems and new antihypertensive agents. Clin. Exp. Hypertens. [A]. **17:** 141-152.

Alpha-2 Adrenergic Receptors

Immunohistochemical Localization and Role in Mediating Inhibition of Adrenergic RVLM Presympathetic Neurons by Catecholamines and Clonidine

R. L. STORNETTA, D. HUANGFU, D. L. ROSIN, K. R. LYNCH, AND P. G. GUYENET

Department of Pharmacology
University of Virginia Health Sciences Center
Charlottesville, Virginia 22908

The rostral ventrolateral medulla (RVLM) is vital for sympathetic vasomotor tone generation. This site also integrates most sympathetic vasomotor reflexes (e.g., baroreflex, chemoreflex, and somatosympathetic pressor reflex; for reviews see refs. 1 and 2). The RVLM contains the C1 group of adrenergic neurons, thus named because they express all the enzymes of the catecholamine biosynthetic pathway including phenylethanolamine *N*-methyl transferase. A large fraction of C1 cells projects monosynaptically to spinal preganglionic neurons (SPGNs) and, as such, qualifies as a presympathetic element.[3–5] These particular C1 cells are thought to provide an excitatory drive to vasomotor SPGNs, thereby contributing to sympathetic tone and blood pressure maintenance. (For reviews see refs. 1 and 2.) Clonidine, an antihypertensive agent with partial α_2-adrenergic receptor (α_2-AR) agonist activity, seems to exert a large part of its sympatholytic effect within the RVLM.[6]

This report discusses the notion that C1 cells, like many other central nervous system (CNS) aminergic neurons (e.g., see ref. 7), may have autoreceptors that mediate some form of collateral or autoinhibition. Evidence that these inhibitory autoreceptors are of the α_{2A}-adrenergic variety is presented. Finally, we conclude that the sympatholytic effect triggered by the presence of clonidine within the RVLM may be due, at least in part, to direct activation of these inhibitory autoreceptors by clonidine. These notions are summarized in schematic form in FIGURE 1.

NEUROPHYSIOLOGICAL IDENTIFICATION OF RVLM PRESYMPATHETIC VASOMOTOR CELLS

The RVLM contains various types of tonically active neurons (reviewed in ref. 1). Vasomotor cells have been defined as tonically active neurons whose pattern of discharge bears striking similarities to that of identified vasoconstrictor sympathetic postganglionic neurons.[1,8,9] These similarities include ongoing activity at rest, exquisite barosensitivity, respiratory inputs, and, more generally, a pattern of evoked discharge that closely mimics that of vasomotor postganglionic cells in response to

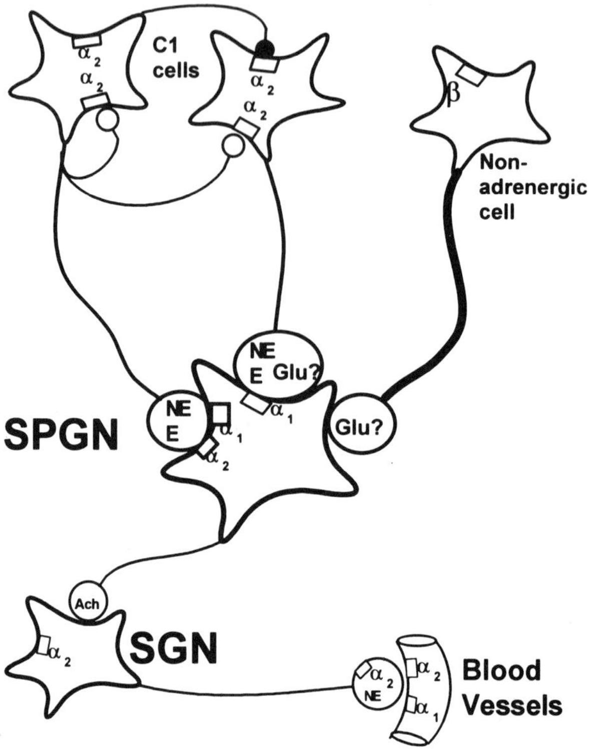

FIGURE 1. Schematic representation of the two types of presympathetic vasomotor cells in the rostral ventrolateral medulla (C1 adrenergic and nonadrenergic). The thickness of the axons symbolizes the relative degree of myelination of the RVLM presympathetic cells. The presumed neurotransmitter content of the two neuronal types is shown in their respective synaptic terminals within the spinal cord. Catecholaminergic receptors present on these cells (α_1-, α_2-, or β) are also shown. Abbreviations: ACh = acetylcholine; E = epinephrine; Glu = glutamate; NE = norepinephrine; SGN = sympathetic postganglionic neuron; SPGN = sympathetic preganglionic neuron.

a large array of peripheral or central stimuli. (For reviews see ref. 1.) The vast majority of RVLM cells with such characteristics project to the vicinity of the intermediolateral cell column in the thoracic spinal cord and are thought to contribute a monosynaptic excitatory drive to preganglionic cells.[1] Because all RVLM spinally projecting cells so far identified have the same characteristic barosensitivity, it is assumed that this functionally identified class of RVLM cells includes C1 adrenergic neurons. This assumption recently received limited confirmation by intracellular recording and staining of these cells *in vivo* (J. Lipski, University of Auckland, New Zealand, personal communication). However, it is increasingly probable that these vasomotor presympathetic cells include a second cell type besides the adrenergic C1 group. These cells are nonaminergic, appear to have intrinsic pacemaker discharges, and do not have α_2-ARs. (For details see ref. 1.)

ANATOMICAL EVIDENCE THAT C1 ADRENERGIC NEURONS EXPRESS α_{2A}-AUTORECEPTORS

Subtype-specific polyclonal antibodies against the two most prevalent subtypes of α_2-ARs in brain (A and C, clones RG20 and RG10) have been raised in our laboratories (Rosin and Lynch) using recombinant proteins as antigens.[10,11] The antibody against the A subtype was raised against a 47 amino acid fragment of the third intracellular loop of the rat α_{2A}-AR. This antibody recognizes specifically the α_{2A}-AR based on the following evidence: (1) COS cells exhibit positive immunohistochemical staining with the α_{2A} subtype antibody after transfection with cDNA for the α_{2A}-AR but not the α_{2B}- (RNG clone) or α_{2C}-AR,[10] (2) the α_{2A}-AR antibody recognizes a broad band that is of the correct molecular weight range for the glycosylated protein on western blots of α_{2A}-AR transfected COS cell membranes (~66 kD), (3) the antibody immunoprecipitates photoaffinity labeled α_{2A}-AR protein from membranes of LL-CPK1 cells transfected with the pig α_{2A} receptor (Lee and Lynch, unpublished data), (4) immunoreactivity detected by histology, western blot, or immunoprecipitation with the α_{2A}-AR antibody is absent when the antibody is preincubated with the fusion protein used to raise it, and (5) the pattern of α_{2A}-AR-like immunostaining correlates well with the distribution of α_{2A}-AR mRNA detected by hybridization histochemistry.[12]

With this antibody, virtually all RVLM C1 adrenergic cells were found to contain specific α_{2A}-AR-like immunoreactivity (α_{2A}-LIR), some in the form of submicron punctata. With confocal microscopy, these punctata were found in an intracellular location, suggesting that the C1 cells synthesize this receptor.[11] To determine if the nonadrenergic vasomotor cells of RVLM also have α_{2A}-ARs, a triple labeling study was undertaken.[11] RVLM cells with axonal projections to the thoracic spinal cord were labeled by retrograde transport of spinally injected fluorescent microbeads. The adrenergic component of the projection was identified by the presence of immunoreactivity to tyrosine hydroxylase. This experiment revealed that over 95% of the C1 adrenergic cells with spinal projection contain α_{2A}-LIR, whereas the noncatecholaminergic part of the projection had undetectable levels of α_{2A}-LIR.

These histological results suggest that (FIG. 1): (1) the C1 adrenergic cells of RVLM, including those with a presympathetic function, express high levels of α_{2A}-ARs, and (2) this type of receptor may not be present on the noncatecholaminergic component of the projection from RVLM to thoracic spinal cord.

The possibility that α_2-ARs of the 2A subtype might be selectively involved in the central sympatholytic effect of clonidine may explain the notoriously low potency of yohimbine and rauwolscine as antagonists of this effect of clonidine in rodents. Because of a single nucleotide mutation, the rodent α_{2A} receptor exhibits a reduced affinity for yohimbine and its diastereoisomer rauwolscine by comparison with the human orthologous receptor.[13] The peculiarly low potency of yohimbine in the rat has frequently been taken as evidence that clonidine's central hypotensive action is mediated by receptors other than α_2-ARs (e.g., NAIBS). This pharmacological oddity of the rodent provides further evidence that clonidine works through α_{2A}-ARs in the RVLM.

NEUROPHYSIOLOGICAL EVIDENCE THAT C1 PRESYMPATHETIC CELLS HAVE INHIBITORY AUTORECEPTORS

As just indicated, evidence indicates that C1 adrenergic cells belong to a class of tonically active barosensitive neurons with identifiable axonal projections to the thoracic spinal cord (vasomotor presympathetic cells). These cells fall into at least two categories. One type discharges at a slow rate of 1-8 spikes per second (mean of 5, under urethane anesthesia), has a mean axonal conduction velocity of less than 2 m/s (range, less than 0.5-3), and is powerfully inhibited by systemic administration of clonidine (1-20 μg/kg). This subclass is likely to be C1 adrenergic cells.[1,14] The other cell type has a much higher discharge rate (mean, around 20 spikes/s), a higher axonal conduction velocity (mean, 3.6 m/s), and is virtually insensitive to intravenous clonidine in the same dose range.[1,14,15]

Presympathetic vasomotor cells of the RVLM, regardless of type, have been subjected to iontophoretic applications of catecholaminergic receptor agonists and antagonists *in vivo* with the following results.[15] Around 35% of the presympathetic cells recorded were inhibited by catecholamines, and these responses had typical α_2-AR characteristics. (For details, see ref. 15.) FIGURE 2 is an example of one slow-firing, barosensitive bulbospinal cell of RVLM that was strongly inhibited by iontophoretic application of the α_2-AR selective agonist α-methylnorepinephrine. The effect of α-methylnorepinephrine was reversibly blocked by iontophoretic application of a selective α_2-AR antagonist, here by RX821002 (methoxyidazoxan). Moreover, as also illustrated in FIGURE 2B, RVLM bulbospinal barosensitive cells inhibited by iontophoretic application of catecholamines were selectively inhibited by small sympatholytic intravenous doses of clonidine (5-20 μg/kg, FIG. 2B1). This figure (panels B1 and B2) also illustrates that *iontophoretic* application of the α_2-AR antagonist methoxyidazoxan could even reverse the inhibitory effect of *systemically* administered clonidine (FIG. 2B1 and B2). FIGURE 2B1 depicts the inhibitory effect of a first dose of clonidine intravenously (5 μg/kg) and the cell's complete recovery after 30 minutes. FIGURE 2B2 illustrates the effect of a second identical dose of clonidine. Note that the inhibitory effect of the drug can then be reversed by iontophoretic application of the α_2-AR antagonist methoxyidazoxan. The effect of methoxyidazoxan is true antagonism, because the same amount of iontophoretic methoxyidazoxan raises the resting firing rate of the cell to a much smaller degree without clonidine (FIG. 2A) than after intravenous administration of the drug (FIG. 2B2). The antagonism of the effect of systemically administered clonidine by an iontophoretically applied antagonist shows that clonidine inhibits RVLM presympathetic cells by activating receptors located on, or near, the target cell. Finally, experiments such as that shown in FIGURE 2A (for a complete study see ref. 15) also suggest that these receptors are predominantly, if not exclusively, α_2-adrenergic.

In brief, the neurophysiological results described in this section also support the model represented in FIGURE 1. However, definitive neurophysiological evidence that C1 adrenergic cells have inhibitory α_2-ARs needs to be obtained by *in vitro* recordings.

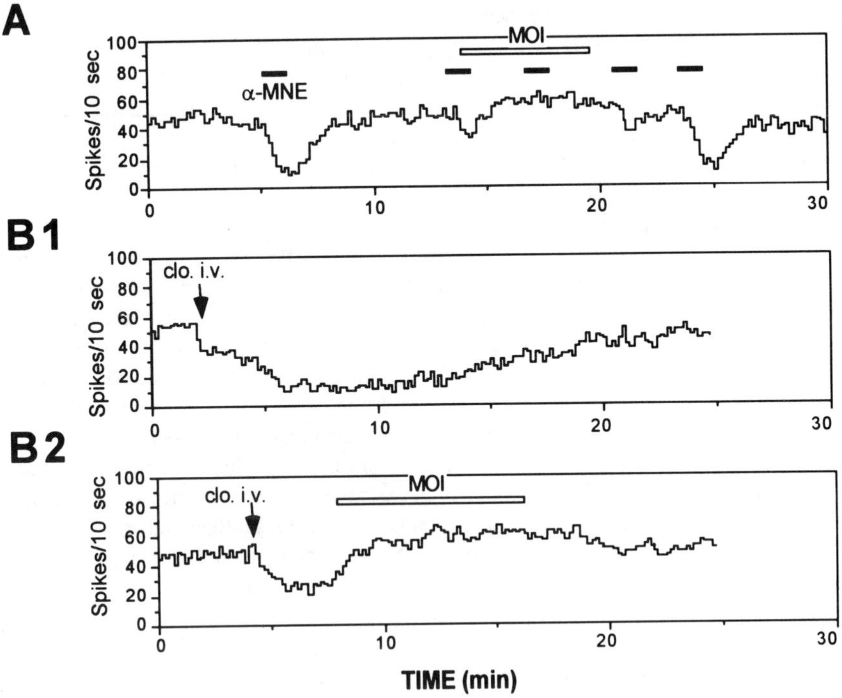

FIGURE 2. Effect of iontophoresis of the α_2-AR antagonist RX821002 (methoxyidazoxan, MOI) on the inhibition of an RVLM vasomotor neuron produced by iontophoretic application of α-methylnorepinephrine (α-MNE, **A**) or intravenous administration of clonidine (5 μg/kg, at *arrows*, **B1** and **B2**). The RVLM cell recorded was barosensitive and antidromically activated from the thoracic spinal cord (axonal conduction velocity 2.7 m/s). α-Methylnorepinephrine (α-MNE, 100 nA) applied at *closed bars* and methoxyidazoxan (MOI) at *open bars.* **A, B1,** and **B2** are consecutive. Recordings were done in a urethane-anesthetized 300–350-g male Sprague-Dawley rat as previously described, using a six-barrel iontophoretic electrode assembly as described in ref. 15. Neuronal activity is represented as integrated rate histograms.

ARE RVLM α_2-ARs OF FUNCTIONAL SIGNIFICANCE IN BLOOD PRESSURE CONTROL?

Bilateral microinjection into RVLM of clonidine or any of a host of substances with α_2-AR agonist properties produces hypotension in the anesthetized rat.[16–18] This hypotension is antagonized by a series of structurally diverse drugs that have α_2-AR antagonist activity as a common pharmacological denominator.[17] Many of these clonidine-like hypotensive agents also have affinity for one or more ''nonadrenergic imidazoline binding sites'' (NAIBS). This has led several investigators (e.g., see ref. 17) to propose that these binding sites rather than α_2-ARs might be responsible for the hypotensive effects of clonidine and related hypotensive agents. These interpretations, to which we do not subscribe for reasons presented in detail elsewhere,[19] are discussed by others in this volume. Proponents of the NAIBS theory generally choose to ignore a major piece of evidence in favor of the α_2-AR hypothesis of clonidine action in RVLM. That is, at least in the rat, injections into RVLM of the selective α_2-AR agonist α-methylnorepinephrine (a catecholamine with near zero affinity for NAIBS) decrease arterial pressure to a degree that is at least equal to that produced by clonidine, and this hypotensive effect is also blocked by α-adrenergic blockers.[20] In addition, α_2-ARs largely predominate in RVLM over NAIBS in terms of binding capacity (at least 75% of bovine medullary binding sites are α_2, and in rat medulla the α_2 binding comprises 100% of the sites[21]).

We therefore consider that the neuronal inhibition and hypotensive effect of clonidine within the RVLM can be explained very satisfactorily by its agonist activity at α_2-ARs located within the RVLM. This interpretation of the microinjection data is congruent with the finer neuroanatomical and neurophysiological evidence described in the two preceding sections. However, these results do not address the larger question of whether these RVLM α_2-ARs are functionally relevant, that is, if they are ever exposed to active concentrations of catecholamines *in vivo*. This appears to be the case based on additional evidence. First microinjections into the RVLM of drugs with α_2-AR antagonist property, such as the selective agent methoxyidazoxan (1 nmol) raise arterial pressure and sympathetic tone in anesthetized rats, that is, they exert effects opposite to those of catecholamines or clonidine (Huangfu and Guyenet, unpublished results, see FIG. 3A). In addition, iontophoretic application of methoxyidazoxan or of other α_2-AR antagonists (idazoxan and piperoxan[15]) raises the discharge rate of C1 type presympathetic cells of RVLM by 10–70% (Huangfu and Guyenet, unpublished results, see FIG. 3B). On the basis of the assumption that these agents block α_2-ARs selectively (methoxyidazoxan microinjection work) or on direct evidence for selectivity in the case of the iontophoretic approach, their excitatory effects suggest that α_2-adrenergic inhibitory receptors must be tonically activated by catecholamines within the RVLM under resting conditions. These catecholamines could originate from C1 cells (norepinephrine or epinephrine) that are tonically active (auto- or collateral inhibition hypothesis of FIG. 1) or from noradrenergic terminals belonging to neurons located elsewhere in the CNS (option not represented in FIG. 1). To distinguish between these two possibilities, rats were treated with bilateral injections of 6-hydroxydopamine into the RVLM, a procedure that destroys noradrenergic terminals but preserves adrenergic cell bodies and dendrites. Microinjection of methoxyidazoxan (1 nmol, bilateral) into the RVLM of 6-hydroxydopamine-treated animals 2 weeks later (urethane anesthesia) produced the same increase in arterial pressure and sympathetic tone as in control rats, suggesting that the endogenous catecholamine that tonically activates RVLM α_2-ARs is released by the C1 adrenergic cells themselves (Huangfu and Guyenet, unpublished results).

Furthermore, the sympathoinhibitory effect of the catecholaminergic uptake inhibitor desmethylimipramine (a commonly used tricyclic antidepressant that produces postural hypotension as a common side effect) also provides evidence suggesting that RVLM α_2-ARs are functionally significant. As indicated in FIGURE 4, the dose-dependent sympathoinhibition produced by desmethylimipramine is reversed by bilateral microinjection of the α_2-AR antagonist methoxyidazoxan into the RVLM in the urethane-anesthetized rat. These results (to be published) are compatible with the possibility that an increase in the extracellular concentration of catecholamine within

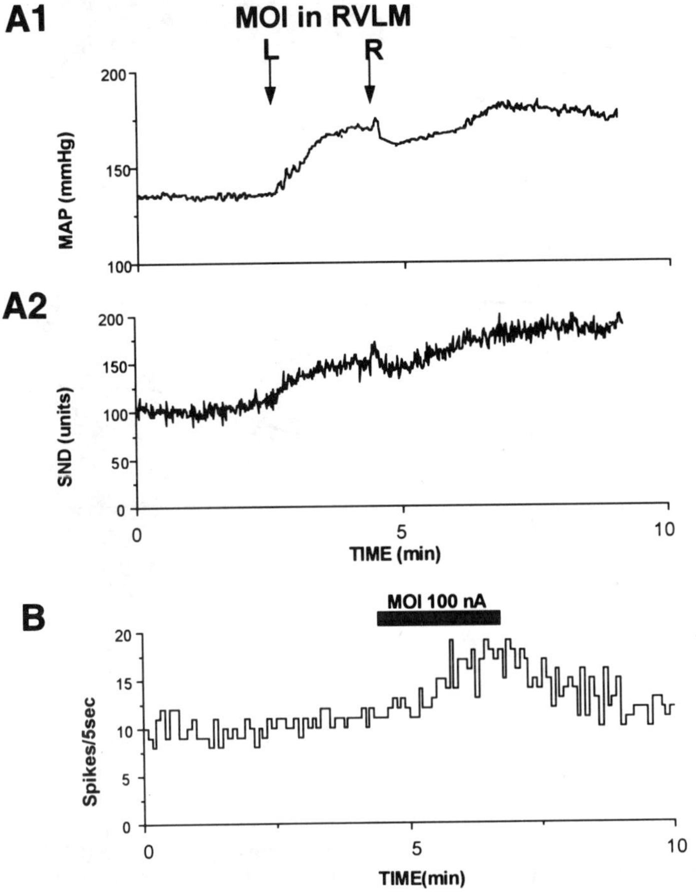

FIGURE 3. (**A**) Effects of right (*R*) and left (*L*) side microinjections of the α_2-AR antagonist methoxyidazoxan (MOI) into RVLM (1 nmol, 50 nL into left, *L,* then right, *R,* side) on mean arterial pressure (MAP, **A1**) and splanchnic sympathetic nerve discharge (SND, units arbitrary, **A2**); baroreceptor-denervated urethane-anesthetized, artificially respirated Sprague-Dawley rat. (**B**) Effect of iontophoretic application of MOI on firing rate of putative C1 adrenergic cell of the RVLM in a baroreceptor intact, urethane-anesthetized Sprague-Dawley rat. Note excitatory effect of MOI in this especially slow cell (2 spikes/s).

the RVLM, triggered by inhibiting reuptake, further activates α_2-ARs, thereby causing the sympathoinhibition that is a well-known effect of desmethylimipramine.

FURTHER EVIDENCE THAT CLONIDINE INHIBITS THE DISCHARGE OF C1 ADRENERGIC CELLS

Several recent studies using c-Fos immunocytochemistry as a marker of cell activity have demonstrated that C1 adrenergic cells are targets of clonidine's actions.

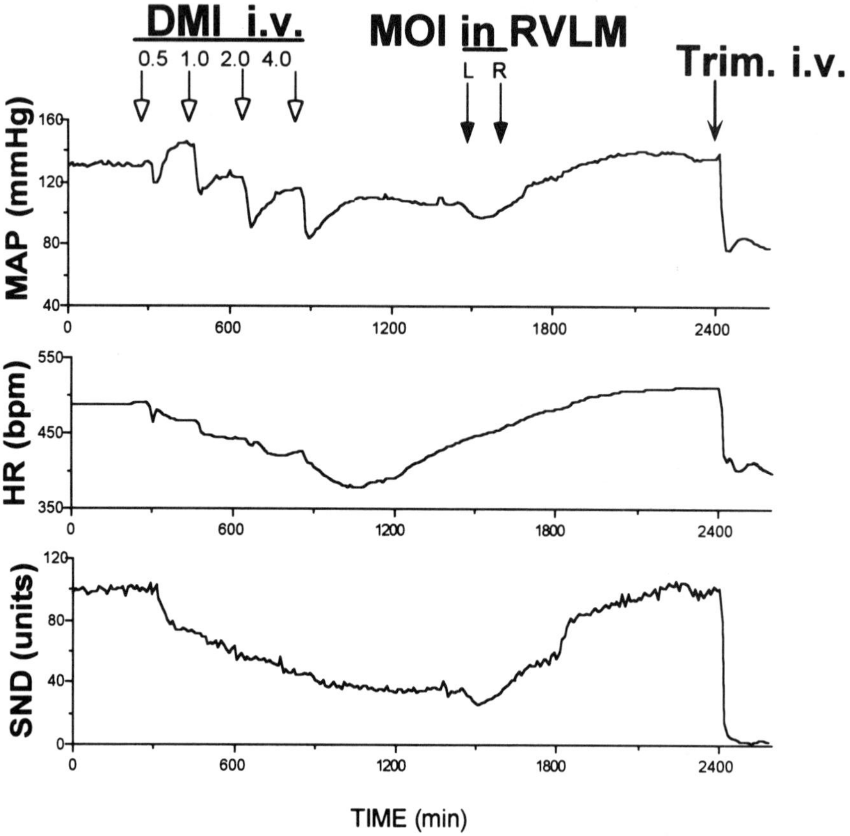

FIGURE 4. Sympathoinhibitory effect of intravenous desmethylimipramine (DMI, cumulative doses indicated in mg/kg) and its reversal by bilateral microinjection of the α_2-AR antagonist methoxyidazoxan (MOI, 1 nmol, 50 nL per side, first left side, *L,* then right side, *R*) into the RVLM. The ganglionic blocker trimethephan (Trim, 300 μl of 500 mg/10 ml) was administered at the end of the experiment to determine zero level of sympathetic nerve discharge (SND, splanchnic nerve). Experiments were performed in barodenervated male Sprague-Dawley rats under urethane anesthesia as previously described.[15] The effect of these pharmacological treatments on heart rate and blood pressure is also represented.

For example, Li and Dampney[22] demonstrated that hypotension induced by nitroprusside infusion activates C1 adrenergic cells (reflected by increased c-Fos immunoreactivity in tyrosine hydroxylase-labeled cells in the RVLM in rabbits). This activation was reversed by intravenous administration of clonidine. Morphine withdrawal is also a potent activator of c-Fos in C1 adrenergic neurons.[23] This activation is also attenuated by systemic clonidine administration.[24] Moreover, the turnover of epinephrine in the CNS is clearly reduced by clonidine.[25] Finally, clonidine greatly reduces the release of DOPAC by C1 adrenergic cells as suggested by electrochemistry *in vivo.*[26]

COLLATERAL AUTOREGULATION OF CNS MONOAMINERGIC CELLS: A MODEL FOR C1 ADRENERGIC CELLS?

The presence of somatodendritic inhibitory autoreceptors on each of the three best studied types of CNS monoaminergic cells (dopaminergic cells of the substantia nigra, noradrenergic cells of the locus coeruleus, and serotonergic neurons of the dorsal raphe) is very well established. (For a review see ref. 27, pages 272, 305, and 359.) Judging from the excitatory effects produced by iontophoretic application of the appropriate antagonists to these autoreceptors *in vivo,* these receptors are of functional significance.[27] The endogenous agonist for these somatic autoreceptors is released either from dendrites (as suggested most convincingly in the case of the dopamine cells of the substantia nigra, (as reviewed in ref. 7) or from axon collaterals.[27] Dendro-dendritic contacts between dopamine cells in the arcuate nucleus of the hypothalamus or in the A10 area[28,29] or between serotenergic neurons in the dorsal raphe[30] may represent the morphological substrate for collateral interactions. However, it is possible that much of the inhibition may be a form of local paracrine regulation (volume transmission) that does not require any special form of synaptic specialization.

Some evidence indicates that the C1 adrenergic cells are organized according to the same scheme as are the other monoaminergic cells. The anatomical evidence that C1 cells synthesize α_{2A}-ARs was presented before in this review. In addition, morphological evidence (immunohistochemistry at the electron microscopic level) for recurrent collateral interactions between C1 cells also exists.[31,32] Synaptic vesicles have also been observed in PNMT-immunoreactive dendrites of the RVLM (T. Milner, personal communication), supporting the possibility of a dendritic release of transmitter by these neurons. Finally, DOPAC is apparently released by C1 adrenergic cell bodies, as suggested by electrochemical detection.[33] The release of this metabolite depends on the activity of the cells, and it is clearly reduced by clonidine. Although DOPAC is not a metabolite of either norepinephrine or epinephrine, its modulated release suggests that other catecholamines may also be locally released by C1 adrenergic cell bodies in an amount that reflects the activity of the cells.

CONCLUSION

The evidence just presented supports the scheme in FIGURE 1. This essentially states that C1 adrenergic cells have a vasomotor sympathoexcitatory function, and like other brain monoaminergic neurons, these cells possess somatodendritic α_2-ARs that mediate some form of collateral inhibition. The sympathoinhibitory effect of clonidine in the RVLM is the result of activation of these autoreceptors and the consequent decrease in the firing rate of these presympathetic cells that set sympathetic tone. We demonstrated the presence of α_{2A}-ARs on these cells by immunohistochemistry and identified the presence of α_2-ARs by neurophysiological approaches *in vivo.* Evidence is also provided that these receptors are functional *in vivo.* There is little doubt that the central sympatholytic action of clonidine involves α_2-AR receptors at several central sites, most notably the spinal cord at the level of preganglionic cells.[34] However, because of their location on cells essential for vasomotor tone generation, the inhibitory α_2-ARs of the RVLM are certainly one important mechanism of the hypotensive action of clonidine and related drugs (FIG. 1).

REFERENCES

1. GUYENET, P. G. 1990. *In* Central Regulation of Autonomic Functions. A. D. Loewy & K. M. Spyer, Eds. :145-167. Oxford University Press. New York.
2. DAMPNEY, R. A. 1994. Progr. Neurobiol. **42:** 197-228.
3. RUGGIERO, D. A., S. L. CRAVO, V. ARANGO & D. J. REIS. 1989. *In* Progress in Brain Research, Vol. 81: The Central Neural Organization of Cardiovascular Control. J. Ciriello, M. M. Caverson & C. Polosa, Eds. :49-79. Elsevier. Amsterdam.
4. ROSS, C. A., D. A. RUGGIERO, T. H. JOH, D. H. PARK & D. J. REIS. 1984. J. Comp. Neurol. **228:** 168-185.
5. MILNER, T. A., S. F. MORRISON, C. ABATE & D. J. REIS. 1988. Brain Res. **448:** 205-222.
6. PUNNEN, S., R. URBANSKI, A. J. KRIEGER & H. N. SAPRU. 1987. Brain Res. **422:** 336-346.
7. CHERAMY, A., V. LEVIEL & J. GLOWINSKI. 1981. Nature. **289:** 537-542.
8. JANIG, W. 1988. Ann. Rev. Physiol. **50:** 525-540.
9. JANIG, W. & E. M. MCLACHLAN. 1992. J. Auton. Nerv. Syst. **41:** 3-13.
10. ROSIN, D. L., D. ZENG, R. L. STORNETTA, F. R. NORTON, T. RILEY, M. D. OKUSA, P. G. GUYENET & K. R. LYNCH. 1993. Neuroscience **56:** 139-155.
11. GUYENET, P. G., R. L. STORNETTA, T. RILEY, F. R. NORTON, D. L. ROSIN & K. R. LYNCH. 1994. Brain Res. **638:** 285-294.
12. SCHEININ, M., J. W. LOMASNEY, D. M. HAYDEN-HIXSON, U. B. SCHAMBRA, M. G. CARON, R. J. LEFKOWITZ & R. T. FREMEAU, JR. 1994. Brain Res. Mol. Brain Res. **21:** 133-149.
13. LINK, R., D. DAUNT, G. BARSH, A. CHRUSCINSKI & B. KOBILKA. 1992. Mol. Pharmacol. **42:** 16-27.
14. HASELTON, J. R. & P. G. GUYENET. 1989. Neuroscience **30:** 199-214.
15. ALLEN, A. M. & P. G. GUYENET. 1993. Am. J. Physiol. **265:** R1065-R1075.
16. ERNSBERGER, P., R. GIULIANO, R. N. WILLETTE, A. R. GRANATA & D. J. REIS. 1988. J. Hypertens. Suppl. **6:** S554-S557.
17. ERNSBERGER, P., R. GIULIANO, R. N. WILLETTE & D. J. REIS. 1990. J. Pharmacol. Exp. Ther. **253:** 408-418.
18. REIS, D. J., P. R. ERNSBERGER & M. P. MEELEY. 1991. *In* Central Neural Mechanisms in Cardiovascular Regulation, vol. 1. G. Kunos & J. Ciriello, Eds. :55-68. Birkhauser. Boston.
19. GUYENET, P. G., K. R. LYNCH, A. M. ALLEN, D. L. ROSIN & R. L. STORNETTA. 1994. *In* Ventral Brainstem Mechanisms and Control. O. Trouth, R. Millis, H. Kiwull-Schone & M. Schlafke, Eds. Marcel Dekker. New York.
20. GRANATA, A. R., Y. NUMAO, M. KUMADA & D. J. REIS. 1986. Brain Res. **377:** 127-146.
21. BRICCA, G., M. DONTENWILL, A. MOLINES, J. FELDMAN, A. BELCOURT & P. BOUSQUET. 1989. Eur. J. Pharmacol. **162:** 1-9.
22. LI, Y. W. & R. A. DAMPNEY. 1994. Neuroscience. **61:** 613-634.
23. STORNETTA, R. L., F. E. NORTON & P. G. GUYENET. 1993. Brain Res. **624:** 19-28.
24. BARABAN, S. C., R. L. STORNETTA & P. G. GUYENET. 1995. Brain Res. In press.
25. SCATTON, B., F. PELAYO, M. L. DUBOCOVICH, S. Z. LANGER & G. BARTHOLINI. 1979. Brain Res. **176:** 197-201.
26. TIBIRIÈA, E., J. FELDMAN, C. MERMET, F. GONON & P. BOUSQUET. 1991. J. Pharmacol. Exp. Ther. **256:** 606-613.
27. COOPER, J. R., F. E. BLOOM & R. H. ROTH. 1991. The Biochemical Basis of Neuropharmacology. 6th Ed. Oxford University Press. New York.
28. PIOTTE, M., A. BEAUDET, T. H. JOH & J. R. BRAWER. 1985. J. Comp. Neurol. **239:** 44-53.

29. BAYER, V. E. & V. M. PICKEL. 1990. J. Neurosci. **10:** 2996-3013.
30. CHAZAL, G. & P. T. OHARA. 1986. J. Neurocytol. **15:** 777-787.
31. MILNER, T. A., V. M. PICKEL, D. H. PARK, T. H. JOH & D. J. REIS. 1987. Brain Res. **411:** 28-45.
32. MILNER, T. A., V. M. PICKEL, S. F. MORRISON & D. J. REIS. 1989. *In* Progress in Brain Research. J. Ciriello, M. M. Caverson & C. Polosa, Eds.: 29-47. Elsevier. Amsterdam.
33. MERMET, C. & L. QUINTIN. 1991. Eur. J. Pharmacol. **204:** 105-107.
34. YOSHIMURA, M., C. POLOSA & S. NISHI. 1989. Prog. Brain Res. **81:** 181-190.

Mechanism of Sympathoinhibition by Imidazolines

BELA SZABO AND ROLF URBAN

Pharmakologisches Institut der Albert-Ludwigs-Universität
Hermann-Herder-Str. 5
D-79104 Freiburg i. Br., Germany

It has long been accepted that the prototype of centrally acting antihypertensive drugs, the α_2-adrenoceptor agonist clonidine, lowers sympathetic tone by activating α_2-adrenoceptors in the central nervous system.[1,2] Initiated by Bousquet[3] in 1984, a competing theory evolved, which we shall call the "imidazoline theory." According to the imidazoline theory, in addition to its affinity for α_2-adrenoceptors, clonidine possesses affinity for a newly discovered class of nonadrenergic radioligand binding sites, the imidazoline$_1$ (I_1) binding sites. Activation of I_1-receptors in the rostral ventrolateral medulla oblongata (RVLM), an important site of action of centrally acting antihypertensive drugs, is postulated to be responsible for the reduction of sympathetic tone by clonidine.[3,4] The present study deals with two relatives of clonidine, the oxazoline rilmenidine and the imidazoline moxonidine, which were recently introduced into the therapy of hypertension. Originally, the hypotensive effect of these drugs, like the effect of clonidine, was attributed to activation of α_2-adrenoceptors.[5,6] With the advent of the imidazoline theory, I_1-receptors were made responsible for the effects of rilmenidine and moxonidine.[7–13]

The main argument for the involvement of I_1-receptors in the sympathoinhibition by rilmenidine and moxonidine is their affinity for I_1-receptors in the RVLM.[10,14–16] In fact, the affinity of rilmenidine for I_1-receptors is 30 times and that of moxonidine 33 times higher than their affinity for α_2-receptors in the bovine medulla oblongata.[10] The imidazoline mode of action is also supported by the results of experiments in anesthetized animals, in which rilmenidine[17–19] and moxonidine[11] were administered into the RVLM, or into its vicinity, the cisterna magna. There are practically no experiments, however, that support involvement of I_1-receptors under therapeutically relevant conditions, that is, after systemic drug administration into conscious animals. Our aim was to characterize the cardiovascular effects of rilmenidine and moxonidine under such conditions, with special emphasis on the receptors involved in sympathoinhibition. We first studied the effects of rilmenidine and the interaction between rilmenidine and the α_2-adrenoceptor antagonists yohimbine and RX821002 [2-(2,3-dihydro-2-methoxy-1,4-benzodioxin-2-yl)-4,5-dihydro-1-imidazole]. In the second series of experiments, rilmenidine was compared with UK 14304 [5-bromo-6-(2-imidazolin-2-ylamino)-quinoxaline], an α_2-adrenoceptor agonist with low affinity for I_1-receptors ($I_1 : \alpha_2$ affinity ratio is 0.01; see ref. 9). Blockade of the effects by SK&F86466 (6-chloro-*N*-methyl-2,3,4,5-tetrahydro-1H-3-benzazepine) was also studied. In the final series of experiments on conscious rabbits, effects of moxonidine were compared with those of UK 14304. Antagonism by yohimbine, given systemically or into the cisterna magna, was also evaluated. The antagonists applied by us, yohimbine, SK&F86466, and RX821002, are selective α_2-blockers with an I_1:α_2 affinity ratio of less than 0.01.[4,20,21] We assumed that if the imidazoline theory is true, the pattern of effects of rilmenidine and moxonidine should differ from those of UK 14304. Furthermore, the effects of UK 14304 should be more susceptible to α_2-adrenoceptor blockade than the effects of rilmenidine and moxonidine.

Theoretically, α_2-adrenoceptor agonists should lower sympathetic tone not only by lowering sympathetic nerve firing through action in the central nervous system. They should also activate presynaptic α_2-autoreceptors at postganglionic sympathetic axons, thus reducing norepinephrine release per action potential.[22] Some experiments were performed in pithed rabbits with electrically stimulated sympathetic outflow to obtain information on the peripheral presynaptic action of rilmenidine and moxonidine.

METHODS

The methods used in conscious rabbits were previously described in detail.[23] Briefly, a bipolar electrode for recording renal sympathetic nerve activity, prepared from teflon-coated stainless steel wire, was implanted into rabbits under halothane anesthesia. The first actual experiment on conscious rabbits was performed 3 days or more after electrode implantation. Up to three additional experiments followed in intervals of 3 days. During the experiment, rabbits were sitting in a plastic box. An ear artery was cannulated for measurement of blood pressure and heart rate and for blood sampling. Blood pressure, heart rate, frequency of compound action potentials of the renal postganglionic sympathetic nerves, and the plasma concentration of norepinephrine and epinephrine were determined. Drugs were injected into an ear vein.

Some experiments were performed on pithed rabbits. The preparation was described previously.[24] The sympathetic outflow was continuously stimulated electrically (2 Hz, 140 mA). Blood pressure, heart rate, and the plasma concentration of norepinephrine were determined. Drugs were injected intravenously.

Values are given throughout as a percentage of the values measured at the beginning of the experiments (PRE values). Comparisons between groups were performed with the Mann-Whitney test or with analysis of variance for repeated measurements. Dose-response curves were compared with the procedure of Motulsky and Ransnas.[25]

RESULTS

Rilmenidine and Its Interaction with Yohimbine and RX821002 in Conscious Rabbits

Saline solution or drugs were injected every 10 minutes. In control animals, which received saline solution five times, all measured parameters remained constant (not shown). Rilmenidine 0.1, 0.3, and 1 mg kg^{-1} was cumulatively injected intravenously. Immediately after its injection, it caused transient hypertension (FIG. 1). A second phase with a long-lasting reduction in blood pressure, renal sympathetic nerve activity, and heart rate followed (FIG. 1). Responses were statistically evaluated 10 minutes after drug administration (FIG. 2). Rilmenidine dose-dependently lowered blood pressure, renal sympathetic nerve activity, heart rate, and the plasma concentration of norepinephrine and epinephrine. Maximum reduction of blood pressure and plasma catecholamines was reached after rilmenidine 0.3 mg kg^{-1}. Heart rate and renal sympathetic nerve activity decreased further after injection of 1 mg kg^{-1} rilmenidine.

In the second part of the experiments, interaction of rilmenidine with the α_2-adrenoceptor antagonist yohimbine was studied. Yohimbine 0.1 and 0.5 mg kg^{-1} was cumulatively injected iv into rabbits that had previously received saline solution or rilmenidine. In saline-treated rabbits, yohimbine did not change blood pressure, heart rate, and renal sympathetic nerve firing (not shown), whereas it slightly increased plasma norepinephrine and markedly increased plasma epinephrine (not shown). When yohimbine was injected in rilmenidine-treated rabbits, it immediately reversed effects of rilmenidine on blood pressure, heart rate, and renal sympathetic nerve firing. (FIGS. 1 and 2). Plasma catecholamines were also increased (FIG. 2). The effects of yohimbine were dose-dependent. After the higher dose, pre-rilmenidine values were reached (cardiovascular parameters) or surpassed (plasma catecholamines).

Another group of conscious rabbits received the α_2-adrenoceptor antagonist RX821002 iv after rilmenidine. The effects of RX821002 0.1 and 0.5 mg kg^{-1} (FIG. 2) were similar to those of yohimbine, that is, blood pressure, heart rate, renal sympathetic nerve activity, and plasma concentration of norepinephrine were dose-dependently increased. Pre-rilmenidine or even higher values were reached after RX821002 0.5 mg kg^{-1}.

Interaction of Rilmenidine and UK 14304 with SK&F86466 in Conscious Rabbits

To further elucidate the involvement of α_2-adrenoceptors in the sympathoinhibition by rilmenidine, in a next series of experiments the effects of rilmenidine were compared with the effects of UK 14304, a "pure" α_2-adrenoceptor agonist. Interaction of both rilmenidine and UK 14304 with SK&F86466 was also studied. SK&F86466 is an α_2-adrenoceptor antagonist with very low affinity for I_1-receptors.

Rilmenidine 0.03-1 mg kg^{-1} and UK 14304 0.001-0.03 mg kg^{-1} were cumulatively injected iv. The pattern of effects of UK 14304 was similar to that of rilmenidine (compare FIGS. 1 and 4): blood pressure increased initially and fell thereafter. Renal sympathetic nerve activity and heart rate fell monophasically. The statistical evaluation of the effects of rilmenidine and UK 14304 (except heart rate), based on measurements taken 10 minutes after drug administration, is shown in FIGURE 3. Rilmenidine and UK 14304 dose-dependently lowered blood pressure, renal sympathetic nerve activity, heart rate, and plasma norepinephrine concentration. ED_{50} values were calculated for each of the effects. The ED_{50} values increased in the same order for both drugs: ED_{50} for plasma norepinephrine fall < ED_{50} for hypotension < ED_{50} for renal sympathoinhibition << ED_{50} for bradycardia. Intravenous pretreatment with SK&F86466 (3 mg

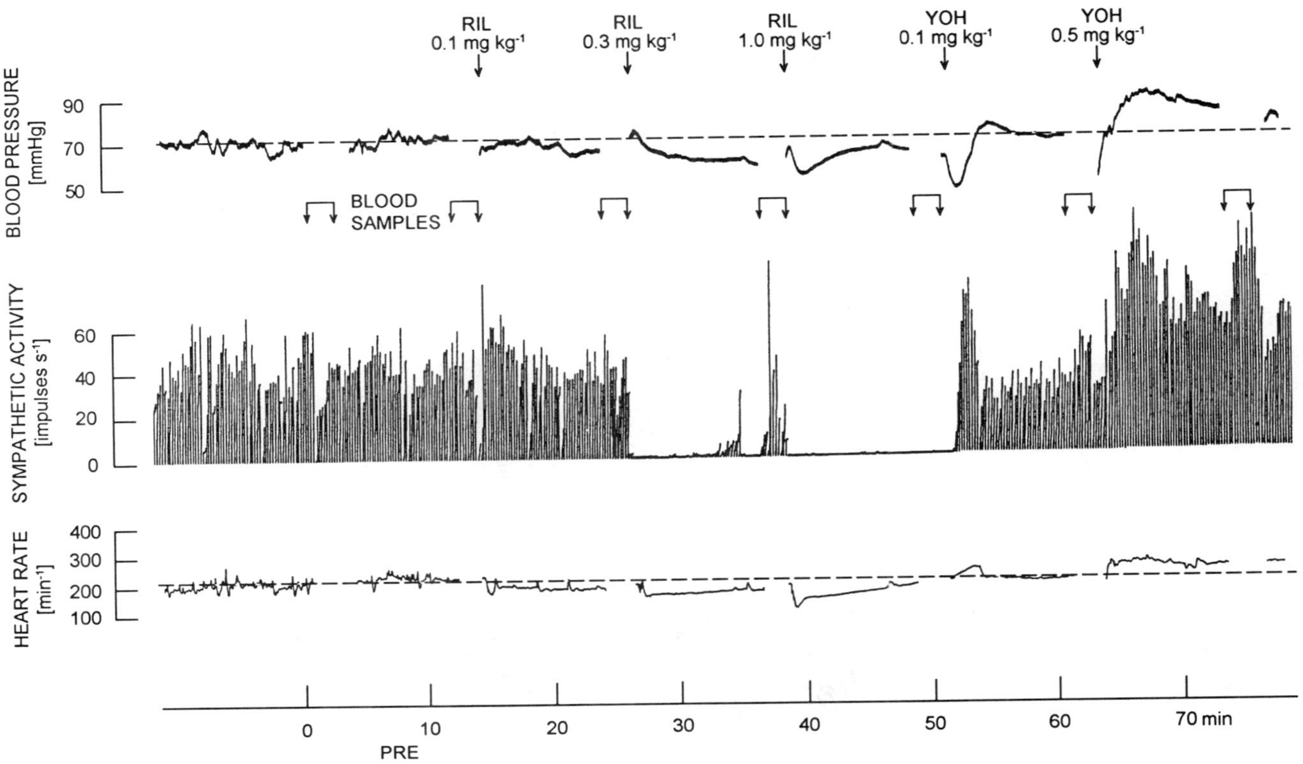

FIGURE 1. Mean arterial pressure, renal sympathetic nerve activity, and heart rate in an experiment in which three increasing doses of rilmenidine (RIL; 0.1, 0.3, and 1 mg kg^{-1} iv) and two increasing doses of yohimbine (YOH; 0.1 and 0.5 mg kg^{-1} iv) were injected into conscious rabbits. Modified from ref. 23.

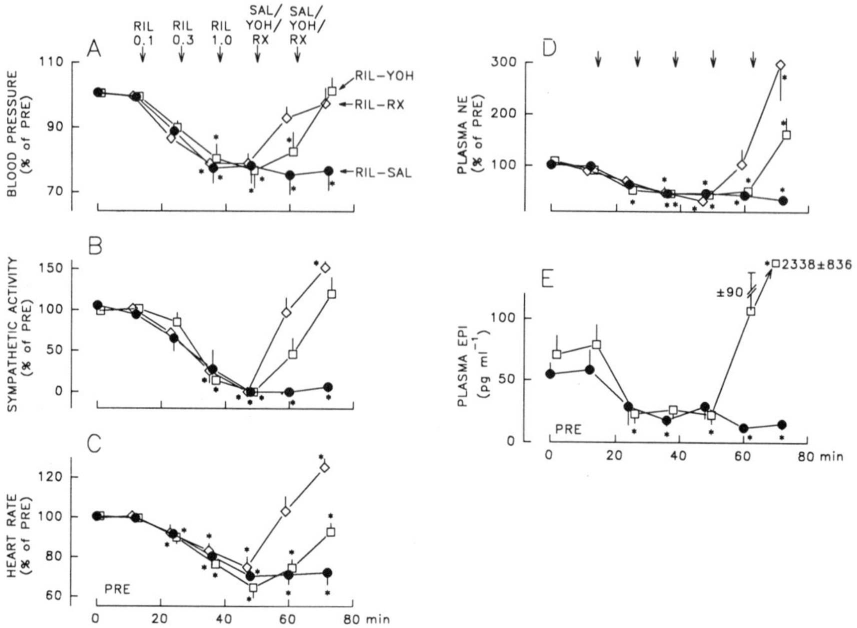

FIGURE 2. Effects of rilmenidine (RIL), saline solution (SAL), yohimbine (YOH), and RX821002 (RX) on mean arterial pressure, renal sympathetic nerve activity, heart rate, and plasma concentration of norepinephrine (NE) and epinephrine (EPI) in conscious rabbits. All rabbits received three increasing doses of RIL (0.1, 0.3, and 1 mg kg^{-1} iv). One group of rabbits thereafter received SAL twice (iv; RIL-SAL). A second group received, after RIL, two increasing doses of YOH (0.1 and 0.5 mg kg^{-1} iv; RIL-YOH), and a third group two increasing doses of RX (0.1 and 0.5 mg kg^{-1} iv; RIL-RX). Except for plasma epinephrine (in absolute units), all values are expressed as a percentage of the respective initial values (PRE). Means ± SEM from 4-5 experiments. Asterisk indicates significant difference from control group of rabbits that received saline solution five times (not shown for sake of clarity). Modified from ref. 23.

kg^{-1} + 1 mg kg^{-1} h^{-1}; bolus + infusion) antagonized the effects of rilmenidine and UK 14304 (FIG. 3) except for bradycardia (not shown). The rightward shifts, by SK&F86466, of the dose-response curves of rilmenidine and UK 14304 were similar.

Interaction of Moxonidine and UK 14304 with Yohimbine in Conscious Rabbits

Cardiovascular effects of moxonidine in conscious rabbits were compared with the effects of UK 14304. Moxonidine 0.003-0.1 mg kg^{-1} and UK 14304 0.001-0.030 mg kg^{-1} were injected cumulatively iv. The pattern of effects of moxonidine (FIG. 4) was similar to those just described for rilmenidine and UK 14304. Immediately after injection moxonidine increased blood pressure. This was followed by a second phase with enduring hypotension, reduction in renal sympathetic nerve firing, and

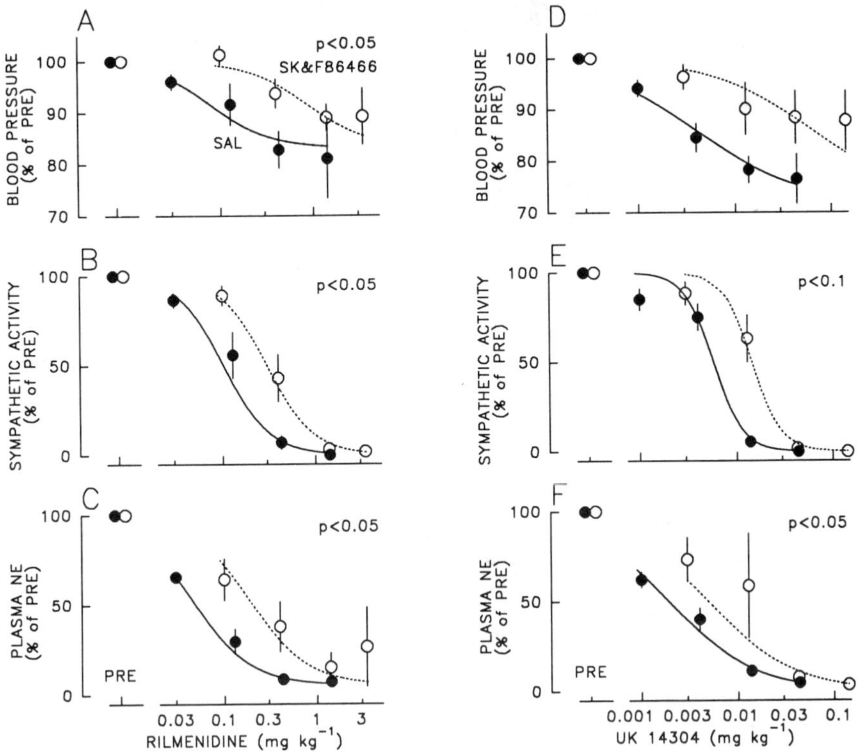

FIGURE 3. Effects of rilmenidine and UK 14304 on mean arterial pressure, renal sympathetic nerve activity, and plasma concentration of norepinephrine (NE) in conscious rabbits. After pretreatment with saline solution (SAL) or SK&F86466 (3 mg kg^{-1} + 1 mg kg^{-1} h^{-1} iv, bolus + infusion), four increasing doses of rilmenidine or UK 14304 were injected iv. All values are expressed as a percentage of the respective initial values (PRE). Means ± SEM from five experiments in each group. The dose response curves obtained in SAL- and SK&F86466-pretreated rabbits were compared for significant difference; results are indicated. Modified from ref. 41.

heart rate. FIGURE 5 shows statistical evaluation of blood pressure changes measured 10 minutes after drug administration. The maximal hypotension, a decrease in blood pressure by 7-9%, appeared after moxonidine 0.03 mg kg^{-1}. The maximal hypotensive effect of UK 14304 with a 16-18% decrease in blood pressure was greater.

FIGURE 5 also shows the interaction of moxonidine and UK 14304 (both iv) with yohimbine. Yohimbine was injected either iv (0.5 mg kg^{-1} + 0.1 mg kg^{-1} h^{-1} or 1 mg kg^{-1} + 0.2 mg kg^{-1} h^{-1}; bolus + infusion) or into the cisterna magna (0.0015 mg kg^{-1}). Intravenous yohimbine attenuated the effects of moxonidine and UK 14304 dose-dependently; the degree of antagonism of the hypotension by moxonidine and by UK 14304 was similar. Intracisternal yohimbine also counteracted the effects of moxonidine and UK 14304. Again, the effects of moxonidine and UK 14304 were similarly antagonized.

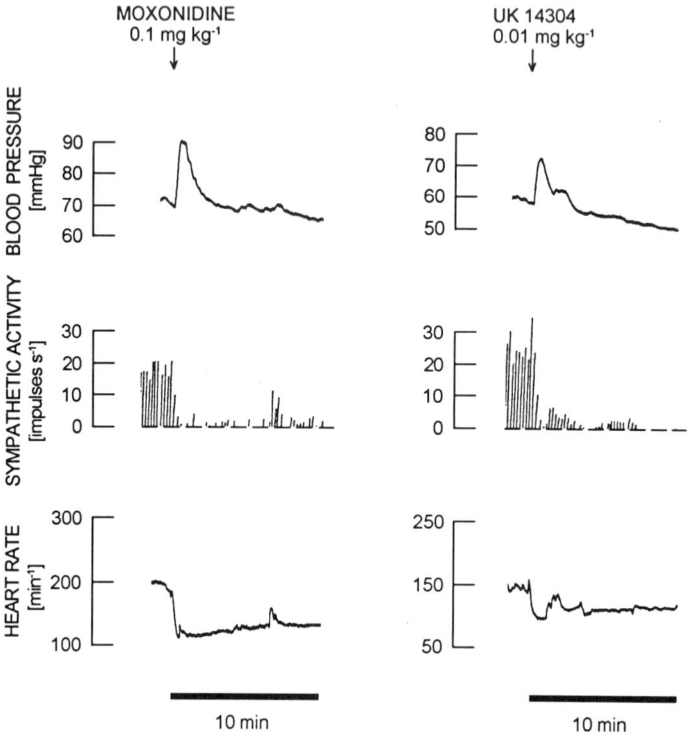

FIGURE 4. Mean arterial pressure, renal sympathetic nerve activity, and heart rate after injection of moxonidine (0.1 mg kg^{-1} iv) or UK 14304 (0.01 mg kg^{-1} iv) into conscious rabbits.

Rilmenidine, Moxonidine, UK 14304, and Yohimbine in Pithed Rabbits

Peripheral effects of rilmenidine, moxonidine, and UK 14304 on presynaptic receptors of postganglionic sympathetic axons were studied in pithed rabbits in which an artificial sympathetic outflow was maintained by electrical stimulation of preganglionic sympathetic nerves at 2 Hz. TABLE 1 shows only changes in plasma norepinephrine, because they reflect peripheral presynaptic drug action. In control rabbits, which received saline solution thrice iv, plasma norepinephrine remained constant. Rilmenidine 0.1-1 mg kg^{-1}, moxonidine 0.01-0.1 mg kg^{-1}, and UK 14304 0.003-0.03 mg kg^{-1} iv, dose-dependently lowered the plasma concentration of norepinephrine. The highest doses inhibited norepinephrine release very markedly, decreasing plasma norepinephrine to approximately 10% of the initial values. The effective doses in pithed rabbits were very similar to the doses that inhibited renal sympathetic nerve firing (see FIG. 3 for rilmenidine and UK 14304) or lowered blood pressure (see FIG. 5 for moxonidine) in conscious rabbits.

Pithed rabbits with electrically-stimulated sympathetic outflow were also used to determine the extent of α_2-adrenoceptor blockade after yohimbine 0.1 and 0.5 mg

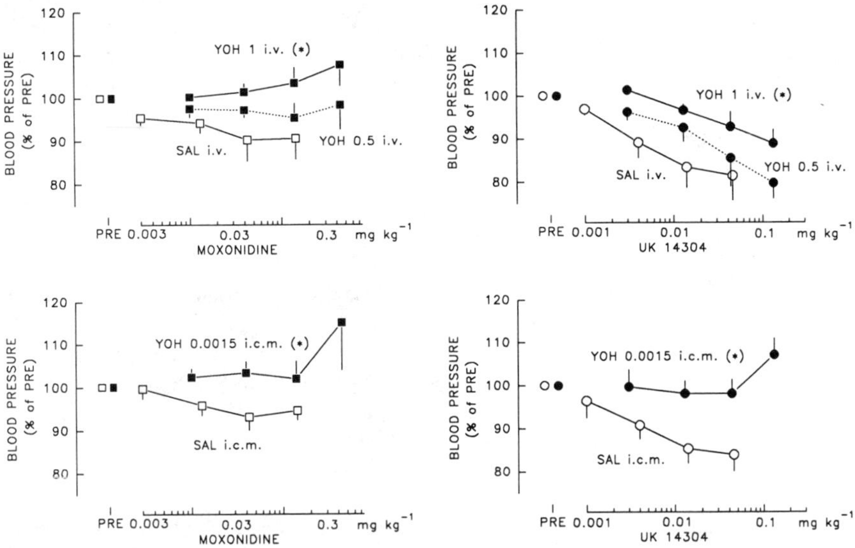

FIGURE 5. Effects of moxonidine, UK 14304, saline solution (SAL), and yohimbine (YOH) on mean arterial pressure in conscious rabbits. After pretreatment with SAL iv, YOH iv (0.5 mg kg^{-1} + 0.1 mg kg^{-1} h^{-1} or 1 mg kg^{-1} + 0.2 mg kg^{-1} h^{-1}), SAL into the cisterna magna (icm), or YOH icm (0.0015 mg kg^{-1}), four increasing doses of moxonidine or UK 14304 were injected iv. All values are expressed as a percentage of the respective initial values (PRE). Means ± SEM from 5-6 experiments. Asterisk indicates significant difference from SAL-pretreated rabbits.

TABLE 1. Effects of Saline Solution, Rilmenidine, Moxonidine, and UK 14304 on Plasma Concentration of Norepinephrine in Pithed Rabbits with Electrically Stimulated (2 Hz) Sympathetic Outflow[a]

	Plasma Norepinephrine (% of PRE)		
Saline (ml kg^{-1})	1	1	1
	90 ± 4%	92 ± 6%	99 ± 10%
Rilmenidine (mg kg^{-1})	0.1	0.3	1
	60 ± 4%[b]	30 ± 5%[b]	13 ± 3%[b]
Moxonidine (mg kg^{-1})	0.01	0.03	0.1
	53 ± 10%[b]	21 ± 7%[b]	10 ± 2%[b]
UK 14304 (mg kg^{-1})	0.003	0.01	0.03
	41 ± 3%[b]	17 ± 3%[b]	5 ± 1%[b]

[a] Saline or three increasing doses of rilmenidine, moxonidine, or UK 14304 were injected intravenously. All values are expressed as a percentage of the respective initial values (PRE). Means ± SEM from 4-6 experiments.

[b] Significantly different from saline-treated rabbits.

kg^{-1} iv, doses that were applied in conscious rabbits. Yohimbine increased the plasma norepinephrine concentration in pithed rabbits (not shown) apparently by interrupting physiological autoinhibition of norepinephrine release mediated by presynaptic α_2-autoreceptors at postganglionic sympathetic axons (see ref. 26). The effect of yohimbine 0.1 mg kg^{-1} was submaximal.

DISCUSSION

Rilmenidine, moxonidine, and UK 14304 lowered sympathetic tone in conscious rabbits, as shown by direct measurement of sympathetic nerve activity and plasma catecholamines. The functional consequence was a decrease in blood pressure and heart rate. Most likely, α_2-adrenoceptors were involved in the reduction of sympathetic tone by all three agonists. α_2-Adrenoceptors played a twofold role. Their activation in the central nervous system lowered sympathetic nerve firing, whereas their activation in the periphery lowered norepinephrine release per action potential from postganglionic sympathetic axons.

To our knowledge, these are the first experiments showing inhibition of sympathetic nerve activity by rilmenidine, moxonidine, and UK 14304 in conscious animals. It was previously shown that sympathetic nerve firing is reduced by rilmenidine in anesthetized dogs[27] and rats[28] and by moxonidine in anesthetized cats.[29] Since clonidine, the prototype of the mixed I_1/α_2-receptor agonists, does not inhibit ganglionic transmission,[30] inhibition of renal sympathetic nerve (a postganglionic nerve) activity reflects sympathoinhibition in the central nervous system. Antagonism of the hypotension by moxonidine by a low dose of yohimbine given into the cisterna magna also points to a central site of sympathoinhibition.

It is unsettled if α_2-adrenoceptors or I_1-receptors are responsible for the central sympathoinhibition by the mixed I_1/α_2-receptor agonists rilmenidine and moxonidine. The *imidazoline mode of action* is supported by the affinity of rilmenidine and moxonidine for I_1-receptors, as mentioned in the Introduction. Results of experiments with antagonists are also compatible with involvement of I_1-receptors. Thus, in experiments in which both rilmenidine and the antagonists were injected into the cisterna magna, idazoxan, which has affinity for I_1-receptors as well as α_2-adrenoceptors, is more effective against rilmenidine than the "pure" α_2-adrenoceptor antagonists yohimbine or RX821002.[17–19] Also compatible with the imidazoline theory is the observation that RX821002 is more potent at antagonizing the hypotension by α-methyldopa (which is the prodrug of the "pure" α_2-adrenoceptor agonist α-methylnorepinephrine) than the hypotension by rilmenidine.[19] Involvement of I_1-receptors in sympathoinhibition by systemically injected rilmenidine is supported only by one study: idazoxan injected into the RVLM more potently antagonized the hypotensive effect of intravenously administered rilmenidine than did SK&F86466, an α_2-adrenoceptor antagonist devoid of affinity for I_1-receptors.[31] Experimental support for involvement of I_1-receptors in the sympathoinhibition by moxonidine is sparse. The hypotensive action of moxonidine injected into the RVLM was more efficiently antagonized by efaroxan, an α_2-adrenoceptor antagonist with affinity for I_1-receptors, than by SK&F86466 (unpublished data; in ref. 11). The *α_2-adrenoceptor mode of action* is supported by the observation that α_2-adrenoceptor antagonists without affinity for I_1-receptors can antagonize the effects of rilmenidine[17,19,27,32] and moxonidine.[33] Sannajust and Head[19] try to fit these observations into the imidazoline theory of mode of action by arguing that rilmenidine (and imidazolines in general) primarily activate I_1-receptors, but activation of α_2-adrenoceptors is necessary for turning this primary event into sympathoinhibition. The antagonism by α_2-antagonists would, hence, be a functional antagonism.

We have observed that the preferential I_1-receptor agonists rilmenidine and moxonidine had a similar pattern of effects than did the α_2-adrenoceptor agonist UK 14304, which lacks affinity for I_1-receptors. Furthermore, three "pure" α_2-adrenoceptor antagonists, yohimbine, SK&F86466, and RX821002, intravenously administered, antagonized the effects of rilmenidine and moxonidine. Doses of the antagonists were kept low in order to selectively inhibit α_2-adrenoceptors without blockade of I_1-receptors. For example, yohimbine 0.1 mg kg^{-1} caused only submaximal blockade of α_2-autoreceptors at postganglionic sympathetic axons, as verified in experiments on pithed rabbits. Hypotension by moxonidine was also antagonized by a very low dose of yohimbine injected into the cisterna magna. The degree of blockade of the effects of rilmenidine and moxonidine by yohimbine, SK&F86466, and RX821002 was similar to the degree of blockade of the effects of the "pure" α_2-adrenoceptor agonist UK 14304. The simplest interpretation of these results is that rilmenidine, moxonidine, and UK 14304 primarily activated α_2-adrenoceptors. We cannot exclude the possibility that the α_2-blockers caused functional antagonism, as mentioned in the previous paragraph. But we can argue that α_2-adrenoceptor antagonists, as applied in this study, do not grossly interfere with sympathetic regulation. Intravenously injected yohimbine and rauwolscine had no major effect on the blood pressure-renal sympathetic nerve activity relationship in anesthetized rabbits,[34] and yohimbine also did not change sympathoinhibition by vasopressin.[23] Most data supporting involvement of I_1-receptors were obtained with drug administered directly into the RVLM or its vicinity. It is possible that after systemic drug administration, and this is the therapeutically relevant application form, the mode of action differs, and α_2-adrenoceptors play the decisive role.

Recently, the "old" α_2-adrenoceptor theory of the mechanism of action of centrally active antihypertensive agents received support in several publications. Hieble and Kolpak[35] studied the antagonism by SK&F86466 and idazoxan against clonidine at doses that were equipotent at blocking α_2-adrenoceptors. From the observation that the two antagonists were equally effective by reversing the hypotensive effect of clonidine, they concluded that α_2-adrenoceptors are the primary targets of clonidine. Guyenet *et al.*[36] recently demonstrated the presence of α_2-adrenoceptors (of the rat α_{2A}-subtype) on spinally-projecting neurons of the RVLM as the neuroanatomical basis for α_2-adrenoceptor-mediated sympathoinhibition. Finally, the spinally-projecting neurons of the RVLM, which were inhibited by systemically or locally (onto the soma of the neuron) applied clonidine, were also inhibited by locally applied α-methyl-norepinephrine, and the inhibitory effects of clonidine were antagonized by locally applied piperoxan, an antagonist without known affinity for I_1-receptors.[37]

Concerning the role of peripheral presynaptic α_2-adrenergic inhibition of norepinephrine release, Kobinger and Pichler[2] in their review of centrally acting hypotensive drugs write that "It remains open whether . . . [the] presynaptic mechanism contributes to the antihypertensive effect of the drugs in patients." In a recent contribution,

Garty *et al.*[38] found only negligible, if any, peripheral presynaptic inhibition after doses of clonidine which already caused central sympathoinhibition. In our experiments, rilmenidine, moxonidine, and UK 14304 markedly lowered the plasma norepinephrine concentration in pithed rabbits in which a constant sympathetic tone was established by electrical stimulation of the preganglionic sympathetic nerves. This is probably the result of activation of presynaptic receptors at postganglionic sympathetic axon terminals, because the prototype of the mixed I_1/α_2-receptor agonists, clonidine, does not cause ganglionic blockade.[30] Presynaptic α_2-autoreceptors were the likely targets of the drugs; a role of the recently described presynaptic imidazoline receptors can be excluded at least for moxonidine (see refs. 39 and 40). Presynaptic inhibition by rilmenidine, moxonidine, and UK 14304 occurred after the same doses in pithed rabbits which inhibited firing of renal sympathetic nerves in conscious rabbits. This has two implications. First, in conscious rabbits central sympathoinhibition and peripheral presynaptic inhibition take place simultaneously, synergistically lowering sympathetic tone. Second, probably the same receptor is responsible for both the peripheral as well as the central effect, the α_2-adrenoceptor.

SUMMARY

The aim of the present study is to characterize the cardiovascular effects of rilmenidine and moxonidine, two recently developed centrally acting antihypertensive drugs. Rilmenidine and moxonidine are α_2-adrenoceptor agonists and, in addition, possess affinity for imidazoline (I_1)-receptors. To determine if α_2- or I_1-receptors are involved in sympathoinhibition, rilmenidine and moxonidine were compared with UK 14304, an α_2-agonist devoid of affinity for I_1-receptors, and antagonism by the "pure" α_2-adrenoceptor antagonists yohimbine, SK&F86466, and RX821002 was studied.

When injected intravenously into conscious rabbits, rilmenidine and moxonidine, on the one hand, and UK 14304, on the other, elicited a similar pattern of effects. Thus, transient hypertension was followed by long-lasting hypotension accompanied by a decrease in heart rate, renal sympathetic nerve firing rate, and plasma norepinephrine concentration. The effects of rilmenidine, moxonidine, and UK 14304 were antagonized by intravenously administered yohimbine, SK&F86466, and RX821002. The effects of moxonidine and UK 14304 were also prevented by yohimbine injected into the cisterna magna. Altogether, the degree of antagonism of the effects of rilmenidine and moxonidine did not differ from the degree of antagonism of the effects of UK 14304. Rilmenidine, moxonidine, and UK 14304 were also given to pithed rabbits in which a constant sympathetic tone was maintained by electrical stimulation of the sympathetic nerves. At doses that caused sympathoinhibition in conscious rabbits, they lowered the plasma norepinephrine concentration markedly.

Our experiments show by direct measurement of sympathetic nerve activity and plasma norepepinephrine concentration that rilmenidine, moxonidine, and UK 14304 cause sympathoinhibition. As a consequence, blood pressure and heart rate decrease. The simplest interpretation of the blockade of central sympathoinhibition by the selective α_2-adrenoceptor antagonists is that rilmenidine, moxonidine, and UK 14304 primarily activated α_2-adrenoceptors. The decrease in plasma norepinephrine in pithed rabbits indicates peripheral presynaptic α_2-adrenoceptor-mediated inhibition of norepinephrine release per action potential from postganglionic sympathetic axons and suggests a contribution of this mechanism to the overall reduction in sympathetic tone.

ACKNOWLEDGMENT

We are grateful to Claudia Schurr for her expert technical assistance.

REFERENCES

1. KOBINGER, W. 1986. Drugs as tools in research on adrenoceptors. Naunyn-Schmiedeberg's Arch. Pharmacol. **332:** 113-123.
2. KOBINGER, W. & L. PICHLER. 1990. Centrally acting drugs (clonidine, methyldopa, guanfacine). *In* Pharmacology of Antihypertensive Therapeutics. G. V. R. Born, P. Cuatrecasas & H. Herken, Eds. **93:** 227-262. Springer-Verlag. Berlin.
3. BOUSQUET, P., J. FELDMAN & J. SCHWARTZ. 1984. Central cardiovascular effects of alpha adrenergic drugs: Differences between catecholamines and imidazolines. J. Pharmacol. Exp. Ther. **230:** 232-236.
4. ERNSBERGER, P., R. GIULIANO, R. N. WILLETTE & D. J. REIS. 1990. Role of imidazole receptors in the vasodepressor response to clonidine analogs in the rostral ventrolateral medulla. J. Pharmacol. Exp. Ther. **253:** 408-418.
5. VAN ZWIETEN, P. A. 1988. Pharmacology of the alpha$_2$-adrenoceptor agonist rilmenidine. Am. J. Cardiol. **61:** 6D-14D.
6. ARMAH, B. I., E. HOFFERBER & W. STENZEL. 1988. General pharmacology of the novel centrally acting antihypertensive agent moxonidine. Arzneimittelforschung **38:** 1426-1434.
7. VERBEUREN, T. J., A. T. DINH XUAN, E. KOENIG-BERARD & P. VITOU. 1990. Rilmenidine. Cardiovasc. Drug Rev. **8:** 56-70.
8. BOUSQUET, P., G. BRICCA, M. DONTENWILL, J. FELDMAN, H. GRENEY, A. BELCOURT, J. STUTZMANN & E. TIBIRICA. 1992. From the α_2-adrenoceptors to the imidazoline preferring receptors. Fundam. Clin. Pharmacol. **6**(Suppl 1): 15s-21s.
9. ERNSBERGER, P. R., K. L. WESTBROOKS, M. O. CHRISTEN & S. G. SCHÄFER. 1992. A second generation of centrally acting antihypertensive agents act on putative I_1-imidazoline receptors. J. Cardiovasc. Pharmacol. **20** (Suppl 4): S1-S10.
10. ERNSBERGER, P., T. H. DAMON, L. M. GRAFF, S. G. SCHÄFER & M. O. CHRISTEN. 1993. Moxonidine, a centrally acting antihypertensive agent, is a selective ligand for I_1-imidazoline sites. J. Pharmacol. Exp. Ther. **264:** 172-182.
11. ERNSBERGER, P., H. L. ELLIOTT, H.-J. WEIMANN, A. RAAP, M. A. HAXHIU, E. HOFFERBER, A. LÖW-KRÖGER, J. L. REID & H.-J. MEST. 1993. Moxonidine: A second-generation central antihypertensive agent. Cardiovasc. Drug Rev. **11:** 411-431.
12. CHRISP, P. & D. FAULDS. 1992. Moxonidine. A review of its pharmacology, and therapeutic use in essential hypertension. Drugs **44:** 993-1012.
13. MOLDERINGS, G. J., M. GÖTHERT, O. CHRISTEN & S. G. SCHÄFER. 1993. Imidazolrezeptoren und Blutdruckregulation. Dtsch. Med. Wschr. **118:** 953-958.
14. BRICCA, G., M. DONTENWILL, A. MOLINES, J. FELDMAN, E. TIBIRICA, A. BELCOURT & P. BOUSQUET. 1989. Rilmenidine selectivity for imidazoline receptors in human brain. Eur. J. Pharmacol. **163:** 373-377.
15. BRICCA, G., J. ZHANG, H. GRENEY, M. DONTENWILL, J. STUTZMANN, A. BELCOURT & P. BOUSQUET. 1993. Relevance of the use of [^{3}H]-clonidine to identify imidazoline receptors in the rabbit brainstem. Br. J. Pharmacol. **110:** 1537-1543.

16. BRICCA, G., H. GRENEY, J. ZHANG, M. DONTENWILL, J. STUTZMANN, A. BELCOURT & P. BOUSQUET. 1994. Human brain imidazoline receptors: Further characterization with [^{3}H]clonidine. Eur. J. Pharmacol.-Mol. Pharmacol. Sec. **266:** 25-33.
17. FELDMAN, J., E. TIBIRICA, G. BRICCA, M. DONTENWILL, A. BELCOURT & P. BOUSQUET. 1990. Evidence for the involvement of imidazoline receptors in the central hypotensive effect of rilmenidine in the rabbit. Br. J. Pharmacol. **100:** 600-604.
18. MAYOROV, D., M. CHERNOBELSKI & O. MEDVEDEV. 1993. Sympathoinhibitory action of rilmenidine in conscious sinoaortically denervated rats. J. Cardiovasc. Pharamcol. **22:** 314-320.
19. SANNAJUST, F. & G. A. HEAD. 1994. Rilmenidine-induced hypotension in conscious rabbits involves imidazoline-preferring receptors. J. Cardiovasc. Pharmacol. **23:** 42-50.
20. ERNSBERGER, P., M. P. MEELEY, J. J. MANN & D. J. REIS. 1987. Clonidine binds to imidazole binding sites as well as α_2-adrenoceptors in the ventrolateral medulla. Eur. J. Pharmacol. **134:** 1-13.
21. BRICCA, G., M. DONTENWILL, A. MOLINES, J. FELDMAN, A. BELCOURT & P. BOUSQUET. 1989. The imidazoline preferring receptor: Binding studies in bovine, rat and human brainstem. Eur. J. Pharmacol. **162:** 1-9.
22. STARKE, K. 1987. Presynaptic α-autoreceptors. Rev. Physiol. Biochem. Pharmacol. **107:** 74-146.
23. SZABO, B., R. URBAN & K. STARKE. 1993. Sympathoinhibition by rilmenidine in conscious rabbits: Involvement of α_2-adrenoceptors. Naunyn-Schmiedeberg's Arch. Pharmacol. **348:** 593-600.
24. SZABO, B., L. HEDLER, K. LICHTWALD & K. STARKE. 1987. ACTH increases noradrenaline release in pithed rabbits with electrically stimulated sympathetic outflow. Eur. J. Pharmacol. **136:** 391-399.
25. MOTULSKY, H. J. & L. A. RANSNAS. 1987. Fitting curves to data using nonlinear regression: A practical and nonmathematical review. Fed. Am. Soc. Exp. Biol. J. **1:** 365-374.
26. SZABO, B., A. SCHRAMM & K. STARKE. 1992. Effect of yohimbine on renal sympathetic nerve activity and renal norepinephrine spillover in anesthetized rabbits. J. Pharmacol. Exp. Ther. **260:** 780-788.
27. LAUBIE, M., J.-C. POIGNANT, J. SCUVEE-MOREAU, H. DABIRE, A. DRESSE & H. SCHMITT. 1985. Pharmacological properties of (*N*-dicyclopropylmethyl) amino-$_2$-oxazoline (S 3341), an α-$_2$-adrenoceptor agonist. J. Pharmacol. (Paris) **16:** 259-278.
28. SANNAJUST, F., C. BARRES, E. KOENIG-BERARD & J. SASSARD. 1992. Sympathoinhibitory effects of rilmenidine may be mediated by sites located below the brainstem. Br. J. Pharmacol. **105:** 535-541.
29. RAMAGE, A. G. & S. J. WILKINSON. 1989. Evidence that different regional sympathetic outflows vary in their sensitivity to the sympathoinhibitory actions of putative 5-HT_{1A} and α_2-adrenoceptor agonists in anaesthetized cats. Br. J. Pharmacol. **98:** 1157-1164.
30. KOBINGER, W. 1967. Über den Wirkungsmechanismus einer neuen antihypertensiven Substanz mit Imidazolinstruktur. Naunyn-Schmiedeberg's Arch. Pharm. Exp. Path. **258:** 48-58.
31. GOMEZ, R. E., P. ERNSBERGER, G. FEINLAND & D. J. REIS. 1991. Rilmenidine lowers arterial pressure via imidazole receptors in brainstem C1 area. Eur. J. Pharmacol. **195:** 181-191.
32. VAN ZWIETEN, P. A., M. J. M. C. THOOLEN, F. A. M. JONKMAN, B. WILFFERT, A. DE JONGE & P. B. M. W. M. TIMMERMANS. 1986. Central and peripheral effects of S 3341 [(*N*-dicyclopropylmethyl)-amino-2-oxazoline] in animal models. Arch. Int. Pharmacodyn. **279:** 130-149.
33. ARMAH, I. B. & W. STENZEL. 1981. BE-5895, a new clonidine-type antihypertensive aminopyrimidine derivative. Naunyn-Schmiedeberg's Arch. Pharmacol. **316:** R42.
34. SZABO, B., L. HEDLER & K. STARKE. 1989. Peripheral presynaptic and central effects of clonidine, yohimbine and rauwolscine on the sympathetic nervous system in rabbits. Naunyn-Schmiedeberg's Arch. Pharmacol. **340:** 648-657.

35. Hieble, J. P. & D. C. Kolpak. 1993. Mediation of the hypotensive action of systemic clonidine in the rat by α_2-adrenoceptors. Br. J. Pharmacol. **110:** 1635-1639.
36. Guyenet, P. G., R. L. Stornetta, T. Riley, F. R. Norton, D. L. Rosin & K. R. Lynch. 1994. Alpha$_{2A}$-adrenergic receptors are present in lower brainstem catecholaminergic and serotonergic neurons innervating spinal cord. Brain Res. **638:** 285-294.
37. Allen, A. M. & P. G. Guyenet. 1993. α_2-Adrenoceptor-mediated inhibition of bulbospinal barosensitive cells of rat rostral medulla. Am. J. Physiol. **265:** R1065-R1075.
38. Garty, M., A. Deka-Starosta, P. Chang, I. J. Kopin & D. S. Goldstein. 1990. Effects of clonidine on renal sympathetic nerve activity and norepinephrine spillover. J. Pharmacol. Exp. Ther. **254:** 1068-1075.
39. Göthert, M. & G. J. Molderings. 1991. Involvement of presynaptic imidazoline receptors in the α_2-adrenoceptor-independent inhibition of noradrenaline release by imidazoline derivatives. Naunyn-Schmiedeberg's Arch. Pharmacol. **343:** 271-282.
40. Molderings, G. J., F. Hentrich & M. Göthert. 1991. Pharmacological characterization of the imidazoline receptor which mediates inhibition of noradrenaline release in the rabbit pulmonary artery. Naunyn-Schmiedeberg's Arch. Pharmacol. **344:** 630-638.
41. Urban, R., B. Szabo & K. Starke. 1994. Is the sympathoinhibitory effect of rilmenidine mediated by alpha-$_2$ adrenoceptors or imidazoline receptors? J. Pharmacol. Exp. Ther. **270:** 572-578.

α-Adrenoceptors and GABA Receptors in the Hypertensive and Hypotensive Actions of Noradrenaline Intrathecally Injected in Anesthetized Rats

M. C. GARCÍA AND S. M. CELUCH

Cátedra de Farmacología
Facultad de Farmacia y Bioquímica (UBA)
and
ININFA-CONICET
Junín 956 5° (1113)
Buenos Aires, Argentina

Evidence indicates that noradrenergic bulbospinal pathways have an inhibitory effect on sympathetic preganglionic neurons (SPN).[1,2] Nevertheless, intrathecally injected noradrenaline increases both blood pressure and heart rate in conscious rats.[3] Moreover, spinally injected noradrenaline induces both inhibitory and excitatory effects on either sympathetic outflow[4] or heart rate[5] of anesthetized rats. These findings suggest that in the spinal cord, noradrenaline can be either an inhibitory or an excitatory neurotransmitter/neuromodulator for SPN related to control of the cardiovascular system. The current study investigates whether intrathecally injected noradrenaline produces an increase or a decrease in blood pressure and heart rate in pentobarbital anesthetized rats. The subtypes of adrenoceptors involved as well as participation of γ-aminobutyric acid (GABA) noradrenaline-induced responses were also analyzed.

METHODS

Wistar rats of either sex (180–230 g) were anesthetized with pentobarbital (40 mg/kg ip). Blood pressure was measured from the right femoral artery by a Statham P23ID transducer and recorded on a Grass 7B polygraph (Quincy, Massachusetts). Heart rate was calculated from the blood pressure record. A femoral vein was cannulated for intravenous drug injection, and a second cannula was positioned at the level of the T12-L1 intervertebral space for intrathecal injection of drugs, as described by Dib.[6] Intrathecal drug administration was always done at a rate not greater than 1 μl/min using a 5-μl Hamilton microsyringe.

Drugs. Drugs included: clonidine HCl (Sigma, USA), noradrenaline bitartrate (Sigma, USA), prazosin HCl (Pfizer, Argentine), yohimbine HCl (Schering; Argentine), bicuculline (RBI, USA), picrotoxin (Sigma, USA), and 2-hydroxy saclofen (RBI, USA). Drug doses refer to the respective free base.

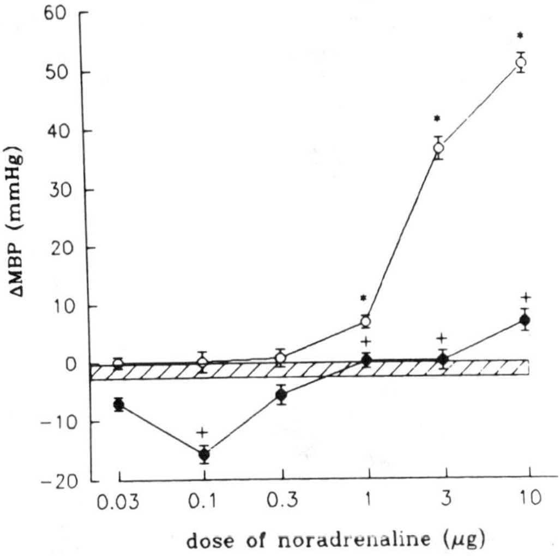

FIGURE 1. Maximal changes in mean blood pressure (ΔMBP, mm Hg) induced by intrathecal injection of noradrenaline in control rats (○) and in rats pretreated with 1 μg intrathecal prazosin (●). Peak pressor effects were obtained 1 minute after injection. Hypotensive responses reached a maximum approximately 15 minutes after injection. Each point represents mean ± SEM for 3-5 rats. *Hatched horizontal bar* represents the effect of 1 μl saline solution (intrathecally; $n = 5$). $^*p < 0.01$ when compared with the corresponding values in the saline-treated group. $^\dagger p < 0.01$ when compared with the control group (Student's t test).

RESULTS

Basal values of mean blood pressure and heart rate in pentobarbital anesthetized rats were 103.8 ± 1 mm Hg and 380 ± 2 beats/min, respectively ($n = 57$). These values were not modified by intrathecally injected saline solution.

Noradrenaline in doses of 0.03-0.3 μg (intrathecally) did not modify mean blood pressure, whereas higher doses (1-10 μg) induced dose-dependent pressor responses (FIG. 1). The α_1-adrenoceptor blocker prazosin (1 μg intrathecally) antagonized the pressor responses induced by noradrenaline and revealed a hypotensive effect with 0.1 μg noradrenaline (FIG. 1). Pretreatment with the α_2-adrenoceptor antagonist yohimbine (1 mg/kg iv) abolished both the hypotensive effect of 0.1 μg noradrenaline and the hypotension caused by the α_2-adrenoceptor agonist clonidine (3 μg intrathecally, TABLE 1). These hypotensive responses were also antagonized by the GABA-A receptor antagonists bicuculline and picrotoxin and the GABA-B receptor antagonist 2-hydroxy saclofen (TABLE 1). Noradrenaline (0.03-10 μg) did not modify heart rate in any of the experimental groups. By contrast, clonidine (3 μg) induced bradycardia (Δ heart rate −56 ± 5 bpm). This effect was antagonized by either yohimbine or GABA-A antagonists (bicuculline and picrotoxin). The GABA-B receptor blocker 2-hydroxy saclofen failed to alter the bradycardia caused by intrathecally injected clonidine (not shown).

TABLE 1. Effects of an α_2-Adrenoceptor Antagonist and of GABA-Receptor Antagonists on the Hypotensive Responses Induced by Intrathecally Injected Noradrenaline and Clonidine

Experimental Group	Noradrenaline (0.1 μg/ith)[a] ΔMBP (mm Hg)[b]	*n*	Clonidine (3 μg/ith) ΔMBP (mm Hg)[b]	*n*
Control	−15.7 ± 1.5	5	−23.3 ± 2.5	6
Yohimbine[c] (1 mg/kg iv)	−1.9 ± 1.7[c]	5	−7.2 ± 2.9[c]	3
Picrotoxin (2.7 μg ith)	−0.4 ± 2.7[c]	4	−8.3 ± 1.7[c]	5
Bicuculline (3.4 μg ith)	−2.9 ± 1.4[c]	4	−7.9 ± 1.4[c]	4
2-Hydroxy saclofen (30 μg ith)	−4.6 ± 1.8[c]	5	−12.6 ± 1.8[c]	5

[a] Animals were pretreated with prazosin (1 μg ith) 5 minutes before the injection of noradrenaline.

[b] Difference between mean blood pressure (MBP) measured immediately before and 15 minutes after intrathecal (ith) injection of noradrenaline or clonidine.

Yohimbine was intravenously injected 60 minutes before the agonists. Picrotoxin, bicuculline, and 2-hydroxy saclofen were intrathecally injected 6 minutes before the agonists. n = number of experiments.

[c] $p < 0.01$ when compared with corresponding control value (analysis of variance followed by Dunnett's test).

DISCUSSION

These results support the idea that in pentobarbital anesthetized rats, noradrenaline possesses both a pressor and a hypotensive effect when injected at the thoracolumbar level of the spinal cord. The pressor effect of high doses of noradrenaline (1-10 μg intrathecally) depended on activation of spinal α_1-adrenoceptors, because it was antagonized by prazosin. *In vitro* assays suggest the presence of excitatory α_1-adrenoceptors on the SPN.[7] Therefore, the pressor effect could be related to a direct action of noradrenaline on SPN involved in the control of vasomotor tone.

In animals pretreated with prazosin, the hypotensive effect induced by low doses of noradrenaline (0.1 μg intrathecally) seems to be mediated by activation of α_2-adrenoceptors, because that effect was prevented by yohimbine. The dose of yohimbine herein employed was also effective in antagonizing the cardiovascular responses to intrathecally injected clonidine, which is known to produce hypotension and bradycardia through activation of spinal α_2-adrenoceptors.[8] The hypotensive responses to noradrenaline and clonidine were antagonized by GABA receptor-blocking drugs. These results could suggest that the hypotension induced by activation of spinal α_2-adrenoceptors is mediated by stimulation of GABAergic neurons in the spinal cord and the subsequent inhibitory action of GABA on SPN. In support of this view, there is evidence of synaptic contacts between GABAergic terminals and SPN,[9,10] but it is not clear if the source of GABAergic innervation is either a spinal interneuron[11,12] or a descending pathway of supraspinal origin.[13] Moreover, in the spinal cord GABA inhibits SPN activity[14,15] and induces a decrease in both blood pressure and heart rate.[16,17] Both GABA-A and GABA-B receptors seem to be involved in the hypotensive responses to intrathecally injected noradrenaline and clonidine, because those effects were antagonized by GABA-A receptor antagonists bicuculline and picrotoxin as well as by the GABA-B receptor antagonist 2-hydroxy saclofen.

Cardiac sympathetic innervation arises mainly from the T2 level of the spinal cord.[5] Therefore, the lack of effects of noradrenaline on heart rate in the present study is possibly related to the low thoracolumbar level site of injection (T12-L1) and the rapid clearance of noradrenaline from the biphase by an active uptake mechanism. By this way, the possibility of diffusing to a higher level of the spinal cord is scarce. Conversely, diffusion of clonidine to higher thoracic spinal levels[18] could account for the bradycardia induced by this drug.

In conclusion, this study suggests that in the rat thoracolumbar spinal cord, noradrenaline induces a predominant pressor effect which is mediated by activation of α_1-adrenoceptors. A hypotensive effect mediated by α_2-adrenoceptors occurs with a low dose of noradrenaline, and it is revealed only when α_1-adrenoceptors are blocked. This hypotensive response could involve activation of a spinal GABAergic neuron.

REFERENCES

1. COOTE, J. H. 1985. J. Auton. Nerv. Syst. **14:** 255-262.
2. ELGHOZI, J. L., G. A. HEAD, C. R. ANDERSON & P. I. KORNER. 1989. Brain Res. **499:** 39-52.
3. GRADIN, K., A. P. NICHOLAS, P. HJEMDAHL, T. SVENSSON & T. HOKFELT. 1992. J. Cardiovasc. Pharmacol. **20:** 367-374.
4. SHI, H., D. I. LEWIS & J. H. COOTE. 1988. J. Auton. Nerv. Syst. **23:** 69-78.
5. SUNDARAM, K., J. MURUGAIAN & H. SAPRU. 1991. Brain Res. **544:** 227-234.
6. DIB, B. 1984. Pharmacol. Biochem. Behav. **20:** 45-48.
7. YOSHIMURA, M., C. POLOSA & S. NISHI. 1987. Brain Res. **414:** 138-142.
8. KUBO, T., J. NAGURA & Y. MISU. 1987. J. Pharmacol. Exp. Ther. **240:** 298-302.
9. BACON, S. J. & A. D. SMITH. 1988. J. Auton. Nerv. Syst. **24:** 97-122.
10. CHALMERS, J., L. ARNOLDA, V. KAPOOR, I. LLEWELLYN-SMITH, J. MINSON & P. PILOWSKY. 1992. J. Hypertens. **10**(suppl. 7): S27-S37.
11. TAPPAZ, M. L., J. A. ZIVIN & I. J. KOPIN. 1976. Brain Res. **111:** 220-223.
12. CALVET, M.-C., C. LEVALLOIS, J. CALVET & J.-M. KAMENKA. 1993. Brain Res. **608:** 299-309.
13. FUJI, K., E. SENBA, S. FUJI, I. NOMURA, J.-Y. WU, Y. UEDA & M. TOHYAMA. 1985. Neuroscience **14:** 881-894.
14. BACKMAN, S. B. & J. L. HENRY. 1983. Brain Res. **277:** 365-369.
15. WU, S. Y. & N. J. DUN. 1992. Brain Res. **572:** 94-102.
16. GORDON, J. 1985. Brain Res. **328:** 165-169.
17. HASSÉSSIAN, H., A. PRAT, J. DE CHAMPLAIN & R. COUTURE. 1991. Eur. J. Pharmacol. **202:** 51-60.
18. MARWAHA, J., J. H. KEHNE, R. L. COMMISSARIS, J. LAKOSKI, W. SHAW & M. DAVIS. 1983. Brain Res. **276:** 379-382.

Imidazoline and Adrenoceptor Involvement in Cardiovascular Responses to Clonidine in Sham and Sinoaortic Denervated Rats

D. RICCI AND C. A. TAIRA

Cátedra de Farmacología
Facultad de Farmacia y Bioquímica
Universidad de Buenos Aires
Junin 956 piso 5to
Buenos Aires, Argentina

The α-adrenergic and imidazoline receptors of the central nervous system have a role in central cardiovascular regulation.[1,2] The intracerebroventricular (icv) administration of clonidine (CLO), an imidazoline, in conscious rats with intact baroreflex induces a pressor response and bradycardia, but in sinoaortic denervated (SAD) rats a transient pressor response is followed by hypotension and bradycardia.[2] This work studies involvement of the central imidazoline receptor and α-adrenoceptor subtypes in the responses to CLO (icv) in sham-operated and SAD rats.

MATERIALS AND METHODS

Wistar rats (200-250 g) were used. Sinoaortic denervation (SAD) was performed according to the method of Krieger.[3] A sham operation was also carried out. Experiments were performed 7 days after the corresponding operation.

Two days before the experiments, a stainless steel guide cannula (22G) was implanted in the left lateral ventricle of rats. Twenty-four hours before the experiments, rats were anesthetized with ether and the right femoral artery was cannulated with a polyethylene cannula (PE50) filled with heparinized saline solution (25 $U{\cdot}ml^{-1}$). The cannula was connected to a Statham Gould P23ID pressure transducer coupled to a Grass 79 D polygraph. Blood pressure was measured in conscious rats and heart rate was counted from blood pressure waves.

Drugs were injected intracerebroventricularly by inserting an injection needle (28G) connected to a Hamilton syringe through the guide cannula. All drugs were dissolved in saline solution, except prazosin and 5-methyl-urapidil which was dissolved in a solution of glucose (5%) and drops of acetic acid (3%). Drugs were injected icv in a volume of 3 μl for 1 minute. The next antagonists were injected icv before CLO: prazosin, yohimbine, idazoxan, chloroethylclonidine, and 5-methyl-urapidil.

Data are expressed as the mean ± SEM. Analysis of variance (ANOVA) and Student's *t* test were used for statistical evaluation of the data.

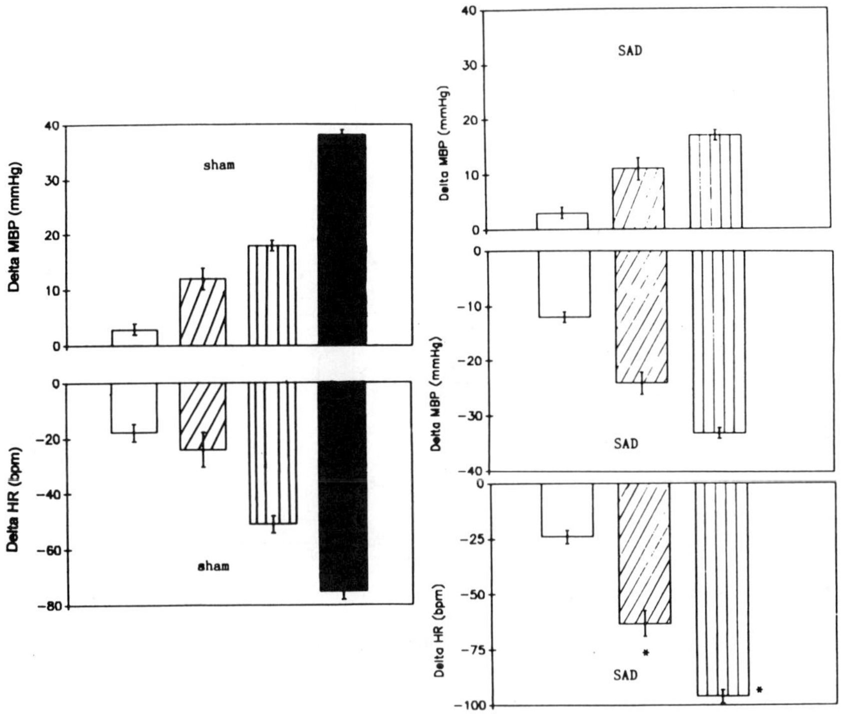

FIGURE 1. Dose-response curve to clonidine (CLO icv) in sham and sinoaortic denervated (SAD) rats. Changes in mean blood pressure (MBP) and heart rate (HR) after clonidine: 1 μg (□), 3 μg (▨), 10 μg (▥), and 30 μg (■); $n = 7$ rats in each group. *$p < 0.05$ vs the corresponding dose from the sham-operated group.

RESULTS AND DISCUSSION

The icv administration of CLO (1-30 μg) induced an increase in mean blood pressure (MBP) associated with bradycardia in conscious sham-operated rats. In SAD rats CLO induced a rise in MBP followed by a fall associated with bradycardia that was significatively greater than that in sham-operated rats. These actions of CLO were dose-dependent (FIG. 1). The results suggest supersensibility of SAD rats to the cardiovascular actions of CLO.

In both groups of rats, neither antagonist altered the basal cardiovascular parameters 15 minutes after administration. Experiments with the antagonists showed that blockade was greater with idazoxan (imidazoline) than with yohimbine (α_2) and 5-methyl-urapidil (α_{1A}) (FIG. 2). On the other hand, prazosin (α_1) and chloroethylclonidine (α_{1B}) did not modify responses to CLO (FIG. 2). The results suggest that cardiovascular responses to centrally injected CLO can be mediated by brain α_{1A} and α_2-adrenoceptors and imidazoline receptors in both groups of rats.

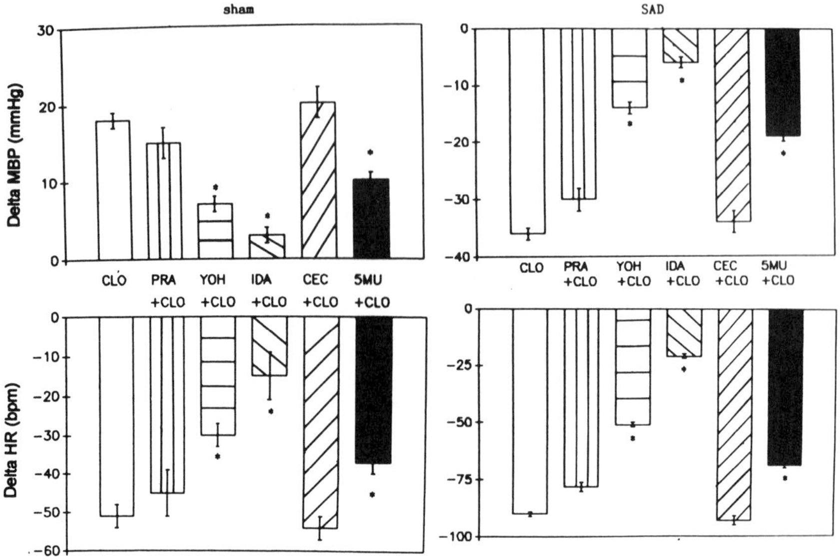

FIGURE 2. Effects of antagonists on cardiovascular responses to clonidine (CLO, 10 μg) in sham and sinoaortic denervated (SAD) rats. The following antagonists were used: idazoxan (IDA, 15 μg), yohimbine (YOH, 30 μg), 5-methyl-urapidil (5 MU, 100 μg), prazosin (PRA, 15 μg), and chloroethylclonidine (CEC, 100 μg). $n = 6$ in each group. *$p < 0.05$ vs the corresponding control group.

REFERENCES

1. Bousquet, P., J. Feldman & S. Schwartz. 1984. Central cardiovascular effects of α adrenergic drugs: Differences between catecholamines and imidazolines. J. Pharmacol. Exp. Ther. **230:** 232-236.
2. Ricci, D., C. A. Taira & M. A. Enero. 1992. Cardiovascular responses of sinoaortic-denervated rats to intracerebroventricular injection of α_1- and α_2-adrenoceptor agonists. Eur. J. Pharmacol. **212:** 195-200.
3. Krieger, E. M. 1964. Neurogenic hypertension in the rat. Circ. Res. **15:** 511-521.

Influence of Imidazolines on Catecholamine Release in Pithed Spontaneously Hypertensive Rats

W. HÄUSER, J. GÜTTING, TH. NGUYEN, AND P. DOMINIAK [a]

Institute of Pharmacology
Medical University of Lübeck
Ratzeburger Allee 160
D-23538 Lübeck, Germany

Imidazoline derivatives have long been considered potent α_2-adrenoceptor agonists or antagonists.[1] The first evidence for different sites of action for phenylethylamines and imidazolines arose from Ruffolo *et al.*[2–4] and Bousquet *et al.*,[5] who reported differences in the responses induced by these substances; however, they were still discussed as having arisen from the effects on α_2-adrenoceptors. Clonidine is known to exert a strong effect on presynaptic α_2-adrenoceptors, which is characterized by significantly reduced norepinephrine release from the sympathetic varicosities.[6–8] It was previously discussed that the reduction of plasma norepinephrine concentration which occurs during clonidine therapy is partly a result of this effect.[9] Several studies have investigated the effects of imidazolines on norepinephrine release of isolated organ preparations.[10–13] However, the effects of imidazolines on norepinephrine release were discussed to be α_2-adrenoceptor mediated and less to imidazoline receptors. The present study investigates the actions of some imidazoline derivatives (clonidine, moxonidine, and rilmenidine) on norepinephrine overflow and epinephrine release in pithed spontaneously hypertensive rats (SHR) during preganglionic electric stimulation of the spinal cord to demonstrate the influence of putative presynaptic imidazoline receptors on catecholamine release. Moreover, the actions of agmatine, a recently discovered endogenous ligand at imidazoline receptors clonidine-displacing substance (CDS), on catecholamine release should be studied.[14]

MATERIALS AND METHODS

Male SHR weighing about 200–250 g were used for the experiments and pithed by a standard method as described elsewhere.[15,16] Polyethylene catheters were inserted into both femoral veins (PE-10) for drug administration and into both carotid arteries (PE-50) for measuring blood pressure and collecting blood samples. Dose response curves of clonidine, moxonidine, and rilmenidine on increases in blood pressure were installed. From the respective curves the EC_{25}, EC_{50}, EC_{75}, and EC_{95} concentrations

[a] To whom correspondence should be addressed.

were calculated and infused into a femoral vein during electrical stimulations of the spinal cord. Agmatine was administered intravenously in doses from 10-5,000 μg/kg*min during stimulations in a separate set of experiments. Control SHR received the same volume of 0.9% sodium chloride. Stimulation of peripheral sympathetic nerves was induced using a steel rod (Ø 1.5 mm) coated with enamel, avoiding the region of the thoracolumbar spinal cord. Sympathetic outflow was induced by preganglionic electric stimulation (10 V, 1 ms, 0.5 Hz, 3 min) and assessed by measuring plasma norepinephrine and epinephrine concentrations at the end of each stimulation period in blood samples obtained from a PE-50 catheter which was inserted into the left carotid artery. Some experiments were performed under α_2-adrenoceptor blockade with rauwolscine (1 and 3 mg/kg iv) and uptake$_1$-inhibition with desipramine (0.5 mg/kg iv). At the end of each stimulation period, blood samples were taken from the left carotid artery, mixed with glutathione/EDTA, and centrifuged at 4°C. Plasma samples were kept frozen at –70°C until analyzed. The plasma catecholamine concentration was assayed according to Erikson and Persson.[17] Data presented in the figures and tables are mean values ± standard error of the mean. Statistical analysis was performed by either ANOVA and U test or Bonferroni test.

RESULTS

Clonidine, moxonidine, and rilmenidine dose-dependently and significantly reduced stimulation-dependent norepinephrine concentrations in the plasma of pithed SHR; moxonidine was the most effective substance in decreasing plasma norepinephrine (FIG. 1). When compared to the respective controls, all substances investigated led to a significant decrease in norepinephrine overflow (FIG. 1). However, stimulation-dependent epinephrine release into plasma remained unchanged with the exception of the highest concentration of the three compounds administered (EC_{95}), resulting in at least a fivefold increase in epinephrine release (not depicted). Inhibition of norepinephrine reuptake by desipramine (0.5 mg/kg iv) and blockade of the α_2-autoreceptor by rauwolscine (1 and 3 mg/kg iv) led to a dose-dependent increase in norepinephrine overflow to about 1,500 pg/ml and 2,000 pg/ml, respectively (FIG. 2). Clonidine, moxonidine, and rilmenidine significantly and dose-dependently reduced stimulation-dependent plasma norepinephrine concentrations when 1 mg respectively 3 mg/kg rauwolscine were administered to the pithed SHR (FIG. 2). Interestingly, rauwolscine 3 mg/kg shifted the dose response curves for the decrease in norepinephrine overflow parallel to higher norepinephrine concentrations when compared to 1-mg curves (FIG. 2). By contrast to norepinephrine, the epinephrine release during electric stimulation and inhibition of both reuptake and α_2-adrenoceptors was enhanced dose-dependently to the imidazoline derivatives injected and to rauwolscine (not depicted).

The clonidine-displacing substance agmatine was investigated in incremental doses with respect to its action on norepinephrine overflow and epinephrine release during electric stimulations of the spinal cord using 0.5, 3.3, and 5.0 Hz. Agmatine failed to change the norepinephrine concentrations in plasma under any experimental condition (not depicted). However, epinephrine release was significantly enhanced

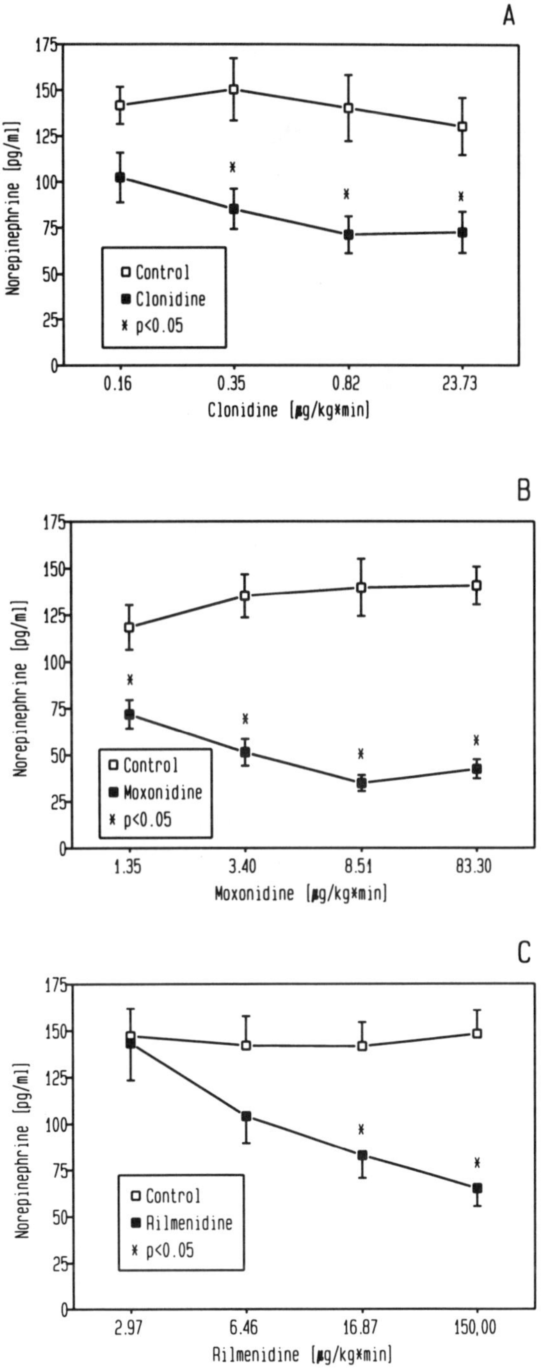

FIGURE 1. Changes in stimulation-dependent plasma norepinephrine concentrations of pithed spontaneously hypertensive rats (SHR) during incremental infusions of the imidazolines clonidine (**A**), moxonidine (**B**), and rilmenidine (**C**). The doses of each imidazoline compound were calculated from the respective dose-response curves on blood pressure increases. The significance ($p < 0.05$) indicates the differences between each imidazoline dose and the respective control value. Stimulation conditions: 10 V, 0.5 Hz, 1 ms, 3 min.

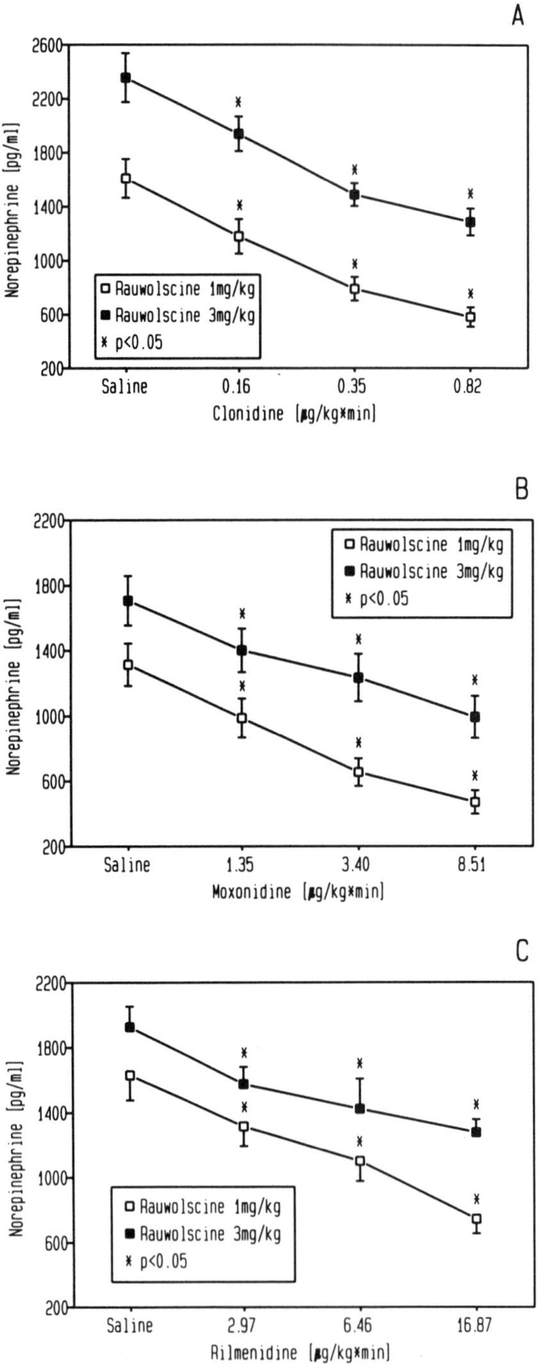

FIGURE 2. Changes in stimulation-dependent plasma norepinephrine concentrations of pithed SHR during incremental infusions of the imidazolines clonidine (**A**), moxonidine (**B**), and rilmenidine (**C**). Stimulation-dependent norepinephrine overflow was measured during inhibition of specific norepinephrine uptake by desipramine (0.5 mg/kg iv) and blockade of the α_2-autoreceptors by rauwolscine, as indicated. The significance (p <0.05) indicates the differences of each norepinephrine concentration from the respective saline values. Between the 1-mg rauwolscine curve and the 3-mg curve of each imidazoline a significant difference could be calculated.

during each stimulation frequency when the highest dose of agmatine (5,000 μg/kg*min) was intravenously administered (not depicted). Lower doses of agmatine did not influence epinephrine release.

DISCUSSION

Clonidine, moxonidine, and rilmenidine are imidazoline respectively oxazoline derivatives and were described as selective agonists at the putative I_1-receptor.[18] However, all three substances also possess graded affinity to the α_2-adrenoceptor, of which clonidine is the most potent agent.[18] With the present experiments we demonstrated that the three compounds dose-dependently diminished norepinephrine overflow in pithed SHR. To decrease the stimulation-dependent norepinephrine plasma concentration of 75 pg/ml doses of 0.35 μg/kg clonidine, 3.4 μg/kg moxonidine and 150 μg/kg rilmenidine were needed. These differences in concentration of the imidazolines used correspond to their affinities to the α_2-adrenoceptor.[18] On the other hand, the α_2-adrenoceptor antagonist rauwolscine increased the stimulation-dependent norepinephrine overflow, which is in agreement with previous reports.[6–8] During α_2-adrenoceptor blockade and inhibition of norepinephrine reuptake with desipramine, the imidazolines further reduced norepinephrine overflow dose-dependently. Similar results were obtained in isolated rabbit pulmonary artery strips and the isolated rat kidney.[10–13] Therefore, a presynaptic action of imidazolines on norepinephrine release, independently of α_2-adrenoceptors, has been concluded. However, the parallel shifting of the dose response curves for decreasing norepinephrine concentrations in plasma to higher levels (FIG. 2) when the α_2-adrenoceptor antagonist rauwolscine was administered in doses of 1 and 3 mg/kg is more responsible to an α_2-agonistic action of clonidine, moxonidine, and rilmenidine on sympathetic nerve endings.[6–8,11,13]

Imidazoline binding sites have been reported on the adrenal medulla, but it is lacking α_2-adrenoceptors.[18] Our results concerning an increase in stimulation-dependent epinephrine release when the highest dose of each imidazoline was administered can support the stimulating effects of those substances on the adrenal medulla.[18] The high concentrations needed for epinephrine release are in agreement with the comparably low affinities of imidazoline compounds on adrenal medullary imidazoline receptors.[19,20]

A stimulating action of agmatine on the adrenal medulla was recently reported.[14] In the same investigation its rather low affinity to imidazoline receptors could also be demonstrated.[14] With the use of a different experimental model our results on epinephrine release during agmatine infusions confirm the previous report. However, without norepinephrine reuptake blockade and α_2-adrenoceptor inhibition, agmatine did not influence stimulation-dependent norepinephrine overflow into plasma.

In conclusion, the present results do not allow the postulation of presynaptically located imidazoline receptors of which a local norepinephrine release could be regulated. The existence of imidazoline receptors on the adrenal medulla mediating epinephrine release, however, could be confirmed.

SUMMARY

The actions of the imidazoline derivatives clonidine, moxonidine, and rilmenidine and of the recently discovered clonidine-displacing substance agmatine on stimulation-induced norepinephrine overflow and epinephrine release were studied in pithed spontaneously hypertensive rats. All three imidazolines dose-dependently decreased norepinephrine overflow and led to an increase in epinephrine release when the highest dose of each compound was injected. The α_2-adrenoceptor antagonist rauwolscine shifted the dose-response curves of plasma norepinephrine concentrations to higher levels. Agmatine did not change norepinephrine overflow but increased epinephrine release into plasma after the highest dose administered. It is concluded that the investigated imidazolines decrease norepinephrine overflow via presynaptic α_2-adrenoceptors, whereas epinephrine release is mediated through putative imidazoline receptors on the adrenal medulla.

REFERENCES

1. KOBINGER, W. & L. PICHLER. 1990. Centrally acting drugs (clonidine, methyldopa, guanfacine). *In* Pharmacology of Antihypertensive Therapeutics. D. Ganten & P. J. Mulrow, Eds.: 227-262. Springer Verlag. Berlin, Heidelberg.
2. RUFFOLO, R. R., J. W. FOWBLE, D. D. MILLER & P. N. PATIL. 1976. Binding of [^{3}H]dihydrozapetine to alpha adrenoceptor-related proteins from rat vas deferens. Proc. Natl. Acad. Sci. USA **73:** 2730-2734.
3. RUFFOLO, R. R., P. J. RICE, P. N. PATIL, A. HAMADA & D. D. MILLER. 1983. Differences in the applicability of the Easson-Stedman hypothesis to the α_1 and α_2 adrenergic effects of phenylethylamines and imidazolines. Eur. J. Pharmacol. **86:** 471-475.
4. RUFFOLO, R. R., B. S. TUROWSKI & P. N. PATIL. 1977. Lack of cross desensitization between structurally dissimilar α-adrenoceptor agonists. J. Pharm. Pharmacol. Commun. **29:** 378-380.
5. BOUSQUET, P., J. FELDMAN & J. SCHWARTZ. 1984. Central cardiovascular effects of alpha adrenergic drugs: Differences between catecholamines and imidazolines. J. Pharmacol. Exp. Ther. **253:** 232-236.
6. STARKE, K. 1977. Regulation of noradrenaline release by presynaptic receptor systems. Rev. Physiol. Biochem. Pharmacol. **77:** 1-124.
7. STARKE, K. 1981. α-Adrenoceptor subclassification. Rev. Physiol. Biochem. Pharmacol. **88:** 199-236.
8. STARKE, K. 1987. Presynaptic α-autoreceptors. Rev. Physiol. Biochem. Pharmacol. **107:** 73-146.
9. BRAVO, E. L., M. D. CRESSMAN & M. A. POHL. 1985. Cardiovascular and neurohumoral effects of long-term oral clonidine monotherapy in essential hypertensive patients. *In* Low Dose Oral and Transdermal Therapy of Hypertension. M. A. Weber, J. I. M. Drayer & R. Kolloch, Eds. Steinkopff. Darmstadt.
10. GÖTHERT, M. & G. J. MOLDERINGS. 1991. Involvement of presynaptic imidazoline receptors in the α_2-adrenoceptor-independent inhibition of noradrenaline release by imidazoline derivatives. Naunyn-Schmiedeberg's Arch. Pharmacol. **343:** 271-282.
11. GÖTHERT, M. & G. J. MOLDERINGS. 1992. Modulation of norepinephrine release in blood vessels: Mediation by presynaptic imidazoline receptors and α_2-adrenoceptors. J. Cardiovasc. Pharmacol. **20**(Suppl. 4): S16-S20.
12. MOLDERINGS, G. J., F. HENTRICH & M. GÖTHERT. 1991. Pharmacological characterization of the imidazoline receptor which mediates inhibition of noradrenaline release in the rabbit pulmonary artery. Naunyn-Schmiedeberg's Arch. Pharmacol. **344:** 630-638.

13. BOHMANN, C., P. SCHOLLMEYER & L. C. RUMP. 1994. Effects of imidazolines on noradrenaline release in rat isolated kidney. Naunyn-Schmiedeberg's Arch. Pharmacol. **349:** 118-124.
14. LI, G., S. REGUNATHAN, C. J. BARROW, J. ESHRAGI, R. COOPER & D. J. REIS. 1994. Agmatine: An endogenous clonidine-displacing substance in the brain. Science **263:** 966-969.
15. GILLESPIE, J. S. & T. C. MUIR. 1967. A method of stimulating the complete sympathetic outflow from the spinal cord to blood vessels in the pithed rat. Br. J. Pharmacol. Chemother. **30:** 78-87.
16. DOMINIAK, P., M. SIMON, A. BLÖCHL & P. BRENNER. 1992. Changes in peripheral sympathetic outflow of pithed spontaneously hypertensive rats after bradykinin and desArg-bradykinin infusions: Influence of converting enzyme inhibition. J. Cardiovasc. Pharmacol. **20**(Suppl. 9): S35-S38.
17. ERIKSON, B. M. & B. A. PERSSON. 1982. Determination of catecholamines in rat heart tissue and plasma samples by liquid chromatography with electrochemical detection. J. Chromatogr. **228:** 143-154.
18. ERNSBERGER, P. R., K. L. WESTBROOKS, M. O. CHRISTEN & S. G. SCHÄFER. 1992. A second generation of centrally acting antihypertensive agents act on putative I_1-imidazoline receptors. J. Cardiovasc. Pharmacol. **20**(Suppl. 4): S1-S10.
19. MOLDERINGS, G. J., D. MOURA, K. FINK, H. BÖNISCH & M. GÖTHERT. 1993. Binding of [^{3}H]clonidine to I_1-imidazoline sites in bovine adrenal medullary membranes. Naunyn-Schmiedeberg's Arch. Pharmacol. **348:** 70-76.
20. REGUNATHAN, S., M. P. MEELEY & D. J. REIS. 1993. Expression of non-adrenergic imidazoline sites in chromaffin cells and mitochondrial membranes of bovine adrenal medulla. Biochem. Pharmacol. **45:** 1667-1675.

Imidazoline Receptors in Vascular Smooth Muscle and Endothelial Cells[a]

SOUNDARARAJAN REGUNATHAN,
CHARLOTTE YOUNGSON, HONG WANG, AND
DONALD J. REIS

Division of Neurobiology
Department of Neurology and Neuroscience
Cornell University Medical College
New York, New York 10021

Clonidine, an imidazoline and α_2-adrenergic partial agonist, contracts vascular smooth muscle.[1] The effect has long been attributed to clonidine's action as an agonist at α_2-adrenergic receptors.[2] However, other related imidazolines, such as tolazoline, cirazoline, and dexmeditomedine,[3,4] irrespective of their actions as agonists or antagonists at α_2-adrenergic receptors, also contract blood vessels. This finding raises the possibility that the vasoconstriction elicited by clonidine is mediated by a nonadrenergic receptor.

One group of receptors that may be relevant are the imidazoline receptors (I-receptors) of either the I_1 or I_2 subclass. Although I_2 receptors have been detected in rabbit urethral smooth muscle,[19] it is not known if they are expressed in vascular smooth muscle and/or endothelium. In the present study, we therefore sought to determine by ligand binding and by immunostaining with a specific antibody to an I-receptor-associated binding protein if smooth muscle cells of rat aorta (ASMC) examined *in vivo* and *in vitro* and/or cultured endothelial cells of bovine pulmonary artery express I-receptors and, if so, to determine their subclass.

MATERIALS AND METHODS

Culturing of Vascular Smooth Muscle and Endothelial Cells

Primary cultures of rat aortic smooth muscle cells were prepared by modification[20] of the method of Meyer-Lehnert and Schrier.[21] Briefly, rats were sacrificed by anesthetic overdose, vessels removed and incubated with collagenase (2 mg/ml) in Dulbecco's modified Eagle's minimum essential medium (DMEM), and the adventitia stripped away. The vessels were cut longitudinally and the intima scraped off. After further collagenase treatment, the remaining tissue was passed 20 times through 18

[a]This research is supported by National Institutes of Health grant HL18974.

and 16 gauge needles. The resulting cell suspension was centrifuged at 500 × *g* for 5 minutes and resuspended in DMEM containing penicillin (100 units/ml), streptomycin (100 μg/ml), 2 mM L-glutamine, and 10% fetal calf serum. The suspended cells were plated in T-75 flasks and cultured until cells reached confluency. The cultured cells were then subcultured in 100-mm plates, and the confluent cells were used for membrane preparation. The viability of the cells was tested by exclusion of trypan blue. Smooth muscle cells were phenotypically defined by immunostaining with anti-smooth muscle actin (Sigma). Cultured smooth muscle cells were usually used for about 10 passages and were replaced by freshly prepared cells. Endothelial cells of bovine pulmonary artery were obtained from American Type Culture Collection and cultured as described in the provided protocol. The homogeneity of the preparation was verified by demonstrating that most cells were immunostained by anti-Factor VIII (Sigma).

Ligand Binding Assays

Membranes from cultured cells were prepared for ligand binding assays by methods described elsewhere.[6,11] Briefly, harvested cells were homogenized in HEPES-sucrose buffer (pH 7.4) with a Teflon-glass homogenizer and centrifuged at 1,000 × *g* for 10 minutes. The pellets were suspended in ice-cold Tris-HCl buffer (pH 7.7) with 5 mM EDTA and homogenized using a Polytron (Brinkman Instruments) for 10 seconds. The homogenate was centrifuged at 40,000 × *g* for 30 minutes; the resulting membrane pellet was washed three times by resuspension in buffer and re-centrifugation. The final washed pellet was suspended in fresh Tris-HCl buffer (pH 7.4) to provide approximately 100 μg protein per assay tube. Binding assays were performed in Tris-HCl buffer (pH 7.7) using 1 nM[^{3}H]*p*-aminoclonidine (PAC), 1 nM [^{3}H]rauwolscine, or 5 nM [^{3}H]idazoxan. Nonspecific binding was defined by 100 μM of cold ligand. Following incubation at 25°C for 30 minutes the reaction was terminated by rapid vacuum filtration over Whatman GF/B filters and washed with 10 ml of ice-cold buffer. Filters were suspended in scintillation fluid and radioactivity counted in a liquid scintillation counter. Binding data were analyzed with LIGAND (Elsevier Biosoft), a parametric nonlinear regression analysis. Membrane protein was assayed by the Coomassie Blue method (Pierce) using bovine serum albumin as standard.

Immunostaining

Cultured ASMC and cryosectioned tissues were immunostained with antibodies to I-receptor-associated protein by the immunoperoxidase method of Hsu *et al.*[22] Polyclonal antibody was prepared in this laboratory in rabbits from an imidazoline receptor-associated binding protein partially purified by idazoxan or *para*-aminoclonidine affinity chromatography. The binding protein binds [^{3}H]idazoxan with comparable kinetic constants to native receptor and the antibody immunoprecipitated receptor protein partially blocks specific binding and immunostains tissues expressing the I-receptor.[12,13]

Briefly, cultures were fixed with 4% paraformaldehyde and blocked with 0.5% bovine serum albumin (BSA) in 0.1M Tris-saline solution. Cultures were then incubated with antiserum (I-receptor protein, 1 : 500 dilution in 0.1% BSA containing Tris-saline solution) at 4°C overnight. Immunoreactivity was visualized by the sequential addition of: (a) biotinylated anti-rabbit IgG, 1:50, for 30 minutes; (b) avidin-biotin complex, 1 : 100, for 30 minutes both in 0.1% BSA containing Tris-saline solution, pH 7.4; and (c) 3,3′-diaminobenzidine, 0.44 mg/100 ml Tris-saline solution with 10 μl of 30% hydrogen peroxide. To permeabilize the tissue, 0.1-0.3% Triton-X was included in these steps to improve penetration of the antibody. Cultures were rinsed in Tris-saline solution between each incubation, dehydrated, and mounted.

To detect I-receptor protein in situ, vessels were removed from rats immediately after sacrifice and immersed in 3.75% acrolein in 2% paraformaldehyde for 30 minutes and cryoprotected in 30% sucrose overnight. The fixed tissues were prepared as 20-μm frozen sections on a freezing microtome (cryostat). Tissue sections were postfixed on the slide for 1 minute with 4% paraformaldehyde, incubated in 1% sodium borohydride in 0.1 M phosphate buffer for 30 minutes, and rinsed in phosphate buffer until the gaseous bubbles disappeared. Sodium borohydride significantly increased the antigenicity of the tissue. To reduce endogenous peroxidase activity, tissue sections were incubated in 4 ml of 30% hydrogen peroxide in 100 ml of methanol followed by incubation in primary antiserum (I-receptor protein, 1 : 2000), and immunoreactivity was visualized as just outlined. Control immunostaining consists of: (a) preimmune serum as a negative control, and (b) anti-smooth muscle α-actin (1 : 1000) antiserum (Sigma) as primary antibody to verify the purity of smooth muscle cells.

Materials

[^{3}H]Idazoxan (40 Ci/mmol) was purchased from Amersham Corporation, and [^{3}H]*p*-aminoclonidine (52 Ci/mmol) and [^{3}H]rauwolscine (80 Ci/mmol) were from New England Nuclear. Cirazoline was a generous gift from Synthelabo (Paris). All other drugs were purchased from Research Biochemical Int. (Natick, Massachusetts).

RESULTS

Aortic Smooth Muscle Cells

In initial experiments, we measured binding of [^{3}H]idazoxan, [^{3}H]PAC, or [^{3}H]rauwolscine to membranes (100 μg protein) prepared from rat ASMCs. Five nM of [^{3}H]ligand was incubated in the presence of 10 μM cold ligand to define nonspecific binding. Over 70% of the binding of each ligand was specific. The specific binding of [^{3}H]rauwolscine and [^{3}H]PAC was completely inhibited by epinephrine (10 μM), indicating that these ligands were binding only to α_2-adrenergic receptors. On the other hand, only about 10% of the specific binding of [^{3}H]idazoxan was inhibited by epinephrine (10 μM). With increasing concentrations of idazoxan, binding of [^{3}H]idazoxan (5 nM) was inhibited with a K_i of about 10 nM, whereas epinephrine

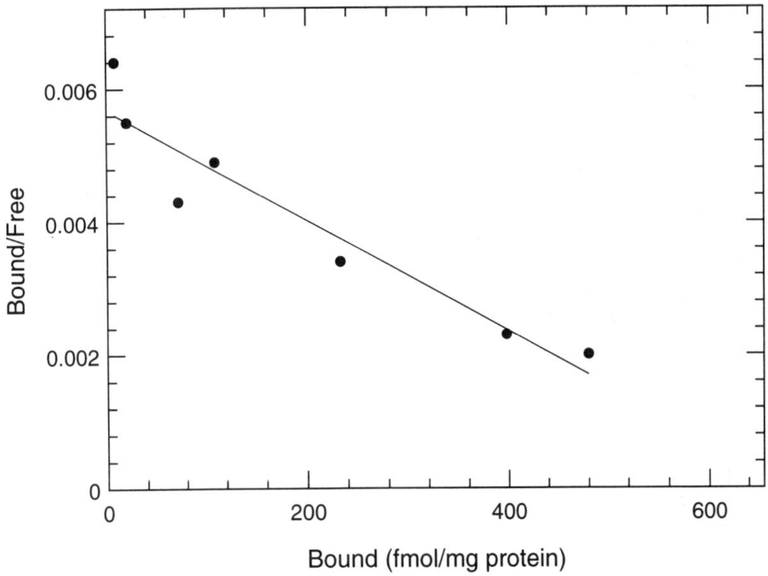

FIGURE 1. Scatchard analysis of the saturation binding of [^{3}H]idazoxan to membranes of aortic smooth muscle cells. Membranes (about 100 μg protein) were incubated with various concentrations of [^{3}H]idazoxan in the presence of 10 μM of epinephrine; nonspecific binding was defined by 10 μM of idazoxan. Values are from one representative experiment repeated with three different cell membrane preparations. The kinetic constants given in the text are from all three experiments analyzed by LIGAND.

was much less potent (K_i > 10 μM) (data not shown). The results suggest that most binding of [^{3}H]idazoxan was to nonadrenergic sites.

Further characterization of the binding site was therefore undertaken using [^{3}H]-idazoxan as ligand. The binding of [^{3}H]idazoxan to membranes of ASMC, performed in the presence of 10 μM epinephrine to mask α_2-adrenergic receptors, was specific and saturable. Scatchard analysis was consistent with a single-site fit (FIG. 1) with a K_D of 8.9 ± 1.2 nM. This high-affinity binding is similar to the affinity of [^{3}H]idazoxan to nonadrenergic sites in other tissues.[11,23,24] The number of nonadrenergic binding sites (B_{max}) was 610 ± 45 fmol/mg protein, about 10-fold higher than that of α_2-adrenergic sites as measured by [^{3}H]rauwolscine (data not shown).

The pharmacological specificity of the binding site was analyzed by examination of the relative potencies of ligands to compete with the binding of [^{3}H]idazoxan (5 nM) to ASMC membranes. The inhibitory effects of cirazoline, naphazoline, clonidine, guanabenz, amiloride, and rauwolscine are depicted in FIGURE 2.

Cirazoline and naphazoline bind with high affinity, whereas clonidine was less potent with K_is > 1 μM. The guanidiniums guanabenz and amiloride also had low affinities for this site in ASMC membranes (data not shown). The Hill coefficients for cirazoline, naphazoline, and clonidine were significantly less than unity, indicating the presence of either multiple binding sites or interactive forms of binding sites as reported in other tissues.[25] Agmatine, an endogenous ligand for I-receptors,[18] also

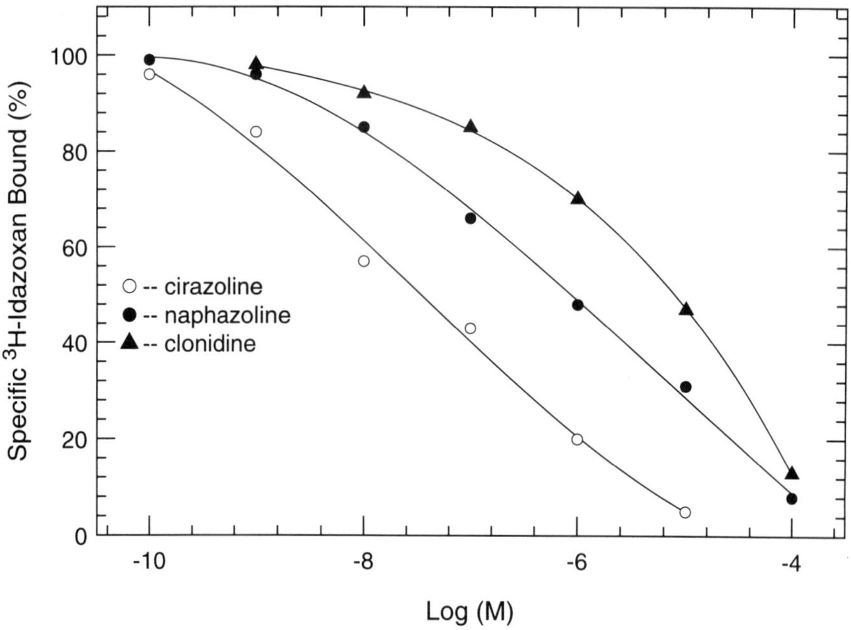

FIGURE 2. Inhibition of [^{3}H]idazoxan binding to aortic smooth muscle cell membranes by various adrenergic agents. Membranes were incubated with 5 nM [^{3}H]idazoxan in the presence of various concentrations of cirazoline, naphazoline, or clonidine. Epinephrine (10 μM) was present in all incubations, and nonspecific binding was defined by 10 μM of idazoxan. Values are from one experiment repeated three times. Inhibition constants (K_i) were calculated from all experiments using the LIGAND program.

completely inhibited binding of [^{3}H]idazoxan to ASMC membranes with a K_i of 240 ± 25 nM and Hill coefficient of 0.64 (data not shown).

These results indicate that the I-receptors of ASMC are of the I_2 subclass (idazoxan-preferring) and differ from the variant of this subclass which exhibits high affinity for synthetic guanidinium compounds and is expressed in rabbit tissue and bovine adrenal chromaffin cells.[11,24]

Endothelial Cells

In contrast to ASMC, [^{3}H]idazoxan binds to α_2-adrenergic and I-receptors in membranes of endothelial cells. As shown in FIGURE 3, epinephrine inhibited specific binding by almost 50% and idazoxan inhibited it completely. Competition of other agents for binding of [^{3}H]idazoxan to nonadrenergic binding demonstrated that it had a pharmacological profile of the I_2 subclass (data not shown).

Immunostaining with Antiserum to I-Receptor Protein

Cultured ASMC and sections of rat carotid arteries were immunostained with a specific polyclonal antibody raised against an I-receptor-associated protein isolated

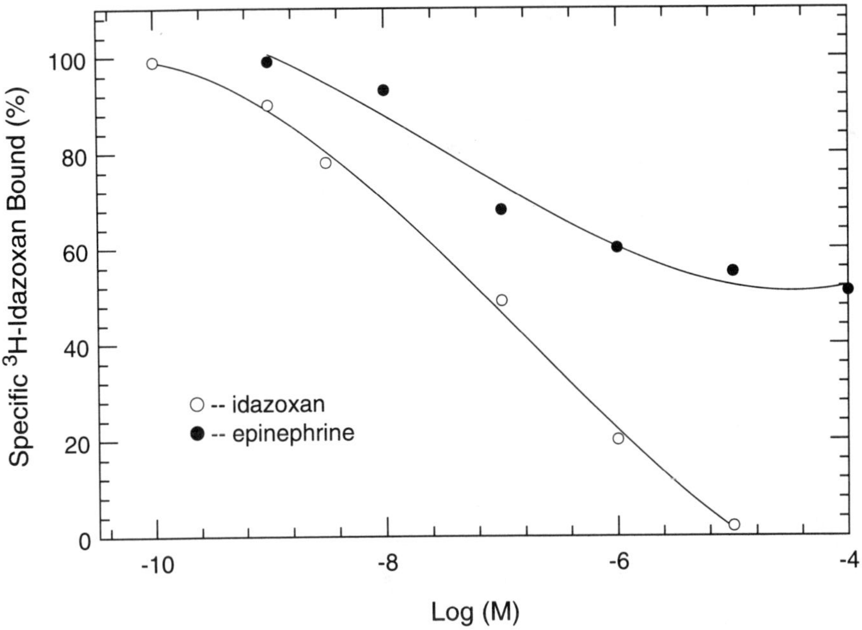

FIGURE 3. Inhibition of [^{3}H]idazoxan binding to endothelial cell membranes by idazoxan and epinephrine. Membranes were incubated with 5 nM [^{3}H]idazoxan in the presence of various concentrations of idazoxan or epinephrine, and nonspecific binding was defined by 10 μM of idazoxan. Values are from one experiment repeated in two different membrane preparations. Inhibition constants (K_i) were calculated from all experiments using the LIGAND program.

from bovine adrenal chromaffin cells[13] (see Methods). Cultured ASMC (FIG. 4A) discrete patches of cells stained with antiserum but not with control preimmune serum (FIG. 4B). Strong immunostaining with antiserum to smooth muscle actin indicated that most of the cells were smooth muscle cells (FIG. 4C). Whereas actin was largely distributed throughout the cytoplasm, I-receptors were concentrated at or around the nucleus. Double-labeling immunofluorescence verified that actin-positive cells also contained I-receptor immunoreactivity (data not shown). In stained sections of excised carotid artery, smooth muscle cells lining arterioles were also positively stained by antireceptor antiserum (FIG. 4A, inset). Thus, rat and bovine receptors appear to share common structural elements; moreover, the receptor-associated protein is expressed in cultured ASMC as well as in smooth muscle cells of arteries in situ.

DISCUSSION

In the present study, we discovered that ASMC of rat express a saturable, specific, and high affinity binding site for [^{3}H]idazoxan. The fact that these cells also bind [^{3}H]rauwolscine indicates that they express α_2-adrenergic receptors. However, after adrenergic receptors were blocked with epinephrine, most [^{3}H]idazoxan binding

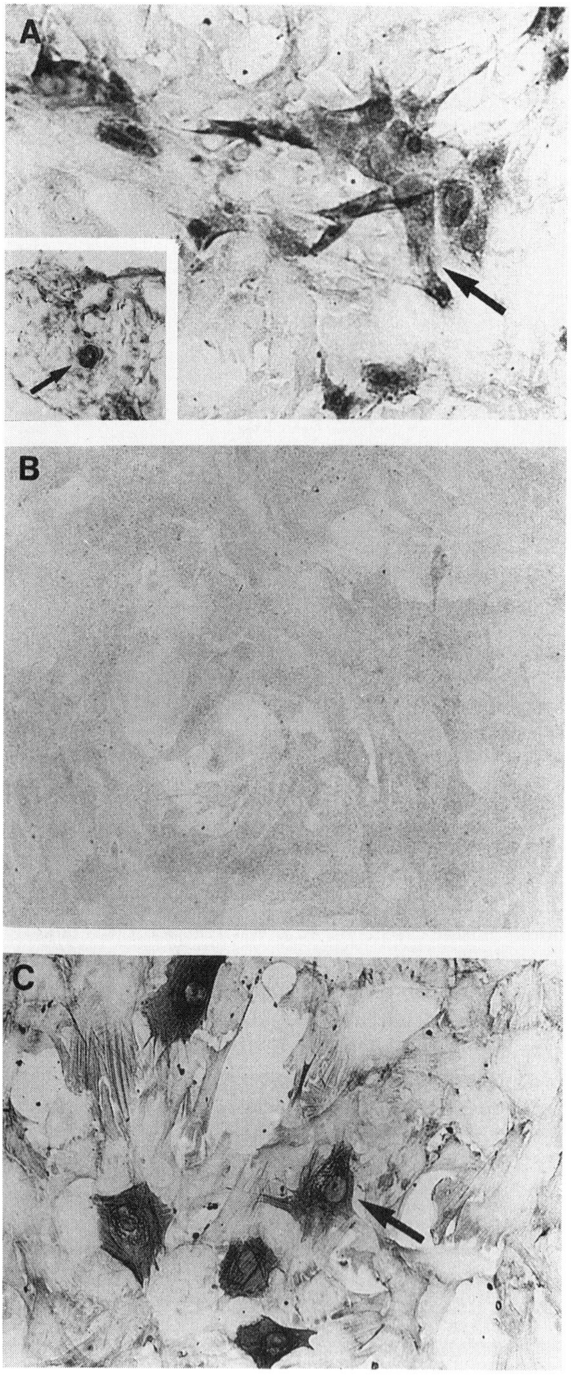

FIGURE 4. Localization of I-receptor protein immunoreactivity in cultured aortic smooth muscle cells and rat carotid artery. Cells or cryosectioned tissue was processed by the immunoperoxidase method with I-receptor antiserum (**A**), control preimmune serum (**B**), or anti-smooth muscle actin antiserum (**C**). An example of immunostaining of an artery with I-receptor antiserum (*Inset in A*). Strong positive immunoreactivity was observed in patches of smooth muscle cells with anti-receptor antiserum, whereas no immunoreactivity was observed with preimmune serum.

remained. Computation indicated that nonadrenergic binding sites were 10 times more abundant than α_2-receptors.

Further characterization of the binding site indicated that it was of the idazoxan-preferring I_2 class of I-receptors. Thus, competitive binding assays, performed in the presence of epinephrine, demonstrated that the ligand was potently inhibited by idazoxan, naphazoline, and cirazoline and only weakly inhibited by clonidine and phentolamine, a profile characteristic of I_2 receptors in various tissues in most species.[8,11,23] The failure of amiloride or guanabenz to interact with [^{3}H]idazoxan binding indicates that the I_2-receptor of ASMC is the kind expressed in brain, kidney, liver, and platelets of various species[8,24–26] and not the one expressed in rabbit tissues.[19,24]

Further evidence that the receptor is not adrenergic is that K^+ but not Na^+ interfered with the binding of [^{3}H]idazoxan to ASMC membranes (data not shown) as others have observed.[11,27] Although in most tissue so far studied I_2 receptors are localized to mitochondrial membranes,[11,28] in this study we did not separate mitochondrial from sarcolemmal membranes, and hence while it is possible the I_2 receptors of ASMC are also mitochondrial, this point has not been proved.

The Hill coefficients (0.5-0.7) for cirazoline, naphazoline, clonidine, and agmatine suggest that [^{3}H]idazoxan may be binding to more than one site in ASMC membranes. The other site is not an α_2-adrenergic receptor because inhibition studies were performed in the presence of epinephrine (10 μM). Whether [^{3}H]idazoxan binds to two populations of I-receptors, as proposed in brain,[26] or alternatively that I-receptors exist in two interconvertible state[25] is not clear.

That ASMC express I-receptors in situ has been demonstrated by immunostaining carotid arterioles of rat with a polyclonal antiserum to an I-receptor binding protein of bovine adrenal chromaffin cells.[12,13] The binding protein, isolated from solubilized receptor using affinity columns to which idazoxan or *p*-aminoclonidine were bound, binds ligands for I-receptors with comparable kinetics to native receptors, failed to bind ligands selective for α_2-adrenergic receptors, immunoprecipitates I-receptors, partially inhibits ligand binding, and cross-reacts with rat tissue as also demonstrated here. The presence of I-receptor immunoreactivity in situ indicates that receptor expression in culture does not result from dedifferentiation of ASMCs but that the I-receptors are a natural constituent of smooth muscle cells in life.

We also discovered that I_2 receptors are expressed in endothelial cells of bovine pulmonary artery. Whether endothelium of all vessels expresses the receptor is not known.

We also observed that agmatine binds to I-receptors of ASMC. The observation is of interest because agmatine is the first native ligand of I-receptors that has been identified. Isolated from brain,[18] it represents the first clonidine-displacing substance (CDS) whose structure has been established. The affinity of agmatine for I-receptors in ASMC is relatively high (K_i 250 ± 25 nM) and comparable to that of other endogenous ligands for vascular receptors; for example, the affinity of catecholamines for α_2-adrenergic receptors is 0.2-1 μM. Although there are a few reports on the effects of imidazoline agents and CDS on the contraction of aortic rings, these studies were not supported by receptor binding studies. This report, therefore, is the first demonstration that vascular smooth muscles express nonadrenergic I-receptors. Indeed, the only other report of the presence of nonadrenergic sites for [^{3}H]idazoxan in smooth muscle cells was performed in rabbit urethra.[19] The functions of I-receptors in vascular smooth muscle cells are not known. Although clonidine, an imidazoline and α_2-adrenergic agonist, contracts vascular smooth muscles,[1] this action may not be entirely related to α_2-adrenergic agonism because idazoxan, an antagonist at α_2-adrenergic receptors, will also potently contract aortic rings (Estrada *et al.*, unpublished data). Moreover, the fact that serum of rat[17] and man[16] contains appreciable amounts of classic CDS and immunoreactive (ir)-CDS, ligands that bind to I-receptors[29] and that constrict aortic rinds[17] by mechanisms independent of α_2-adrenergic receptors,[3] suggests that vascular I-receptors may contribute to vasoconstriction. On the other hand, agmatine also present in serum[30] is relatively inert in blood vessels (Estrada *et al.*, unpublished data) even though it is stored in vascular smooth muscle cells and in endothelium where it appears to be synthesized (unpublished data). Thus, while agmatine appears to have elements in common with other endothelially derived vascular mediators, its precise function in vascular biology is obscure.

SUMMARY

We sought to determine if smooth muscle and endothelial cells of blood vessels express imidazoline receptors. Membranes of cultured smooth muscle cells specifically bind with high affinity to α_2-adrenergic ligands, $[^3H]$*p*-aminoclonidine, $[^3H]$rauwolscine, and $[^3H]$idazoxan. All of $[^3H]$rauwolscine and $[^3H]$*p*-aminoclonidine but less than 10% of $[^3H]$idazoxan binding was displaced by 10 μM epinephrine, indicating a nonadrenergic binding site for $[^3H]$idazoxan. $[^3H]$Idazoxan binding was inhibited with a rank order of potency: cirazoline > idazoxan > naphazoline >> guanabenz > amiloride > clonidine = phentolamine. Agmatine, an endogenous ligand for I-receptors, inhibited binding with a K_i of 240 ± 25 nM. The binding of $[^3H]$idazoxan to membranes of pulmonary artery endothelial cells was to both α_2-adrenergic and imidazoline receptors. Cultured smooth muscle cells, as well as rat carotid arterioles, were specifically immunostained by antibodies to an I-receptor-associated protein.

We conclude that vascular smooth muscle and endothelial cells express not only α_2-adrenergic receptors but also I-receptors of the I_2 subclass with high affinity for agmatine. Since serum contains an endogenous ligand for I-receptors, possibly agmatine, the results suggest the presence of a novel receptor mechanism on vascular smooth muscle which may regulate vascular tone.

ACKNOWLEDGMENT

The authors thank Sonja Bramwell for technical support.

REFERENCES

1. Decker, N., J. D. Erhardt, G. Leclerc & J. Schwartz. 1984. Post-junctional α-adrenoceptors: α_1- and α_2-subtypes in rat vasculature *in vitro* and *in vivo*. Naunyn-Schmiedeberg's Arch Pharmacol. **326:** 1-6.
2. Flavahan, N. A., H. Shimokawa & P. M. Vanhoutte. 1989. Pertussis toxin inhibits endothelium-dependent relaxations to certain agonists in porcine coronary arteries. J. Physiol. (Lond.) **408:** 549-560.

3. SHAMS, G., B. V. VENKATARAMAN, A. HAMADA, D. D. MILLER, P. N. PATIL & D. R. FELLER. 1991. Diversity of the pharmacological actions of some tolazoline analogues in human platelets and rat aorta. Eur. J. Pharmacol. **199:** 315-323.
4. KARLSSON, B. R., M. FORSMAN, O. K. ROALD, M. S. HEIER & P. A. STEEN. 1990. Effect of dexmedetomidine, a selective and potent α_2-agonist, on cerebral blood flow and oxygen consumption during halothane anesthesia in dogs. Anesth. Analg. **71:** 125-129.
5. BOUSQUET, P., J. FELDMAN & J. SCHWARTZ. 1984. Central cardiovascular effects of alpha adrenergic drugs: Differences between catecholamines and imidazolines. J. Pharmacol. Exp. Ther. **230:** 232-236.
6. ERNSBERGER, P. R., M. P. MEELEY, J. J. MANN & D. J. REIS. 1987. Clonidine binds to imidazole binding sites as well as α_2-adrenoceptors in the ventrolateral medulla. Eur. J. Pharmacol. **134:** 1-14.
7. BRICCA, G., M. DONTENWILL, A. MOLINES, J. FELDMAN, A. BELCOURT & P. BOUSQUET. 1989. The imidazoline preferring receptor: Binding studies in bovine, rat and human brainstem. Eur. J. Pharmacol. **162:** 1-9.
8. PARINI, A., I. COUPRY, R. M. GRAHAM, I. UZIELLI, D. ATLAS & S. M. LANIER. 1989. Characterization of an imidazoline guanidinium receptive site distinct from the α_2-adrenergic receptor. J. Biol. Chem. **264:** 11874-11878.
9. MICHEL, M. C. & P. A. INSEL. 1989. Are there multiple imidazoline binding sites? Trends Pharmacol. Sci. **10:** 342-344.
10. MICHEL, M. C. & P. R. ERNSBERGER. 1992. Keeping an eye on the I site: Imidazoline-preferring receptors. Trends Pharmacol. Sci. **13:** 369-370.
11. REGUNATHAN, S., M. P. MEELEY & D. J. REIS. 1993. Expression of nonadrenergic imidazoline sites in chromaffin cells and mitochondrial membranes of bovine adrenal medulla. Biochem. Pharmacol. **45:** 1667-1675.
12. WANG, H., S. REGUNATHAN, M. P. MEELEY & D. J. REIS. 1992. Isolation and characterization of imidazoline receptor proteins from bovine adrenal chromaffin cells. Mol. Pharmacol. **42:** 792-801.
13. WANG, H., S. REGUNATHAN, D. A. RUGGIERO & D. J. REIS. 1993. Production and characterization of antibodies specific for the imidazoline receptor protein. Mol. Pharmacol. **43:** 509-515.
14. DIAMANT, S., A. ELDOR & D. ATLAS. 1987. A low molecular weight brain substance interacts, similarly to clonidine, with α_2-adrenoceptors of human platelets. Eur. J. Pharmacol. **144:** 247-255.
15. ERNSBERGER, P., M. P. MEELEY & D. J. REIS. 1988. An endogenous substance with clonidine-like properties: Selective binding to imidazole sites in the ventrolateral medulla. Brain Res. **441:** 309-318.
16. SYNETOS, D., V. G. MANOLOPOULOS, D. ATLAS & E. PIPILI-SYNETOS. 1991. Human plasma-derived material with clonidine displacing substance (CDS)-like properties contracts the isolated rat aorta. J. Auton. Pharmacol. **11:** 343-351.
17. MEELEY, M. P., M. L. HENSLEY, P. ERNSBERGER, D. FELSEN & D. J. REIS. 1992. Evidence for a bioactive clonidine-displacing substance in peripheral tissues and serum. Biochem. Pharmacol. **44:** 733-740.
18. LI, G., S. REGUNATHAN, C. J. BARROW, J. ESHRAGHI, R. COOPER & D. J. REIS. 1994. Agmatine is an endogenous clonidine-displacing substance in brain. Science **263:** 966-969.
19. YABLONSKY, F. & J. P. DAUSSE. 1991. Non-adrenergic binding sites for the α_2-antagonist [^{3}H]idazoxan in the rabbit urethral smooth muscle. Biochem. Pharmacol. **41:** 701-707.
20. ERLINGE, D., H. YOO, L. EDVINSSON, D. J. REIS & C. WAHLESTEDT. 1993. Mitogenic effects of ATP on vascular smooth muscle cells vs. other growth factors and sympathetic cotransmitters. Am. J. Physiol. **265** (Heart Circ. Physiol. 34): H1089-H1097.
21. MEYER-LEHNERT, H. & R. W. SCHRIER. 1989. Potential mechanism of cyclosporin A-induced vascular smooth muscle contraction. Hypertension **13:** 350-361.

22. Hsu, S. M., L. Raine & H. Fanger. 1981. Use of avidin-biotin-peroxidase complex (ABC) in immunoperoxidase techniques: A comparison between ABC and unlabeled antibody (PAP) procedures. J. Histochem. Cytochem. **29:** 557-580.
23. Zonnenschein, R., S. Diamant & D. Atlas. 1990. Imidazoline receptors in rat liver cells: A novel receptor or subtype of α_2-adrenoceptors? Eur. J. Pharmacol. **190:** 203-215.
24. Coupry, I., D. Atlas, P. A. Podevin, I. Uzielli & A. Parini. 1989. Imidazoline-guanidinium receptive sites in renal proximal tubule: Asymmetric distribution, regulation by cations and interaction with an endogenous clonidine displacing substance. J. Pharmacol. Exp. Ther. **252:** 293-299.
25. Wikberg, J. E. S., S. Uhlen & V. Chhajlani. 1992. Evidence that drug binding to non-adrenergic [^{3}H]-idazoxan binding sites (I-receptors) occurs to interacting or interconvertible affinity forms of the receptor. Pharmacol. Toxicol. **70:** 208-219.
26. Bricca, G., H. Greney, J. Dontenwill-Kieffer, J. Zhang, A. Belcourt & P. Bousquet. 1993. Heterogeneity of the specific imidazoline binding of [^{3}H]-idazoxan in the human cerebral cortex. Neurochem. Int. **22:** 153-163.
27. Atlas, D. 1991. Clonidine-displacing substance (CDS) and its putative imidazoline receptor. New leads for further divergence of α_2-adrenergic receptor activity. Biochem. Pharmacol. **41:** 1541-1549.
28. Tesson, F., C. Prip-Buus, A. Lemoine, J.-P. Pegorier & A. Parini. 1991. Subcellular distribution of imidazoline-guanidinium-receptive sites in human and rabbit liver: Major localization to the mitochondrial outer membrane. J. Biol. Chem. **260:** 155-160.
29. Reis, D. J., G. Li & S. Regunathan. 1995. Endogenous ligands of imidazoline receptors: Classic and immunoreactive clonidine displacing substance (CDS) and agmatine. Proceedings of the II International Symposium on Imidazoline Receptors. Ann. N.Y. Acad. Sci. This volume.
30. Raasch, W., S. Regunathan, G. Li & D. J. Reis. 1995. Agmatine, the bacterial amine, is widely distributed in mammalian tissues. Life Sci. In press.

The Relationship between the Adrenoceptor and Nonadrenoceptor-Mediated Effects of Imidazoline- and Imidazole-Containing Compounds[a]

J. C. McGRATH,[b] C. M. BROWN,[c] C. J. DALY,[b]
D. KENDALL,[d] A. MacKINNON,[c] D. J. MILLER,[b]
M. NAGADEH,[b] A. O'DOWD,[b] J. J. O'DOWD,
D. PINTHONG,[d] G. SINGH,[d] A. G. B. TEMPLETON,[b]
AND V. G. WILSON[d]

[b]*Institute of Physiology*
University of Glasgow
Glasgow G12 8QQ, Scotland

[c]*Syntex Research Centre*
Edinburgh, EH14 4AP, Scotland

[d]*Department of Physiology and Pharmacology*
University of Nottingham
Nottingham NG7 2UH, UK

Over the last 15 years our laboratories have been studying the pharmacological effects of imidazoline- and imidazole-containing compounds with a view to understanding the biological processes with which they interact. This has extended to processes in which (1) they are experimental surrogates for endogenous compounds, particularly as activators or ligands for α-adrenoceptors, and (2) these compounds or close relatives might themselves be involved endogenously, particularly the endogenous imidazole-containing dipeptides. During this period came the discovery of tissue extracts containing clonidine-displacing substance (CDS) and of binding sites for imidazoline and related compounds which are separate from α-adrenoceptors. This provided further facets to our analyses: First, could there be new receptors to consider in explaining the pharmacological actions of imidazolines which never quite parallel those of catecholamines at α-adrenoceptors? Second, in our analysis of tissue extracts, could the imidazole compounds in which we were interested exert actions through some novel cell surface receptors and also could our tissue extracts contain substances

[a]We are grateful to the Medical Research Council, British Heart Foundation, Wellcome Trust, and Smith Kline Foundation for financial support. The Glasgow Clinical Research Initiative Laboratory is supported by the Medical Research Council and is a member of the EU Biomed Projects, EureCa, and EURAD.

akin to CDS or imidazoline-receptor ligands which might account for the biological actions that we were assaying?

IMIDAZOLINES AND SMOOTH MUSCLE PHARMACOLOGY

In the pharmacology of smooth muscle and of adrenoceptors, subtle differences among otherwise similar actions have often been found between the effects of compounds that are related to the natural agonists at these receptors, the catecholamines, that is, phenylethylamine derivatives, and imidazolines or structurally related compounds. Given the disparity in structures between these two major classes of compounds and the desire to explain drug-receptor interactions in chemical terms, analysis and speculation have followed two general lines.

1. The structure-activity relationships for particular effects (These were generally classical pharmacological studies in which the cellular effect per se was not a relevant factor; it is implicit that the cellular responses involved within the assay of drug effects are identical throughout and are mediated by a single type of receptor, usually considered an "adrenoceptor" because the endogenous catecholamines are active) were pursued broadly along the lines of the Easson-Steadman hypothesis for three point attachment to the receptor, with the general conclusion that different logics seem to apply to imidazolines and phenylethylamines, in particular with regard to stereo selectivity.[12] A corollary of this is that for these two sets of compounds, the attachment to the receptor is different and a general assumption is that within each set it is probably similar.

2. Comparisons were made of the characteristics of the responses obtained in functional preparations when employing the two groups as agonists. Differences were commonly found between the two groups either in the "nature" of the response in a given preparation or in the correlation of some function of response amongst preparations. Two alternative substrands of analysis follow from this:

A. That the two groups of compounds can initiate different responses even though a single receptor is involved,[3] or

B. That different receptors mediating two responses are involved, which thus have different agonist profiles.[4]

It is a small step from 2B to a third class of experiment, the demonstration, during assay of adrenoceptor ligand binding characteristics, of binding sites with a preference for imidazoline-containing molecules and with some characteristics that separate them from adrenoceptors. The issue then arises of differences in response to imidazolines and phenylethylamines being explained by their own classes of "physiological" receptors, and with this comes the implication of, and hence the search for, the "physiological" agonist for the imidazoline receptor(s).[5–10]

The implications of 2A are more difficult to unravel, because the possibility of a heterogeneous receptor population is difficult to exclude absolutely when the responses of heterogeneous tissue samples are involved.

Each of these classes of experimental endeavor has been fruitful in unearthing new aspects of the biology of receptor-effector coupling and in predicting new receptors without, it may be fair to say, producing a fundamental explanation for the difference in receptor attachment/biology between imidazolines and phenylethylamines. Indeed, it could be said that it is the promiscuity of agonists such as oxymetazoline and adrenaline that has led to the subclassifications so far.

CONTRIBUTIONS OF IMIDAZOLINES TO ADRENOCEPTOR SUBCLASSIFICATION

The subdivision of α-adrenoceptors into α_1 and α_2 (originally pre- and postjunctional at sympathetic nerve terminals) depended on the discovery of both agonists and antagonists with selectivity between the subtypes. On the agonist side this was almost entirely due to the development of compounds similar to clonidine. The subsequent discovery of postjunctional α_2-adrenoceptors was entirely due to pharmacological investigation of the agonist effects of these prejunctionally active compounds, including the imidazolines clonidine and oxymetazoline and their analogs xylazine and guanabenz, which unexpectedly produced postjunctional responses.[11,12] Thus, the ability to demonstrate the two major subtypes of α-adrenoceptor by pharmacological means, rather than the previous, more tenuous, anatomical location, was dependent on the use of imidazoline-containing compounds or analogs as agonists. Essentially this was a process of clarifying inconsistencies in the interactions of agonist and antagonists acting at "adrenoceptors."

However, it must be stressed that all of these imidazoline agonists possess agonism, to some degree, at α_1-adrenoceptors, so discrimination of their effects at particular subtypes requires the use of selective antagonists such as prazosin (α_1) or rauwolscine (α_2). This is illustrated below using the example of UK-14304. Even in the first illustration of postjunctional α_2-adrenoceptors,[11,12] although the "inconsistency" of vascular responses to (previously) presumed α_2-adrenoceptor agonists could be explained by a combination of α_1- and α_2-adrenoceptor agonism, this conclusion could not be reached in other smooth muscle preparations, such as anococcygeus and vas deferens, in which the agonist profiles did not fit current concepts. Subsequent study of responses in these other preparations led to the first postulated subdivision of α_1-adrenoceptors into α_{1a} and α_{1b} on the basis of a correlation between the nature of the response, particularly involvement of Ca^{2+} ions, and the ability to produce agonism with both imidazolines and phenylethylamines (α_{1a}) or only phenylethylamines (α_{1b}).[13,14,15] This second order subclassification of α_1-adrenoceptors was subsequently put on a firmer footing by ligand binding studies in which one particular imidazoline, oxymetazoline, figured largely.[16] Interestingly, oxymetazoline also figures largely in the second order classification of α_2-adrenoceptors into α_{2A} and α_{2B}, which relies entirely on binding studies.[17] Indeed, oxymetazoline and another agonist featuring widely in adrenoceptor classification, UK-14304, both show relative selectivity for the α_{2A}-adrenoceptor subtype (TABLE 1 and FIG. 1).

The imidazoline UK-14304 has been used for many years as a selective agonist for α_2-adrenoceptor subtypes.[18] Although the demonstration of postjunctional α_2-adrenoceptors in isolated vascular smooth muscle preparations is difficult, particularly in arteries, clear evidence exists for the presence of postjunctional α_2-adrenoceptors as well as α_1-adrenoceptors on the smooth muscle of many vascular preparations.[19] In many cases, responses are small, but intervention by raising the tone or inhibition of nitric oxide production can enhance or "uncover" responses to α_2-agonists in

TABLE 1. Affinity Values (pIC_{50}) for Commonly Used Imidazoline Ligands against the α_2-Adrenoceptor Ligand [^{3}H]RS-15385-197

	UK-14304	Oxymetazoline
α_{2A} (human platelet)	7.9	8.5
α_{2B} (neonatal lung)	7.0	6.9
α_1 (rat cortex)	6.4	7.7

certain blood vessels.[20,21] This is not an effect exclusive to α_2-adrenoceptors but influences various receptors, particularly when the agonist has low efficacy, giving scope for potentiation. This phenomenon can be shown with many imidazoline agonists and with a wide variety of modulating factors including Ca^{2+} channel modulators, angiotensin, and blood gases.[22–24] Another limitation on the use of imidazoline agonists, even when they are as selective as UK-14304, is that selectivity is not an absolute; there is always some α_1-adrenoceptor agonism. Thus, in tissues in which the α_1-adrenoceptor system is efficiently coupled to contraction, at least part of its action can be mediated through α_1-adrenoceptors. These factors can be well illustrated in tissues with relatively homogeneous receptor populations. The rat isolated common carotid artery provides a good example.

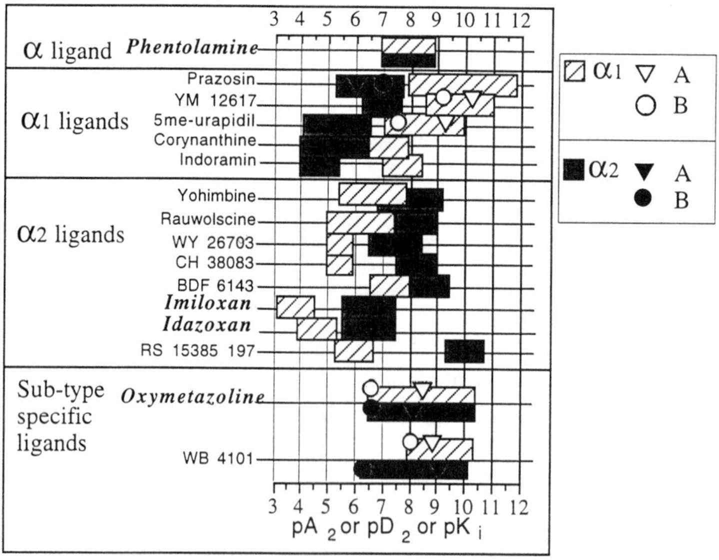

FIGURE 1. Subclassification of α-adrenoceptors by functional ligands. Key imidazolines are shown in bold.

In rings of rat carotid artery, UK-14304 produces a concentration-dependent contraction with a maximum 23% of that to noradrenaline. Removal of the endothelium increased UK-14304's maximum response to 47%, and treatment with the NO synthase inhibitor L-NAME (100 μM) potentiated the maximum response to 73% of noradrenaline's maximum. This illustrates that, as in other preparations, the contractile effect of this imidazoline is under the tonic influence of constitutive NOS, the constant background release of NO inhibiting the response to the agonist.

Pharmacological investigation of the contraction to UK-14304 in this preparation showed clearly that contractile responses were produced via α_1- but not α_2-adrenoceptors, being blocked by low concentrations of the α_1-antagonist prazosin but only by high, nonselective concentrations of the α_2-antagonist rauwolscine. The logic of UK-14304 being a ligand for α_1-adrenoceptors was further tested by assessing UK-14304 as an antagonist versus the α_1-adrenoceptor agonist phenylephrine. UK-14304 produced concentration-related antagonism with an estimated dissociation constant of 10 μM; this is consistent with its binding affinity at α_1 sites in rat cortex of 0.5 μM (TABLE 1), binding estimates routinely being higher. Thus, it was possible with this preparation both to show the potential for uncovering functional responses to imidazolines and to illustrate that UK-14304's affinity for α_1-adrenoceptors results in it being a partial agonist, displaying both agonist and antagonist actions.

These observations on UK-14304 in the rat isolated common carotid artery illustrate two key issues in dealing with imidazoline pharmacology. First, even this compound, which is much more potent at α_2-adrenoceptors than at α_1, has sufficient affinity and efficacy at α_1 that its pharmacology at this receptor can be clearly described in tissue where α_2 is not present to complicate the picture. Given the high degree of synergism between α-adrenoceptors and related receptor types,[19–24] great caution is required in interpreting pharmacological phenomena with such compounds as due to a novel receptor rather than to a complex interaction at several receptors. Perhaps the more unexpected outcome is that its affinity for α_1 could cause it to attenuate, rather than potentiate, a response to α_1-adrenoceptor agonists in some circumstances. Second, the influence of NO illustrates how the rather fragile "modulatory" action of this imidazoline can be moderated by other factors and might therefore be missed in some circumstances.

Contributions to the two orders of subclassification of α-adrenoceptors have thus been made by imidazolines, sometimes as agonists and sometimes as affinity ligands. In practice, the tendency towards agonism has limited the use of imidazoline-like compounds as functional antagonists, with the possible exception of some benzodioxan-imidazoline combinations such as idazoxan or imiloxan. It is thus of further interest, though pharmacologically complicated, that idazoxan should enable the identification of a further class of binding site, separate from adrenoceptors, that is, the "imidazoline" or "imidazoline-preferring" receptor(s) (to be discussed).

CROSS-PHARMACOLOGY: ACTIONS OF ADRENOTROPIC DRUGS AT SMOOTH MUSCLE NONADRENOCEPTOR SITES

Imidazoline derivatives were reported to elicit nonadrenoceptor responses in several peripheral preparations, such as pancreatic β-cells,[25] adrenal chromaffin cells,[26,27] and rat isolated portal vein.[28] However, it is debatable if these represent functional correlates for nonadrenoceptor, imidazoline binding sites. Satake *et al.*[29] reported that the imidazoline derivative phentolamine can induce transient contractions of the isolated detrusor muscle of the guinea pig bladder and presented evidence to discount a role for α_2-adrenoceptors, histamine H_1 and H_2 receptors, muscarinic receptors, and dopamine receptors. To investigate this we compared the action of phentolamine in the guinea pig isolated bladder with that of cirazoline and idazoxan, two agents known to possess high affinity for nonadrenoceptor, imidazoline binding sites.

We confirmed that phentolamine can elicit transient contractions of longitudinal strips of the guinea pig isolated bladder and demonstrated that cirazoline produces similar effects. However, these responses were qualitatively different from those produced by KCl (ref. 29 and the present study), making it unlikely that the effect of the imidazoline derivatives is due to membrane depolarization followed by opening of voltage-dependent Ca^{2+} channels.

It has been suggested that nonadrenoceptor, imidazoline binding sites may allow an interaction with K^+ channels.[30] Significantly, high concentrations of many imidazoline derivatives are reported to inhibit ATP-sensitive K^+ channels in both vascular[28] and nonvascular tissues[25,31] and, in the case of the rat portal vein, to produce qualitatively similar effects to those reported in the guinea pig isolated bladder.[28,32,33] However, several observations in the present study argue against involvement of nonadrenoceptor, imidazoline binding sites in the effect of phentolamine on the guinea pig isolated bladder. First, the concentration of phentolamine required to produce 50% of the maximum contractile responses, approximately 10 μM, is 30-fold greater than the K_i values at nonadrenoceptor, imidazoline binding sites.[34] Second, cirazoline, a ligand with a high affinity for nonadrenoceptor, imidazoline binding sites ($K_i \times 2$ nM[34,35]), was significantly less potent than phentolamine at producing contractions in the guinea pig isolated bladder. Third, idazoxan, the agent most commonly used to define nonadrenoceptor, imidazoline binding sites, was practically devoid of activity in this system and failed to modify the action of phentolamine. Because idazoxan is also a potent α_2-adrenoceptor antagonist, this latter observation provides further evidence against a role for α_2-adrenoceptors in the spasmogenic action of phentolamine (see also ref. 29).

Thus, although many of our findings are consistent with an effect of phentolamine and cirazoline on K^+ channels, in this case these do not appear to be linked to nonadrenoceptor, imidazoline binding sites.

IMIDAZOLE-CONTAINING DIPEPTIDES: RELATIONSHIP TO CLONIDINE DISPLACING SUBSTANCE AND IMIDAZOLINE RECEPTORS

The search for an endogenous ligand for nonadrenoceptor, imidazoline binding sites has resulted in the isolation and, in some cases, partial purification, primarily from brain, of a substance known as "clonidine-displacing substance" (CDS). The exact chemical nature of CDS is still to be determined, but it was originally characterized by its ability to bind to and displace [^{3}H]clonidine and its analog [^{3}H]*p*-aminoclonidine (PAC) from α_2-adrenoceptor binding sites in brain cerebral cortex membranes.[36,37] Later studies demonstrated that brain-derived CDS was also able to recognize nonadrenoceptor, imidazoline binding sites labeled by either [^{3}H]PAC[38] or [^{3}H]idazoxan.[39] More recent evidence indicates that CDS may be located in several peripheral tissues.

Because the method of preparation of the CDS-containing extract is similar to that used in the extraction of endogenous histidyl compounds from brain and other tissues,[40] it seemed possible that these compounds could be present in CDS-containing extracts and account for the activity detected in radioligand binding assays.

Clonidine-displacing substance extracted from brain was reported to have several biological effects including contraction of smooth muscle in rat gastric fundus,[41] stimulation of catecholamine release from adrenal chromaffin cells,[42] potentiation of ADP-induced aggregation in human platelets,[43] and inhibition of the twitch response in rat vas deferens.[44] The latter two effects were yohimbine sensitive and presumed to be mediated via α_2-adrenoceptors. However, in some cases reports of putative biological actions have been contradictory. In 1986 Bousquet and coworkers[45] reported that their CDS preparation had a hypertensive effect when applied to the cat ventrolateral medulla, whereas Meeley *et al.*[37] reported that their preparation had a hypotensive effect when injected centrally in rats. Furthermore, Felsen *et al.*[41] found that CDS had no effect on vascular smooth muscle, whereas Synetos *et al.*[46] reported that a human plasma-derived CDS-like material had contractile activity on rat isolated aortic smooth muscle.

One explanation for the divergence of these results could be variation in the composition of the CDS extract prepared by the two groups. Although the initial extraction method of both groups is very similar (methanolic extraction of a boiled aqueous homogenate), Atlas and Burstein[36] employed further purification steps to give a "partially purified" CDS. It is reasonable to postulate that the crude extract may contain a variety of endogenous brain compounds. In particular, the method employed has some similarities to that used in our own work for the extraction of endogenous histidyl compounds from rat brain[40] which involved extraction in 80% ethanol/H_2O at 60°C. Carnosine (β-alanyl-L-histidine) and homocarnosine (g-aminobutyryl-L-histidine) have both been described in mammalian brain where they may function as neurotransmitters or neuromodulators in specific regions.[47,48] In addition, some evidence indicates that carnosine (and possibly other related compounds) may have effects on vascular smooth muscle.[49,50]

With these factors in mind, we analyzed such extracts for their chemical composition and their biological effects. We analyzed crude, methanolic extracts from both bovine brain and lung, the latter extract having previously been shown to contain a noncatecholamine substance capable of recognizing α_2-adrenoceptor and nonadrenoceptor, imidazoline binding sites,[51] to determine if any of these substances are present and, therefore, could contribute to the nonadrenoceptor actions of CDS-containing extracts. The α_2-adrenoceptor and nonadrenoceptor, imidazoline binding sites employed are those described by Atlas and Burstein[36] and Vigne *et al.*,[9] respectively.

Extracts from both bovine lung and brain, prepared by a method common to that described by Atlas and Burstein[36] and Ernsberger *et al.*,[52] contain a substance(s) capable of interacting with [^{3}H]clonidine binding sites on bovine cerebral cortex membranes (α_2-adrenoceptors[53]) and [^{3}H]idazoxan binding sites on porcine renal cortex membranes (nonadrenoceptor, imidazoline binding sites[9,54]). The extracts exerted similar potency at each site, consistent with the reported activity of the novel endogenous material CDS.[55,56]

The CDS described here appears to be different from that reported in human serum.[46] Principally, little CDS activity was detected with either the brain extract or lung extract (unpublished observations) following solid phase extraction with a Sep-Pak C18 column and elution by 35% propan-1-ol in distilled water. The significance of the observation is not clear, but may indicate the existence of more than one substance capable of interacting with both α_2-adrenoceptor and nonadrenoceptor, imidazoline binding sites.

The crude, methanolic CDS extract from bovine brain, similar to that used in some functional studies, contains several biologically active substances including histidine and the histidyl derivatives carnosine and homocarnosine. A similarly prepared extract of bovine lung, while containing virtually no catecholamines, contains histamine, imidazole-4-acetic acid, histidine, and the histidyl dipeptide anserine. Both extracts contain several *N*-acetylated histidyl compounds in lesser quantities and several other peaks which remain unidentified. The finding that homocarnosine levels in the brain extract exceed that of carnosine is consistent with that reported earlier for rat brain.[57] A number of histidyl dipeptides such an anserine and homocarnosine and their acetylated derivatives have been found in muscular tissue, notably heart and skeletal muscle,[40,58] and the results obtained in this study extend this to a nonmuscular organ, that is, the lung.

With the exception of histamine, none of the histidyl-containing compounds tested, carnosine, *N*-acetyl carnosine, homocarnosine, *N*-acetyl homocarnosine, histidine, *N*-acetyl histidine, anserine, and imidazole-4-acetic acid, interacted with either the α_2-adrenoceptor or the nonadrenoceptor, imidazoline binding site. This is consistent with an earlier report with fewer histidyl dipeptides and a different nonadrenoceptor, imidazoline binding assay,[5] [^{3}H]*para*-aminoclonidine binding to bovine ventrolateral medulla membranes, now recognized to represent a different subtype of nonadrenoceptor, imidazoline binding sites from that employed in this study.[59] Nonetheless, their presence has obvious implications for the interpretation of results from studies carried out with crude extract, particularly those aimed at assessing the effect of CDS on the central control of blood pressure and direct effects on smooth muscle.

Du Vigneau and Hunt,[49] for example, reported that carnosine produced a stereospecific (L-carnosine only) depressor effect when injected intravenously in anesthetized cats; it was not clear if this was mediated centrally or peripherally. Melville *et al.*[50] repeated this observation, but found that when the same protocol was used with anesthetized or pithed rats carnosine produced a pressor response, suggesting a peripheral site of action. Further experiments with isolated rings of rabbit saphenous vein revealed that carnosine, over a concentration range of 1 nmol/l to 1 μmol/l, induced contractions which showed marked tachyphylaxis. In addition, this effect was associated with significant attenuation of contractions mediated via the selective α_2-adrenoceptor agonist UK-14304. It is unlikely, however, that this effect is due to specific blockade of α_2-adrenoceptors, as we demonstrated that carnosine does not recognize α_2-adrenoceptor binding sites on the bovine cerebral cortex membranes even up to a concentration of 300 μmol/l. Other actions of histidyl dipeptides that could influence smooth muscle function include calcium sensitization, as reported in skeletal muscle,[60] antioxidant properties,[61] and of buffering intracellular pH.[62] The possibility of a central effect of carnosine on blood pressure cannot be excluded, but it has never been investigated. Significantly, Granata and Reis[63] reported that imidazole-4-acetic acid, the oxidation product of histamine and a major contaminant in crude lung extract, produced a hypotensive effect when microinjected into the ventrolateral medulla in the rat. The ventrolateral medulla is the site previously shown to be responsive to synthetic imidazolines[64] and to brain-derived CDS[37,45] but not to catecholamines.

We detected the presence of large quantities of histamine in bovine lung extract. Although this cannot account for the inhibitory effect of the extract at either the α_2-adrenoceptor or nonadrenoceptor, imidazoline binding site, it does pose considerable problems for functional experiments. For example, we observed that lung extract, but not brain extract, can elicit contractions of vascular and intestinal smooth muscle that are sensitive to the histamine H_1-receptor antagonist mepyramine. Clearly, any further functional experiments with extracts is dependent upon prior removal of histamine, imidazole-4-acetic acid, and the histidyl dipeptides or manipulation of conditions to exclude their effects. With this point in mind, we attempted to use solid phase extraction, as described by Synetos *et al.*,[46] to separate CDS activity and the histidyl dipeptides but were unsuccessful; the Sep-Pak C18 failed to retain any of these compounds.[65]

In summary, available evidence suggests that CDS, as assessed by its ability to interact with α_2-adrenoceptor and nonadrenoceptor, imidazoline binding sites, is a novel substance that is not synonymous with any known neurotransmitter or mediator. However, when extracts of limited purity are used, we cannot exclude the possibility that other biologically active compounds may be present and may, at least in part, be responsible for some of the observed properties of extract. Their presence might also explain some discrepancies between the observed biological actions of crude and semipurified extracts. The properties of the active principle in CDS, if indeed it is a single substance, and its relevance to cardiovascular and other systems will only be fully elucidated when it has been isolated and structurally characterized.

NONADRENOCEPTOR IMIDAZOLINE BINDING SITES

Direct evidence for an imidazoline-preferring binding site came from studies of binding of the α_2-adrenoceptor agonist [^{3}H]*p*-aminoclonidine in membranes prepared from the bovine ventrolateral medulla. It was found that only 70% of the sites labeled were competed for by noradrenaline and other catecholamines, whereas compounds containing the imidazoline moiety inhibited the remaining noradrenaline-insensitive binding.[5] The imidazoline binding sites are localized in the medulla region as membranes from the frontal cortex are entirely sensitive to noradrenaline. The rank order of affinity for displacing drugs is clonidine = rilmenidine (6 nM) > phentolamine (42 nM) >> guanabenz (10,000 nM).[6] It appears that central imidazoline sites labeled by [^{3}H]*p*-aminoclonidine may be partly involved in the hypotension observed with some imidazoline and oxazoline compounds.

A second subtype of imidazoline binding sites (I_2), pharmacologically distinct from the [^{3}H]*p*-aminoclonidine site, were identified using the potent α_2-adrenoceptor

TABLE 2. Binding Affinities ($pIC_{50}s$)

Tissue	Idazoxan	Clonidine	Amiloride	Guanabenz
[^{3}H]*p*-Aminoclonidine				
bovine brain	–	8.2	<5	<5
[^{3}H]Clonidine human				
NRL	7.5	7.0	. . .	. . .
[^{3}H]Idazoxan				
Kidney	8.8	6.5	7.4	8.7
Rat brain	8.4	6.3	6.3	8.2

Abbreviations: NRL = nucleus reticularis lateralis.

antagonist radioligand [^{3}H]idazoxan. This subtype has been reported in a number of tissues, most notably rabbit, rat, pig, and human kidney[6,7,9] and in the CNS rat cortex.[10] Again, additional sites labeled by [^{3}H]idazoxan were not competed for by catecholamines but showed high affinity for several compounds such as cirazoline, guanabenz, and amiloride (TABLE 2). In autoradiographic studies [^{3}H]idazoxan binding in the rat brain is restricted to the circumventricular organs and some areas of the hypothalamus. The most convincing functional effect for the [^{3}H]idazoxan site is a marked effect on eating which is not related to any α_2- or α_1-adrenoceptor blockade.[66] Idazoxan itself is hyperphagic, but it is still not clear if it acts as an agonist or antagonist at the I_2 site. Other studies have shown additional effects of idazoxan which cannot be adequately explained by α_2-adrenoceptor blockade, including blockade of nonadrenergic-noncholinergic relaxations of the rat anococcygeus muscle[67] and inhibition of noradrenaline release in rabbit aorta and pulmonary artery.[68]

CONCLUSIONS

We set out to answer two general questions. Could novel receptors explain the pharmacological actions of imidazolines, which never quite parallel those of catecholamines at α-adrenoceptors? In parallel with other laboratories we were able to distinguish "imidazoline" binding sites in various tissues including the CNS and kidney. This, however, has not led to any new insights into adrenoceptor/smooth muscle pharmacology except, perhaps, to provide a series of negatives for the nonadrenoceptor actions of phentolamine and some related compounds. There remain, nevertheless, many observations on the agonist effects of imidazolines that are only partly explained by the action at α-adrenoceptors.

Second, in our analysis of tissue extracts, (1) could the imidazole compounds in which we were interested exert actions through some novel cell surface receptors? (2) Could our tissue extracts contain substances akin to CDS or imidazoline-receptor ligands that might account for the biological actions that we were assaying?

The answers to both parts of this question seem to be positive. With respect to cell surface receptors, it seems that our imidazole-containing dipeptides have synergistic actions with other agents. In the second part, it is clear that the tissue extracts that have been prepared by us and others contain many biologically active compounds, rendering bioassay complex. However, it also seems clear that although we can identify a CDS in extracts, the imidazole-containing dipeptides are not good CDSs.

Overall we conclude that the biological actions of this family of compounds remain of great interest, but we do not believe that the imidazoline-preferring binding sites as currently described provide the answer to the biological effects of imidazolines in excitable peripheral tissue, which remain to be elucidated. A general theme that recurs in all this work, both with identified compounds and with uncharacterized extracts, is that biological activity tends to be ''synergistic'' rather than ''full activation,'' pointing to a need for subtlety in its assay and also for a modulatory rather than a ''controlling'' physiological role for these compounds or those for which they are surrogates.

SUMMARY

This article brings together work on imidazoline or imidazole-containing compounds concerned with the pharmacology of α-adrenoceptors, principally on smooth muscle, to illustrate how imidazolines have contributed to the subclassification of α-adrenoceptors and how, against this background, attempts have been made to use this knowledge to uncover ''nonadrenoceptor''-mediated biological effects of previously uncharacterized compounds, notably imidazole-containing dipeptides and ''clonidine displacing substance'' (CDS). Recent data are included on (1) the pharmacology of UK-14304, (2) nonadrenoceptor actions of phentolamine, (3) the pharmacology of tissue extracts containing imidazole-containing dipeptides and CDS activity, and (4) ligand binding data at I_1 and I_2 sites.

REFERENCES

1. Ruffolo, R. R. 1983. Structure-activity relationships of alpha-adrenoceptor agonists. *In* Adrenoceptors and Catecholamine Action, Part B. G. Kunos, Ed. John Wiley & Sons, Inc.
2. Ruffolo, R. R. & J. E. W. Waddell. 1982. Receptor interactions of imidazolines: α-adrenoceptors of rat and rabbit aortae differentiated by relative potencies, affinities and efficacies of imidazoline agonists. Br. J. Pharmacol. **77:** 169-176.
3. McGrath, J. C. 1982. Is there more than one α_1-adrenoceptor or is this the wrong question? Br. J. Pharmacol. **76:** 210P.
4. MacDonald, A. & J. C. McGrath. 1982. Characterization and location of α-adrenoceptors in the rat vas deferens. Br. J. Pharmacol. **76:** 211P.
5. Ernsberger, P., M. P. Meeley, J. J. Mann & D. J. Reis. 1987. Clonidine binds to imidazoline binding sites as well as α_2-adrenoceptors in the ventrolateral medulla. Eur. J. Pharmacol. **134:** 1-13.
6. Michel, M. C. & P. A. Insel. 1989. Are there multiple imidazoline binding sites? Trends Pharmacol. Sci. **10:** 342-344.
7. Hamilton, C. A., J. R. Reid & M. A. Yakuba. 1988. [^{3}H]-Yohimbine and [^{3}H]-idazoxan bind to different sites on rabbit forebrain and kidney membrane. Eur. J. Pharmacol. **146:** 345-348.
8. Michel, M. C., O.-E. Brodde, B. Schnepel, J. Behrendt, R. Tschada, H. J. Motulsky & P. A. Insel. 1989. [^{3}H]-Idazoxan and some other α_2-adrenergic drugs also bind with high affinity to a non-adrenergic site. Mol. Pharmacol. **35:** 324-330.

9. VIGNE, P., M. LAZDUNSKI & C. FRELIN. 1989. Guanabenz, guanachlor, guanoxan and idazoxan bind with high affinity to a non-adrenergic sites in pig kidney membranes. Eur. J. Pharmacol. **160:** 295-298.
10. BROWN, C. M., A. C. MACKINNON, J. C. MCGRATH, M. J. SPEDDING & A. T. KILPATRICK. 1990. Heterogeneity of α_2-adrenoceptors in rat cortex but not human platelet can be defined by 8-OH-DPAT, RU 24969 and methysergide. Br. J. Pharmacol. **99:** 481-486.
11. DOCHERTY, J. R., J. C. MCGRATH & A. MACDONALD. 1979. Further subclassification of α-adrenoceptors in the cardiovascular system, vas deferens and anococcygeus of the rat. Br. J. Pharmacol. **67:** 421P.
12. DOCHERTY, J. R. & J. C. MCGRATH. 1980. A comparison of pre- and post-junctional potencies of several α-adrenoceptor agonists in the cardiovascular system and anococcygeus of the rat. Evidence for two types of post-junctional α-adrenoceptor. Naunyn-Schmiedeberg's Arch. Pharmacol. **312:** 107.
13. MCGRATH, J. C. 1982. Evidence for more than one type of post-junctional α-adrenoceptor. Biochem. Pharmacol. **31:** 467-484.
14. MCGRATH, J. C. 1982. Is there more than one α_1-adrenoceptor or is this the wrong question? Br. J. Pharmacol. **76:** 210P.
15. MCGRATH, J. C. 1985. α-Adrenoceptor agonists and the Ca^{2+}-dependence of smooth muscle contraction: Evidence for subtypes of receptors or for agonist-dependent differences in the agonist-receptor interaction. Clin Sci. **86**(Suppl 10): 55s-63s.
16. MORROW, A. L. & I. CREESE. 1986. Characterization of α_1-adrenergic receptor subtypes in rat brain: A reevaluation of [^{3}H]-WB-4101 and [^{3}H]prazosin binding. Mol. Pharmacol. **29:** 321-330.
17. BYLUND, B. D. 1988. Subtypes of α_2-adrenoceptors: Pharmacological and molecular biological evidence converge. Trends Pharmacol. Sci. **9:** 356-361.
18. CAMBRIDGE, D. 1981. UK-14304, a potent and selective α_2-agonist for the characterisation of α-adrenoceptor subtypes. Eur. J. Pharmacol. **72:** 413-415.
19. WILSON, V. G., C. M. BROWN & J. C. MCGRATH. 1991. Are there more than two types of α-adrenoceptors involved in physiological processes? Exp. Physiol. **76:** 317-346.
20. MACLEAN, M. R., K. M. MCCULLOCH & J. C. MCGRATH. 1993. Effects of α_1-adrenoceptor stimulation on responses to noradrenaline in the rat pulmonary artery. Br. J. Pharmacol. **108:** 155-161.
21. TEMPLETON, A. G. B., J. MACMILLAN, J. C. MCGRATH, N. D. STOREY & V. G. WILSON. 1989. Evidence for a prazosin-resistant, rauwolscine-sensitive adrenoceptor-mediated pressor responses to NA and UK-14304 in the isolated perfused vascular bed of the rat tail. Br. J. Pharmacol. **97:** 563-571.
22. GRANT, T. L., J. C. MCGRATH & J. O'BRIEN. 1985. The influence of blood gases on α_1- and α_2-adrenoceptor-mediated pressor responses in the pithed rat. Br. J. Pharmacol. **86:** 69-77.
23. DUNN, W. R., C. J. DALY, J. C. MCGRATH & V. G. WILSON. 1991. A comparison of the effects of angiotensin II and Bay K 8644 on responses to noradrenaline mediated via postjunctional α_1- and α_2-adrenoceptors in rabbit isolated blood vessels. Br. J. Pharmacol. **103:** 1475-1483.
24. MCGRATH, J. C., C. M. BROWN & V. G. WILSON. 1989. α-Adrenoceptors: A critical review. Med. Res. Rev. **9:** 407-532.
25. JONAS, J. C., T. D. PLANT & J. C. ONENQUIN. 1992. Imidazoline antagonists of α_2-adrenoceptors increase insulin release *in vitro* by inhibiting ATP-sensitive K^+ channels in pancreatic channels. Br. J. Pharmacol. **107:** 8-14.
26. OHARA-IMAIZUMI, M., Y. MIYAKAWA & K. KUMAKURA. 1988. Pertussis toxin attenuates clonidine inhibition of catecholamine release in adrenal chromaffin cells. Neurosci. Letts. **93:** 294-299.
27. OHARA-IMAIZUMI, M. & K. KUMAKURA. 1992. Effects of imidazoline compounds on catecholamine release in adrenal chromaffin cells. Cell. Mol. Neurobiol. **12:** 273-283.

28. IBBOTSON, T., G. EDWARDS & A. H. WESTON. 1993. Antagonism of levocromakalin by imidazoline- and guanidine-derivatives in rat portal vein: Involvement of the delayed rectifier. Br. J. Pharmacol. **110:** 1556-1564.
29. SATAKE, N., S. SHIBATA & S. UEDA. 1984. Phentolamine-induced rhythmic contractions in bladder detrusor muscle of guinea-pig. Br. J. Pharmacol. **83:** 965-971.
30. ATLAS, D. 1991. Clonidine-displacing substance (CDS) and its putative imidazoline receptor: New leads for further divergence of α_2-adrenergic receptor activity. Biochem. Pharmacol. **41:** 1541-1549.
31. DUNNE, M. J. 1991. Block of ATP-regulated potassium channels by phentolamine and other α-adrenoceptor antagonists (in RINm5F cells). Br. J. Pharmacol. **103:** 1847-1850.
32. DOWNING, O. A., W. STEIN, K. A. WILSON & V. G. WILSON. 1988. A comparison of the action of noradrenaline and UK-14304 in the longitudinal smooth muscle of the rat isolated portal vein: No evidence for a population of post-junctional α_2-adrenoceptors. Naunyn-Schmiedeberg's Arch. Pharmacol. **338:** 239-245.
33. SCHWIETERT, R., D. WILHELM, B. WILLFERT & P. A. VAN ZWIETEN. 1992. The effect of some alpha-adrenoceptor antagonists on spontaneous myogenic activity in the rat portal vein and putative involvement of ATP-sensitive K^+ channels. Eur. J. Pharmacol. **211:** 87-95.
34. YABLONSKY, F. & J. P. DAUSSE. 1991. Non-adrenergic binding sites for the α_2-antagonist [^{3}H]-idazoxan in the rabbit urethral smooth muscle. Biochem. Pharmacol. **41:** 701-707.
35. HAMILTON, C. A. 1992. The role of imidazoline receptors in blood pressure regulation. Pharmacol. Ther. **54:** 231-248.
36. ATLAS, D. & Y. BURSTEIN. 1984. Isolation and partial purification of a clonidine-displacing endogenous brain substance. Eur. J. Biochem. **144:** 287-293.
37. MEELEY, M. P., P. R. ERNSBERGER, A. R. GRANATA & D. J. REIS. 1986. An endogenous clonidine-displacing substance from bovine brain: Receptor binding and hypotensive actions in the ventrolateral medulla. Life Sci. **38:** 1119-1126.
38. ERNSBERGER, P., M. P. MEELEY & D. J. REIS. 1988. An endogenous substance with clonidine-like properties: Selective binding to imidazole sites in the ventrolateral medulla. Brain Res. **441:** 309-318.
39. PARINI, A., I. COUPREY, R. M. GRAHAM, I. UZIELLI, D. ATLAS & S. M. LANIER. 1989. Characterisation of an imidazoline/guanidinium receptive site distinct from the α_2-adrenergic receptor. J. Biol. Chem. **264:** 11874-11878.
40. O'DOWD, J. J., M. T. CAIRNS, M. TRAINOR, D. J. ROBINS & D. J. MILLER.1990. Analysis of carnosine, homocarnosine, and other histidyl derivatives in rat brain. J. Neurochem. **55:** 446-452.
41. FELSEN, F., P. ERNSBERGER, M. P. MEELEY & D. J. REIS. 1987. Clonidine displacing substance is biologically active on smooth muscle. Eur. J. Pharmacol. **142:** 453-455.
42. REGUNATHAN, S., M. P. MEELEY & D. J. REIS. 1992. Clonidine-displacing substance from bovine brain binds to imidazoline receptors and releases catecholamines in adrenal chromaffin cells. Mol. Pharmacol. **40:** 884-888.
43. DIAMANT, S., A. ELDOR & D. ATLAS. 1987. A low molecular weight brain substance interacts, similarly to clonidine, with α_2-adrenoceptors of human platelets. Eur. J. Pharmacol. **144:** 247-255.
44. DIAMANT, S. & D. ATLAS. 1986. An endogenous brain substance, CDS (clonidine-displacing substance), inhibits the twitch response of rat vas deferens. Biochem. Biophys. Res. Comm. **134:** 184-190.
45. BOUSQUET, P., J. FELDMAN & D. ATLAS. 1986. An endogenous non-catecholamine clonidine antagonist increases mean arterial blood pressure. Eur. J. Pharmacol. **124:** 167-170.

46. Synetos, D., V. G. Manolopoulos, D. Atlas & E. Pipili-Synetos. 1991. Human plasma-derived material with clonidine displacing substance (CDS)-like properties contracts the isolated rat aorta. J. Auton. Pharmacol. **11:** 343-351.
47. Margolis, F. L., M. Grillo, T. Kawano & A. I. Farbman. 1985. Carnosine synthesis in olfactory tissue during ontogeny: Effect of exogenous β-alanine. J. Neurochem. **44:** 1459-1464.
48. Mitchell, R., G. Grieve, R. Dow & G. Fink. 1983. Endogenous GABA receptor ligands in hypophyseal portal blood. Neuroendocrinology **37:** 169-176.
49. du Vigneaud, V. & M. Hunt. 1936. The synthesis of d-carnosine, the enantiomorph of the naturally occurring form, and a study of its depressor effect on the blood pressure. J. Biol. Chem. **115:** 93-100.
50. Melville, C. A., M. Trainor, J. C. McGrath, C. Daly, J. J. O'Dowd & D. J. Miller. 1990. Carnosine shows α_2-adrenoceptor agonist and antagonist properties in saphenous vein isolated from rabbit. J. Physiol. **427:** 29P.
51. Hussain, J. F., D. A. Kendall & V. G. Wilson. 1993. Species-selective binding of [^{3}H]-idazoxan to α_2-adrenoceptors and non-adrenoceptor, imidazoline binding sites in the central nervous system. Br. J. Pharmacol. **109:** 831-838.
52. Ernsberger, P., M. P. Meeley, J. J. Mann & D. J. Reis. 1987. Clonidine binds to imidazole binding sites as well as α_2-adrenoceptors in the ventrolateral medulla. Eur. J. Pharmacol. **134:** 1-13.
53. Hussain, J. F., C. M. Brown, D. A. Kendall, A. C. Mackinnon & V. G. Wilson. 1991. A crude extract of bovine lung displaces ligands from α_2-adrenoceptors and non-adrenoceptor imidazoline binding sites. Br. J. Pharmacol. **104:** 80P.
54. Hussain, J. F., D. A. Kendall & V. G. Wilson. 1992. A possible linkage of non-adrenoceptor, imidazoline binding sites labelled by [^{3}H]-idazoxan to K^+ channels and G proteins. Br. J. Pharmacol. **107:** 323P.
55. Atlas, D. 1991. Clonidine-displacing substance (CDS) and its putative imidazoline receptors. Biochem. Pharmacol. **41:** 1541-1549.
56. Hamilton, C. A. 1992. The role of imidazoline receptors in blood pressure regulation. Pharmacol. Ther. **54:** 231-248.
57. O'Dowd, J. J., D. J. Robbins & D. J. Miller. 1988. Detection, characterisation, and quantification of carnosine and other histidyl derivatives in cardiac and skeletal muscle. Biochim. Biophys. Acta **967:** 241-249.
58. O'Dowd, A., J. J. O'Dowd, J. M. O'Dowd, N. McFarlane, H. Abe & D. J. Miller. 1992. Analysis of novel imidazoles from isolated perfused rabbit heart by two high-performance liquid chromatographic methods. J. Chromatog. Biomed. Appl. **577:** 347-353.
59. Michel, M. C. & P. Ernsberger. 1992. Keeping an eye on the I site: Imidazoline-preferring receptors. TIPS **13:** 369-370.
60. Lamont, C. & D. J. Miller. 1992. Calcium sensitizing action of carnosine and other endogenous imidazoles in chemically-skinned striated muscle. J. Physiol. **454:** 421-434.
61. McFarlane, N., J. McMurray, J. J. O'Dowd, H. J. Dargie & D. J. Miller. 1991. Synergism of histidyl dipeptides as antioxidants. J. Mol. Cell. Cardiol. **23:**1205-1207.
62. Crush, K. G. 1970. Carnosine and related substances in animal tissues. Comp. Biochem. Physiol **34:** 3-30.
63. Granata, A. R. & D. J. Reis. 1987. Hypotension and bradycardia elicited by histamine into the C1 area of the rostral ventrolateral medulla. Eur. J. Pharmacol. **136:** 157-162.
64. Bousquet, P., J. Feldman & J. Schwartz. 1984. Central cardiovascular effects of alpha adrenergic drugs: Differences between catecholamines and imidazolines. J. Pharmacol. Exp. Ther. **230:** 232-236.
65. Singh, G., J. F. Hussain, A. MacKinnon, C. M. Brown, D. A. Kendall & V. G. Wilson. 1995. Evidence for the presence of a non-catecholamine, clonidine-displacing substance

in crude, methanolic extracts of bovine brain and lung. Naunyn-Schmiedeberg's Arch. Pharmacol. **351:** 17-26.

66. Jackson, H. C., I. J. Griffin & D. J. Nutt. 1991. The effects of idazoxan and other α_2-adrenoceptor antagonists on food and water intake in the rat. Br. J. Pharmacol. **104:** 258-262.

67. Ramagopal, M. V. & H. J. Leighton. 1989. Analysis of the presence of postjunctional α_2 adrenoceptors in the rat anococcygeus muscle. J. Pharmacol. Exp. Ther. **250:** 492-499.

68. Gothert, M. & G. J. Molderings. 1991. Involvement of presynaptic imidazoline receptors in the α_2-adrenoceptor-independent inhibition of noradrenaline release by imidazoline derivatives. Naunyn-Schmiedeberg's Arch. Pharmacol. **343:** 271-282.

Effect of the Imidazoline Receptor Ligand Moxonidine on Human Platelet Aggregation

SALLY A. WIEST AND MITCHELL I. STEINBERG [a]

Lilly Research Laboratories
Eli Lilly and Company
Indianapolis, Indiana 46285

Recently it was shown that moxonidine binds with high affinity (0.7 nM) to sites on human platelet plasma membranes.[1] These sites appear analogous to imidazoline type 1 (I_1) sites described in the bovine, rat, and human brain stem.[2–4] Stimulation of I_1 receptors in the central nervous system reduces sympathetic outflow[5] mediating the antihypertensive effects of moxonidine.[6,7] The function(s) of I_1 receptors on the platelet membrane, however, is not known. Therefore, we investigated the effect of moxonidine on platelet function and agonist-mediated aggregation and ATP release.

Blood from healthy, human volunteers was collected by venipuncture and mixed with heparin (60 ml of blood added to 0.4 ml of heparinized saline solution [4 mg heparin/ml saline]). Platelet-rich plasma (PRP) was prepared by centrifuging whole blood (150 × *g*), and aggregation studies were performed on PRP that was continuously stirred at 37°C. A typical tracing comparing the change in optical density after the addition of epinephrine or moxonidine is illustrated in FIGURE 1A. Moxonidine caused a concentration-dependent aggregation of platelets at concentrations from 10^{-7} to 10^{-4} M that was similar in magnitude and potency to the aggregation response to the α_2 agonist epinephrine (FIG. 1B). Furthermore, aggregation to both epinephrine and moxonidine was completely blocked by the α_2 antagonist yohimbine (10^{-5} M) (FIG. 1B). Thus, both epinephrine and moxonidine induce platelet aggregation by binding to α_2 adrenergic sites present on platelet plasma membranes as previously reported.[1,8,9]

The α_2-mediated agonist effect of moxonidine was blocked by yohimbine (10^{-5} M), and the ability of moxonidine to influence agonist-induced platelet function through I_1-mediated interaction was evaluated. Agonists including ADP, arginine vasopressin, collagen, thromboxane agonist U46619, and platelet-activating factor were added to PRP 1 minute after exposure to moxonidine (10^{-8} or 10^{-5} M). In the presence of yohimbine, maximal and submaximal aggregation induced by all agonists tested, as well as shape change to low concentrations of U46619 and collagen, was unaffected by moxonidine (FIG. 2).

Adenosine diphosphate and collagen-induced ATP secretion was measured simultaneously with platelet aggregation. The luciferin/luciferase reaction was used to

[a] Address for correspondence: Mitchell Steinberg, Ph.D., Cardiovascular Research, MC 304, Lilly Research Laboratories, Lilly Corporate Center, Indianapolis, Indiana 46285.

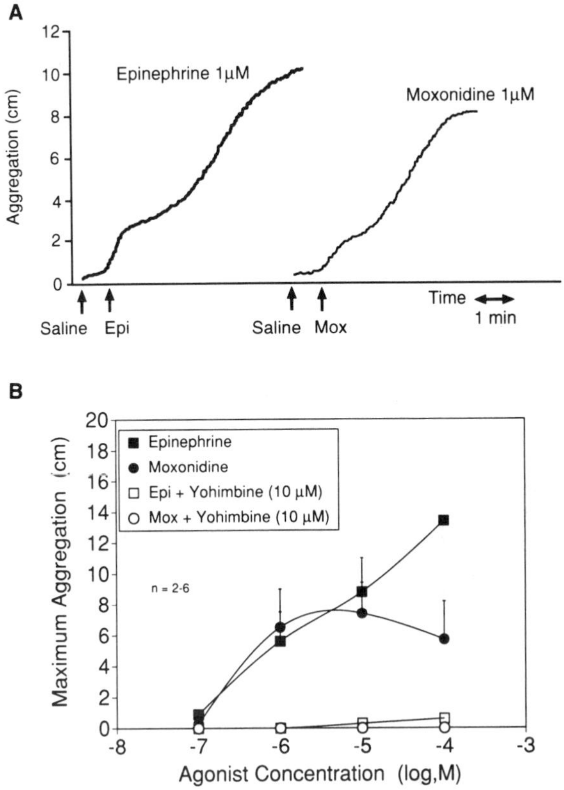

FIGURE 1. (**A**) Typical tracings of stirred human platelets showing change in optical density after the addition of epinephrine (10^{-6} M) or moxonidine (10^{-6} M). Aggregation induced by both epinephrine and moxonidine was biphasic and proceeded immediately, without shape change, upon the addition of agonist. (**B**) Concentration-response curve for moxonidine- and epinephrine-induced α_2-mediated platelet aggregation. The α_2 antagonist yohimbine (10^{-5} M) completely blocked epinephrine as well as moxonidine-induced aggregation.

measure ATP release, and 0.2 nmol of ATP was used to quantitate the response. In the presence of yohimbine, ATP release from platelet-dense granules in response to submaximal (10^{-6} M) and maximal (10^{-5} M) aggregatory concentrations of ADP and a nonaggregatory concentration of collagen (0.2 μg/ml) was not affected by moxonidine (10^{-7} to 10^{-4} M). Control ATP release induced by ADP (10^{-6} and 10^{-5} M) and collagen (0.2 μg/ml) was 20.0 ± 0.2, 300 ± 10, and 13 ± 2 pmol/1.5 × 10^8 platelets, respectively (n = 3).

In conclusion, moxonidine appears to induce human platelet aggregation by an α_2-like adrenergic mechanism. A specific I_1 receptor-mediated effect of moxonidine on platelet aggregation or the release reaction to various agonists could not be demonstrated. However, I_1 receptors might still play a role in platelet physiology as suggested by the following observations: (1) I_1 sites are upregulated in platelets from

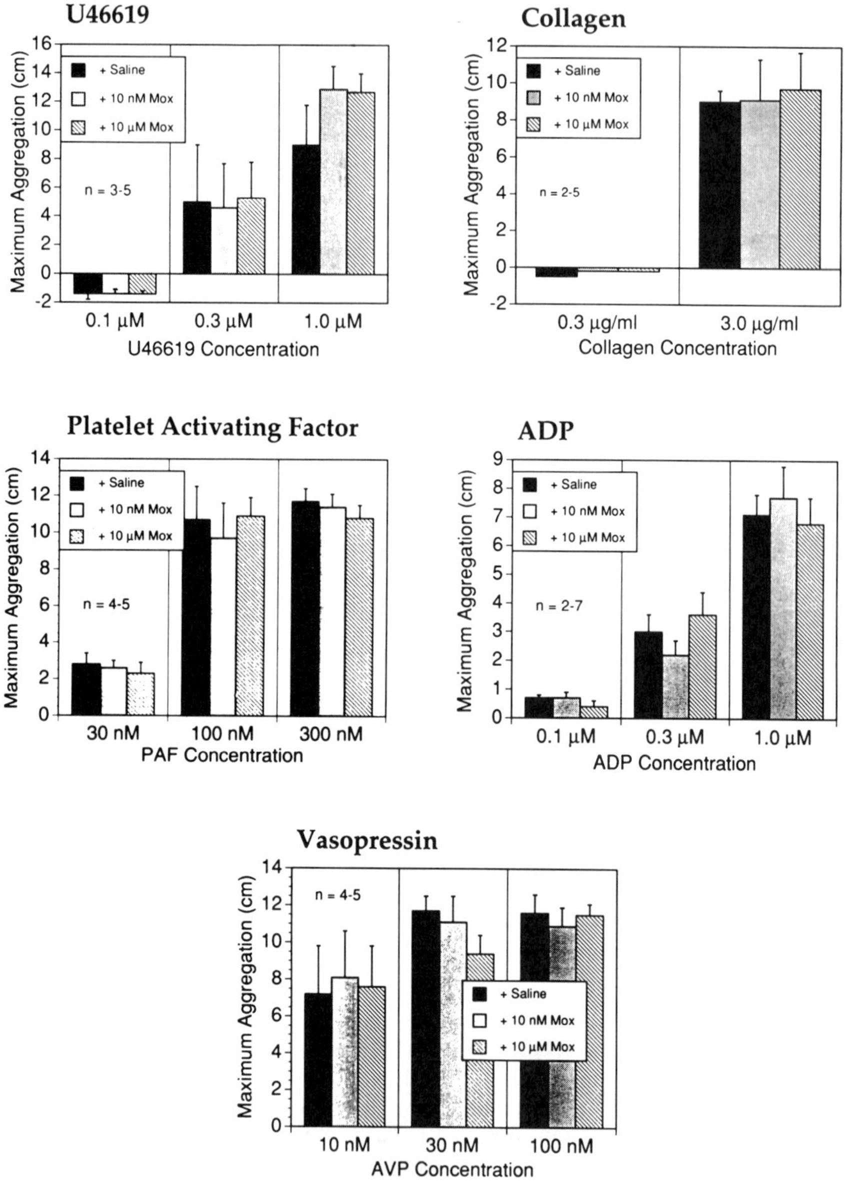

FIGURE 2. Agonist-mediated platelet aggregation is not influenced by moxonidine in the presence of yohimbine (10^{-5} M). Moxonidine at low or high concentrations did not affect the ability of ADP, U46619, arginine vasopressin, collagen, or platelet-activating factor to induce concentration-dependent platelet aggregation. At low concentrations of U46619 and collagen, shape change alone could be evaluated and is represented as negative aggregation values. High concentrations of agonists represent irreversible aggregation, and intermediate concentrations of ADP and U46619 represent reversible aggregation.

depressed patients and are normalized after successful antidepressant treatment.[10] (2) Moxonidine inhibits prostaglandin E_1-stimulated platelet adenylate cyclase activity similar to epinephrine; however, an imidazoline-mediated effect of moxonidine on platelet inhibitory pathways cannot be ruled out.[8] (3) Finally, the ability of clonidine to modulate serotonin metabolism in the rat central nervous system via a nonadrenergic mechanism could have potential relevance to platelet serotonin turnover.[1,11]

REFERENCES

1. Piletz, J. E. & K. Sletten. 1993. Nonadrenergic imidazoline binding sites on human platelets. J. Pharmacol. Exp. Ther. **267:** 1493-1502.
2. Ernsberger, P., M. P. Meeley, J. J. Mann & D. J. Reid. 1987. Clonidine binds to imidazoline binding sites as well as α_2-adrenoceptors in the ventrolateral medulla. Eur. J. Pharmacol. **134:** 1-13.
3. Ernsberger, P., T. H. Damon, L. M. Graff, S. G. Schafer & M. O. Christen. 1993. Moxonidine, a centrally acting antihypertensive agent, is a selective ligand for I_1-imidazoline sites. J. Pharmacol. Exp. Ther. **264:** 172-182.
4. Bricca, G., M. Dontenwill, A. Molines, J. Feldman, A. Belcourt & P. Bousquet. 1989. The imidazoline preferring receptor: Binding studies in bovine, rat and human brainstem. Eur. J. Pharmacol. **162:** 1-9.
5. Bousquet, P., J. Feldman & J. Schwartz. 1984. Central cardiovascular effects of α-adrenergic drugs: Differences between catecholamines and imidazolines. J. Pharmacol. Exp. Ther. **230:** 232-236.
6. Armah, B. I., E. Hofferber & W. Stenzel. 1988. General pharmacology of the novel centrally acting antihypertensive agent moxonidine. Arzneim. Forsch./Drug Res. **38:** 1426-1434.
7. Ernsberger, P. R., K. L. Westbrooks, M. O. Christen & S. G. Schafer. 1992. A second generation of centrally acting antihypertensive agents act on putative I_1-imidazoline receptors. J. Cardiovasc. Pharmacol. **20:** S1-S10.
8. Bergerhausen, J. 1985. Moxonidine (BE 5895), a full agonist at human platelet α_2-adrenoceptors. Naunyn-Schmiedebergs' Arch. Pharmacol. (Suppl.) **329:** R80.
9. Petrusewicz, J. & R. Kaliszan. 1991. Human blood platelet alpha adrenoceptor in view of the effects of various imidazol(in)e drugs on aggregation. Gen. Pharmacol. **22:** 819-823.
10. Piletz, J. E., A. Halaris, A. Saran & M. R. Marler. 1991. Desipramine lowers ^{3}H-*para*-aminoclonidine binding in platelets of depressed patients. Arch. Gen. Psychiatry **48:** 813-820.
11. Aslanian, V. & B. Renaud. 1989. Changes in serotonin metabolism in the rat raphe magnus and cardiovascular modifications following systemic administration of clonidine and other central alpha$_2$-agonists: An *in vivo* voltammetry study. Neuropharmacology **28:** 387-395.

Role of Imidazoline Receptors in Halothane-Epinephrine Arrhythmias[a]

YUKIO HAYASHI,[b,c] TAKAHIKO KAMIBAYASHI,[d] ATSUSHI YAMATODANI,[d] MASAKAZU KURO,[b] AND IKUTO YOSHIYA [d]

[b]*Department of Anesthesiology*
National Cardiovascular Center
Suita, Osaka 565, Japan

[d]*Department of Anesthesiology and Medical Physics*
Osaka University Faculty of Medicine
Suita, Osaka 565, Japan

Halothane, a volatile anesthetic agent still applied in clinical settings, is well known to sensitize the myocardium to the arrhythmogenic effect of epinephrine.[1,2] Traditionally, changes in hemodynamic variables, such as hypertension and tachycardia, were considered important in triggering arrhythmias.[3] After the recent development of selective adrenergic antagonists, the role of adrenergic receptors in the cardiovascular system involved in the genesis of halothane-epinephrine arrhythmias has been well explored. Thus far, we and others have implicated myocardial α_1- and β_1-adrenoceptors in the genesis of arrhythmias and indirect protective action of β_2-adrenoceptors.[4–9] Besides the cardiovascular system, several studies have indicated that the central nervous system (CNS) plays an important role in the onset of other types of arrhythmias,[10–13] and our recent report also suggests that the CNS modulates halothane-epinephrine arrhythmias.[14]

Alpha$_2$-adrenoceptors are widely distributed throughout the CNS and are involved in several physiological processes including sedation, analgesia, and hemodynamic control[15,16]; therefore, α_2-adrenoceptor agonists have been extensively applied to clinical settings in anesthesiology.[16] Recent investigations have documented that some compounds, such as clonidine, besides being able to bind to α_2-adrenoceptors, also functionally bind to nonadrenergic receptor, that is, the imidazoline receptor (IR), and some functions that were classically attributed to activation of α_2-adrenoceptors may be mediated by IRs.[17–20] Clonidine has been reported to inhibit the onset of digitalis-induced arrhythmias,[21,22] and recently we documented that dexmedetomidine, a selective α_2-agonist,[23,24] dose-dependently prevented halothane-epinephrine

[a]These studies were supported by a grant-in-aid for scientific research (B) from the Minister of Education, Science, and Culture in Japan.

[c]Address for correspondence: Yukio Hayashi, MD, Department of Anesthesiology, National Cardiovascular Center, 5-7-1, Fujishiro-dai, Suita, Osaka 565, Japan.

arrhythmias.[25] However, involvement of IRs in the receptor mechanism of the α_2-agonist- induced antiarrhythmic property is obscure.

In this paper we review several studies that we previously reported of a dog model of halothane-epinephrine arrhythmias[25–27] and demonstrated the contribution of IRs in central modulation of the genesis of halothane-epinephrine arrhythmias.

MATERIALS AND METHODS

Studies were conducted under guidelines provided by the Animal Care Committee of Osaka University Faculty of Medicine. Adult mongrel dogs of either sex, weighing 7.3-13.0 kg, were used. Whenever different experiments were performed with the same dog, more than 7 days elapsed between experiments. The same dog was not used more than once in the same experimental group. Two arrhythmogenic doses obtained in the same dog in separate experiments were considered two different observations. Anesthesia was induced with halothane alone and maintained at an end-tidal concentration of 1.3%, which was monitored continuously by an anesthetic gas analyzer (model AA-102-30-00, Datex, Helsinki, Finland). The trachea of each dog was intubated with a cuffed tracheal tube and the lungs were mechanically ventilated. The end-tidal CO_2 concentration was continuously monitored with an expired gas monitor (model 1H 21A, Minato, Osaka, Japan) and maintained at a level of 35-40 mm Hg. A heating lamp and circulating water blanket were used to maintain the esophageal temperature at 36.5-38.5°C.

Lead II of the electrocardiogram was monitored continuously. A femoral artery catheter was inserted for both pressure monitoring and blood gas and serum electrolyte sampling. A right femoral vein was cannulated for administration of both drugs and lactated Ringer's solution, which was infused at a rate of 10 $ml \cdot kg^{-1} \cdot hr^{-1}$. Serum K^+ was maintained between 3.5 and 4.5 $mEq \cdot l^{-1}$ by infusing K^+ at a rate of 1-10 $mEq \cdot hr^{-1}$. Arterial pH, oxygen tension (PaO_2), and serum Na^+ were maintained within the ranges of 7.35 and 7.45, 85 and 100 mm Hg, and 135 and 150 $mEq \cdot l^{-1}$, respectively.

The arrhythmia threshold was achieved when four or more premature ventricular contractions occurred within 15 seconds. The arrhythmogenic dose of epinephrine was defined as the smallest dose that produced arrhythmias. According to the method we previously reported,[28] the arrhythmogenic dose of epinephrine was determined with logarithmically spaced infusions of epinephrine lasting 3 minutes. Recovery periods of 10-30 minutes were allowed between infusions until the hemodynamic parameters (arterial blood pressure and heart rate) returned to basal levels. Infusion was started at a minimum dose of 0.67 $\mu g \cdot kg^{-1} \cdot min^{-1}$, and the dose was increased by $e^{0.4}$ ($e = 2.72$) until the arrhythmia threshold was achieved. If arrhythmias occurred at one of these doses, a smaller arrhythmic dose, divided by $e^{0.2}$, was tested. When the criterion for arrhythmogenic dose had been satisfied, an arterial blood sample was collected to measure the plasma concentration of epinephrine. Epinephrine was determined in a fully automated high-performance liquid chromatography-fluorometric system (model HLC-8030 Catecholamine Analyzer, Tosoh, Tokyo, Japan) using a diphenylethylenediamine condensation method.[29] This assay method has a limit of sensitivity of 10 $pg \cdot ml^{-1}$ for epinephrine. Inter- and intra-assay variations were less than 3%.

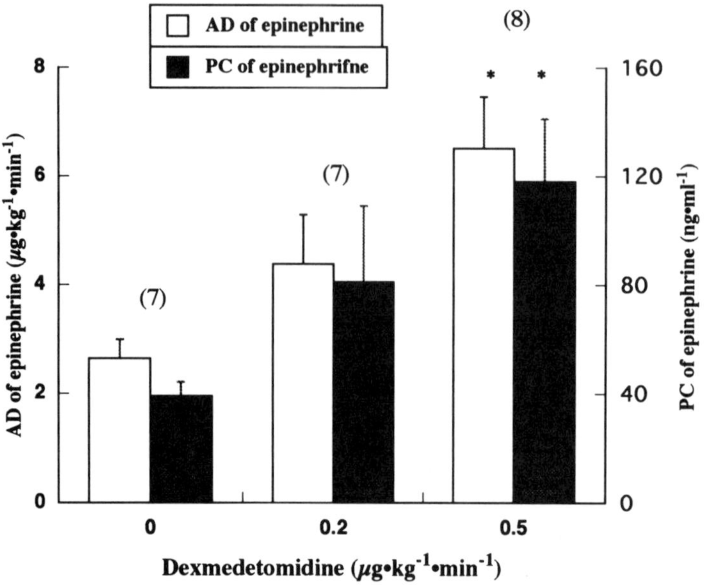

FIGURE 1. Arrhythmogenic doses (AD) and plasma concentrations (PC) of epinephrine in the presence of dexmedetomidine 0, 0.2, and 0.5 $\mu g \cdot kg^{-1} \cdot min^{-1}$ during halothane anesthesia in intact dogs. *$p < 0.05$ compared with the 0 dose.

At first, the arrhythmogenic dose of epinephrine was determined in the presence of dexmedetomidine 0, 0.2, and 0.5 $\mu g \cdot kg^{-1} \cdot min^{-1}$, which was administered intravenously. Second, to elucidate which receptor mechanism was involved in the antiarrhythmic effect of dexmedetomidine in the CNS, we examined whether two kinds of α_2 antagonists, that is, atipamezole, which has an imidazole structure and may have an affinity to IRs,[19,30] and L-659,066, a nonimidazoline compound,[31] inhibited the effects of dexmedetomidine. In this protocol, antagonists were administered into the cisterna magna through a 20 G needle percutaneously. The doses of each antagonist were determined to roughly share equieffective potency according to previous data.[32,33] In a third group of experiments, we investigated the effect of atipamezole and L-659,066 on the arrhythmogenic dose of epinephrine in the absence of dexmedetomidine. As atipamezole can cross the blood-brain barrier,[30] we administered the drug either intravenously or into the cisterna magna and compared the effect between the two different routes of drug administration. On the other hand, we only tested the effect of central administration of L-659,066, because it does not penetrate the blood-brain barrier.[31] These antagonists were administered 30 minutes before starting epinephrine infusions.

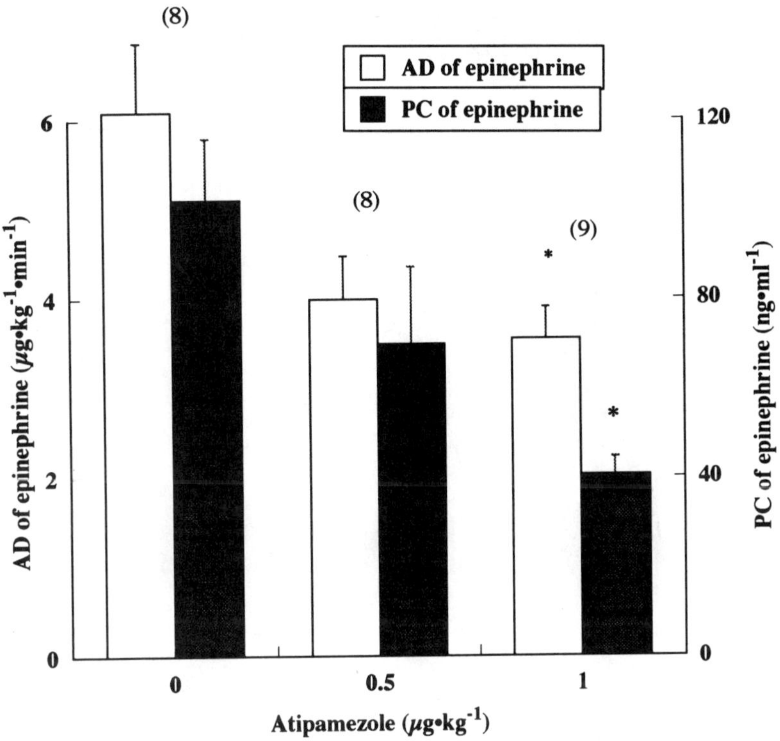

FIGURE 2. Effect of atipamezole on the arrhythmogenic threshold of epinephrine in the presence of dexmedetomidine, 0.5 $\mu g \cdot kg^{-1} \cdot min^{-1}$, during halothane anesthesia in dogs (mean ± SEM; number of observations is shown in parentheses). $*p < 0.05$ compared with dexmedetomidine alone (dose 0). AD = arrhythmogenic dose; PC = plasma concentration.

The data were expressed as mean ± SEM. The results of multiple groups were analyzed by one-way analysis of variance, and comparisons between groups were assessed by Scheffe's test. The two groups were compared by Student's *t* test. We considered $p < 0.05$ to be statistically significant.

RESULTS

Dexmedetomidine significantly increased the arrhythmia thresholds of epinephrine in a dose-dependent fashion, achieving statistical significance at 0.5 $\mu g \cdot kg^{-1} \cdot min^{-1}$ (FIG. 1). Therefore, we used the highest dose (0.5 $\mu g \cdot kg^{-1} \cdot min^{-1}$) of dexmedetomidine in the second experiment. The effects of central administration of atipamezole and L-659,066 on the antiarrhythmic action of dexmedetomidine are shown in FIGURES 2 and 3, respectively. Atipamezole significantly inhibited the antiarrhythmic effect of dexmedetomidine, whereas the effect of L-659,066 was not significant.

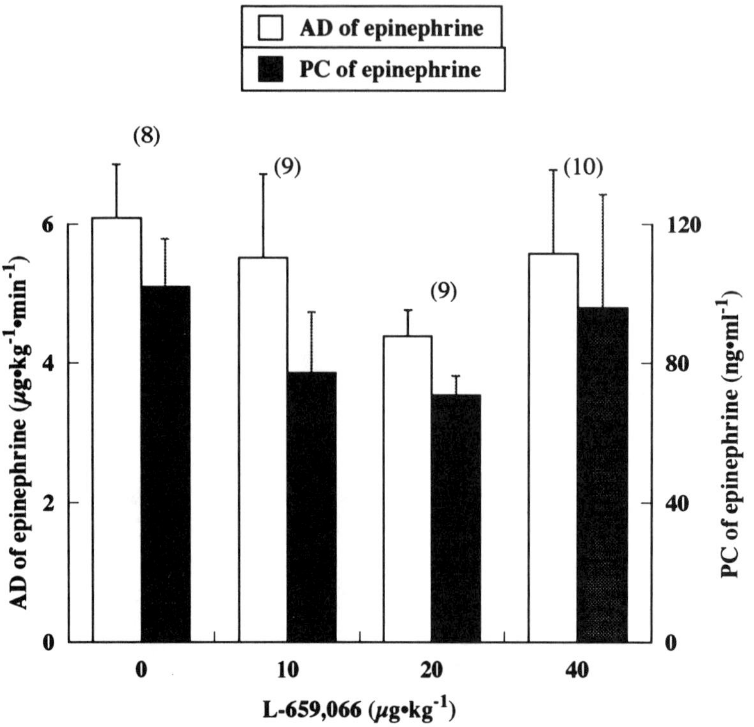

FIGURE 3. Effect of L-659,066 on the arrhythmogenic threshold of epinephrine in the presence of dexmedetomidine, 0.5 $\mu g \cdot kg^{-1} \cdot min^{-1}$, during halothane anesthesia in dogs (mean ± SEM; number of observations is shown in parentheses). AD = arrhythmogenic dose; PC = plasma concentration.

Furthermore, both intravenous and central treatment of atipamezole potentiated epinephrine-induced arrhythmias during halothane anesthesia, whereas L-659,066 given to the cisterna magna did not change the thresholds despite slight potentiation of the arrhythmias (Figs. 4-6).

DISCUSSION

The results demonstrate that dexmedetomidine, a selective α_2 agonist, prevented the arrhythmogenic effect of epinephrine during halothane anesthesia and this response was blocked by central administration of atipamezole, but not by L-659,066. Furthermore, atipamezole alone significantly potentiated the arrhythmias, whereas the effect of L-659,066 was insignificant.

Recent radioligand binding studies indicate that selective α_2-adrenergic legends bind to nonadrenergic receptors, namely, IRs.[19,34,35] Before these studies, Bousquet *et al.*[17] first suggested that the hypotensive effect of clonidine, previously believed

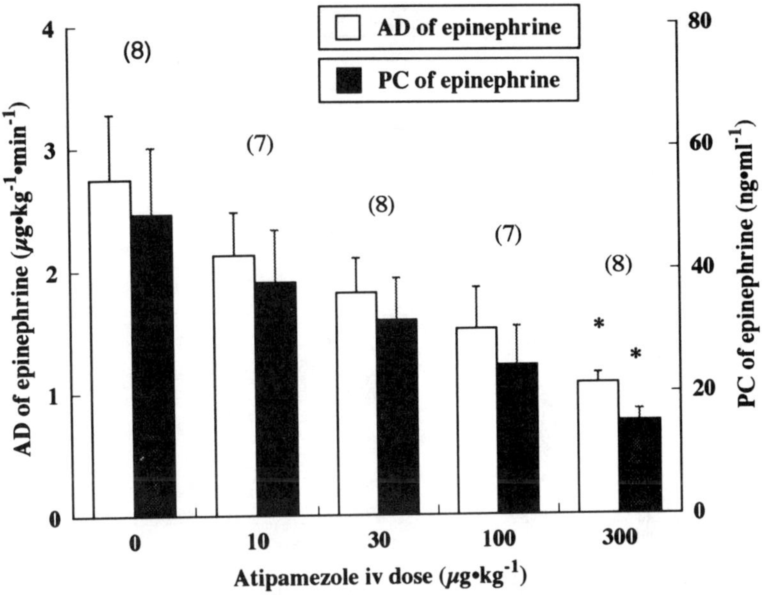

FIGURE 4. Arrhythmogenic doses (ADs) and plasma concentrations (PCs) of epinephrine in the presence of atipamezole administered intravenously (iv) during halothane anesthesia in dogs (mean ± SEM; number of observations is shown in parentheses). *$p < 0.05$ compared with the control value (dose 0). (Reproduced with permission from Hayashi *et al.*[26])

to be exerted through activation of α_2-adrenoceptors, was mediated through another type of receptor which had a specific affinity for drugs with an imidazoline structure. This idea was confirmed by their later reports.[20,36] In addition, recent studies documented that IRs are also responsible for several pharmacological properties of α_2 agonists.[37,38] Because dexmedetomidine has an imidazoline structure and binds to IR,[19,23,30] it is important to clarify which receptors are more responsible for its antiarrhythmic property. The present results, showing that atipamezole provoked significant inhibition, whereas L-659,066 did not produce a significant blockade despite exerting minor inhibition, indicate that the imidazoline receptors are more important than the α_2-adrenoceptors in the dexmedetomidine-induced antiarrhythmic property.

In the third separate group of experiments, atipamezole enhanced epinephrine-induced arrhythmias during halothane anesthesia. Conversely, although L-659,066 produced minor potentiation of the arrhythmias, its effect was also insignificant. As just discussed, this difference between atipamezole and L-659,066 regarding potentiation of arrhythmias may also be due to blockade of IRs. Furthermore, we examined the effect of both intravenous and central administration of atipamezole in the third study. The different routes of drug administration may be helpful in elucidating the site of action for a particular response. When systemically administered, atipamezole may exert arrhythmogenic action through either a central or a

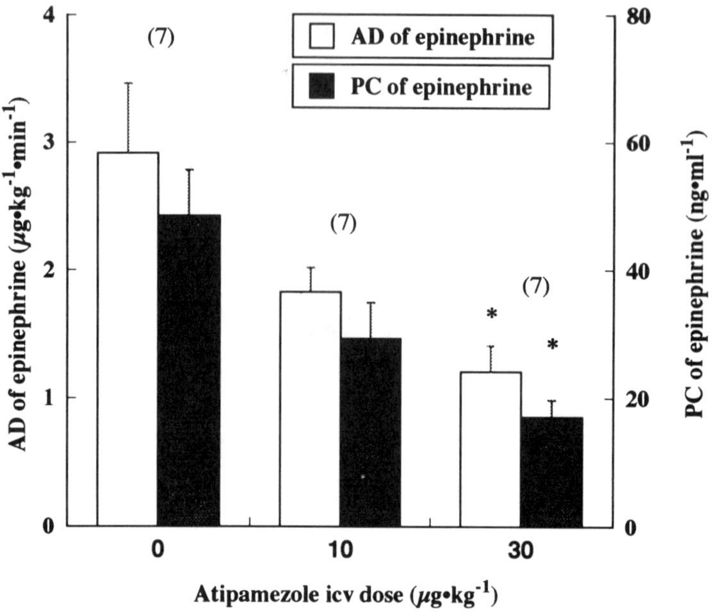

FIGURE 5. Arrhythmogenic doses (ADs) and plasma concentrations (PCs) of epinephrine in the presence of atipamezole administered intracerebroventricularly (icv) during halothane anesthesia in dogs (mean ± SEM; number of observations is shown in parentheses). $^*p < 0.05$ compared with the control value (dose 0). (Reproduced with permission from Hayashi *et al.*[26])

peripheral mechanism, because it can cross the blood-brain barrier.[30] However, the present finding that a smaller dose was equally potent in the case of central administration indicated a central site of action. Accordingly, both activation and blockade of IRs in the CNS modulate the genesis of halothane-epinephrine arrhythmias. Here, since L-659,066 exerted some proarrhythmic action in this experiment despite statistical insignificance, we may deduce that α_2-adrenoceptors in the CNS are also partly involved in the central modulation of the onset of arrhythmias. Again however, our data do not deny the concept that IRs played a more important role than α_2-adrenoceptors in the central modulation of halothane-epinephrine arrhythmias.

The precise site in the CNS involved in controlling vulnerability to halothane-epinephrine arrhythmias has not been elucidated. The nucleus reticularis lateralis in the ventrolateral medulla is rich in IRs and is a site of the hypotensive action of clonidine.[20,34,36] This brain area is connected functionally with the nucleus tractus solitarius, which governs autonomic control, including vagal activity.[39] Activation of IRs may decrease vasomotor activity, presumably leading to increased vagal outflow.[34,36] Actually, stimulation of vagal activity was reported to protect the genesis of halothane-epinephrine arrhythmias.[3,40] On the contrary, blockade of IRs by atipamezole may remove vagal stimulation, resulting in the opposite situation.

In conclusion, IRs in the CNS modulate the genesis of epinephrine-induced arrhythmias during halothane anesthesia in dogs, that is, prevention and potentiation of arrhythmias were induced by activation and blockade of IRs, respectively.

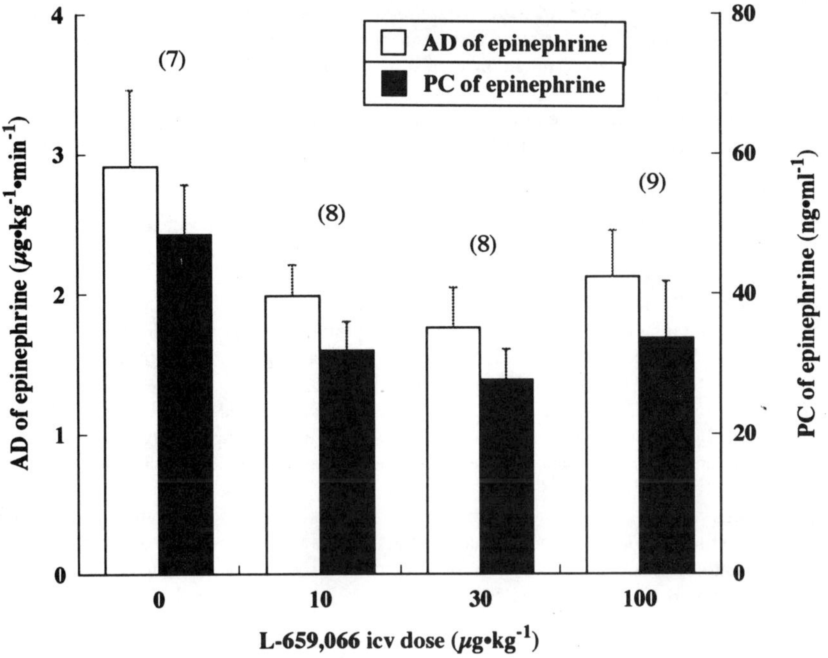

FIGURE 6. Arrhythmogenic doses (ADs) and plasma concentrations (PCs) of epinephrine in the presence of L-659,066 administered intracerebroventricularly (icv) during halothane anesthesia in dogs (mean ± SEM; number of observations is shown in parentheses). (Reproduced with permission from Hayashi *et al.*[26])

REFERENCES

1. Atlee, J. L. & Z. J. Bosnijak. 1990. Mechanisms for cardiac dysrhythmias during anesthesia. Anesthesiology **72:** 347-374.
2. Raynold, A. K. 1984. On the mechanism of myocardial sensitization to catecholamines by hydrocarbon anesthetics. Can. J. Physiol. Pharmacol. **62:** 183-198.
3. Zink, J., B. I. Sayniuk & P. E. Dressel. 1975. Halothane-epinephrine-induced cardiac arrhythmias and the role of heart rate. Anesthesiology **43:** 548-555.
4. Sharma, P. L. 1966. Effect of propranolol on catecholamine-induced arrhythmias during nitrous oxide-halothane anaesthesia in the dog. Br. J. Anaesth. **38:** 871-877.
5. Maze, M. & C. M. Smith. 1983. Identification of receptor mechanism mediating epinephrine-induced arrhythmias during halothane anesthesia in the dog. Anesthesiology **59:** 322-326.
6. Maze, M., E. Hayward, Jr. & D. Gaba. 1985. Alpha$_1$-adrenergic blockade raises epinephrine-arrhythmia threshold in halothane-anesthetized dogs in a dose-dependent fashion. Anesthesiology **63:** 611-615.
7. Hayashi, Y., K. Sumikawa, C. Tashiro & I. Yoshiya. 1988. Synergistic interaction of α_1 and β adrenoceptor agonists on induction arrhythmias during halothane anesthesia in dogs. Anesthesiology **68:** 902-907.
8. Hayashi, Y., K. Sumikawa, M. Kuro, K. Fukumitsu, C. Tashiro & I. Yoshiya. 1991. Role of β_1- and β_2-adrenoceptors in the mechanism of halothane myocardial sensitization in dogs. Anesth. Analg. **72:** 435-439.

9. Hayashi, Y., K. Sumikawa, T. Kamibayashi, A. Yamatodani, T. Mammoto, M. Kuro & I. Yoshiya. 1992. Selective $beta_1$ and $beta_2$ adrenoceptor blockade on epinephrine-induced arrhythmias in halothane anaesthetized dogs. Can. J. Anaesth. **39:** 873-876.
10. Garan, H., J. N. Ruskin & W. J. Powell. 1981. Centrally mediated effect of phenytoin on digoxin-induced ventricular arrhythmias. Am. J. Physiol. **241:** H67-H72.
11. DeSilva, R. A., R.L. Verrier & B. Lown. 1978. The effects of psychological stress and vagal stimulation with morphine on vulnerability to ventricular fibrillation (VF) in the conscious dog. Am. Heart J **95:** 197-203.
12. Furey, S. A., III & M. N. Levy. 1983. Interactions among heart rate, autonomic activity, and arterial pressure upon the multiple repetitive extrasystole threshold in the dog. Am. Heart J. **106:** 1112-1120.
13. Corr, P. B., K. A. Yamada & F. X. Witkowski. 1986. Mechanism controlling cardiac autonomic function and their relation to arrhythmogenesis. *In* The Heart and Cardiovascular System. H. A. Fozzard, E. Haber, R. A. Jennings, A. M. Katz & H. E. Morgan, Eds.: 1343-1403. Raven Press. New York.
14. Hayashi, Y., T. Kamibayashi, K. Sumikawa, A. Yamatodani, M. Kuro & I. Yoshiya. 1991. Phenytoin prevents epinephrine-induced arrhythmias through central nervous system in halothane-anesthetized dogs. Res. Comm. Chem. Pathol. Pharmacol. **74:**59-70.
15. Unnerstall, J. R., T. A. Kopajtic & M. J. Kuhar. 1984. Distribution of α_2 agonist binding sites in the rat and human central nervous system: Analysis of some functional, anatomic correlates of the pharmacologic effects of clonidine and related adrenergic agents. Brain Res. Rev. **7:** 69-101.
16. Hayashi, Y. & M. Maze. 1993. $Alpha_2$ adrenoceptor agonists and anaesthesia. Br. J. Anaesth. **71:** 108-118.
17. Bousquet, P., J. Feldman & J. Schwartz. 1984. Central cardiovascular effects of alpha adrenergic drugs: Differences between catecholamines and imidazolines. J. Pharmacol. Exp. Ther. **230:** 232-236.
18. Lehmann, J., E. Koening-Berard & P. Vitou. 1989. The imidazoline-preferring receptor. Life Sci. **45:** 1609-1615.
19. Wikberg, J. E. S. & S. Uhlen. 1990. Further characterization of the guinea pig cerebral cortex idazoxan receptor: Solution, distinction from the imidazole site, and demonstration of cirazoline as an idazoxan receptor-selective drug. J. Neurochem. **55:** 192-203.
20. Tibirica, E., J. Feldman, C. Mermet, F. Gonon & P. Bousquet. 1991. An imidazoline-specific mechanism for the hypotensive effect of clonidine: A study with yohimbine and idazoxan. J. Pharmacol. Exp. Ther. **256:**606-613.
21. Gillis, R. A., R. A. Dionne & F. G. Standaert. 1972. Suppression by clonidine (St-155) of cardiac arrhythmias induced by digitalis. J. Pharmacol. Exp. Ther. **182:** 218-226.
22. Chen, S. A., R. H. Liu, H. Ting, M. S. Chang, B. N. Chiang & J. S. Kuo. 1991. Termination of digitalis-induced ventricular tachycardias by clonidine involved central α_2 adrenoceptors in cats. Br. J. Pharmacol. **103:** 1114-1118.
23. Savola, J.-M., H. Ruskoaho, J. Puurunen, J. S. Salonen & N. T. Karki. 1986. Evidence for medetomidine as selective and potent agonist at alpha-2 adrenoceptor. J. Auton. Pharmacol. **5:** 275-284.
24. Virtanen, R., J.-M Savola, V. Saano & L. Nyman. 1988. Characterization of the selectivity, specificity and potency of medetomidine as an α_2 adrenoceptor agonist. Eur. J. Pharmacol. **150:** 9-14.
25. Hayashi, Y., K. Sumikawa, M. Maze, A. Yamatodani, T. Kamibayashi, M. Kuro & I. Yoshiya. 1991. Dexmedetomidine prevents epinephrine-induced arrhythmias through stimulation of central α_2 adrenoceptors in halothane-anesthetized dogs. Anesthesiology **73:** 113-117.

26. HAYASHI, Y., T. KAMIBAYASHI, M. MAZE, A. YAMATODANI, K. SUMIKAWA, M. KURO & I. YOSHIYA. 1993. Role of imidazole-preferring receptors in the genesis of epinephrine-induced arrhythmias in halothane-anesthetized dogs. Anesthesiology **78:** 524-530.
27. KAMIBAYASHI, T., MAMMOTO, T., HAYASHI, Y., SUMIKAWA, K., YAMATODANI, A. & I. YOSHIYA. 1993. A role of imidazoline preferring receptors in antiarrhythmic effects of dexmedetomidine on halothane-epinephrine arrhythmias in dogs. Anesthesiology **79:** A684.
28. HAYASHI, Y., K. SUMIKAWA, A. YAMATODANI, T. KAMIBAYASHI, M. KURO & I. YOSHIYA. 1991. Myocardial epinephrine sensitization with subanesthetic concentration of halothane in dogs. Anesthesiology **74:** 134-137.
29. NOHTA, H., A. MITSUI & Y. OHKURA. 1987. Spectrofluorimetric determination of catecholamines with 1,2-diphenylethlenediamine. Anal. Chim. Acta **165:** 171-175.
30. VIRTANEN, R., J.-M. SAVOLA & V. SAANO. 1989. Highly selective and specific antagonism of central and peripheral α_2-adrenoceptors by atipamezole. Arch. Int. Pharmacodyn. Ther. **297:** 190-204.
31. CLINESCHMIDT, B. V., D. J. PETTIBONE, V. J. LOTTI, H. B. HUCKER, B. M. SWEENEY, D. R. REISS, E. D. LIS, J. R. HUFF & J. VACCA. 1988. A peripherally acting alpha-2 adrenoceptor antagonist: L-659,066. J. Pharmacol. Exp. Ther. **245:** 32-40.
32. DOXEY, J. C., A. G. ROACH & C. F. C. SMITH. 1983. Studies on RX 781094: A selective, potent and specific antagonism of $alpha_2$-adrenoceptors. Br. J. Pharmacol. **78:** 489-505.
33. DOZE, V. A., B.-X. CHEN & M. MAZE. 1989. Dexmedetomidine produces a hypnotic-anesthetic action in rats via activation of central $alpha_2$ adrenoceptors. Anesthesiology **71:** 171-175.
34. ERNSBERGER, P., R. GIULIANO, N. WILLETTE & D. J. REIS. 1990. Role of imidazole receptors in the vasodepressor response to clonidine analogs in the rostral ventrolateral medulla. J. Pharmacol. Exp. Ther. **253:** 408-418.
35. KAMISAKI, Y., Y. ISHIKAWA, H. TAKAO, H. OMODANI, N. KUNO & T. ITOH. 1990. Binding of [^{3}H] *p*-aminoclonidine to two sites: Distribution of imidazoline binding sites in rat brain. Brain Res. **514:** 15-21.
36. TIBRICA, E., J. FELDMAN, C. MERMET, L. MONASSIER, F. GONON & P. BOUSQUET. 1991. Selectivity of rilmenidine for the nucleus reticularis lateralis, a ventrolateral medullary structure containing imidazoline-preferring receptors. Eur. J. Pharmacol. **209:** 213-221.
37. MAIESE, K., L. PEK, S. B. BERGER & D. J. REIS. 1992. Reduction in focal cerebral ischemia by agents acting at imidazole receptors. J. Cereb. Blood Flow Metab. **12:** 53-63.
38. POTTER, D. & M. J. OGIDIGBEN. 1991. Medetomidine-induced alterations of intraocular pressure and contraction of the nictitating membrane. Invest. Ophthalmol. Visual Sci. **32:** 2799-2805.
39. ROSS, C. A., D. A. RUGGIERO & D. S. REIS. 1985. Projections from the nucleus tractus solitarii to the rostral ventrolateral medulla. J. Comp. Neurol. **242:** 511-534.
40. WAXMAN, M. B., A. D. SHARMA, J. ASTA, D. A. CAMERON & R. W. WALD. 1989. The protective effect of vagus nerve stimulation on catecholamine-halothane-induced ventricular fibrillation in dogs. Can. J. Physiol. Pharmacol. **67:** 801-809.

Antiarrhythmic Effect of the Selective I_1-Imidazoline Receptor Modulator Moxonidine on Ouabain-Induced Cardiac Arrhythmia in Guinea Pigs

H.-J. MEST, P. THOMSEN, AND A. RAAP

Department of Pharmacology
Beiersdorf-Lilly GmbH
Wiesingerweg 25
20253 Hamburg, Germany

Enhanced susceptibility to arrhythmia under enhanced sympathetic tone is associated with an alteration in electrophysiological parameters expressed by increased pacemaker activity, increased asynchronism in the recovery of excitability, a shortened refractory period, and a lowered diastolic threshold.[1] Enhanced sympathetic tone can be responsible for disturbances in cardiac rhythm under specific circumstances. The use of drugs to reduce sympathetic stimulation or blocking adrenoceptors is well known antiarrhythmic treatment. Beta-receptor blocking agents are established as so-called "class II" antiarrhythmics, as defined by Vaughan-Williams. The relevance of α_1-, α_2-, and imidazoline receptors in cardiac arrhythmia is still under discussion. The reduction in centrally mediated sympathetic tone induces antiarrhythmic effects. Central modulation of imidazoline receptors seems responsible for the reduced sympathetic tone demonstrated by moxonidine.[2] It remains to be seen if the centrally stimulated α_2-adrenoceptors (e.g., by clonidine) are relevant in the reduction of sympathetic tone. On the other hand, the existence of α_1- and α_2-adrenoceptors in the sinoatrial node of the rat had been identified by Dashwood and Spyer.[3] Alpha$_1$-adrenoceptors can mediate chronotropic responses in the heart,[4,5] and α_2-adrenoceptors can modulate the response to α_1-adrenoceptor activation.[6]

Recent investigations have shown that the α_1-adrenergic system is included in arrhythmogenesis especially in the ischemic heart[7,8] and in ventricular tachycardia.[9] The relevance of ischemia-induced changes in α_1-receptor density in connection with arrhythmia and inotropy is controversial.[10–13] Participation of imidazoline receptors in the central regulation of sympathetic tone is relevant in the blood pressure lowering effect of imidazoline receptor modulators such as moxonidine or clonidine. The possible existence of imidazoline receptors in cardiac tissue and their relevance in heart function are still unknown. The present study investigates the influence of moxonidine-induced attenuation of sympathetic tone on experimentally induced cardiac arrhythmia and the influence of different imidazoline receptor-modulating substances and α_1- and α_2-antagonists on heart rhythm.

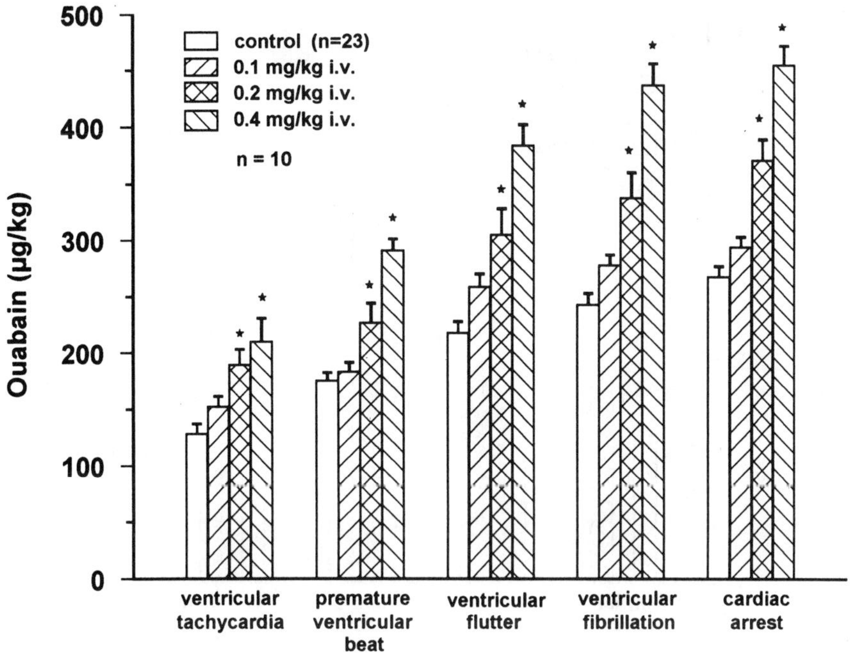

FIGURE 1. Moxonidine in a dose ranging between 0.1 and 0.4 mg/kg induced an antiarrhythmic effect on ouabain-induced cardiac arrhythmia in the guinea pig. The threshold doses of ouabain needed to induce a different kind of cardiac arrhythmia were enhanced in a dose-dependent manner (*$p < 0.05$).

MATERIAL AND METHODS

Ouabain-Induced Cardiac Arrhythmia in the Guinea Pig

Guinea pigs weighing 300-450 g were anesthetized with urethane (1.4 g/kg ip). After recording a control ECG, test substances were given intravenously 2 minutes before a continuous infusion of ouabain (10 µg/kg/min) in the right jugular vein. The ECG was recorded continuously until the death of the animal using an Einthoven II lead. The threshold dose of ouabain needed to induce ventricular tachycardia (VT), premature ventricular beats (PVB), ventricular flutter (VF), ventricular fibrillation (FIB), and asystole (A) was determined and expressed in µg/kg. The amount of ouabain infused was calculated against the infusion time. A decrease in the threshold dose of ouabain needed to induce a different cardiac arrhythmia indicated an arrhythmogenic effect and an increase in the threshold dose indicated an antiarrhythmic effect.

Aconitine-Induced Arrhythmia in Auricle Preparation

Guinea pigs weighing 300-430 g were sacrificed by cervical dislocation, and the heart was rapidly removed. Auricles were prepared in an organ bath with Tyrode's

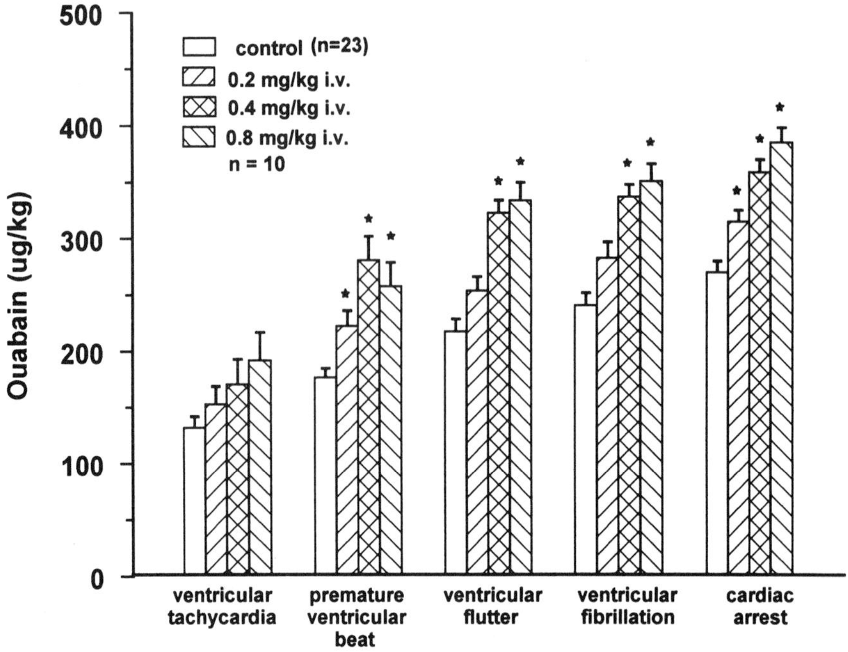

FIGURE 2. Clonidine in a dose ranging between 0.2 and 0.8 mg/kg induced an antiarrhythmic effect on ouabain-induced cardiac arrhythmia in the guinea pig (*$p < 0.05$).

solution. The auricle was dissected and mounted in an organ bath containing modified Tyrode's solution composed of (mM): NaCl (119.8), Kcl (5.4), $CaCl_2$ (1.8), $MgCl_2$ (1.05), $NaHCO_3$ (22.6), NaH_2PO_4 (0.42), ascorbinic acid (0.28), Na_2 EDTA (0.05), and glucose (5.0). The bath solution (35 °C) was continuously oxygenated (with 95% O_2, 5% CO_2) and kept at pH 7.4. Spontaneous beating of the auricle was measured with a Statham transducer (Gould, Cleveland, Ohio), exerting an initial tension of 1.0 g. The rate and force of the beating auricle were recorded by a linearcorder (Watanabe, Tokyo, Japan). Premature beats were registered when the electrical signal came 30% earlier than the normal distance between two beats. Cardiac arrhythmia was induced by 100 nM of aconitine. The number of premature beats, increase in frequency, and time until the start of premature beats were registered. Test substances were given 20 minutes before aconitine application.

Substances

The following substances were used: Ouabain (Merck, Darmstadt, Germany), aconitine (Sigma, Deisenhofen, Germany), efaroxan (Sigma, Deisenhofen, Germany), idazoxan (a gift of Reckitt & Colman, Hull, UK), SKF 86466A (a gift of Smith Kline Beecham Pharmaceuticals, King of Prussia, PA), propranolol (synthesized by

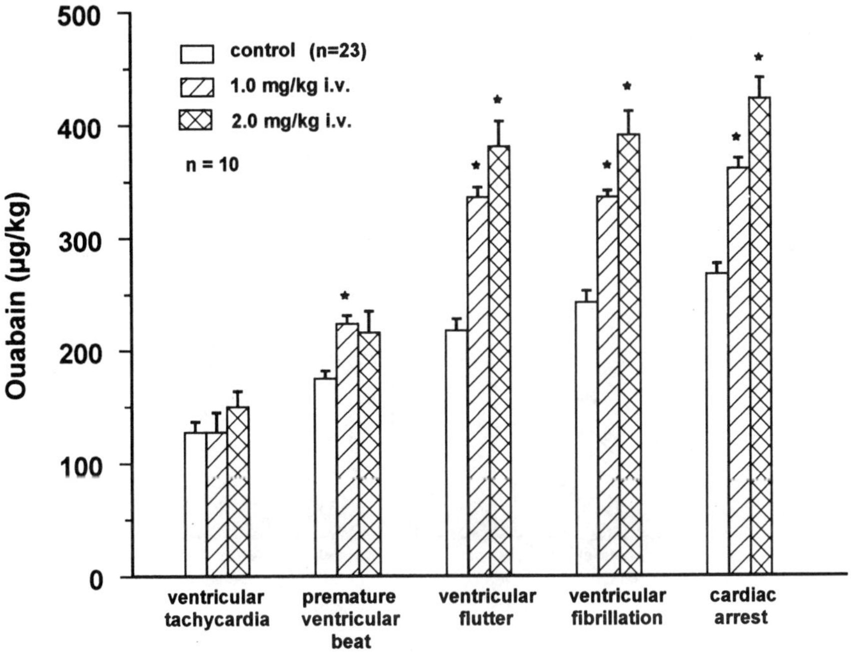

FIGURE 3. The β-blocking agent propranolol demonstrated an antiarrhythmic effect on this method (*$p < 0.05$).

Beiersdorf-Lilly GmbH, Hamburg, Germany), and clonidine (synthesized by Beiersdorf-Lilly GmbH, Hamburg, Germany).

Statistics

Data are expressed as the mean ± SEM and analyzed by unpaired Student's *t* test with Bonferoni correction. Differences were considered significant at $p < 0.05$.

RESULTS

Effect of Moxonidine and Clonidine on Ouabain-Induced Cardiac Arrhythmia

FIGURES 1 and 2 demonstrate the antiarrhythmic effects of moxonidine and clonidine in a dose-dependent manner. The antiarrhythmic effect is expressed as an enhanced threshold dose of ouabain necessary to induce different disturbances in cardiac rhythm. Standard antiarrhythmic propranolol in an approximately 10-fold higher dosage increased the ouabain threshold dose of all kinds of arrhythmias in the same manner (FIG. 3). The moxonidine-induced increase in threshold dose within

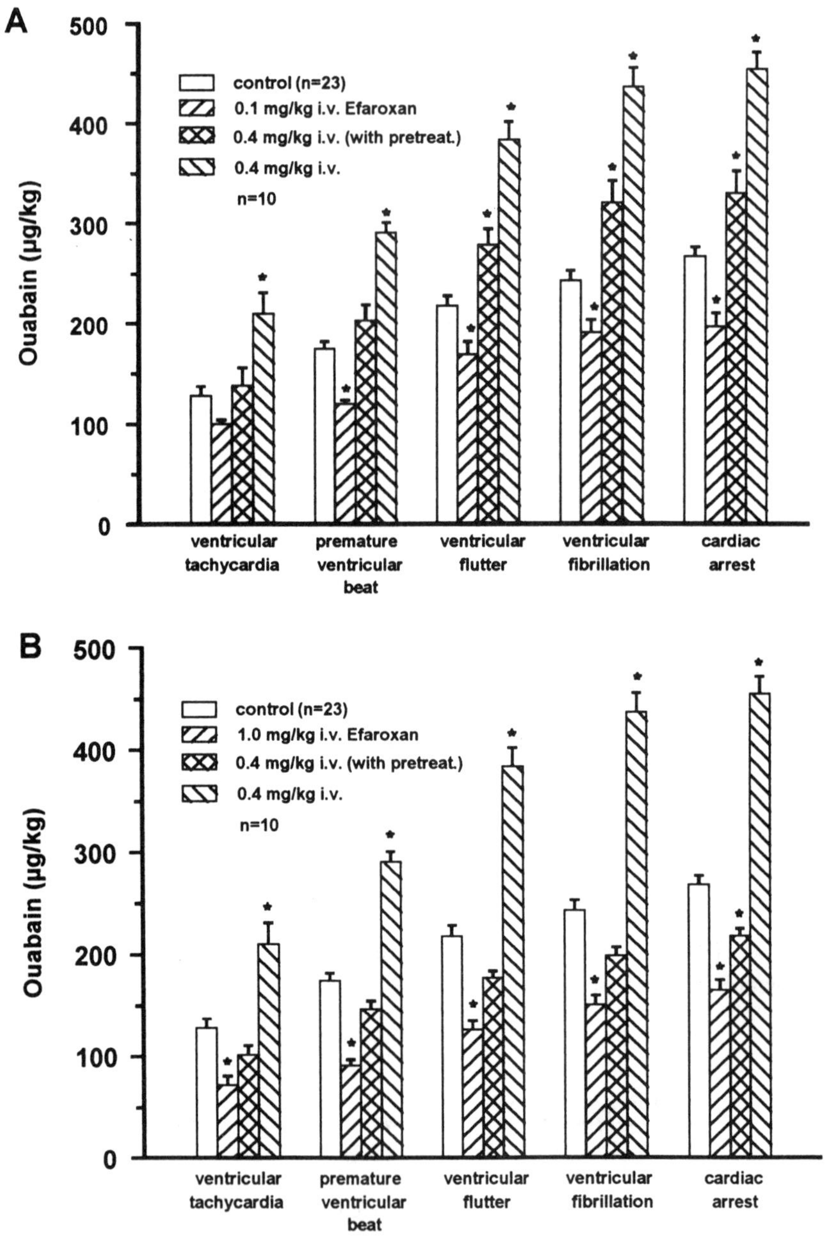

FIGURE 4. The antiarrhythmic effect of moxonidine is reduced by the imidazoline receptor antagonist efaroxan. Efaroxan, in a dose of 0.1 mg/kg (**A**) and 1.0 mg/kg (**B**), alone demonstrates an arrhythmogenic effect on ouabain-induced cardiac arrhythmia in the guinea pig ($*p < 0.05$).

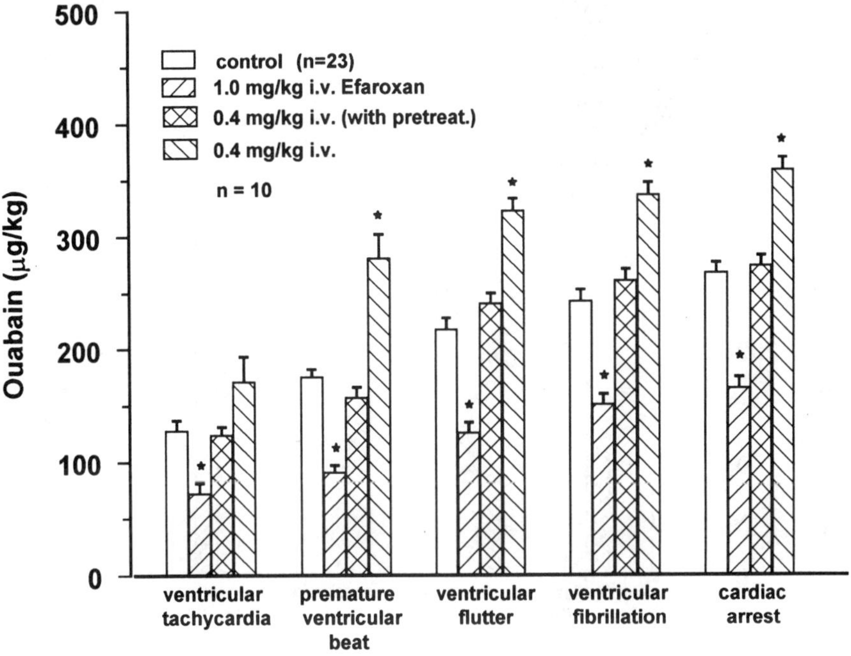

FIGURE 5. Efaroxan (pretreatment, 1.0 mg/kg iv) abolished the antiarrhythmic effect of clonidine in this method too (*$p < 0.05$).

PVB, VF, and FIB was between 66% and 80%. A double dose of clonidine increased the ouabain threshold dose between 44% and 53%. Propranolol induced an increase in the ouabain threshold dose by between 23% and 75%.

Influence of the I_1-Modulator Efaroxan

Efaroxan in a dose of 0.1 and 1 mg/kg iv had a strong arrhythmogenic effect expressed by a maximum reduction in threshold dose within PVB, VF, and FIB of between 38% and 48%. The effect of moxonidine as well as clondine was abolished by efaroxan. Efaroxan in a dose of 1 mg/kg reduced the moxonidine effect between 49% and 54%, and compared to control values the threshold dose of ouabain was also reduced between 16% and 19% (FIG. 4). The clonidine effect was reduced between 26% and 39%, and in comparison with control values the enhanced threshold dose induced by clonidine was equalized by efaroxan (FIG. 5.)

Influence of the I_1-Modulator and α_2-Adrenergic Receptor Antagonist Idazoxan

Idazoxan reduced the ouabain threshold dose to the same extent as did efaroxan (between 36% and 45%). The moxonidine effect was absolutely abolished by idazoxan

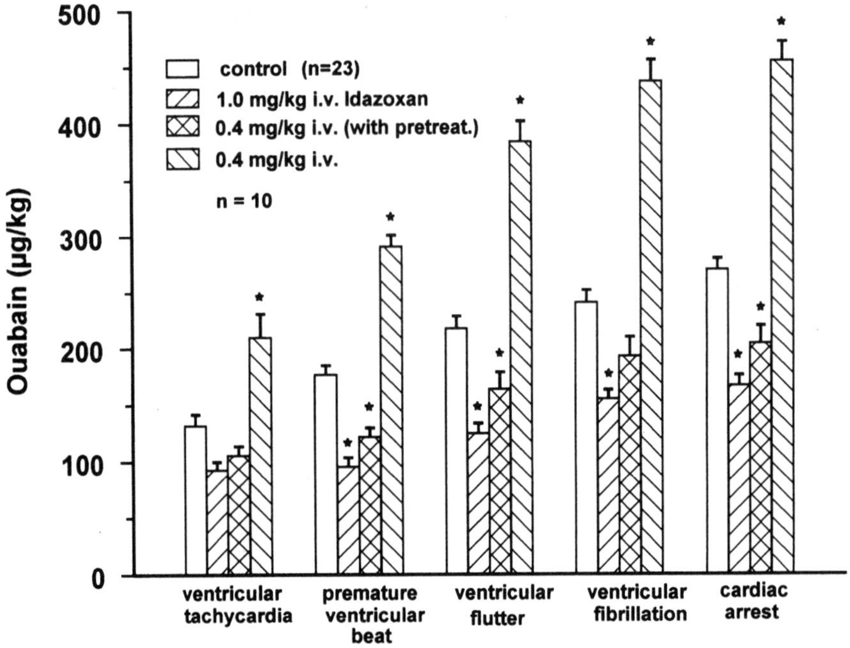

FIGURE 6. The imidazoline and α_2-adrenoceptor antagonist idazoxan (pretreatment, 1.0 mg/kg iv) reduced the antiarrhythmic effect of moxonidine on ouabain-induced arrhythmia in the guinea pig. Idazoxan alone had an arrhythmogenic effect (*$p < 0.05$).

and the threshold doses dropped between 21% and 30% under the control level. The moxonidine effect was reduced by half by idazoxan (FIG. 6). The effect of clonidine was also abolished by idazoxan, and the threshold dose of ouabain corresponds to the control level (FIG. 7).

Influence of the α_2-Adrenergic Receptor Antagonist SKF 86466

SKF 86466 reduced the threshold dose of ouabain necessary to induce cardiac arrhythmias between 27% and 36%. The enhanced threshold dose by moxonidine was nearly completely inhibited by SKF 86466 (FIG. 8). On the other hand, the clonidine effect on the ouabain threshold dose was not influenced by SKF 86466 (FIG. 9).

Influence of the α_1-Adrenergic Receptor Antagonist Prazosin

Prazosin in a dose of 1 mg/kg iv had no influence on the ouabain threshold dose. The moxonidine-induced increase in the threshold dose was partially reduced (FIG. 10). The clonidine effect was not changed by prazosin (FIG. 11).

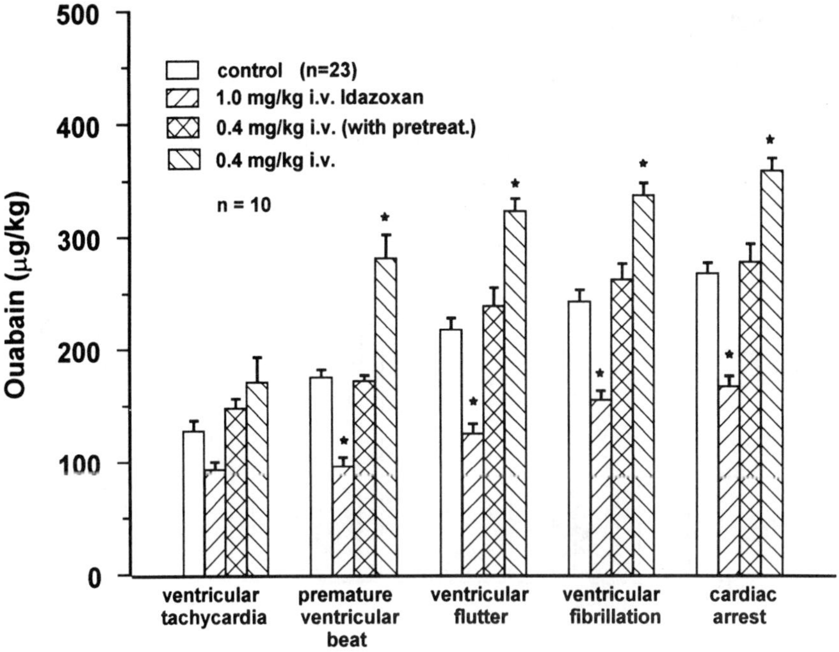

FIGURE 7. Idazoxan (pretreatment, 1.0 mg/kg iv) inhibited the antiarrhythmic effect of clonidine on ouabain-induced cardiac arrhythmia in the guinea pig (*$p < 0.05$).

Effect of Moxonidine on Aconitine-Induced Arrhythmia in the Guinea Pig Auricle

Moxonidine in a concentration of 10 and 100 nM reduced the number of extrasystoles induced by aconitine in an auricle preparation. The 1 μM concentration of moxonidine showed no statistically significant antiarrhythmic effect (FIG. 12).

DISCUSSION

The reduction in centrally mediated sympathetic tone by moxonidine or clonidine induced antiarrhythmic effects on ouabain-induced cardiac arrhythmias in guinea pigs, because the sympathetic nervous system can be involved in the pathogenesis of disturbances in cardiac rhythm. Adrenergic receptor blockade has been beneficial in cardiac arrhythmias; β-blocking agents especially were effective as "class II" antiarrhythmics according to Vaughan-Williams. The role of postsynaptic α_1- and α_2-adrenoceptors for cardiac rhythm is still under discussion. On the other hand, it is more logical to reduce sympathetic tone by centrally active substances. The antiarrhythmic effect of clonidine seemed to be induced by stimulation of presynaptic α_2-adrenergic receptors, reducing sympathetic tone. Moxonidine was 20-fold less

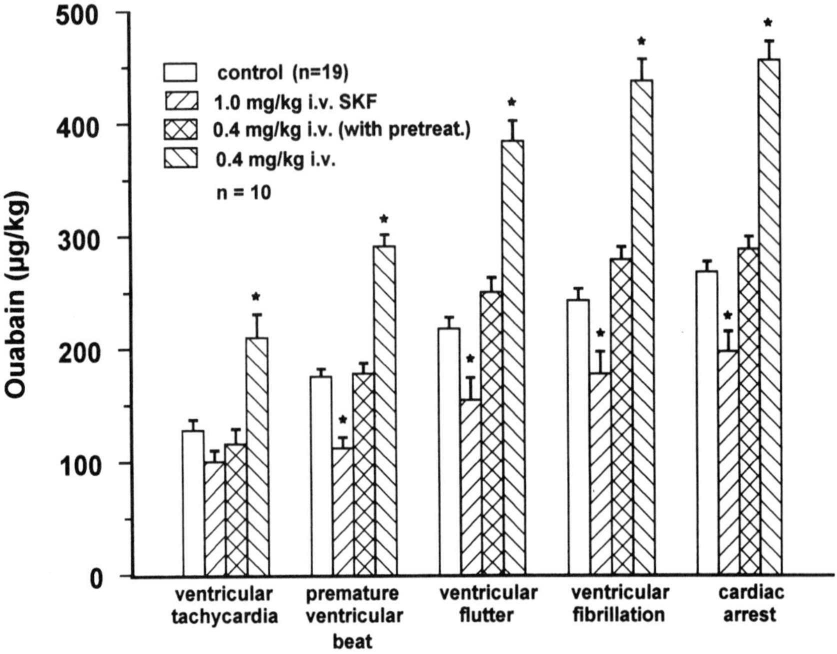

FIGURE 8. The α_2-adrenoceptor antagonist SKF 86466-A (pretreatment, 1.0 mg/kg iv) also induced an arrhythmogenic effect. The antiarrhythmic effect of moxonidine was reduced by SKF 86466 on ouabain-induced arrhythmia in the guinea pig (*$p < 0.05$).

potent on α_2-adrenergic receptors compared to clonidine,[14] but the antiarrhythmic effect of moxonidine was significantly more pronounced than was that of clonidine. Recent evidence suggests that a novel receptor may be involved in centrally mediated sympathetic tone.[15–17] The first direct evidence for the nonadrenergic action of imidazolines was reported by Bousquet *et al.*,[18,19] who used microinjection to administer clonidine directly into their site of action in the ventrolateral medulla. Cirazoline, an imidazoline that is an α_2-adrenergic antagonist, showed clonidine-like vasodepressor action. These findings suggest that the vasodepressor action of clonidine was mediated by a novel receptor specific for imidazolines.[20,21]

The antiarrhythmic effect of moxonidine and clonidine is induced by a reduction in sympathetic tone mediated by modulation of imidazoline receptors. I_1-imidazoline receptors could be responsible for centrally mediated effects.[2,14] The role of α_2-adrenoceptors in the antiarrhythmic effect of moxonidine and clonidine is still unknown. Efaroxan, an antagonist to moxonidine and clonidine concerning to I_1-imidazoline receptors, abolished the antiarrhythmic effects of both drugs. Efaroxan alone in a dose of 1 mg/kg reduced the threshold dose of ouabain needed to induce cardiac arrhythmias demonstrating an arrhythmogenic effect. The selectivity ratio of efaroxan to imidazoline receptors versus α_2-adrenoceptors is approximately 50:1.[2] Idazoxan is an antagonist to imidazoline and α_2-adrenoceptors. Idazoxan induced similar effects

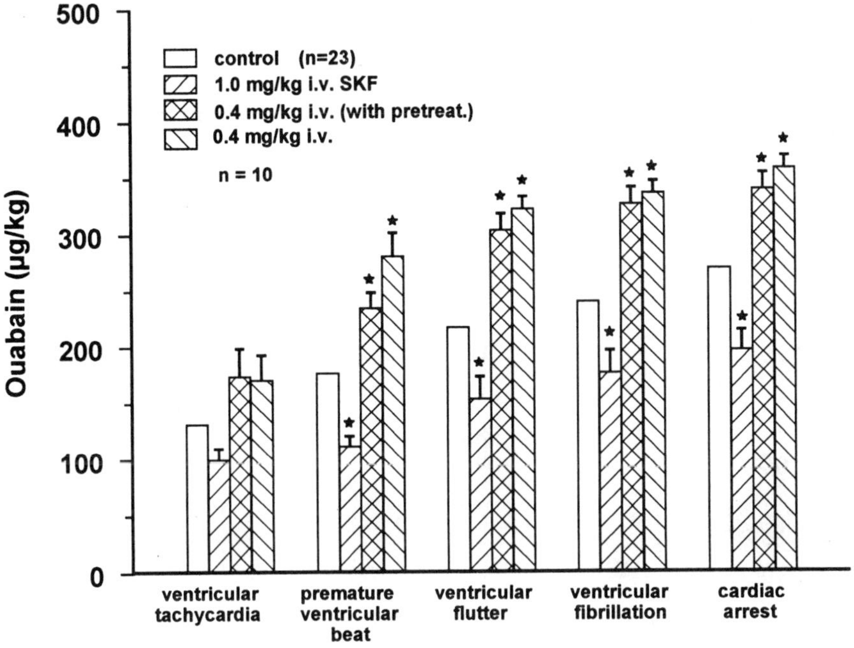

FIGURE 9. SKF 86466-A (pretreatment, 1.0 mg/kg iv) did not influence the antiarrhythmic effect of clonidine in this method (*$p < 0.05$).

as did efaroxan. On the other hand, SKF 86466 is about three times more selective for α_2-receptors than for imidazoline receptors. SKF 86466 could also reduce the threshold dose of ouabain, signaling an arrhythmogenic effect. The antiarrhythmic effect of moxonidine was reduced by SKF 86466, whereas the effect of clonidine was unchanged by pretreatment with SKF 86466. The antiarrhythmic effect of moxonidine and clonidine could be mediated by modulation of imidazoline receptors. In this case, clonidine should be the most potent substance, but moxonidine was more potent but the affinity to imidazoline receptors is about half this. In this respect the role of pre- and postsynaptic α_1- and α_2-adrenoceptors is still unknown. Clonidine has 20-fold greater binding to α_2-adrenoceptors as does moxonidine. The relevance of α_2-binding for the antiarrhythmic effect seems very low, because the antiarrhythmic effect of clonidine was smaller than that of moxonidine and the clonidine effect was unchanged by an α_2-antagonist. On the other hand, the role of peripheral imidazoline receptors as well as α_1- and α_2-adrenoceptors in cardiac tissue is also unknown.[22] The antiarrhythmic effect of moxonidine on aconitine-induced arrhythmia in auricle preparation demonstrated a peripheral mediated effect. In this method the antiarrhythmic effect of moxonidine disappeared in a higher dose like classic antiarrhythmics, but moxonidine induced no arrhythmogenic effect. It is unknown which receptors are responsible for this effect. Moxonidine is also involved in activation of peripheral vascular postsynaptic α_1- and α_2-adrenoceptors.[23–25] The possible role of α-adrenocep-

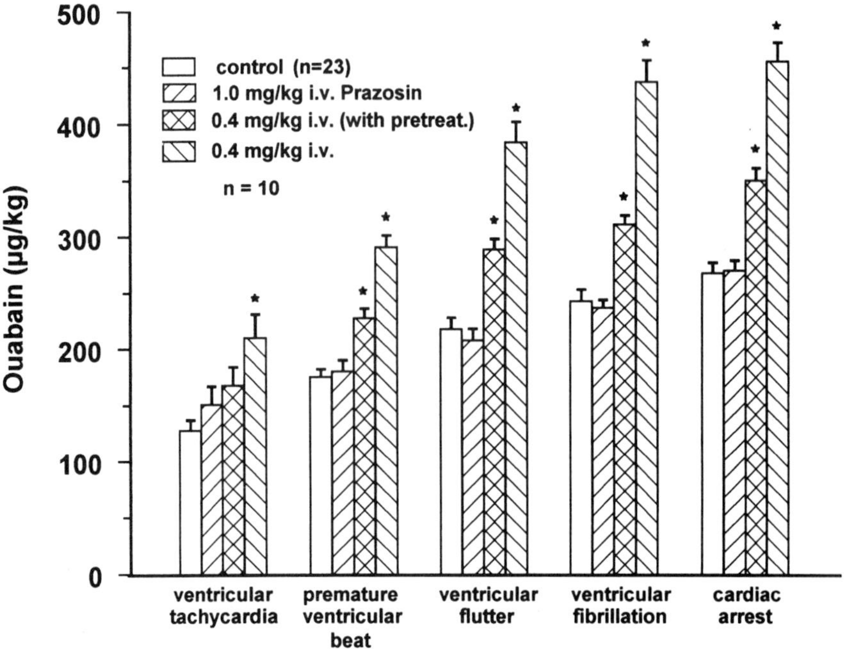

FIGURE 10. The α_1-blocking agent prazosin had no influence on the threshold dose of ouabain needed to induce cardiac arrhythmias. Prazosin reduced the antiarrhythmic effect of moxonidine ($*p < 0.05$).

tors was just discussed. An unspecific effect cannot exclude from possible mechanism. There are grounds for believing that there are further mechanisms responsible for the antiarrhythmic effect of moxonidine and clonidine. It is well accepted that the antiarrhythmic effect of both drugs is at least partly mediated by a reduction in sympathetic tone via central action. Imidazoline-preferring rather than α_2-adrenergic receptors appear responsible for the antiarrhythmic property of moxonidine and clonidine.[26] The antiarrhythmic effect of moxonidine was also seen in ischemia- and reperfusion-induced cardiac arrhythmias.[27] Independent of the mode of action, the antiarrhythmic effect of moxonidine is an additional beneficial effect of this drug.

SUMMARY

Moxonidine is a centrally acting antihypertensive agent with potent action on I_1-imidazoline receptors. Moxonidine as an SIR modulator elicits a persistent reduction in circulating levels of epinephrine, demonstrating a reduction in sympathetic tone.

In the first experiment the threshold dose of ouabain needed to induce ventricular arrhythmia and asystole was determined in guinea pigs, and the influence of moxonidine was tested. In a dose range of 0.1–0.4 mg/kg body weight iv, moxonidine

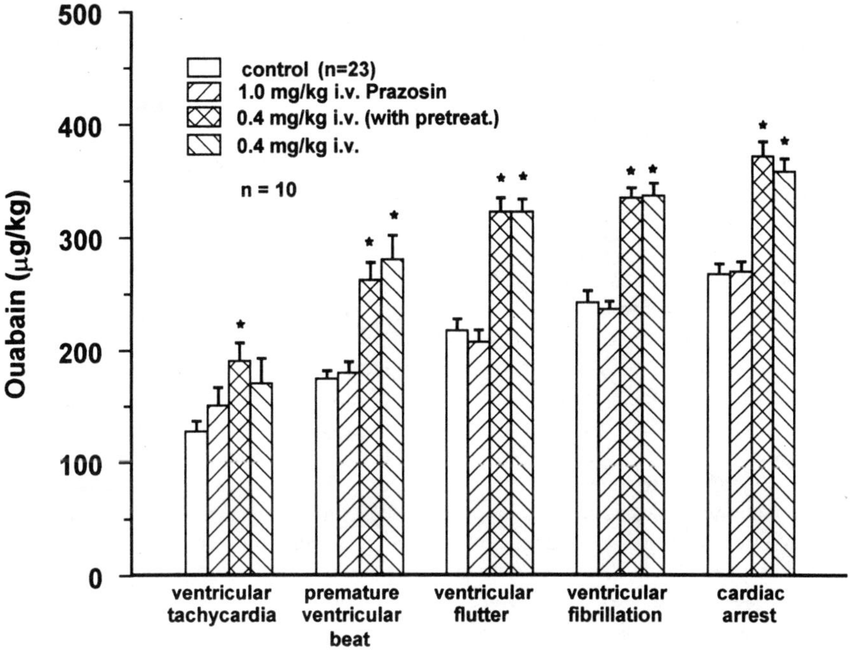

FIGURE 11. The antiarrhythmic effect of clonidine was unchanged by pretreatment with prazosin (1.0 mg/kg iv) on ouabain-induced arrhythmia in the guinea pig (*$p < 0.05$).

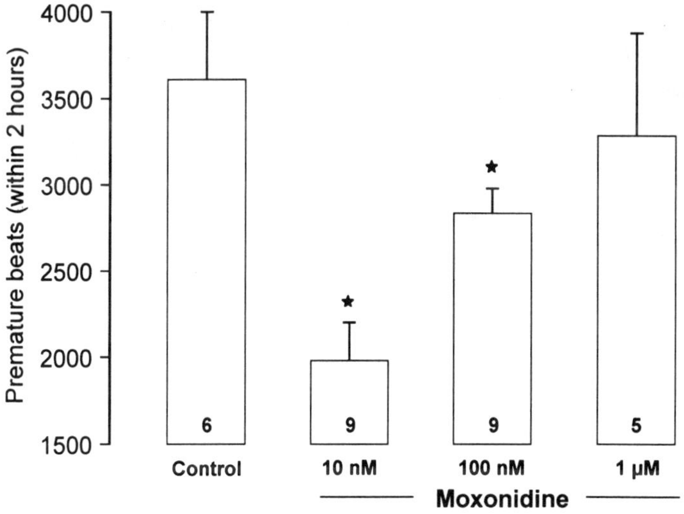

FIGURE 12. Moxonidine given in a small dose inhibited extrasystoles induced by aconitine in an auricle preparation of the guinea pig (*$p < 0.05$).

increased the threshold dose needed to induce ventricular tachycardia, premature ventricular beats, ventricular flutter, ventricular fibrillation, and asystole. The effect was dose-dependent and statistically significant. Clonidine, in a dose range of 0.2-0.8 mg/kg body weight iv, also increased the threshold dose of ouabain necessary to induce different cardiac rhythm disturbances. Moxonidine was more effective than clonidine. Pretreatment with the α_2-receptor and I_1-receptor-influencing substances efaroxan, idazoxan, and SKF 86466 attenuated the effect of moxonidine and clonidine. Efaroxan, idazoxan, or SKF 86466 alone reduced the threshold dose of ouabain necessary to induce cardiac arrhythmia as a sign for arrhythmogenic effects. The α_1-receptor antagonist prazosin had no influence on ouabain-induced arrhythmia. Pretreatment with prazosin reduced the moxonidine but not the clonidine effect. In the second experiment the influence of moxonidine on aconitine-induced extrasystoles (ES) in the spontaneously beating guinea pig auricle was investigated. Moxonidine in a dose of 10^{-7}-10^{-8} M reduced the number of ES. A 10-fold higher dose had no influence on ES number. The beta-blocking agent propranolol showed antiarrhythmic effects in both methods. The ouabain-induced cardiac arrhythmia is associated with increased sympathetic tone on central stimulation. The reduced sympathetic tone by centrally acting moxonidine via imidazoline receptors seems responsible for the antiarrhythmic effect of this drug. Clonidine also reduced the sympathetic tone via imidazoline receptors. The selectivity of clonidine to imidazoline receptors is less pronounced than is that of moxonidine. The interaction of moxonidine with imidazoline receptors is not clear. The possible interaction between imidazoline and α-adrenoceptors in relation to the antiarrhythmic effect of moxonidine or clonidine is also unknown. Modulation of imidazoline receptors by moxonidine could be an agonistic effect or an antagonism to an endogenous agonistic or antagonistic substance and vice versa.

REFERENCES

1. SZEKERES, L. & GY. J. PAPP. 1971. Experimental cardiac arrhythmias and antiarrhythmic drugs. Akadémiai Kiaó. Budapest.
2. ERNSBERGER, P., H. L. ELLIOTT, H.-J. WEIMANN, A. RAAP, M. A. HAXHIU, E. HOFFERBER, A. LÖW-KRÖGER, J. L. REID & H.-J. MEST. 1993. Moxonidine: A second-generation central antihypertensive agent. Cardiovasc. Drug Rev. **11:** 411-431.
3. DASHWOOD, M. R. & K. M. SPYER. 1986. Autoradiographic localisation of α-adrenoceptors, muscarinic acetylcholine receptors and opiate receptors in the heart. Eur. J. Pharmacol. **127:** 279-282.
4. FLAVAHAN, N. A. & J. C. MCGRATH. 1981. α_1-adrenoceptors can mediate chronotropic responses in the rat heart. Br. J. Pharmacol. **73:** 586-588.
5. SCHOLZ, H., R. BRÜCKNER, A. MÜGGE & C. REUPCKE. 1986. Myocardial α-adrenoceptors and positive inotropy. J. Cell Cardiol. **18:** 79-87.
6. RAND, M. J. & L. H. TUNG. 1987. Cardiac α_2-adrenoceptors. *In* Cardiovascular Pharmacology '87. Results, Concepts and Perspectives. J. Gy Papp, Ed.: 575-581. Akadémiai Kiadó. Budapest.
7. KURZ, T., K. A. YAMADA, S. D. DATORRE & P. B. CORR. 1991. Alpha$_1$-adrenergic system and arrhythmias in ischemic heart disease. Eur. Heart J. **12:** (Suppl.): 88-98.
8. MOLINA-VIAMONTE, V., E. P. ANYUKHOVSKY & M. R. ROSEN. 1991. An alpha$_1$-adrenergic receptor subtype is responsible for delayed after depolarizations and triggered activity during simulated ischemia and reperfusion of isolated canine Purkinje fibers. Circulation **84:** 1732-1740.

9. HASUMI, M., M. HIROE, N. FUJITA, S. HASUMI, T. TOYOSAKI, H. NAGAO, S. HOSODA, T. KASAJIMA & M. SEKIGUCHI. 1990. Significant myocardial pathology and increase of α_1-adrenergic receptor number affecting the arrhythmogenic condition in cases with ventricular tachycardia. Heart Vessels **5** (Suppl): 31-36.
10. CORR, P. B., G. P. HEATHERS & K. A. YAMADA. 1989. Mechanisms contributing to the arrhythmogenic influence of α_1-adrenergic stimulation in the ischemic heart. Am. J. Med. **87:** 19-25.
11. MANNING, A. S., R. CROME, N. KLEIN, D. J. HEARSE & M. E. MAGUIRE. Dissociation between reperfusion induced arrhythmias and increases in ventricular α_1-receptor density in the anaesthetized rat. Cardiovasc. Res. **23:** 852-858.
12. BRÜCKNER, R., A. MÜGGE & H. SCHOLZ. 1985. Existence and functional role of α_1-adrenoceptors in the mammalian heart. J. Mol. Cell Cardiol. **17:** 639-645.
13. BENEEY, B. G. 1993. Functions of myocardial alpha-adrenoceptors. J. Autonom. Pharmacol. **13:** 351-372.
14. ERNSBERGER, P., T. H. DAMON, L. M. GRAFF, S. G. SCHÄFER & M. O. CHRISTEN. 1993. Moxonidine, a centrally acting antihypertensive agent, is a selective ligand for I_1-imidazoline sites. J. Pharmacol. Exp. Ther. **264:** 172-182.
15. KARPPANEN, H. 1981. Interrelationships between clonidine and histaminergic mechanisms. Trends Pharmacol. Sci. **2:** 35-37.
16. KARPPANEN, H., I. PAAKKARI & P. PAAKKARI. 1977. Further evidence for central histamine H_2-receptor involvement in the hypotensive effect of clonidine in the rat. Eur. J. Pharmacol. **42:** 299-302.
17. KARPPANEN, H., I. PAAKKARI, P. PAAKKARI, R. HUOTARI & A. L. ORMA. 1976. Possible involvement of central histamine H_2-receptors in the hypotensive effect of clonidine. Nature **259:** 587-588.
18. BOUSQUET, P., J. FELDMAN, R. BLOCH & J. SCHWARTZ. 1984. Central effects of alpha-adrenergic drugs: Differences between catecholamines and imidazolines. J. Pharmacol. Exp. Ther. **230:** 232-236.
19. BOUSQUET, P. & J. SCHWARTZ. 1983. Alpha-adrenergic drugs: Pharmacological tools for the study of the central nervous system. Biochem. Pharmacol. **32:** 1459-1465.
20. HEAD, G. A. & F. SANNAJUST. 1992. Importance of imidazoline receptors in the cardiovascular responses to clonidine and rilmenidine in conscious rabbits. Fundam. Clin. Pharmacol. **6** (Suppl. 1): 31-34.
21. REIS, D. J., S. REGUNATHAN, H. WANG, D. L. FEINSTEIN & M. P. MEELEY. 1992. Imidazoline receptors in the nervous system. Fundam. Clin. Pharmacol. **6:** (Suppl. 1): 23-29.
22. FUDER, H. & P. SCHWARZ. 1993. Desensitization of inhibitory prejunctional alpha-2-adrenoceptors and putative imidazoline receptors on rabbit heart sympathetic nerves. Naunyn-Schmiedeberg's Arch. Pharmacol. **348:** 127-133.
23. ARMAH, B. I. 1988. Unique presynaptic $alpha_2$-receptor selectivity and specificity of the antihypertensive agent moxonidine. Arzneim. Forsch. **38:** 1435-1442.
24. DOCHERTY, J. R., J. W. CONSTANTINE & K. STARKE. 1981. Smoothe muscle of rabbit aorta contains alpha-1 but not alpha-2 adrenoceptors. Naunyn-Schmiedeberg's Arch. Pharmacol. **317:** 5-7.
25. TOWART, R. J. 1982. A new method for investigating the isometric contractile responses of small blood vessels. Br. J. Pharmacol. **75:** 105P.
26. HAYASHI, Y., T. KAMIBAYASHI, M. MAZE, A. YAMATODANI, K. SUMIKAWA, M. KURO & I. YOSHIYA. 1993. Role of imidazoline-preferring receptors in the genesis of epinephrine-induced arrhythmias in halothane-anesthetized dogs. Anesthesiology. **78:** 524-530.
27. LEPRAN, I. & J. GY. PAPP. 1994. Effect of moxonidine on arrhythmias induced by coronary artery occlusion and reperfusion. J. Cardiovasc. Pharmacol. In press.

Rilmenidine and the High Risk Hypertensive Patient

JOHN H. LARAGH

The New York Hospital-Cornell Medical Center
New York, New York 10021

That the central nervous system is a major player in the pathophysiology of hypertension has always been a prime consideration in the diagnosis and treatment of hypertension, even though many of the processes that fueled and invoked its participation were, and still remain, mysteries. Early neuropharmacological approaches to treatment depended on ganglionic blockers, but their radical side effects made their use appropriate only for the most seriously ill patients. No drug strategy against the central nervous system was benign enough to respond adequately to the increasing evidence that treatment could reduce blood pressure and lower its morbid and mortal consequences in patients with milder forms of hypertension.

I remember my excitement when rauwolfia and reserpine were introduced. Their actions, believed due to depletion of catecholamines and 5-hydroxytryptamine stores in the brain, restrained peripheral vascular resistance, and, if administered in large doses, could bridle the heart rate, myocardial contractile function, and cardiac output. However, that excitement was considerably calmed when it was found that even though the side effects were milder than those of the ganglionic blockers, these alkaloids still induced enough sedation, depression, and blunted mentation to severely limit their use.

Enthusiasm returned with the advent of methyldopa, clonidine, and its analogs. These α_2-agonists, by quieting brain sympathetic outflow, appeared to offer a more selective strategy. Encouragingly, their action slowed the heart rate, reduced peripheral resistance, and markedly suppressed plasma renin activity. But discouragingly, these positive effects were countered by such negative ones as dry mouth, depression, impaired mentation, and a palpably blunted facial expression. Helpful in the short term, they were not appropriate for long-term treatment of the alert and vigorous people who make up the largest portion of hypertensive patients and whose lives, in quality and duration, could be measurably enhanced by more benign therapy.

In fact, reasonably kindly alternatives were already in the market. These were the new beta blockers, ACE inhibitors, alpha blockers, and calcium antagonists. Properly employed, they promised finer discrimination of specific physiologic dysfunctions that stoke the hypertensions: excessive fluid volume accumulation and excessive renin-angiotensin secretion. They also added the virtue of marksmanship: they appeared to address specific, primary locations in the causal pathways. At this point in time they are the mainstay in antihypertensive treatment.

Nevertheless, the search for a relatively benign, selective, and effective means of controlling sympathetic outflow continued, for it could not be ignored that sympathetic outflow is involved in most (and perhaps all) hypertensive states. The search has been fruitful. Certain areas in the brain stem contain recently identified noradrenergic receptors different from the traditional α_2-receptors.[1,2] Avidly sensitive to the imidazoline portion of molecular structures, even more so than the α_2-receptors, they may not actually be engaged by methyldopa or clonidine.

Moreover, pharmacologic science, rapidly reaching beyond theory to fact, has already developed relatively inert compounds rich in imidazoline-like structures (IPRs). The newest of these is rilmenidine. By usurping the functions of norepinephrine, the natural cerebral transmitter for sympathetic nervous system outflow, these imidazoline-rich compounds can tame sympathetic outflow, restrain heart rate and vasomotor influences, and reduce the negative quality-of-life effects induced by earlier agents acting on the CNS. The possibilities opened by this development are important not only for extending the treatment options available in the clinic, but also for introducing new research options in the laboratory, where we still struggle for greater understanding of the mechanisms of hypertension.

It should be remembered, however, that there is no single cause of hypertension and no single, all-encompassing treatment. We made that mistake long ago with diuretics. The last couple of decades have revealed the interactions of many factors in the hypertensive cascade; we are now better equipped to determine primary causes and to address them with specific treatments. And we have learned that the selection of treatment strategy in individual patients is determined not only by attention to primary cause but also by the degree of disability and by weighing the risks of hypertension against those posed by treatment.

THE LESSONS OF MALIGNANT HYPERTENSION

Thirty-five years ago malignant hypertension was a major cause of sickness and death among hypertensive patients, but it is now rather rare thanks to what we have learned since. Indeed, its very extremity together with its swift onset and conclusion made it a revealing accelerative model for studying the hypertensive mechanisms that for other patients developed more slowly and obscurely. Of particular interest to us were the dramatic extremities of endocrine derangements in this deadly illness (FIG. 1).

Thus, in 1960, studying 30 patients with malignant hypertension, we were able to demonstrate that every one of them had massive oversecretion of aldosterone as a direct consequence of renin and angiotensin release by damaged kidneys.[3–5] Twelve years later we showed that a beta-adrenergic blocking drug produced a dramatically steep decline in renin secretion with remission of the hypertension.[6,7] Continuing study showed that beta blockers maintain a renin level lowered by at least 75% as long as they are given.[8,9] So reliable is this effect that at our Center we identify patients who are taking beta blockers by simply doing a renin assay; renin levels in such patients are always very suppressed.

What is the value of so reducing the renin level? Dialysis experience reveals that when both kidneys, the source of plasma renin, are removed in patients with malignant hypertension, arteritis heals, blood pressure returns to normal, and the viciousness of the disease virtually evaporates.[10] This has also been demonstrated in animal models. In normal levels the vasoconstrictive physiologic assignment of the hormone renin supports life by supporting the circulation; in abnormal levels, providing exces-

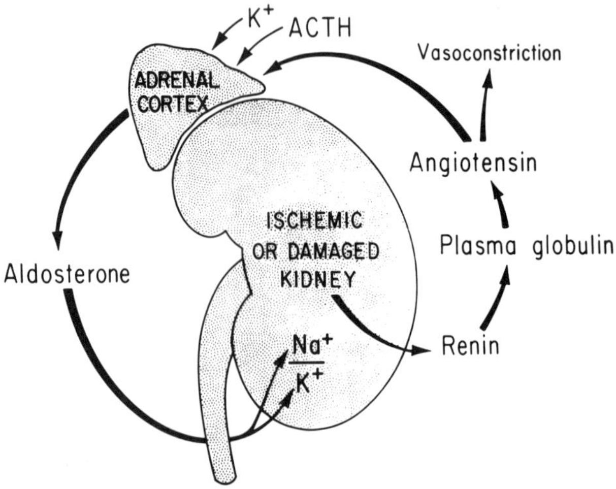

FIGURE 1. The renal-adrenal axis in malignant hypertension. The renal-adrenal axis is a hormonal cascade involving renal renin secretion, angiotensin formation, and aldosterone release for regulation of sodium and potassium balance and blood pressure. The interaction is depicted as it was first discovered in studies of patients with malignant hypertension in whom, because of defective feedback, it is involved in causation. Accordingly, there is massive excess of both renin and aldosterone that cannot turn itself off, that is, the vicious cycle or ''vascular crisis'' of malignant hypertension. Antirenin system drugs or binephrectomy corrects the disorder. The sympathetic nervous system is also a major factor in amplifying renin secretion.

sive vasoconstriction, renin can be destructive. Renin is not the only culprit in this regard; catecholamine excess in pheochromocytoma, insulin deficiency, and parathyroid hormone excess, all can produce severe vascular disease in the heart, brain, and kidneys. Indeed, whenever a cause for hypertensive or normotensive vascular disease has been found, it has proved to be hormonal or humoral.

Clearly, severe hypertension is associated with marked deformation of the renin system and, presumably, the automatic nervous system, with great overactivity of CNS outflow. It is also clear by now that the physiopathic relationships revealed in the study of malignant hypertension are also seen, but with lesser intensity, in milder forms of hypertension.[11]

HIGH RISK HYPERTENSIVE PATIENTS

There is good evidence that plasma renin excess causes human vascular disease manifesting heart attack, stroke or kidney failure. In 1972 we completed a large 8-year study of untreated patients divided according to low, medium, and high plasma values assessed by new, greatly improved assay methods. Significantly, no drugs were available at that time to contaminate the study. We found that even though the low-renin patients had the highest blood pressures and were older by 10 years than their medium- or high-renin counterparts, none of them had a heart attack or stroke *despite their more severe hypertension.* However, among the medium- and high-renin patients there was an 11-14% incidence of stroke, heart attack, and death.[11] This convinced us that an overactive renin system was a large risk factor in vascular disease.

Renin-induced damage is visible. In a study by Giese,[12] a normal rat mesenteric artery 3 hours after a mildly depressor infusion of angiotensin II clearly showed stenotic beading of the arterial tree with poststenotic dilatation, a classic feature of angiotensin. After an injection of India ink, which blinds the endothelium and perfectly defines its normal architecture, massive diapedeses, disruption of the entire architecture of the arteriole, and the beginning of ischemic injury can be observed.

A study by Gavras *et al.*[13] demonstrates the advent of an infarcted area in a rabbit myocardium after a 3-hour mildly pressor angiotensin infusion. Giacomelli's group[14] provided an electron microscopic view of the myocardium of rats given a pressor infusion of angiotensin II for 3 hours. Vacuoles and blood formation were be observed in the arterioles, and condensation of chromatin and bleb formation were noticeable in three of the myofibrils. The investigators concluded that these angiotensin-induced changes were exactly like those of human myocardial infarction in incipiency.

In 1991 we published an 8-year prospective trial of 1,717 patients with mild hypertension who were treated with the then traditional stepped-care regimen.[15] FIGURE 2 shows that a high plasma renin value enhances the cardiac risk imposed by hyperlipidemia, diabetes, or tobacco use more than threefold. But more significantly, in the large subgroup developing heart attacks in the absence of the aforementioned traditional risk factors, plasma renin was an *even more potent predictor of risk:* absolutely no heart attacks occurred in 241 consecutive such low-renin patients, whereas the high-renin subjects exhibited a more than sevenfold increase in the heart attack rate.

Another persuasive demonstration of renin involvement begins in Tobian's studies in stroke-prone hypertensive rats, which develop a lacunar ischemic infarct similar to that seen in the most common form of human stroke. He showed that very few strokes occurred in the first 12 weeks of life if the rats were fed a normal diet, but when fed a high-salt diet, almost all were dead of stroke by 12 weeks. However, he learned that when potassium is included in the diet, impressive protection from stroke occurs even though blood pressure is not reduced.[16–18] Following this evidence and knowing that dietary potassium intake can markedly suppress renin levels, we repeated Tobian's study, assaying plasma renin on each rat at weeks 6, 8, and 12. We found that those destined to die of stroke had enormously higher pre-stroke renin levels. Verifying Tobian's observations we showed that dietary potassium markedly suppressed renin levels and protected the animals.[19]

At our center, Drs. James and Pickering, using ambulatory blood pressure monitoring to study the 24-hour variations in blood pressure, found that in untreated patients the higher the renin levels the higher was the variability in blood pressures.[20] Others have shown that risk in hypertension is associated with fluctuations and flings in blood pressure and that reduction of risk is associated with damped blood pressure. The blood pressure of low-renin patients is generally damped.

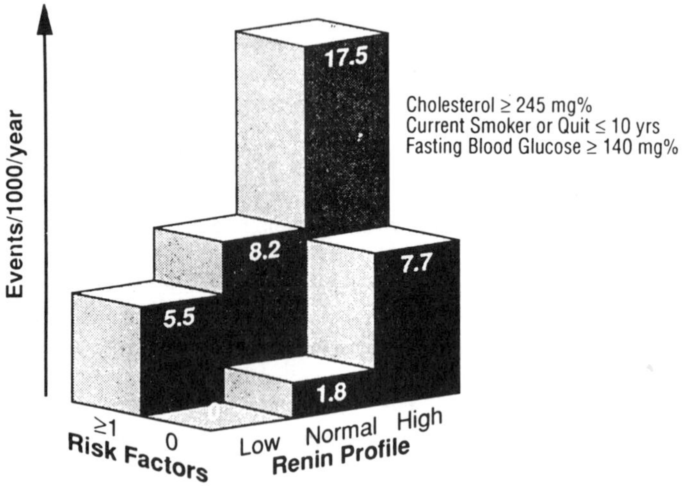

FIGURE 2. The Alderman study of 1,700 patients, redrawn. Plot of the incidence of myocardial infarction related to the baseline plasma renin level among 1,717 patients with mild hypertension followed for 8 years. The back row presents those 997 patients who had one or more known risk factors (cholesterol, smoking, or diabetes). The front row presents the heart attack rates for those patients who had no such risk factors. The plasma renin level appears as a powerful, continuous, and most pervasive risk factor, multiplying the heart attack rate more than threefold among those who had other risk factors and more than sevenfold among those without the other risk factors. No heart attacks occurred in 241 such low-renin patients, suggesting that the renin level is a continuous variable. These data are further supported by animal studies describing myocardial, cerebral, and renin lesions after injection of renin or angiotensin. Redrawn from Alderman *et al.*[15] (These data extend the previous report of Brunner, Laragh *et al.* 1972. N. Engl. J. Med. **286:** 441-449.)

RENIN AND THE NERVOUS SYSTEM

As I and my colleagues at The New York Hospital-Cornell Medical Center try to understand the regulation of renal renin secretion, we are increasingly convinced from evidence just noted that the nervous system plays a major role. Jon Blumenfeld of our group has shown that renin secretion is almost entirely shut off when the nervous system is inhibited by spinal anesthesia in patients undergoing surgery.[21] What is more, under these conditions kidney renin secretion cannot be increased by stimuli that normally increase its production.

Certain commonly used antihypertensive medications, such as the beta blockers, work by blocking autonomic nervous signals to the kidneys. When the sympathetic nervous system is reacting normally, adrenergic blockade effectively lowers renin secretion. But although beta blockers are effective in some patients, they are less so, or not effective at all, in others in whom sympathetic controls are malfunctioning. This information is useful in identifying patients without effective neural drive who cannot be treated with drugs that block neural signals to the kidney.

DEALING WITH RISK IN HYPERTENSION

Up to about 4 or 5 years ago the medical community, acting on the evidence furnished by the historic VA study showing that blood pressure alone can and does reduce risk, has dealt with risk in hypertension by employing measures, sometimes draconian, designed to lower blood pressure by any means. However, we now know that not all hypertensions are alike in either their pathophysiologic origins or degree of risk. An energetic pharmacological community has provided us with several classes of drugs targeted to specific causes and known risks. ACE inhibitors and beta blockers have significant and consistent effects, curtailing plasma renin activity–ACE inhibitors by blocking angiotensin II formation and beta blockers by markedly reducing the renal secretion of renin. Acting preponderantly on the opposing limb of the pressure equation–the sodium-volume factor–are the calcium antagonists, diuretics, and alpha blockers.

If the clinical experience with rilmenidine continues its current positive performance, it may very well turn out to have virtually universal applications in the treatment of hypertension. An effective centrally acting drug, it can enhance natriuresis, thus addressing the sodium-volume component, and it can simultaneously suppress the humoral factors, including renal renin secretion, that work the vasoconstrictive side of the equation. The neurohumoral domain over which imidazoline-preferring receptors preside also has jurisdiction over heart rate as well as the beta and neural pathways that modulate vasomotor tone.

Unique, possibly global, antihypertensive potentials are suggested by such centrally acting drugs. Reduction of central sympathetic outflow could produce balanced inhibition of alpha- and beta-adrenergic tone. Reduced peripheral resistance is produced by reducing alpha tone, which in turn reduces plasma renin and at the same time unloads the vasculature and relieves cardiac work by vasodilating the arterial tree. Reduced beta tone turns off renin secretion, thereby further vasodilating the arterial tree; it also reduces cardiac contractility and rate, which in turn reduces shear stress on the major arterial vessels and thus the likelihood of vascular injury over the long term. With cardiac work reduced, protection is provided against myocardial ischemia and infarction (beta blockers are so far the only form of therapy for which myocardial infarction protection has been convincingly demonstrated). Additional benefits of lowered beta tone are an antiarrhythmic effect, an increased fibrillation threshold from beta blockade, a protection from arrhythmias associated with catecholamine surge, and also an antiarrhythmic facilitation of vagal tone.

This new clonidine-displacing substance and others like it are bringing us to the point where specific loci in the brain can be addressed selectively to produce a constellation of desirable responses not achievable by any other single agent. If the experience bears up and if the drowsiness and mental sedation that accompany the older, less selective centrally acting drugs can be bypassed or dose-adjusted to tolerable levels, we may be very near to being able to mount a central assault against hypertension and its effects with a single agent.

Future research with this drug type or its clinical use should be accompanied, I believe, by monitoring of plasma renin and catecholamine levels, two simple measurements that can tell the researcher or clinician whether or not central sympathetic outflow is reduced and for how long.

REFERENCES

1. REIS, D. J., D. A. RUGGIERO & S. F. MORRISON. 1989. The C1 area of the rostroventrolateral medulla oblongata. A critical brainstem region for control of resting and reflex integration of arterial pressure. Am. J. Hypertens. **2:** 368S-374S.
2. BOUSQUET, P., J. FELDMANN, E. TIBIRICA *et al.* 1992. Imidazoline receptors: A new concept in central regulation of the arterial blood pressure. Am. J. Hypertens. **5:** 47S-50S.
3. LARAGH, J. H., S. ULICK, V. JANUSZEWICZ, O. B. DEMING, W. G. KELLY & S. LIEBERMAN. 1960. Aldosterone secretion and primary and malignant hypertension. J. Clin. Invest. **39:** 1091-1106.
4. LARAGH, J. H., M. ANGERS, W. G. KELLY & S. LIEBERMAN. 1960. Hypotensive agents and pressor substances. The effect of epinephrine, norepinephrine, angiotensin II and others on the secretory rate of aldosterone in man. JAMA **174:** 234-240.
5. LARAGH, J. H. 1960. The role of aldosterone in man: Evidence for regulation of electrolyte balance and arterial pressure by renal-adrenal system which may be involved in malignant hypertension. JAMA **174:** 293-295.
6. BÜHLER, F. R., J. H. LARAGH, L. BAER, E. D. VAUGHAN, JR. & H. R. BRUNNER. 1972. Propranolol inhibition of renin secretion. A specific approach to diagnosis and treatment of renin-dependent hypertensive diseases. N. Engl. J. Med. **287:** 1209-1214.
7. LARAGH, J. H. 1976. Modern system for treating high blood pressure based on renin profiling and vasoconstriction-volume analysis: A primary role for beta blocking drugs such as propranolol. Am. J. Med. **61:** 797-810.
8. BÜHLER, F. R., J. H. LARAGH, E. D. VAUGHAN, JR., H. R. BRUNNER, H. GAVRAS & L. BAER. 1973. Antihypertensive action of propranolol. Specific antirenin responses in high and normal renin forms of essential, renal, renovascular and malignant hypertension. Am. J. Cardiol. **32:** 511-522.
9. MAN IN'T VELD, A. J. & A. H. VAN DEN MEIRACKER. 1990. Effects of antihypertensive drugs on cardiovascular hemodynamics *In* J. H. Laragh & B. M. Brenner, (Eds.): 2117-2130. Raven Press, Ltd. New York.
10. LARAGH, J. H., L. BAER, H. R. BRUNNER, F. R. BÜHLER, J. E. SEALEY & E. D. VAUGHAN, JR. 1972. Renin, angiotensin and aldosterone system in pathogenesis and management of hypertensive vascular disease. Am. J. Med. **52:** 633-652.
11. BRUNNER, H. R., J. H. LARAGH, L. BAER *et al.* 1972. Essential hypertension: Renin and aldosterone, heart attack and stroke. N. Engl. J. Med. **286:** 441-449.
12. GIESE, J. 1973. Renin, angiotensin and hypertensive vascular damage: A review. Am. Heart. J. **55:** 315-332.
13. GAVRAS, H., J. J. BROWN, A. F. LEVER, R. F. MACADAM & J. I. ROBERTSON. 1971. Acute renal failure, tubular necrosis and myocardial infarction induced in the rabbit by intravenous angiotensin II. Lancet **2:** 19.
14. GIACOMELLI, F., P. ANVERSA & J. WEINER. 1976. Effect of angiotensin-induced hypertension on rat coronary arteries and myocardium. Am. J. Pathol. **84:** 111-138.
15. ALDERMAN, M. H., S. MADHAVAN, W. L. OOI, H. COHEN, J. E. SEALEY & J. H. LARAGH. 1991. Association of the renin-sodium profile with the risk of myocardial infarction in patients with hypertension. N. Engl. J. Med. **324:** 1098-1104.
16. TOBIAN, L. J., J. LANG, K. ULM, L. WOLD & J. IWAI. 1985. Potassium reduces cerebral hemorrhage and death rate in hypertensive rats, even when blood pressure is lowered. Hypertension **7**(Suppl I): I110-I114.
17. SEALEY, J. E., I. CLARK, M. B. BULL & J. H. LARAGH. 1970. Potassium balance and the control of renin secretion. J. Clin. Invest. **49:** 2119-2127.
18. BRUNNER, H. R., L. BAER, J. E. SEALEY, J. G. LEDINGHAM & J. H. LARAGH. 1970. The influence of potassium administration and of potassium deprivation on plasma renin in normal and hypertensive subjects. J. Clin. Invest. **49:** 2128-2138.
19. VOLPE, M., M. J. F. CAMARGO, F. B. MUELLER *et al.* 1990. Relation of plasma renin to end-organ damage and to protection of K^+ feeding in stroke-prone hypertensive rats. Hypertension **15:** 318-326.

20. JAMES, G. D., T. G. PICKERING & J. H. LARAGH. 1991. Ambulatory blood pressure variation is related to plasma renin activity in borderline hypertensive men. Am. J. Hypertens. **4:** 525-528.
21. ZAYAS, V. M., J. D. BLUMENFELD, B. BADING *et al.* 1993. Adrenergic regulation of renin secretion and renal hemodynamics during deliberate hypotension in man. Am. J. Physiol. **265:** F686-F692.

I_1 Agents: A New Approach to the Treatment of Hypertension

M. DUBAR[a] AND G. PILLION

Institut de Recherches Internationales Servier
6 place des Pléiades
92415 Courbevoie Cedex, France

The recent observation, that "there may be some form of structure-activity relationship which would indicate the existence, in this particular region of the medulla oblongata, of sites preferring the imidazoline structure," published only 10 years ago,[1] led to the definition of a new receptor system whose therapeutic applications are of growing importance. Since then, fundamental and applied clinical research has contributed to the establishment of a new generation of antihypertensive agents that act selectively on I_1-imidazoline receptors. Rilmenidine represents the prototypic "I_1 agent." Its high selectivity for I_1 receptors (see further) means that this drug has been extensively studied by pharmacologists and clinicians to demonstrate the validity of the imidazoline concept and the reliability of this original therapeutic approach.

From all existing possibilities for treating essential hypertension, no single pharmacological class of agents meets the needs of the heterogeneous population of hypertensive patients, especially those with concomitant cardiovascular complications or coexisting disease. Moreover, failure of current pharmacologic therapy to consistently lower the risks of cardiovascular and renal morbidity and mortality has dampened research into new strategies of treatment. The rationale, the mechanism of antihypertensive action, and the therapeutic profile of I_1 agents are viewed hereafter through extended accumulated experience with rilmenidine.

THE I_1-IMIDAZOLINERGIC SYSTEM

Evidence from both binding and functional studies at the central nervous system and peripheral levels could give the notion that the "I_1-imidazolinergic" system is particularly involved in the regulation of cardiovascular function. This notion is important to consider in pharmacology in which a description of a specific link between a receptor system and a defined function is often a prerequisite for new therapeutic approaches benefiting from optimal targeting and less undesirable effects. Because the α_2-adrenergic system influences many central functions such as vigilance, glucose metabolism, and behavior, the advent of agents selective for I_1 receptors offers a more specific approach to lowering blood pressure with less interference

[a] Address for correspondence: Michel Dubar, I.R.I.S., 6, place des Pléiades, 92415 Courbevoie Cedex, France.

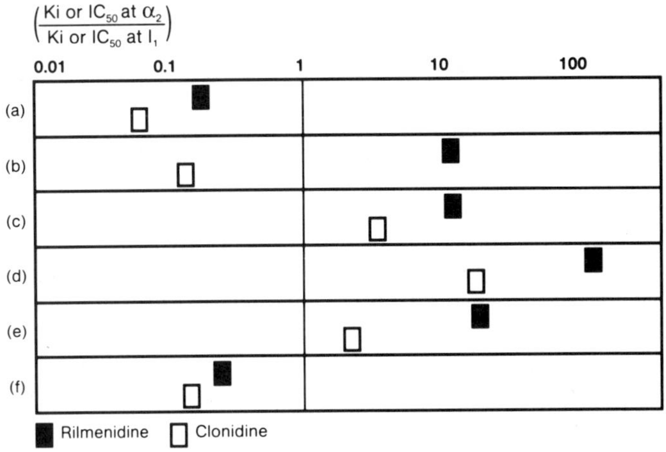

FIGURE 1. Comparison of selectivity ratios for I_1- versus α_2-adrenergic sites between clonidine and rilmenidine. Data represent the ratio of affinity constants (K_i or IC_{50}) at α_2- versus I_1-binding sites as obtained by different groups. Differences in experimental binding conditions, ligands, tissues, and species employed can largely account for differences in the respective affinities observed between groups. From: (a) Bricca *et al.*[5]; (b) Bricca *et al.*[6]; (c) Gomez *et al.*[7]; (d) Hamilton *et al.*[8]; (e) Ernsberger *et al.*[9]; and (f) De Vos *et al.*[10]

from other physiological functions, and it therefore represents real progress in antihypertensive therapy.

Central Nervous System Level

The existence of imidazoline-preferring binding sites, classified later into I_1 subgroup, was first reported in the rostro-ventrolateral medulla (RVLM) of bovine brain by Reis *et al.*[2] and was later reported in human brain in the corresponding region (nucleus reticularis lateralis, NRL) by Bousquet *et al.*[3] Since that time, many experiments have been carried out using not only different ligands and binding conditions but also membranes prepared from different areas of various species. (For a review, see ref. 4.) Perhaps the most common feature of all these binding studies is the constant observation that clonidine and rilmenidine have a high affinity for these sites. Overall, rilmenidine has proven to be the most selective drug for I_1-binding sites relative to α_2-adrenergic sites (FIG. 1 and refs. 5-10). By contrast, numerous studies have clearly established that rilmenidine demonstrates a low affinity for the I_2 subtype of imidazoline binding sites.[11–13]

Some important additional information was gained recently by the presentation of brain autoradiography of I_1 sites using [^{3}H]clonidine in humans, enabling a comparison of regional distributions of α_2-adrenergic and idazoxan binding sites.[10] The following conclusions have been drawn from these experiments: (1) I_1 and idazoxan sites are distinct molecular entities, although possibly associated physically in some regions (i.e., NRL, striatum). This would explain the ability of idazoxan to displace the binding of [^{3}H]clonidine in the latter regions. (2) In certain brain areas, I_1- and α_2-adrenergic sites are detected separately, and a lack of correlation (r = 0.26) is reported between the two respective distributions. Consequently, these two sites do not seem to be associated in human brain. (3) Whereas α_2-adrenergic and I_1 sites were colocalized in animal preparations of RVLM,[5,14] the corresponding area of human brain is enriched in I_1 sites labeled by [^{3}H]clonidine.[6,10]

A high density of nonadrenergic clonidine sites[10] or rilmenidine sites[15] was found in brain areas that do not participate directly in cardiovascular control such as the cerebral cortex and striatum. However, the nature of the nonadrenergic binding sites revealed by the latter radiolabeled ligand is a matter of debate. Nevertheless, the functional role of the I_1 sites found outside the medulla oblongata is not clear and remains to be investigated.

To further elucidate the role of I_1 sites within the brain, three different techniques have been used successfully in *in vivo* studies: microinjection, electrochemical, and [^{14}C]2-deoxyglucose autoradiography techniques.

Microinjection Techniques. Rilmenidine was injected bilaterally into three areas in the medulla of anesthetized Sprague-Dawley rats.[7] Only the rilmenidine injection into the rostro-ventrolateral part of the medulla induced hypotension. The response was preferentially antagonized by idazoxan, an α_2-adrenergic antagonist with a high affinity for I_1 receptors, rather than by SKF 86466, an equipotent α_2-adrenergic antagonist with no affinity for I_1 receptors, both administered peripherally. By contrast, microinjection of rilmenidine was ineffective in the nucleus tractus solitarii (NTS), a site of action of α_2-adrenergic agonists, and in the caudal part of the ventrolateral medulla.

Later, Head *et al.*[16] extended these observations in anesthetized rabbits in which blood pressure and renal sympathetic nerve activity were recorded. Using microinjections, they confirmed that the RVLM but not the NTS (unpublished results) was involved in the sympathoinhibitory and hypotensive effects of rilmenidine. In further studies using intracisternal administration in conscious rabbits, they assessed the preferential antagonism of rilmenidine using idazoxan, indicating involvement of I_1 receptors, whereas hypotension induced by clonidine was more efficiently blocked by 2-methoxy idazoxan, an imidazoline demonstrating selective antagonism for α_2-adrenoceptors.[16] It is important to note that the doses of antagonists used in the later experiments were based on an equi-effective antagonism against the action of alpha-methyldopa, a catecholamine selective for α_2-adrenoceptors. These combined results indicate that the main action of rilmenidine is primarily localized in the RVLM where the resulting sympathoinhibition is mainly dependent on interaction with I_1 sites.

Electrochemical Techniques. Other *in vivo* experiments were performed in normotensive rats using differential voltammetry, an electrochemical technique. Cardiovascular parameters were studied along with neuronal metabolic activity of different areas by recording DOPAC, with a specific electrode probe, released from noradrenergic cell bodies. The effects of rilmenidine and azepexole, an α_2-adrenergic agonist, were compared in the NRL and locus coeruleus, a region rich in α_2 adrenoceptors.[17] Moderate doses of rilmenidine (0.3 and 1.5 mg/kg) given systemically decreased blood pressure by −15 ± 5% and −25 ± 7%, respectively. When the activity of noradrenergic neurons was recorded within the areas just mentioned, a dose-dependent reduction in catechol oxidation peak was found in the NRL, whereas doses 50 times higher were necessary to significantly inhibit neuronal activity in the locus coeruleus. Furthermore, azepexole showing low affinity for I_1 sites induced a comparable reduction in blood pressure, but hypotension was not accompanied by a DOPAC peak height decrease in the NRL.

Previous experiments[18] have shown that pretreatment with intracisternally injected idazoxan can prevent the hypotensive effect of intravenously administered clonidine as well as inhibit the metabolic activity of NRL neurons. This pretreatment did not prevent a reduction in the activity of locus coeruleus-catecholaminergic neurons induced by larger doses of clonidine. By contrast, yohimbine, a reference antagonist for α_2 adrenoceptors, could prevent the effects of clonidine only in the locus coeruleus region, the site responsible for the most frequent clinical side effect of clonidine, that is, sedation. The preferential antagonism by idazoxan of the effects of clonidine on blood pressure and NRL neuronal activity indicates that the mechanism of action occurs via I_1 receptors.

Using an alternative technique, Pineda *et al.*[19] recently demonstrated that rilmenidine (ED_{50} = 1.5 mg/kg) and clonidine (ED_{50} = 0.7 mg/kg) can even stimulate the firing rate of locus coeruleus noradrenergic neurons by a non-α_2 adrenoceptor mechanism that contrasted with inhibition of the activity of the same neurons by α_2 agonists.

To summarize, the hypotensive and sedative effects of clonidine or related drugs were shown to be separate both by their receptor mechanisms and by their sites of action. I_1 agents may act selectively within the NRL of the medulla oblongata to induce sympathoinhibition without interfering with α_2-adrenergic-mediated functions.

[^{14}C] 2-Deoxyglucose Autoradiography Technique. This study evaluated changes in local glucose utilization within the CNS of conscious spontaneously hypertensive rats induced by rilmenidine or BHT 933, a reference α_2 adrenoceptor agonist, at doses that produced comparable hypotension. Neuroanatomical changes in the metabolic activity of neuronal cells were determined using this powerful technique[20] to gain valuable information about the functional site of action of the centrally acting drugs, independent of their receptor mechanism. Autoradiography of the brain after rilmenidine treatment revealed significant reductions in local glucose utilization in limited areas of the brain stem and spinal cord that are normally associated with cardiovascular control (FIG. 2). None of the other brain regions was affected by rilmenidine, and no significant changes in local glucose utilization were present in BHT 933-treated rats. The reported mapping of α_2- and I_1-binding sites in brain coupled with the data just cited provide evidence for the involvement of I_1 receptors in the effects of rilmenidine on specialized areas in central cardiovascular control mechanisms, with no reported effects on other parts of the brain at hypotensive doses.[21]

The Peripheral Level

In the periphery, the search for I_1 binding sites has not been systematic in all organs. Only renal, adrenal, and platelet determinations of I_1 sites will be given attention here, because an attempt has been made to link these organs with a functional activity.

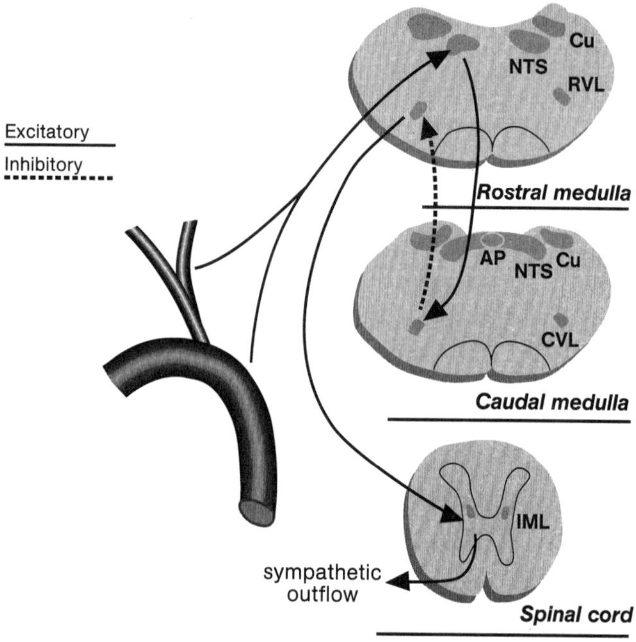

FIGURE 2. Schematic representation of brain areas where rilmenidine (1 mg/kg subcutaneously) induces a significant reduction in glucose utilization as assessed by the [^{14}C]2-deoxyglucose technique. The baroreceptor sympathetic reflex pathway is superimposed to indicate the close relations between rilmenidine-sensitive neurons and bulbospinal sympathetic areas associated with cardiovascular control. (From Browne and Macrae.[21]) Abbreviations: AP = area postrema; Cu = cuneate nucleus; CVL = caudo-ventrolateral reticular nucleus; IML = intermedio-lateral cell column; NTS = nucleus tractus solitarii; RVL = rostro-ventrolateral reticular nucleus.

In kidney, both binding (refs. 9 and 22 and Parini *et al.*, this volume) and autoradiography techniques[15,23] demonstrated the existence of a subpopulation of I_1- or nonadrenergic [^{3}H]rilmenidine-binding sites whose exact localization, however, is still a matter of debate (cortex versus medulla). A series of *in vivo* studies in uninephrectomized rats also suggested involvement of local I_1 receptors in eliciting diuresis consecutive to an increase in osmolar clearance.[24,25] The consequences of such modifications in renal function will be discussed, but it is noteworthy that α_2 agonists increase water clearance by inhibiting vasopressingergic effects in rat kidney and can mediate antinatriuresis in the human as previously reported.[26]

The presence of I_1 binding sites was shown in adrenal medullary cell membranes[9,27] but not in primary cultures of chromaffin cells using [^{3}H]*p*-aminoclonidine ligand.[28] Although the designation of these nonadrenergic imidazoline binding sites to the I_1-subtype described in brain must still be confirmed, nonadrenergic receptors have been proposed to inhibit evoked catecholamine release[29,30] in isolated adrenal cells and to increase the content of mRNA for PNMT[31] or to modulate second messenger systems[32] in chromaffin cell culture following clonidine application. Moreover, the adrenal medulla has a particularly high concentration of clonidine-displacing substance (CDS), the putative endogenous ligand for imidazoline receptors.[33,34] This observation suggests that an endogenous imidazolinergic system is normally important for this organ's function. Nevertheless, the exact role of I_1 receptors in adrenal glands has not yet clearly been established. Whatever the origin of the mechanism of action, centrally by sympathoinhibition or locally, it should be emphasized that imidazolines and related derivatives can decrease adrenal activity and hence plasma catecholamines, whereas other antihypertensive drugs are neutral (β-blockers, ACE inhibitors) or tend to increase them (α_1 blockers, calcium antagonists, and vasodilators).[35]

Imidazoline receptors are also reported in human platelets,[36–38] and both intracellular I_2-binding sites and I_1-sites on the plasma membrane have been described. Although the binding characteristics of platelet I_1 sites differ to some extent (GTP dependence, for instance) from those established in the brain, some nonadrenergic effects of imidazoline derivatives attributable to I_1 receptors have been reported. Of primary importance in the cardiovascular field are antiaggregatory properties to activated human blood platelets[39,40] or the inhibitory effect of certain imidazoline compounds on adenylate cyclase activity.[41] However, these results are inconclusive with regard to involvement on an imidazoline receptor mechanism. A parallel observation was that I_1 sites were higher on platelets in depressed patients.[37]

Finally, recent findings have revealed the existence of peripheral presynaptic nonadrenregic receptors that may inhibit the release of noradrenaline in nerve terminals in rabbit pulmonary artery,[42] in rabbit heart,[43] and in rat isolated kidney.[44] Again, whether these receptors belong to the I_1 subgroup remains to be established, because a distinct profile of activity of imidazoline compounds from centrally I_1-mediated effects is actually described.

EFFECTS OF I_1 AGENTS ON BLOOD PRESSURE CONTROL SYSTEMS

Determinants of Blood Pressure Control Over Time

Any deviation from a basal level of blood pressure is detected and then corrected by the interaction of the central and peripheral regulatory mechanisms. In Guyton's classic concept of regulation of arterial pressure, the circulation's regulatory mechanisms can roughly be divided into two successive temporal sessions to oppose a sustained change in blood pressure (FIG. 3).[45]

At first, a rapid phase is mainly dependent on baroreceptor mechanisms of various origins, including the classic sinoaortic baroreceptors as well as other nonarterial baroreceptors.[46] As the CNS can integrate a variety of stimuli to finely regulate blood pressure,[47] baroreceptor mechanisms are especially important for the moment to moment stabilization of blood pressure, but are subject to being reset with time to the prevailing level of blood pressure in the presence of a sustained stimulus.

The following period (long-term phase) consists of the participation of hormonal mechanisms (adrenaline, vasopressin, renin/angiotensin II, and aldosterone) minutes after pressure stimuli and the renal response governed by the pressure-natriuresis

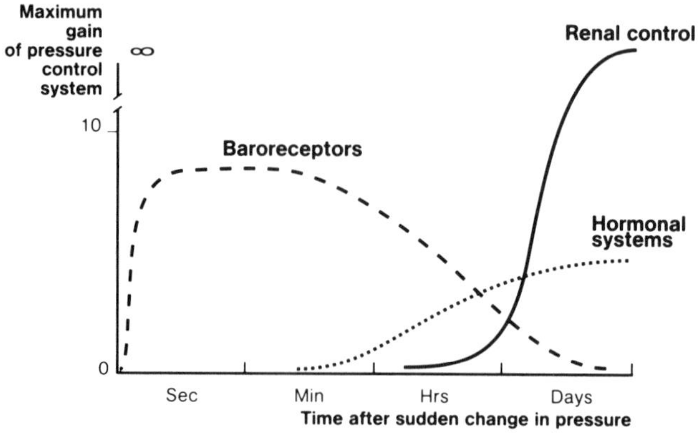

FIGURE 3. Degree of response of three major arterial pressure control mechanisms over time following a sudden change in blood pressure (based on Guyton *et al.*[45]).

system. These two pressure control systems have the common property of being relatively resistant to adaptation. The renal response has the additional property of an infinite gain, thus permitting complete restoration of blood pressure to within closely defined limits.[45]

Evidence that central influences on circulation may last much longer than the rapid phase was recently reviewed and integrated into a wider theory on the etiology of essential hypertension.[48] The ability of the CNS to determine the long-term blood pressure level has been a controversial issue. In brief, confusion may come from the concept that the kidney-fluid volume mechanism of the pressure control system detects and controls blood pressure at a fixed level, optimal for survival (e.g., renal controllers), but this concept ignores other systems that can influence blood pressure at any time and possibly on a long-term basis. Consistent examples from *in vivo* models (reviewed in refs. 47 and 48) and from particular clinical situations support the contribution of the CNS in long-term control oi the circulation, even when a clear primary initiating cause is present such as renovascular hypertension in which sympathetic efferent nerve activity is involved in the maintenance of hypertension.[49]

Moreover, because CNS influence and fluid volume regulation are so intertwined and nervous system activity is difficult to assess experimentally, especially at the kidney level,[50] the precise role of the CNS in both hormonal and renal control systems of blood pressure needs thorough investigation. The CNS via the renal nerves is critically important in adjusting fluid volume because of its action on different parameters of renal function (renal blood flow, glomerular filtration rate, renal tubular sodium, and water reabsorption) and the renin/angiotensin II secretion rate.[51] Recently, sympathoinhibitory action on renal nerve activity by certain antihypertensive agents was shown to preserve renal blood flow while reducing blood pressure in conscious SHRs.[52]

Consequently, because nervous and endocrine factors can modulate the sensitivity of the pressure-diuretic system, it can be speculated that the role of neural mechanisms may exceed the rapid phase of control of blood pressure by driving, through the

TABLE 1. Effect of Rilmenidine (1 to 2 mg/day, 8 wk) on Neurohormonal System in Patients (n = 19) Treated for Mild to Moderate Hypertension[a]

	Standardized Conditions of Tests					
Plasma hormones	At Rest		Intellectual Stress		After Walking	
Noradrenaline	−26%	p <0.01	−28%	p = 0.02	−14%	NS
Adrenaline	−11%	NS	−9%	NS	−23%	NS
Plasma renin activity	−43%	p <0.05	−43%	NS	−44%	p = 0.02
Aldosterone	−9%	NS	−16%	NS	−27%	NS

[a] Based on Zech *et al.*[57]

sympathetic/parasympathetic balance, the various processes regulating blood pressure on a long-term basis.

Short-Term Action of I_1 Agents

The primary mechanism underlying the hypotensive effects of I_1 agents such as rilmenidine is consecutive to withdrawal of sympathetic tone, possibly associated with restoration of baroreflex function altered in hypertension and thereby with a reduction in humoral secretion of vasoconstrictive hormones.

The sympathoinhibitory action of rilmenidine is well documented in various species of anesthetized or conscious animals[53–56] using different approaches. A limited number of studies have been carried out in hypertensive patients to assess the sympathoinhibitory action of rilmenidine. Indeed, evaluation of such action necessitates the use of recent techniques such as noradrenaline spillover, microneurography, and pharmacological blockade, each of which has its own limitations.[50] By power spectral analysis of short-term circulatory rhythms of borderline hypertensive patients, Elghozi *et al.* showed a significant reduction in the 0.1-Hz peak (Mayer's waves) of standing systolic arterial pressure fluctuations recorded with a Finapres device following acute treatment with rilmenidine (1 mg) compared with placebo. Data suggest a reduction in the sympathetic component of blood pressure variability (unpublished results).

This profile of action is also reflected by a significant decrease in plasma norepinephrine before and after treatment both at rest or after intellectual stress following 8 weeks of treatment by rilmenidine in patients with essential hypertension (n = 19). A sustained decrease (over 40%) in plasma renin activity was reported in the same study, with a slight but insignificant change in plasma aldosterone levels (TABLE 1).[57] These results parallel those found in animals[53,58,59] with the addition of a marked reduction in plasma vasopressin in SHRs, but not in normotensive rats after 12 weeks of treatment (0.6 mg/kg).[59]

Furthermore, tests of autonomic function have been realized in normotensive volunteers before and after rilmenidine treatment in conditions of both low and high sympathetic tone.[60] Briefly, data indicated that after an oral dose of 2 mg rilmenidine, cardiac baroreflex sensitivity increased as already reported with classic centrally-acting agents.[61] In tests in which sympathetic tone was enhanced (standing for 3 minutes), rilmenidine blunted the reflex increase, whereas when activity of the sympathetic system was low, the drug enhanced vagal tone (standing for 5 seconds but not a hand-grip test and Valsalva maneuver).[60]

A comparable set of experiments designed to compare the effects of rilmenidine and atenolol on autonomic function were carried out recently in patients with mild to moderate essential hypertension following 4 weeks of treatment (Reid *et al.*, this volume). Overall results indicated that rilmenidine did not interfere with physiologic cardiovascular responses during testing.

Whereas only the cardiac baroreflex can generally be studied in humans by the ramp method,[46] it is evident that the autonomic nervous system supplies each organ through separate pathways with distinct patterns of reflex activity.[50] Therefore, the effects of rilmenidine on baroreflex control of renal sympathetic nerve activity, as well as cardiac baroreflex, was investigated in conscious rabbits.[56] Experiments showed that rilmenidine treatment at hypotensive doses both shifted the sympathetic activity baroreflex curves to lower blood pressure values and decreased the maximal and minimal sympathetic activity levels available, reducing the baroreflex range by 40%. Conversely, although cardiac sympathetic tone is reduced and baroreflex curve is shifted to the left, its range is preserved in the same conditions. Hence, the specific response of rilmenidine on renal sympathetic efferent activity could account for preservation of renal circulatory function and natriuretic capacity under transient fluctuations in blood pressure (stress response, for example) or hypertension.

The available evidence indicates that the short-term antihypertensive effect of rilmenidine and possibly I_1 agents is produced by central sympathoinhibition, which extends in distinct degrees to vital organs, and less reflex stimulation in neurohormonal mechanisms accompanied by restoration of baroreflex function and a respect of autonomic functions.

This profile confers favorable clinical consequences, as illustrated by the absence of orthostatic hypotension[62,63] and the preserved adaptation to exercise response (Reid *et al.*, this volume) following rilmenidine treatment.

Long-Term Action of I_1 Agents

A unique feature of I_1 agents lies in their possible ability to modulate renal function by two separate mechanisms, that is, central and local, that act in concert to oppose sodium retention. At the end of the 1980s, imidazolines and related derivatives were shown to inhibit the Na^+/H^+ exchanger by an adrenergic-independent mechanism in the renal proximal tubule of the rabbit kidney.[64] The specific action of these compounds, as opposed to activation of the antiport elicited by α_2 agonists, was suggested first to be mediated by renal idazoxan-sensitive imidazoline receptors which have been characterized in both rabbit[65] and human kidney.[66] Recent studies examined the role of renal I_1-subtype receptors in mediating a natriuretic response further to the administration of rilmenidine or other I_1 agents in a 1 kidney-anesthetized preparation.[24,25] Moreover, their existence in kidney was confirmed by different groups.[9,15,22,23] Recently, Parini *et al.* (this meeting) demonstrated the coexpression of I_1 and I_2 sites in the proximal renal tubule. From these findings and in the light of evidence of dysfunction of the Na^+/H^+ antiport in hypertension,[67] the role of each I-subtype receptor in inhibiting the Na^+/H^+ exchanger must be reevaluated to specify the mechanism of action of imidazolines and related derivatives on renal function.

The influence of sympathetic renal nerves on renal function has been amply demonstrated.[51] As their activity is subjected to rilmenidine's modulation, Kline and Cechetto[68] evaluated the CNS contribution of hypotensive doses of rilmenidine on renal function in anesthetized Wistar rats with innervated and contralateral denervated kidneys. With renal perfusion pressure being kept at pretreatment levels, rilmenidine increased renal blood flow and osmolar clearance, mainly through enhancement of sodium excretion, and decreased water clearance in the innervated kidney. These data were coupled with a marked reduction in renal sympathetic efferent activity. In these experimental conditions, only minor changes were seen in the denervated kidney. Overall, these elements suggest that both local and central I_1 receptors are susceptible to the diuretic and natriuretic effects of rilmenidine found under experimental conditions.

In essential hypertension, the kidney-fluid volume control system is chronically altered.[69] As a consequence, higher renal arterial pressure is necessary for sodium excretion, allowing a steady-state fluid balance to be achieved.[47,51,70] Furthermore, overactivity in renal sympathetic efferent activity is found in hypertension in various animal models and in humans.[51,70,71] In this situation, the pressure-natriuresis relation has to be reset, and an effective antihypertensive agent must shift the renal function curve back towards normal by either directly (i.e., affecting tubular reabsorption) and/or indirectly (i.e., changing neural or hormonal influences) increasing the ability of the kidney to excrete sodium. As for I_1 agents, their combined effects on renal function may contribute to major efficacy in the long term.

This view is illustrated by our clinical experience with rilmenidine. Efficacy (reviewed in ref. 72) has been assessed in various populations of hypertensive patients (middle aged and elderly persons and patients with renal failure) and for long periods of treatment (up to 24 months). Sodium retention and weight gain described with chronically administered α_2 agonist antihypertensive drugs were absent with rilmenidine; a reduction in weight, a reliable marker of water-sodium retention, was reported in two separate studies of mild to moderate essential hypertension.[73,74]

Acting homeostatically through a unique receptor system on the central, hormonal, and renal pressure-control mechanisms, I_1 agents are potentially able to meet the criteria of efficacy and tolerability required for current standard antihypertensive therapy as assessed in the different populations of hypertensive patients.

THERAPEUTIC HIGHLIGHTS OF I_1 AGENTS

To achieve the status of a first-line antihypertensive drug, I_1 agents must demonstrate efficacy and acceptability and should reduce cardiovascular risk factors to lower the morbidity and mortality associated with essential hypertension.

Rilmenidine compares favorably in terms of both efficacy and tolerance to reference drugs such as clonidine, methyldopa, hydrochlorothiazide, or atenolol in double-blind, randomized parallel design studies. (For reviews, see refs. 72 and 75.) The disadvantages of classic centrally-acting agents (namely, sedation, dry mouth, and

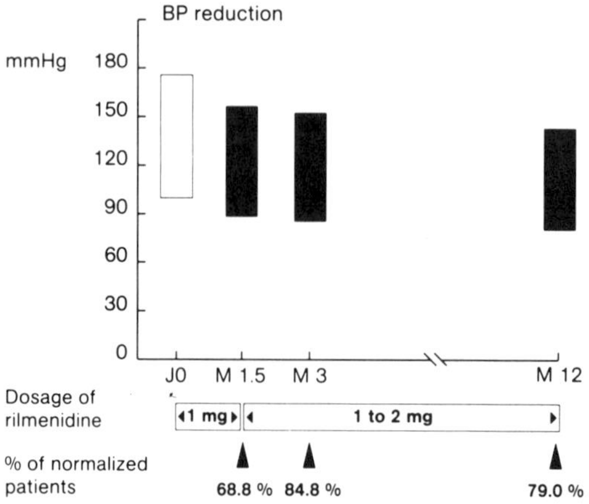

FIGURE 4. Efficacy of rilmenidine administered as monotherapy during 1 year of treatment (1 to 2 mg according to the therapeutic schedule) in patients with mild to moderate hypertension (n = 2,635), taking into account the type of population.[78])

withdrawal reactions), in particular, are limited to a frequency of no real clinical significance.[76] Also, minimal metabolic disturbances (in plasma electrolytes, glucose, lipids, and uric acid) were confirmed in a large population study using rilmenidine for 2 years.[75] The latter result contrasts with the known deleterious metabolic effects of diuretics or β-blockers.[77]

A similar profile was found in a recent trial designed to evaluate rilmenidine in a general practice population (2,635 hypertensive patients enrolled) including populations at risk after 12 months of treatment.[78] This pharmaco-epidemiologic study best exemplifies efficacy (FIG. 4) and safety (TABLE 2) in daily therapeutic prescription, as monotherapy or combination therapy in patients receiving multiple concomitant treatments. Around 80% of patients affected by mild to moderate hypertension, including specific populations prone to cardiovascular complications, normalized blood pressure by taking rilmenidine 1-2 mg/day as monotherapy. The frequency of complaints during follow-up of this heterogeneous population (TABLE 2) does not vary to a great extent between events or according to specific populations.[78] Therefore, the acceptability of rilmenidine is satisfactory considering that adverse events were recorded whatever the cause (imputation was not evaluated).

Although systolic pressure is a good risk indicator of cardiovascular events, including stroke, congestive heart failure, and coronary heart diseases, therapeutic reduction of blood pressure is not completely successful in reducing coronary morbidity or mortality and sudden cardiac death.[79,80] Consequently, new emphasis has been placed on pressure-independent risk factors that can possibly be linked to sympathetic nervous system activity.[79,81]

Left Ventricular Hypertrophy

Development of left ventricular hypertrophy (LVH) in hypertension results from multifactorial processes, including mechanical factors (blood pressure/volume over-

TABLE 2. Distribution of Adverse Events (incidence >2.5%), Irrespective of Their Imputability, Reported during 1 Year of Treatment with Rilmenidine (1 to 2 mg/day, monotherapy or combination therapy) in a General Practice Population of Hypertensive Patients (n = 2,635)[a]

	Mouth Dryness	Asthenia	Drowsiness	Dizziness
General Population				
of which				
Populations at Risk:	5.8%	4.6%	3.5%	3.5%
Age ≥ 60 yr	6.7%	4.8%	6.4%	3.6%
Diabetes	7.4%	6.4%	3.6%	3.8%
Renal insufficiency	4.3%	5.2%	3.7%	2.4%
Angina pectoris	7.3%	6.3%	3.6%	3.3%
Heart failure	4.8%	3.6%	3.4%	7.1%
Dyslipidemia	6.4%	4.9%	3.0%	4.1%

[a] Reproduced with permission from Pillion *et al.*[78]

load), neurohumoral disturbances (activation of renin/angiotensin II and noradrenergic systems), and abnormal cardiac cell trophic mechanisms.[80] In contrast to diuretics, arterial vasodilators and β-blockers with intrinsic sympathomimetic activity that activate neurohumoral processes, drugs reducing sympathetic tone and/or the renin-angiotensin II system produce a consistent and rapid reversal of left ventricular mass.[77,80,82] These results suggest that rapid regression of LVH may result from additional mechanisms to hemodynamic changes alone that accompany treatment by diuretics and arterial vasodilators.[82] Left ventricular hypertrophy regression has been assessed in hypertensive patients after 1 year of treatment with rilmenidine.[83] By comparison with data obtained in a recent metaanalysis of 109 studies,[84] rilmenidine can significantly reduce left ventricular mass to a level comparable to that produced by ACE inhibitors and greater than that produced by calcium antagonists and β-blocking agents.[83] This finding was associated with improvement in brachial artery compliance that persisted 1 month after withdrawal of treatment. Cardiovascular structural changes after chronic rilmenidine administration were previously demonstrated in DOCA-salt hypertensive rats (reduction in left ventricular collagen mass).[85] The counterpart of left ventricular remodeling was suggested in the aforementioned clinical study to be represented by a progressive reduction in plasma ANF. This response occurred independently of blood pressure normalization after 1 year of rilmenidine treatment.[83]

Ventricular Arrhythmias

Patients with LVH have a higher prevalence of premature ventricular arrhythmias (VA) than do hypertensive subjects. Ventricular arrhythmias are particularly life-threatening in patients with coronary artery diseases under psychological stress, leading to ventricular fibrillation and sudden death. It is likely that both the genesis of VA and its clinical complications under acute emotional stress are attributable, at least in part, to a high background level of cardiac sympathetic tone.[86]

Using the halothane-epinephrine arrhythmia model in dogs, Hayashi *et al.*[87] suggested the involvement of central nonadrenergic imidazoline receptors in the antiarrhythmic properties of atipemazole, an α_2 adrenoceptor agonist imidazoline compound. The same group has investigated the cardioprotective effects of rilmenidine in a similar model. Rilmenidine (1, 3, and 10 μg/kg iv) prevented the genesis of halothane-epinephrine arrhythmias. Following antagonistic experiments employing idazoxan or rauwolscine, a reference α_2-antagonist, by the intracisternal route, the results indicated that central I_1- rather than α_2-receptors were involved in the rilmenidine response. Therefore, I_1 agents could exert protective effects against the development of arrhythmias by acting through medulla I_1 receptors to reduce cardiac sympathetic activity.

X Syndrome-Risk Factors

The triad of metabolic abnormalities, namely, hyperinsulinemia, insulin resistance, and hyperglycemia, contributes to the development of hypertension and atherosclerosis.[88] These risk factors, with the addition of hypertension and dyslipidemia, have been referred to by Reaven[89] as syndrome X. Further, epidemiological evidence indicates that these factors are strongly involved in the pathogenesis of atherosclerosis and coronary heart disease.[90]

These hemodynamic and metabolic disturbances were reproduced in normotensive rats by feeding them a fructose-enriched diet for 4 weeks.[91] Contrasting with the results obtained after clonidine treatment,[91] rilmenidine (1 mg/kg/day in drinking water) was shown recently to reverse hypertension as does clonidine, but also fructose-induced overweight and peripheral and hepatic insulin resistance (Penicaud *et al.*, unpublished results). The resultant increase in triglyceridemia under a fructose regimen was also reduced to normal levels by rilmenidine treatment.

Whether insulin resistance and hypertension are the consequence of sympathetic overactivity remains debatable. However, numerous experimental findings support this view.[81,86,90] A recent study provided indirect evidence of abnormal sympathetic activity in hypertensive patients receiving insulin.[92] Consequently, I_1 agents may demonstrate a beneficial influence in hypertensive populations at risk, besides ACE inhibitors and α_1-blockers that have been reported to improve insulin sensitivity when used as antihypertensive drugs.[88]

Moreover, phase III clinical studies have confirmed the safety and the excellent biochemical acceptability of rilmenidine on glucidic and lipidic metabolisms in hypertensive patients with concomitant diseases including insulin-dependent or -independent diabetes, renal failure,[72] and dyslipidemia.[93] Studies showed that the treatment may even have some slight beneficial effects on the blood lipid profile.

CONCLUSION

The primary goal of an antihypertensive drug is to reduce the cardiovascular morbidity and mortality associated with high blood pressure, and treatment with

TABLE 3. Reported Effects of Different Antihypertensive Pharmacological Classes on Cardiovascular Risk Factors[a]

Pressure-Independent Risk Factors	I_1 Agents (Expected Profile)	Diuretics	α_1-Blocker	β-Blocker	Calcium Antagonist	ACE Inhibitor
Left ventricular hypertrophy	+	–	+	+/–	+	+
Ventricular arrhythmias	+	–	+	+	+	+
Dyslipidemia	NC/+	–	+	–	NC	NC
Insulin resistance	+	–	+	–	NC	+
Hyperinsulinemia	NC	–	+	–	NC	+
Hyperuricemia	NC	–	NC	–	NC	NC

NOTE: + = positive effects; – = negative effects; NC = no change.
[a] Based on Houston.[77]

current classes of antihypertensive drugs has failed to show the expected reduction in the risk of coronary heart disease.[94] As pressure-independent risk factors for cardiovascular events have been identified, differences in the effects of antihypertensive drugs on these risk factors may explain in part their disappointing results in the prevention of cardiovascular mortality. It is likely that certain antihypertensive agents induce potentially negative changes on pressure-independent risk factors and affect the benefit of lowering arterial blood pressure. Based on the remarkable profile of the effects of I_1 agents on the risk factors that are expected, as just mentioned (TABLE 3), and on the proposed evidence of sympathetic hyperactivity as the primary mechanism responsible for cardiovascular events,[79,81,86] I_1 agents could constitute an especially commendable therapeutic class in the management of hypertension. Furthermore, by acting homeostatically on the whole mechanism of blood pressure control systems, the unique mode of action of I_1 agents can be considered the most comprehensive and physiological way to treat essential hypertension. The intrinsic therapeutic qualities of rilmenidine, such as efficacy, safety, and ease of use, has taught us the advantages of this new generation of antihypertensive agents. All the elements just presented suggest that I_1 agents have the potential to fulfill the criteria required for first-line antihypertensive treatment. Long-term controlled clinical trials, however, are necessary to clearly establish the treatment benefit of I_1 agents in cardiovascular diseases associated with hypertension.

ACKNOWLEDGMENTS

The authors gratefully acknowledge Dr. R. L. Kline and Dr. G. A. Head for careful revision of the text and A. Mannion for excellent secretarial assistance.

REFERENCES

1. Bousquet, P., J. Feldman & J. Schwartz. 1984. J. Pharmacol. Exp. Ther. **230:** 232-236.
2. Ernsberger, P., M. P. Meeley, J. J. Mann & D. J. Reis. 1987. Eur. J. Pharmacol. **134:** 1-13.
3. Bricca, G., M. Dontenwill, A. Molines, J. Feldman, A. Belcourt & P. Bousquet. 1989. Eur. J. Pharmacol. **162:** 1-9.
4. Hieble, J. P. & R. R. Ruffolo. 1992. Fundam. Clin. Pharmacol. **6** (suppl. 1): 7s-13s.
5. Bricca, G., J. Zhang, H. Greney, M. Dontenwill, J. Stutzmann, A. Belcourt & P. Bousquet. 1993. Br. J. Pharmacol. **110:** 1537-1543.
6. Bricca, G., H. Greney, J. Zhang, M. Dontenwill, J. Stutzmann, A. Belcourt & P. Bousquet. 1994. Eur. J. Pharmacol.—Mol. Pharmacol. Sect. **266:** 25-33.
7. Gomez, R. E., P. Ernsberger, G. Feinland & D. J. Reis. 1991. Eur. J. Pharmacol. **195:** 181-191.
8. Hamilton, C. A., M. A. Yakuba, E. Jardine & J. L. Reid. 1991. J. Auton. Pharmacol. **11:**277-283.
9. Ernsberger, P., T. H. Damon, L. M. Graff, S. G. Schafer & M. O. Christen. 1993. J. Pharmacol. Exp. Ther. **264:** 172-182.
10. De Vos, H., G. Bricca, J. De Keyser, J. P. De Backer, P. Bousquet & G. Vauquelin. 1994. Neuroscience **59:** 589-598.
11. Bricca, G., H. Greney, M. Dontenwill-Kieffer, J. Zhang, A. Belcourt & P. Bousquet. 1993. Neurochem. Int. **22:** 153-163.
12. Miralles, A., G. Olmos, M. Sastre, F. Barturen, I. Martin & J. A. Garcia-Sevilla. 1993. J. Pharmacol. Exp. Ther. **264:** 1187-1197.
13. Renouard, A., P. Widdowson & A. Cordi. 1993. Br. J. Pharmacol. **109:** 625-631.
14. Ernsberger, P., R. Giuliano, N. Willette & D. J. Reis. 1990. J. Pharmacol. Exp. Ther. **253:** 408-418.
15. King, P. R., S. Suzuki, T. I. Nero, A. Gunglach & W. J. Louis. 1993. J. Hypertens. **11** (suppl. 5): S160-S161.
16. Head, G. A., S. J. Godwin & F. Sannajust. 1993. J. Hypertens. **11** (Suppl. 5): S322-S323.
17. Tibirica, E., J. Feldman, C. Mermet, L. Monassier, F. Gonon & P. Bousquet. 1991. Eur. J. Pharmacol. **209:** 213-221.
18. Tibirica, E., J. Feldman, C. Mermet, F. Gonon & P. Bousquet. 1991. J. Pharmacol. Exp. Ther. **256:**606-613.
19. Pineda, J., L. Ugedo & J. A. Garcia-Sevilla. 1993. Naunyn-Schmiedeberg's Arch. Pharmacol. **348:** 134-140.
20. Sokoloff, L. 1981. J. Cereb. Blood Flow Metab. **1:** 7-36.
21. Browne, S. E. & I. M. Macrae. 1994. Brain Res. **666:** 216-222.
22. Hamilton, C. A., E. Jardine & J. L. Reid. 1993. Eur. J. Pharmacol. **243:** 95-97.
23. Ernsberger, P., G. Feinland, M. P. Meeley & D. J. Reis. 1990. Am. J. Hypertens. **3:** 90-97.
24. Smyth, D. D., P. Li, D. E. Blandford & S. B. Penner. 1992. J. Pharmacol. Exp. Ther. **261:** 1080-1086.
25. Penner, S. B. & D. D. Smyth. 1994. Br. J. Pharmacol. **112:** 200-206.
26. Grabie, M., P. Nussbaum & S. Golfarb. 1980. Clin. Pharmacol. Ther. **27:** 522-527.
27. Molderings, G. J., D. Moura, K. Fink, H. Bonisch & M. Gothert. 1993. Naunyn-Schmiedeberg's Arch. Pharmacol. **348:** 70-76.
28. Reis, D. J., S. Regunathan, H. Wang, D. L. Feinstein & M. P. Meeley. 1992. Fundam. Clin. Pharmacol. **6** (Suppl. I): 23s-29s.
29. Powis, D. A. & P. F. Baker. 1986. Mol. Pharmacol. **29:** 134-141.
30. Ohara-Imaizumi, M. & K. Kumakura. 1992. Cell. Mol. Neurobiol. **12:** 273-283.
31. Evinger, M. J., P. Ernsberger, T. H. Joh & D. J. Reis. 1989. Soc. Neurosci. Abstr. **15:** 500.

32. REGUNATHAN, S., M. J. EVINGER, M. P. MEELEY & D. J. REIS. 1991. Biochem. Pharmacol. **42:** 2011-2018.
33. HENSLEY, M. L., M. P. MEELEY, P. M. MCCAULEY, P. ERNSBERGER & D. J. REIS. 1989. Am. J. Hypertens. **2:** 917-919.
34. MEELEY, M. P., M. L. HENSLEY, P. ERNSBERGER, D. FELSEN & D. J. REIS. 1992. Brain Res. **582:** 261-267.
35. IZZO, J. L. 1989. Am. J. Hypertens. **2:** 305S-312S.
36. MICHEL, M. C., J. W. REGAN, M. A. GERHARDT, R. R. NEUBIG, P. A. INSEL & H. J. MOTULSKY. 1990. Mol. Pharmacol. **37:** 65-68.
37. PILETZ, J. E. & K. SLETTEN. 1993. J. Pharmacol. Exp. Ther. **267:** 1493-1502.
38. ZONNENSCHEIN, R., S. DIAMANT & D. ATLAS. 1990. Eur. J. Pharmacol. **190:** 203-215.
39. PETRUSEWICZ, J. & R. KALISZAN. 1991. Gen. Pharmacol. **22:** 819-823.
40. SCHAMS, G., B. V. VENKATARAMAN, A. HAMADA, D. D. MILLER, P. N. PATIL & D. R. FELLER. 1991. Eur. J. Pharmacol. **199:** 315-323.
41. FERRY, N., D. HENRY, E. BATTAIS, A. MARY, C. BONNE & J. HANOUNE. 1986. Biochem. Pharmacol. **35:** 1511-1516.
42. MOLDERINGS, G. J., F. HENTRICH & M. GOTHERT. 1991. Naunyn-Schmiedeberg's Arch. Pharmacol. **344:** 630-638.
43. FUDER, H. & P. SCHWARZ. 1993. Naunyn-Schmiedeberg's Arch. Pharmacol. **348:** 127-133.
44. BOHMANN, C., P. SCHOLLMEYER & L. C. RUMP. 1994. Naunyn-Schmiedeberg's Arch. Pharmacol. **349:** 118-124.
45. GUYTON, A. C., J. E. HALL, T. E. LOHMEIER, T. E. JACKSON & P. R. KASTNER. 1981. Fed. Proc. **40:** 2252-2256.
46. KORNER, P. I. & G. A. HEAD. *In* Central Neural Mechanisms of Blood Pressure Regulation. G. Kunos & J. Ciriello, Eds. Birkhauser. Boston, Basel, Berlin.
47. COWLEY, A. W. 1992. Physiol. Rev. **72:** 231-300.
48. JULIUS, S. 1988. J. Hypertens. **6:** 177-185.
49. MATHIAS, C. J. 1991. Hypertension **18** (Suppl. III): III22-III30.
50. ESLER, M. 1993. Pharmacol. Toxicol. **73:** 243-253.
51. DIBONA, G. F. 1992. Hypertension **19** (Suppl. I): I28-I35.
52. TAKISHITA, S., H. MURATANI, N. KAWAZOE, Y. KIMURA, M. TOZAWA & K. FUKIYAMA. 1994. Hypertension **23** (Suppl. I): I97-I101.
53. LAUBIE, M., J. C. POIGNANT, J. SCUVEE-MOREAU, H. DABIRE, A. DRESSE & H. SCHMITT. 1985. J. Pharmacol. **16:** 259-278.
54. VALET, P., M. A. TRAN, C. DAMASE-MICHEL, G. DE SAINT-BLANQUAT, L. DANG TRAN, F. ANGLADET, E. KOENIG-BERARD & J. L. MONTASTRUC. 1988. J. Auton. Pharmacol. **8:** 319-326.
55. MAYOROV, D., M. CHERNOBELSK & O. MEDVEDEV. 1993. J. Cardiovasc. Pharmacol. **22:** 314-320.
56. SANNAJUST, F. & G. HEAD. 1993. J. Hypertens. **11** (Suppl. 5): S328-S329.
57. ZECH, P. & N. POZET. 1986. 11th scientific meeting of the International Society of Hypertension, Heidelberg.
58. SANNAJUST, F., C. JULIEN, C. BARRES, C. CERUTTI, E. KOENIG-BERARD & J. SASSARD. 1989. Clin. Exp. Pharmacol. Physiol. **16:** 837-848.
59. GHAEMMAGHAMI, F., J. IBANEZ, G. GEELEN, M. VINCENT, J. FRUTOSO & C. GHARIB. 1990. J. Cardiovasc. Pharmacol. **15:** 68-74.
60. MCKAIGUE, J. P. & D. W. G. HARRON. 1992. Clin. Pharmacol. Ther. **52:** 511-517.
61. BADOER, E., G. A. HEAD & P. I. KORNER. 1983. J. Cardiovasc. Pharmacol. **5:** 760-767.
62. GALLEY, P., G. MANCIET, J. L. HESSEL & J. P. MICHEL. 1988. Am. J. Cardiol. **61:** 86D-90D.
63. WILKINSON, R., S. MANSY, C. CORCORAN & C. T. DOLLERY. 1982. Arch. Mal. Coeur (suppl.): 31-38.
64. BIDET, M., P. POUJEOL & A. PARINI. 1990. Biochem. Biophys. Acta **1024:** 173-178.

65. LACHAUD, V., I. COUPRY, R. A. PODEVIN, J. P. DAUSSE, E. KOENIG & A. PARINI. 1988. J. Hypertens. **6** (Suppl. 4): S511-S513.
66. LACHAUD, V., R. A. PODEVIN, Y. CHRETIEN & A. PARINI. 1991. Eur. J. Pharmacol. **206:** 23-31.
67. ROSSKOPF, D., R. DUSING & W. SIFFERT. 1993. Hypertension **21:** 607-617.
68. KLINE, R. L. & D. F. CECHETTO. 1993. J. Pharmacol. Exp. Ther. **233:** 1556-1562.
69. GUYTON, A. C. 1991. Hypertension **18** (Suppl. III): III49-III53.
70. WYSS, J. M., S. OPRIL & W. SRIPAIROJTHIKOON. 1992. Can. J. Physiol. Pharmacol. **70:** 759-770.
71. ESLER, M., C. FERRIER, G. LAMBERT, G. EISENHOFER, H. COX & G. JENNINGS. 1991. Hypertension **17** (Suppl. III): III29-III35.
72. LAURENT, S. & M. SAFAR. 1992. Am. J. Hypertens. **5:** 99S-105S.
73. DOLLERY, C. T. *et al.* 1990. Curr. Ther. Res. **47:** 194-221.
74. DALLOCCHIO, M., P. GOSSE, J. P. FILLASTRE, G. GROLLIER, P. MORAND, M. DRICI & C. CORCORAN. 1991. Presse Méd. **20:** 1265-1271.
75. BEAU, B., F. MAHIEUX, M. PARAIRE, S. LAURIN, B. BRISGAND & P. VITOU. 1988. Am. J. Cardiol. **61:** 95D-102D.
76. HARRON, D. W. G. 1992. Am. J. Hypertens. **5:** 91S-98S.
77. HOUSTON, M. C. 1989. Am. Heart J. **117:** 911-950.
78. PILLION, G., B. FEVRIER, P. CODIS & D. SCHUTZ. 1994. Am. J. Cardiol. **74:** 58A-65A.
79. ROSENMAN, R. H. 1989. Am. J. Hypertens. **2:** 313S-338S.
80. GILES, T. D. 1992. Hypertension **19:** I124-I129.
81. JULIUS, S. 1993. Hypertension **21:** 886-893.
82. TAMARGO, J., E. DELPON & C. VALENZUELA. 1993. Eur. Heart. J. **14** (Suppl J): 102-106.
83. TRIMARCO, B., G. ROSIELLO, D. SARNO, F. LAMENZA, S. RUBATTU, N. DELUCA & M. VOLPE. 1994. Am. J. Cardiol. **74:** 36A-42A.
84. DAHLOF, B., K. PENNERT & L. HANSSON. 1992. Am. J. Hypertens. **5:** 95-110.
85. CALLENS-EL AMRANI, F., F. PAOLAGGI & B. SWYNGHEDAUW. 1989. J. Hypertension **7:** 947-954.
86. ESLER, M., G. LAMBERT & G. JENNINGS. 1990. J. Hypertens. **8** (Suppl. 7): S53-S57.
87. MAMMOTO, T., T. KAMIBAYASHI, Y. HAYASHI, K. TAKADA, A. YAMATODANI & I. YOSHIYA. 1993. ASA Congress, Washington.
88. SOWERS, J. R., P. R. STANDLEY, J. L. RAM, S. JACOBER, L. SIMPSON & K. ROSE. 1993. Am. J. Hypertens. **6:** 260S-270S.
89. REAVEN, G. M. 1988. Diabetes **37:** 1595-1607.
90. DONNELLY, R. & M. C. CONNELL. 1992. Clin. Sci. **83:** 265-275.
91. HWANG, I. S., H. HO, B. B. HOFFMAN & G. M. REAVEN. 1987. Hypertension **10:** 512-516.
92. LEMBO, G., R. NAPOLI, B. CAPALDO, V. RENDINA, G. IACCARINO, M. VOLPE, B. TRIMARCO & L. SACCA. 1992. J. Clin. Invest. **90;** 24-29.
93. SCEMAMA, M., B. FEVRIER, I. BEUCLER, F. DAIROU & P. UNGER. 1993. XIII Journées de l'hypertension arterielle. Paris.
94. Joint National Committee on Detection, Evaluation, and Treatment of High Blood Pressure (JNC V). Fifth Report. 1993. Arch. Intern. Med. **153:** 154-183.

Why Imidazoline Receptor Modulator in the Treatment of Hypertension?

S. G. SCHÄFER,[a] E. C. KAAN,[a] M. O. CHRISTEN,[b] A. LÖW-KRÖGER,[c] H.-J. MEST,[c] AND G.-J. MOLDERINGS[d]

[a]*Solvay Pharma Germany*
Hanover, Germany

[b]*Solvay France*
Suresnes, France

[c]*BDF-Lilly*
Hamburg, Germany

[d]*University of Bonn*
Bonn, Germany

Growing evidence indicates that the sympathetic nervous system is implicated in the pathogenesis of hypertension.[1,2] Moreover, an overactive sympathetic nervous system leading to excess of circulating and tissue catecholamines may be an independent risk factor for further cardiovascular complications. A high sympathetic drive seems to directly promote cardiac and vascular smooth muscle hypertrophy,[3] causes platelet aggregation and enhances thrombosis, accelerates the development of atherosclerosis,[4] and facilitates underperfusion of the kidney.[5,6] Also, the incidence of arrhythmias and sudden cardiac death seems to be increased by high sympathetic tone.[7] Finally, catecholamines may also have an impact on metabolic status, raising total cholesterol, triglycerides, and insulin while decreasing HDL cholesterol and causing insulin resistance and glucose intolerance.[8] Therefore, an important aspect of overall treatment of hypertensive patients is not only to lower blood pressure per se, but also to reduce sympathetic overactivity; in particular, an effective antihypertensive agent should also reduce cardiac hypertrophy, because epidemiological studies have indicated that cardiac hypertrophy is an independent risk factor for cardiovascular mortality in hypertensive patients.

Centrally acting antihypertensive drugs of the first generation such as α-methyldopa, guanfacine, guanabenz, and clonidine are well known agents in the treatment of moderate to severe forms of hypertension. These agents lower blood pressure by modulating the activity of neurons in the brain stem, resulting in peripherally reduced sympathetic tone. Their antihypertensive action is accompanied by a slowed heart rate and marked suppression of plasma renin activity. The latter is an attractive feature, as this parameter is associated with subsequent development of ventricular hypertrophy. Furthermore, clonidine is an effective antihypertensive agent in patients with renal failure, including those who are undergoing chronic hemodialysis.

Finally, centrally acting drugs have the advantage that their use avoids activation of compensatory reflex mechanisms in hypertensive patients.

However, there are problems with the centrally acting antihypertensive drugs of the first generation, inasmuch as they all have a relatively high incidence of side effects in common such as dry mouth, sedation, reduced vigilance, impaired mood, and rebound hypertension. For this reason, despite their attractive hemodynamic, endocrine, and metabolic effects, these drugs are no longer recommended for first-line use and are reserved for a special indications or when other agents fail. To improve antihypertensive treatment, a more specific drug was needed to lower blood pressure without the aforementioned adverse drug reactions. The recent discovery of I_1-imidazoline receptors with pharmacological properties distinct from those of α_2-adrenoceptors was an important step forward. Because it became possible to discriminate between the pharmacological mechanisms involved in the therapeutically desired hypotensive effect and the major side effect (sedation) of centrally acting antihypertensive agents, further development should lead to nonsedating antihypertensive drugs.

ROLE OF I_1-IMIDAZOLINE RECEPTORS IN BLOOD PRESSURE CONTROL

It has been established that a clearly defined region in the brain stem, termed the rostral ventrolateral medulla (RVLM), contains a subset of adrenergic neurons (C1 group) that are tonically active in maintaining a resting level of arterial pressure and integrating many cardiovascular reflexes.[9,10] These sympathoexcitatory neurons descend into the spinal cord to innervate monosynaptically preganglionic sympathetic neurons of the intermediolateral columns. The RVLM itself is innervated by multiple inputs from other areas of the brain and spinal cord. Among these are dense projections from the nucleus tractus solitarii and from a sympathoinhibitory area in the caudal ventrolateral medulla. Mainly through these pathways the RVLM is engaged in the mediation of a number of reflex responses (baroreceptor reflex and reflex elevations of arterial pressure associated with pain, exercise, and emotional behavior). Compounds that bind with high affinity and selectivity to the cells in this brain stem area might be able to modulate blood pressure very effectively. A new class of receptors, termed I_1-imidazoline receptors, was recently identified on these cells.

Evidence has been established that the I_1-imidazoline receptors in the C1 area of the RVLM are the predominant site of the hypotensive action of imidazoline-like compounds such as moxonidine and rilmenidine. The sedative effect of the clonidine-like drugs, on the other hand, has been attributed to an action within the nucleus locus coeruleus, a noradrenergic structure of the reticular system, which is directly involved in the regulation of the sleep-wake cycle. Activation of α_2-adrenoceptors by clonidine in the locus coeruleus leads to inhibition of noradrenergic neurons in this brain area, resulting in sedation.

The hypothesis that the I_1-imidazoline receptor is the first rank target of centrally active imidazoline antihypertensive drugs to lower blood pressure is supported by several observations (indicated below). It should be noted, however, that adrenergic receptors also contribute to blood pressure reduction,[11–13] but regulation may be complex; for example, α_2-adrenoceptors in the spinal cord or nucleus tractus solitarii appear to be involved in the regulatory process.[13–16]

TABLE 1. Comparison of the Selectivity of First- and Second-Generation Centrally Acting Antihypertensive Agents for I_1-Imidazoline over α_2-Adrenergic Binding Sites in the Bovine Rostral Ventrolateral Medulla (modified from Ernsberger[63])[a]

	K_I at L_1-Sites (nM)	K_I at α_2-Sites (nM)	Selectivity Ratio (α_2-I_1)
	First-Generation Agents		
Guanabenz	>10,000	7.2 ± 0.6	0.0007
Guanfacine	2,500 ± 190	2.3 ± 0.7	0.0009
Lofexidine	1.9 ± 0.7	6.9 ± 0.7	3.6
Clonidine	1.0 ± 0.3	3.8 ± 0.1	3.8
	Second-Generation Agents		
Rilmenidine	5.3 ± 2.6	113 ± 23	21
Moxonidine	2.3 ± 0.5	76 ± 8	33

[a] K_I values are ligand concentration in nM ± SEM and represent the combined analysis using computerized nonlinear curve-fitting of a saturation experiment in quadruplicate using eight concentrations of [^{3}H]clonidine, together with three to five competition experiments using three different ligands. A four-ligand two-component model was used; imidazoline and nonimidazoline compounds were analyzed simultaneously to isolate the I_1-imidazoline and α_2-adrenergic components of binding. Nonspecific binding was defined by using 10 μM of phentolamine of BDF 6143.

IMIDAZOLINE RECEPTOR BINDING

Radioligand binding studies with [^{3}H]clonidine, [^{3}H]*para*-aminoclonidine, and [^{3}H]rauwolscine and membranes prepared from tissue of the RVLM from various species including man show that cells in this brain region are endowed with I_1-imidazoline receptors and α_2-adrenoceptors.[17,18] In human RVLM I_1-imidazoline receptors exceed α_2-adrenoceptors by a factor of about 10.[17,18] Therefore, compounds with high selectivity and affinity to this receptor represent real progress in the treatment of hypertension. In radioligand binding studies moxonidine was 33-fold more selective for I_1-imidazoline recognition sites than for α_2-adrenoceptors, whereas clonidine had a selectivity ratio (affinity to α_2-adrenoceptor over affinity to I_1-imidazoline sites) only of about 4 (TABLE 1). Thus, at low dosage, moxonidine should primarily stimulate I_1-imidazoline receptors in the RVLM, yielding a decrease in arterial blood pressure.

Indeed, after oral doses of moxonidine 0.2 mg, peak plasma concentrations between 4 and 6 nmol/l were detected in man after 0.6–1.1 hour which closely correspond to the binding affinity of moxonidine for I_1-imidazoline sites but fall below the range needed to occupy α_2-adrenoceptors (TABLE 1).

Because clonidine is considerably less selective for I_1-imidazoline receptors (TABLE 1), the compound should activate at an equihypotensive dose both receptor systems (I_1-imidazoline receptors in the RVLM *and* α_2-adrenoceptors in the locus coeruleus); therefore, the hypotensive effect of clonidine is often accompanied by increased sedation.

MICROINJECTION OF IMIDAZOLINES INTO THE RVLM

Microinjection of moxonidine, rilmenidine, clonidine, and related compounds into the left and right RVLM of normotensive and spontaneously hypertensive rats,[19–22] cats,[23] and rabbits[24] induced a dose-dependent decrease in arterial pressure. Administration of 4 nmol of moxonidine 1.5 mm dorsal or 2 mm caudal to the RVLM had no effect on blood pressure or heart rate, demonstrating that the effect of moxonidine was specific to the region of the RVLM in the brain stem.[25,26] By contrast, bilateral injection of catecholamines into the RVLM only caused a small reduction in blood pressure or was even without effect.

INTERACTION EXPERIMENTS WITH I_1- AND α_2-RECEPTOR ANTAGONISTS

Results of interaction experiments in spontaneous hypertensive rats with receptor antagonists (idazoxan, efaroxan, yohimbine, and SKF 86466) confirmed the involvement of imidazoline receptors in the RVLM in the blood pressure-lowering effect of the imidazolines. Microinjection of a low dose of idazoxan, an antagonist at both α_2-adrenoceptors and imidazoline receptors, or of efaroxan into the RVLM could reverse the hypotension elicited by intracisternal application of clonidine, moxonidine, or rilmenidine[19,20,24,27] and by systemic application of moxonidine and clonidine[21,22,27]; by contrast, intracisternal administration of the selective α_2-adrenoceptor antagonists SKF 86466 and yohimbine had no effect on drug-induced hypotension.

Electrochemical *in vivo* investigations determining the metabolic activity of noradrenergic neurons further support the new concept of blood pressure regulation. Metabolism in noradrenergic neurons in the RVLM was dose-dependently inhibited by clonidine and rilmenidine when intravenously administered.[21,28] This inhibitory effect on neuronal activity could be correlated with the hypotensive effect of the drugs. By contrast, within the locus coeruleus (site of sedative activity), which is densely endowed with α_2-adrenoceptors, both agents inhibited neuronal activity only at doses 25-50 times higher than those required to lower arterial pressure. Similarly, the imidazoline receptor antagonist idazoxan, in low doses, prevented both the hypotensive effect and the inhibition of neuronal activity of clonidine and rilmenidine within the RVLM, but it was ineffective in counteracting the inhibitory effect of higher doses of clonidine and rilmenidine in the locus coeruleus.[21,28] By contrast, the inhibitory effect of the imidazolines in the RVLM was unaffected by yohimbine (a selective α_2-adrenoceptor antagonist), whereas their effect on locus coeruleus neurons was inhibited by yohimbine.

CORRELATION BETWEEN HYPOTENSIVE POTENCY AND BINDING AFFINITY

A significant correlation was established between the decrease in blood pressure elicited by microinjection of a fixed 1-nmol dose of the respective agents into the

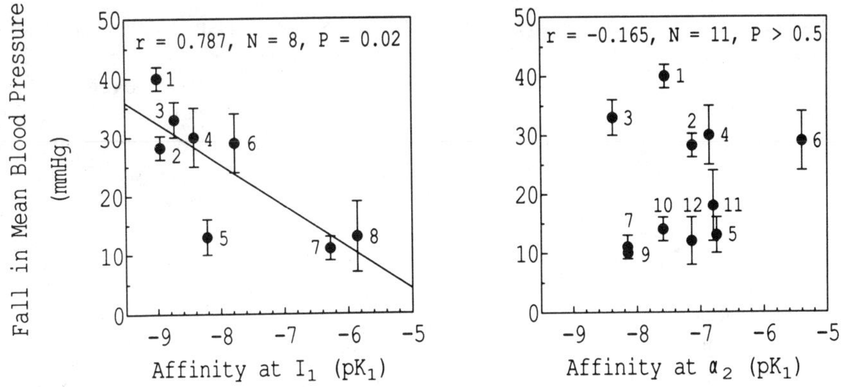

FIGURE 1. Relation between vasodepressor potency and binding affinity at I_1-imidazoline bindings sites (*left panel*) and at α_2-adrenoceptors (*right panel*). *Abscissa:* Affinity expressed as the logarithm of the K_i value (pK_i) at I_1-imidazoline binding sites (*left panel*) and at α_2-adrenoceptors (*right panel*). *Ordinate:* Maximum decrease in blood pressure in mm Hg elicited by bilateral microinjection of the test compound (1 nmol/l) in the rostral ventrolateral medulla in urethane-anesthetized, paralyzed, artificially ventilated normotensive rats. (**1**) clonidine, (**2**) moxonidine, (**3**) *p*-NH_2-clonidine, (**4**) oxymetazoline, (**5**) rilmenidine, (**6**) cimetidine, (**7**) naphazoline, (**8**) imidazol-4-acetic acid, (**9**) guanabenz, (**10**) adrenaline, (**11**) noradrenaline, and (**12**) α-methylnoradrenaline; *r* = correlation coefficient (modified from ref. 63).

RVLM of normotensive rats and the binding affinity of the drugs at I_1-imidazoline binding sites in bovine RVLM.[19,20] By contrast, no correlation was found between binding affinity at α_2-adrenoceptors and vasodepressor potency within the RVLM (FIG. 1). Moveover, the blood pressure-lowering potency of guanfacine and guanabenz is substantially inferior to that of clonidine and moxonidine, even though both guanabenz and guanfacine have equal or even greater affinity for α_2-adrenoceptors than does clonidine. The affinity of moxonidine, moreover, is again 70-fold higher than that of clonidine. Finally, no relation between the decrease in blood pressure elicited by each compound and the efficacy at α_2-adrenoceptors could be determined.

EFFECT OF IMIDAZOLINES' VIGILANCE AND THE RESPIRATORY SYSTEM

The consequences of activating the imidazoline or α_2-receptor population are reflected by the results of psychometric studies in man. Moxonidine and clonidine differ from each other distinctly on the behavioral level. In objective psychometric tests moxonidine hardly differed from placebo, whereas clonidine reduced performance immediately and markedly.[29,30] With the use of various EEG parameters, vigilance was consistently reduced by clonidine, but was only marginally, if at all, affected by moxonidine. On the level of subjective perception, moxonidine did not differ from placebo, whereas clonidine led to strong perceived tiredness and decreased activity. Moreover, moxonidine was without effect on reaction time and driving performance.[31] In this context, the results of recent experimental studies suggest that activation of imidazoline receptors on noradrenergic neurons in the locus coeruleus, in contrast to stimulation of somatoinhibitory α_2-adrenoceptors in this area, *increased* neuronal activity and thereby might enhance vigilance.[32] Thus, moxonidine and rilmenidine at the low blood pressure-lowering dose, by contrast to clonidine, might predominantly stimulate I_1-imidazoline receptors also in the locus coeruleus and might activate only a few of the inhibitory α_2-adrenoceptors on locus coeruleus neurons, thus eliciting no substantial depression of these neurons.

Moreover, structures near the ventral medullary surface are influential in shaping the pattern of hypoglossal nerve activity and in maintaining balanced activity of upper airway and chest wall muscles.[33,34] Specifically, the response of the hypoglossal nerve supplying dilator muscles of the upper airway was strongly inhibited by clonidine but almost unaffected by moxonidine.[35] In hypertensive patients with sleep-related respiratory disorders (sleep apnea), centrally acting antihypertensive drugs of the first generation might elicit deleterious effects by depressing respiratory activity, which may contribute to sleep apnea.[36] Moxonidine, in therapeutic doses, had little or no depressive effect on respiration,[35] most likely because of its selectivity for I_1-imidazoline receptors over α_2-adrenoceptors, suggesting that this drug may be of particular value in hypertensive patients with sleep-related respiratory disorders.

IMIDAZOLINE RECEPTORS IN THE PERIPHERY

Peripheral presynaptic inhibitory imidazoline receptors on sympathetic nerves innervating smooth muscle cells of the blood vessels might also be involved in blood pressure modulation. Activation of presynaptic imidazoline receptors causes inhibition of the release of the vasoconstrictor noradrenaline.[37,38] Furthermore, it has repeatedly been demonstrated that several imidazolines inhibit catecholamine release from the adrenal medulla.[39] Both mechanisms lead to a reduction in sympathetic tone. The centrally mediated blood pressure-lowering effect of antihypertensive drugs with an imidazoline moiety could be supported by these mechanisms.

RENAL EFFECTS OF IMIDAZOLINES

I_1- and I_2-imidazoline binding sites have been detected in the kidney of various species, including man (for a review, see ref. 40), and the presence of the putative endogenous ligand of imidazoline receptors, endazoline, has been demonstrated in rat renal tissue. In rat kidney, activation of imidazoline receptors leads to remarkable natriuresis, probably by blocking Na^+/H^+-antiporter in the renal tubules (FIG. 2; refs. 41–45). Urine flow rate was increased secondary to an increase in solute excretion, and this effect was independent of the renal action of vasopressin. Clonidine mainly increased free water clearance which was completely attenuated by a selective V_2 vasopressin receptor antagonist, and only at higher concentrations was an increase in osmolar clearance (i.e., natriuresis) found.[45,46]

These findings are compatible with the existence of two independent renal targets, namely, α_2-adrenoceptors and I_1-imidazoline receptors. In this context, it is of interest that renal I_1-imidazoline recognition sites are upregulated early in the development

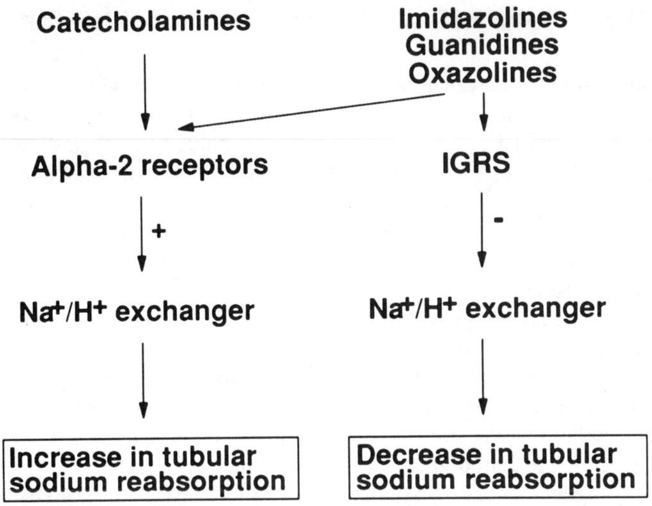

FIGURE 2. Effect of α_2-adrenoceptor (alpha-2 receptor) and imidazoline receptor (IGRS) stimulation on the Na^+/H^+-antiporter inhuman renal proximal tubule (modified from ref. 43).

of hypertension in an angiotensin II infusion model,[47] consistent with a role in blood pressure regulation. Recently, it was also shown that the number of imidazoline sites in rabbit kidney may be modified during chronic treatment with centrally acting antihypertensive drugs and that this effect is dependent on the drugs' affinity for imidazoline receptors.[48]

METABOLIC EFFECTS OF IMIDAZOLINES

A large percentage of hypertensive patients show concomitant disorders of glucose metabolism.[49] Increased sympathetic tone is associated with mobilization of glucose through inhibition of insulin release from pancreatic β-cells, reduced glucose transporter translocation and diminished tyrosine kinase activity of the insulin receptor, increased glycogenolysis and gluconeogenesis in the liver, and increased lipolysis.[8] Direct insulinotropic or insulin-like effects of imidazolines and agmatine (the putative endogenous ligand at imidazoline receptors) have also been described.[50,51] Therefore, compounds that reduce sympathetic outflow without inherent peripheral sympaticomimetic activity can be assumed to have beneficial effects on glucose-intolerant or insulin-resistant states. Moxonidine, as a selective compound for imidazoline I_1 receptors, could therefore be particularly suitable for the treatment of hypertension associated with derangements of glucose metabolism. In a retrospective analysis of fasting plasma glucose values (FPG) in 228 patients treated for 6 weeks with moxonidine 0.1, 0.2, or 0.4 mg bid or a placebo, a trend towards dose-dependent antihyperglycemic activity of moxonidine was observed.[52] Further prospective studies, however, are required to investigate the metabolic effects of moxonidine in patients with pathological plasma glucose or lipid values.

ENDOGENOUS LIGAND AT IMIDAZOLINE RECEPTORS

The physiological relevance of imidazoline receptors in blood pressure regulation is emphasized by the isolation of an endogenous ligand at these receptors, a noncatecholamine of low molecular weight and not susceptible to peptide digestion, which was termed endazoline. The affinity of endazoline for I_1-imidazoline recognition sites is about 30 times higher than that for α_2-adrenoceptors.[53–55] By contrast to catecholamines, endazoline cross-reacts with clonidine antibodies.[56–58] Although the exact chemical structure of endazoline is unknown, it can be determined in most tissues and in blood serum by radioimmune techniques.[54,59] The level of endazoline in blood serum is mainly due to endazoline release from the adrenal medulla.[59] Thus, endazoline may act as a circulating hormone with far-reaching response in multiple target organs. Changes in the concentration of this putative endogenous ligand of imidazoline receptors might be linked directly to the genesis and severity of some forms of essential hypertension.[55,60] In preliminary investigations the concentrations of endazoline in the blood serum of patients with nontreated hypertension were considerably higher than those in normotensives subjects.[61,62] It should be emphasized that endazoline seems to be the physiological correlate for the imidazoline receptors. Endazoline is the endogenous I_1-imidazoline receptor agonist that seems to be one of the mediators centrally regulating blood pressure. If endazoline is responsible for a blood pressure increase, moxonidine should be classified as an antagonist. However, if endazoline decreases blood pressure physiologically, moxonidine should be described as an agonist. Therefore, moxonidine could inhibit the putative receptor like a blocking agent, whereas enhanced sympathetic tone is induced by endazoline. In the "black box of essential hypertension," the balance between blood pressure-elevating and blood pressure-decreasing mediators may be a relevant pathophysiological problem. It is reasonable to assume that endazoline could be a blood pressure-upregulating mediator, and in that case moxonidine could be a potential antagonist. Thus, antihypertensive therapy with imidazoline derivatives might be a physiological therapeutic regimen for some forms of essential hypertension.

CONCLUSION

Moxonidine and rilmenidine represent a new generation of centrally acting antihypertensive drugs that decrease arterial blood pressure via a completely new mechanism of action (FIG. 3). The drugs bind more selectively to I_1-imidazoline receptors than to α_2-adrenoceptors in the RVLM (TABLE 1). Ample evidence indicates that these I_1-imidazoline receptors may be essentially responsible for regulating arterial blood pressure, whereas central α_2-adrenoceptors mediate the side effects of centrally acting antihypertensive drugs such as sedation, bradycardia, or dry mouth. By contrast to clonidine, a first-generation agent, moxonidine and rilmenidine did not reduce heart rate. Moreover, rebound hypertension has not been observed in preclinical or clinical trials with moxonidine. The decrease in plasma catecholamine concentrations, plasma renin activity, and other hormones involved in hemodynamic regulation clearly demonstrates the decrease in central nervous sympathetic activity. In particular, reduction in plasma renin activity and, consequently, in angiotensin II concentrations might

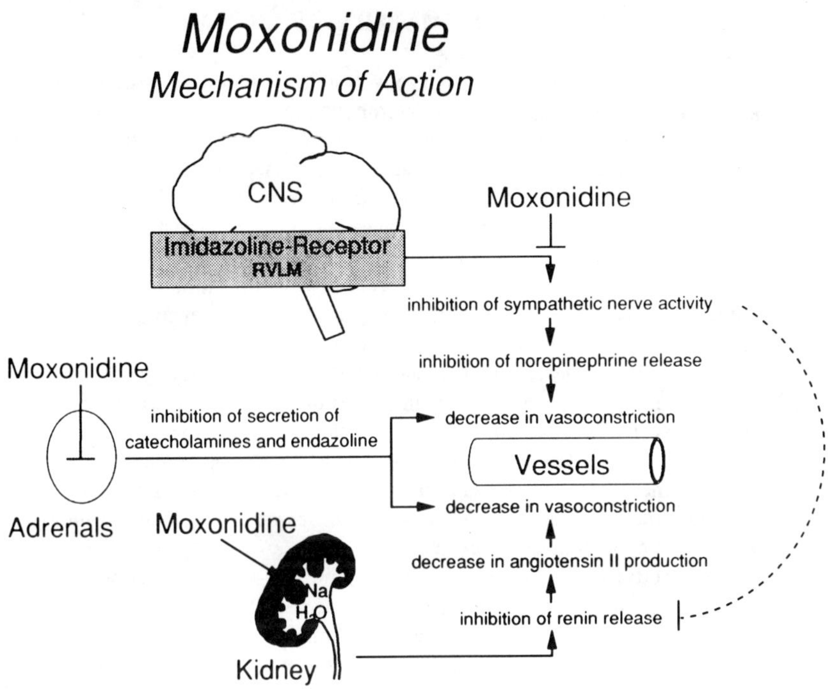

FIGURE 3. Central and peripheral mechanisms by which imidazoline receptor agonists might reduce arterial blood pressure. CNS = central nervous system (modified from ref. 64).

prevent sodium and water retention (FIG. 3). Furthermore, renal imidazoline receptors also are potential targets for rilmenidine and moxonidine to mediate a therapeutic effect, because renal imidazoline receptors seem to be responsible for increased sodium excretion (FIGS. 2 and 3), an additional benefit in the treatment of hypertension. Activation of presynaptic inhibitory imidazoline receptors as well as imidazoline recognition sites that inhibit release of catecholamines and endazoline from adrenal medulla might also contribute to the blood pressure-lowering effect of the second-generation antihypertensive drugs (FIG. 3).

Thus, second-generation antihypertensive drugs moxonidine and rilmenidine represent antihypertensive drugs that should have a promising benefit-to-risk ratio because of their high selectivity for I_1-imidazoline receptors. The drugs launched in France and Germany, respectively, and registered in the EU (moxonidine) will overcome the shortcomings that have kept centrally acting antihypertensive drugs from being first-line treatment.

Because the more favorably profile is due to moxonidine's selectivity for I_1-imidazoline receptors (moxonidine is the most selective and most potent agent at I_1-imidazoline receptors described to date), the development of moxonidine and rilmenidine represents a real advance in the treatment of hypertension through selective modulation of CNS pathways that regulate vasomotor tone. Hypertension can be treated by moxonidine by acting selectively at the I_1-imidazoline receptor.

SUMMARY

The influence of the sympathetic nervous system on blood pressure control was impressively demonstrated in 1940 by bilateral excision of sympathetic nerve fibers. Thereafter, the first generation of drugs lowering blood pressure by central modulation of the sympathetic outflow through α_2-adrenoceptor for stimulation, such as α-methyldopa, guanabenz, clonidine, and guanfacine, were marketed. However, these compounds were often tolerated poorly, because they caused orthostatic hypotension, sedation, tachycardia or bradycardia, dry mouth, and reduced cardiac output. The mode of action of the second generation centrally acting antihypertensive drugs moxonidine and rilmenidine is different from that of the first generation compounds (e.g., clonidine). Contrary to clonidine, the newer drugs bind more selectively to I_1-imidazoline receptors rather than to α_2-adrenoceptors where first-generation drugs act. The high affinity and selectivity of these two drugs for this recently discovered new receptor class make it possible to discriminate between I_1-imidazoline receptor-mediated blood pressure lowering, on the one hand, and α_2-adrenoceptor-mediated side effects, on the other. Discrimination of the two effects was substantiated either by studies using moxonidine alone or in interaction experiments with I_1-imidazoline receptor or α_2-adrenoceptor antagonists.

The high selectivity of moxonidine at the I_1-imidazoline receptor allows discrimination between α_2-adrenoceptors and I_1-imidazoline receptors and is reflected in man by the relatively low incidence of adverse drug events during moxonidine treatment.

Concentration of endazoline, a specific mediator of I_1-imidazoline receptors, is elevated in some patients with essential hypertension. Modulation of I_1-imidazoline receptors by moxonidine could be interpreted as antagonism with regard to the endogenous agonistic effect of the endogenous "transmitter" endazoline. On the other hand, moxonidine acted directly as an agonist at the putative I_1-imidazoline receptor. Therefore, to clear the ground, characterization as well as physiological function of the mediator for imidazoline receptors seems essential. The therapeutic relevance of using drugs selective for I_1-imidazoline receptors for blood pressure reduction in hypertensive patients is substantiated by the finding that in human rostral ventrolateral medulla (RVLM), which is essential in central blood pressure regulation, the relation between α_2-adrenoceptors and I_1-imidazoline receptors is about one to ten (1 : 10). Reduction of a long-lasting sympathetic overdrive may avoid the deteriorating effects on the heart and peripheral circulation. These recent findings give a rational explanation for the very low incidence of sedation and the absence of respiratory depression, orthostatic hypotension, and rebound hypertension that banned the former central acting antihypertensive drugs from first-line treatment despite the advantages of central mediated blood pressure control.

REFERENCES

1. JULIUS, S. & A. B. WEDER. 1989. Brain and the regulation of blood pressure: A hemodynamic perspective. Clin. Exp. Hypertens. **11**(Suppl. 1): 1-19.
2. FERRARIO, C. M. & D. B. AVERILL. 1991. Do primary dysfunctions in neural control of arterial pressure contribute to hypertension? Hypertension **18**(Suppl.): 138-151.
3. LONG, C. S., K. KARIYA, L. KARNS & P. C. SIMPSON. 1991. Sympathetic activity: Modulator of myocardial hypertrophy. J. Cardiovasc. Pharmacol. **17**(Suppl. 2): S20-S24.

4. KAPLAN, J. R., K. PETTERSSON, S. B. MANUCK & G. OLSSON. 1991. Role of sympathoadrenal medullary activation in the initiation and progression of atherosclerosis. Circulation **84**(Suppl. 6): VI23-VI32.
5. ERNSBERGER, P. & D. C. U'PRICHARD. 1987. Modulation of agonist and antagonist interactions at kidney α_1-adrenoceptors by nucleotides and metal ions. Eur. J. Pharmacol. **133:** 165-176.
6. DIBONA, G. F. 1992. Sympathetic neural control of the kidney in hypertension. Hypertension **19**(Suppl. 1): 128-135.
7. ROSEN, M. R., P. DANILO, JR., R. B. ROBINSON, A. SHAH & S. F. STEINBERG. 1988. Sympathetic neural and α-adrenergic modulation of arrhythmias. Ann. N.Y. Acad. Sci. **533:** 200-209.
8. RUPP, H. & R. JACOB. 1993. Antihypertensiva bei Hyperinsulinämie: Durch Sympathikusregulation ans Ziel. Therapiewoche **43:** 1686-1693.
9. REIS, D. J., S. M. MORRISON & D. RUGGIERO. 1988. The C1 area of the brainstem in tonic and reflex control of blood pressure. Hypertension **11**(2 Pt 2): I8-I13.
10. REIS, D. J., D. A. RUGGIERO & S. F. MORRISON. 1989. The C1 area of the rostral ventrolateral medulla oblongata. A critical brainstem region for control of resting and reflex integration of arterial pressure. Am. J. Hypertens. **2**(12 Pt 2): 363S-374S.
11. SANNAJUST, F., C. BARRES, E. KOENIG-BERARD & J. SASSARD. 1992. Sympathoinhibitory effects of rilmenidine may be mediated by sites located below the brainstem. Br. J. Pharmacol. **105:** 535-541.
12. TIBIRICA, E., J. FELDMAN & P. BOUSQUET. 1992. Contribution of catecholaminergic neurons of the dorsomedial and ventrolateral medulla oblongata to the hypotensive effect of clonidine in the spontaneously hypertensive rats: *In vivo* voltammetric studies. J. Hypertens. **10:** 1327-1334.
13. SZABO, B., R. URBAN & K. STARKE. 1993. Sympathoinhibition by rilmenidine in conscious rabbits: Involvement of α_2-adrenoceptors. Fundam. Clin. Pharmacol. **7:** 382.
14. KOBINGER, W. 1978. Central alpha-adrenergic systems as targets for hypotensive drugs. Rev. Physiol. Biochem. Pharmacol. **81:** 39-100.
15. BOUSQUET, P. & J. SCHWARTZ. 1983. Alpha-adrenergic drugs: Pharmacological tools for the study of the central nervous system. Biochem. Pharmacol. **32:** 1459-1465.
16. SVED, A. F., K. TSUKAMOTO & A. M. SCHREIHOFER. 1992. Stimulation of α_2-adrenergic receptors in nucleus tractus solitarius is required for the baroreceptor reflex. Brain Res. **576:** 297-303.
17. BRICCA, G., M. DONTENWILL, A. LONIES, J. FELDMAN, A. BELCOURT & P. BOUSQUET. 1989. The imidazoline preferring receptor: Binding studies in bovine, rat and human brainstem. Eur. J. Pharmacol. **162:** 1-9.
18. BRICCA, G., M. DONTENWILL, A. MOLINES, J. FELDMAN, E. TIBIRICA, A. BELCOURT & P. BOUSQUET. 1989. Rilmenidine selectivity for imidazoline receptors in human brain. Eur. J. Pharmacol. **163:** 373-377.
19. ERNSBERGER, P., R. GIULIANO, R. N. WILLETTE & D. J. REIS. 1990. Role of imidazole receptors in the vasodepressor response to clonidine analogs in the rostral ventrolateral medulla. J. Pharmacol. Exp. Ther. **253:** 408-418.
20. GOMEZ, R. E., P. ERNSBERGER, G. FEINLAND & D. J. REIS. 1991. Rilmenidine lowers arterial pressure via imidazole receptors in brainstem C1 area. Eur. J. Pharmacol. **195:** 181-191.
21. TIBIRICA, E., J. FELDMAN, C. MERMET, F. GONON & P. BOUSQUET. 1991. An imidazoline-specific mechanism for the hypotensive effect of clonidine: A study with yohimbine and idazoxan. J. Pharmacol. Exp. Ther. **256:** 606-613.
22. TIBIRICA, E., J. FELDMAN, C. MERMET, L. MONASSIER, F. GONON & P. BOUSQUET. 1991. Selectivity of rilmenidine for the nucleus reticularis lateralis, a ventrolateral medullary structure containing imidazoline-preferring receptors. Eur. J. Pharmacol. **209:** 213-221.

23. Bousquet, P., J. Feldman & J. Schwartz. 1984. Central cardiovascular effects of alpha adrenergic drugs: Differences between catecholamines and imidazolines. J. Pharmacol. Exp. Ther. **230:** 232-236.
24. Feldman, J., E. Tibirica, G. Bricca, M. Dontenwill, A. Belcourt & P. Bousquet. 1990. Evidence for the involvement of imidazoline receptors in the central hypotensive effect of rilmenidine in the rabbit. Br. J. Pharmacol. **100:** 600-604.
25. Haxhiu, M. A., I. Dreshai, B. Erokwu & P. Ernsberger. 1992a. Vasodepression elicited in hypertensive rats and normal cats by the selective I_1-imidazoline agonist moxonidine administered into the rostral ventrolateral medulla (RVLM). Fundam. Clin. Pharmacol. **6:** 58s.
26. Haxhiu, M. A., I. Dreshaj, B. Erokwu, S. G. Schäfer, M. O. Christen & P. R. Ernsberger. 1992. Vasodepression elicited in hypertensive rats by the selective I_1-imidazoline agonist moxonidine administered into the rostral ventrolateral medulla. J. Cardiovasc. Pharmacol. **20**(Suppl. 4): 11-15.
27. Haxhiu, M. A., I. Dreshaj, S. G. Schäfer & P. Ernsberger. 1993. Selective antihypertensive action of moxonidine is mediated mainly by I_1-imidazoline receptors in the rostral ventrolateral medulla. J. Cardiovasc. Pharmacol. In press.
28. Mermet, C. & L. Quintin. 1991. Effect of clonidine on catechol metabolism in the rostral ventrolateral medulla: An *in vivo* electrochemical study. Eur. J. Pharmacol. **204:** 105-107.
29. Ziegler, G. & A. Gemeinhardt. 1988. ZNS-Effekte von Moxonidin, einem neuen zentralnervös wirkenden Antihypertensivum, im Vergleich zu Clonidin und Placebo. Klin. Wochenschr. **66**(Suppl. XIII): 41-42.
30. Hermann, W. M. 1989. Pharmaco-EEG study and psychometric study with moxonidine versus clonidine and placebo in normotensives to determine the sedative and/or performance-decreasing potential. Controlled, randomized, double-blind study in a threefold crossover design. Kali-Chemie AG, Report No. K.220.5002.
31. Schmidt, U., H. Frerick, K. Kraft, N. Schenk & A. Löw-Kröger. 1992. Hypertension: A possible risk in road traffic. J. Cardiovasc. Pharmacol. **20**(Suppl. 4): S50-S56.
32. Pineda, J., L. Ugedo & J. A. Garcia-Sevilla. 1993. Stimulatory effects of clonidine, cirazoline and rilmenidine on locus coeruleus noradrenergic neurones: Possible involvement of imidazoline-preferring receptors. Naunyn-Schmiedeberg's Arch. Pharmacol. **348:** 134-140.
33. Haxhiu, M. A., J. Mitra, E. van Lunteren, E. N. Bruce & N. S. Cherniack. 1984. Hypoglossal and phrenic responses to cholinergic agents applied to the ventral medullary surface. Am. J. Physiol. **247:** R939-R944.
34. Haxhiiu, M. A., J. Mitra, E. van Lunteren, N. R. Prabhakar, E. N. Bruce & N. S. Cherniack. 1986. Responses of hypoglossal and phrenic nerves to decreased respiratory drive in cats. Respiration **50:** 130-138.
35. Haxhiu, M. A., I. Dreshaj, S. G. Schäfer & P. Ernsberger. 1993. The I_1-imidazoline agonist moxonidine is a selective central antihypertensive: lack of depressant action on central respiratory drive. Cardiovasc. Drugs Ther. **7**(Suppl. 2): 425.
36. Strohl, K. P., N. S. Cherniack & B. Gothe. 1986. Physiological basis for therapy of sleep apnea. Am. Rev. Resp. Dis. **134:** 791-802.
37. Göthert, M. & G. J. Molderings. 1991. Involvement of presynaptic imidazoline receptors in the α_2-adrenoceptor-independent inhibition of noradrenaline release by imidazoline derivatives. Naunyn-Schmiedeberg's Arch. Pharmacol. **343:** 271-282.
38. Molderings, G. J., F. Hentrich & M. Göthert. 1991. Pharmacological characterization of the imidazoline receptor which mediates inhibition of noradrenaline release in the rabbit pulmonary artery. Naunyn-Schmiedeberg's Arch. Pharmacol. **344:** 630-638.
39. Ohara-Imaizumi, M. & K. Kumakura. 1992. Effects of imidazole compounds on catecholamine release in adrenal chromaffin cells. Cell. Mol. Neurobiol. **12:** 273-283.

40. MOLDERINGS, G. J., M. C. MICHEL, M. GÖTHERT, O. CHRISTEN & S. G. SCHÄFER. 1992. Imidazolrezeptoren: Angriffsort einer neuen Generation von antihypertensiven Arzneim. Dtsch. Med. Wochenschr. **117:** 67-71.
41. ARMAH, B. L., E. HOFFERBER & W. STENZEL. 1988. General pharmacology of the novel centrally acting antihypertensive agent moxondine. Drug Res. **38:** 1426-1434.
42. BIDET, M., P. POUJEOL & A. PARINI. 1990. Effect of imidazolines on Na^+ transport and intracellular pH in renal proximal tubule cells. Biochim. Biophys. Acta **1024:** 173-178.
43. LIMON, I., I. COUPRY, F. TESSON, V. LACHAUD-PETTITI & A. PARINI. 1992. Renal imidazoline-guanidinium receptive site. A potential target for antihypertensive drugs. J. Cardiovasc. Pharmacol. **20**(Suppl.): S21-S23.
44. SMYTH, D. D., D. R. ALLAN, P. LI & S. B. PENNER. 1992. Direct renal actions of moxonidine are distinct from renal α_2-adrenoceptors. Fundam. Clin. Pharmacol. **6:** 50s.
45. ALLAN, D. R., S. B. PENNER & D. D. SMYTH. 1993. Renal imidazoline preferring sites and solute excretion in the rat. Br. J. Pharmacol. **108:** 870-875.
46. BLANDFORD, D. E. & D. D. SMYTH. 1991. Potentiation of the effects of renal α_2-adrenoceptor stimulation by indomethacin. Can. J. Physiol. Pharmacol. **69:** 1196-1203.
47. ERNSBERGER, P., C. SIMS & J. G. DOUGLAS. 1991. I_1-imidazoline sites in rat renal medulla: Binding properties and up-regulation by subpressor angiotensin infusion. Hypertension **18:** 380.
48. HAMILTON, C. A., E. JARDINE & J. L. REID. 1993. Down-regulation of imidazoline sites in rabbit kidney. Eur. J. Pharmacol. **243:** 95-97.
49. ZAVARONI, I., S. MAZZA, E. DALL'AGLIO, P. GASPARINI, M. PASSERI & S. M. REAVAN. 1992. Prevalence of hyperinsulinaemia in patients with high blood pressure. J. Int. Med. **231:** 235-240.
50. SCHULZ, A. & A. HASSELBLATT. 1989. Dual action of clonidine on insulin release: Suppression, but stimulation when α_2-adrenoceptors are blocked. Naunyn-Schmiedeberg's Arch. Pharmacol. **340:** 712-714.
51. WEITZEL, G., R. RENNER & H. GUGLIELMI. 1971. Insulinähnliche Aktivität von Arginylverbindungen in vitro. Hoppe-Seyler's Z. Physiol. Chem. **352:** 1617-1630.
52. KAAN, E. C., R. BRÜCKNER & D. ZIEGLER. 1994. Patent application P 4423177.6.
53. ERNSBERGER, P., M. P. MEELEY & D. J. REIS. 1988. An endogenous substance with clonidine-like properties: Selective binding to imidazole sites in the ventrolateral medulla. Brain Res. **441:** 309-318.
54. ATLAS, D. 1991. Clonidine-displacing substance (CDS) and its putative imidazoline receptor. Biochem. Pharmacol. **41:** 1541-1549.
55. REIS, D. J., S. REGUNATHAN & M. P. MEELEY. 1992. Imidazoline receptors and clonidine-displacing substance in relationship to control of blood pressure, neuroprotection, and adrenomedullary secretion. Am. J. Hypertens. **5:** 51S-57S.
56. ATLAS, D. & Y. BURSTEIN. 1984. Isolation and partial purification of a clonidine-displacing endogenous brain substance. Eur. J. Biochem. **144:** 287-293.
57. DONTENWILL, M., G. BRICCA, A. MOLINES, A. BELCOURT & P. BOUSQUET. 1987. A polyclonal antibody raised against clonidine: A model for the specific imidazoline receptor. Eur. J. Pharmacol. **137:** 143-144.
58. MEELEY, M. P., A. C. TOWLE, P. ERNSBERGER & D. J. REIS. 1988. A specific antiserum recognizes clonidine-displacing substance: Implications for the structure of the brain's own clonidine. Neurosci. Lett. **84:** 84-90.
59. MEELEY, M. P., M. L. HENSLEY, P. ERNSBERGER, D. FELSEN & D. J. REIS. 1992. Evidence for a bioactive clonidine-displacing substance in peripheral tissues and serum. Biochem. Pharmacol. **44:** 733-740.
60. BOUSQUET, P., J. FELDMAN, E. TIBIRICA, G. BRICCA, A. MOLINES, M. DONTENWILL & A. BELCOURT. 1989. New concepts on the central regulation of blood pressure. Am. J. Med. **87:**(Suppl. 3C): 10S-13S.

61. KREISBERG, G. A., D. DIAMANT, Y. Z. DIAMANT & D. ATLAS. 1987. Raised levels of an endogenous nonadrenergic substance in the serum of pregnancy-induced hypertension patients. Isr. J. Med. Sci. **23:** 1194-1197.
62. DONTENWILL, M., A. MOLINES, A. VERDUN, G. BRICCA, A. BELCOURT & P. BOUSQUET. 1992. A circulating imidazoline-like substance crossreacts with anti-clonidine antibodies; high levels in hypertensive patients. Fundam. Clin. Pharmacol. **6:**(Suppl. 1): 49s.
63. ERNSBERGER, P. R., K. L. WESTBROOKS, M. O. CHRISTEN & S. G. SCHÄFER. 1992. A second generation of centrally acting antihypertensive agents act on putative I_1-imidazoline receptors. J. Cardiovasc. Pharmacol. **20**(Suppl. 4): S1-S10.
64. MOLDERINGS, G. J., M. GÖTHERT, O. CHRISTEN & S. G. SCHÄFER. 1993. Imidazolrezeptoren und Blutdruckregulation—Hohe Rezeptorselektivität von Moxonidin. Dtsch. Med. Wochenschr. **118:** 953-958.

Clinical Pharmacology of Drugs Acting on Imidazoline and Adrenergic Receptors

Studies with Clonidine, Moxonidine, Rilmenidine, and Atenolol

JOHN L. REID,[a] VSEVOLOD PANFILOV, GRAHAM MacPHEE, AND HENRY L. ELLIOTT

Division of Clinical Pharmacology
Department of Medicine and Therapeutics
University of Glasgow
Glasgow, Scotland

Hypertension, or raised blood pressure, is one of the established risk factors for atherosclerosis and an important predictor of death and disability from coronary artery disease and stroke.[1] Considerable progress has been made in recent years in understanding the mechanisms of blood pressure regulation. However, except in a minority of patients with well characterized secondary hypertension, it is not yet possible to identify the factors underlying raised blood pressure. There are clearly genetic and environmental factors. Raised blood pressure is likely to include functional disturbances in the renin-angiotensin-aldosterone system, responsiveness to sodium intake, and central or peripheral disturbances in autonomic function.

Regardless of the mechanism of elevated blood pressure in essential hypertension, increasingly convincing evidence indicates that reduction in blood pressure, even in the elderly and in those with mild to moderate hypertension, substantially reduces stroke risk and significantly reduces coronary heart disease sequelae.[2] Blood pressure can be reduced by a range of drugs with widely differing mechanisms of action. There is to date no good evidence that the benefit of blood pressure reduction is limited to the effects of a particular group of drugs.[3] It is the ultimate aim of antihypertensive therapy, however, to identify the mechanisms that cause and maintain high blood pressure and to develop treatment strategies that specifically and selectively interfere with these mechanisms.

Drugs acting on the sympathetic nervous system either peripherally (ganglion blockers) or centrally (reserpine) were among the earliest effective treatments of hypertension. Subsequently these approaches were refined by the development of adrenergic neuron blockers and $alpha_1$ and beta receptor antagonists to inhibit peripheral sympathetic function.

[a] Address for correspondence: Dr. J. L. Reid, Department of Medicine and Therapeutics, Western Infirmary, Glasgow, G11 6NT, Scotland.

Alpha-methyldopa and clonidine represented parallel approaches to improving centrally acting antihypertensive therapy.[4] Although these approaches resulted in more consistent decreases in blood pressure, treatment with these agents was associated with frequent complaints of central nervous system side effects of sedation and drowsiness together with dry mouth and stuffy nose. At least with clonidine, these adverse effects were dose (and plasma concentration) related and in many patients they limited the clinical usefulness of centrally acting drugs.[5] Several hypotheses were proposed in the 1960s and 1970s to explain the actions of methyldopa. Until recently it was assumed that alpha-methyldopa, through its metabolite alpha-methylnorepinephrine, acted like clonidine as an $alpha_2$ adrenoceptor agonist in the central nervous system.[6] Experimental studies in animals and clinical studies in man suggested that sedation and dry mouth were a consequence of interaction with an $alpha_2$ receptor similar to one that lowered blood pressure. Recent work based on radioligand binding suggested that in addition to the $alpha_2$ adrenoceptor an additional binding site preferred by imidazoline compounds might be involved in the central regulation of blood pressure.[7] Evidence for imidazoline-preferring receptors and the possibility of an endogenous ligand for these receptors has been reviewed elsewhere.[7,8] There is evidence of functional as well as binding differences between "classic" $alpha_2$ adrenoceptors and these imidazoline binding sites. These findings have led to the hypothesis that imidazoline compounds may be useful antihypertensive drugs that lower blood pressure without the same level of central side effects as experienced with $alpha_2$ adrenoceptor agonists.

We have had an opportunity in recent years to study two new imidazoline compounds, rilmenidine and moxonidine, in both normal subjects and in patients with essential hypertension. We compared these new compounds with clonidine and with the $beta_1$ selective adrenoceptor antagonist atenolol. Our studies support the suggestion that imidazoline compounds differ from drugs acting directly on adrenoceptors.

CLINICAL PHARMACOLOGY OF CLONIDINE

Clonidine was introduced into hypertension treatment in the late 1960s. It was proposed to lower blood pressure by a central sympathoinhibitory effect. Even in early clinical studies, however, its use was associated with increased frequency of drowsiness, sedation, and complaints of dry mouth. Some years after its introduction we undertook a comprehensive series of investigations into the clinical pharmacology of clonidine in normotensive subjects[9] and in patients with essential hypertension.[10] Our aim was not only to define the effects of acute and chronic dosing on blood pressure and heart rate but also to measure in quantitative terms the major side effects of sedation and dry mouth. Together with the measurement of plasma concentrations of clonidine we aimed to define the concentration effect relationship and to determine both the mechanism of action and factors determining response. In the course of these studies we confirmed the likely sympathoinhibitory mechanism by showing that in normal and hypertensive subjects a decrease in blood pressure was accompanied by a decrease in circulating plasma norepinephrine.[11] The central site of action was supported by studies in patients with specific neurological deficit including a group with high cervical cord transection[12] and another group with a range of autonomic neuropathies.[13]

In both normal subjects[9] and patients,[10] both sedation and dry mouth were closely and steeply related to plasma drug concentration. Within the same concentration range, blood pressure was related to concentration in a more complex manner. There was a positive linear relation, lower drug levels, and apparent conversion to a negative relation at the highest levels studied. This "bell-shaped" relation was observed in both normal and hypertensive subjects. It was suggested that although the initial depressor effect was related to central $alpha_2$ agonist stimulation, the later high concentration pressor effect was the result of peripheral postjunctional $alpha_2$ receptor activation.[10] Whatever the mechanism, these studies suggested that clonidine had a relatively low therapeutic index and confirmed that in many patients adverse effects were likely to occur within the therapeutic dose range. These studies supported the view that clonidine would likely be most useful at lower dose levels, and the relatively short duration of effect stimulated work on alternative formulations including transdermal delivery.

We used the strategy developed to assess clonidine to examine the newer imidazoline derivatives.

CLONIDINE AND MOXONIDINE IN NORMAL SUBJECTS

Moxonidine was developed as a centrally acting antihypertensive drug with features different from those of clonidine. Early clinical studies suggested that its hypotensive response was similar to that of clonidine but it had significantly fewer adverse effects, particularly sedation and dry mouth.[14] More recently moxonidine was shown to interact preferentially with imidazoline binding sites.[15] We undertook a double-blind controlled trial with clonidine 200 μg, moxonidine 200 μg orally, and placebo in a balanced Latin Square design in nine healthy, normotensive male subjects. Hemodynamic, behavioral, and autonomic effects were measured and subjective side effects and drug plasma levels were quantified. The results of this study were published in full.[16] In these salt-replete normal subjects both moxonidine and clonidine significantly reduced both supine and standing systolic and diastolic blood pressure when compared to placebo. Clonidine, however, had a small but consistently significantly greater effect on blood pressure than did moxonidine. Sedation was assessed with a visual analog scale, and compared to placebo clonidine caused a marked sedative effect that was maximal at 2 hours and lasted more than 8 hours. Moxonidine did not have any statistically significant effect on sedation. Similarly, salivary flow assessed by weighed cotton rolls was consistently and significantly reduced by clonidine than by either placebo or moxonidine. Psychomotor evaluation of critical flicker fusion threshold and choice reaction time showed substantial and consistent effects of clonidine, whereas moxonidine did not differ from placebo. The results of this study were consistent with the view that moxonidine could have a different profile of effects on sedation and dry mouth. It is unfortunate that although previous clinical studies suggested that 200 μg of clonidine and moxonidine were equivalent in terms of their ability to lower blood pressure,[15,17] this finding was not confirmed in normotensive subjects in the present study.

RILMENIDINE AND ATENOLOL IN ESSENTIAL HYPERTENSION

Rilmenidine is also a new drug that apparently acts via the central nervous system and interacts preferentially with imidazoline receptors.[18] Early clinical pharmacological studies in normal subjects confirmed its blood pressure lowering activity and suggested that at least compared to clonidine side effects of sedation and dry mouth might be less prominent.[19–21] Clinical trials comparing it to placebo and to other centrally acting drugs confirm its efficacy and improved patient tolerability. We recently completed a double-blind cross-over comparison of rilmenidine and atenolol on autonomic function, mental stress, and dynamic exercise in patients with mild to moderate essential hypertension. Twelve male patients were recruited and all completed a cross-over study of 4 weeks of atenolol (50 or 100 mg) and 4 weeks of rilmenidine (1 or 2 mg). Each active treatment period was preceded by 4 weeks of treatment with an identical placebo respectively as a run-in and wash-out period. Patients were studied on four occasions at the end of each placebo treatment period and after 4 weeks of active atenolol or rilmenidine treatment. The aim of the study was to compare observations at the end of the active treatment periods.

Both rilmenidine and atenolol were well tolerated and no patients dropped out of the study. They both caused similar and significant decreases in supine and standing systolic and diastolic blood pressure. Tests of parasympathetic and sympathetic autonomic function as well as isometric and dynamic (bicycle) exercise and mental stress were evaluated. Rilmenidine had no effect on any of the autonomic tests or stresses examined. Atenolol had predictable beta adrenoceptor antagonist actions in limiting the heart rate increase on exercise and modifying the heart rate response to the Valsalva maneuver. From this study we conclude that rilmenidine is an effective, well tolerated antihypertensive drug that does not interfere with physiological responses to physical, mental stress, or autonomic regulation.

ROLE OF IMIDAZOLINE AGENTS IN CLINICAL PRACTICE

Centrally acting antihypertensive drugs offer a potentially attractive strategy for treatment of hypertension. Not only are they effective and safe but they may also, at least in some patients, modify an important underlying mechanism for hypertension. The disadvantage of this class of drugs until now has been the relatively high frequency of centrally mediated symptomatic side effects which appeared as an inevitable accompaniment to the blood pressure reduction. In recent years there has been an increasing understanding of the potential importance of imidazoline binding sites compared to classical alpha$_2$ adrenoceptors. These developments resulted from *in vitro* radioligand binding studies of alpha adrenoceptor heterogeneity. However, there is now increasing support for a distinct functional role of these imidazoline-preferring receptors. Clinical studies with imidazoline derivatives lend support to the view that the actions of imidazolines can be clinically distinguished from those of adrenoceptor agonists and antagonists. Imidazolines lower blood pressure, and evidence to date is consistent with the less frequent appearance and severity of central side effects of drowsiness, sedation, and also dry mouth. If these experiences can be confirmed in further clinical practice, centrally acting imidazoline antihypertensive drugs could become an important new choice in the therapy of hypertension.

SUMMARY

Centrally acting antihypertensive drugs are recognized to be safe and effective treatment for high blood pressure. Centrally mediated side effects, such as sedation, are commonly dose- and treatment-limiting events. Imidazoline-preferring receptors, while functionally similar to $alpha_2$ adrenoceptors, are distinguishable not only on the basis of *in vitro* radioligand binding but also *in vivo* in terms of side effects. Drugs with an imidazoline structure lower blood pressure but are less likely to impair psychomotor function. A placebo-controlled study compared moxonidine 0.1 mg with clonidine 0.1 mg orally in nine normal subjects. Both active drugs lowered blood pressure compared to placebo (clonidine more than moxonidine). However, psychomotor function and self-scored sedation and dry mouth were significantly affected only by clonidine. In a long-term (4 weeks) double-blind cross-over study in essential hypertension, rilmenidine was well tolerated and had similar effects to those of atenolol on erect and supine blood pressure. Rilmenidine had no effect on a wide range of autonomic and psychomotor tests or on responses to mental or physical stress. Atenolol, by contrast, had the predicted effects of a beta adrenoceptor antagonist on heart rate during exercise and the Valsalva maneuver. Imidazoline-preferring drugs offer a new and realistic approach to antihypertensive therapy with blood pressure reduction not limited by marked sedation within the therapeutic dose range.

ACKNOWLEDGMENTS

The studies described in this review were undertaken in collaboration with clinical, nursing, and technical colleagues at the Department of Clinical Pharmacology, Royal Postgraduate Medical School, London, the Department of Materia Medica, Stobhill Hospital, Glasgow, and the Department of Medicine and Therapeutics, Western Infirmary, Glasgow. Mary Wood kindly typed the manuscript.

REFERENCES

1. MacMahon, S., R. Peto, J. Cutler, R. Collins, P. Sorlie, J. Neaton *et al.* 1990. Blood pressure, stroke and coronary heart disease. Part 1. Prolonged differences in blood pressure: Prospective observational studies corrected for the regression dilution bias. Lancet **335:** 765-774.
2. Collins, R., R. Peto, S. MacMahon, P. Hebert, N. H. Fiebach, K. A. Eberlein *et al.* 1990. Blood pressure, stroke and coronary heart disease. Part 2. Short-term reductions in blood pressure: Overview of randomised drug trials in their epidemiological context. Lancet **335:** 827-838.
3. Sever, P., G. Beevers, C. Bulpitt, A. Lever, L. Ramsay, J. Reid *et al.* 1993. Management guidelines in essential hypertension: Report of the Second Working Party of the British Hypertension Society. Br. Med. J. **306:** 983-987.

4. Reid, J. L. 1979. Central action of antihypertensive drugs. *In* Drug Therapeutics. K. Melmon, Ed.: 135-150. Elsevier. New York.
5. Reid, J. L. 1981. The clinical pharmacology of clonidine and related central antihypertensive agents. Br. J. Clin. Pharmacol. **12:** 295-302.
6. Reid, J. L. & H. L. Elliott. 1988. Methyldopa. *In* Handbook of Hypertension. A. E. Doyle, Ed. **3:** 103-124. Elsevier. Amsterdam.
7. Bousquet, P., J. Feldman & J. Schwartz. 1984. Central cardiovascular effects of alpha adrenergic drugs: Differences between catecholamines and imidazolines. J. Pharmacol. Exp. Ther. **230:** 232-236.
8. Ernsberger, P., M. P. Meeley & D. J. Reis. 1988. An endogenous substance with clonidine-like properties. Brain Res. **441:** 309-318.
9. Dollery, C. T., D. S. Davies, G. H. Draffan, H. J. Dargie, C. R. Dean, J. L. Reid, R. A. Clare & S. Murray. 1976. Clinical pharmacology and pharmacokinetics of clonidine. Clin. Pharmacol. Ther. **19:** 11-18.
10. Wing, L. M. H., J. L. Reid, D. S. Davies, H. J. Dargie & C. T. Dollery. 1977. Pharmacokinetic and concentration effect relationships of clonidine in essential hypertension. Eur. J. Clin. Pharmacol. **12:** 463-469.
11. Wing, L. M. H., J. L. Reid, C. A. Hamilton, P. S. Sever, D. S. Davies & C. T. Dollery. 1977. Effects of clonidine on biochemical indices of sympathetic function and plasma renin activity in normotensive man. Clin. Sci. Mol. Med. **53:** 45-53.
12. Reid, J. L., L. M. H. Wing, C. J. Mathias, H. L. Frankel & E. Neill. 1977. The central hypotensive effect of clonidine: Studies in tetraplegic subjects. Clin. Pharmacol. Ther. **21:** 375-381.
13. Reid, J. L., K. K. Tangri & L. M. H. Wing. 1977. The central hypotensive action of clonidine and propranolol in animals and man. Progr. Brain Res. **47:** 369-384.
14. Hamilton, C. A. 1994. Chemistry, mode of action and experimental pharmacology of moxonidine. *In* The I_1 Imidazoline Receptor. P. A. van Zwieten, C. A. Hamilton & B. N. C. Pritchard, Eds. 1-23. Royal Society of Medicine Services. London.
15. Planitz, V. 1984. Crossover comparison of moxonidine and clonidine in mild to moderate hypertension. Eur. J. Clin. Pharmacol. **27:** 147-152.
16. Macphee, G. J. A., C. A. Howie, H. L. Elliott & J. L. Reid. 1992. A comparison of the haemodynamic and behavioural effects of moxonidine and clonidine in normotensive subjects. Br. J. Clin. Pharmacol. **33:** 261-267.
17. Ernsberger, P., H. L. Elliott, H.-J. Weismann, A. Raag, M. A. Haxhiu, E. Hofferber, A. Low-Kroger, J. L. Reid & H.-J. Mest. 1993. Moxonodine: A second generation central antihypertensive drug. Cardiovasc. Drug Rev. **11:** 411-431.
18. Verbeuren, T. J., A. Tuan Dinh Xuan, E. Konig-Berard & P. Viton. 1990. Rilmenidine. Cardiovasc. Drug Rev. **8:** 56-70.
19. Weerasuriya, K., E. Shaw & P. Turner. 1984. Preliminary clinical pharmacological studies with S3341, a new hypotensive agent: A comparison with clonidine in normal males. Eur. J. Clin. Pharmacol. **27:** 281-286.
20. Dollery, C. T., D. S. Davies, J. Duchier, B. Pannier & M. E. Safar. 1988. Dose and concentration effect relations for rilmenidine. Am. J. Cardiol. **61:** 60D-66D.
21. McKaigue, J. P. & D. W. G. Harron. 1992. The effects of rilmenidine on tests of autonomic function in humans. Clin. Pharmacol. Ther. **52:** 511-517.

Central Nervous System Norepinephrine Turnover in Essential Hypertension[a]

GAVIN W. LAMBERT,[b,c] CLAUDIA FERRIER,[c]
DAVID M. KAYE,[c] GARRY L. JENNINGS,[c]
VICTOR KALFF,[d] MICHAEL J. KELLY,[d]
HELEN S. COX,[c] ANDREA G. TURNER,[c] AND
MURRAY D. ESLER[c]

[c]*Human Autonomic Function Laboratory*
and
Alfred Baker Medical Unit
Baker Medical Research Institute
Commercial Road
Prahran Vic 3181, Australia

[d]*Division of Nuclear Medicine*
Alfred Hospital
Commercial Road
Prahran Vic 3181, Australia

The impetus for examining sympathetic nervous system activity in human hypertension came with von Euler's characterization of norepinephrine as the sympathetic postganglionic neurotransmitter[1] and his subsequent demonstration of elevated urinary excretion of norepinephrine in some hypertensive patients.[2] A variety of techniques, including clinical microneurography, power spectral analysis of circulatory rhythms, and regional norepinephrine spillover rate determinations have since been developed and applied to the assessment of sympathetic function in essential hypertension.[3–6] These methods have provided evidence, particularly in younger patients, of activation of sympathetic outflows to the kidneys, heart, and skeletal muscle. The central nervous system mechanisms controlling this increased firing in efferent postganglionic sympathetic fibers remain to be unequivocally elucidated, but given that in younger hypertensive subjects cerebrospinal fluid norepinephrine levels are elevated,[7–9] activation of central noradrenergic neurons has been implicated.[7–10] The demonstration of a relationship between norepinephrine levels in cerebrospinal fluid and blood pressure[11]

[a]This work is supported by a National Health and Medical Research Council Grant to the Baker Medical Research Institute. Gavin Lambert is the recipient of a Dora Lush National Health and Medical Research Council Postgraduate Scholarship.

[b]Address for correspondence: Gavin W. Lambert, Human Autonomic Function Laboratory, Baker Medical Research Institute, Commercial Road, Prahran Vic 3181, Australia.

lends credence to the concept of a direct role in blood pressure regulation by central noradrenergic nuclei.

Reciprocal aberrations in central nervous system noradrenergic and dopaminergic neuronal activity in essential hypertension, with noradrenergic activity being elevated and dopaminergic activity being diminished, were described by DeQuatro *et al.*[8] and more recently by Lambert *et al.*[12] The hypotensive effects of clonidine in human subjects have been attributed to diminished central nervous system norepinephrine release combined with concomitant enhancement of central dopaminergic activity.[13] Central monoaminergic systems are recognized as being of importance in the regulation of cardiovascular reflexes.[14] Stimulation of pressor noradrenergic neurons in a variety of brain areas, including the A5 region, locus coeruleus, and hypothalamus, increases sympathetic outflow.[15–18] Abnormal function in central nervous system catecholaminergic neurons has been suspected in animal models of hypertension.[19,20] Indeed, Qualy and Westfall[21] recently demonstrated increased neuronal release of norepinephrine in the hypothalamic paraventricular nucleus in spontaneously hypertensive rats, with increased transmitter release being most prominent during the phase of development of the disease in younger animals.[21] Ferrier *et al.*,[22] using direct blood sampling from the internal jugular veins combined with cerebral blood flow scans, found an elevated release of norepinephrine from subcortical brain regions in essential hypertensive patients.

Studies using internal jugular vein sampling demonstrated the release of norepinephrine from subcortical brain regions.[22–24] In the present study we measured the internal jugular venous overflow, in hypertensive patients and normotensive controls, of norepinephrine and its major lipophilic central nervous system metabolites 3,4-dihydroxyphenylglycol (DHPG) and 3-methoxy-4-hydroxyphenylglycol (MHPG). We combined bilateral jugular vein sampling with cerebral blood flow scans to ascertain which areas of the human brain are responsible for the overflow of these neurochemicals into the peripheral circulation and to determine whether central nervous system dysfunction of noradrenergic neurons exists in essential hypertension. We measured concurrently heart rate, blood pressure, and the rate of total body norepinephrine spillover to plasma in order to investigate possible relationships between cerebral neuronal systems, peripheral hemodynamics, and sympathetic nervous outflow.

METHODS

Subjects

Sixteen subjects with mild to moderately severe essential hypertension (aged 43 ± 3 years) and 25 healthy, normotensive volunteers (aged 38 ± 4 years) participated in this study. Hypertensive subjects were recruited from medical outpatient clinics of the Alfred and Baker Medical Unit. Renal vein catheterization and bilateral renal vein plasma renin activity determinations confirmed the absence of renal artery stenosis. Other forms of secondary hypertension were excluded by appropriate clinical testing. The majority of patients were newly diagnosed, whereas the remainder were treated with conventional antihypertensive regimens comprising β-adrenoreceptor blocking drugs, thiazide diuretics, and angiotensin-converting inhibitors either singly or in combination. Clinical blood pressure in all patients was consistently greater than 150/90 mm Hg prior to the initiation of any therapy. Medications were withdrawn 4 weeks prior to the day of study. The healthy subjects were recruited by local advertisement from the general community and were reimbursed a nominal amount for their time and efforts. All subjects underwent a comprehensive clinical and physical examination to screen for any previously undiagnosed medical conditions prior to their entry into the experimental protocol. Exclusion criteria included a history of major illness, cardiovascular disease, current drug medication, and previous psychiatric illness. All healthy volunteers were nonsmokers. Screening procedure entailed a complete blood examination including white cell differential analysis, serum biochemistry, and tests for previous exposure to the hepatitis B and human immunodeficiency viruses.

General Procedure

All studies were performed with subjects in the supine position. Caffeinated beverages, alcohol, and tobacco smoking were prohibited for the 12 hours preceding the catheter study. Subjects had a standardized light breakfast, typically consisting of fruit juice and toast, at 8:00 AM the day of the study. In our experience this level of food intake does not significantly influence plasma neurochemical concentrations.[25] All subjects gave written informed consent for their participation in the study which was approved by the Alfred Hospital Ethics Review Committee.

Blood samples were obtained from central venous and arterial catheters percutaneously inserted under local anesthesia according to our established protocols.[23,24,26] With fluoroscopic control, a 7F type CCS-7U coronary sinus thermodilution catheter (Webster Laboratories, California) was positioned in the internal jugular vein beyond the mandibular angle, upstream to points of entry of veins draining the face and neck to minimize any contamination of the cerebral venous effluent. Catheter position was verified with 2 ml of radiopaque contrast medium (Omnipaque, Winthrop Pharmaceuticals, New York). The catheter was used for sampling internal jugular vein blood and for determining jugular blood flow by thermodilution. Blood flows were converted to plasma flow using the subjects' hematocrit. When possible, bilateral jugular vein sampling was achieved by repositioning this catheter to lie in the opposite internal jugular vein or, more commonly, by percutaneous placement of a second catheter. Throughout the course of the catheter study tritiated norepinephrine (0.7 μCi/min 1-[7-^{3}H] norepinephrine, New England Nuclear, Boston, Massachusetts) was infused for determining norepinephrine plasma kinetics and for ascertaining the rate of spillover of norepinephrine from the brain.[23,27,28] In all studies, blood samples for biochemical analysis were immediately placed in chilled tubes containing anticoagulant and antioxidant and placed on ice. At the completion of the catheter study blood samples were centrifuged (800 *g* for 30 min at 4°C) and the plasma stored at –80°C until assayed. Arterial hematocrits were determined for each subject.

Cerebral Blood Flow Scans

Cerebral blood flow scans were performed in the Alfred Hospital Division of Nuclear Medicine according to our previously described methods.[22,26] Briefly, 2-3 ml of blood was obtained from the subjects and the red cells were labeled with technetium. The radioactively tagged red blood cells were then reinjected into the subject and the scintigraphic image was monitored. This procedure enabled the detection of the internal jugular vein which formed the continuation of the superior sagittal sinus. By inference the route of drainage of the inferior and straight sinuses, carrying the bulk of subcortical venous return,[29] was then ascertained.

Plasma Catecholamine and MHPG Determination

The catechols dihydroxyphenylalanine (DOPA), norepinephrine, and DHPG were extracted from plasma with alumina adsorption, separated by high performance liquid chromatography (HPLC), and the amounts quantified by electrochemical detection as previously described.[30,31] Fractions of eluant leaving the electrochemical cell were collected for measurement of tritium-labeled norepinephrine by liquid scintillation spectroscopy.

The plasma concentration of MHPG was determined as previously described.[12,32] The MHPG was extracted from ultrafiltered plasma by vigorous shaking with sequanal grade ethyl acetate followed by drying under gaseous nitrogen. The MHPG was then reconstituted in mobile phase and quantified by HPLC with electrochemical detection.

Calculations

After steady-state arterial plasma tracer concentrations of [^{3}H]norepinephrine had been reached, the overall release rate into plasma of endogenous norepinephrine was determined by isotope dilution according to methods developed in our laboratory.[27,33,34]

$$\text{Total Spillover Rate} = \frac{[^3\text{H}]\ \text{Catechol Infusion Rate}}{\text{Plasma Catechol Specific Radioactivity}}$$

The estimated central nervous system norepinephrine turnover was calculated by summing the respective overflows from the brain of norepinephrine, DHPG, and MHPG. Venoarterial plasma concentration differences combined with an appropriate internal jugular vein flow measurement were used, according to the Fick principle, to determine metabolite overflows from the brain.

The spillover of norepinephrine from the brain to plasma was calculated from the product of the venoarterial difference in plasma norepinephrine concentration and the plasma flow with an adjustment allowing for the fractional extraction of tritium-labeled norepinephrine across the brain during a constant-rate infusion of radiolabeled norepinephrine.[28,33,35] At steady state:

$$\text{Internal Jugular Spillover} = [(NE_{ven} - NE_{art}) + (NE_{art} \times NE_{ex})] \times \text{Plasma Flow}$$

where NE_{ven} and NE_{art} are the venous and arterial norepinephrine concentrations, respectively, and NE_{ex} is the fractional extraction of tritiated norepinephrine at steady state in a single passage through the brain.

TABLE 1. Hemodynamic Profile of Normotensive and Hypertensive Subjects[a]

Hemodynamic Measurements	Normotensive Subjects	Hypertensive Subjects
Systolic blood pressure (mm Hg)	138 ± 3	162 ± 7*
Diastolic blood pressure (mm Hg)	70 ± 2	84 ± 4*
Mean arterial pressure (mm Hg)	93 ± 2	110 ± 4*
Heart rate (beats/min)	61 ± 2	78 ± 4*

[a] Intraarterial blood pressure and heart rate measurements recorded in cardiac catheterization laboratory. * $p < 0.001$ significantly greater than corresponding value in normotensive subjects.

The overflow of MHPG and DHPG from the brain were determined according to the following formula:

$$\text{Overflow} = (\text{Conc}_v - \text{Conc}_a) \times \text{BF}$$

where Conc_v and Conc_a are the matching venous and arterial concentrations, respectively, and BF is the internal jugular vein blood flow. As there is no evidence of extraction of the metabolites from blood during transcerebral passage,[36] net overflow was calculated without recourse to isotope dilution methodology. Blood flow is used in the calculation of overflow, as MHPG and DHPG rapidly and equally distribute between red cells and plasma.[26]

Statistical Analysis

All results, unless otherwise specified, are expressed as mean ± standard error of the mean (SEM). Between-group comparisons were made using one-way analysis of variance. Relationships between continuous variables were examined by the least squares method of linear regression. Analysis of covariance was used to compare the slopes and intercepts of the relation for two continuous variables between groups. The null hypothesis was rejected at $p < 0.05$.

RESULTS

Intraarterial blood pressure and heart rate on the day of the catheter procedure were significantly elevated in the hypertensive patients (TABLE 1). The plasma concentration profile of the neurochemicals examined differed between the hypertensive patients and their normotensive counterparts (TABLE 2). The elevation in arterial plasma concentration of DOPA, norepinephrine, and MHPG, by 12, 33, and 21% respectively in the hypertensive group, was presumably indicative of sympathetic nervous system activation. The plasma concentration of the intraneuronally produced norepinephrine metabolite DHPG was identical in the two groups studied. Resting heart rate was not significantly related to the arterial levels of DOPA, norepinephrine, DHPG, or

TABLE 2. Arterial Plasma Neurochemical Concentrations[a]

Arterial Plasma Concentrations (nmol/L)	Normotensive Subjects	Hypertensive Subjects
DOPA	6.4 ± 0.2	7.1 ± 0.3*
Norepinephrine	1.2 ± 0.1	1.6 ± 0.2*
DHPG	5.3 ± 0.2	5.7 ± 0.3
MHPG	15.9 ± 0.8	18.1 ± 1.6*

[a] Arterial plasma concentrations of 3,4-dihydroxyphenylalanine (DOPA), norepinephrine, 3,4-dihydroxyphenylglycol (DHPG), and 3-methoxy-4-hydroxyphenylglycol (MHPG) in normotensive and hypertensive subjects. * $p < 0.05$, significantly higher than corresponding value in normotensive subjects.

MHPG. The arterial DOPA level was related to the mean arterial blood pressure ($p < 0.05$, y = 5.1x + 65.2, $r = 0.36$). Such a relationship did not occur for any of the other neurochemicals examined.

In the hypertensive subjects, total body norepinephrine spillover to plasma was increased by 43% (4.39 ± 0.61 nmol/min vs 3.08 ± 0.24 nmol/min, $p < 0.05$). Indicative of a link between blood pressure regulation and sympathetic nervous activity, the integrated neuronal firing rate, as indicated by total body norepinephrine spillover to plasma, was significantly related to the intraarterially determined blood pressure ($p < 0.01$, FIG. 1). Analysis of covariance indicated that the slope of the relationship was similar in the two groups studied. Total body norepinephrine spillover was not related to resting heart rate but was significantly related to arterial norepinephrine ($p < 0.001$, y = 0.2x + 2.5, $r = 0.78$) and DHPG ($p < 0.02$, y = 0.5x + 0.6, $r = 0.38$) concentrations.

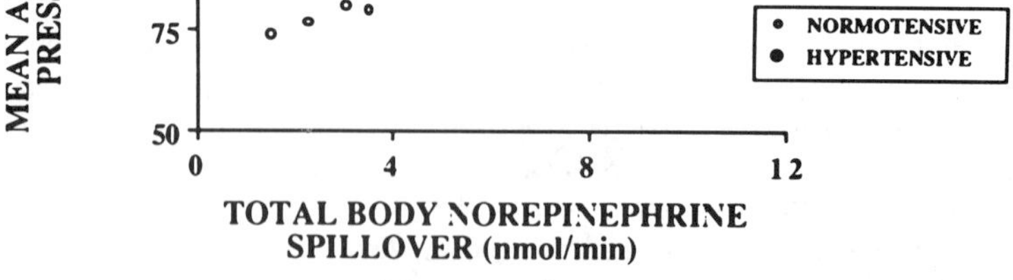

FIGURE 1. Relationship between mean arterial blood pressure and total body norepinephrine spillover in normotensive and hypertensive subjects combined. **$p < 0.01$, y = 3.8x + 85, $r = 0.50$.

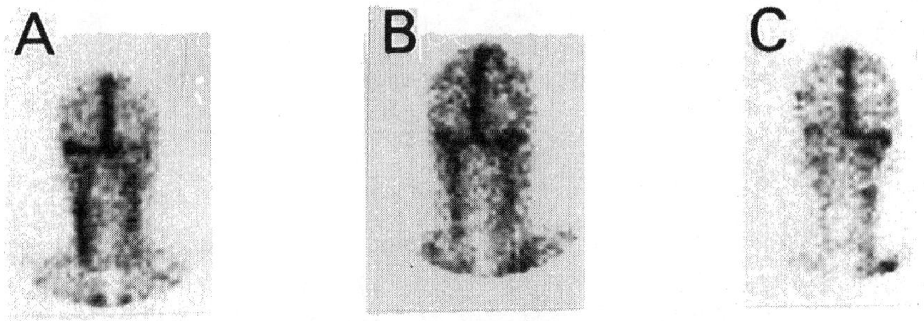

FIGURE 2. Posterior views of dynamic cerebral blood flow scans showing the drainage route of the superior sagittal sinus. (**A**) Drainage of superior sagittal sinus (carrying the bulk of cortical venous return) into the left transverse sinus and left internal jugular vein. (**B**) Nonlateralizing scan showing symmetrical drainage of superior sagittal sinus into left and right internal jugular veins. (**C**) Predominant drainage of superior sagittal sinus into right transverse sinus and right internal jugular vein.

Cerebral blood flow scans were performed in 11 of the hypertensive subjects studied. The route of superior sagittal sinus drainage, carrying the bulk of cortical venous return, was unequivocally identified as being either to the right or to the left internal jugular vein in 7 cases (FIG. 2). The cerebral sinus drainage of the remaining subjects displayed a symmetrical, or nonlateralizing pattern. Healthy subjects were preselected on the basis of their cortical drainage being identifiably either to the right or to the left internal jugular vein.

Activation of central nervous system noradrenergic neurons was evident in hypertension with the estimated total brain norepinephrine turnover in these subjects being over three times that of the corresponding value in the normotensive subjects ($p < 0.01$, FIG. 3). Further investigation revealed that this increased norepinephrine turnover in the brain was confined to subcortical brain regions (FIG. 3). In those subjects in whom simultaneous bilateral jugular vein catheterization was successfully performed, the estimated total brain norepinephrine turnover (combined right jugular and left jugular venous overflows of norepinephrine, DHPG, and MHPG) was significantly related to both heart rate ($p <0.01$, $y = 6.5x + 58.1$, $r = 0.79$) and blood pressure ($p <0.001$, $y = 11.9x + 127.9$, $r = 0.89$ for systolic blood pressure and $p <0.01$, $y = 5.7x + 68.1$, $r = 0.66$ for diastolic blood pressure), but only weakly with total body norepinephrine spillover ($p = 0.09$, $y = x + 3.1$, $r = 0.44$).

Compared to their normotensive counterparts the hypertensive subjects had an over fivefold increase in the estimated turnover of norepinephrine in subcortical brain regions (1,940 ± 397 vs 359 ± 200 pmol/min, $p <0.01$). Estimated cortical norepinephrine turnover in healthy volunteers was 710 ± 220 pmol/min and was not significantly different from that of hypertensive patients (587 ± 305 pmol/min). Subcortical norepinephrine turnover was significantly related to the rate of norepinephrine spillover to plasma and to heart rate and blood pressure (FIG. 4). Cortical norepinephrine turnover was not related to the level of sympathetic nervous activation, heart rate, or blood pressure.

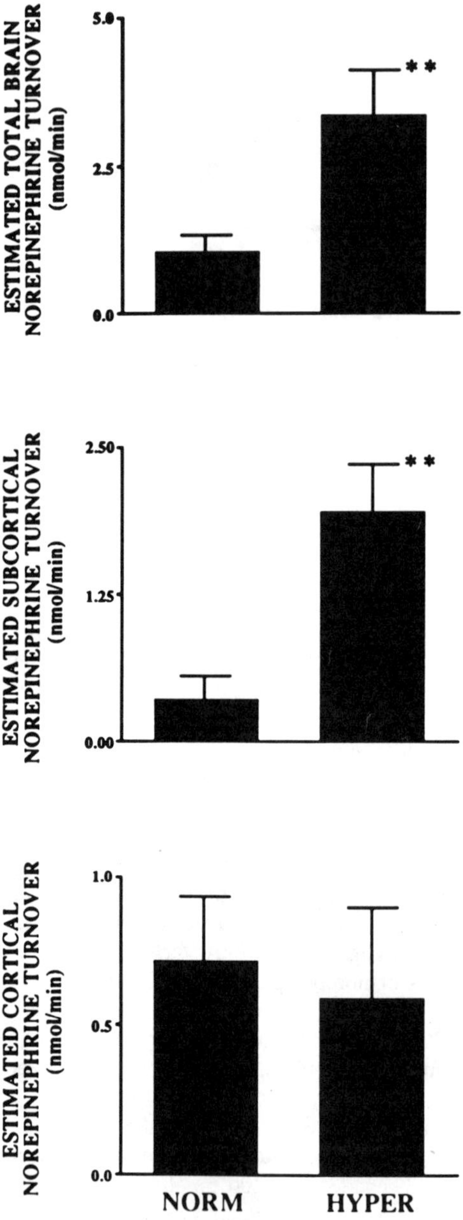

FIGURE 3. Norepinephrine turnover, as estimated from combined overflows of norepinephrine, 3-methoxy-4-hydroxyphenylglycol, and 3,4-dihydroxyphenylglycol, in whole brain (*top*), subcortical regions (*middle*), and cortical regions (*bottom*) in normotensive (NORM) and hypertensive (HYPER) patients. Values shown are mean ± SEM. $^{**}p < 0.01$, significantly greater than corresponding value in normotensive subjects.

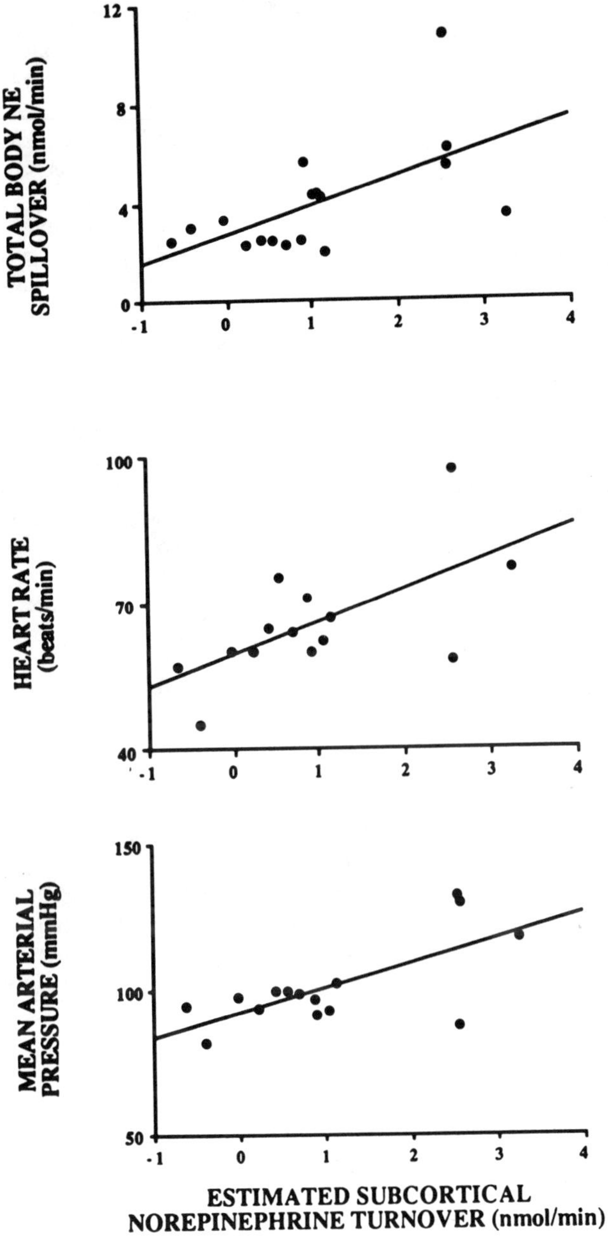

FIGURE 4. Relation between subcortical norepinephrine turnover, as estimated from combined overflows of norepinephrine, 3-methoxy-4-hydroxyphenylglycol, and 3,4-dihydroxyphenylglycol, and total body norepinephrine spillover to plasma (*top,* $^{**}p = 0.01$, $y = x + 2.7$, $r = 0.60$), heart rate (*middle,* $^{**}p = 0.01$, $y = 7x + 59$, $r = 0.63$), and mean arterial blood pressure (*bottom,* $^{**}p < 0.01$, $y = 9x + 92$, $r = 0.68$).

DISCUSSION

This study reinforces the previous findings of this laboratory demonstrating enhanced subcortical noradrenergic activity and its link with sympathetic activation in essential hypertension. Using radioenzymatic and HPLC methodologies we have now studied the release of norepinephrine and metabolites into the cerebrovascular circulation in 35 hypertensive subjects, over 60 healthy volunteers, 25 patients with chronic congestive heart failure, and 9 patients with chronic postural hypotension, in 5 of whom the diagnosis of idiopathic peripheral autonomic insufficiency was confirmed.[12,22–24,37–39] The demonstration of similar cerebral norepinephrine overflows in healthy subjects and in patients with autonomic failure, who had existing biochemical evidence of almost complete postganglionic sympathetic denervation,[40] suggested that the norepinephrine emanated from central noradrenergic neurons and not from cerebrovascular sympathetic nerves. The inability of ganglionic blockade, with trimethaphan, to reduce cerebral norepinephrine overflow in hypertensive subjects[24] underpins the concept of a central origin of the norepinephrine found in these studies. Cerebral norepinephrine efflux per se in fact underestimates the degree of cerebral noradrenergic neuronal activation in hypertension, because barriers to norepinephrine overflow from the brain, whether they be metabolic, structural, or a combination of both, may exist.[41–43]

MHPG is the predominant norepinephrine metabolite released from the brain into the bloodstream in primates[44,45] and, being a neutral metabolite, readily diffuses through tissue membranes.[46,47] Previous reports from our laboratory have indicated that the glycol metabolites of noradrenaline are readily released into the cerebral venous effluent in hypertensive and normotensive subjects alike, with the overflow in hypertensive subjects being appreciably greater than that in normotensive subjects.[12,22] Facilitated washout, with higher internal jugular blood flows, can be excluded as a possible mechanism for this increased washout of norepinephrine and its metabolites given that subcortical blood flows are not increased in essential hypertension.[12] These results support previous findings of laboratories demonstrating a degree of central nervous system noradrenergic activation in essential hypertension.[7–10]

The existence of a relation between central nervous system norepinephrine turnover and resting, supine blood pressure is in agreement with earlier studies which assessed central noradrenergic activity by measuring cerebrospinal fluid levels of norepinephrine and MHPG[8] or norepinephrine alone.[7,9,10] Contrary to the observations of these investigators, we found no relation of the arterial concentration of norepinephrine to blood pressure and heart rate.

The relevance of activation of brain noradrenergic neurons in the hypertensive setting remains uncertain. Abundant evidence exists that central noradrenergic neuronal systems participate in the regulation of sympathetic outflow and autonomic cardiovascular control. The link between subcortical norepinephrine turnover and total body norepinephrine spillover that we describe adds further credence to the concept of forebrain noradrenergic neurons acting in a sympathoexcitatory fashion. This is not to deny the existence of sympathoinhibitory, depressor noradrenergic neurons in the brain stem.[48] Previous reports from our laboratory have illustrated the dependence of sympathetic outflow on cerebral noradrenergic activity[24,34] and have provided evidence of a relation between subcortical noradrenergic neuronal activity and renal sympathetic activation.[22] Interestingly, Eide and colleagues[7] previously demonstrated a relation between cerebrospinal fluid norepinephrine levels and plasma renin activity. These results are supported by experiments conducted in animals in which stimulation of noradrenergic pressor regions of the hypothalamus and amygdala increase renal nerve firing and renal vascular resistance.[16,49] Also of note is the recent demonstration by Qualy and Westfall[21] of increased overflow of norepinephrine from the paraventricular hypothalamic nucleus in spontaneously hypertensive rats. Given that several CNS noradrenergic nuclei, including those of the locus coeruleus, the lateral and paraventricular nuclei of the hypothalamus, the A5 region, and the amygdala, are unequivocally sympathoexcitatory[15–18,21,49,50] and that using techniques involving the retrograde transynaptic transport of pseudorabies virus, a number of brain regions have been identified as innervating sympathetic preganglionic outflow at all thoracolumbar levels,[51,52] it is pertinent to consider from which brain noradrenergic nuclei the increased internal jugular vein overflow of norepinephrine and metabolites found in our hypertensive patients might derive. The topographical limitations of our technique preclude precise anatomical interpretation of our findings, so that animal experimentation must be relied upon for the identification of the probable neuronal cell groups activated in human hypertension.

There exist 15 catecholaminergic neuronal groups in the reticular formation. The cell bodies of the noradrenergic groups, designated A1-7, are confined to medullary and pontine parts of the brain stem but exhibit complex ascending and descending projections in addition to local destinations in the brain stem.[29] With no less than 50% of brain norepinephrine being located in the locus coeruleus, or A6 group,[53] it seems probable that this brain region may provide a major input to our overflow measurements. Because profuse branching of both ascending and descending noradrenergic fibers exists, locus coeruleus activation will impart effects over broad brain regions. Descending A6 fibers are joined by some A7 neurons and provide collaterals to numerous bulbar nuclei, together forming the ventrolateral ceruleospinal tract, while a substantial bundle joins the superior cerebellar peduncle to form the ceruleocerebellar tract.[29] Ascending fibers form the dorsal tegmental norepinephrine bundle and traverse the midbrain through the lateral hypothalamic part of the diencephalon as part of the medial forebrain bundle. The dorsal noradrenergic pathway projects to the frontal cortex, hippocampus, amygdaloid complex, thalamus, cerebellum, lateral hypothalamic area, and the periventricular, supraoptic, and paraventricular nuclei. The nucleus coeruleus receives intrinsic brainstem input from other reticular neuronal systems, in particular B5-8 serotonergic projections.[29] Functionally, the A6 region and its hypothalamic and amygdala projections are involved in behavioral responses associated with autonomic activation.[53] Activation of the A6 region might possibly be linked to previous behavioral observations in hypertensive patients, such as the linking of suppression of hostility with sympathetic activation.[54]

In another series of experiments conducted in our laboratory we examined central nervous system neuronal activation in patients with severe congestive heart failure, who constitute another clinical model of sympathetic nervous system activation.[39] This syndrome of congestive cardiac failure is characterized by diminished pumping performance of the heart, complex neural pathophysiology, and substantial morbidity and mortality.[39,55–57] It is generally acknowledged that sympathetic nervous system stimulation in patients with heart failure is commonly present to the detriment of the failing myocardium,[55,56,58] predisposing even in mild failure to lethal ventricular arrhythmias.[59] In our patients the levels of release of norepinephrine and its lipophilic metabolites DHPG and MHPG from the brain into the internal jugular vein indicated the presence of increased central nervous system norepinephrine turnover in cardiac failure. A link between cardiac sympathetic activation, pulmonary artery pressures, and central nervous system noradrenergic activation has been proposed.[39] Whether elevated pulmonary pressures, in fact, provide the hemodynamic signal for increased sympathetic outflow to the myocardium remains to be established. Cardiopulmonary volume receptor afferents do project to the noradrenergic nuclei of the locus coeruleus, and the firing rate of locus coeruleus neurons is changed by alterations in cardiopulmonary pressures.[50,60] As noted previously, the A6 region is also closely linked with behavioral responses involving autonomic activation.[53] This convergence of inputs to the locus coeruleus may be pertinent in the context of reports, indicating that experimental hypothalamic stimulation can induce cardiac arrhythmias[61,62] and that psychological stress is a common antecedent of potentially fatal ventricular fibrillation in those with underlying heart disease.[63,64]

Central monoaminergic pathways are implicated in the maintenance of circulatory control and the regulation of sympathetic outflow from the brain. The existence of a blood-brain barrier to the outward flux of monoamines from the brain to the bloodstream would hinder any attempt to study central noradrenergic activity using our techniques. In this study, along with norepinephrine, we measured the overflow from the brain of the major lipophilic metabolites of norepinephrine, DHPG and MHPG, thereby circumventing interpretational problems arising from the possible existence of a diffusion barrier to monoamines. In accordance with previous reports from our group,[12,22,24] we found an increased turnover of norepinephrine in the forebrain in essential hypertension. Such findings are supported by the demonstration of increased norepinephrine concentration in the cerebrospinal fluid of hypertensive patients.[7–10] Elevated subcortical noradrenergic activity has been implicated as playing a possible sympathoexcitatory role in the regulation of sympathetic preganglionic neurons of the thoracolumbar cord in essential hypertension,[22] a proposal further supported by this study which demonstrated the existence of a correlation between subcortical norepinephrine turnover and total body norepinephrine spillover to plasma, resting blood pressure, and heart rate. We propose that activation of the A6 region may underly the increased subcortical noradrenergic activity in hypertensive patients and could possibly account for the reported association of behavioral phenomena, of which one is the suppression of hostility,[54] with human hypertension.

ACKNOWLEDGMENTS

The authors wish to thank Elizabeth Dewar, Leonie Johnston, and Kaye Varcoe for their expert assistance in the research catheter laboratory and Dr. Kenneth Kirk of the National Institutes of Health, Bethesda, Maryland, for synthesizing and donating the internal standards used in this study. The photographic expertise of Neil Potter is gratefully acknowledged.

REFERENCES

1. VON EULER, U. S. 1946. A specific sympathetic ergone in adrenergic nerve fibres (sympathin) and its relation to adrenaline and noradrenaline. Acta Physiol. Scand. **12:** 73-97.
2. VON EULER, U. S., S. HELLNER & A. PURKHOLD. 1954. Excretion of noradrenaline in the urine in hypertension. Scand. J. Clin. Lab. Invest. **6:** 54-59.
3. ESLER, M., G. JENNINGS, B. BIVIANO, G. LAMBERT & G. HASKING. 1985. Mechanism of elevated plasma noradrenaline in the course of essential hypertension. J. Cardiovasc. Pharmacol. **8:** 539-544.
4. GUZZETTI, S., E. PICCALUGA, R. CASATI, S. CERUTTI, F. LOMBARDI, M. PAGANI & A. MALLIANI. 1988. Sympathetic predominance in essential hypertension: A study employing spectral analysis of heart rate variability. J. Hypertens. **6:** 711-717.
5. ANDERSON, E. A., C. A. SINKEY, W. J. LAWTON & A. L. MARK. 1989. Elevated sympathetic nerve activity in borderline hypertensive humans: Evidence from direct intraneural recordings. Hypertension **14:** 177-183.
6. YAMADA, Y., E. MIYAJIMA & O. TOCHIKUBO. 1989. Age-related changes in muscle sympathetic nerve activity in essential hypertension. Hypertension **13:** 870-877.
7. EIDE, I., R. KOLLOCH, V. DEQUATTRO, L. MIANO, R. DUGGER & J. VAN DER MEULEN. 1979. Raised cerebrospinal fluid norepinephrine in some patients with primary hypertension. Hypertension **1:** 255-260.
8. DEQUATRO, V., P. SULLIVAN, R. MINAGAWA, I. KOPIN, J. BORNHEIMER, A. FOTI & R. BARNDT. 1984. Central and peripheral noradrenergic tone in primary hypertension. Fed. Proc. **43:** 47-51.
9. KAWANO, Y., K. FUKIYAMA, Y. TAKEYA, I. ABE & T. OMAE. 1984. Catecholamines, angiotensin II and sodium concentrations in cerebrospinal fluid in young men with borderline hypertension. Clin. Exp. Hypertens.—Theory Prac. **A6:** 1131-1145.
10. LAKE, C. R., H. G. GULLNER, R. J. POLINSKY, M. H. EBERT, M. G. ZIEGLER & F. C. BARTTER. 1981. Essential hypertension: Central and peripheral norepinephrine. Science **211:** 955-957.
11. ZIEGLER, M. G., C. R. LAKE, J. H. WOOD & B. R. BROOKS. 1980. Relationship between cerebrospinal fluid norepinephrine and blood pressure in neuroleptic patients. Clin. Exp. Hypertens. **2:** 995-1008.
12. LAMBERT, G. W., C. FERRIER, D. M. KAYE, V. KALFF, M. J. KELLY, H. S. COX, A. G. TURNER, G. L. JENNINGS & M. D. ESLER. 1994. Monoaminergic neuronal activity in subcortical brain regions in essential hypertension. Blood Pressure **3:** 55-66.
13. MARTIN, P. R., M. H. EBERT, E. K. GORDON, M. LINNOILA & I. J. KOPIN. 1984. Effects of clonidine on central and peripheral catecholamine metabolism. Clin. Pharmacol. Ther. **35:** 322-327.
14. PALKOVITS, M. 1981. Neuropeptides and biogenic amines in central cardiovascular control mechanisms. *In* Central Nervous System Mechanisms in Hypertension. J. P. Buckley & C. M. Ferrario, Eds. F3-F8. Raven Press. New York.
15. VAN HUYSSE, J. W. & S. L. BEALER. 1991. Central nervous system norepinephrine release hypotension and hyperosmolality in conscious rats. Am. J. Physiol. **260:** R1071-R1076.
16. HUANGFU, D., N. KOSHIYA & P. GUYENET. 1991. A5 noradrenergic unit activity and sympathetic nerve discharge in rats. Am. J. Physiol. **261:** R393-R402.
17. HUANGFU, D., L.-J. HWANG, T. A. RILEY & P. GUYENET. 1992. Splanchnic nerve responses to A5 area stimulation in rats. Am. J. Physiol. **263:** R437-R446.
18. SINGEWALD, S. & A. PHILIPPU. 1993. Catecholamine release in the locus coeruleus is modified by experimentally induced changes in haemodynamics. Naunyn-Schmiedeberg's Arch. Pharmacol. **347:** 21-27.
19. HOWES, L. G. 1984. Central catecholaminergic neurons and spontaneously hypertensive rats. J. Auton. Pharmacol. **4:** 207-217.
20. TIBIRICA, E., J. FELDMAN & P. BOUSQUET. 1992. Contribution of catecholaminergic neurons of the dorsomedial and ventrolateral medulla oblongata to the spontaneously hypertensive rats: *In vivo* voltammetric studies. J. Hypertens. **10:** 1327-1334.

21. QUALY, J. M. & T. C. WESTFALL. 1993. Age-dependent overflow of endogenous norepinephrine from paraventricular hypothalamic nucleus of hypertensive rats. Am. J. Physiol. **265:** H39-H46.
22. FERRIER, C., G. L. JENNINGS, G. EISENHOFER, G. LAMBERT, H. COX, V. KALFF, M. KELLY & M. D. ESLER. 1993. Evidence for increased noradrenaline release from subcortical brain regions in essential hypertension. J. Hypertens. **11:** 1217-1227.
23. ESLER, M. D., G. L. JENNINGS & G. W. LAMBERT. 1988. Release of noradrenaline into the cerebrovascular circulation in patients with primary hypertension. J. Hypertens. **6:** 493-496.
24. FERRIER, C., M. D. ESLER, G. EISENHOFER, B. G. WALLIN, M. HORNE, H. S. COX, G. LAMBERT & G. JENNINGS. 1992. Increased norepinephrine spillover into the jugular veins in essential hypertension. Hypertension **19:** 62-69.
25. LAMBERT, G. W., G. EISENHOFER, G. L. JENNINGS & M. D. ESLER. 1993. Regional homovanillic acid production in humans. Life Sci. **53:** 63-75.
26. LAMBERT, G. W., G. EISENHOFER, H. S. COX, M. HORNE, V. KALFF, M. KELLY, G. L. JENNINGS & M. D. ESLER. 1991. Direct determination of homovanillic acid release from the human brain, an indicator of central dopaminergic activity. Life Sci. **49:** 1061-1072.
27. ESLER, M., G. JACKMAN, A. BOBIK, D. KELLEHER, G. JENNINGS, P. LEONARD, H. SKEWS & P. KORNER. 1979. Determination of norepinephrine apparent release rate and clearance in humans. Life Sci. **25:** 1461-1470.
28. ESLER, M., G. JENNINGS, P. KORNER, P. BLOMBERY, N. SACHARIAS & P. LEONARD. 1984. Measurement of total and organ-specific norepinephrine kinetics in humans. Am. J. Physiol. **247:** E21-E28.
29. GRAY, H. 1980. Angiology and neurology. *In* Gray's Anatomy, 36th Ed. P. L. Williams & R. Warwick, Eds.: 738-751. Churchill Livingston. Edinburgh.
30. EISENHOFER, G., D. S. GOLDSTEIN, R. STULL, T. KEISER, T. SUNDERLAND, D. L. MURPHY & I. KOPIN. 1986. Simultaneous liquid chromatographic determination of 3,4-dihydroxyphenylglycol, catecholamines, and 3,4-dihydroxyphenylalanine in plasma and their responses to inhibition of monoamine oxidase. Clin. Chem. **31:** 2030-2033.
31. MEDVEDEV, O. S., M. D. ESLER, J. A. ANGUS, H. S. COX & G. EISENHOFER. 1990. Simultaneous determination of plasma noradrenaline and adrenaline kinetics, responses to nitroprusside-induced hypotension and 2-deoxyglucose-induced glucopenia in the rabbit. Naunyn-Schmiedeberg's Arch. Pharmacol. **341:** 192-199.
32. EISENHOFER, G., D. S. GOLDSTEIN, T. G. ROPCHAK, H. Q. NGUYEN, H. R. KEISER & I. J. KOPIN. 1988. Source and physiological significance of plasma 3,4-dihydroxyphenylglycol and 3-methoxy-4-hydroxyphenylglycol. J. Auton. Nerv. Syst. **24:** 1-14.
33. ESLER, M., G. JENNINGS, P. KORNER, I. WILLETT, F. DUDLEY, G. HASKING, W. ANDERSON & G. LAMBERT. 1988. Assessment of human sympathetic activity from measurements of norepinephrine turnover. Hypertension (Dallas) **11:** 3-20.
34. ESLER, M., G. JENNINGS, G. LAMBERT, I. MEREDITH, M. HORNE & G. EISENHOFER. 1990. Overflow of catecholamine neurotransmitters to the circulation: Source, fate, and functions. Physiol. Rev. **70:** 963-985.
35. ESLER, M. D., G. L. JENNINGS, P. LEONARD, N. SACHARIAS, F. BURKE, J. JOHNS & P. BLOMBERY. 1984. Contribution of individual organs to total noradrenaline release in humans. Acta Physiol. Scand. **527:** 11-16.
36. GOLDSTEIN, D. S., R. O. CANNON, A. QUYYUMI, P. CHANG, M. DUNCAN, J. E. BRUSH, JR. & G. EISENHOFER. 1991. Regional extraction of circulating norepinephrine, DOPA, and dihydroxyphenylglycol in humans. J. Auton. Nerv. Syst. **34:** 17-36.
37. ESLER, M., G. JENNINGS, G. LAMBERT & P. KORNER. 1987. Local autonomic activity in primary hypertension. *In* The Early Pathogenesis of Primary Hypertension. A. Hofman, D. E. Grobbe & M.A.D.H. Schalekamp, Eds.: 141-155. Excerpta Medica. Amsterdam.

38. Esler, M., C. Ferrier, G. Lambert, G. Eisenhofer, H. Cox & G. Jennings. 1991. Biochemical evidence of sympathetic hyperactivity in human hypertension. Hypertension **17:** III 29-III 35.
39. Kaye, D. M., G. W. Lambert, J. Lefkovits, M. Morris, G. Jennings & M. D. Esler. 1994. Neurochemical evidence of cardiac sympathetic activation and increased central nervous system norepinephrine turnover in congestive heart failure. J. Am. Coll. Cardiol. **23:** 570-578.
40. Meredith, I. T., M. D. Esler, H. S. Cox, G. W. Lambert, G. L. Jennings & G. Eisenhofer. 1991. Biochemical evidence of sympathetic denervation of the heart in pure autonomic failure. Clin. Auton. Res. **1:** 187-194.
41. Glowinski, J., A. Cheramy, R. Romo & L. Barbeito. 1988. Presynaptic regulation of dopaminergic transmission in the striatum. Cell. Molec. Neurobiol. **8:** 7-17.
42. Pardridge, W. 1983. Brain metabolism: A perspective from the blood-brain barrier. Physiol. Rev. **63:** 1481-1535.
43. Kopin, I. J. 1985. Catecholamine metabolism: Basic aspects and clinical significance. Pharmacol. Rev. **37:** 333-364.
44. Karoum, F., J. Moyer-Schwing, S. G. Potkin & R. J. Wyatt. 1977. Presence of free, sulfate, and glucuronide conjugated 3-methoxy-4-hydroxy phenylglycol (MHPG) in human brain, cerebrospinal fluid and plasma. Brain Res. **125:** 333-339.
45. Elsworth, J. D., R. H. Roth & D. E. Redmond, Jr. 1983. Relative importance of 3-methoxy-4-hydroxyphenylglycol and 3,4-dihydroxyphenylglycol as norepinephrine metabolites in rat, monkey, and humans. J. Neurochem. **41:** 786-793.
46. Kessler, J. A., J. D. Fenstermacher & C. S. Patlak. 1976. 3-Methoxy-4-hydroxyphenylethyleneglycol (MHPG) transport from the spinal cord during spinal subarachnoid perfusion. Brain Res. **102:** 131-141.
47. Trendelenburg, U., H. Bonisch, K. H. Graefe & M. Henseling. 1980. The rate constants for the efflux of metabolites of catecholamines and phenylethylamines. Pharmacol. Rev. **31:** 179-203.
48. Chalmers, J. & P. Pilowsky. 1991. Brainstem and bulbospinal neurotransmitter systems in the control of blood pressure. J. Hypertens. **9:** 675-694.
49. Koepke, J. P., S. Jones & G. F. DiBona. 1987. A2 adrenoreceptors in amygdala control renal sympathetic nerve activity and renal function in conscious spontaneously hypertensive rats. Brain Res. **404:** 80-88.
50. Elam, M., T. H. Svensson & P. Thoren. 1985. Differentiated cardiovascular afferent regulation of locus coeruleus and sympathetic nerves. Brain Res. **358:** 77-84.
51. Strack, A. M., W. B. Sawyer, J. H. Hughes, K. B. Platt & A. D. Loewy. 1989. A general pattern of CNS innervation of the sympathetic outflow demonstrated by transneuronal pseudorabies viral infections. Brain Res. **491:** 156-162.
52. Strack, A. M., W. B. Sawyer, K. B. Platt & A. D. Loewy. 1989. CNS cell groups regulating the sympathetic outflow to adrenal gland as revealed by transneuronal cell body labelling with pseudorabies virus. Brain Res. **491:** 274-296.
53. Foote, S. L., F. E. Bloom & G. Aston-Jones. 1983. Nucleus coeruleus: New evidence of anatomical and physiological specificity. Physiol. Rev. **63:** 844-914.
54. Esler, M., S. Julius, A. Zweifler, O. Randall, E. Harburg, H. Gardiner & V. DeQuattro. 1977. Mild high-renin essential hypertension: A neurogenic human hypertension? N. Engl. J. Med. **296:** 405-411.
55. Cohn, J., T. Levine, M. Olivari, V. Garberg, D. Tura, G. Francis, A. Simon & T. Rector. 1984. Plasma norepinephrine as a guide to prognosis in patients with chronic congestive heart failure. N. Engl. J. Med. **311:** 819-823.
56. Hasking, G. J., M. D. Esler, G. L. Jennings, D. Burton & P. I. Korner. 1986. Norepinephrine spillover to plasma in patients with congestive heart failure: Evidence of increased overall and cardiorenal sympathetic nervous activity. Circulation **73:** 615-621.

57. THAMES, M. D., T. KINUGAWA, M. L. SMITH & M. E. DIBNER-DUNLAP. 1993. Abnormalities of baroreflex control in heart failure. J. Am. Coll. Cardiol. **22:** 56A-60A.
58. LEIMBACH, W. N., G. B. WALLIN, R. G. VICTOR & P. E. AYLWARD. 1986. Direct evidence from intraneural recordings for increased central sympathetic outflow in patients with heart failure. Circulation **73:** 913-919.
59. MEREDITH, I. T., A. BROUGHTON, G. L. JENNINGS & M. D. ESLER. 1991. Evidence for a selective increase in resting cardiac sympathetic activity in some patients suffering sustained out of hospital ventricular arrhythmias. N. Engl. J. Med. **325:** 618-624.
60. ELAM, M., T. YAO, H. SVENSSON & P. THOREN. 1984. Regulation of locus coeruleus neurons and splanchnic, sympathetic nerves by cardiovascular afferents. Brain Res. **290:** 281-287.
61. HOCKMAN. L. H., H. P. MAUCK & E. C. HOFF. 1966. Changes resulting from cerebral stimulation: A spectrum of ventricular arrhythmia of sympathetic origin. Am. Heart J. **71:** 695-700.
62. VERRIER, R. L., A. CALVERT & B. LOWN. 1975. Effect of posterior hypothalamic stimulation on the ventricular fibrillation threshold. Am. J. Physiol. **225:** 923-927.
63. ENGEL, G. L. 1971. Sudden and rapid death during psychological stress: Folk lore or folk wisdom? Ann. Intern. Med. **74:** 771-782.
64. TRICHOPOULOS, D., K. KATSOUYANNI, X. ZAVITSANOS, A. TZONOU & P. DALLA-VORGIA. 1983. Psychological stress and fatal heart attack: The Athens (1981) earthquake natural experiment. Lancet **1:** 441-444.

Subject Index

Index of Contributors